T0140337

dvances in Intelligent Systems and Computing" contains publications
plications, and design methods of Intelligent Systems and Intelligent
irtually all disciplines such as engineering, natural sciences, computer
on science, ICT, economics, business, e-commerce, environment,
e science are covered. The list of topics spans all the areas of modern
tems and computing such as: computational intelligence, soft comput-
neural networks, fuzzy systems, evolutionary computing and the fusion
ligms, social intelligence, ambient intelligence, computational neuro-
cial life, virtual worlds and society, cognitive science and systems,
nd Vision, DNA and immune based systems, self-organizing and
tems, e-Learning and teaching, human-centered and human-centric
ecommender systems, intelligent control, robotics and mechatronics
man-machine teaming, knowledge-based paradigms, learning para-
ine ethics, intelligent data analysis, knowledge management, intelligent
igent decision making and support, intelligent network security, trust
, interactive entertainment, Web intelligence and multimedia.

cations within "Advances in Intelligent Systems and Computing" are
oceedings of important conferences, symposia and congresses. They
icant recent developments in the field, both of a foundational and
haracter. An important characteristic feature of the series is the short
time and world-wide distribution. This permits a rapid and broad
on of research results.

**g: The books of this series are submitted to ISI Proceedings,
dex, DBLP, SCOPUS, Google Scholar and Springerlink ****

nation about this series at http://www.springer.com/series/11156

Advances in Intelligent S

Volume 938

Series Editor

Janusz Kacprzyk, Systems Research Institu
Warsaw, Poland

Advisory Editors

Nikhil R. Pal, Indian Statistical Institute, Ko
Rafael Bello Perez, Faculty of Mathematics,
Universidad Central de Las Villas, Santa Cla
Emilio S. Corchado, University of Salamanc
Hani Hagras, Electronic Engineering, Univer
László T. Kóczy, Department of Automation
Gyor, Hungary
Vladik Kreinovich, Department of Computer
at El Paso, El Paso, TX, USA
Chin-Teng Lin, Department of Electrical Eng
Tung University, Hsinchu, Taiwan
Jie Lu, Faculty of Engineering and Informatio
University of Technology Sydney, Sydney, N!
Patricia Melin, Graduate Program of Compute
of Technology, Tijuana, Mexico
Nadia Nedjah, Department of Electronics Engin
Rio de Janeiro, Brazil
Ngoc Thanh Nguyen, Faculty of Computer Sci
Wrocław University of Technology, Wrocław,
Jun Wang, Department of Mechanical and Auto
The Chinese University of Hong Kong, Shatin,

The series "
on theory, a
Computing.
and informa
healthcare, I
intelligent s
ing includin
of these par
science, art
Perception
adaptive sy
computing,
including I
digms, mac
agents, inte
manageme

The pub
primarily I
cover sign
applicable
publicatio
disseminat

** Index
EI-Comp

More info

Zhengbing Hu · Sergey Petoukhov ·
Ivan Dychka · Matthew He
Editors

Advances in Computer Science for Engineering and Education II

 Springer

Editors
Zhengbing Hu
School of Educational Information
Technology
Central China Normal University
Wuhan, Hubei, China

Ivan Dychka
"Igor Sikorsky Kiev Polytechnic Institute"
National Technical University of Ukraine
Kiev, Ukraine

Sergey Petoukhov
Mechanical Engineering Research Institute
of the Russian Academy of Sciences
Moscow, Russia

Matthew He
Halmos College of Natural Sciences
and Oceanography
Nova Southeastern University
Ft. Lauderdale, FL, USA

ISSN 2194-5357   ISSN 2194-5365   (electronic)
Advances in Intelligent Systems and Computing
ISBN 978-3-030-16620-5   ISBN 978-3-030-16621-2   (eBook)
https://doi.org/10.1007/978-3-030-16621-2

Library of Congress Control Number: 2018942171

This Springer imprint is published by the registered company Springer Nature Switzerland AG
The registered company address is: Gewerbestrasse 11, 6330 Cham, Switzerland

Preface

Thanks to computer science, modern engineering and educational technologies receive radically new opportunities. These include the possibility of performing computational experiments, detailed visualization of the objects being constructed and studied, distance learning, fast retrieval of information in huge databases, creation of artificial intelligence systems, etc. Therefore, the problems of computer science and its applications in engineering and education are at the center of attention of governments and scientific–technological communities. Accordingly, higher education institutions face the pressing tasks of educating new generations of specialists who can effectively use and further develop achievements of computer science and their applications.

By these reasons, the National Technical University of Ukraine "Igor Sikorsky Kyiv Polytechnic Institute" and the International Research Association of Modern Education and Computer Science (RAMECS) jointly organized the Second International Conference on Computer Science, Engineering and Education Applications (ICCSEEA2019), January 26–27, 2019, Kiev, Ukraine. ICCSEEA2019 brought together the top researchers from different countries around the world to exchange their research results and address open issues in computer science, engineering, and education applications.

The best contributions to the conference were selected by the program committee for inclusion in this book out of all submissions.

Our sincere thanks and appreciation to the board members as listed below:

Michael Zgurovsky, Ukraine
Yurii Yakymenko, Ukraine
Oleksandr Pavlov, Ukraine
Ivan Dychka, Ukraine
Valeriy Zhuykov, Ukraine
Pavlo Kasyanov, Ukraine
M. He, USA
Georgy Loutskii, Ukraine
Felix Yanovsky, Ukraine

Igor Ruban, Ukraine
Volodymyr Tarasenko, Ukraine
Prof. Janusz Kacprzyk, Poland
Prof. E. Fimmel, Germany
PhD G. Darvas, Hungary
Dr. K. Du, China
Dr. Oleksii K. Tyshchenko, Ukraine
Prof. N. A. Balonin, Russia
Prof. S. S. Ge, Singapore
Prof. A. U. Igamberdiev, Canada
Prof. A. V. Borisov, Russia
Dr. X. J. Ma, China
Prof. S. C. Qu, China
Prof. Y. Shi, USA
Dr. J. Su, China
Dr. O. K. Ban, USA
Prof. J. Q. Wu, China
Prof. A. Sachenko, Ukraine
Prof. Q. Wu, China
Prof. Z. W. Ye, China
Prof. C. C. Zhang, Taiwan
Prof. O. R. Chertov, Ukraine
Prof. N. D. Pankratova, Ukraine.

Finally, we are grateful to Springer-Verlag and Janusz Kacprzyk as the editor responsible for the series "Advances in Intelligent System and Computing" for their great support in publishing this conference proceedings.

January 2019   Zhengbing Hu
                                                              Sergey Petoukhov
                                                                 Ivan Dychka
                                                                 Matthew He

Organization

Honorary Chairs

Michael Zgurovsky NASU Academician, Rector of the National Technical University of Ukraine "Igor Sikorsky Kyiv Polytechnic Institute," Kyiv, Ukraine

Yurii Yakymenko NASU Academician, First Vice Rector of Igor Sikorsky Kyiv Polytechnic Institute, Kyiv, Ukraine

Chairs

Oleksandr Pavlov National Technical University of Ukraine "Igor Sikorsky Kiev Polytechnic Institute," Kyiv, Ukraine

Ivan Dychka National Technical University of Ukraine "Igor Sikorsky Kiev Polytechnic Institute," Kyiv, Ukraine

Valeriy Zhuykov National Technical University of Ukraine "Igor Sikorsky Kiev Polytechnic Institute," Kyiv, Ukraine

Pavlo Kasyanov National Technical University of Ukraine "Igor Sikorsky Kiev Polytechnic Institute," Kyiv, Ukraine

General Co-chairs

M. He Nova Southeastern University, USA

Georgy Loutskii National Technical University of Ukraine "Kyiv Polytechnic Institute," Kyiv, Ukraine

Felix Yanovsky IEEE Fellow, Chair of IEEE Ukraine Section, National Aviation University, Kyiv, Ukraine

Igor Ruban Kharkiv National University of Radio Electronics, Kharkiv, Ukraine

Volodymyr Tarasenko National Technical University of Ukraine "Igor Sikorsky Kyiv Polytechnic Institute," Kyiv, Ukraine

International Program Committee

Janusz Kacprzyk	Systems Research Institute, Polish Academy of Sciences
E. Fimmel	Institute of Mathematical Biology of the Mannheim University of Applied Sciences, Germany
G. Darvas	Institute Symmetrion, Hungary
K. Du	National University of Defense Technology, China
Oleksii K. Tyshchenko	Kharkiv National University of Radio Electronics, Kharkiv, Ukraine
N. A. Balonin	Institute of Computer Systems and Programming, St. Petersburg, Russia
S. S. Ge	National University of Singapore, Singapore
A. U. Igamberdiev	Memorial University of Newfoundland, Canada
A. V. Borisov	Udmurt State University, Russia
X. J. Ma	Huazhong University of Science and Technology, China
S. C. Qu	Central China Normal University, China
Y. Shi	Bloomsburg University of Pennsylvania, USA
J. Su	Hubei University of Technology, China
O. K. Ban	IBM, USA
J. Q. Wu	Central China Normal University, China
A. Sachenko	Ternopil National Economic University, Ternopil, Ukraine
Q. Wu	Harbin Institute of Technology, China
Z. W. Ye	Hubei University of Technology, China
C. C. Zhang	Feng Chia University, Taiwan
O. R. Chertov	National Technical University of Ukraine "Igor Sikorsky Kyiv Polytechnic Institute," Kyiv, Ukraine
N. D. Pankratova	National Technical University of Ukraine "Igor Sikorsky Kiev Polytechnic Institute," Kyiv, Ukraine

Steering Chairs

Vadym Mukhin	National Technical University of Ukraine "Kyiv Polytechnic Institute," Kyiv, Ukraine
Z. B. Hu	Central China Normal University, China

Local Organizing Committee

Oleg Barabash	State University of Telecommunication, Kyiv, Ukraine
Dmitriy Zelentsov	Ukrainian State University of Chemical Technology, Ukraine
Yuriy Kravchenko	Taras Shevchenko National University of Kyiv, Ukraine

Viktor Vishnevskiy	State University of Telecommunication, Kyiv, Ukraine
Igor Parhomey	National Technical University of Ukraine "Kyiv Polytechnic Institute," Kyiv, Ukraine
Elena Shykula	Kyiv State Maritime Academy, Kyiv, Ukraine
Alexander Stenin	National Technical University of Ukraine "Kyiv Polytechnic Institute," Kyiv, Ukraine
Sergiy Gnatyuk	National Aviation University, Kyiv, Ukraine
Yaroslav Kornaga	National Technical University of Ukraine "Kyiv Polytechnic Institute," Kyiv, Ukraine
Yevgeniya Sulema	National Technical University of Ukraine "Igor Sikorsky Kyiv Polytechnic Institute," Kyiv, Ukraine
Andrii Petrashenko	National Technical University of Ukraine "Igor Sikorsky Kiev Polytechnic Institute," Kyiv, Ukraine

Publication Chairs

Z. B. Hu	Central China Normal University, China
Sergey Petoukhov	Mechanical Engineering Research Institute of the Russian Academy of Sciences, Moscow, Russia
Ivan Dychka	National Technical University of Ukraine "Igor Sikorsky Kiev Polytechnic Institute," Kyiv, Ukraine
Oksana Bruy	National Technical University of Ukraine "Igor Sikorsky Kiev Polytechnic Institute," Kyiv, Ukraine

Contents

Perfection of Computer Algorithms and Methods

Computer Science and Education

Computer Science for Manage of Natural and Engineering Processes

Malware Detection Using Artificial Neural Networks

Ivan Dychka[1] , Denys Chernyshev[2] , Ihor Tereikovskyi[1(✉)] ,
Liudmyla Tereikovska[2] , and Volodymyr Pogorelov[3]

[1] National Technical University of Ukraine "Igor Sikorsky Kyiv Polytechnic
Institute", Kyiv, Ukraine
dychka@scs.ntu-kpi.kiev.ua, terejkowski@ukr.net
[2] Kyiv National University of Construction and Architecture, Kyiv, Ukraine
{taqm, tereikovskal}@ukr.net
[3] National Aviation University, Kyiv, Ukraine
volodymyr.pogorelov@gmail.com

Abstract. This paper deals with improvement of malware protection efficiency. The analysis of applied scientific researches devoted to creation of malware protection systems suggests that the improvement of mathematical tools using modern neural network models based on deep neural networks is a promising trend in the development of malware detection systems. Also, the results of analysis have determined the need to create a development method for the deep neural network architecture suitable for use within the modern malware detection means. As part of the study, a method for developing a deep neural network architecture designed to detect malicious software has been suggested. In contrast to the existing methods, it helps avoid long-term numerical experiments to determine the expediency of application of the neural network model and optimize its structural parameters during the development. At the same time, multiple experiments conducted using Microsoft BIG-2015 malware database have shown that the method constructs a neural network model that provides a detection error commensurate with the error of modern malware detection systems. Prospective research is related to the adaptation of the suggested method for the application of deep neural networks in behaviour analysers.

Keywords: Malware · Detection · Neural network model ·
Deep neural network · Neural network architecture

1 Introduction

According to many reputable studies, malware has been among the most dangerous threats in both general and special-purpose modern computer systems for quite a long time [1, 5, 10, 11]. This is suggested by well-known cases of successful cyber-attacks implemented using various malicious programs [12, 15]. A number of experts are occupied with the development of appropriate security tools, but the problem is still far from being resolved. The main difficulties arise when detecting the malware as it penetrates the computer system. For this purpose, modern protection systems use

© Springer Nature Switzerland AG 2020
Z. Hu et al. (Eds.): ICCSEEA 2019, AISC 938, pp. 3–12, 2020.
https://doi.org/10.1007/978-3-030-16621-2_1

solutions related to the artificial neural networks (ANN) theory on a large scale. The future prospects of this trend are confirmed by separate successful applications of ANN in computer virus detection means (open source code ClamAV antivirus, Deep Instinct startup) and a large number of relevant theoretical and practical works, which are summarized in [7, 9, 16]. At the same time, insufficient detection accuracy and poor adaptability to operating conditions, as well as the closed nature of the solutions used limits their scope considerably, while the constant progress of the neural networks theory indicates that significant improvement of proven neural network tools (NNT) for malware detection is possible. This explains the relevance of research for improvement of the existing NNT, which, through the use of modern theoretical solutions, would allow for effective malware detection.

2   Analysis of Literature Sources in the Field of Research

The main objective of the ANN use in malware protection means is malware detection based on the generalization of controlled parameters as shown in case studies [13, 14, 16]. At the same time, the neural network malware detection process typically involves the evaluation of a set of controlled parameter values by the neural network. If an ANN estimation is within a certain range, then malware is considered to be detected, and in case it is out of the range, it is assumed that there is no malware in the computer system. According to the information protection NNT development methodology described in [9], the main trends for increasing the efficiency of such tools are associated with the adaptation of the neural network model (NNM) type and parameters to the expected conditions of use, which are primarily determined by the used set of input parameters.

For example, in [2, 3] the methods for determining the program code fragments to list and assess the ANN input parameter values used in the malware detection systems and in anti-virus protection systems are described. Also, an approach for NNM application based on Kohonen self-organizing topographic map is described. The choice of the NNM type is stated to be due to the implementation of comparative numerical experiments. The training time has been used as a comparison standard. In [9], the description and results of experiments in which ANN was used for malware detection on the basis of a two-layer perceptron are provided. At the same time, neither NNM parameters nor the training procedure have been optimized.

In [5], an approach has been suggested for the determination of NNM input parameters intended for malware detection based on the obfuscated code analysis. The approach assumes the use of theoretical solutions for deobfuscation in order to optimize the code. A procedure for software code deobfuscation using a value/state dependency graph has also been developed. It has been established that the developed procedure represents the functional semantics of the tested programs in the form of a graph. As a result, neural network malware detection on the basis of its implementation semantics has become possible.

[2, 3, 7, 13] offer a computer virus detection system based on the neural network analysis of normalized signatures.

The detection accuracy is declared to be within 80–91%. The possibility of polymorphic viruses detection is indicated. The main difference between the results in

[2, 3, 7, 13] is the use of different approaches for the preliminary processing of ANN input parameters. It should be noted that in [2, 3, 7, 13], no mechanisms for optimizing the two-layer perceptron structure or forming the training sample are given. Also, there are certain doubts as to the expediency of using NNM based on obsolete ANN, including the two-layer perceptron and Kohonen topographic map. It should be noted that modern NNTs are based on NNM of a deep neural network (DNN) type [4, 6–9].

For example, in [8] DNN with autoencoder pre-education has been used. The network consists of 8 layers with 30 neurons each. To obtain a set of DNN input data, a specially designed method for automatic generation of malware signatures has been used. The results of numerical experiments with the detection accuracy reaching 98% are given. However, [8] states that DNN is taught based on unlabelled data only with the help of the autoencoder mechanism. This somewhat reduces the reliability of the results, since it is considered that in order to ensure high detection accuracy, it is also necessary to predict DNN training using the algorithm for reverse error distribution based on labelled training data.

In order to extend the results of the analysis, academic and research papers devoted to the NNT assessment of information systems security parameters have also been considered. For instance, in [4] a basic set of criteria for the effectiveness of the NNM type used to evaluate security parameters has been formed. The ways to expand this set have been specified. In addition, the procedure for NNT development to estimate the security parameters of information system resources, which can be used in the construction of malware detection systems, has been developed in the paper. Also, a method to assess the NNT detection effectiveness for Internet-oriented cyber-attacks, which can also include Internet-oriented malware, has been developed.

In [4], NNM to recognize network cyber-attacks on information resources has been developed. The expediency of DNN application is shown. This is due to the fact that this type of the neural network model is characterized by high learning ability, high computing capabilities and high adaptability to the application environment. The results of experiments based on network cyber-attack detection are shown, with signatures listed in the NSL-KDD database.

Also, in related works [1–12, 15, 16], the method for adapting DNN parameters to the conditions of computer virus recognition was not found.

The analysis suggests that the use of DNN-based neural network detection devices is a promising area for improving the effectiveness of malware protection systems. At the same time, the set of input DNN parameters depends on the features of the detection system. For a behaviour analyser, the list is defined by a set of signs of potentially dangerous functions calls in the application interface of the operating system. For an antivirus scanner, the list of features may correlate with malware signatures. A conclusion can be made that in most relevant academic and research papers, no theoretical justification of using DNN in malware detection means, as well as the justification of DNN architectural parameters adaptation to the expected application conditions, is available.

Therefore, the purpose of this study is to ensure the effectiveness of malware detection systems by adapting the type of deep neural network architecture and architectural parameters to the expected conditions of use.

3 Improvement of Methodological Framework

In accordance with the generally accepted methodology for the development of information protection NNT, elements of the methodological framework for DNN architectural parameters adaptation to the expected application conditions have been developed during the first stage of research.

DNN Type Use Acceptability Principle. Among the available types, i-th DNN type (**DNN**$_i$) is included in the set of admissible types (**DNN**$_{avl}$), provided that its main characteristics ($Q(\mathbf{DNN}_i)$, $\tau(\mathbf{DNN}_i)$) meet the requirements for the allowed time (τ_{avl}) and the permissible resource intensity of NNT construction (Q_{avl}):

$$\text{if } (Q(\mathbf{DNN}_i) \leq Q_{avl}) \& (\tau(\mathbf{DNN}_i) \leq \tau_{avl}) \rightarrow \mathbf{DNN}_i \in \mathbf{DNN}_{avl}. \tag{1}$$

DNN Effectiveness Calculation Principle. The effectiveness of the i-th DNN type (**DNN**$_i$) correlates to the extent, to which this type of DNN meets the basic functional requirements described using the efficiency criteria. The following expression is used to quantify the effectiveness:

$$V(\mathbf{DNN}_i) = \sum_{k=1}^{K} \alpha_k H_k(\mathbf{DNN}_i) \tag{2}$$

where $V(\mathbf{DNN}_i)$ is the value of the efficiency function, $H_k(\mathbf{DNN}_i)$ is the value of the k-th criterion for DNN with the i-th architecture, $\alpha \in [0\dots1]$ is the weight factor of the k-th efficiency criterion, and K is the number of criteria.

DNN Effectiveness Estimation Principle. Among the admissible types, i-th DNN type (**DNN**$_i$) is the most effective, provided that its efficiency function (V_i) has the maximum value:

$$\max_{V_i} = \{V_1, \ V_2, ..V_I\}. \tag{3}$$

The development of the above principles allowed us to suggest a *model for determining the DNN effective types*:

$$\mathbf{DNN}_{ent} \rightarrow \mathbf{DNN}_{avl} \rightarrow \mathbf{DNN}_{eff}$$

where $\mathbf{DNN}_{ent}$ is a set of available DNN types, $\mathbf{DNN}_{avl}$ is a set of admissible DNN types, and $\mathbf{DNN}_{eff}$ is a set of effective DNN types.

Based on theoretical developments related to ANN, the following has been defined:

$$\mathbf{DNN}_{ent} = \{dnn_1, dnn_2, dnn_3, dnn_4\}$$

where dnn_1 is fully connected DNNs for which no pretraining procedure is provided, dnn_2 is fully connected DNNs for which the pretraining procedure is used, dnn_3 is convolutional neural networks (CNN) with direct signal propagation, dnn_4 is recurrent CNN (RCN).

The admissibility condition for dnn_1 and dnn_4 DNN types is defined:

$$if\left(20N_x\left(\vartheta_w + 0,2\lambda\left(N_x + N_y\right)\right) \leq \tau_{avl}\right) \rightarrow \{dnn_1, dnn_4\} \in \mathbf{DNN_{avl}} \qquad (4)$$

where N_x, N_y is the number of DNN input and output parameters, ϑ_w is the average time required to create one case study with the expected output signal, λ is the duration of one training iteration.

The admissibility of dnn_3 DNN type is defined by the following expression:

$$if\left(N_x\left(\vartheta_w + 0,2\lambda\left(N_x + N_y\right)\right) \leq \tau_{avl}\right) \rightarrow dnn_3 \in \mathbf{DNN_{avl}}. \qquad (5)$$

The admissibility of dnn_2 type is defined by the following expression:

$$\begin{aligned} if\big(22, 2N_x\left(\vartheta_n + 0,01\lambda\,N_x\left(N_x + N_y\right)\right) \\ + 200N_x\left(\vartheta_m + 0,2\lambda\left(N_x + N_y\right)\right)\big) \leq \tau_{avl} \rightarrow dnn_2 \in \mathbf{DNN_{avl}} \end{aligned} \qquad (6)$$

where ϑ_n is the average time required to create one case study without the output signal expectation.

Also, the results of [4, 9] have been used to suggested the set of criteria for the effectiveness of the DNN type criteria correlating with the basic requirements for the NNM in the MS recognition task, which is presented in Table 1. The suggested efficiency criteria are dimensionless. The list may be further modified in accordance with the specific conditions of the malware detection task.

Table 1. Criteria for the NNM type effectiveness.

Criterion	Requirement
H_1	Ability to use labelled case studies
H_2	Ability to use unlabelled case studies
H_3	Adaptation for training
H_4	Adaptation for training by individual parts
H_5	Training stability
H_6	Training time minimization
H_7	Computing power maximization
H_8	Ability to take into account the topology of the data analysed
H_9	Decision speed maximization
H_{10}	Adaptability to the dynamic data rows analysis

Similarly to [4, 9], it has been assumed that the values of the suggested criteria can vary within the range of 0 to 1. At the same time, the value of the k-th criterion for the i-th DNN architecture is 1 if the corresponding k-th requirement is fully provided in this architecture, and 0 if it is not provided. See Table 2 for the calculated criteria values for the included DNN.

Table 2. Values of effectiveness criteria for tested DNN types.

Criterion	dnn_1	dnn_2	dnn_3	dnn_4
H_1	1	1	1	1
H_2	0	1	0	0
H_3	0	1	0	0
H_4	0	1	0	0
H_5	1	0.5	1	0.5
H_6	1	0.5	0.5	0.5
H_7	1	0.5	0.5	0.5
H_8	0	0	1	1
H_9	1	1	1	1
H_{10}	0	0	0	0.5

4   Method for DNN Architecture Adaptation

Integration of the common methodology for the development of neural network information security tools with the principles, criteria and model for determining the effective DNN types suggests a method for DNN architecture adaptation to the conditions of the malware detection task that consists of 5 stages.

Step 1.   Formalization of the detection conditions in order to obtain the numerical values of the parameters used in expressions (1–6).

Step 2.   Determination of the expediency of use of the DNN neural network model. In order to do this, the mathematical tools (1, 4–6) should be used.

Step 3.   Determination of the significance of efficiency criteria in Table 2.

Step 4.   Definition of the most effective type of DNN NNM architecture using expressions (2, 3).

Step 5.   Determination of the architectural parameters of the most effective type. It is expedient to use the results of [9, 16] to determine the parameters of fully connected DNNs, and the results of [14, 15] to determine the parameters of CNN and RCN.

The input data for the method are parameters that are characterized by the expected conditions for the DNN use in the malware detection tools, and the output is the type and parameters of the DNN architecture. Let us consider the use of the suggested method on a specific example of the DNN architecture adaptation to the following conditions of application:

– The detection system refers to the commonly used hardware and software;
– NNM is used to recognize Windows-based computer viruses;
– The NNM input is represented by information received during test files scanning;
– The permissible time for NNT creation is 1 month (2,592,000 s);
– Microsoft BIG-2015 computer virus database is used for NNM training and testing.

It should be noted that BIG-2015 Database contains sample signatures of 9 computer viruses, which are characterized in Table 3. The total number of examples is 10,868.

Table 3. BIG-2015 Database characteristics.

Virus name	Number of case studies	Virus type (Microsoft classification)
Ramnit	1541	Worm
Lollipop	2478	Adware
Kelihos_ver3	2942	Backdoor
Vundo	475	Trojan
Simda	42	Backdoor
Tracur	751	TrojanDownloader
Kelihos_ver1	398	Backdoor
Obfuscator.ACY	1228	Any kind of obfuscated malware
Gatak	1013	Backdoor

BIG-2015 Database is generated using the Interactive DisAssembler software package, which removes the metadata related to the Assembler language instructions, the registers contents, and the data and functions imported from DLL from the binary file. The Flirt technology application to the disassembled code detects potentially dangerous functions in partition management, file management, registry operation, system information use etc. The list of such functions in [5, 9] contains 300 names in the first approximation. Based on the scoping of the application conditions, we could proceed to the implementation of the DNN development method.

Step 1. Since the use of available databases is foreseen for the formation of a training sample, it is assumed that the case studies are already formed in the estimation calculations. That is $\vartheta_w = \vartheta_n = 0$. At the same time, $\tau_{avl} = 2,592,000$ s, and the number of case studies P = 10,868. It is assumed that the NNM input parameters correlate with the set of all potentially dangerous functions of the Windows operating system, and the initial parameters are related to the names of the viruses presented in the BIG-2015 Database. Therefore, $N_x = 300$, and $N_y = 9$. Also, expert evaluation suggested that the duration of a single training iteration does not exceed 0.01 s, i. e. *λ = 0.001 s.*

Step 2. By substituting the obtained $N_x, N_y, \vartheta_w, \vartheta_n$ and λ values in (1, 4–6) expressions, the following result is obtained: *τ(dnn₁, dnn₄) = 3708 s,* $\tau(dnn_3) = 185$ s, $\tau(dnn_2) = 98818$ s. Since all these values are less than τ_{avl}, it is considered that the expediency of all DNN types using has been proven.

Step 3. Evaluation of the significance of each effectiveness criterion presented in Table 2, has been implemented with the help of the expert method of pair comparison. The obtained results are shown in Table 4.

Step 4.   By substituting the data from Table 4 for each DNN type into expression (3), the followins values are obtained: $V(dnn_1) = 0.75$, $V(dnn_2) = 0.535$, $V(dnn_3) = 0.725$, $V(dnn_4) = 0.64$. Expression (4) determines that the most effective type is dnn_1.

Step 5.   With known values of N_x and N_y, the dnn_1 basic architectural parameters are the number of hidden neural layers, the number of neurons in each hidden layer, and the type of activation function. Three-layer perceptron with the number of hidden layers $K_h = 2$ has been chosen as dnn_1 basic option

Table 4. Weight factors of the DNN types effectiveness criteria.

α_1	α_2	α_3	α_4	α_5	α_6	α_7	α_8	α_9	α_{10}
0.1	0.02	0.03	0.02	0.2	0.15	0.2	0.15	0.1	0.03

According to the recommendations in [4, 9], the architectural parameters of such a DNN are defined by the following expression:

$$N_h = Round \left(\frac{\sqrt{P \times N_x}}{N_y \times K_h} \right) \tag{7}$$

where N_h is the number of neurons in each hidden layer, K_h is the number of hidden layers; $f(z_k) = max(0, z_k)$ is activation function of the hidden and output layer neurons, where z_k is the total input signal of the k-th neuron in the hidden or output layer. By substituting the known P, N_x, N_y and K_h values into expression (7), $N_h = 135$ is obtained.

Following the determination of architectural parameters we could proceed to the development of relevant software. This was done using the Python programming language and the TensorFlow library (developed by Google). The experiments have been performed on a personal computer (AMD FX-9800P (2.7–3.6 GHz)/RAM 8 GB/HDD 1 TB/AMD Radeon RX 540, 2 GB) with OS Windows 10.

The training had 100 stages. After about 90 training stages, the training error stabilized at the level of 0.01. Subsequently, test samples not used in the training have been submitted to the DNN input from the BIG-2015 Database. Figure 1 shows recognition errors for different viruses.

Analysis of Fig. 1 indicates that the biggest recognition error is typical of Simda, Tracur, and Vundo viruses. This can be explained by a small number of case studies that correspond to these viruses. At the same time, the average error of recognition for all virus types is 0.036. It should also be noted that due to the use of the suggested method during NN development, we managed to avoid long-term numerical experiments aimed at determining the expediency of its application and at optimizing its structural parameters. Considering that the recognition error achieved corresponds to the error of modern antivirus tools [1, 2, 5, 7, 11, 12], this indicates the effectiveness of the suggested solutions.

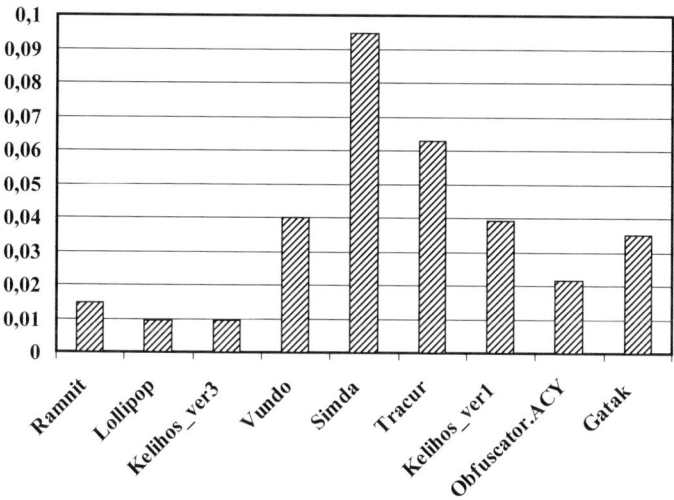

Fig. 1. Recognition error per sample.

5   Conclusions

It has been shown that improving the mathematical support of malware detection systems using modern neural network models based on deep neural networks is a promising area of malware detection systems development. The need for creating a development method for this model, which is adapted to the conditions of application as an anti-virus means, has been determined. The method for developing a deep neural network architecture designed to detect malware has been suggested. In contrast to the existing methods, this method allows avoiding the long-term numerical experiments to determine the expediency of the neural network model application and optimize its structural parameters during its development. At the same time, numerical experiments using the Microsoft BIG-2015 computer virus database show that the method allows for construction of a neural network model that provides a recognition error commensurate with the error of modern computer virus detection systems. Further research is related to the adaptation of the suggested method for the deep neural networks application in behaviour analysers.

References

1. Ahmadi, M., Ulyanov, D., Semenov, S., Trofimov, M., Giacinto, G.: Novel feature extraction, selection and fusion for effective malware family classification. In: Proceedings of the Sixth ACM Conference on Data and Application Security and Privacy (CODASPY 2016), New York, NY, USA, pp. 183–194 (2016)
2. Akhmetov, B.A., Lakhno, V.B., Akhmetov, B.C., Alimseitova, Z.: Development of sectoral intellectualized expert systems and decision making support systems in cybersecurity. In: Advances in Intelligent Systems and Computing, 2nd Computational Methods in Systems and Software (CoMeSySo), vol. 860, pp. 162–171 (2018). https://doi.org/10.1007/978-3-030-00184-1_15

3. Artemenko, A.V., Golovko, V.A.: Analysis of neural network computer virus recognition methods. Molodezhnyiy innovatsionnyiy forum «INTRI», GU «BelISA», Minsk, Belorus, 239 p. (2010). (in Russian)
4. Asiru, O.F., Dlamini, M.T., Asiru, J.M.B.: Application of artificial intelligence for detecting derived viruses. In: 16th European Conference on Cyber Warfare and Security (ECCWS 2017), University College Dublin, Dublin, Ireland, 29–30 June 2017, pp. 217–227 (2017)
5. Bapiyev, I.M., Aitchanov, B.H., Tereikovskyi, I.A., Tereikovska, L.A., Korchenko, A.A.: Deep neural networks in cyber attack detection systems. Int. J. Civil Eng. Technol. (IJCIET) **8**(11), 1086–1092 (2017)
6. Dychka, I., Tereikovskyi, I., Tereikovska, L., Pogorelov, V., Mussiraliyeva, S.: Deobfuscation of computer virus malware code with value state dependence graph. In: Advances in Intelligent Systems and Computing, pp. 370–379 (2018). https://doi.org/10.1007/978-3-319-91008-6
7. Falaye, A.A., Oluyemi, E.S., Victor, A.N., Uchenna, U.C., Ogedengbe, O., Ale, S.: Parametric equation for capturing dynamics of cyber attack malware transmission with mitigation on computer network. Int. J. Math. Sci. Comput. (IJMSC) **3**(4), 37–51 (2017). https://doi.org/10.5815/ijmsc.2017.04.04
8. Hu, Z., Tereykovskiy, I.A., Tereykovska, L.O., Pogorelov, V.V.: Determination of structural parameters of multilayer perceptron designed to estimate parameters of technical systems. Int. J. Intell. Syst. Appl. (IJISA) **9**(10), 57–62 (2017). https://doi.org/10.5815/ijisa.2017.10.07
9. Korchenko, A., Tereykovskiy, I., Karpinskiy, N., Tynymbayev, S.: Neural network models, methods and tools for assessing the security parameters of internet-oriented information systems. In: TOV NashFormat, Kiev, Ukraine, 275 p. (2016). (in Russian)
10. Lakhno, V., Malyukov, V., Parkhuts, L., Buriachok, V., Satzhanov, B., Tabylov, A.: Funding model for port information system cyber security facilities with incomplete Hacker. J. Theor. Appl. Inf. Technol. **96**(13), 4215–4225 (2018)
11. Rudenko, O.H., Bodianskyi, Ye.: Artificial neural networks, Kharkiv, Kompaniia SMIT, 404 p. (2016). (in Ukranian)
12. Shah, S., Jani, H., Shetty, S., Bhowmick, K.: Virus detection using artificial neural networks. Int. J. Comput. Appl. **84**(5), 17–23 (2013)
13. Sharma, S.: Design and implementation of malware detection scheme. Int. J. Comput. Netw. Inf. Secur. (IJCNIS) **10**(8), 58–66 (2018). https://doi.org/10.5815/ijcnis.2018.08.07
14. Sujyothi, A., Acharya, S.: Dynamic malware analysis and detection in virtual environment. Int. J. Modern Educ. Comput. Sci. (IJMECS) **9**(3), 48–55 (2017). https://doi.org/10.5815/ijmecs.2017.03.06
15. Tahir, R.: A study on malware and malware detection techniques. Int. J. Educ. Manag. Eng. (IJEME) **8**(2), 20–30 (2018). https://doi.org/10.5815/ijeme.2018.02.03
16. Tereykovska, L., Tereykovskiy, I., Aytkhozhaeva, E., Tynymbayev, S., Imanbayev, A.: Encoding of neural network model exit signal, that is devoted for distinction of graphical images in biometric authenticate systems. News Natl. Acad. Sci. Repub. Kaz. Ser. Geol. Tech. Sci. **6**(426), 217–224 (2017)

Applying Wavelet Transforms for Web Server Load Forecasting

Zhengbing Hu[1] ⓘ, Ihor Tereikovskyi[2] ⓘ, Liudmyla Tereikovska[3] ⓘ,
Mikola Tsiutsiura[3] ⓘ, and Kostiantyn Radchenko[2](✉) ⓘ

[1] School of Educational Information Technology,
Central China Normal University, Wuhan, China
hzb@mail.ccnu.edu.cn

[2] National Technical University of Ukraine "Igor Sikorsky Kyiv Polytechnic
Institute", Kyiv, Ukraine
terejkowski@ukr.net, radche00l@gmail.com

[3] Kyiv National University of Construction and Architecture, Kyiv, Ukraine
tereikovskal@ukr.net, teodenor@gmail.com

Abstract. The study focuses on increasing the effectiveness of web server load forecasting systems, which are utilized for technical state diagnostics and ensuring data security of distributed computer systems and networks. The analysis of applied research papers has shown that a promising way of web server load forecasting systems development is improving their mathematical background by using modern frequency-time signal analysis methods based on the wavelet transformation theory. It has been established that the challenges of using the wavelet transformation theory are primarily related to the choice of basis wavelet type, parameters of which shall be adapted to the application conditions in a particular forecasting system. A new basis wavelet type selection method, which is most effective for web server load parameters forecasting, has been proposed. This method is based on a series of conditions and criteria to achieve significant effectiveness for the given forecasting task by choosing a basis wavelet type. Also, simulation modelling based on the web server request statistics collected by the authors in a Ukrainian university has shown that the method allows selecting the basis wavelet type, which ensures the approximation error level similar to that of the modern web server load forecasting systems. The possibility to avoid long-term simulation modelling typically used for basis wavelet type selection is a significant advantage of the proposed method for wavelet model development. Prospects for further research are related to the refinement of the effectiveness criteria calculation process and improvement of the proposed method by developing a basis wavelet parameters calculation procedure.

Keywords: Wavelet model · Forecasting · Web server · Load ·
Mother wavelet

© Springer Nature Switzerland AG 2020
Z. Hu et al. (Eds.): ICCSEEA 2019, AISC 938, pp. 13–22, 2020.
https://doi.org/10.1007/978-3-030-16621-2_2

1 Introduction

Due to the intensive growth of distributed computer network systems, the challenge of ensuring their operation reliability becomes more and more important and urgent as an increasing number of such systems are being integrated in various branches of human activity. The structure of virtually all modern computer network systems includes one or more servers to ensure integration with the global computer network, Internet. Typical services provided by web servers include WWW, FTP and e-mail. Practical experience has shown repeated disruptions of networked computer systems operation due to web server software failures, which are both due to excessive server loads and successful DDoS attacks [2, 4, 6]. It is obvious that the prevention of such violations is possible by developing effective tools for web server load forecasting tools. Many researchers [1, 3, 8, 9, 16] suggest that the effectiveness of such tools cannot be increased without developing new methods and models to create sufficiently reliable forecasts of operational parameters.

It must be noted that the created forecasts are used as a base for both network resources planning and determining the control actions applied to them. Therefore, the computer networks operation efficiency directly depends on the reliability of operational parameters forecasting methods and models, the development of which is a general subject of this article.

2 Analysis of Literature Sources in the Field of Research

Based on the general model development methodology for evaluating performance parameters of web-oriented information systems, it has been established that the architecture of this model directly depends on many parameters that determine web server load. Considering the typical web server tasks, as well as the network communication protocols used, three groups of parameters can be defined, each characterizing the utilization of:

- Server computer hardware resources.
- Operating system resources.
- Network resources.

As a rule, all aforementioned groups are registered in operation via commonly used software and hardware. The earlier studies [2, 4, 12, 14] suggested forecasting these parameters over time using forecast models described by analytical expressions as follows:

$$Z = \alpha + \beta t, \tag{1}$$

$$Z = \sum_{h=0}^{H} \left(\alpha_h + t^h \right), \tag{2}$$

$$Z = \alpha \exp(\lambda t), \tag{3}$$

$$Z_{t+1} = \frac{Z_t + Z_{t-1} + \ldots + Z_{t-N+1}}{N}, \tag{4}$$

$$A_{t+1} = \omega \times Z_t + (1 - \omega) \times A_{t-1} \tag{5}$$

where Z is the parameter value, t is time, α, β are regression coefficients, K is the approximating polynomial power, α_i is the i-th coefficient of the approximating polynomial, λ is the coefficient, Z_t is the parameter value at time t, N is the sliding window size, A_t is the smoothed parameter value at time t, and w is the exponential smoothing coefficient.

These models have names as follows: 1 - linear, 2 - polynomial, 3 - exponential, 4 - sliding average, 5 - exponential smoothing. Their main advantage is the mathematical apparatus simplicity. At the same time, the practical use has revealed a major drawback of these models. In the case of a long-term load forecast, models (1–5) do not adequately factor in the non-stationary and multi-periodic nature of typical time dependencies of most web server operational parameters [2, 4, 12, 14].

In addition, the analysis of models (1–5) suggests with a high degree of certainty that certain periodic components occur at time $t \neq 0$. This fact, as well as conclusions [4, 6], indicate that the forecast model for web server load parameters shall adequately include the non-stationary time-frequency nature of these parameters.

According to a number of studies [6, 8, 9, 16], this requirement can be fulfilled by applying elements of the wavelet transform theory to the forecast model.

Studies [8, 9] propose a prediction method based on the integration of wavelet analysis and neural network in order to improve the accuracy of web traffic predicates. A non-linear and non-stationary time series of web traffic have been decomposed and then rebuilt into several branches using the wavelet method. These branches were forecasted by neural networks, respectively, and the final value was a combination of these forecasted results. The theoretical analysis and experimental results show that the wavelet analysis can decompose the initial traffic series into several time series, which have simpler frequency components and are easier to forecast. Therefore, it has been shown that the developed method has higher forecast accuracy than conventional forecasting approaches.

Study [16] presents a server load forecasting model based on the analysis of wavelet packets. The model assumes that server load time series is decomposed and reconstructed using wavelet packet analysis to obtain multiple server branches with the same historical series length. The proposed method has been proven experimentally to be superior to the conventional forecasting approach.

An approach for predicting short-term load was developed in [6]. This approach uses vector machines (LS-SVM) and the wavelet transform (WT) theory. The historical time series is analysed using wavelet transforms with the results being forecasted using a single LS-SVM predictor. The new forecast model combines the advantages of WT and LS-SVM. As compared to other predictors, this forecasting model is declared to have a greater generalizing ability and higher accuracy.

The conducted analysis suggests that application of wavelet transform theory is a promising approach for increasing effectiveness of web server load forecasting systems. However, it can be argued that modern wavelet systems pay insufficient attention to questions related to mother wavelet parameters selection. Also in related works the method of choosing the most effective base wavelet was not found [1–16]. At the same time, the wavelet transform theory suggests that the wrong choice of these parameters can greatly reduce the effectiveness of the forecasting system. Therefore, the **goal of the study** is to develop a new basis wavelet type selection method, which is most effective for web server load parameters forecasting.

3   Basic Principles of the Method

Similarly to the well-known methods for developing tools for computer system parameters forecasting [1, 2, 13, 15], the basis wavelet type selection method being developed is based on the following provisions:

- Criteria for selecting the most effective basis wavelet type should reflect the extent of its suitability for the given forecasting task.
- The k-th criterion for determining the most effective basis wavelet type means the degree, in which this type of wavelet fulfils the k-th requirement of the forecasting task.
- Calculation of the i-th basis wavelet type effectiveness can be represented in the following form:

$$R_i = \sum_{k=1}^{K} \alpha_k r_k(i) \tag{6}$$

where R_i is the index of the integral effectiveness of the i-th basis wavelet type, $r_k(i)$ is the k-th effectiveness criterion of the i-th type of the mother wavelet, α_k is the weight coefficient of the k-th effectiveness criterion, K is the number of effectiveness criteria.

- The most effective basis wavelet type is determined by the following expression:

$$R_{eff} = \max\{R\}_I \tag{7}$$

where $\{R\}_I$ is the set of parameters of basis wavelet types integral effectiveness, I is the cardinal of the set.

4   Effectiveness Criteria for Mother Wavelets

Stating the basic provisions of the basis wavelet selection method development allows proceeding to the next stage of research, which is the development of a set of effectiveness criteria used in expression (6). For this purpose, the characteristics of the most proven basis wavelet types have been analysed with the results listed in Table 1. The analysis is based on theoretical studies [5, 7, 10–12].

Table 1. Tested basis wavelet types.

Basis wavelet type name	Basic wavelet type analytic form	Basis type
WAVE wavelet	$\psi(t) = -t \exp(-\frac{t^2}{2})$	Real continuous
MHAT	$\psi(t) = (1 - t^2) \exp(-\frac{t^2}{2})$	
Nth order Gaussian	$\psi_n(t) = (-1)^n \frac{\partial^n}{\partial t^n} (\exp(-\frac{t^2}{2}))$	
DOG wavelet	$\psi(t) = \exp(-\frac{|t|^2}{2}) - \frac{1}{2}\exp(-\frac{|t|^2}{8})$	
LP wavelet	$\psi(t) = (\pi t)^{-1}(\sin(2\pi t) - \sin(\pi t))$	
Daubechies wavelet	$\psi(t) = \sqrt{2} \sum_k g_k \phi(2t - k)$	
HAAR	$\psi(t) \geq \begin{cases} 1, & 0 \leq t < 1/2 \\ -1, & 1/2 \leq t < 1 \\ 0, & t < 0, t \geq 1 \end{cases}$	Real discrete
FHAT	$\psi(t) \geq \begin{cases} 1, & |t| \leq 1/3 \\ -1/2, & 1/3 < |t| \leq 1 \\ 0, & |t| > 1 \end{cases}$	
Morlet	$\psi(t) = \exp(ik_0 t) \exp(-\frac{t^2}{2})$	Complex
Paul	$\psi(t) = \Gamma(n+1) \frac{i^n}{(1-it)^{n+1}}$	

As a result of the analysis, a number of effectiveness criteria have been developed in the first approximation (see Table 2).

Table 2. Basis wavelet type effectiveness criteria.

Criterion	Description
r_1	Redundant information present in wavelet coefficients
r_2	No redundant information present in wavelet coefficients
r_3	Fast wavelet transform can be applied
r_4	Infinite regularity present
r_5	Random regularity present
r_6	Basis function is symmetrical
r_7	Basis function is asymmetrical
r_8	Basis function is orthogonal
r_9	Scaling function present
r_{10}	Full signal recovery is possible
r_{11}	Basis function is compact
r_{12}	Basic function geometry is similar to analysed process geometry

By analogy with [13, 15], it was assumed that the values of the proposed criteria from r_1 to r_{11} in the first approximation can be estimated on a two-point discrete scale. In this case, the value of the k-th criterion for i-th basis wavelet type is equal to 1, if the corresponding k-th requirement is fully fulfilled, and equals 0 if not fulfilled. In the future, the criteria can be calculated more accurately in the range from 0 to 1. See Table 3 for the calculated criteria values for tested basis wavelet types.

Table 3. Criteria values for tested basis wavelet types.

Basis wavelet type	r_1	r_2	r_3	r_4	r_5	r_6	r_7	r_8	r_9	r_{10}	r_{11}
WAVE wavelet	0	1	0	1	0	1	0	0	0	1	0
MHAT	0	1	0	1	0	1	0	0	0	1	0
Nth order Gaussian	0	1	0	0	1	0	1	0	0	0	0
DOG wavelet	0	1	0	0	1	0	1	0	0	1	0
LP wavelet	0	1	0	0	1	0	1	0	0	0	0
Daubechies wavelet	1	0	1	0	1	0	1	1	1	1	1
HAAR	1	0	1	0	1	0	1	1	1	1	1
FHAT	1	0	1	0	1	1	0	0	0	1	0
Morlet	0	1	0	1	0	1	0	0	0	1	0
Paul	0	1	0	1	0	1	0	0	0	1	0

It should be noted that a priori it is difficult to determine the value of the efficiency criterion r_{12}, which determines the similarity of the basis function geometry compared to the analysed process geometry. Therefore, Table 3 doesn't contain this value.

5   Basis Wavelet Type Selection Method

The use of the developed principles for determining the most effective basis wavelet type and the results of the research related to the creation of effective neural network models for parameter estimation made it possible to propose a basis wavelet type selection method for web server load forecasting.

Step 1.   Formalization of the forecasting task conditions in order to determine the type and values of parameters that describe web server load.

Step 2.   Determining the value of r_{12} criterion for each of the tested basis wavelet types. The determination involves estimating the similarity of the basis wavelet shape geometry and the forecasted process geometry.

Step 3.   Determining the significance of each of the effectiveness criteria presented in Table 2. This may be performed via expert evaluation.

Step 4.   Calculation of the effectiveness of each basis wavelet type. Expression (6) shall be used.

Step 5.   Determining the most effective basis wavelet type using expression (7).

Let us consider using the proposed method on a specific basis wavelet type selection example for web server load forecasting.

Step 1.   Let us evaluate the web server load using the request statistics (graph in Fig. 1). Therefore, the forecasting function is described by expression $K = f(t)$, where

K is the number of web server requests during the day, and t is the time. It should be noted that the statistics collected by the authors corresponds to the number of web server requests at a Ukrainian university in 2018.

Step 2.   For each of the tested basis wavelet types, the r_{12} criterion value is obtained by comparing the geometry of the corresponding function with the request statistics graph shown in Fig. 1. The calculated values are listed in Table 4.

Table 4. r_{12} criterion values for tested basis wavelet types.

Basis type	WAVE wavelet	MHAT	Nth order Gaussian	DOG wavelet	LP wavelet
r_{12} value	0.4	0.5	0.5	0.5	0.5
Basis type	FHAT	HAAR	Daubechies wavelet-4	Morlet	Paul
r_{12} value	0.5	0.3	0.9	0.3	0.5

Step 3.   The significance of each effectiveness criteria in the given task was evaluated using the expert method of pair comparison. The results are listed in Table 5.

Table 5. Weighting factors of basis wavelet type effectiveness criteria.

r_1	r_2	r_3	r_4	r_5	r_6	r_7	r_8	r_9	r_{10}	r_{11}	r_{12}
0.15	0.01	0.07	0.06	0.06	0.06	0.06	0.06	0.06	0.15	0.06	0.2

Step 4.   The integral effectiveness indicator for each of the tested basis wavelet types is calculated using the data from Tables 3, 4 and 5 in expression (6). The results are listed in Table 6.

Table 6. Integral effectiveness indicator values for tested basis wavelet types.

Basis type	Wave wavelet	MHAT	Nth order Gaussian	DOG wavelet	LP wavelet
R value	0.36	0.38	0.23	0.38	0.23
Basis type	FHAT	HAAR	Daubechies wavelet-4	Morlet	Paul
R value	0.59	0.73	0.85	0.34	0.38

Step 5.   Expression (7) indicates that Daubechies wavelet-4 is the most effective wavelet.

Following determination of the most effective basis wavelet type, simulation modelling was conducted to verify the obtained solutions by checking the statistical data approximation accuracy and the possibility of using wavelet coefficients to determine the local characteristics of the load process.

Calculations were performed using Mathcad. The results of the analysis are presented in Fig. 1 as wavelet spectrum graphs.

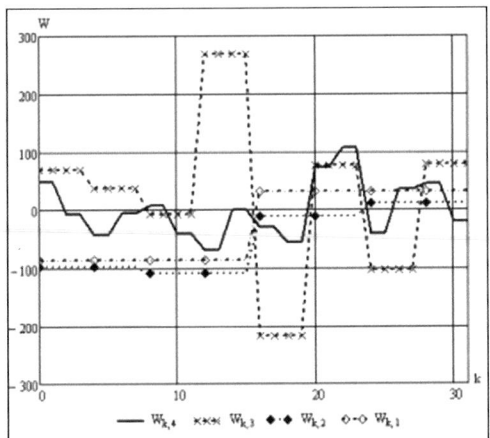

Fig. 1. The results of the analysis of tested wavelet types. Web server traffic wavelet spectrum

The analysis of data shown in Fig. 1 indicates that the analysed process has 4 periodic components, which are equal to 2, 4, 8 and 16 days. The third component seems to occur approximately on the 14–16th day of operation, and the rest do not have clear time localization.

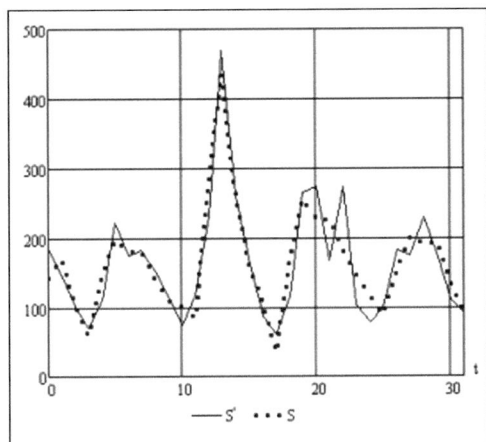

Fig. 2. The results of the analysis of tested wavelet types. Graphs of the original (S) and recovered signal (S')

In addition, the inverse discrete wavelet transform of the signal was performed using a set of wavelet coefficients. Figure 2 shows graphs of the recovered and the original signal.

A sufficiently high similarity of the indicated graphs can be noted. Additionally, the average approximation error does not exceed 5%. At the same time, the average approximation error of other basis wavelet types is approximately 10%. It can be noted that, as shown by studies [6, 8, 9, 16], the approximation error for such statistics is 5–7%.

Therefore, the results of the experiments confirm the possibility of applying the developed method, the prospects for improvement of which lie in refining the calculation of effectiveness criteria and their weights. In addition, the proposed method can be further elaborated in basis wavelet parameters calculation.

6 Conclusions

It has been shown that a promising way of web served load forecasting systems development is improving their mathematical background by using modern frequency-time signal analysis methods based on the wavelet transformation theory. It has been established that the challenges of using the wavelet transformation theory are primarily related to the choice of basis wavelet type, parameters of which shall be adapted to the application conditions in a particular forecasting system. A new basis wavelet type selection method, which is most effective for web server load parameters forecasting, has been proposed. This method is based on a series of conditions and criteria used to achieve significant efficiency for a given forecasting task by choosing a basis wavelet type. In addition, simulation modelling based on the web server request statistics collected by the authors in a Ukrainian university, was used to show that the method allows selecting the basis wavelet type, which ensures the approximation error level similar to that of the modern web server load forecasting systems. The possibility to avoid long-term simulation modelling typically used for basis wavelet type selection is a significant advantage of the proposed method for wavelet model development. Prospects for further research are related to the refinement of the efficiency criteria calculation process and improvement of the proposed method by developing a basis wavelet parameters calculation procedure.

References

1. Bapiyev, I.M., Aitchanov, B.H., Tereikovskyi, I.A., Tereikovska, L.A., Korchenko, A.A.: Deep neural networks in cyber attack detection systems. Int. J. Civil Eng. Technol. (IJCIET) **8**(11), 1086–1092 (2017)
2. Chen, Q.-S., Zhang, X., Xiong, S.-H., Chen, X.-W.: Short-term power load forecasting with least squares support vector machines and wavelet transform. In: International Conference on Machine Learning and Cybernetics, vol. 3, pp. 1425–1429 (2008)
3. Dychka, I., Tereikovskyi, I., Tereikovska, L., Pogorelov, V., Mussiraliyeva, S.: Deobfuscation of computer virus malware code with value state dependence graph. In: Advances in Intelligent Systems and Computing, pp. 370–379 (2018). https://doi.org/10.1007/978-3-319-91008-6

4. Hu, Z., Tereykovskiy, I.A., Tereykovska, L.O., Pogorelov, V.V.: Determination of structural parameters of multilayer perceptron designed to estimate parameters of technical systems. Int. J. Intell. Syst. Appl. (IJISA) **9**(10), 57–62 (2017). https://doi.org/10.5815/ijisa.2017.10.07

5. Kavitha, K.J., Shan, B.P.: Reversible joint watermarking for medical images and videos. Int. J. Eng. Manuf. (IJEM) **8**(5), 10–21 (2018). https://doi.org/10.5815/ijem.2018.05.02

6. Pereberin, A.V.: On the classification of wavelet transforms. Vyich. met. programmirovanie **2**(3), 15–40 (2001). (in Russian)

7. Quadry, K.M., Govardhan, A., Misbahuddin, M.: A novel approach of kurtosis based watermarking by using wavelet transformation. Int. J. Image Graph. Sig. Process. (IJIGSP) **10**(7), 42–50 (2018). https://doi.org/10.5815/ijigsp.2018.07.05

8. Savakar, D.G., Pujar, S.: Digital image watermarking using DWT and FWHT. Int. J. Image Graph. Sig. Process. (IJIGSP) **10**(6), 50–67 (2018). https://doi.org/10.5815/ijigsp.2018.06.06

9. Starck, J.-L., Murtagh, F., Bijaoui, A.: Image and Data Analysis: The Multiscale Approach. Cambridge University Press, Great Britain, 307 p. (1998)

10. Steinbuch, M., van de Molengraft, M.J.G.: Wavelet Theory and Applications: A Literature Study. Eindhoven University of Technology, Netherlands, 39 p. (2005)

11. Takore, T.T., Kumar, P.R., Devi, G.L.: A new robust and imperceptible image watermarking scheme based on hybrid transform and PSO. Int. J. Intell. Syst. Appl. (IJISA) **10**(11), 50–63 (2018). https://doi.org/10.5815/ijisa.2018.11.06

12. Tereykovska, L., Tereykovskiy, I., Aytkhozhaeva, E., Tynymbayev, S., Imanbayev, A.: Encoding of neural network model exit signal, that is devoted for distinction of graphical images in biometric authenticate systems. News Natl. Acad. Sci. Repub. Kaz. Ser. Geol. Tech. Sci. **6**(426), 217–224 (2017)

13. Yakuben, M.B.: Detection of cyber attacks by searching for anomalies based on probabilistic and verification modelling. Shtuchnyj intelekt **3**, 679–687 (2005). (in Russian)

14. Yang, Z.: Research on server load prediction based on wavelet packet theory. In: First IEEE International Symposium on Information Technologies and Applications in Education, pp. 610–613 (2007). https://doi.org/10.1109/isitae.2007.4409360

15. Yao, S., Hu, C., Peng, W.: Server load prediction based on wavelet packet and support vector regression. In: International Conference on Computational Intelligence and Security, vol. 2, pp. 1016–1019 (2006)

16. Yao, S., Hu, C., Sun, M.: Prediction of web traffic based on wavelet and neural network. In: 6th World Congress on Intelligent Control and Automation, vol. 1, pp. 4026–4028 (2006)

On the Calculation by the Method of Direct Linearization of Mixed Oscillations in a System with Limited Power-Supply

Alishir A. Alifov[(✉)]

Mechanical Engineering Research Institute of the Russian Academy of Sciences, 4, Maly Kharitonievskiy Pereulok, Moscow 101990, Russia
a.alifov@yandex.ru

Abstract. The calculation procedure using the method of direct linearization of mixed self-oscillations, forced and parametric oscillations in a system with limited power-supply is considered. On the basis of this method, the equations of non-stationary and stationary motions are derived. Using these equations and the Routh-Hurwitz criteria, the conditions for the stability of steady-state oscillations are obtained. Calculations were carried out to obtain information on the amplitude-frequency dependence and stability of oscillations in order to compare with the results obtained by the known methods of nonlinear mechanics. These calculations show that the results based on the method of direct linearization are qualitatively completely similar to the results obtained using the well-known methods of nonlinear mechanics, there are only fairly small quantitative differences. At the same time, in contrast to the known methods of nonlinear mechanics, the use of the method of direct linearization is quite simple, it takes much less time and labor spent.

Keywords: Mixed oscillations · Self-oscillations · Forced oscillations · Parametric oscillations · Limited excitation · Imperfect energy source · Method · Direct linearization

1   Introduction

The models of linear and nonlinear oscillatory systems, methods and technics of their calculation are considered in many papers, for example, [1–18]. As is known from the theory of oscillations, oscillatory processes are divided into 4 types: free oscillations; forced oscillations; parametric oscillations; self-oscillations. The mixed oscillations (4 classes) representing the "mixture" of these types of vibrations are more complex than the first ones. The most complex class of mixed oscillations, both in mathematical and physical terms, is "mixed self-oscillations, forced oscillations, parametric oscillations". This class, as shown in [1, 2], is considered in a very small number of papers, and in the case of an ideal source of energy. These mixed oscillations, both in the case of an ideal and nonideal energy source, are studied in [1].

The well-known Sommerfeld effect, experimentally discovered at the beginning of the last century (1902), led to the emergence of a new direction in the theory of oscillations - the theory of oscillatory systems with limited power-supply or non-ideal

© Springer Nature Switzerland AG 2020
Z. Hu et al. (Eds.): ICCSEEA 2019, AISC 938, pp. 23–31, 2020.
https://doi.org/10.1007/978-3-030-16621-2_3

sources of energy. Its systematic basis was laid by the works of Kononenko, V.O., who considered all the above types of oscillations, taking into account the limited power of the energy source [3]. This direction of the theory of oscillations has been further developed in the papers of Kononenko, V.O. and his followers, and at present it has acquired particular relevance in connection with energy and environmental problems.

The main method for studying oscillatory systems with limited excitation is the Krylov-Bogolyubov method for constructing approximate analytical solutions of non-linear differential equations with a small non-linearity. Both this method and other well-known methods for studying nonlinear systems (energy balance, harmonic lin-earization, etc.) require significant labor spent, time, etc. These disadvantages are significantly reduced using the methods of direct linearization described in [4–8] which significantly ease the calculation of the parameters of various technical objects. In this context, we note the work [9], in which, with reference to the works [10–13], it is indicated that one of the main problems of nonlinear system dynamics is the high labor spent for analyzing connected oscillator networks, which play an important role in biology, chemistry, physics, electronics, neural networks, etc. The purpose of the article is to develop the procedure for calculating nonlinear oscillatory systems with limited excitation based on direct linearization methods. Its structure includes the following sections: system model; replacing nonlinear functions with linear ones using the direct linearization method; solving linearized equations; stability conditions for stationary motions; calculations; conclusion.

2   System Model

Consider mixed self-oscillations, forced and parametric oscillations in a system with limited excitation using the example of a model (see Fig. 1) of a friction self-oscillating system [1, 3], which is under external influences: the driving force $\lambda \sin v_2 t$ and parametric $bx \cos v_1 t$ disturbance. The functioning of the system is supported by the engine (energy source) with the torque characteristic of the $M(\dot{\varphi})$. Self-oscillations are caused by the nonlinear friction force $T(U)$, depending on the relative velocity $U = r\dot{\varphi} - \dot{x}$, at the point of contact of the body of mass m with the tape.

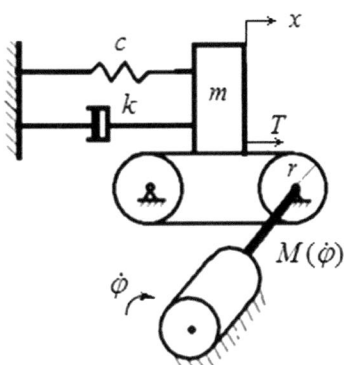

Fig. 1. System model

The equations of motion of the system are

$$m\ddot{x} + k\dot{x} + cx + F(x) = T(U) + \lambda \sin v_2 t - bx \cos v_1 t,$$
$$J\ddot{\varphi} = M(\dot{\varphi}) - rT(U) \tag{1}$$

where k is the damping coefficient, c is the stiffness coefficient, $F(x)$ is the nonlinear part of the spring force of the spring, λ and v_2 is the amplitude and frequency of the driving force, b and v_1 respectively, the modulation depth and frequency of the parametric effect ($b \ll c$), J is the total moment of inertia of engine parts rotating at a velocity $\dot{\varphi}$, r is the radius of the point of application of the friction force $T(U)$.

3   Replacing Nonlinear Functions with Linear Linearization Method

Nonlinear characteristics of forces in practice are described in most cases by polynomial functions. In this regard, we represent the polynomial functions of nonlinear forces

$$F(x) = \sum_s \gamma_s x^s, \ T(U) = R\bar{f}(U), \bar{f}(U) = \text{sgn}U + \sum_i \delta_i U^i. \tag{2}$$

Here $\gamma_s = const$, $\delta_i = const$, $s = 2, 3, 4, \ldots, i = 1, 2, 3, \ldots, R$ is the normal reaction force, $\bar{f}(U)$ is the coefficient of friction, sgn $U = 1$ at $U > 0$ and sgn $U = -1$ at $U < 0$. Since $U = r\dot{\varphi} - \dot{x}$, then we transform the expression $\bar{f}(U)$ and write in the form $\bar{f}(U) = \text{sgn } U + f(\dot{x}), f(\dot{x}) = \sum_{n \geq 0} \alpha_n \dot{x}^n$, where the coefficients α_n depend on δ_n and $\dot{\varphi}$.

Using the direct linearization method [5], we now replace functions (2) with linear functions

$$F_*(x) = B_F + k_F x, \ f_*(\dot{x}) = B_f + k_f \dot{x}. \tag{3}$$

The linearization coefficients B_F, k_F, B_f, k_f included in (3) are determined by the expressions

$$B_F = \sum_s \gamma_s N_s a^s, \ \ s = 2, 4, 6, \ldots (s \text{ is even number}), \tag{4}$$

$$k_F = \sum_s \gamma_s \bar{N}_s a^{s-1}, \ \ s = 3, 5, 7, \ldots (s \text{ is odd number}),$$

$$B_f = \sum_n \alpha_n N_n v^n, \ \ n = 0, 2, 4, \ldots (n \text{ is even number}),$$

$$k_f = \sum_n \alpha_n \bar{N}_n v^{n-1}, \ \ n = 1, 3, 5, \ldots (n \text{ is odd number}).$$

Here $a = \max|x|$, $\upsilon = \max|\dot{x}|$, $N_s = (2r+1)/(2r+1+s)$, $\bar{N}_s = (2r+3)/$
$(2r+2+s)$, $N_n = (2r+1)/(2r+1+n)$, $\bar{N}_n = (2r+3)/(2r+2+n)$. Regardless of
the value of the parameter r, on which the accuracy of linearization depends, there are
$N_s = 1$, $N_n = 1$ for $s = n = 0$ and $\bar{N}_s = 1$, $\bar{N}_n = 1$ for $s = n = 1$. Although the value
of r can be different, it is enough to choose from the interval $(0, 2)$, as shown in [5].
Equations (1) in view of (3) take the form

$$m\ddot{x} + k_0\dot{x} + (\bar{c} + b \cos v_1 t) x = B + \lambda \sin v_2 t + R \operatorname{sgn} U,$$
$$J\ddot{\varphi} = M(\dot{\varphi}) - r_0 R(\operatorname{sgn} U + B_f + k_f\dot{x})$$

(5)

where $k_0 = k - Rk_f$, $\bar{c} = c + k_F$, $B = RB_f - B_F$.

4 Equation Solutions

To reflect the captured (resonant) oscillations caused by the simultaneous fulfillment of
certain relations between the frequency of self-oscillations and the frequencies of both
effects, the condition of synchronism of the resonant frequencies of external and
parametric effects was introduced in [1]. Since, in practice, the main interest is the main
resonances from the influence of both effects, in accordance with this condition, we
have for the main resonances $v_2 = v_1/2$, for which we give solutions (6a, 6b).

In accordance with the method of replacing variables with averaging [5], we have
$x = a \cos \psi$, $\psi = pt + \xi$, $p = v_1/2$, and we can immediately write down solutions for
determining the non-stationary values of the amplitude a and the phase ξ of the
oscillations:

(a) $u \geq ap$

$$\frac{da}{dt} = -\frac{k_0 a}{2m} - \frac{1}{2pm}(\lambda \cos \xi - \frac{ba}{2}\sin 2\xi),$$
$$\frac{d\xi}{dt} = \frac{\omega_0^2 - p^2}{2p} + \frac{k_F}{2pm} + \frac{1}{4pam}(2\lambda \sin \xi + ba \cos 2\xi);$$

(6a)

(b) $u < ap$

$$\frac{da}{dt} = -\frac{a}{2m}\left[k_0 + \frac{4R}{\pi a^2 p^2}\sqrt{a^2 p^2 - u^2}\right] - \frac{1}{2pm}(\lambda \cos \xi - \frac{ba}{2}\sin 2\xi),$$
$$\frac{d\xi}{dt} = \frac{\omega_0^2 - p^2}{2p} + \frac{k_F}{2pm} + \frac{1}{4pam}(2\lambda \sin \xi + ba \cos 2\xi)$$

(6b)

where $\omega_0^2 = c/m$.

Under the conditions $\dot{a} = 0$, $\dot{\xi} = 0$ from (6a, 6b), the equations for calculating the
stationary values of a and ξ follow. Since in the resonance region the frequency
difference $\omega_0 - p$ is rather small, it is possible to accept $(\omega_0^2 - p^2)/2p \approx \omega_0 - p$. As

shown in [1] and other works of the author analytically and by modeling on an analog computer, the characters of $x(t)$, $\dot{x}(t)$ solutions at $u \geq ap$ and $u < ap$ speeds are different. The amplitude in the case of $u < ap$ and small external influences can be calculated by the approximate equality $ap \approx u$ or $a\omega_0 \approx u$.

The procedure for applying direct linearization methods for calculating oscillatory systems with limited excitation, described in [7, 8], based on extracting the main part of the Ω solution and discarding small vibrational components ε_{vibr} in the expression $\dot{\varphi} = \Omega + \varepsilon_{vibr}$, allows us to find the equations from (5)

(a) $u \geq ap$

$$\frac{du}{dt} = \frac{r_0}{J}\left[M\left(\frac{u}{r}\right) - r_0 R(1 + B_f)\right]; \tag{7a}$$

(b) $u < ap$

$$\frac{du}{dt} = \frac{r_0}{J}\left[M\left(\frac{u}{r}\right) - r_0 R(1 - B_f) - \frac{r_0 R}{\pi}(3\pi - 2\psi_*)\right], \tag{7b}$$

whence for stationary movements we have the relation

$$M(u/r) - S(u) = 0, \tag{8}$$

wherein
(a) $u \geq ap$ $S_+(u) = r_0 R(1 + B_f)$;
(b) $u < ap$ $S_-(u) = r_0 R\left[(1 - B_f) + \pi^{-1}(3\pi - 2\psi_*)\right]$

where $u = r_0 \Omega$, $\psi_* = 2\pi - \arcsin(u/ap)$.

The expression $S(u)$ reflects the load on the energy source and in the case of $u < ap$ is determined taking into account the approximate equality $ap \approx u$. The stationary values of the velocity u are determined by the intersection points of the curves $M(u/r)$ and $S(u)$.

5   Stability Conditions for Stationary Solutions

To derive the stability conditions for stationary motions, we make up the equations in variations for (6a, 6b) and (7a, 7b) and use the Routh-Hurwitz criteria, with the result that we get the following conditions:

$$D_1 > 0, D_3 > 0, D_1 D_2 - D_3 > 0 \tag{9}$$

where

$$D_1 = -(b_{11} + b_{22} + b_{33}),$$
$$D_2 = b_{11}b_{33} + b_{11}b_{22} + b_{22}b_{33} - b_{23}b_{32} - b_{12}b_{21},$$
$$D_3 = b_{11}b_{23}b_{32} + b_{12}b_{21}b_{33} - b_{11}b_{22}b_{33}.$$

In the case of $u \geq ap$ we have

$$b_{11} = \frac{r_0}{J}(Q - r_0 R \frac{\partial B_f}{\partial u}), \; b_{12} = -\frac{r_0^2 R}{J}\frac{\partial B_f}{\partial a}, b_{21} = -\frac{a}{2m}\frac{\partial k_0}{\partial u},$$

$$b_{22} = -\frac{1}{2m}(k_0 + a\frac{\partial k_0}{\partial a}) + \frac{b}{4pm}\sin 2\xi, \; b_{23} = \frac{1}{2pm}(\lambda \sin \xi + ab \cos 2\xi), \quad (10a)$$

$$b_{32} = \frac{1}{2pm}(\frac{\partial k_F}{\partial a} - \frac{\lambda}{a^2}\sin \xi), b_{33} = \frac{1}{2pma}(\lambda \cos \xi - ab \sin 2\xi)$$

in the case of $u < ap$, the expressions b_{23}, b_{32}, b_{33} do not change, only change

$$b_{11} = \frac{r_0}{I}\left[Q - r_0 R\frac{\partial B_f}{\partial u} - \frac{2r_0 R}{\pi\sqrt{a^2p^2 - u^2}}\right], b_{12} = -\frac{Rr_0^2}{J}\left[\frac{\partial B_f}{\partial a} + \frac{2u}{\pi a\sqrt{a^2p^2 - u^2}}\right],$$

$$b_{21} = -\frac{a}{2m}\left[\frac{\partial k_0}{\partial u} - \frac{4uR}{\pi a^2p^2\sqrt{a^2p^2 - u^2}}\right],$$

$$b_{22} = -\frac{1}{2m}(k_0 + a\frac{\partial k_0}{\partial a} + \frac{4Ru^2}{\pi a^2p^2\sqrt{a^2p^2 - u^2}}) + \frac{b}{4pm}\sin 2\xi$$

$$\quad (10b)$$

where $Q = \frac{d}{du}M(\frac{u}{r})$.

In (8), the values of a, ξ and Ω, determined from the equations of stationary motions and the partial derivatives of $\partial k_f / \partial u$, $\partial B_f / \partial u$, are calculated with the known specific form of the function $T(U)$.

In the case of a widespread and used characteristic of a self-oscillation causing nonlinear friction force

$$\bar{f}(U) = \text{sgn} \, U - \delta_1 U + \delta_3 U^3 \quad (11)$$

taking into account the fact that with averaging $\dot{\varphi} = \Omega$, $u = r\Omega$, we have

$$f(\dot{x}) = \sum_{n=0}^{3} \alpha_n \dot{x}^n \quad \text{where} \quad \alpha_0 = \delta_0 + \delta_1 u + \delta_2 u^2 + \delta_3 u^3, \quad \alpha_1 = -(\delta_1 + 2\delta_2 u + 3\delta_3 u^2),$$

$\alpha_2 = \delta_2 + 3\delta_3 u$, $\alpha_3 = -\delta_3$, and the partial derivatives in (10a, 10b)

$$\frac{\partial B_f}{\partial u} = \frac{\partial \alpha_0}{\partial u} + N_2(ap)^2\frac{\partial \alpha_2}{\partial u}, \frac{\partial k_f}{\partial u} = ap\frac{\partial \alpha_1}{\partial u} + \bar{N}_3(ap)^3\frac{\partial \alpha_3}{\partial u},$$

$$\frac{\partial \alpha_0}{\partial u} = \delta_1 + 2\delta_2 u + 3\delta_3 u^2, \frac{\partial \alpha_1}{\partial u} = -2(\delta_2 + 3\delta_3 u), \frac{\partial \alpha_2}{\partial u} = 3\delta_3, \quad (12)$$

$$\frac{\partial \alpha_3}{\partial u} = 0, k_0 + a\frac{\partial k_0}{\partial a} = k - R(\alpha_1 + 3\alpha_3\bar{N}_3 p^2 a^2).$$

Note that in calculating $\partial B_f / \partial u$, only even powers of n and, respectively, α_0, α_2 are taken into account, and in calculating $\partial k_f / \partial u$, odd powers of n and, respectively, α_1, α_3

are taken into account. As shown in [14], the dependence of the form (11) was also observed when considering the problem of measuring friction forces in a space experiment.

6 Calculation

To obtain information on the characteristics of stationary modes, calculations were performed in case of characteristic (11) with the following parameters: $\omega_0 = 1c^{-1}$, $m = 1\,\text{kgf} \cdot c^2 \cdot cm^{-1}$, $\lambda = 0.02\,\text{kgf}$, $b = 0.07\,\text{kgf} \cdot cm^{-1}$, $R = 0.5\,\text{kgf}$, $k = 0.02\,\text{kgf} \cdot c \cdot cm^{-1}$, $\delta_1 = 0.84\,c \cdot cm^{-1}$, $\delta_3 = 0.18\,c^3 \cdot cm^{-3}$, $r_0 = 1\,cm$, $I = 1\,\text{kgf} \cdot c \cdot cm^2$. Depending on the linearization accuracy parameter r, the values of the coefficients N_2 and $\bar{N}_3$ are as follows: $N_2 = 0.6$ $(r = 1)$, $\bar{N}_3 = 0.75$ $(r = 1.5)$.

Figure 2 shows some results of calculations in the case of $u = 1.16\,cm \cdot c^{-1}$ and the linear characteristic of the elastic force ($k_F \equiv 0$), where Fig. 2b reflects the dependence of the amplitude on the velocity u for $p = 2$. Self-oscillations in the velocity range $u = u_0 \geq 1.217\,cm \cdot c^{-1}$ do not exist. The amplitude of self-oscillations (a_s) is indicated in Fig. 2a with a line with a dot. The stable and unstable amplitudes of oscillations with an ideal source of energy are shown in Fig. 2a, respectively, the solid and dashed curves. With a non-ideal source of energy, oscillations are stable within shaded sectors. For the sake of brevity, these sectors are shown on the amplitude curve, although they should be indicated on the load curve $S(u)$. Depending on the steepness, the characteristic of the energy source and the direction in which its velocity changes, a number of interesting dynamic phenomena arise in the system, similar to those described in [1].

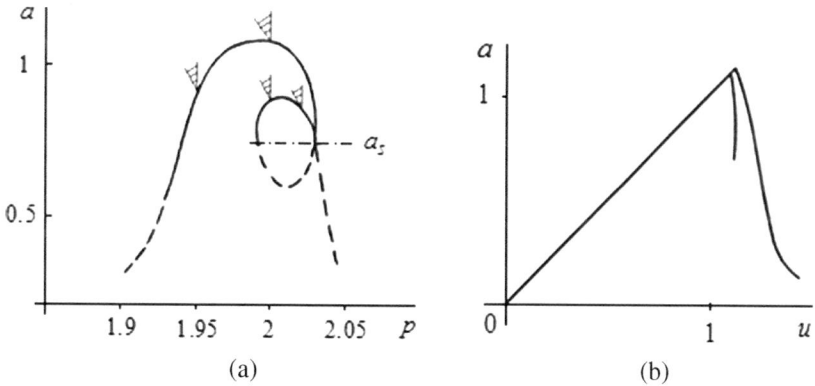

(a)   (b)

Fig. 2. Dependences of amplitude on frequency (p) and speed (u)

7 Conclusions

We considered the procedure for applying the methods of direct linearization to calculate the most complex class of mixed oscillations (the interaction of self-oscillations, forced oscillations and parametric oscillations) with a non-ideal source of energy that supports oscillations. The results on the basis of the direct linearization method demonstrate that they are similar to those obtained using the well-known methods of nonlinear mechanics. In the well-known methods of studying nonlinear systems, the form of the solution is usually given, and if a nonlinear characteristic of a complex type is used, the derivation of the final calculated ratios becomes a big challenge. This causes great difficulty in practical calculations, because the ratios obtained as a result of calculations depend on the type of characteristic. As can be seen from the above results, this disadvantage is absent in the direct linearization method, which allows to obtain the final calculated ratios, regardless of the specific type of nonlinear characteristic. And also, in comparison with the known methods for analyzing nonlinear systems, methods of direct linearization by several orders of magnitude reduce labor and time spent in calculations, which significantly increases their efficiency in practice. Its application is quite simple, which is especially valuable for calculating real objects for various purposes.

References

1. Alifov, A.A., Frolov, K.V.: Interaction of Nonlinear Oscillatory Systems with Energy Sources. Hemisphere Publishing Corporation. Taylor & Francis Group, New York (1990)
2. Tondl, A.: On the interaction between self-exited and parametric vibrations. National Research Institute for Machine Design Bechovice. Series: Monographs and Memoranda, no. 25 (1978)
3. Kononenko, V.O.: Vibrating Systems with Limited Power-Supply. Iliffe, London (1969)
4. Alifov, A.A.: Methods of Direct Linearization for Calculation of Nonlinear Systems. RCD, Moscow (2015). (in Russian)
5. Alifov, A.A.: Method of the direct linearization of mixed nonlinearities. J. Mach. Manuf. Reliab. **46**(2), 128–131 (2017). https://doi.org/10.3103/S1052618817020029
6. Alifov, A.A.: About some methods of calculation nonlinear oscillations in machines. In: Proceedings International Symposium of Mechanism and Machine Science, Izmir, Turkey, 5–8 October 2010, pp. 378–381 (2010)
7. Alifov, A.A.: About application of methods of direct linearization for calculation of interaction of nonlinear oscillatory systems with energy sources. In: Proceedings of the Second International Symposium of Mechanism and Machine Science (ISMMS – 2017), Baku, Azerbaijan, 11–14 September 2017, pp. 218–221 (2017)
8. Alifov, A.A., Farzaliev, M.G., Jafarov, E.N.: Dynamics of a self-oscillatory system with an energy source. Russ. Eng. Res. **38**(4), 260–262 (2018). https://doi.org/10.3103/S1068798X18040032
9. Gourary, M.M., Rusakov, S.G.: Analysis of oscillator ensemble with dynamic couplings. In: The Second International Conference of Artificial Intelligence, Medical Engineering, Education, AIMEE 2018, pp. 150–160 (2018)
10. Acebrón, J.A., et al.: The Kuramoto model: a simple paradigm for synchronization phenomena. Rev. Mod. Phys. **77**(1), 137–185 (2005)
11. Bhansali, P., Roychowdhury, J.: Injection locking analysis and simulation of weakly coupled oscillator networks. In: Li, P., et al. (eds.) Simulation and Verification of Electronic and Biological Systems, pp. 71–93. Springer, Dordrecht (2011)
12. Ashwin, P., Coombes, S., Nicks, R.J.: Mathematical frameworks for oscillatory network dynamics in neuroscience. J. Math. Neurosci. **6**(2), 1–92 (2016)

13. Ziabari, M.T., Sahab, A.R., Fakhari, S.N.S.: Synchronization new 3D chaotic system using brain emotional learning based intelligent controller. Int. J. Inf. Technol. Comput. Sci. (IJITCS) 7(2), 80–87 (2015). https://doi.org/10.5815/ijitcs.2015.02.10

14. Bronovec, M.A., Zhuravljov, V.F.: On self-oscillations in systems for measuring friction forces. Izv. RAN, Mekh. Tverd. Tela 3, 3–11 (2012). (in Russian)

15. Karabutov, N.: Structural identification of nonlinear dynamic systems. Int. J. Intell. Syst. Appl. (IJISA) 7(9), 1–11 (2015). https://doi.org/10.5815/ijisa.2015.09.01

16. Karabutov, N.: Frameworks in problems of structural identification systems. Int. J. Intell. Syst. Appl. (IJISA) 9(1), 1–19 (2017). https://doi.org/10.5815/ijisa.2017.01.01

17. Da-Xue, C., Liu, G.H.: Oscillatory behavior of a class of second-order nonlinear dynamic equations on time scales. Int. J. Eng. Manuf. (IJEM) 1(6), 72–79 (2011). https://doi.org/10.5815/ijem.2011.06.11

18. Wang, Q., Fu, F.: Numerical oscillations of runge-kutta methods for differential equations with piecewise constant arguments of alternately advanced and retarded type. Int. J. Intell. Syst. Appl. (IJISA) 3(4), 49–55 (2011). https://doi.org/10.5815/ijisa.2011.04.07

Dynamic Model of a Stepping Robot for Arbitrarily Oriented Surfaces

Mikhail N. Polishchuk[ID] and Volodymyr V. Oliinyk[✉][ID]

National Technical University of Ukraine "Igor Sikorsky Kyiv Polytechnic Institute", Peremohy Avenue 37, Kyiv 03056, Ukraine
oliinyk.volodymyr@gmail.com

Abstract. Climbing robots for vertical and inclined surfaces are rather new modification of mobile robots used to perform contact technological operations while simultaneously overcoming the forces of gravity. In this paper, we propose new approach to increase energy efficiency of such robots by equipping them with elastic elements used as energy accumulators. This approach to the design and control of mobile robot allows using accumulated energy of motion when servomotors are turned off. So main feature of the proposed dynamic model is transformation of accumulated potential energy into kinetic energy of motion. The simulations of proposed model prove the concept of significant decrease of energy consumption and show the influence of stiffness parameter of elastic elements on beneficial effect of its use.

Keywords: Mobile robot · Stepping robot · Climbing robot ·
Vertical movement robot · Energy recovery · Dynamic model

1 Introduction

Stepping robots for arbitrarily orientated surfaces, including vertical walls (also known as climbing robots) are rather new modification of mobile robots used to perform technological operations for monitoring industrial facilities, installation and dismantling of building structures, repair and preventive maintenance of their components etc. Such robots are especially useful in extreme conditions of man-made disasters that are dangerous for human stay.

Important feature of wall climbing robots is the retention system designed to keep the robot on a surface overcoming the gravitational force. It has effect on design parameters calculation increasing importance of dynamic model of such robots and significance of robot weight. Gravitational component of the dynamic load can be reduced by optimization of climbing robot energy efficiency lowering weight of the drives and power sources. Such optimization is the main goal of this research.

Our solution provides significant energy savings thanks to accumulation of potential energy in elastic elements at the first step of movement and its conversion into kinetic energy of motion with the engine off at each second step.

The fundamental contributions of this work are:

© Springer Nature Switzerland AG 2020
Z. Hu et al. (Eds.): ICCSEEA 2019, AISC 938, pp. 32–42, 2020.
https://doi.org/10.1007/978-3-030-16621-2_4

- New design of climbing mobile robot with energy accumulators is proposed and its dynamic model is presented.
- Proposed idea of elastic elements embedded in actuators provides significant energy efficiency.
- Influence of elastic energy accumulators' stiffness is studied and valuable practical results are received.

Rest of the paper is organized as follows: in Sect. 2, brief survey of recent achievements in mobile robotics for vertical movement is made. Fundamentally new construction of stepping robot is proposed in Sect. 3 followed by description of this robot dynamic model in Sect. 4. Section 5 contains experimental results of elastic energy accumulators simulations and discussions. The paper is concluded in Sect. 6.

2 Related Work

The newest designs of vertical movement robots without dynamic analysis of their movement process are described in articles [1, 2]. In papers [3, 4] vacuum grippers of wall climbing robots were considered but without calculations of the device. Calculations of mechanical grippers of mobile robot are given in [5]. Paper [6] contains description of rope climbing robot for video surveillance. The need for such guides (ropes) complicates the design of the robot, and limits its technological capabilities. Original mobile rescue robot is proposed in [7]. However, the wheel transmission of the specified robot limits its movement only to horizontal or slightly inclined surfaces. Symbiosis of mobile robot and aircraft for extinguishing fires is described in the article [8]. The mathematical model of this robot, unfortunately, is limited to matrices of angular coordinates and does not display its dynamic parameters. Dynamic model of a mobile robot is most fully represented in the research [9] but only in relation to vacuum coupling with a vertical surface. In contrast to the studies above, works [10, 11] propose the principles of integration of energy storage devices to robot's actuators, which increases the robot's energy efficiency. However, those studies are limited to static models, without taking dynamic parameters into account. Thus, the task of designing a stepping robot dynamic model with energy recovery is most relevant.

3 Stepping Robot Principle

Taking into account the fundamental difference between the robot [12] and similar solutions, namely its ability to accumulate potential energy in the first stage of the movement and convert it into kinetic energy in the second stage, let us first consider the principle of the robot action. Figure 1 shows three stages A, B, C of the stepping robot movement along the vertical wall. The robot itself includes a housing 1 with a rotating actuator 2 located on it. The actuator is connected to pedipulators (stepping mechanisms) 3 via the transmission. They are equipped with elastic elements 4, which are used for potential energy accumulation in the first half of the movement cycle (the first half of the robot's step).

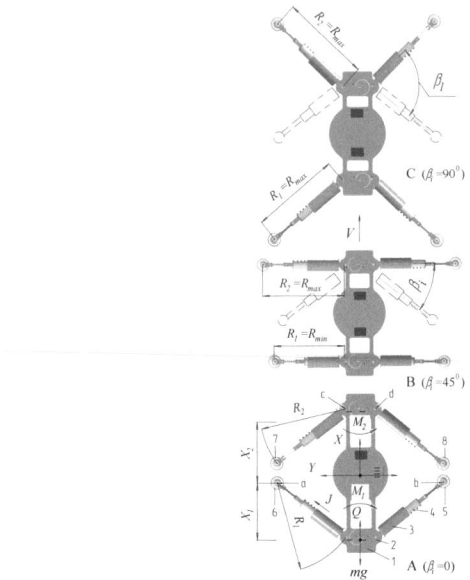

Fig. 1. Diagram of the robot movement along the vertical wall

The grippers 5…8 keep the robot on vertical or arbitrarily oriented surface. They can be implemented in different ways, i.e. as mechanical, vacuum, electromagnetic or adhesive devices. The type of gripper is determined by the surface of movement and the robot's technological functions. The grippers 5, 6 and 7, 8 are enabled alternately in each half of the robot movement cycle (step). When grippers 5 and 6 are attached to the surface of the movement, drive 2 rotates pedipulators 3 by a certain degree β_i around corresponding points 'a' and 'b' with a varying radius $R_1 = Var$. In this case, the elastic elements 4 are compressed, accumulating the potential energy $U = \frac{jx^2}{2}$ (where: j is stiffness of the elastic element, N/m; x is value of the spring deformation).

Simultaneously, when the grippers 7 and 8 are disabled, a similar drive rotates this pair of pedipulators around points 'c' and 'd' with the radius $R_2 = Const$. R_2 is constant because these grippers are disabled. Robot body 1 is moved along the distance X_1 with the speed V, so the robot passes the first half of its movement cycle distance. In the 'B' state, when $\beta_i = 45°$, the elastic elements 4 reach maximum compression, accumulating maximum amount of potential energy.

During the second half of the movement cycle, the drive 2 is disabled and the robot moves under the influence of elastic elements 4 straightening. When distance X_2 is covered, the grippers 7 and 8 are turned on and therefore become attached to the surface of movement, while the grippers 5 and 6 are released. Then full cycle of such longitudinal movement is repeated.

Thus, on the second half of the movement cycle mobile robot moves using energy of elastic elements deformation accumulated during the first stage of the movement. As a result, energy resources are saved for main technological operations.

4 Stepping Robot Dynamic Model

Initially, we will build dynamic model of the stepping robot. It will allow us to define objective functions and proceed with parametric optimization using a method that was improved by the authors.

To describe the robot movement, we use the Lagrange equations of the second kind:

$$\frac{d}{dt}\left(\frac{\partial T}{\partial \dot{q}_i}\right) - \left(\frac{\partial T}{\partial q_i}\right) = Q_{q_i}, \quad i = 1 \ldots k \tag{1}$$

where: k is degree of freedom of a mechanical system; q_i are generalized coordinates; $\dot{q}_i$ are generalized velocities; $T(q_i, \dot{q}_i)$ is kinetic energy of mechanical system, which is a function of generalized coordinates and generalized velocities; Q_{q_i} is generalized force, which matches the generalized coordinate q_i. Let us determine potential and kinetic energy of robot movement in different stages, depending on the design parameters. Then we can get differential equations of a robot motion on the first and second parts of the movement cycle (step) and make dependencies between the parameters of the robot transmission.

In current model, the system has two degrees of freedom. We choose rotation angle of the supporting leg $q_1 = \beta_1$ and the rotation angle of free leg $q_2 = \beta_2$ as generalized coordinates. Then $\dot{\beta}_1, \dot{\beta}_2$ are the generalized velocities. Let us name the leg attached to the surface of movement the 'supporting' leg. Then the 'free' leg is the one not attached to the surface.

The robot body travels distance s that can be described with the following formula:

$$s = R_2 \cos 45°(1 - tg(45° - \beta_1)), \quad 0 \leq \beta_1 \leq 90° \tag{2}$$

Then velocity of the robot's body is

$$V = \frac{ds}{dt} = \frac{R_2 \cos 45°}{\cos^2(45° - \beta_1)}\dot{\beta}_1; \quad (\cos 45° = \sin 45° = \frac{\sqrt{2}}{2}) \tag{3}$$

In addition, equation for the kinetic energy T_k of body will be

$$T_k = \frac{mV^2}{2} = \frac{mR_2^2}{4\cos^4(45° - \beta_1)}(\dot{\beta}_1)^2 \tag{4}$$

where m is mass of robot body.

Let us consider the mass of each leg m_1 to be evenly distributed along the robot. Free leg performs a plane parallel motion: it moves forward with the body with velocity V and rotates with angular velocity around the fixing point mounted to the body (points 'c' and 'd' in Fig. 1). Speed projections of arbitrary point with coordinates x_1, y_1 can be represented as follows

$$V_{x_1} = \dot{\beta}_2 y \sin(45° - \beta_2); \quad V_{y_1} = V + \dot{\beta}_2 y \cos(45° - \beta_2). \tag{5}$$

Kinetic energy of robot's free leg is

$$T_1 = \frac{1}{2} \int (V_{x_1}^2 + V_{y_1}^2) dm. \tag{6}$$

We substitute the expressions for the velocity projections and $dm = m_1 dy / R_2$ in (6) and perform integration:

$$T_1 = \frac{m_1}{2} \left(V^2 + VR_2 \dot{\beta}_2 \cos(45° - \beta_2) + \frac{1}{3} (\dot{\beta}_2 R_2)^2 \right). \tag{7}$$

We then substitute expression (3) for the velocity V and get the final formula for finding kinetic energy of the free leg:

$$T_1 = \frac{m_1 R_2^2}{2} \left(\frac{(\dot{\beta}_1)^2}{2 \cos^4(45° - \beta_1)} + \frac{\dot{\beta}_1 \dot{\beta}_2 \sqrt{2} \cos(45° - \beta_2)}{2 \cos^2(45° - \beta_1)} + \frac{1}{3} (\dot{\beta}_2)^2 \right). \tag{8}$$

Supporting leg, having its grips attached to the surface, performs rotational movement with angular velocity $\dot{\beta}_1$. At the same time, the mutual displacement of the leg parts due to the spring deformation may be neglected. We can get the expression for kinetic energy of supporting leg from the formula (7) by substituting robot linear velocity $V = 0$ and angular velocities of the pedipulators $\dot{\beta}_2 = \dot{\beta}_1$:

$$T_2 = \frac{m_1 R_2^2}{6} (\dot{\beta}_1)^2. \tag{9}$$

Then total kinetic energy T during two halves of the robot movement cycle will be:

$$T = \frac{R_2^2}{2} \left(\frac{(2m_1 + m)(\dot{\beta}_1)^2}{2 \cos^4(45° - \beta_1)} + \frac{m_1 \dot{\beta}_1 \dot{\beta}_2 \sqrt{2} \cos(45° - \beta_2)}{\cos^2(45° - \beta_1)} + \frac{2m_1}{3} ((\dot{\beta}_2)^2 + (\dot{\beta}_1)^2) \right) \tag{10}$$

To get the differential equations of robot movement during both parts of the cycle, we first construct an equation for the generalized forces. In order to do this, we find the partial derivatives of the kinetic energy that are contained in Eq. (1):

$$\frac{\partial T}{\partial \dot{\beta}_1} = \frac{R_2^2}{2}\left(\frac{(2m_1+m)\dot{\beta}_1}{\cos^4(45°-\beta_1)} + \frac{m_1\dot{\beta}_2\sqrt{2}\cos(45°-\beta_2)}{\cos^2(45°-\beta_1)} + \frac{4m_1}{3}\dot{\beta}_1\right)$$

$$\frac{\partial T}{\partial \beta_1} = -\frac{R_2^2\sin(45°-\beta_1)}{2}\left(\frac{2(2m_1+m)(\dot{\beta}_1)^2}{\cos^5(45°-\beta_1)} + \frac{2m_1\dot{\beta}_1\dot{\beta}_2\sqrt{2}\cos(45°-\beta_2)}{\cos^3(45°-\beta_1)}\right)$$

$$\frac{\partial T}{\partial \dot{\beta}_2} = \frac{R_2^2}{2}\left(\frac{m_1\dot{\beta}_1\sqrt{2}\cos(45°-\beta_2)}{\cos^2(45°-\beta_1)} + \frac{4m_1}{3}\dot{\beta}_2\right)$$

$$\frac{\partial T}{\partial \beta_2} = \frac{R_2^2}{2}\left(\frac{m_1\dot{\beta}_1\dot{\beta}_2\sqrt{2}\sin(45°-\beta_2)}{\cos^2(45°-\beta_1)}\right)$$

(11)

Additionally, it is recommended to calculate total derivatives with respect to time. After that, generalized forces Q_{q_i} can be found using

$$Q_{q_i} = \frac{\delta A_{q_i}}{\delta_{q_i}}$$

(12)

where δ_{q_i} is possible increment of the generalized coordinate q_i; δA_{q_i} is possible work of forces that influence the mechanical system during the corresponding movement.

In our case, we use possible increment δ_{β_1} of rotation angle β_1 of mobile robot leg and resulting possible increment δ_s of displacement s. This is the distance on which drive forces perform possible work

$$\delta A_{\beta_1} = \left(\frac{2M_1 i}{nz} - 2J\sin(45° - \beta_1) - (m+2m_1)g\sin(\gamma)\right)\delta s$$
$$- m_1 g\sin(\gamma)R_2\cos(45° - \beta_1)\delta\beta_1$$

(13)

where M_1 is pedipulator(leg) drive torque, N/m; i is transmission ratio; n, z are gear module and teeth number respectively; m, m_1 are mass of robot body and single leg mass respectively. The force of pedipulator's elastic element is equal to:

$$J = P_{\min} + jR_2\left(1 - \frac{\cos 45°}{\cos(45° - \beta_1)}\right); 0 \le \beta_1 \le 90°.$$

(14)

Taking into account Eq. (2), we will get the increment of the distance s:

$$\delta_S = \frac{R_2\sin 45°}{\cos^2(45° - \beta_1)}\delta_{\beta_1}.$$

(15)

Therefore, generalized force is equal to

$$Q_{\beta_1} = Q_2 + Q_1$$

(16)

where driving force on the second half of robot movement:

$$Q_2 = \left(\frac{2M_1 i}{nz}\right)\frac{R_2\sin 45°}{\cos^2(45° - \beta_1)}$$

(17)

and on the first half:

$$Q_1 = \frac{-(2J \sin(45° - \beta_1) + (m + 2m_1)g \sin\gamma)R_2 \sin 45°}{\cos^2(45° - \beta_1)} - m_1 g \sin\gamma R_2 \cos(45° - \beta_1).$$

(18)

Let us express the possible increment δ_{β_2} of the turn angle β_2. As the result, we will get the possible increment $\delta\varphi = \delta\beta_2 i$, where i is transmission ratio and β_2 is rotation angle of the link to which the torque M_2 is applied (Fig. 1). This is the distance where momentum M_2 performs possible work $\delta A_{\beta_2} = M_2 i \delta\beta_2$. So generalized force will be:

$$Q_{\beta_2} = M_2 i - m_1 g \sin(\gamma)R_2 \cos(45° - \beta_2).$$

(19)

Next, we put expressions for the derivatives and generalized forces to (1). We leave the summands containing second derivatives with respect to the rotation angles in the left side and transfer other summands to the right side of expression (1). As a result, we get differential equations for the first part of the mobile robot movement cycle:

$$\frac{R_2^2}{2}\left(\frac{(2m_1 + m)\ddot{\beta}_1}{\cos^4(45° - \beta_1)} + \frac{m_1\ddot{\beta}_2\sqrt{2}\cos(45° - \beta_2)}{\cos^2(45° - \beta_1)} + \frac{4m_1}{3}\ddot{\beta}_1\right) = Q_2 + Q_1 + \frac{R_2^2}{2}A + \frac{\partial T}{\partial\beta_1}$$

$$\frac{R_2^2}{2}\left(\frac{m_1\ddot{\beta}_1\sqrt{2}\cos(45° - \beta_2)}{\cos^2(45° - \beta_1)} + \frac{4m_1}{3}\ddot{\beta}_2\right) = M_2 i + \frac{R_2^2}{2}B + \frac{\partial T}{\partial\beta_2}$$

(20)

$$0 \leq \beta_1 \leq \beta_o$$

where β_o is extreme value of β_1, when the momentum M_1 still has effect. Then:

$$L_1 = R_2 \cos 45°[(1 - tg(45° - \beta_o))] \text{ and } L_2 = R_2 \cos 45°[(1 + tg(45° - \beta_o))] \quad (21)$$

where L_1 is part of the distance travelled under the influence of drive force F_1 and L_2 is distance covered under the influence of elastic element decompression F_2.

The value of the angle β_o depends on the momentum M_1 and the stiffness j of elastic element. It is selected using simulations with the goal to make velocity V close to zero at the end of supporting leg rotation. Otherwise, there will be unnecessary energy expenditure and the leg can significantly hit movement limiter. In our solution, the angle is equal to $\beta_o = \pi/6$.

We can get the differential equations for the robot movement on the second part of the cycle from (20). In order to do this, we put the value $M_1 = 0$ into the expression for generalized force Q_{β_1} when the leg rotation angle β_1 is within the limits $\beta_o < \beta_1 \leq 90°$:

$$\frac{R_2^2}{2}\left(\frac{(2m_1 + m)\ddot{\beta}_1}{\cos^4(45° - \beta_1)} + \frac{m_1\ddot{\beta}_2\sqrt{2}\cos(45° - \beta_2)}{\cos^2(45° - \beta_1)} + \frac{4m_1}{3}\ddot{\beta}_1\right) = Q_1 + \frac{R_2^2}{2}A + \frac{\partial T}{\partial\beta_1}$$

$$\frac{R_2^2}{2}\left(\frac{m_1\ddot{\beta}_1\sqrt{2}\cos(45° - \beta_2)}{\cos^2(45° - \beta_1)} + \frac{4m_1}{3}\ddot{\beta}_2\right) = M_2 i + \frac{R_2^2}{2}B + \frac{\partial T}{\partial\beta_2}$$

(22)

Those differential equations can be solved with the numerical method that requires expressing them as a system of first order equations and solving with respect to derivatives for the first and second stages of pedipulators movement.

5   Simulation Results

The system of differential equations was solved using numerical fourth order Runge-Kutta method and simulated in the MATLAB environment. The results are shown below. As it was mentioned above, pedipulators use elastic elements (Fig. 1, item 4) for accumulation of potential energy in the first half of the step. The main characteristic of elastic elements is their stiffness j - parameter that defines the compression force of these elements and the accumulated potential energy in the first half of the pedipulator step. Figure 2(a) shows that this parameter has more significant effect on the robot displacement $s(t)$ in the second half of the step than in the first half when the drive is disabled and accumulated potential energy is converted into the kinetic energy of motion. Similarly, the stiffness of elastic elements affects the angular movement of pedipulators (Fig. 3b).

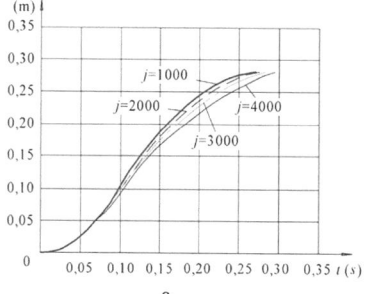
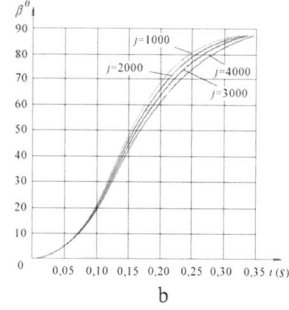

Fig. 2. Distance-time (a) and angular movement (b) graphs for different stiffness j (N/m) of the elastic elements

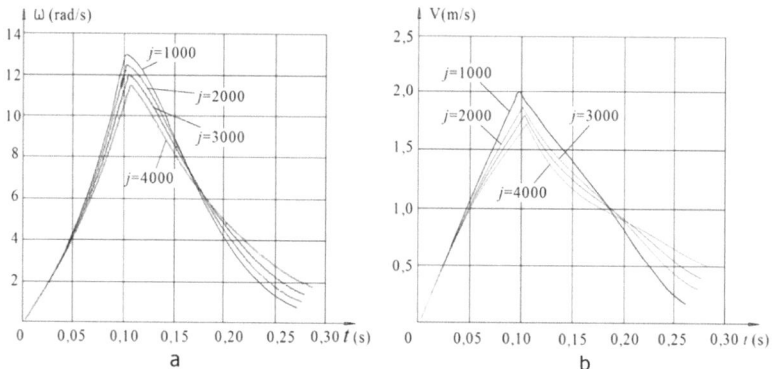

Fig. 3. Angular (a) and linear (b) velocity graphs of the pedipulator, j (N/m) - stiffness of elastic elements

Analysis of the graphs in Fig. 2 shows that it is advisable to increase stiffness in order to get an increase of kinetic energy of robot motion even despite corresponding increase of counteraction to the drive in the first half of the step and resulting decrease of its efficiency. This drawback can be compensated by an increase in transmission ratio of the robot drive. The impact of the stiffness j of potential energy storage devices on the robot angular velocity ω and the linear velocity V is shown in Fig. 3. Provided graphs allow to state that the dominant effect of stiffness on angular and linear velocities has place in the second half of the step. Moreover, the maximum stiffness value $j = 4000$ N/m has increasing effect at the end of the process of potential energy transformation into kinetic energy of motion.

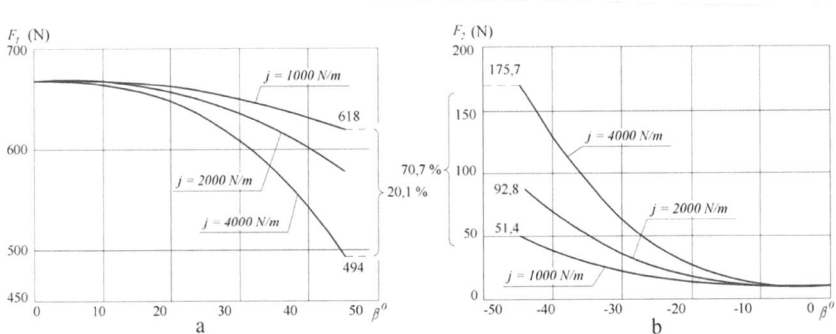

Fig. 4. Driving force F_1 of the robot when the engine is turned on (a) and the force F_2 when the engine is off and the robot moves under the influence of elastic elements (b)

Simulation results on Fig. 4 demonstrate typical influence of elastic elements on moving forces. Both drive force F_1 with enabled motor and force of elastic element decompression F_2 when the motor is turned off are decreasing during corresponding movement stage. However, force F_2 receives much higher relative gain compared to F_1 for higher values of stiffness providing more overall energy efficiency.

So, the main practical results of stepping robot movement simulation are as follows:

1. It is recommended to increase the stiffness of elastic elements that will result in higher kinetic energy of robot motion. Increasing counteraction to the drive in the first half of the step and corresponding drive efficiency loss can be compensated by bigger ratio of the drive transmission.
2. The impact from high values of stiffness of elastic elements increases upon completion of accumulated potential energy usage process in the second stage of the step.
3. The limitation on rotation angle in the first half of the movement cycle is introduced in differential equations of the robot movement. The result is extreme angle value at which the pedipulator torque still has effect. The value of this angle is selected by simulations to ensure that the robot linear speed at the end of supporting leg rotation is close to zero. It allows reducing energy costs and avoiding hitting the leg limiter.

4. Equations (16, 17, 18, 19, 21) can be directly used for calculation of design parameters of similar stepping robots.

In practice, in order to reduce dynamic loads during transient processes, it is recommended to use dampers – devices that absorb and convert kinetic energy of impact into the heat energy of fluid (typically).

6   Conclusions

This work presents new design of climbing robot with embedded energy accumulators in a form of elastic elements. Important part of the paper is dedicated to dynamic model of such type of robots. Proposed model provides engineering technique for calculation of dynamic loads and main design parameters. Application of potential elastic elements energy accumulators integrated in actuators of the stepping robot can significantly reduce the total power requirements of the robot (up to 30–40% depending on other design parameters). This is confirmed by the simulations provided in this work as well as by the qualification technical expertise (Patent UA 111021). Additionally, some valuable practical results of simulation were received including recommendation of using highest possible values of elastic elements stiffness.

References

1. Dethe, R.D., Jaju, S.B.: Developments in wall climbing robots: a review. Int. J. Eng. Res. Gen. Sci. **2**(3), 33–42 (2014)
2. Cardinaels, R., et al.: GECO: the development of a wall-climbing robot. In: 20th International Conference on Climbing and Walking Robots and the Support Technologies for Mobile Machines (CLAWAR 2017), Porto, Portugal, 11–13 September 2017 (2017)
3. Apostolescu, T.C., et al.: Development of a climbing robot with vacuum attachment cups. In: International Conference on Innovations, Recent Trends and Challenges in Mechatronics, Mechanical Engineering and New High-Tech Products Development – MECAHITECH 2011, vol. 3 (2011)
4. Fujita, M., Ikeda, S., Fujimoto, T., Shimizu, T., Ikemoto, S., Miyamoto, T.: Development of universal vacuum gripper for wall-climbing robot. Adv. Robot. **32**(6), 283–296 (2018)
5. Lam, T.L., Xu, Y.: Tree Climbing Robot: Design, Kinematics and Motion Planning. Springer, Heidelberg (2012)
6. Zafar, K., Hussain, I.M.: Rope climbing robot with surveillance capability. Int. J. Intell. Syst. Appl. (IJISA) **5**(9), 1–9 (2013). https://doi.org/10.5815/ijisa.2013.09.01
7. Moniruzzaman, M., Zishan, M.S.R., Rahman, S., Mahmud, S., Shaha, A.: Design and implementation of urban search and rescue robot. Int. J. Eng. Manuf. (IJEM) **8**(2), 12–20 (2018). https://doi.org/10.5815/ijem.2018.02.02
8. Alshbatat, A.I.N.: Fire extinguishing system for high-rise buildings and rugged mountainous terrains utilizing quadrotor unmanned aerial vehicle. Int. J. Image Graph. Sig. Process. (IJIGSP) **10**(1), 23–29 (2018). https://doi.org/10.5815/ijigsp.2018.01.03
9. Dyshenko, V.S.: Research on dynamics of the mobile robot for vertical surfaces. Ph.D. thesis, Kursk State Technical University (2006). (in Russian)

10. Polishchuk, M., Opashnianskyi, M., Suyazov, N.: Walking mobile robot of arbitrary orientation. Int. J. Eng. Manuf. (IJEM) **8**(3), 1–11 (2018). https://doi.org/10.5815/ijem.2018.03.01
11. Yampolskiy, L., Polishchuk, M., Persikov, V.: Method and device for movement of pedipulators of walking robot. UA Patent 111021, 10 March 2016
12. Polishchuk, M.N., Oliinyk, V.V.: Mobile climbing robot with elastic energy accumulators. Mech. Adv. Technol. **1**(82), 116–122 (2018). https://doi.org/10.20535/2521-1943.2018.82.115920

The Cryptography of Elliptical Curves Application for Formation of the Electronic Digital Signature

Olexander Belej[(⊠)] ⓘ

Lviv Polytechnic National University,
5 Mytropolyt Andrei Street, Building 4, Room 324, Lviv 79000, Ukraine
Oleksandr.I.Belei@lpnu.ua

Abstract. Today, commercial structures can no longer run their business without using the Internet and cloud technologies. Electronic transactions conducted through the global Internet play an increasing role in the modern global economy, and their importance is rapidly increasing every year. However, the reality is the Internet is quite vulnerable in the terms of threats of most diverse types. Facing reality makes commercial organizations increasingly focus on Web security. In this article we are focuses on encryption of Web pages using encryption based on elliptical curves. We examined the algorithm of this approach and engagements are possible protecting information by transforming data from a distortion by an attacker. Even a simple transformation of information is a very effective means, making it possible to hide its meaning from the majority of unskilled offenders. We also considered an algorithm of constructing an electronic-digital message signature based on encryption using elliptic curves.

Keywords: Cryptography · Elliptic curve · Algorithm · Digital signature · Message · Security key

1 Introduction

Most products and standards that use public key cryptography for encryption and authentication are based on the RSA algorithm [12]. However, the number of key bits required for reliable data protection when using RSA has increased dramatically in recent years, which has led to a corresponding increase in load on systems using RSA. The attractiveness of the approach based on elliptic curves in comparison with RSA is that using elliptic curves provides an equivalent level of protection with a much smaller number of digits, resulting in reduced processor utilization. At the same time, the degree of trust in cryptographic methods using elliptic curves is not as high as the degree of trust in RSA.

Neil Koblitz proposed using algebraic properties of elliptic curves in cryptography [8]. The role of the main cryptographic operation is performed by the operation of scalar multiplication of a point on an elliptic curve by a given integer, determined through the operations of addition and doubling of points of an elliptic curve. The latter, in turn, are performed on the basis of the operation of addition, multiplication and inversion in a finite field, over which the curve is considered [8].

© Springer Nature Switzerland AG 2020
Z. Hu et al. (Eds.): ICCSEEA 2019, AISC 938, pp. 43–57, 2020.
https://doi.org/10.1007/978-3-030-16621-2_5

The attractiveness of the approach based on the elliptic curves in comparison with RSA is that using an elliptic curve provides an equivalent level of protection with a significantly smaller number of discharges. This helps reduce CPU usage. Cryptography using RSA is used more often than cryptography using elliptic curves, and the degree of confidence, respectively, is much higher. But, at equal lengths of keys, the computational effort required for using RSA and cryptography based on elliptic curves does not vary greatly. Thus, in comparison with RSA at equal levels of protection, the clear computational advantage belongs to cryptography based on elliptic curves with shorter key lengths.

Of particular interest to the cryptography of elliptic curves is due to the advantages that its use in wireless communications gives - high speed and short key length. Asymmetric cryptography is based on the complexity of solving some mathematical problems. When using algorithms on elliptic curves, it is assumed that there are no sub-exponential algorithms for solving discrete logarithm problems in groups of their points. The order of the group of points determines the complexity of the task [4].

In the article [17] describes an im-proved method of the Lopez-Dahab-Montgomery (LD-Montgomery) scalar point multiplication in terms of working with binary elliptic curves. This algorithm is used to compute point multiplication results of the curves over binary Galois Fields featuring experimental results based on different scalars [17].

In [14] are using Matrix Array Symmetric Key (MASK) for the key generation and Chaos based approach for the image encryption. In this approach, author has adapted the concept of partial encryption of image pixels instead of complete encryption so that in case of arrack, intruder can be confused with the partial encrypted image. That concept is evaluated based on the parameters of Information Entropy, Elapsed Time, Precision, Recall and F-Measure [14].

In [5] the main encryption stages of C-GET are chaotic map functions, fuzzy logic and genetic operations. For testing C-GET, digital images are used because they become an important resource of communication. Experimental results show that C-GET technique has multilayer protection stages against various attacks and a powerful security based on the multi-stages, multiple parameters, fuzzy logic and genetic operations [5].

In [10] public key cryptographic schemes are vastly used to ensure confidentiality, integrity, authentication and nonrepudiation. The implementation of SSC to secure different recent communication technologies such as cloud and fog computing is on demand due to the assorted security services, digital signature and data integrity. In that paper provide a systematic review of SSC public key cryptosystem to help crypto-designers to implement SSC efficiently and adopt it in hardware or software-based applications [11].

In [7] are writing about the cloud computing features aim at providing security and confidentiality to its customers. The business officials have gained maximum profit using the cloud environment. Therefore the critical data moving to and from the cloud server must be secured. The privacy preservation and cloud security issue is the most serious issue that we are facing in today's world in cloud era [7].

In other studies [2, 9, 11, 13], this deals with cloud and security data not disclosed security data and emails using elliptic curve algorithms.

As an international standard, the American digital signature algorithm on elliptic curves (ECDSA) is adopted. This standard uses elliptic curves over the field of characteristic 2. However, cryptographically resistant curves over the field of such a characteristic are relatively small. Therefore, we consider the electronic signature on elliptic curves defined over the field of higher performance [1, 4].

2   Elliptic Curve Cryptography in SSL and RSA

The most widely used cryptographic method in SSL implementations is the RSA algorithm. However, alternative ECC (Elliptic Curve Cryptography) is becoming increasingly attractive from the practical side. Developed in 1985 by Victor Miller and Neal Koblitz, it gradually gains acceptance in cryptography [3]. NIST (National Institute of Standards and Technology) has already recommended the government of USA to use ECC technology in solving problems in the field of information security.

   In the case of cryptography using elliptic curves, one has to deal with a reduced form of an elliptic curve, which is defined over a finite field. Of particular interest for cryptography is an object called an elliptic group modulo p, then p is a prime number. The elliptic curve over a finite field is given by the equation:

$$y^2 = x^3 + a \cdot x + b \pmod p \tag{1}$$

The main arithmetic operation in ECC is the operation of scalar multiplication of curve points, which allows determining the point $Q = k \times p$ (the point P multiplied by the integer k turns into the point Q). Scalar multiplication is performed by several combinations of addition and doubling of points of an elliptic curve. The reliability and crypto resistance of elliptical cryptography is based on the difficulty of solving the ECDLP (Elliptic Curve Discrete Logarithm Problem) problem, the essence of which consists in finding an integer k from known points P and $Q = k \times p$. In addition to the equation, an important parameter of the curve is the base (generating) point G, which is selected for each curve separately. The secret key in accordance with the ECC technology is a large random number k, and the reported public key is the product of k by the base point G.

   Elliptic curves are so called simply because they are described by cubic equations similar to those used to calculate the curve of an ellipse (Fig. 1).

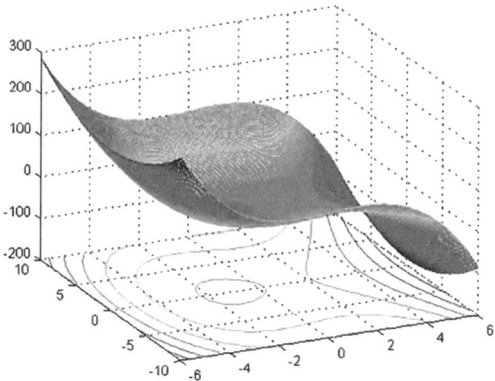

Fig. 1. Spatial graph of the elliptic curve $y^2 = x^3 - 5 \cdot x + 1$.

   Not every curve provides the required cryptographic strength, and for some of them the ECDLP task is solved quite effectively. Since an unsuccessful choice of a curve may lead to a decrease in the level of security provided, standardization organizations select entire blocks of curves that have the necessary reliability. The use of standardized curves is also recommended because better compatibility between different implementations of information security protocols becomes possible.

Diffie-Hellman key exchange (ECDH) and the digital signature algorithm on the Elliptic Curve (ECDSA) are similar in essence to the Diffie-Hallman and RSA algorithms, respectively [6]. Key exchange can be performed as follows. First, a large prime number p and the parameters of the curve equation are selected. This defines the group of points at which the base point G is selected. When choosing G, it is important that the smallest value of n, at which $nG = O$, be a very large prime number. The curve equation and the point G are known to all participants. The exchange of keys between users A and B can be carried out as follows:

1. Party A selects an integer nA less than n. This number will be the private key of participant A. Then participant A generates the public key PA = $nA|G$. The public key is a certain point of the curve.
2. Similarly, participant B chooses the private key nB and calculates the public key PB.
3. Member A generates a secret key K = $nA \times PB$, and participant B generates a secret key K = $nB \times PA$.

Formulas for secret keys give the same result:

$$nA \times PB = nA \times (nB \times G) = nB \times (nA \times G) = nB \times PA. \tag{2}$$

This scheme to crack, the enemy will have to solve the discrete logarithm problem on the curve, which is assumed to be an intractable problem. In the case of the handshake protocol, several messages are generated exchanged between the client and the server. In Fig. 2 shows the messaging scheme required, when establishing a logical connection.

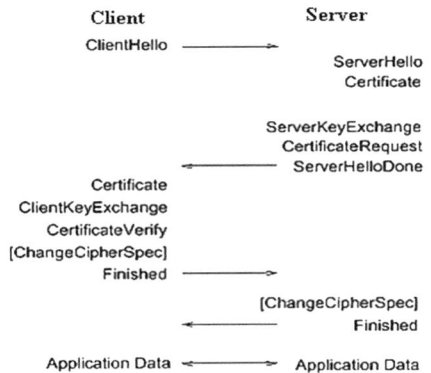

Fig. 2. The algorithm of the handshake's protocol.

Separately, the 160-bit and 224-bit ECC keys provide the same levels of protection as the 1024-bit and 2048-bit RSA keys, respectively. Shorter key lengths lead to faster calculations, lower power consumption, and less memory and peripheral loads. All this makes the ECC technology indispensable for small devices and allows you to facilitate the work of servers aimed at ensuring the security of network connections.

3 Digital Signature Algorithm Based on a Group of Points on an Elliptic Curve

An electronic digital signature for a message is a number depending on the message itself and on a secret key known only to the signatory of the subject. At the same time, the signature should be easily verified without knowing the secret key. In the event of a dispute arising from the refusal of the signer of the fact of signing a certain message or attempting to fake the signature, a third independent party should be able to resolve the dispute [15].

The use of digital signature allows solving the following tasks: to authenticate the source of the message; establish the integrity of the message; ensure that it is impossible to refuse the signature of a specific message.

A common practice is to generate an EDS not for the message itself, but for its hash image with an appropriate choice of the hash function.

In standard of ECDSA is uses elliptic curves over a field of characteristic with a value of 2. However, cryptographically resistant curves over a field of such a characteristic are relatively small. Therefore, we consider EDS on elliptic curves given over a field of greater characteristic [16].

The choice of a curve and a point on it implies the solution of a number of auxiliary problems. First of all, it is counting the number of points on the curve. If N is the number of points on (F_p), then the following conditions must be met:

$$p + 1 - 2\sqrt{p} \leq N \leq p + 1 + 2\sqrt{p}, \; G \in E(F_p) \Rightarrow N \cdot G = O. \tag{3}$$

Thus, in order to weed out the extra numbers from the interval $(p + 1 - 2\sqrt{p}, p + 1 + 2\sqrt{p})$, we can check condition (3) for different points of G. The only remaining number will be the desired order of the curve.

There are several methods for optimizing the determination of the order of a curve, the method of large-small steps, the Shuf-method. You can get acquainted with them from the books given in the list of references [7].

To obtain a cryptographically secure EDS system, the following conditions must be met:

- $N \neq p$ and $N \neq p + 1$, where N is the order of the curve;
- $p^k \neq 1 \pmod{n}$ for all $k = 1, \ldots, C$, where C is so large that it is impossible to calculate the discrete logarithm in $\mathbb{F}_{p^c}$ in a reasonable time;
- Currently, where $C = 20$ value is considered sufficient.

The order of the point G used in the system of digital signature should be a prime number n, $n > \max\{2^{160}, 4\sqrt{p}\}$.

A possible way to protect against known attacks and against possible attacks for special classes of curves that may be detected in the future is to choose the curve E in a random way so that the specified conditions are met.

After the order N of the curve is determined, it is required to find a large prime divider n of the order of the curve. Such a divisor may not exist, and then it will be necessary to repeat the procedure of selecting a curve until all the required conditions are fulfilled. The search for the number n may require both the factorization of the number N and the proof of the simplicity of the number n.

Point G can be selected as follows. Find a random point $G' \in$ (Fp) and calculate $G = \frac{N}{n} \cdot G'$. If $G \neq O$, then the required point is found, if it is $G = O$, then choose another point G'.

The described parameters may be common to all users. To generate and verify signatures, individual user parameters are also required - these are secret and public keys. The signature key (secret key) is a random number d, $0 < d < n$. The signature verification key (public key) is the point of the elliptic curve $Q = d \cdot G$. The EDS algorithm also uses a hash function, denoted by h.

Signature generation (r, s) is executed by the following algorithm:

– Choose a random number k in the interval $[1, n - 1]$.
– Calculate $(x, y) = k \cdot G$.
– Calculate $r = x \bmod n$.
– If $r = 0$, then go back to step 1.
– Calculate $z = k^{-1} \bmod n$.
– Calculate $e = h(m)$.
– Calculate $s = z(e + dr) \bmod n$.
– If $s = 0$, then go back to step 1.
– Print a pair (r, s) - a signature to m.

The signature verification algorithm takes several steps:

– If at least one of the conditions is violated $1 \leq r \leq n - 1$, $1 \leq s \leq n - 1$, then the signature is fake and the algorithm is completed.
– Calculate $e = h(m)$.
– Calculate $v = s^{-1} \bmod n$.
– Calculate $u_1 = ev \bmod n$.
– Calculate $u_2 = rv \bmod n$.
– Calculate $X = u_1 \cdot G + u_2 \cdot Q = (x, y)$.
– If $r = x \bmod n$, then the signature is valid, otherwise the signature is fake.

The proof of the correctness of the generation algorithm and the signature verification algorithm is very simple and is provided further.

4 Formation of the Electronic Digital Signature

Before embarking on the implementation of the algorithm of digital signature, we write a class to work with the points of an elliptic curve.

To create a digital signature, we use a large number of d, which is a permanent secret key of the scheme, and will be known only to the signer. In order to calculate the signature of a message M, using the NIST algorithm, we take the next steps:

Step 1.   Calculate hash messages M: $H = h(M)$;
Step 2.   Calculate the integer α, the binary representation of which is H;

Step 3.   Define $e = \alpha mod n$, if $e = 0$, set $e = 1$;
Step 4.   Generate a random number k satisfying the condition $0 < k < n$;
Step 5.   Calculate the point of an elliptic curve $C = k \times G$;
Step 6.   Determine $r = xCmod n$, where xC is the x-coordinate of point C. If $r = 0$, then return to step 4;
Step 7.   Calculate the value $s = (rd + ke)mod n$. If $s = 0$, then go back to step 4;
Step 8.   Return the value of $r||s$ as a digital signature

To implement all these steps, we have written the function SignGen:

```
public string SignGen(byte[] h, BigInteger d)

{

BigInteger alpha = new BigInteger(h);

BigInteger e = alpha % n;

if (e == 0) e = 1;

BigInteger k = new BigInteger();

ECPoint C=new ECPoint();

BigInteger r=new BigInteger();

BigInteger s = new BigInteger();

do

{

do

{

k.genRandomBits(n.bitCount(), new Random());

} while ((k < 0) || (k > n));

C = ECPoint.multiply(k, G);

r = C.x % n;

s = ((r * d) + (k * e)) % n;

} while ((r == 0)||(s==0));

string Rvector =padding(r.ToHexString(),n.bitCount()/4);

stringSvector=padding(s.ToHexString(), n.bitCount()/4);

return Rvector + Svector;

}
```

In this part of our code, the filling function complements the hexadecimal representation of the numbers r and s of the length of the module p, so that when checking the signature can be analyzed by the recipient.

To verify the signature, we use the Q point that satisfies our equality $Q = d \times G$. The point Q is the public key of the scheme and may be known to any verifier. Our proposed signature verification process is based on the following algorithm:

Step 1.   Using the received signature, restore the numbers r and s. If the inequalities $0 < r < n$ and $0 < s < n$ are not met, then return "the signature is not true";

Step 2.   Calculate the message hash $M: H = h\ (M)$;

Step 3.   Calculate the integer α, the binary representation of which is H;

Step 4.   Define $e = \alpha mod n$, if $e = 0$, set $e = 1$;

Step 5.   Calculate $v = (e - 1) mod n$;

Step 6.   Calculate the values of $z_1 = s \times v \times mod n$ and $z_1 = -r \times v \times mod n$;

Step 7.   Calculate the point of an elliptic curve $C = z_1 \times G + z_2 \times Q$;

Step 8.   Determine $R = xc\ mod\ n$, where xc is the x-coordinate of point C;

Step 9.   If $R = r$, then the signature is correct. Otherwise, the signature is not accepted

For a better understanding of the algorithm, we describe the signature verification process in the form of formulas:

$$C_v = z_1 \times G + z_2 \times Q = svG \text{ - } rvdG = se^{-1}G - re^{-1}dG$$
$$= (rd + ke)e^{-1}G - re^{-1}dG = kG = C_s \qquad (4)$$

As we see, at the verification stage, we get the same point $C = k \times G$ as when signing up. We will represent the SignVer that performs the verification as follows:

```
public bool SignVer(byte[] H, string sign, ECPointQ)

{

string Rvector = sign.Substring(0, n.bitCount()/4);

string Svector=sign.Substring(n.bitCount()/4, n.bitCount()/4);

BigInteger r = new BigInteger(Rvector, 16);

BigInteger s = new BigInteger(Svector, 16);

if ((r<1)||(r>(n-1))||(s<1)||(s>(n-1)))
```

```
return false;

BigInteger alpha = new BigInteger(H);

BigInteger e = alpha % n;

if (e == 0)

e = 1;

BigInteger v = e.modInverse(n);

BigInteger z1 = (s * v) % n;

BigInteger z2 = n + ((-(r * v)) % n);

this.G = GDecompression();

ECPoint A = ECPoint.multiply(z1, G);

ECPoint B = ECPoint.multiply(z2, Q);

ECPoint C = A + B;

BigInteger R = C.x % n;

if (R == r)

return true;

else

return false;

}
```

With GDecompression (), we unpack encrypted digital messages. Below we offer a DSGost class that implements digital signing and message verification using the algorithm NIST:

```
private void ECTest()

{

BigInteger p=newBigInteger("627710173538668076383578
9423206666416083908700390324961279", 10);

BigInteger a = new BigInteger("-3", 10);

BigInteger b=newBigInteger("64210519e59c80e70fa7e9
ab72243049feb8deecc146b9b1", 16);

byte[] xG = FromHex-
StringToByte("03188da80eb03090f67cbf20eb43a18800f4ff0afd82ff1012");

BigInteger n = new BigInteger("ffffffffffffffffffffffff99def836146bc9b1b4d22831",
16);

DSGost DS = new DSGost(p, a, b, n, xG);

BigInteger d=DS.GenPrivateKey(192);

ECPoint Q = DS.GenPublicKey(d);

GOST hash = new GOST(256);

byte[] H = hash.GetHash(Encoding.Default.GetBytes("Message"));

string sign = DS.SignGen(H, d);

bool result = DS.SignVer(H, sign, Q);

}
```

The main question that arises when we are offered a "ready" curve, or is it really generated in a random manner. This question, as a rule, is relevant to most crypto-graphic algorithms. It is possible that the proposed algorithm has some property that allows the "crack" of the cryptosystem, and the person who developed the curve may later access the information encrypted using this algorithm. The problem of proof of the absence of special properties of the algorithm is reduced to the problem of such proof for randomly selected parameters. In particular, for the curve recommended by us above, we will have to prove the randomness of the choice of parameter b. The solution to this problem can be shown in a certain way. Let $h(x)$ be a cryptographically stable hash function. To generate the number b, first select the number s, and then calculate $b = h(s)$ and give both the numbers: b and s. If our hash function satisfies all the requirements of stability, then b can not have any predefined properties, and we can safely use it. The number s in our case will be a "certificate", which confirms the "cleanliness" of the number. The above-discussed algorithm based on the elliptic curve has a similar certificate based on the SHA-1 hash function, and the standard defines the procedure for using this certificate. Thus, we can be sure that the algorithm is actually generated by chance.

5 Result and Discussion

It is important to note that the ECDLP solution may not exist. This is due to the fact that the group of points on an elliptic curve is not always a cyclic group. However, it can always be represented as a result of the interaction of two cyclic groups. As for DLP, we believe that its solution always exists. This is due to the fact that the multiplicative field group is cyclic. To study cryptographic stability, we need to understand the complexity of the algorithm. Basically, the complexity of the algorithm is the number of arithmetic operations performed by them. We represent the complexity of the function of the length of the input signal, that is, the number of bits n needed to write the input data. If this function is a polynomial of n, then the algorithm has a polynomial complexity. If this function looks like e^{Cn}, $c = const$, then the algorithm has exponential complexity.

$$L_p(\mu, c) = \exp(c(Lnp)^{\mu}(LnLnp)^{1-\mu}). \tag{5}$$

When $\mu = 0$, this function is polynomial in Lnp, $\mu = 1$, we consider it exponential. We will consider the behavior of this function at $0 < \mu < 1$ as sub-exponential.

The best currently known DLP solution algorithm in F_p has complexity $Lp(\frac{1}{3}, c_0)$, $c_0 \approx (\frac{64}{9})^{\frac{1}{3}} \approx 1,92$. This algorithm was proposed by Shirokuars and implemented as a program by Weber. The basic idea is to modify a sifting algorithm for a numerical field. Now we note that the best among the currently known algorithms for solving the ECDLP has the complexity $O\sqrt{p}$ of addition operations in the group $\langle E(F_p, +) \rangle$.

We can analyze the difference in key lengths for the same level of cryptographic security in the final fields (DLP) and on elliptic curves (ECDLP). We have $C_{ECDLP}(n) \approx C_{DLP}(l)$, therefore is $n = 2(\log_2(\exp(c_1 l^{\frac{1}{3}}(\ln(l \ln 2))^{\frac{2}{3}})))$ (Fig. 3).

From the above calculations, it can be seen that:

$$C_{ECDLP}(n) \approx C_{DLP}(s, n), \quad s = O\left(\frac{n^2}{(\ln 2)^2}\right), \tag{6}$$

where ECDLP is considered over the field F_p, and DLP over the field $F_p S$. Consequently, there was an increase in complexity (and as a result of the system's cryptographic strength) on the nonlinear multiplier $s = O\left(\frac{n^2}{(\ln 2)^2}\right)$ only due to the transition to the language of elliptic curves. It is worth noting that elliptic curves are a universal means of generalization, this means that most of the existing cryptosystems can be

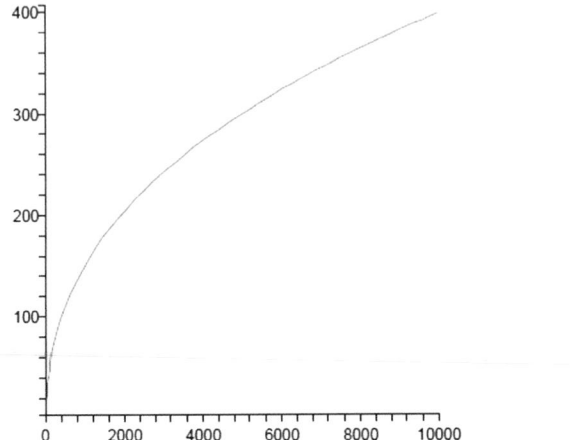

Fig. 3. The ratio of key lengths in cryptosystems over finite fields (horizontal axis) and cryptosystems based on elliptic curves (vertical axis) having the same security level.

transferred to elliptic curves, as a result of which we will get a significant gain in cryptoresistance. Note that the transition to elliptic curves is somewhat equivalent to taking an extension of a field of degree s (Fig. 4).

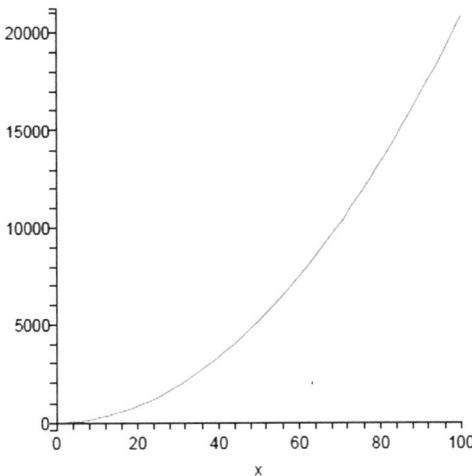

Fig. 4. Field expansion ratio (vertical axis), bit length (horizontal axis).

It should be noted that the operation in the fields with such a large expansion is impossible for modern requirements of cryptography (where speed and the ability of the entire system to work in real time). Based on the foregoing, the advantage of using elliptic curves is obvious. We get a serious level of cryptographic security using keys that are significantly shorter than we could have remained in the framework of finite fields and their extensions.

Below are the conclusions regarding the key length ratio in standard systems? It should be noted that, according to the RSA standard, the key length in DLP systems similar to the El-Gamal system is assumed to be 512 bits. With the same level of cryptographic security when switching to elliptic curves, the key length is about 100 bits (Table 1).

Table 1. The ratio of key is lengths in standard RSA and DLP systems.

The name of cryptosystem	Key length in the case of DLP, bits	Key length in case of ECDLP, bits
El-Gamal system	512	112
Mesci-Omura system	512	112
Diffie-Hellman system	512	112
El-Gamal system	1024	152
Mesci-Omura system	1024	152
Diffie-Hellman system	1024	152
El-Gamal system	2048	206
Mesci-Omura system	2048	206
Diffie-Hellman system	2048	206

Taking into account the standard for digital signature DSA (RSA), we give the gain in the transition to elliptic curves (ECDSA) in this case (Table 2).

Table 2. The gain in the transition to elliptic curves, bits.

The system based on elliptic curve	The system based on RSA/DSA
106	512
132	768
160	1024
224	2048

As we can see from the last table, the 1024-bit DSA digital signature scheme can be easily replaced with the 160-bit digital signature scheme on the ECDSA elliptic curves. In this case, there is a serious decrease in key size.

The performance of computing devices has recently been taken to be assessed at MIPS (Million Instruction Per Second): $1\ MIPS = 10^6$ $operations/s$. MIPS year essentially characterizes the complexity of the algorithm, which requires the annual operation of the computer to open the corresponding cipher. With respect to the elliptic curves, the performance of 1 MIPS corresponds to approximately $4*10^4$ $operations$ of addition of the curve per second, since the key length significantly exceeds the length of a data unit. The stability of cryptography algorithms is usually evaluated in MIPS years. In other words, resilience is the number of years of uninterrupted work required by a computer with a capacity of 1 MIPS to crack this cipher. In this regard, it seems appropriate to give the MIPS characteristics in the latter case (Table 3).

Table 3. The performance of computing devices in standard RSA and DLP systems.

Time to crack MIPS, years	RSA/DSA key size	Key size ECC	The ratio of the lengths of keys RSA/DSA vs ECC
10^4	512	106	5:1
10^8	768	132	6:1
10^{11}	1024	160	7:1
10^{20}	2048	210	10:1

6 Conclusions

We consider the construction of an electronic digital signature with the help of cryptography of elliptic curves. We can say that this method is more simple compared to others and much more economical than computing resources.

Cryptography of elliptical curves is one of the basic technologies for constructing the public key infrastructure. To withstand existing algorithms for solving the problem of discrete logarithms on an elliptic curve, an elliptic curve must meet certain conditions. Such curves are usually called cryptographically strong. One of the known problems of elliptic cryptography is the generation of strong elliptic curves over finite fields of simple order.

Our future research will focus on building more other encryption algorithms based on elliptic curves. Also, special attention will be paid to considering possible criminal access to the database and to calculate the reliability of the implementation of the ECDLP in the operation of cloud storage data for enterprises of corporate type.

References

1. Petrov, A.A., Karpinski, M., Petrov, O.S.: Development of methodological basis of management of information protection in the segment of corporate information systems. In: 18th International Multidisciplinary Scientific Geoconference: Informatics, SGEM 2018. Informatics, go Informatics and Remote Sensing, Conference Proceedings, STEF92 Technology, no. 2.1, vol. 18, pp. 317–324 (2018)
2. Kamboj, D., Gupta, D.K., Kumar, A.: Efficient scalar multiplication over elliptic curve. Int. J. Comput. Netw. Inf. Secur. (IJCNIS) 8(4), 56–61 (2016). https://doi.org/10.5815/ijcnis. 2016.04.07
3. Federal Information Processing Standards Publication 46-2. Data Encryption Standard (DES). NIST, US Department of Commerce, Washington D.C. (1993)

4. FIPS PUB 186-4. Federal Information Processing Standards Publication. Digital Signature Standard (DSS). National Institute of Standard and Technology (2013)
5. Mousa, H.M.: Chaotic genetic-fuzzy encryption technique. Int. J. Comput. Netw. Inf. Secur. (IJCNIS) **10**(4), 10–19 (2018). https://doi.org/10.5815/ijcnis.2018.04.02
6. Jacobson, M., Menezes, A.: Solving Elliptic Curve Discrete Logarithm Problems Using Weil Descent. University of Illinois (2001)
7. Agarkhed, J., Ashalatha, R.: Security and privacy for data storage service scheme in cloud computing. Int. J. Inf. Eng. Electron. Bus. (IJIEEB) **9**(4), 7–12 (2017). https://doi.org/10.5815/ijieeb.2017.04.02
8. Koblitz, N.: Random Curves. Journeys of a Mathematician, pp. 312–313. Springer, Heidelberg (2009). ISBN 9783540740780
9. Akomolafe, O.P., Abodunrin, M.O.: A hybrid cryptographic model for data storage in mobile cloud computing. Int. J. Comput. Netw. Inf. Secur. (IJCNIS) **9**(6), 53–60 (2017). https://doi.org/10.5815/ijcnis.2017.06.06
10. Al-Haija, Q.A., Asad, M.M., Marouf, I.: A systematic expository review of schmidt-samoa cryptosystem. Int. J. Math. Sci. Comput. (IJMSC) **4**(2), 12–21 (2018). https://doi.org/10.5815/ijmsc.2018.02.02
11. Alimoradi, R.: A study of hyperelliptic curves in cryptography. Int. J. Comput. Netw. Inf. Secur. (IJCNIS) **8**(8), 67–72 (2016). https://doi.org/10.5815/ijcnis.2016.08.08
12. Rivest, R.L.: The RC5 Encryption Algorithm. Crypto bytes, RSA Laboratories, vol. 1, no. 1, pp. 9–11 (1995)
13. Gaur, T., Sharma, D.: A secure and efficient client-side encryption scheme in cloud computing. Int. J. Wirel. Microwave Technol. (IJWMT) **6**(1), 23–33 (2016). https://doi.org/10.5815/ijwmt.2016.01.03
14. Kumar, T., Chauhan, S.: Image cryptography with matrix array symmetric key using chaos based approach. Int. J. Comput. Netw. Inf. Secur. (IJCNIS) **10**(3), 60–66 (2018). https://doi.org/10.5815/ijcnis.2018.03.07
15. Olorunfemi, T.O.S., Alese, B.K., Falaki, S.O., Fajuyigbe, O.: Implementation of elliptic curve digital signature algorithms. J. Softw. Eng. **1**, 1–12 (2007)
16. Wiener, M.J., Zuccherato, R.J.: Faster attacks on elliptic curve cryptosystems. In: Selected Areas in Cryptography. Lecture Notes in Computer Science, vol. 1556, pp. 252–266. Springer, Heidelberg (1999)
17. Hu, Z., Dychka, I., Onai, M., Ivaschenko, M., Su, J.: Improved method of López-Dahab-Montgomery scalar point multiplication in binary elliptic curve cryptography. Int. J. Intell. Syst. Appl. (IJISA) **10**(12), 27–34 (2018). https://doi.org/10.5815/ijisa.2018.12.03

Mathematical Model of Heat Exchange for Non-stationary Mode of Water Heater

Igor Golinko and Iryna Galytska[✉]

National Technical University of Ukraine
"Igor Sikorsky Kyiv Polytechnic Institute",
Peremogy Street 37, Kyiv 03056, Ukraine
irinagalicka@gmail.com

Abstract. The dynamical model of heat exchange for a water heater with lumped parameters, which can be used for synthesis of control systems by inflowing-exhaust ventilation installations, or industrial complexes of artificial microclimate, is considered. A mathematical description that represents the dynamical properties of a water heater concerning the main channels of regulation and perturbation is presented. Numerical simulation of transient processes for the VEZA VNV 243.1 heater according to the influence channels was carried out. The resulting dynamical model of a water heater can be the basis for the synthesis of automatic control systems and simulation of transients. A significant advantage of a mathematical model in the state space is the possibility of synthesis and analysis of a multidimensional control system.

Keywords: Dynamical model · State space · Water heater

1 Introduction

When heating the buildings, water and air systems are most often used. Air heating systems have been used recently and have proven themselves as high-speed, with a small specific capital cost. Air heating systems use electric heaters with a distributed automatic control system [1]. For air heating of commercial and business centers, warehouses and industrial buildings, centralized ventilation and conditioning systems are used, where the main equipment is a water heater. The achievement of high performance indicators of industrial ventilation and conditioning systems involves the development of adequate mathematical models of water heaters and their control methods. The use of numerical methods and computer technology made it possible to substantially elaborate the mathematical description of heat exchangers. In developing a mathematical model, the task is to determine the limits of its detalization. The dynamical model should be simple for its research and synthesis of the control system, and also take into account the features of heat exchange.

Researchers use mathematical models with lumped [2, 3] and distributed [4, 5] parameters to simulate the dynamical processes of heat exchange devices. Models with lumped parameters are simpler in the calculations and make it possible to obtain an analytical solution, models with distributed parameters claim to provide a more precise mathematical description. An analytical modeling of heat exchangers with distributed

© Springer Nature Switzerland AG 2020
Z. Hu et al. (Eds.): ICCSEEA 2019, AISC 938, pp. 58–67, 2020.
https://doi.org/10.1007/978-3-030-16621-2_6

parameters is a rather complicated task, transcendental functions appear in the solution [4]. For such tasks numerical methods of the solution are used in practice.

Considering surface heat exchangers used in industrial ventilation and air conditioning systems, it is necessary to take into account that the air is intensively mixed with the fans of the air conditioner. Practical studies of non-stationary heat exchange characterize the process as clearly aperiodic, which with sufficient accuracy described by transfer functions of the second order. Thus, a mathematical model with lumped parameters for a water cooler will be perfectly acceptable.

2 Using Dynamic Models

Typically, static models of process devices are used to design and calculate the power of a particular unit. Dynamic models of process devices are used for the synthesis and analysis of control systems. When designing and optimizing any control system, the structure and parameters of the controller are primarily dependent on the dynamic behavior of a controlled plant. The choice of whether a different law of control is determined by the dynamic properties of a controlled plant. The choice of a law of control is determined by the dynamic properties of a controlled plant. In automation systems, developers use: classical PI controllers, PID controllers; adaptive controllers [6, 7]; controllers with fuzzy logic [8, 9]; controllers with fractional derivatives [10]; intelligent controllers [11]; controllers based on neural networks and many other types of controllers. In their research, experts compare the operation of control systems with different types of controllers and draw conclusions about the quality of management. In the study of control systems with different types of controllers unchanged in the system there is a controlled plant, which characterizes the dynamic behavior of the technological apparatus and formalized in the form of differential equations controlled plant. Thus, an adequate mathematical model of a controlled plant is the basis for the synthesis and analysis of a qualitative control system.

The purpose of the publication is to develop a mathematical model of the heat transfer process for a water heater in the state space, which will allow the analysis of dynamic characteristics of water heaters of industrial air conditioners. An additional requirement is the convenience of using the resulting model in the MatLAB environment.

3 Dynamical Model of Water Heater

The following simplifications were made during the development of a dynamic model of a water heater: heat exchange with the environment is absent, since the thermal losses of modern heaters do not exceed 5%; the model contains three main dynamic elements with lumped parameters (water, heat exchange surface and air); the physical properties of the material flows and the heat transfer surface are brought to the averaged values of the working range. The calculation scheme of the water heater is shown in Fig. 1.

On the calculation scheme, the following notation is taken. The heat-carrier is water with a consumption $G_W(t)$, the temperature of the heat-carrier at the heater input is $\theta_{W0}(t)$, at the heater output is $\theta_W(t)$. The heater contains n tubes with the length H. The heat-carrier heats the heat exchange tubes, the average temperature of which is $\theta_M(t)$. The air flow comes into the heater $G_A(t)$ uncrossing the heat-carrier flow. The input air temperature is $\theta_{A0}(t)$, the output air temperature is $\theta_A(t)$. The geometric dimensions L, C, H are the depth, width and height of a water heater.

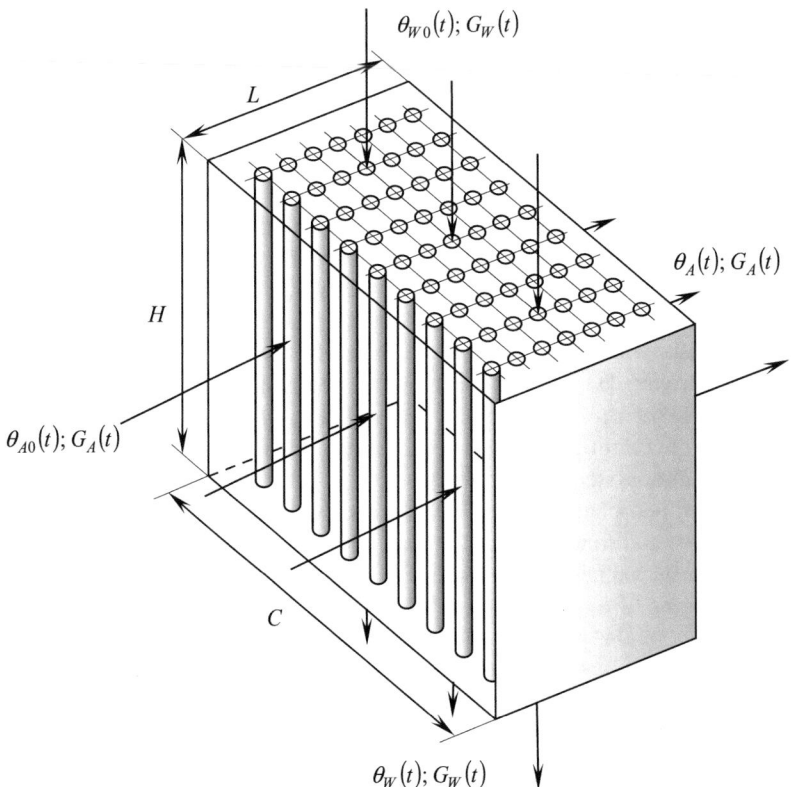

Fig. 1. Schematic diagram for simulation

Let's consider the thermal balance in dynamics for each dynamic element.
The thermal balance for the heat-carrier flowing in the tubes is:

$$G_W c_W (\theta_{W0} - \theta_W) - \alpha_0 F_0 (\theta_W - \theta_M) = M_W c_W \frac{d\theta_W}{dt}, \qquad (1)$$

where c_W is heat capacity of the heat-carrier; α_0 is coefficient of heat transfer between the coolant and the inner surface of the pipes; F_0 is the area of the internal surface of all the pipes, $F_0 = n L \pi d_0$; d_0 is the internal diameter of the pipes; M_W is the mass of the heat-carrier in all tubes of the heater, $M_W = n \rho_W L \pi d_0^2 / 4$; ρ_W is the density of the heat-carrier.

After simplification and linearization, Eq. (1) will take the form:

$$T_W \frac{d \Delta \theta_W}{dt} + \Delta \theta_W = k_0 \Delta \theta_{W0} + k_1 \Delta \theta_M + k_2 \Delta G_W, \tag{2}$$

where $K_W = c_W G_W + \alpha_0 F_0$; $T_W = \frac{c_W M_W}{K_W}$; $k_0 = \frac{c_W G_W}{K_W}$; $k_1 = 1 - k_0$; $k_2 = \frac{c_W (\theta_{W0} - \theta_W)}{K_W}$.

Heat balance for the heat-exchanger surface of the heater:

$$\alpha_0 F_0 (\theta_W - \theta_M) - \alpha_1 F_1 (\theta_M - \theta_A) = M_M c_M \frac{d \theta_M}{dt}, \tag{3}$$

where c_M is heat capacity of the metal; α_1 is coefficient of heat transfer between the outer surface of pipes and the air; F_1 is heat exchange area of the external surface of the heater; M_M is mass of metal of heat exchange surface.

After simplifying Eq. (3) we obtain the differential equation in increments:

$$T_M \frac{d \Delta \theta_M}{dt} + \Delta \theta_M = k_3 \Delta \theta_W + k_4 \Delta \theta_A, \tag{4}$$

where $K_M = \alpha_0 F_0 + \alpha_1 F_1$; $T_M = \frac{c_M M_M}{K_M}$; $k_3 = \frac{\alpha_0 F_0}{K_M}$; $k_4 = 1 - k_3$.

Heat balance for airspace of the heater:

$$G_A c_A (\theta_{A0} - \theta_A) - \alpha_1 F_1 (\theta_M - \theta_A) = M_A c_A \frac{d \theta_A}{dt}, \tag{5}$$

where c_A is air heat capacity; M_A is mass of air in volume $H \times L \times C$ of a water heater.

After simplification and linearization, Eq. (5) will take the form:

$$T_A \frac{d \Delta \theta_A}{dt} + \Delta \theta_A = k_5 \Delta \theta_{A0} + k_6 \Delta \theta_M + k_7 \Delta G_A, \tag{6}$$

where $K_A = c_A G_A + \alpha_1 F_1$; $T_A = \frac{c_A M_A}{K_A}$; $k_5 = \frac{c_A G_A}{K_A}$; $k_6 = 1 - k_5$; $k_7 = \frac{c_A (\theta_{A0} - \theta_A)}{K_A}$.

Equations (2), (4) and (6) describe the behavior of the dynamical elements in the water heater. The mathematical model of the heat transfer of the water heater is represented by a system of ordinary differential equations:

$$\begin{cases} T_W \frac{d \Delta \theta_W}{dt} + \Delta \theta_W = k_0 \Delta \theta_{W0} + k_1 \Delta \theta_M + k_2 \Delta G_W; \\ T_M \frac{d \Delta \theta_M}{dt} + \Delta \theta_M = k_3 \Delta \theta_W + k_4 \Delta \theta_A; \\ T_A \frac{d \Delta \theta_A}{dt} + \Delta \theta_A = k_5 \Delta \theta_{A0} + k_6 \Delta \theta_M + k_7 \Delta G_A. \end{cases} \tag{7}$$

We represent (7) as the system of algebraic equations in the Laplace space.

$$\Delta\theta_W(T_W p + 1) = k_0\,\Delta\theta_{W0} + k_1\,\Delta\theta_M + k_2\,\Delta G_W; \tag{8}$$

$$\Delta\theta_M(T_M p + 1) = k_3\,\Delta\theta_W + k_4\,\Delta\theta_A; \tag{9}$$

$$\Delta\theta_A(T_A p + 1) = k_5\,\Delta\theta_{A0} + k_6\,\Delta\theta_M + k_7\,\Delta G_A. \tag{10}$$

From (8) we find $\Delta\theta_W(p)$; from (9) determine $\Delta\theta_M(p)$, taking into account $\Delta\theta_W(p)$; the determined $\Delta\theta_M(p)$ we substitute in (10).

After grouping similar and simplifying we obtain:

$$\begin{aligned}
\Delta\theta_A = \frac{1}{a_3 p^3 + a_2 p^2 + a_1 p + 1} & \left[\left(b_2\,p^2 + b_1\,p + b_0 \right)\Delta\theta_{A0} \right.\\
& \left. + \left(b_5\,p^2 + b_4\,p + b_3 \right)\Delta G_A + b_6\,\Delta\theta_{W0} + b_7\,\Delta G_W \right],
\end{aligned} \tag{11}$$

where

$$a_1 = \frac{T_W + T_M + T_A - k_1 k_3 T_A - k_4 k_6 T_W}{1 - k_1 k_3 - k_4 k_6}; \quad a_2 = \frac{T_W T_M + T_W T_A + T_M T_A}{1 - k_1 k_3 - k_4 k_6};$$

$$a_3 = \frac{T_W T_M T_A}{1 - k_1 k_3 - k_4 k_6}; \quad b_0 = \frac{k_5 (1 - k_1 k_3)}{1 - k_1 k_3 - k_4 k_6}; \quad b_1 = \frac{k_5 (T_W + T_M)}{1 - k_1 k_3 - k_4 k_6};$$

$$b_2 = \frac{k_5 T_W T_M}{1 - k_1 k_3 - k_4 k_6}; \quad b_3 = \frac{k_7 (1 - k_1 k_3)}{1 - k_1 k_3 - k_4 k_6}; \quad b_4 = \frac{k_7 (T_W + T_M)}{1 - k_1 k_3 - k_4 k_6};$$

$$b_5 = \frac{k_7 T_W T_M}{1 - k_1 k_3 - k_4 k_6}; \quad b_6 = \frac{k_0 k_3 k_6}{1 - k_1 k_3 - k_4 k_6}; \quad b_7 = \frac{k_2 k_3 k_6}{1 - k_1 k_3 - k_4 k_6}.$$

Applying Laplace's inverse transformation, one can find an analytical solution (11) to the main channels of regulation and perturbation.

The mathematical model (7) in the state space will take the form:

$$\mathbf{X}' = \mathbf{AX} + \mathbf{BU}, \tag{12}$$

where

$$\mathbf{X}' = \begin{bmatrix} \Delta\theta'_A \\ \Delta\theta'_M \\ \Delta\theta'_W \end{bmatrix}; \quad \mathbf{A} = \begin{bmatrix} -1/T_A & k_6/T_A & 0 \\ k_4/T_M & -1/T_M & k_3/T_M \\ 0 & k_1/T_W & -1/T_W - 1/T_W \end{bmatrix}; \quad \mathbf{X} = \begin{bmatrix} \Delta\theta_A \\ \Delta\theta_M \\ \Delta\theta_W \end{bmatrix};$$

$$\mathbf{B} = \begin{bmatrix} k_5/T_A & k_7/T_A & 0 & 0 \\ 0 & 0 & 0 & 0 \\ 0 & 0 & k_0/T_W & k_2/T_W \end{bmatrix}; \quad \mathbf{U} = \begin{bmatrix} \Delta\theta_{A0} \\ \Delta G_A \\ \Delta\theta_{W0} \\ \Delta G_W \end{bmatrix}.$$

4   Structural Scheme of the Heater's Mathematical Model

The analysis of the mathematical model (11) allows us to obtain a structural scheme of the water heat transfer apparatus, as shown in Fig. 2.

The temperature of the air $\Delta\theta_A(t)$ at the outlet of the heater varies depending on the temperature $\Delta\theta_{A0}(t)$ and air consumption $\Delta G_A(t)$ at the inlet of the device, as well as the temperature $\Delta\theta_{W0}(t)$ and heat-carrier consumption supplied to the device. The control influence can be carried out by two control channels, in practice only one is used.

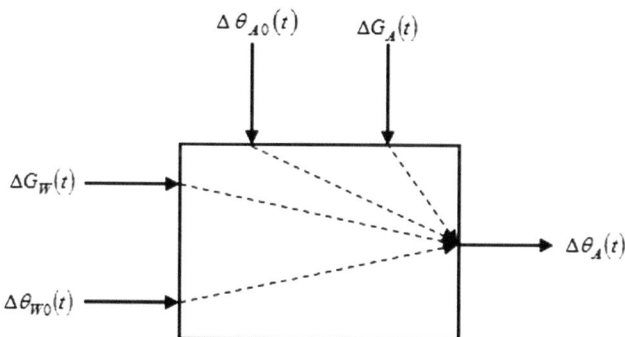

Fig. 2. Structural scheme of the heater's mathematical model

For example, a change in the heat load of the heater can be made by changing the consumption $\Delta G_W(t)$, or by changing the temperature $\Delta\theta_{W0}(t)$ of the direct heat-carrier. The control channel is determined by the technological connection of the heater. One-way regulating valve is used to change the heat-carrier consumption $\Delta G_W(t)$. The change in the temperature of the heat-carrier at the inlet to the heater $\Delta\theta_{W0}(t)$ is ensured by the use of a three-way regulating valve with the mixing of the return heat-carrier $\Delta\theta_W(t)$ to the direct $\Delta\theta_W(t)$ through the three-way valve. Wile the heat-carrier consumption $\Delta G_W(t)$ remains unchanged. Other channels of influence will be perturbation channels.

5   Simulation of the Dynamic Mode of the Water Heater

According to the dynamical model (12), let's carry out the simulation of the transition processes for the VEZA VNV 243.1 water heater [12]. Table 1 shows the thermophysical parameters for the VEZA VNV 243.1 water heater.

Table 1. Thermophysical parameters of the VEZA VNV 243.1 water heater

Parameter name	Marking	Numerical value	Dimension
Dimensions of the water heater	$H \times L \times C$	$1570 \times 1.965 \times 0.18$	m
Water consumption	G_W	2.44	kg/sec
Water density	ρ_W	986	kg/m^3
Water heat capacity	c_W	4185	J/(kg °C)
Mass of water in the heater	M_W	18.54	kg

(*continued*)

Table 1. (*continued*)

Parameter name	Marking	Numerical value	Dimension
Area of the inner surface of heat transfer	F_0	6.54	m^2
Heat transfer coefficient for the inner surface	α_0	800	$W/(m^2 \; °C)$
Metal tubing density	ρ_M	3334.8	kg/m^3
Metal heat capacity	c_M	866.5	$J/(kg \; °C)$
Weight of heat exchanger tubing	M_M	21	kg
Heat exchange area of the external surface	F_1	274	m^2
Coefficient of heat transfer for the external surface	α_1	80	$W/(m^2 \; °C)$
Air consumption through the heater	G_A	16.67	kg/sec
Air density	ρ_A	1.2	kg/m^3
Air heat capacity	c_A	1010	$J/(kg \; °C)$
Mass of air in the heater	M_A	2.76	kg
Input water temperature	θ_{W0}	90	°C
Output water temperature	θ_W	70	°C
Input air temperature	θ_{A0}	0	°C
Output air temperature	θ_A	18	°C

The calculation of parameters for models (11) and (12) of the VEZA VNV 243.1 water heater was carried out in the MatLAB environment.

We have the following numerical values of the matrices for the model (12):

$$\mathbf{A} = \begin{bmatrix} -13.9033 & 7.8634 & 0 \\ 1.2046 & -1.4922 & 0.2875 \\ 0 & 0.0674 & -0.199 \end{bmatrix}; \; \mathbf{B} = \begin{bmatrix} 6.0399 & -6.5217 & 0 & 0 \\ 0 & 0 & 0 & 0 \\ 0 & 0 & 0.13 & 1.08 \end{bmatrix}.$$

Simulation modeling of the dynamic mode for the water heater was carried out in the Simulink MatLAB environment using the State Space functional block. The results of transients simulation by perturbation channels are shown in Fig. 3. The results of transients simulation by possible control channels are presented in Fig. 4.

The results of the simulation can be summarized as follows. The inertia of the regulation and perturbation channels is insignificant compared to the inertia of the temperature sensor and the actuating mechanism. For these reasons, when designing the system for controlling water heaters, it is necessary to take into account the inertial properties of the temperature sensor and the actuating mechanism. The inertia of the regulation channels is greater in comparison with the perturbation channels, which is explained by the heat transfer through the metal heat exchange tubes, which must be warmed up. The resulting transient processes for the channels of perturbation and regulation have an aperiodic nature without delay. By the nature of the transients, the PI controller should be recommended for controlling the water heat exchanger. Given the low inertia of transient processes and the absence of delay, a proportional part of the regulator will quickly move the apparatus to a new operating mode, and the integral part will ensure accurate monitoring of the task of the regulator. According to practical recommendations, the use of PID controllers does not justify itself for the controlling of high-speed objects. The dynamical model of the water heater is represented by

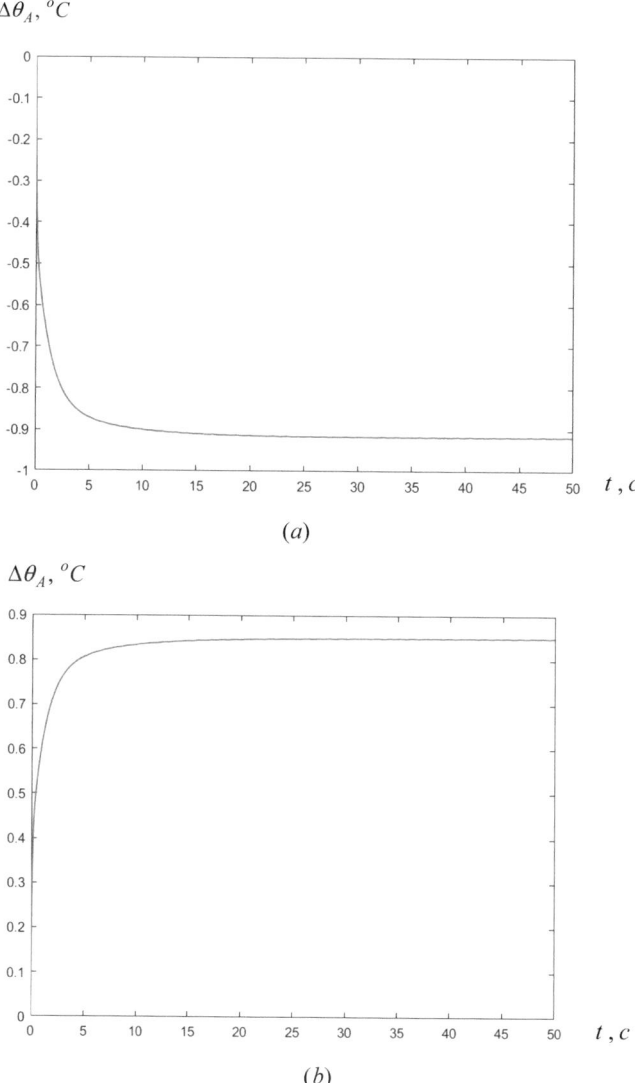

Fig. 3. Graphs of the resulting transient processes for the perturbation channels: (a) $\Delta\theta_{A0} \rightarrow \Delta\theta_A$, $\Delta\theta_{A0} = 1°C$; (b) $\Delta G_{A0} \rightarrow \Delta\theta_A$, $\Delta G_{A0} = 1\,кг/с$

equivalent dependencies (11) and (12), which allow us to qualitatively assess the transition processes for heaters of industrial ventilation and conditioning systems. The choice between dependencies (11) or (12) is determined by the synthesis methods of control systems as well as the preferences of a researcher. In the general case, determining the adequacy for the obtained models is an incorrect task.

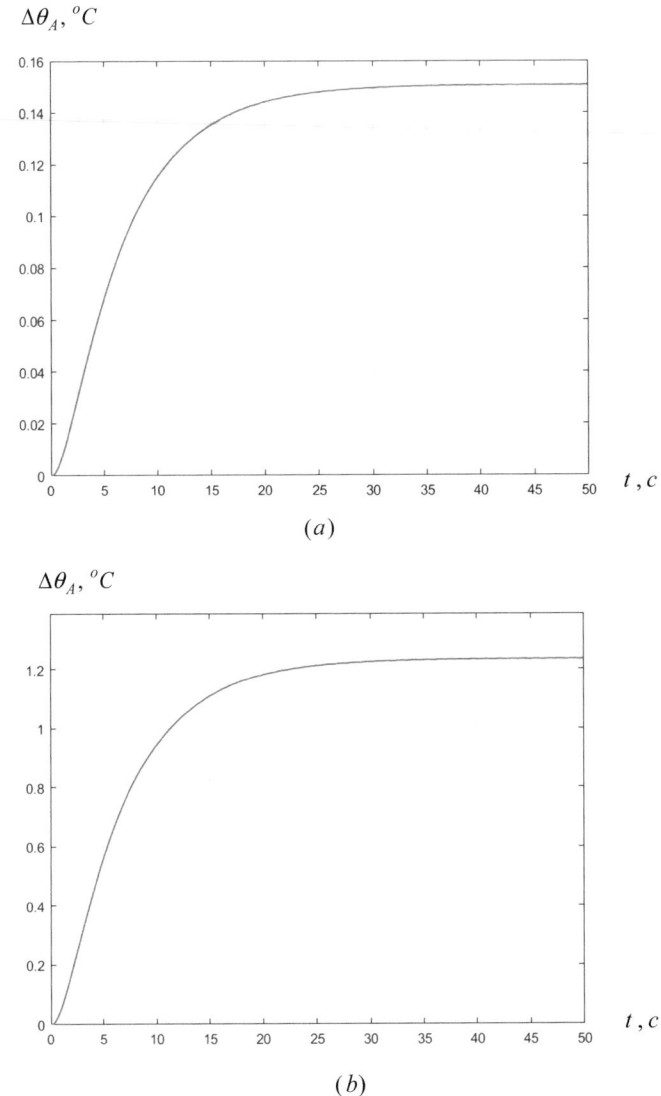

(a)

(b)

Fig. 4. Graphs of the resulting transient processes for the regulation channels: (a) $\Delta\theta_{W0} \rightarrow \Delta\theta_A$, $\Delta\theta_{W0} = 1\,°C$; (b) $\Delta G_{W0} \rightarrow \Delta\theta_A$, $\Delta G_{W0} = 1\,\kappa\varepsilon/c$

It is possible to speak about the adequacy only for a particular object and the specific conditions of the technological process with verification of conformity on the basis of experimental research. This article does not address this problem.

The values of the matrices **A** and **B** of the model (12) were calculated for the Table 1. The thermophysical values for the model (12) are determined with high accuracy from reference books, except for the coefficients of heat transfer. The heat transfer coefficients depend on many factors, they can be determined with sufficient accuracy on the basis of experimental studies for the specific equipment. In order to obtain adequate quantitative dynamical characteristics of the heater, it is necessary to additionally carry out experimental studies of the surface heat exchanger and adapt the mathematical model to the specific operating conditions.

References

1. Tkachov, V., Gruhler, G., Zaslavski, A., Bublikov, A., Protsenko, S.: Development of the algorithm for the automated synchronization of energy consumption by electric heaters under condition of limited energy resource. Eastern Eur. J. Enterp. Technol. **8**(92), 51–61 (2018)
2. Vychuzhanin, V.V.: Mathematical models of non-stationary modes of air processing in the central ACS. Bull. Odessa Natl. Marit. Univ. **23**, 172–185 (2007)
3. Golinko, I.M.: Non-stationary model of heat and mass transfer for a water cooler. In: 12th International Scientific and Practical Conference "Modern Problems of Scientific Support of Power Engineering", Kyiv, Ukrainian, p. 137 (2014)
4. Golinko, I.M., Kubrak, N.A.: Modeling and optimization of control systems. Ruta, Kamyanets-Podilskii, Ukrainian (2012)
5. Golinko, I.M., Ladanuk, A.P., Koshelieva, L.D.: Dynamic model of the heat mode of the calorifier. Inf. Technol. Comput. Eng. **3**(16), 59–63 (2009)
6. Singhal, P., Agarwal, S.K., Kumar, N.: Advanced adaptive particle swarm optimization based SVC controller for power system stability. Int. J. Intell. Syst. Appl. (IJISA) **7**(1), 101–110 (2015). https://doi.org/10.5815/ijisa.2015.01.10
7. Puangdownreong, D.: Multiobjective multipath adaptive tabu search for optimal PID controller design. Int. J. Intell. Syst. Appl. (IJISA) **7**(8), 51–58 (2015). https://doi.org/10.5815/ijisa.2015.08.07
8. Misra, Y., Kamath, H.R.: Design algorithm and performance analysis of conventional and fuzzy controller for maintaining the cane level during sugar making process. Int. J. Intell. Syst. Appl. (IJISA) **7**(1), 80–93 (2015). https://doi.org/10.5815/ijisa.2015.01.08
9. Yazdanpanah, A., Piltan, F., Roshanzamir, A., Mirshekari, M., Mozafari, N.G.: Design PID baseline fuzzy tuning proportional-derivative coefficient nonlinear controller with application to continuum robot. Int. J. Intell. Syst. Appl. (IJISA) **6**(5), 90–100 (2014). https://doi.org/10.5815/ijisa.2014.05.10
10. Soukkou, A., Belhour, M.C., Leulmi, S.: Review, design, optimization and stability analysis of fractional-order PID controller. Int. J. Intell. Syst. Appl. (IJISA) **8**(7), 73–96 (2016). https://doi.org/10.5815/ijisa.2016.07.08
11. Lakshmi, K.V., Srinivas, P., Ramesh, C.: Comparative analysis of ANN based intelligent controllers for three tank system. Int. J. Intell. Syst. Appl. (IJISA) **8**(3), 34–41 (2016). https://doi.org/10.5815/ijisa.2016.03.04
12. Nikolaeva, K.A., Golinko, I.M.: Dynamic model of a heater in the state space. In: 14th International Scientific and Practical Conference "Modern Problems of Scientific Support of Power Engineering", Kyiv, Ukrainian, p. 28 (2016)

Applying Recurrent Neural Network for Passenger Traffic Forecasting

Zhengbing Hu[1], Ivan Dychka[2], Liubov Oleshchenko[2(✉)],
and Sergiy Kukharyev[2]

[1] School of Educational Information Technology,
Central China Normal University, Wuhan, China
hzb@mail.ccnu.edu.cn
[2] National Technical University of Ukraine
"Igor Sikorsky Kyiv Polytechnic Institute", Kyiv, Ukraine
dychka@pzks.fpm.kpi.ua,
oleshchenkoliubov@gmail.com, s.a.kukharev@gmail.com

Abstract. The article represents the analysis of neural networks that can be used to predict passenger traffic between cities. Passenger data nonstationary timetable is considered. A class of recurrent neural networks (RNN) have also been considered, among which the expediency of using the Long Short-Term Memory (LSTM) neural network for analysis and prediction of passenger traffic on the interurban route investigated is selected and substantiated. The stages of the research are represented. The data of the Ukrainian motor transport enterprise for 2007–2015 were used for the experiment. The study uses static methods for predicting the moving average, exponential smoothing of Holt-Winters, linear and logarithmic trends for verification and comparison of forecast accuracy with various methods.

Keywords: Passenger traffic · Forecasting · Non-stationary time series · Motor transport enterprises · Neural network · Recurrent neural networks · Long Short-Term Memory

1 Introduction

The main factor affecting the efficiency of traffic management in intercity communication is the speed of processing and obtaining data on all passengers who use the transport services at a given time. In order to analyze data and make operational forecasts, it is necessary to know the dynamics of changes in the value of passenger flow. Therefore, there is a need to create software that will collect data on passengers, make forecasts, and then, based on these forecasts, optimally use the vehicles of motor transport enterprises (MTE).

In the context of rising fuel prices, there is a need for the use of information technologies (IT) aimed to optimize the work of motor transport enterprises. The emergence of the problem of rational use of vehicles of MTE is due to the increase of mobility of the population in the regions as a result of the establishment in the East of

© Springer Nature Switzerland AG 2020
Z. Hu et al. (Eds.): ICCSEEA 2019, AISC 938, pp. 68–77, 2020.
https://doi.org/10.1007/978-3-030-16621-2_7

Ukraine and the search for places for work, and, thus, this requires an increase in the number of vehicles to meet the growing demand for passenger traffic.

An analysis of available work has shown that at present in Ukraine there is insufficiently developed IT base to improve the quality of passenger service provided that the expenses of the MTE are minimized, as well as the planning and maintenance of vehicles in conditions of daily, weekly and seasonal changes in passenger transport demand. There are no technologies for automated decision-making on choosing the optimal distribution of vehicles referred to the possible value of passenger traffic.

The task of maximizing the profit ($P \rightarrow \max$) of the MTE is reduced to the task of minimizing the costs of servicing vehicles ($C \rightarrow \min$) with restrictions on the number of passengers carried in one hour, the total number of seats in buses, as well as the average interval of time of departure of vehicles [11]. Solving this task requires the exact value of the passenger flow determination for each subsequent hour, based on the study of statistical data for the previous period.

1.1 Passenger Flows Forecasting as a Non-stationary Time Series

Passenger traffic is variable at different time intervals and has its own AM-frequency characteristics depending on the chosen route, as well as time of day, day of the week and season. They tend to change in time depending on trends in population mobility in the region.

Figure 1 presents the daily fluctuations of passenger traffic between Ukrainian cities of Chernihiv and Kyiv in the forward and reverse directions.

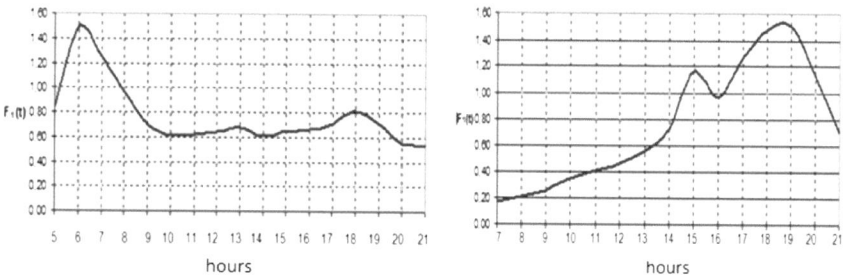

Fig. 1. Daily changes of passenger flow, route "Chernihiv-Kyiv" and "Kyiv-Chernihiv".

Adaptive methods of short-term forecasting, based on the theory of time series, are used to predict socio-economic processes. The main ones are methods of moving average, exponential smoothness of Holt-Winters, method of smoothing errors (Trigg method), Trigg-Lich methods and Chow method [7, 8, 12].

Data on the number of passengers at regular intervals can be expressed as a dynamic number Y_t, where $t \in [0, T]$ and can be represented in an additive form [13]:

$$Y_t = U_t + V_t + E_t + Z_t + \gamma_t \tag{1}$$

where U_t is the trend of the dynamic series, regular component characterizing the general tendency; V_t is seasonal component, in the general case it is a cyclic component; E_t is random component; Z_t is a component that provides the parallelism of the elements of a dynamic series; γ_t is the control component, by which the members of a row are influenced in order to shape its desired trajectory in the future.

For a specific computation of the component U_t, V_t, E_t the series Y_t use the filter-this component of the series. If you want to evaluate the trend component in a seasonal manner, i.e., a smoothing process. The seasonal component V_t is characterized by the duration of the period of seasonal fluctuations, their amplitude, the expansion of maxima and minima in time. A prerequisite for evaluating the periodic component is the exclusion of a regular Y_t of the regular component (trend U_t). Forecasting of seasonal processes is based on the decomposition of a dynamic series. It is assumed that in the future the tendency and the same nature of fluctuations will continue. Under these conditions, the forecast for any month (quarter), determined by the method of trend extrapolation, is adjusted by the seasonal index.

The dynamics of passenger traffic indicators does not have a clear-cut trend of development. Due to the constant redistribution of the influence of factors that shape the process dynamics, the intensity of the dynamics, frequency and amplitude of oscillations can be changed.

Statistical methods of prediction of time series can not predict a significant and uneven increase in passenger traffic on holidays, which are unevenly distributed over the seasons, so we will use RNN that lack such a disadvantage.

1.2 Data Model

Before we begin the forecasting of the time series, we analyze and process the data on passenger traffic. Passenger data is stored on a remote cloud service, which prevents them from being destroyed and simplifies the data integration process. The data is synchronized in a .xls file. For the experiment, we have a set of time series data, which provides average information about passenger traffic between Ukrainian cities of Chernihiv and Kyiv. This data set contains passenger information for the period of 2007–2015.

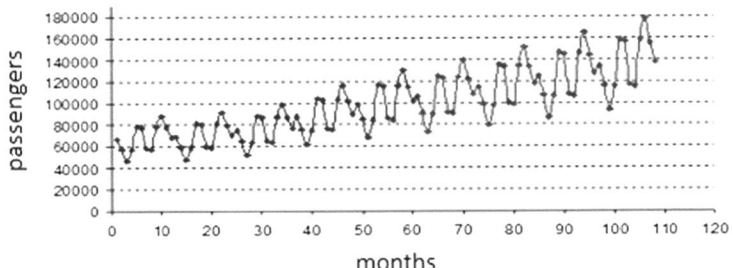

Fig. 2. Number of passengers on the route "Kyiv-Chernihiv" for 2007–2015

To implement time analysis, we remove data from the cloud and begin the process of forecasting. First of all, we will build a schedule based on passenger traffic data for 2007–2011 (Fig. 2). In our experiment, we use 56% of the data for training the neural network and 44% of the data for testing.

1.3   Objective

The main objective of the research is to predict the passenger traffic on the long-distance route using the technology of recurrent neural network. The object of the study is statistical data on passenger traffic and data on the rolling stock of MTE on the long-distance route. The subject of the study is the methods of processing data of passengers and solving optimization problems by means of the use of neural networks.

2   Related Works

The neural network model that has been developed to forecast passenger traffic flows includes artificial neural networks. The number of the neural networks corresponds to the forecasted level of passenger traffic. Time Series data is large in volume, highly dimensional and continuous updating. Time series data analysis for forecasting, is one of the most important aspects of the practical usage [6, 16].

[14] comparison of hidden Markov model, XGBoost(decision tree based boosting), recurrent neural network (RNN) and kind of RNN known as long short-term memory (LSTM) is done.

In [9] feed forward neural network using back propagation algorithm and Levenberg-Marquardt training function has been suggested. However, proposed artificial neural network model showed optimistic results for forecasting two months duration only.

3   Choosing a Neural Network to Predict Non-stationary Rows of Passenger Statistics

Time range is a sequence of vectors $x(t)$, $t = 0, 1, \ldots n$, where t is time passed. We will consider only sequences of numerical values, namely the number of passengers.

The value of x will be chosen to give a series of discrete data values, evenly distributed over time, by days, weeks, and months. The method of the variable window helps us to create a training set of data, for example: $x(1)$, $x(2)$, $x(3) \ldots x(n)$, which is a vector of time series. Table 1 represents the creation of displaying the input and output data set after using a slider window, the size of which is d.

Table 1. Display of input and output data set

Data entry	Goals
x(1), x(2), x(3), ... x(d)	x(d + 1)
x(2), x(3), x(4) ... x(d + 1)	x(d + 2)
...	...
x(n − d − 1), x(n − d), x(n − d + 1) ... x(n − 1)	x(n)

The neural network of direct propagation includes the input layer, the hidden layers and the output layer. In a neural network of direct propagation, all neurons and input blocks are connected to the anterior neuron and are not connected with the neurons, which are located in the same layer and in the previous one [2]. Elman Recurrent Networks are periodically repeated [16]. They are similar to direct propagation networks, but they have feedback in hidden layers.

Time rows can have noises, this causes problems when trying to predict. Reducing the degree of error allows the use of direct collection networks, but it significantly increases not only the complexity of the structure itself, but also the time of its learning.

The use of the Elman recurrent network allows us to solve the pro-grenade problem even in very noisy time series (this is especially important for business). In the general case, this neural network is a structure of three layers, as well as a set of additional elements (inputs). The back links go from the hidden layer to these elements; each bond has a fixed-weight weight equal to one. In each time interval, the input data is distributed by the neurons in the forward direction; then they apply a training rule. Due to fixed feedbacks, contextual elements always keep a copy of hidden layer values for the previous step (since they are sent backwards before the application of the rule). Thus, the noise of the time series is gradually offset, and with it the error is minimized. We obtain a forecast which in the general case will be more accurate than the result of the classical approach, which is confirmed experimentally in the work [10].

Theoretically there will be many hidden layers, but practically you may usually use one hidden layer, which is proved in [3, 10, 15]. The number of units in the input layer is equal to the window size. We can't define how many neurons we invest in the hidden layer, because the more coming-bathrooms neurons, the more calculations. Therefore, the optimal number of hidden neurons will be different for each dataset.

As the activation function in this case uses a simple difference-Aries nonlinear logistic function:

$$f(x) = \frac{1}{1 + e^{-x}}. \tag{2}$$

In training, we use the method of back propagation. The method of reciprocal distribution is one of the simplest and most common methods of controlled learning of multilayer neural networks [13]. The basic approach to learning is to start working with an unprepared network, present a tutorial in the inner layer, transmit signals through the network, and define output in the output layer. Here, these exits are compared with the target values; any difference corresponds to a mistake. This error or function criterion is a scalar weight function that is minimized when the network outputs match the desired results.

We consider the learning error in the sample as the square input units sum and the difference between the desired output value t_k given by the teacher and the actual output z_k:

$$J(w) = \frac{1}{2} \sum_{k=1}^{c} (t_k - z_k)^2 = \frac{1}{2} \|t - z\|^2 \tag{3}$$

Here t and z are the target and network source vectors with lengths c and w, which predefined all scales in the network.

The rule of reverse propagation training is based on the gradient run-off. Scales are not initiated by random values, and then they change in the direction that will reduce the error:

$$\Delta w = -\eta \frac{\partial J}{\partial w} \tag{4}$$

where η is the learning speed, indicating the relative magnitude of changes in scales. This repetitive algorithm requires the adoption of a weight vector on iteration m and updating it as:

$$w(m+1) = w(m) + \Delta w(m) \tag{5}$$

where m indexes the partial model of representation.

In order to solve the problem of passenger traffic forecasting, we chose a RNN, namely, long short-term memory (LSTM), because the input data for the task is a time series [1].

A RNN is well suited for predicting time-based data, since it takes into account the "experience" that was obtained earlier. The main re-weight of LSTM is the presence of nodes that are allocated to store values for long periods of time. This is due to the fact that such nodes do not use activation functions that do not change the data.

In RNN, the links between the elements form a directed sequence. This allows you to process a series of events in time or sequential spatial chains. Unlike multi-threaded perceptron networks, recurrent networks can use their internal memory to handle sequences of any length.

Like most RNN, the LSTM network is universal in the sense that, with a sufficient number of network elements, it is capable of performing any computation that a conventional computer is capable of, this requires a matrix of weights that can be considered as a program. Unlike the classical RNN, the LSTM-network is well-suited for training on tasks of processing and forecasting of time series in cases where important events are separated by time lags with an indefinite three-valiancy and boundaries. The LSTM has the form of a chain of repetitive modules of the neural network. The LSTM has four hidden layers and uses sigmoidal-like function and hyperbolic tangent as an activation function.

Neural network training is carried out with the help of reverse distribution in time, for which an iterative gradient descent is used that changes each weight coefficient in proportion to its derivative in relation to the error.

4 Method for Traffic Flow Forecasting Using RNN-LSTM

Before we begin the forecasting of the time series, we analyze and process the data on passenger traffic. Passenger data is stored on a remote cloud service, which prevents them from being destroyed and simplifies the data integration process. The data is synchronized in a .xls file. For the experiment, we have a set of time series data, which provides average information about passenger traffic between Ukrainian cities of Chernihiv and Kyiv. This data set contains passenger information for the period of 2011–2015.

For the time series of passenger traffic using LSTM, we perform the following sequence of steps.

Step 1.   Collect data for training.

Step 2.   Preparation and organizing of data.

Step 3.   Select the topology of the neural network (number of layers and the presence of feedback).

Step 4.   Experimental selection of characteristics of the neural network.

Step 5.   Empirical selection of training parameters.

Step 6.   Teaching the neural network.

Step 7.   Check the training for the adequacy of the task.

Step 8.   Adjust the parameters taking into account the previous step, the final training.

Step 9.   Verbalize the neural network (description using several algebraic or logic functions) for further use.

To implement prediction based on the neural network LSTM used high-level programming language Python and open Keras library for processing passenger data [5].

The entry of passenger data is organized according to the size of the lot, the number of time steps and the hidden size. The input is sent to the LSTM client layer in two layers. Output data is passed to a layer that has an activation tied to it. This output is compared to the training data for each batch. Learning data in this case are inputs x, which have passed one stage, that is, at each step of time the model tries to predict the next value of the sequence.

To get the prediction in the correct sequence, for each unique value in the body a unique whole index is assigned. The output file is re-indexed into a list of these unique integers, this allows the use of passenger data in the neural network for forecasting.

During the training of neural networks, we transmit data to them in small parts (mini-parts). Keras library uses the *fit_generator* function, which automatically pulls out training data from a pre-installed Python iterator or generator object and introduces them into a model, performs gradient steps, maintains a log of accuracy and callback performance [3–5].

The first layer on the network is a layer of insertion of records. It is converts entries (reference to integers into data) into weight vectors. This layer takes the number of passengers as the first argument, and then the size of the resulting embedded vector as the next argument.

In order to test the trained model of LSTM, we compare the projected trips with actual trips in the training set and testing set.

The model reboots from the trained data. Then a cycle of extracting fictitious data from the generator is created - in order to control where the data set indicates comparative sentences.

From the generator data, the number to the expected trips is returned. The relative forecasting error for these neural networks at the adaptation stage was less then 5%.

Table 2 shows results of comparison actual and predicted trips after training on a set of the passenger traffic.

Table 2. Results of forecasting after training

Actually values	Predicted values (100 epochs)	Predicted values (400 epochs)
20	20	20
32	28	32
50	46	48
60	62	60
100	100	100
34	70	32
38	38	38

The completion of the learning algorithm optimizes synoptic weighted connections to establish a reliable relationship between inputs and outputs. During the testing and validation phase, these free options remain unchanged, while new inputs are served on the network to make a series of outputs. These outputs are compared with a set of test data corresponding to the actual results obtained. If the actual output differs from the set of test outputs above the threshold value of the error, then you need to adjust the training and prepare the neural network again.

Compare methods of forecasting the moving average, Holt-Winters exponential smoothing, linear and logarithmic trends [8, 11] and LSTM for forecasting passenger traffic between cities of Chernihiv and Kyiv for 2012–2015, using data from 2007–2011. Table 3 shows the accuracy of the forecast obtained by these methods.

Table 3. Indicators of forecast accuracy using different methods

Forecasting method	Accuracy of forecast
Moving average	0.961
Holt-Winters exponential smoothing	0.933
Linear trend	0.930
Logarithmic trend	0.901
Long short-term memory (LSTM)	0.983

5 Conclusions

With the use of the LSTM neural network, the accuracy of passenger traffic modeling will be increased up to several percents in the future, which allows the carrier to reduce the cost of servicing the rolling stock MTE. The method allows to predict the number of passengers hourly for any day. The model achieves its maximum accuracy if you anticipate not less than 10 days in advance and every week to provide up-to-date data and study the model. The proposed forecasting passenger traffic method using neural networks can reduce the cost of passenger traffic and improve the quality of transport services on the routes studied. The obtained refined predictive values on the studied routes allow us improve the work of the MTE, in particular, when planning the modes of movement and the choice of transport vehicles for the transportation of passengers.

References

1. Al-Maqaleh, B.M., Al-Mansoub, A.A., Al-Badani, F.N.: Forecasting using artificial neural network and statistics models. Int. J. Educ. Manag. Eng. (IJEME) **6**(3), 20–32 (2016). https://doi.org/10.5815/ijeme.2016.03.03
2. Geraldine Bessie Amali, D., Dinakaran, M.: A new quantum tunneling particle swarm optimization algorithm for training feedforward neural networks. Int. J. Intell. Syst. Appl. (IJISA) **10**(11), 64–75 (2018). https://doi.org/10.5815/ijisa.2018.11.07
3. Cyprich, O., Konečný, V., Kilianová, K.: Short-term passenger demand forecasting using univariate time series theory. Promet - Traffic Transp. **25**(6), 533–541 (2013)
4. Graves, A., Mohamed, A.-R., Hinton, G.: Speech recognition with deep recurrent neural networks. arXiv preprint arXiv:1303.5778 (2013)
5. Greff, K., Srivastava, R.K., Koutník, J., Steunebrink, B.R., Schmidhuber, Jü.: LSTM: A Search Space Odyssey. arXiv preprint arXiv:1503.04069 (2015)
6. Hu, Z., Bodyanskiy, Y.V., Tyshchenko, O.K., Boiko, O.O.: Adaptive forecasting of non-stationary nonlinear time series based on the evolving weighted neuro-neo-fuzzy-ANARX-model. Int. J. Inf. Technol. Comput. Sci. (IJITCS) **8**(10), 1–10 (2016). https://doi.org/10.5815/ijitcs.2016.10.01
7. Khashei, M., Bijari, M.: A new class of hybrid models for time series forecasting. Expert Syst. Appl. **39**, 4344–4357 (2012)
8. Medvedev, M.G., Oleshchenko, L.M.: The optimal control models of interurban bus transport. Electron. Control Syst. **1**(39), 85–90 (2014)
9. Mishra, N., Soni, H.K., Sharma, S., Upadhyay, A.K.: Development and analysis of artificial neural network models for rainfall prediction by using time-series data. Int. J. Intell. Syst. Appl. (IJISA) **10**(1), 16–23 (2018). https://doi.org/10.5815/ijisa.2018.01.03
10. Nielsen, M.A.: Neural Networks and Deep Learning [Electronic], Determination Press (2015). http://michaelnielsen.org
11. Oleshchenko, L., Grabariev, A.: Technology of prediction of variable value of passenger traffic using dynamic gravity model. Econ. Manag. **3**(67), 118–124 (2015)
12. Oleshchenko, L., Mykolaichyk, V.: Neural network for optimization of intercity passenger transportations. In: XI Conference of Young Scientists "Applied Mathematics and Computing" (PMK-2018-2), National Technical University of Ukraine "Igor Sikorsky Kyiv Polytechnic Institute", Kyiv, Ukraine, 14–16 November 2018, pp. 222–226 (2018)

13. Pawade, D., Sakhapara, A., Jain, M., Jain, N., Gada, K.: Story scrambler - automatic text generation using word level RNN-LSTM. Int. J. Inf. Technol. Comput. Sci. (IJITCS) **10**(6), 44–53 (2018). https://doi.org/10.5815/ijitcs.2018.06.05

14. Sharma, M., Tomer, M.S.: Predictive analysis of RFID supply chain path using long short term memory (LSTM): recurrent neural networks. Int. J. Wirel. Microw. Technol. (IJWMT) **8**(4), 66–77 (2018). https://doi.org/10.5815/ijwmt.2018.04.05

15. Sutskever, I., Vinyals, O., Le, Q.V.: Sequence to sequence learning with neural networks. arXiv preprint arXiv:1409.3215 (2014)

16. Yongchun, L.: Application of Elman neural network in short-term load forecasting. In: International Conference on Artificial Intelligence and Computational Intelligence, pp. 141–144 (2010). https://doi.org/10.1109/aici.2010.153

Thread Pool Parameters Tuning Using Simulation

Inna V. Stetsenko$^{(\boxtimes)}$ ⓘ and Oleksandra Dyfuchyna ⓘ

Igor Sikorsky Kyiv Polytechnic Institute,
37 Prospect Peremogy, Kiev 03056, Ukraine
stiv.inna@gmail.com, sashadif@gmail.com

Abstract. The problems of multithreaded programming are discussed. One way to improve the quality of writing a parallel program is to use a model which can represent the process of parallel computing in details. In this research construction of model of the multithreaded algorithm is carried out using Petri-object simulation technology grounded on stochastic Petri net and object-oriented approach. Synchronous and asynchronous performance can be reproduced in model according to the resources availability. The parallel computation is reproduced with taking into account not only formal rules of threads interaction but also the time delays of program instructions. Due to that simulation results are close to the real which has been obtained by launching a multithreaded program. In particular, the model of a thread pool as one of the most efficient high-level tools of parallel programming is developed. The experimental results show speedup dependence on algorithm complexity, computing resources and parameters of the thread pool. This dependence is revealed both in multithreaded program computation and in the model. Hence, simulation can be used for testing a parallel program with given parameters and resources.

Keywords: Modeling · Multithreaded programming · Stochastic Petri net · Java

1 Introduction

Multithreaded programming is a popular technique of modern software development. It is widely used in web applications, graphics, and big data computing because it makes program performance much faster. However, multithreaded program development often requires valuable efforts to achieve effective implementation of an algorithm.

The main problems of parallel programming are deadlock, starvation, livelock and memory consistency error. In work [1] they are discussed in details. Deadlock occurs when threads are waiting for each other. When a thread can not catch resource for computing because other threads occupy them, the starvation has happened. If a thread can not progress in their running although it hasn't blocked it means livelock situation. Memory consistency error can occur when threads concurrently modify shared data. These problems can appear only if the conflict between threads has happened. In addition, through the stochastic behaviour of parallel computing, the problems in a program can stay unrevealed long time. When a correct program has been modified

© Springer Nature Switzerland AG 2020
Z. Hu et al. (Eds.): ICCSEEA 2019, AISC 938, pp. 78–89, 2020.
https://doi.org/10.1007/978-3-030-16621-2_8

especially extended by new classes it can destruct the parallel algorithm performance and this kind of error is hard to catch.

Another problem, that the testing of parallel programs should be performed on different computer platforms to guarantee its correctness and efficiency. It takes many resources and a lot of time. Software developers point on the necessity to create tools for improving the process of debugging multithreaded program because existing tools are effective only for debugging and performance optimization of single-threaded software systems [2].

Related works The practice of multithreaded programming with the use of high-level tools shows that their effectiveness has a strong dependence on many parameters such as resources number, tasks number and its computational complexity. So the goal of this research is the development of tool which could help to define the correctness of using a thread pool and estimate the speedup reached by using parallelism. The tool is based on the simulation model of parallel algorithm which reproduces the computational process in details.

The model which is developed provides the estimation of the time performance which depends on parameters of algorithm and characteristics of equipment. It gives us a way to estimate the speedup without testing in real time. Also, it provides an opportunity to debug parallel algorithm using visualization of computational process.

Section 1 introduce to the problems of a multithreaded program development. The second section describes related works. The next section represents the foundations of Petri-object approach which is used for simulation. Section 4 describes the java thread pool and demonstrates the experimental results that prove the speedup strong dependence on the parameters of thread pool and computational complexity. The fifth section contains the model description of a multithreaded program that uses a thread pool. Experimental results of simulation are given in the next section. It shows that the model accuracy is not exceeded 10%. The last section summaries the article and gives the perspective of future research.

2 Related Works

Different parallel programming technologies applied for multicores computer systems were compared in work [3] from the point of matrix multiplication time performance. The testing of parallel program using data and communication flow analysis is considered in work [4]. The algorithm for tracing program execution helps to explore interprocess communication. The technique which is proposed can be applied only for an MPI program.

The platform for testing java parallel algorithm using hardware with 60 cores is proposed in work [5] and the numbers of experimental results for different wild known algorithms are given. The execution time and speedup for all algorithms depend on number of threads using for algorithm implementation.

In work [6] the timed automata are used for modeling thread pool. The model shows the valuable reducing of the deadline of tasks in case of parallel threads. The computing time for task performance is provided by the scheduler in this model. Work

[7] discusses the effective scheduling of tasks in a parallel program grounded on genetic algorithm.

The partial problem of parallel programming is explored by researchers and they can not be united in a simple way. Thus, for example, a method for static detection of deadlocks is proposed in [8]. Moreover, special tools have been developed for detecting deadlock [9] or memory consistency errors [10] in a parallel program. However, the development of the tool which is suitable for the testing of all existing problems is the complicated task through the absence of a united approach to investigate them.

In this research, we consider Petri-object simulation models as a base for creating such tool. The significant difference from previous research is that the parallel computation reproduces not only in steps but also in time with taking into account the time delays of program instructions. It is important through the threads interaction. Also, the impact of available resources is taken into account in the model which is developed.

3 Petri-Object Simulation

Construction of Petri-object model is grounded on stochastic Petri net and object-oriented technology. Mathematical description of stochastic Petri net with multichannel transitions is represented by the set of places $\mathbf{P} = \{P\}$, the set of transition $\mathbf{T} = \{T\}$, $\mathbf{P} \cap \mathbf{T} = \varnothing$, the sets of input and output arcs $\mathbf{A} \subseteq (\mathbf{P} \times \mathbf{T} \cup \mathbf{T} \times \mathbf{P})$, the set of arc multiplicity $\mathbf{W} : \mathbf{A} \to \mathrm{N}$, the set of parameters of transition's priority and probability $\mathbf{K} = \{(c_T, b_T) | T \in \mathbf{T}, c \in N, b \in [0; 1]\}$, set of non-negative values defined by determine or stochastic value with given distribution for transition time delay $\mathbf{R} : T \to \Re_+$:

$$Net = (\mathbf{P}, \mathbf{T}, \mathbf{A}, \mathbf{W}, \mathbf{K}, \mathbf{R}). \tag{1}$$

Petri-object is an object that inherited class *PetriSim* in which the algorithm of Petri net functioning is defined:

$$\text{PetriObject} \overset{inherit}{\to} \text{PetriSim}. \tag{2}$$

Petri-object model is a model that aggregate Petri-objects:

$$\text{PetriObjectModel} = \cup_j O_j, \, O_j \overset{inherit}{\to} \text{PetriSim}. \tag{3}$$

There are two ways to connect Petri-objects. The first one is to share place between two or more objects. The second one is to pass marker from the transition of one Petri-object to one or more places of Petri-objects. Both types of connections provide that the dynamics of model described by the Petri net obtained as union of nets of all its Petri-objects [11]. Therefore the simulation algorithm is realized like the algorithm of stochastic Petri net. It's the main Petri-object model distinguish of other known combination object-oriented technology and Petri net, for example [12], that not only for the objects its dynamics described by Petri net but the dynamics of model also described by Petri net. In addition, Petri-object model provides the simulation algorithm that has less complexity than stochastic Petri net.

The Petri-object model is constructed in four stages. In the first stage, the Petri-nets which describe the dynamics of the elements of the model are compounded. In the second stage, the classes of Petri-objects as the object with the dynamics given by Petri net are created. In the third stage, the Petri-objects are created and the connections between them are set. In the fourth stage, the Petri-object model is compounded with the given set of Petri-object. If the classes of Petri-objects have been constructed for a specific area they can be reused for construction of a model. Then the time to create a model considerably reduced.

4   Java Thread Pool

Thread pool is the most efficient tool for implementing multithreaded algorithms. It is used by many popular programming languages such as Java, C++, C#, Python, etc. Being a high-level tool of multithreading it provides convenience, speed, and ease of use. It allows reducing the overhead of creating a thread and as a result, increases the efficiency of using resources. In [13] general problems of using a thread pool are discussed. In [14] the author points on the importance of tuning the right size of a thread pool.

A thread pool reuses its threads to execute current tasks. Since the thread already exists when the request on execution arrives, the delay produced by thread creation is missed. It's the main advantage of using the pool that instead of creating a new thread, any new task is served by a free thread from the pool.

Thread pool consists of worker threads which are able to do the computing work. The work for computing in thread can be represented only by the object of class that implements Runnable or Callable interface. Such objects are called 'tasks' according to Java documentation [15]. In order for worker threads to start computing, tasks must be loaded to the pool. After that with the help of ExecutorService tasks can be executed by the thread pool. Thus, working with thread pool programmer should create pool than load to the pool tasks that have been created earlier, after that stop the loading and than wait for finish of computing all tasks.

The following code represents thread pool creation and execution in Java language:

```
ExecutorService executor = Executors.newFixedThreadPool(2);
for (int j = 0; j < n; j++) {
executor.execute(new TaskCounter(new Counter()));
}
executor.shutdown();
executor.awaitTermination(50, TimeUnit.MINUTES);
```

Experimental results show that the speedup obtained by parallelism strongly depends on the computational complexity of an algorithm and its parameters such as the number of threads which are used. Figure 1 depicted the changing of speedup for different complexity of computation (from 10^3 operations to 10^9 operations) when the number of threads and the number of tasks increase. If the number of operations less than 10^6 the overhead on creating the resources for computing is more than reduction obtained by the parallel implementation of the algorithm. So in this case parallelism is not effective (we see on figure speedup less than 1.0). The increasing of speedup to nearly 2.0 can be observed if only the number of operations is much more than 10^9 and the number of threads is not less than two. However, the value of speedup more than 2.0 cannot be achieved through the limitation of computing resources. Thus it's a big question for every parallel algorithm what parameters provide its effective performance. In order to answer this question the model of thread pool was created. This model allows varying the number of threads and tasks in thread pool with taking into account the computer resources. Instead of running a multithreaded program on different computers many times, thread pool model can be simulated. In that case, finding the right parameters for effective program performance is simplified, and the time and resources costs are reduced.

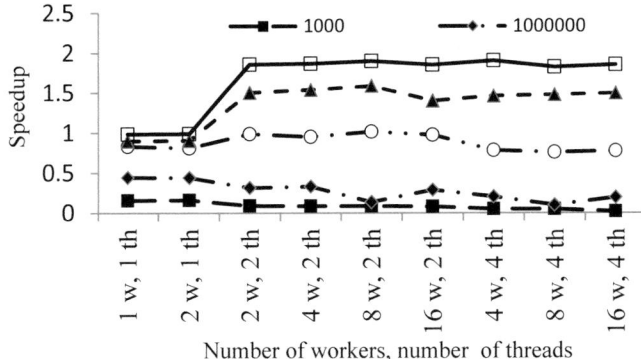

Fig. 1. Speedup dependence on number of threads (th) and tasks (w) for different complexity of computation by the results of running multithreaded program on dual core processor

5   Model Description

A detail description of the basic instructions of the multithreaded program by the fragments of stochastic Petri net has been given in our previous work [16]. We implement them for constructing the model which simulate thread pool known as one of the high-level tool of multithreading. It describes the thread pool loaded with tasks which are running the given method of given class. For simplicity, suppose that the given method is increasing the value which is contained in the class named *Counter*. According to the objects of program the model is combined of Petri-objects '*Main*','*ThreadPool*', '*Task*', '*Counter*' connected with each other by shared places.

Simplest instruction (as add or multiply numbers) of a program is represented by transition with time delay. Repeating instructions can be represented by the same transition. However, the condition of repeating should be added. Therefore the cycle of *n* simple instructions is represented by two transitions, one of which has a bigger priority (Fig. 2). So the second transition can be fired when the first one can't be. Notice that the transition with higher priority is depicted with the bigger width.

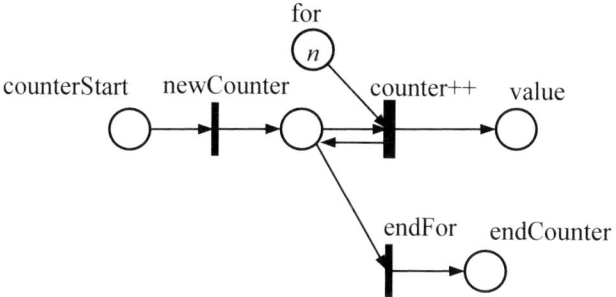

Fig. 2. Petri-object *'Counter'* without synchronization

Two cases can be considered: task runs asynchronous or synchronous. The Petri-object *'Counter'* with synchronization mechanism is depicted in Fig. 3. In both nets, there is the same construction with an alternative to events. The event of incrementing *'counter++'*, as it has higher priority value, will be performed for n times (mark in place *'for'*), each time the result of the operation will be saved as a marking of place *'value'*. After that, the *'endFor'* event is occurring.

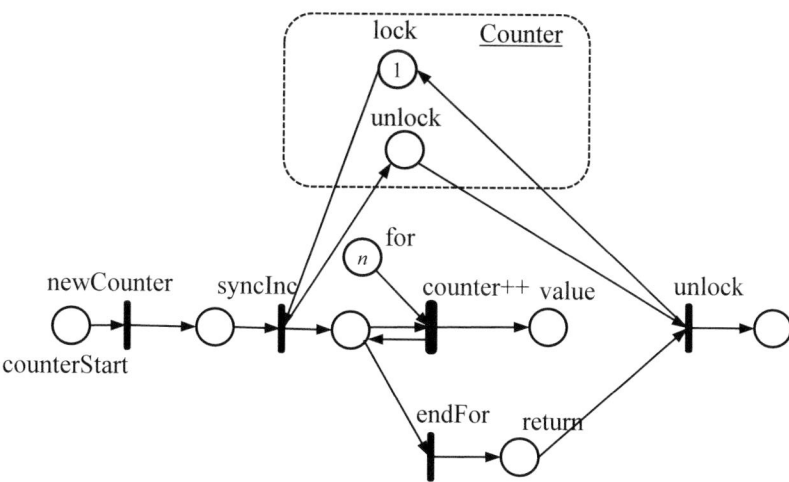

Fig. 3. Petri-object *'Counter'* with synchronization

As it has been already mentioned, Petri net for synchronous counter contains the locking mechanism - place *'lock'* in the figure. That means that before the increment method could be executed the lock object should be captured. It is demonstrated by the transition 'syncInc' and 'unlock', and by the places 'lock' and 'unlock'.

It all starts with the Petri-object '*ThreadPool*' connected to the '*Main*' Petri net through the places '*poolStart*' and '*poolEnd*' (Fig. 4). The place 'cores' simulates the cores of processor as a resource for computing. It is shared with all objects of model. Moreover, all events in model use this place like '*Main*'-object but we will skip the corresponding elements to facilitate the perception of Petri nets.

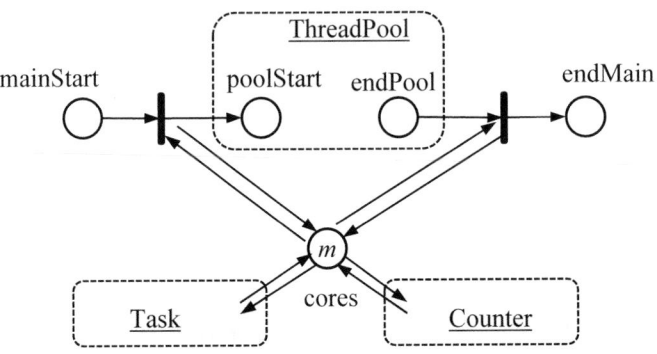

Fig. 4. Petri-object '*Main*'

The net of Petri-object '*ThreadPool*', in its turn, has a connection with Petri-object '*Task*'. It contains shared places '*taskStart*' and '*endTask*' which belong to '*Task*' (Fig. 5). The transition '*newThreadPool*' represents the event of initialization of thread pool. The output multiplicity arc with number k sets the number of threads in the pool. Moving forward, there is an alternative to events. The conflict of transitions '*execute*' and '*shutdown*' is solved with the help of the priority parameter. '*Execute*' transition has higher priority (it is depicted with bigger width) than '*shutdown*', so the event '*execute*' which stands for the tasks loading to the pool will be performed first as long as a condition for input will be true. The marking of place '*numTasks*' matches the number of tasks in the thread pool. When all the tasks are loaded to the thread pool the event '*shutdown*' is executed.

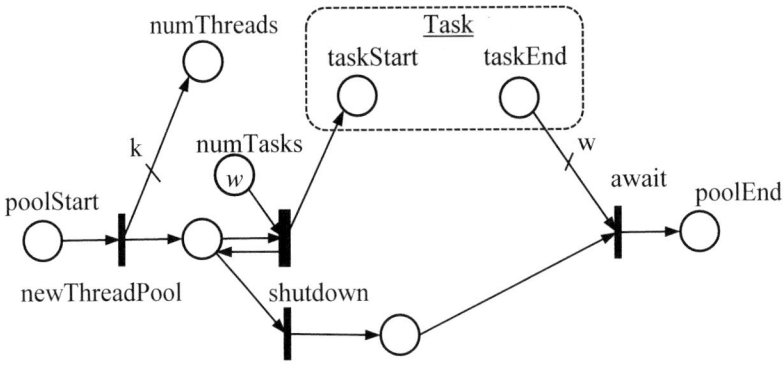

Fig. 5. Petri-object '*ThreadPool*'

As soon as a task is loaded to the pool it starts to do the work. This process is described by the Petri-object 'Task' (Fig. 6). In general, any *Runnable*-object with computing algorithm in its method *run()* can be performed by the task. In our case, the task running the method of class *Counter* which is connected to the 'Task' via shared places 'counterStart' and 'endCounter'. The task of counter has been chosen because of the simplicity of computing algorithm which is well parallelizable since it has no sequential instructions.

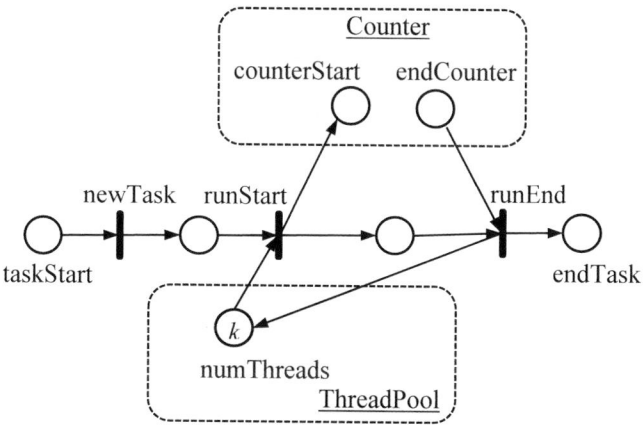

Fig. 6. Petri-object 'Task'

The end of simulation happens when the event 'endMain' of Petri-object 'Main' has fired and one token in place 'endMain' has appeared. The moment of this event is registered as the multithreaded program completion time. This time is considered as the result of simulation in following experiments.

6   Experimental Results

The values of time delay for events in the model have been estimated by the statistical study of the corresponding values during the computing on dual core processor. There are program instructions that need much more time for their completion than other. For example, the time for creating a thread pool is approximately 20 times bigger than the time for creating a new object which, in turn, is 30000 times bigger than the time of the simplest arithmetic operation.

In experiment the speedup which is achieved in multithreaded program with different parameters of thread pool is researched. Changing computational complexity of algorithm which is parallelized from 10^3 to 10^9 we can see the sharp increasing of speedup to approximately 2.0 only for the complexity more than 10^8. If complexity is less than 10^6, the speedup is less than 1.0 which means that using pool in this case is ineffective (Fig. 7). Comparison of the result of simulation and the result which has been obtained when the program is running on dual core processor confirms the correctness of the first one. The model in a right way revealed the limitation of computing resources because of the speedup reaches value 2.0. Notice that this value is equal to the number of cores. The searching of thread pool parameters provided the maximum speedup is presented in Table 1. If the complexity of an algorithm is less than 10^6 the using of a thread pool is not recommended because of the low level of speedup which can be achieved. In the case of an algorithm complexity greater than 10^6 the maximum speedup is achieved when 2 tasks and 2 threads in a thread pool.

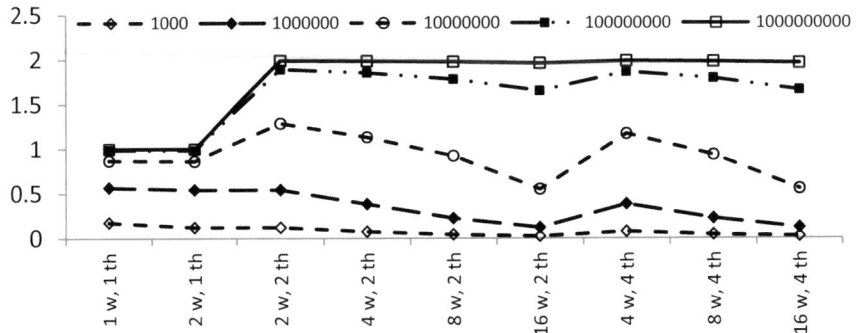

Fig. 7. The speedup dependence on parameters of thread pool (th - threads, w - tasks).

Table 1. The parameters of thread pool (th - threads, w - tasks) to achieve maximum speedup.

Complexity	Parameters								
	1 w, 1 th	2 w, 1 th	2 w, 2 th	4 w, 2 th	8 w, 2 th	16 w, 2 th	4 w, 4 th	8 w, 4 th	16 w, 4 th
1000	**0.18**	0.12	0.12	0.07	0.04	0.02	0.07	0.04	0.02
1000000	**0.57**	0.54	0.54	0.38	0.22	0.12	0.38	0.22	0.12
10000000	0.87	0.86	**1.29**	1.13	0.92	0.55	1.17	0.93	0.55
100000000	0.98	0.98	**1.89**	1.85	1.78	1.65	1.86	1.79	1.66
1000000000	1.00	1.00	**1.99**	1.98	1.98	1.96	1.99	1.98	1.96

The speedup measured in multithreaded program and in its model if thread pool uses 2 threads and 2 tasks depicted in Fig. 8. When the complexity of computation is huge (10^9) the model exactly reproduces the speedup. In other case, the model shows as a rule the bigger speedup than the multithreaded program. It can be explained by the fact that we do not include in model the uses of system resources by other processes. In our future version of model it can be considered as the corresponding event that will be added to the model.

The accuracy of speedup simulation is represented in Table 2. The value of accuracy is calculated as a difference between speedup obtained in model and in real multithreaded program expressed as a percentage of last value. The number of workers *w* and the numbers of threads *th* are varied. The increasing of absolute value of time performance obtained in model and in real multithreaded program for the 2 tasks and 2 threads in thread pool is represented in Fig. 9.

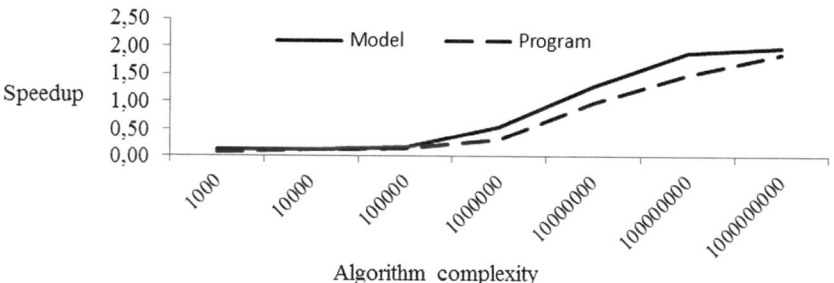

Fig. 8. The comparison of speedup in multithreaded program and in its model

Table 2. Accuracy of simulation result (109 computational complexity).

Number of tasks (w) and threads (th)	1 w, 1 th	2 w, 1 th	2 w, 2 th	4 w, 2 th	8 w, 2 th	16 w, 2 th	4 w, 4 th	8 w, 4 th	16 w, 4 th
Program	0.985	0.990	1.859	1.869	1.901	1.856	1.907	1.828	1.855
Model	0.998	0.998	1.989	1.984	1.976	1.977	1.985	1.977	1.960
Accuracy	1%	1%	7%	6%	4%	7%	4%	8%	6%

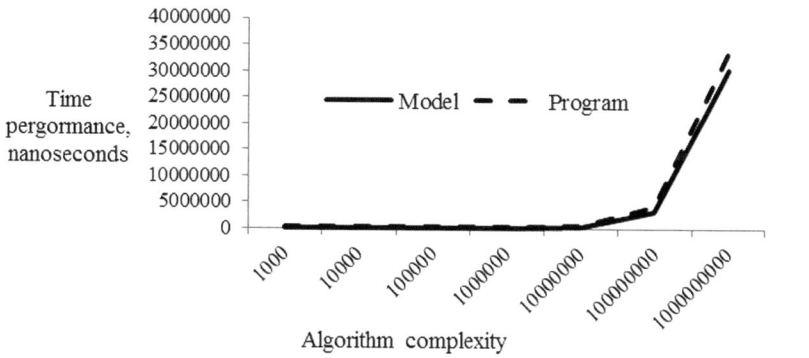

Fig. 9. The comparison of time performance in multithreaded program and in its model if 2 tasks and 2 threads are used

7   Conclusions

The main result we obtained is that the Petri-object model of thread pool reproduces the impact of resources which are used for computing. Although the number of threads in the pool is increased, the time performance of the multithreaded program is not increased through the lack of resources. The tasks loaded to the pool are performed concurrently but not simultaneously. Therefore Petri-object simulation of the multi-threaded program can be used for investigating the parameters used in the program and searching the best ones for the biggest effect of multithreaded tools. Future research will help to give common recommendation for choosing the size of pool in accordance to the complexity of computation and system resources.

References

1. Friesen, J.: Java Threads and the Concurrency Utilities. Apress, New York (2015)
2. Trümper, J., Bohnet, J., Döllner, J.: Understanding complex multithreaded software systems by using trace visualization. In: SOFTVIS 2010 Proceedings of the 5th International Symposium on Software Visualization, pp. 133–142. ACM, Salt Lake City (2010)
3. Ashraf, M.U., Fouz, F., Eassa, F.A.: Empirical analysis of HPC using different programming models. Int. J. Mod. Educ. Comput. Sci. (IJMECS) 8(6), 27–34 (2016). https://doi.org/10.5815/ijmecs.2016.06.04
4. Elnashar, A.I., El-Zoghdy, S.F.: An algorithm for static tracing of message passing interface programs using data flow analysis. Int. J. Comput. Netw. Inf. Secur. (IJCNIS) 7(1), 1–8 (2015). https://doi.org/10.5815/ijcnis.2015.01.01
5. Malinowski, A.: Modern platform for parallel algorithms testing: Java on Intel Xeon Phi. Int. J. Inf. Technol. Comput. Sci. (IJITCS) 7(9), 8–14 (2015). https://doi.org/10.5815/ijitcs.2015.09.02
6. De Boer, F., Grabe, I., Jaghoori, M., Stam, A., Yi, W.: Modeling and analysis of thread-pools in an industrial communication platform. In: Breitman, K., Cavalcanti, A. (eds.) International Conference on Formal Engineering Methods ICFEM 2009: Formal Methods and Software Engineering, pp. 367–386. Springer, Heidelberg (2009)
7. Singh, R.: An optimized task duplication based scheduling in parallel system. Int. J. Intell. Syst. Appl. (IJISA) 8(8), 26–37 (2016). https://doi.org/10.5815/ijisa.2016.08.04
8. Owe, O., Yu I.C.: Deadlock detection of active objects with synchronous and asynchronous method calls. In: Norsk Informatikkonferanse (NIK) OPJ/PKP, Halden, Norway (2014)
9. Software Verify LTD. Smarter tools for better software. https://www.softwareverify.com/thread-analysis-deadlock-detection.php. Accessed 6 Nov 2018
10. Chen, Z., Dinan, J., Tang, Z., Balaji, P., Zhong, H., Wei, J., Huang, T., Qin, F.: MC-checker: detecting memory consistency errors in MPI one-sided applications. In: SC 2014, Institute of Electrical and Electronics Engineers (IEEE), New Orleans, Louisiana, USA (2014)
11. Stetsenko, I.V.: Theoretical foundations of Petri-object modeling of systems. Math. Mach. Syst. 4, 136–148 (2011). (in Russian)
12. Pujari, S., Mukhopadhyay, S.: Petri net: a tool for modeling and analyze multi-agent oriented systems. Int. J. Intell. Syst. Appl. (IJISA) 4(10), 103–112 (2012). https://doi.org/10.5815/ijisa.2012.10.11

13. Goetz, B.: Java theory and practice: thread pools and work queues. Thread pools help achieve optimum resource utilization. https://www.ibm.com/developerworks/library/j-jtp0730/j-jtp0730-pdf.pdf. Accessed 6 Nov 2018
14. Pepperdine, K.: Tuning the size of your thread pool. InfoQ. https://www.infoq.com/articles/Java-Thread-Pool-Performance-Tuning. Accessed 06 Nov 2018
15. Java™ Platform, Standard Edition 8 API Specification. Class ThreadPoolExecutor. https://docs.oracle.com/javase/8/docs/api/java/util/concurrent/ThreadPoolExecutor.html. Accessed 06 Nov 2018
16. Stetsenko, I.V., Dyfuchyna, O.: Simulation of multithreaded algorithms using Petri-object models. In: Hu, Z., Petoukhov, S., Dychka, I., He, M. (eds.) International Conference on Computer Science, Engineering and Education Applications ICCSEEA 2018: Advances in Intelligent Systems and Computing, vol. 754, pp. 391–401. Springer, Cham (2019)

Hierarchical Coordination Method of Inter-area Routing in Backboneless Network

Oleksandr Lemeshko(ID), Olena Nevzorova(✉)(ID),
Andriy Ilyashenko(ID), and Maryna Yevdokymenko(ID)

Kharkiv National University of Radio Electronics, Kharkiv 61000, Ukraine
oleksandr.lemeshko.ua@ieee.org,
{olena.nevzorova,marina.ievdokymenko}@nure.ua,
andy.ilyashenko@gmail.com

Abstract. In this paper, a flow-based decomposition model and a two-level hierarchical coordination method of inter-area routing in a backboneless network are proposed. In the framework of the proposed method of inter-area routing at the lower level there is a calculation of routes for individual areas, and the upper level task is the coordination of the lower-level solutions to provide connectivity of inter-area paths (multipath) both in its structure and in the nature of the users' traffic balancing on it. The results of numerical calculations confirmed the adequacy of the proposed model and the convergence of the method of inter-area routing to optimal solutions for a finite number of iterations.

Keywords: Inter-area routing · Backboneless network · Hierarchical routing · Goal coordination principle · Iteration

1 Introduction

In modern telecommunications networks, an important problem is the provision of the required Quality of Service (QoS) for which the functionality of all seven levels of the OSI (Open Systems Interconnection) model is involved [1–3]. The main means of providing QoS on the Network Level it can be attributed the technology of traffic management, the key of which, by rights, are the routing protocols [4–6]. With the growth of the territorial distribution of modern networks, the number of network devices, supported services and users in general, tasks related to increasing the scalability of routing solutions are emerging. At present, in order to solve these problems, in practice, protocols of hierarchical routing are actively applied, for example, OSPF, IS-IS, BGP in IP/MPLS networks, PNNI in ATM technology [4, 7, 8].

A feature of the operation of these hierarchical routing protocols is that two types of areas such as backbone Area (Area 0) and non-backbone Area (NBA) are distinguished in the network structure [9, 10]. To increase the scalability of the network, routing tasks in these areas are solved quite autonomously, which helps limit the amount of information transmitted about the state of areas and reduce also the size of the routing tables. Within the framework of this research direction, quite intensive scientific and applied attachments are being conducted [11–22]. However, all traffic between NBAs is

© Springer Nature Switzerland AG 2020
Z. Hu et al. (Eds.): ICCSEEA 2019, AISC 938, pp. 90–102, 2020.
https://doi.org/10.1007/978-3-030-16621-2_9

necessarily routed through the backbone routers, which severely reduces the functionality of hierarchical routing solutions on the network.

Generally, to improve the efficiency and flexibility of hierarchical routing solutions, it is advisable to abandon the restrictions associated with the mandatory passage of all inter-area routes through the backbone, i.e. make all network areas functionally equal. Then, in general, desired paths may pass through an arbitrary set of areas. However, the requirements for the order of solving inter-area routing problems are contradictory. On the one hand, to increase the scalability of routing in the network as a whole, the solutions of these private tasks should be as autonomous as possible. But in order to increase the efficiency of the network and ensure the coherence of the inter-area routes, the solutions to these problems must be maximally coordinated. This cannot be realized without corresponding re-examination and improvement of mathematical models and methods of hierarchical routing.

At the same time, the basis for such an improvement may be the flow-based routing solutions proposed in [11, 12, 14–16] and focused on the implementation of hierarchical-coordinating routing in the network. Hierarchical coordination routing along with the increasing of scalability of the network ensures the efficiency of solutions close to centralized routing. This is achieved, as shown in [11, 12, 14–16], by introducing a two-level functional hierarchy, where different hierarchical levels are responsible for solving intra- and inter-area routing problems. At the same time, connectivity of inter-area routes is provided by introduction of coordination of lower-level solutions (area level) from the upper level (network level). This approach is oriented, on the one hand, to providing a higher scalability of the network as a whole, and on the other hand to the coordinated solution of tasks within and inter-area routing, taking into account the real topology of individual areas and the network as a whole, not aggregated.

2   Flow-Based Model of Inter-area Routing in Backboneless Network

Let us the telecommunication network structure has been described by oriented graph $G = (M, \mathrm{E})$ which consists of both the set of graph vertices $M = \{M_i, i = \overline{1,m}\}$ modeling network routers; and the set of graph arcs $E = \{E_{i,j}, i,j = \overline{1,m}, i \neq j\}$ modeling links connecting routers in network. Let denote K like the set of flows circulated in network and based on this, denote $|K| = \tilde{K}$ like the cardinality of the set K describes the total number of flows in network. Denote by λ^k the rate of kth packet flow $(k \in K)$ which measured in packet per second (1/s). Also denote by s^k and d^k the router-sender and destination router of kth flow respectively and they must be included to different areas.

In order to develop the decomposition model of inter-area routing in multi-area network it was supposed that the telecommunication network consists of N interconnected subnetworks named areas. Then every separate pth area in network can be described by the subgraph $G^p = (M^p, \mathrm{E}^p)$ of graph G, which by the analogy with graph G consists of set vertices $M^p = \{M_i^p; i = \overline{1,m_p}\}$ modeling the pth area routers, where m_p is the total number of routers in pth area, and the set of links $E^p = \left\{ E_{i,j}^p; \ i,j = \overline{1,m_p}, \ i \neq j \right\}$ connecting routers in pth area. Also in model two subsets: E_{in}^p $(E_{in}^p = E^p, \ E_{in}^p = \left\{ E_{i,j}^p; \ M_j \in M^p, M_i \in M^q, p \neq q \right\})$ the subset of graph arcs modeling networks links the traffic may incoming through to pth area; and subset E_{out}^p $(E_{out}^p = E^p, \ E_{out}^p = \left\{ E_{i,j}^p; \ M_i \in M^p, M_j \in M^q, p \neq q \right\})$ modeling network links, through which the traffic may outgo from pth area, are introduced.

Agree that boundary between areas (decomposition of network for areas) passes through the links, which is implemented in protocol IS-IS [4]:

$$E^p \cap E^q \neq 0 \text{ and } M^p \cap M^q = 0.$$

i.e. the link may connect routers belonging different areas.

Also let us using $\varphi_{i,j}^p$ as bandwidth of the link $E_{i,j}$, which measured in packet per second (1/s). The link $E_{i,j}$ is between routers M_i and M_j, and at least one of them must be included to the pth area. For solving the task of hierarchical inter-area routing for every pth area it is necessary to calculate routing variables $x_{i,j}^{p,k}$ which determine the fraction of the rate of kth flow in link $E_{i,j}^p \in E^p$.

Implementation of the single-path routing strategy is based on fulfilment of following condition:

$$x_{i,j}^{p,k} \in \{0,1\}, \tag{1}$$

and for implementation of the multipath routing strategy the routing variables have to satisfy the next constraints:

$$0 \leq x_{i,j}^{p,k} \leq 1. \tag{2}$$

The novelty of the inter-area routing model comparted to model presented in [11, 12, 14] is the improved conditions of flow conservation both for transit areas and for domains that include the source router and the destination router. The condition of flow conservation have to be fulfilled for every kth flow and every pth area router to provide connectivity of calculated inter-area routes in network.

If the kth flow was generated in pth area $(M_i^p = s^k)$, the condition of flow conservation for this area take follows:

$$\begin{cases} \sum\limits_{E_{i,j}^p \in E^p} x_{i,j}^{p,k} = 1; \ \ if \ M_i^p = s^k; \\ \sum\limits_{E_{i,j}^p \in E^p} x_{i,j}^{p,k} - \sum\limits_{E_{j,i}^p \in E^p} x_{j,i}^{p,k} = 0; \ \ if \ M_i^p \neq s^k; \\ \sum\limits_{E_{i,j}^p \in E_{out}^p} x_{i,j}^{p,k} = 1; \end{cases} \tag{3}$$

and the first condition of the system (3) covers all the border routers the kth flow incoming through to the pth area; the second condition is introduced for those routers of the pth area, which are transit for the kth flow; and for ensuring that the entire kth flow transferred from the pth area to the other areas it is introduced into the model the third condition of system (3) [16].

If the destination router of kth packets flow is included to pth area the condition of flow conservation take following form:

$$
\begin{cases}
\sum_{E^p_{i,j} \in E^p} x^{p,k}_{i,j} - \sum_{E^p_{j,i} \in E^p} x^{p,k}_{j,i} = 0; \ \ if \ M^p_i \neq d^k; \\
\sum_{E^p_{j,i} \in E^p} x^{p,k}_{j,i} = 1; \ \ if \ M^p_i = d^k; \\
\sum_{E^p_{i,j} \in E^p_{in}} x^{p,k}_{j,i} = 1;
\end{cases}
\tag{4}
$$

and the first condition of the system (4) covers all the routers of the pth area, which are transit for the kth flow; the second condition is introduced for those routers the kth flow outgoing through from the pth area; and similarly, for the pth area which is a receiver, into the model the condition ensured that the kth flow completely incoming to the pth area from other areas is introduced by the third condition of system (4) [16].

If pth area is the transit for kth flow the condition of flow conservation take following form:

$$
\begin{cases}
\sum_{E^p_{i,j} \in E^p} x^{p,k}_{i,j} - \sum_{E^p_{j,i} \in E^p} x^{p,k}_{j,i} = 0; \ \ if \ M^p_i \in M^p; \\
\sum_{E^p_{i,j} \in E^p_{in}} x^{p,k}_{i,j} = \sum_{E^p_{i,j} \in E^p_{out}} x^{p,k}_{j,i}.
\end{cases}
\tag{5}
$$

And it is important that conditions (1)–(5) must be carried out for every kth flow separately.

Also the important part of the model of inter-area routing is the fulfilment of conditions of link overload prevention:

$$
\sum_{k \in K} \lambda^k x^{p,k}_{(i,j)} \leq \varphi^p_{i,j}, \ p = \overline{1, N}.
\tag{6}
$$

Routing variables $x^{p,k}_{i,j}$ are the coordinates of corresponding routing vectors $\vec{x}^k_p$, which with structural decomposition have to make the functional decomposition. Due to the distributed calculation of the vector $\vec{x}^k_p$ with coordinates $x^{p,k}_{i,j}$ within each area in network, it is necessary to provide inter-area interaction of routes passing through the multiple areas. For this it is proposed to introduce the condition of inter-area interaction represented by the following expression:

$$\vec{x}_p^k = C_{q,p}^k \vec{x}_q^k, \ p,q = \overline{1,n}, \ p \neq q, \ k \in K, \tag{7}$$

where $C_{q,p}^k$ is the matrix of interaction between the area pth and area qth; $C_{p,q}^k$ is the matrix of interaction between the area qth and area pth. Elements of matrixes $C_{q,p}^k$ and $C_{p,q}^k$ have been selected according to that coordinates of vectors $\vec{x}_p^k$ and $\vec{x}_q^k$ responsible for order of routing in pth and qth areas may be identical.

3   Development of Hierarchical Coordination Method of Inter-area Routing in Backboneless Network

Proposed hierarchical coordination method of routing in multi-area network is based on solving the optimization task of calculation of vector $\vec{x}_p^k$ of routing variables ($p = \overline{1,N}, k \in K$) with fulfilment of constraints (1)–(7). And next objective function has to be minimized:

$$\mathbf{min} \, F, \quad F = \sum_{p \in N} \sum_{k \in K} (\vec{x}_p^k)^t H_p^k \vec{x}_p^k, \tag{8}$$

where H_p^k is the diagonal matrix of weight coefficients and its coordinates are routing metrics of pth area links; $[\cdot]^t$ is transpose function of vector (matrix).

To make routing variables take properties of hierarchical routing during solving optimization problem with minimization of constraints (1)–(7) the goal coordination principle [23, 24] is used and turned to unconditional extremum problem:

$$\mathbf{min} \, F = \max_{\mu} L, \tag{9}$$

and the Lagrangian L has to be maximize by vectors of Lagrange multipliers $\vec{\mu}$:

$$L = \sum_{p=1}^{N} \sum_{k \in K} (\vec{x}_p^k)^t H_p^k \vec{x}_p^k + \sum_{p=1}^{N} \sum_{\substack{q=1 \\ q \neq p}}^{N} \sum_{k \in K} \vec{\mu}_{p,q}^k (\vec{x}_p^k - C_{q,p}^k \vec{x}_q^k), \tag{10}$$

where $\vec{\mu}_{p,q}$ is the subvector of vector $\vec{\mu}$ assigned to each of the vector-matrix conditions of interaction between the area pth and area qth represented by (7).

Within the principle of goal coordination the vector of Lagrange multipliers $\vec{\mu}$ is calculated on upper level and known for lower level, and based on this (10) can be represented as follows:

$$L = \sum_{p=1}^{N} L_p, \tag{11}$$

$$L_p = \sum_{k \in K} (\vec{x}_p^k)^t H_p^k \vec{x}_p^k + \sum_{\substack{q=1 \\ p \neq q}}^{N} \sum_{k \in K^+} \vec{\mu}_{p,q}^k \vec{x}_p^k - \sum_{\substack{q=1 \\ p \neq q}}^{N} \sum_{k \in K^-} \vec{\mu}_{q,p}^k C_{p,q}^k \vec{x}_p^k, \tag{12}$$

where K_p^+ is the subset of flows incoming to the pth area; K_p^- is the subset of flows outgoing from the pth area $(K_p^+, K_p^- \in K)$.

Thus, within the proposed two-level method, separate inter-area routing tasks are proposed to be distributed at different levels of the hierarchy. That is, it is advisable to solve tasks of calculation of routing variables described by vector $\vec{x}_p^k$ $(p = \overline{1, N}, k \in K)$ at the lower level, and the task of the providing the fulfilment of conditions of inter-area interconnections (7) transfer to the upper level to the network coordinator.

The results of solving the problems of inter-area interaction are collected by the network coordinator (the upper level of the hierarchy), where they are analyzed and coordinated by calculating (correcting) the Lagrange multiplier vectors during the following gradient iteration procedure:

$$\vec{\mu}_{p,q}^k(\alpha + 1) = \vec{\mu}_{p,q}^k(\alpha) + \nabla \vec{\mu}_{p,q}^k, \tag{13}$$

where α is a number of coordination iteration; $\nabla \vec{\mu}_{pq}^k$ is the gradient of function (11) calculated according to the one received from the lower level solutions of routing tasks $\vec{x}_p^{k*}$ $(p = \overline{1, N}, k \in K)$. This gradient is calculated as:

$$\nabla \vec{\mu}_{p,q}^k(x)\big|_{x=x^*} = \vec{x}_p^k - C_{q,p} \vec{x}_q^k, \tag{14}$$

With approaching coordinates of gradient $\nabla \vec{\mu}_{p,q}^{k_r}$ to zero, the fulfilment of condition of inter-area interconnection (7) will be provided. The total optimum is achieved when $\nabla \vec{\mu}_{p,q}^{k_r}$ approaching to the zero.

4 Numerical Research of Hierarchical Coordination Method of Inter-area Routing in Backboneless Network

In order to confirm the efficiency of the proposed hierarchical coordination method of routing in multi-area network, as well as to verify the adequacy and effectiveness of the solutions obtained, the research of the method for various variants of telecommunications network structures was conducted. For example, in this paper the solution of proposed method considered for network structure shown in Fig. 1.

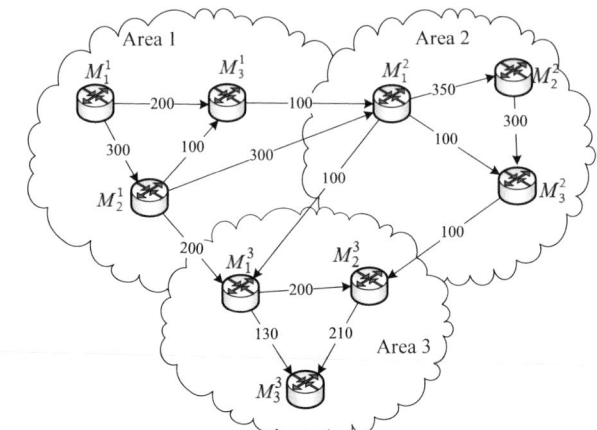

Fig. 1. Telecommunication network structure

The researched network consists of three directly connected to each other areas named Area 1, Area 2 and Area 3. Each area contains three routers presented by sets $M^1 = \{M_1^1, M_2^1, M_3^1\}$, $M^2 = \{M_1^2, M_2^2, M_3^2\}$, $M^3 = \{M_1^3, M_2^3, M_3^3\}$ of routers in Area 1, Area 2 and Area 3 respectively. The boundary between areas (decomposition of the network into subnets or areas) passes through the links. Let the flow incoming to the network through the router M_1^1 (router-sender) located in the Area 1, and also outgoing from network through the router M_3^3 (destination router) located in Area 3. The link bandwidth is represented on links in Fig. 1.

During the research, the rate of packet flow varied from 10 to 300 1/s. For example, let us consider the variant of research when flow rate is equaled 100 1/s. In Fig. 2 there is represented the initial solution of the inter-area routing task, i.e. before the coordination process begins.

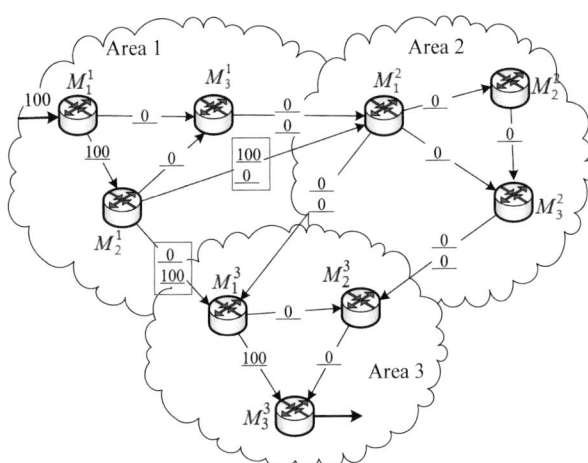

Fig. 2. Initial route calculation for transmission of flow with rate equal 100 1/s

In Fig. 2 on links the flow rate is represented, but in the links participating in the inter-area interaction, there is representation of fraction, where the numerator indicates the calculation performed in the area from which the flow is outgoing, and in the denominator the similar decision made in the area the flow is incoming to.

In view of the fact that the calculation of routes in different areas was made independently of each other, then the condition of inter-area interaction (7) was not provided (in Fig. 2 links where the condition (7) are not fulfilled highlighted in red). According to the Area 1 calculations the flow was sent through the link connecting the router M_2^1 of the first area and router M_1^2 of the second area. In turn, according to the Area 3 calculations, the flow may incoming through the link connecting the router M_2^1 of the first area and the router M_1^3 of third area. The initial calculation of routes in each of the areas is dictated by the fact that they are, on the one hand, the most productive, and on the other hand, contain a minimum number of hops, since the used criterion is additive.

Provided coordination of routing solutions obtained in individual areas and aimed at ensuring the fulfillment of the conditions of inter-area interaction (7) at the upper level of the hierarchy of the proposed method leads to the fact that the connectivity of a single calculated inter-area route was ensured already after the second coordinating iteration (Fig. 3).

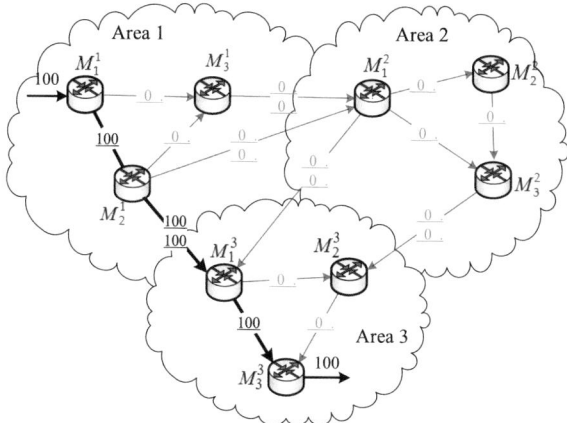

Fig. 3. A coordinated solution of the inter-area routing task of flow with rate of 100 1/s obtained after second iteration

This situation was also observed with an increase the flow rate to 130 1/s, and was justified by the implementation of single-path routing, and convergence was carried out in two iterations. Further increase of the rate of same flow with the same initial data of the network structure and the link bandwidth (Fig. 1), the convergence of the coordinating procedure was provided already after the sixth iteration, and the routing solutions had a multipath routing strategy character. In Fig. 4 shows the initial solution of the inter-area routing task with a packet flow rate of 300 1/s.

The designation of the data represented in the links coincides with that shown in Fig. 2. Route solutions in each area had a multi-path routing strategy character; none of the links can cope this load. However, due to the lack of coordination of route solutions obtained from individual areas, the connectivity of inter-area routes was not provided (Fig. 4).

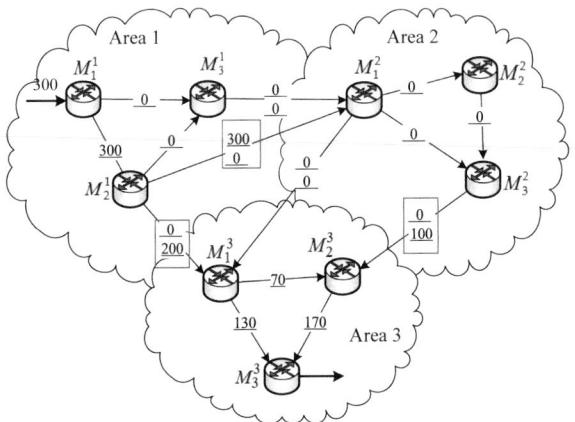

Fig. 4. Initial solution of inter-area routing of flow with rate 300 1/s

Figure 5 shows an intermediate solution of hierarchical inter-area routing task, obtained after the third coordinating iteration. Within the framework of this route solution, the conditions of inter-area interaction (10) are still not satisfied, but there is a smaller discrepancy in the results of calculations obtained in different area in network, compared with the initial solution (Fig. 4).

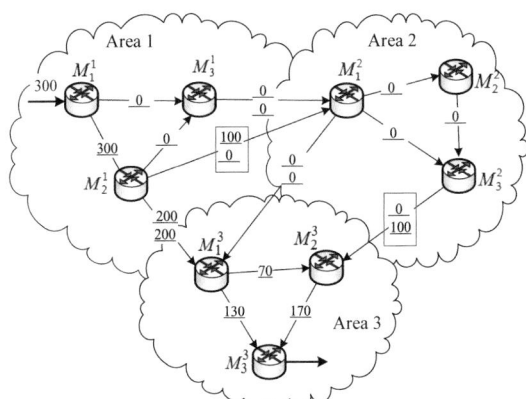

Fig. 5. The solution of the inter-area routing task of flow with rate 300 1/s obtained after the third iteration

Figure 6 presents the resulting coordinated solution of the hierarchical inter-area routing task obtained after the sixth coordinating iteration of the proposed method, leading to the fulfillment of the conditions for inter-area interaction (10). The increase in the number of iterations of the coordinating procedure to obtain the desired solution with increasing flow rate is caused by the necessity to implement a multi-path routing strategy both within and between areas, which implies an expansion of the number of possible solutions to the task (Fig. 7).

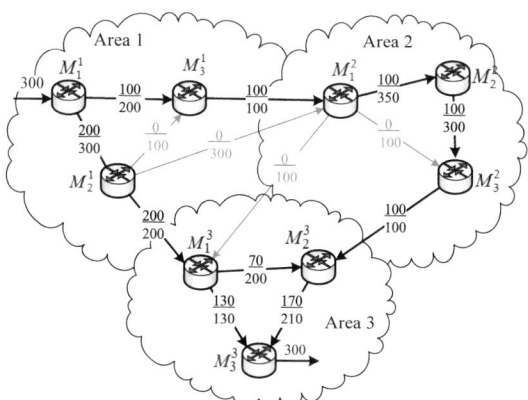

Fig. 6. A coordinated solution of the inter-area routing task of flow with rate of 300 1/s obtained after the sixth iteration

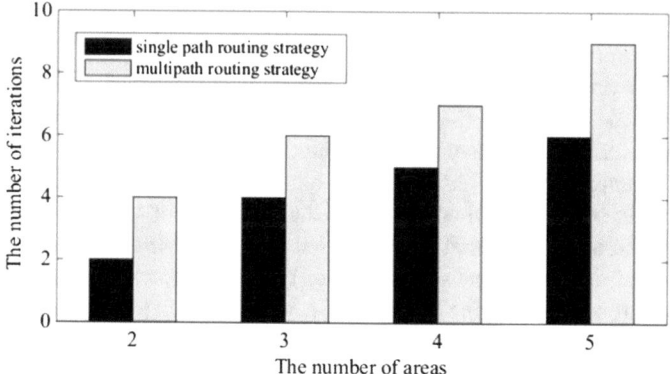

Fig. 7. Dependence number of iterations on number of areas in network and the routing strategy

In the conditions of increasing the number of possible variants of calculated inter-areas routes, the number of coordination iterations (11)–(12) increased. The results of the study confirmed the convergence of the method to an optimal solution by a finite number of iterations. Using the routing of a single route, the number of iterations of coordination varied from 1–2 to 6–7 depending on the number of areas in the network. The use of multipath routing has led to an increase in the number of iterations of the coordination procedure by approximately 1.5–2 times compared to routing decisions along a single path.

5 Conclusions

The decomposition flow-based routing model in a backboneless network is proposed in the paper. The novelty of the model is, firstly, the modification of the conditions of flow conservation for three types of areas: the sender, the receiver, and the set of transit areas; secondly, the introduction of the conditions of inter-area interaction in its structure to ensure the connectivity of inter-area routes both in terms of their structure and the nature of load balancing across them. The model is oriented to the implementation of both single and multipath routing with the support of routing metrics used in existing routing protocols. A feature of the proposed model is also that during the decomposition of the network, the boundary between areas passes through links, as it is realized, for example, in the IS-IS protocol. In addition, the network does not allocate a backbone, and an inter-area route can pass, in general, through an arbitrary set of domains.

Using the proposed model, the routing task was reduced to the optimization task of the linear programming class. Its solution, using the of goal coordination principle, determined the content of the proposed method of hierarchical coordination routing in a multi-area telecommunications network. The method is based on the introduction of a two-level hierarchy of calculations: the lower level is responsible for the calculation of intra-area routes, and the upper level is responsible for coordinating decisions of the lower level aimed at ensuring the connectivity of the calculated inter-area routes by fulfilling the introduced conditions for inter-area interaction. The rate of convergence of the method to optimal values was determined by the number of iterations of the coordination procedure (13)–(14) and, in turn, influenced on the efficiency and the volumes of information transmitted over hierarchical levels about the state of the network. The application of the method is aimed at increasing the scalability of the terminal routing solutions without reducing the efficiency of the network as a whole.

The research of the proposed method based on determining the degree of influence of the structural and functional parameters of the network on the convergence of the coordination procedure and the method as a whole is conducted in this work. The results of the research showed that the main factors influencing on the number of coordinating iterations in the method include the number of areas in the network, its congestion, and the implementation of a multipath routing strategy. The paper shows that, depending on the growth of network traffic, the number of iterations increases on average from 2 to 4 times.

References

1. Marsic, I.: Computer Networks: Performance and Quality of Service. Rutgers University (2013)
2. Lemeshko, O.V., Yeremenko, O.S.: Dynamics analysis of multipath QoS-routing tensor model with support of different flows classes. In: Proceedings of the 2016 International Conference on Smart Systems and Technologies (SST), pp. 225–230. IEEE (2016). https://doi.org/10.1109/SST.2016.7765664
3. Daradkeh, Y.I., Kirichenko, L., Radivilova, T.: Development of QoS methods in the information networks with fractal traffic. Int. J. Electron. Telecommun. **64**(1), 27–32 (2018). https://doi.org/10.24425/118142
4. Cisco Networking Academy (ed.): Routing Protocols Companion Guide. Pearson Education, London (2014)
5. Tkachov, V.M., Savanevych, V.Ye.: Method for transfer of data with intermediate storage. In: Proceedings of the 2014 IEEE First International Scientific-Practical Conference «Problems of Infocommunications. Science and Technology» (PICS&T-2014), pp. 105–106. IEEE (2014). https://doi.org/10.1109/INFOCOMMST.2014.6992315
6. Moza, M., Kumar, S.: Improving the performance of routing protocol using genetic algorithm. Int. J. Comput. Netw. Inf. Secur. (IJCNIS) **8**(7), 10–16 (2016). https://doi.org/10.5815/ijcnis.2016.07.02
7. Af-pnni-0055.000: Private Network-Network Interface. Specification Version 1.0 (PNNI 1.0) (1996)
8. Osunade, O.: A packet routing model for computer networks. Int. J. Comput. Netw. Inf. Secur. (IJCNIS) **4**(4), 13–20 (2012). https://doi.org/10.5815/ijcnis.2012.04.02
9. Wright, B.: Inter-area routing, path selection and traffic engineering. White paper. Data Connection Limited (2003)
10. Wójcik, R., Domżał, J., Duliński, Z., Rzym, G., Kamisiński, A., Gawłowicz, P., Jurkiewicz, P., Rząsa, J., Stankiewicz, R., Wajda, K.: A survey on methods to provide interdomain multipath transmissions. Comput. Netw. **108**, 233–259 (2016). https://doi.org/10.1016/j.comnet.2016.08.028
11. Nevzorova, Ye.S., Arous, K.M., Salakh, M.T.R.: Method for hierarchical coordinated multicast routing in a telecommunication network. Telecommun. Radio Eng. **75**, 1137–1151 (2016). https://doi.org/10.1615/TelecomRadEng.v75.i13.10
12. Lemeshko, O., Yeremenko, O., Nevzorova, O.: Hierarchical method of inter-area fast rerouting. Transp. Telecommun. J. **18**(2), 155–167 (2017). https://doi.org/10.1515/ttj-2017-0015
13. Sharma, G., Singh, M., Sharma, P.: Modifying AODV to reduce load in MANETs. Int. J. Mod. Educ. Comput. Sci. (IJMECS) **8**(10), 25–32 (2016). https://doi.org/10.5815/ijmecs.2016.10.04
14. Lemeshko, O., Nevzorova, O., Vavenko, V.: Hierarchical coordination method of inter-area routing in telecommunication network. In: Proceedings of the International Conference Radio Electronics and Info Communications (UkrMiCo), Kyiv, Ukraine, pp. 1–4 (2016). https://doi.org/10.1109/UkrMiCo.2016.7739626
15. Rajasekaran, K., Balasubramanian, K.: Energy conscious based multipath routing algorithm in WSN. Int. J. Comput. Netw. Inf. Secur. (IJCNIS) **8**(1), 27–34 (2016). https://doi.org/10.5815/ijcnis.2016.01.04

16. Lemeshko, O., Ilyashenko, A., Nevzorova, O., Mal-allah, A.M.: Method of segment hierarchical coordination routing in multi-area network. In: Proceedings of the 2nd International Conference on Advanced Information and Communication Technologies (AICT), Lviv, Ukraine, pp. 262–265 (2017). https://doi.org/10.1109/AIACT.2017.8020115

17. Lemeshko, A.V., Evseeva, O.Y., Garkusha, S.V.: Research on tensor model of multipath routing in telecommunication network with support of service quality by greate number of indices. Telecommun. Radio Eng. **73**(15), 1339–1360 (2014). https://doi.org/10.1615/TelecomRadEng.v73.i15.30

18. Zaman, R.U., Begum, S.S., Khan, K.U.R., Reddy, A.V.: Efficient adaptive path load balanced gateway management strategies for integrating MANET and the internet. Int. J. Wirel. Microw. Technol. (IJWMT) **7**(2), 57–75 (2017). https://doi.org/10.5815/ijwmt.2017.02.06

19. Yeremenko, O.S., Lemeshko, O.V., Nevzorova, O.S., Hailan A.M.: Method of hierarchical QoS routing based on the network resource reservation. In: Proceedings of the 2017 IEEE First Ukraine Conference on Electrical and Computer Engineering (UKRCON), pp. 971–976. IEEE (2017). https://doi.org/10.1109/UKRCON.2017.8100393

20. Lemeshko, O., Yeremenko, O.: Dynamic presentation of tensor model for multipath QoS-routing. In: Proceedings of the 2016 IEEE 13th International Conference on Modern Problems of Radio Engineering, Telecommunications and Computer Science (TCSET), pp. 601–604. IEEE (2016). https://doi.org/10.1109/TCSET.2016.7452128

21. Alinaghian, M., Kalantari, M.R., Bozorgi-Amiri, A., Raad, N.G.: A novel mathematical model for cross dock open-close vehicle routing problem with splitting. Int. J. Math. Sci. Comput. (IJMSC) **2**(3), 21–31 (2016). https://doi.org/10.5815/ijmsc.2016.03.02

22. Sahana, S.K., Al-Fayoumi, M., Mahanti, P.K.: Application of modified ant colony optimization (MACO) for multicast routing problem. Int. J. Intell. Syst. Appl. (IJISA) **8**(4), 43–48 (2016). https://doi.org/10.5815/ijisa.2016.04.05

23. Mesarovic, M.D., Macko, D., Takahara, Y.: Theory of Hierarchical, Multilevel, System. Academic Press, New York and London (1970)

24. Singh, M.G., Titli, A.: Systems: Decomposition, Optimization and Control. Pergamon, Oxford (1978)

On Optimality Conditions for Job Scheduling on Uniform Parallel Machines

Volodymyr Popenko[1] , Maiia Sperkach[1(✉)] ,
Olena Zhdanova[1] , and Zbigniew Kokosiński[2]

[1] National Technical University of Ukraine "Igor Sikorsky Kyiv Polytechnic Institute", 37 Prosp. Peremohy, Kyiv, Ukraine
VolodP@ukr.net, sperkachmaya@gmail.com,
zhdanova.elena@hotmail.com
[2] Faculty of Electrical and Computer Engineering,
Cracow University of Technology, 24 Warszawska Str., Cracow, Poland
zk@pk.edu.pl

Abstract. We study three similar problems of scheduling of unrelated jobs on uniform parallel machines each having a distinct optimization criterion. In the first problem the criterion is the makespan minimization. In the second one the goal is to create a schedule, where the minimum of the completion times of the last jobs on the parallel machines is maximized (machine covering problem). In the third one the goal is to create a schedule with maximally uniform distribution of jobs among the machines. We propose the sufficient conditions of schedule optimality for these problems. First, optimality criteria for the analyzed problems were transformed into functions of makespan's lower boundary deviation. This allows to define auxiliary optimization problems of the mixed-integer programming problems class. The objective of these auxiliary problems is to determine a perfect schedule - the one that gives the perfect value of the corresponding source criterion for the given volume of jobs. Perfect value allows us to determine the sufficient conditions of schedule optimality for all three problems.

Keywords: Schedule · Uniform parallel machines · Makespan ·
Machine covering problem · Auxiliary optimization problem ·
Sufficient conditions of optimality

1 Introduction

Models and methods of scheduling theory are widely used in many areas of activity, starting from manufacturing and service industries to cluster and parallel computing.

One of the typical areas where these methods are used is optimization of some process that involves execution of some set of tasks. Such optimization minimizes the usage of resources involved in the process. In this work we consider three practically meaningful problems of uniform parallel machines scheduling.

Consider a set of jobs, $J = \{1, 2, \ldots, j, \ldots, n\}$, and a number of machines, m. The machines operate simultaneously, are interchangeable, and may differ in job processing

© Springer Nature Switzerland AG 2020
Z. Hu et al. (Eds.): ICCSEEA 2019, AISC 938, pp. 103–112, 2020.
https://doi.org/10.1007/978-3-030-16621-2_10

efficiency. Machines can be sorted in the decreasing order according to the speed of job processing. The resulting order is the same for all the jobs. For any machine i, $i = 1, \ldots, m$, the duration of processing job j equals $k_i p_j$, where: p_j is the duration of processing job j on the reference machine and $k_i > 0$ is a performance coefficient (if $k_i < 1$ then machine i is more efficient than a reference machine with $k_i = 1$; similarly, if $k_i > 1$ – then machine is less efficient than the reference machine). Let us assume that all the jobs from J come simultaneously at time zero, and the processing of each job runs without interruptions until completion. All machines start working at time zero and operate without downtime.

> Problem 1. Find a schedule, where the maximum jobs completion time (C_{max}) is minimized (makespan minimization, $Qm\|C_{max}$).
> Problem 2. Find a schedule, where the minimum of the completion times of the last jobs on the parallel machines is maximized (this problem is known as machine covering problem).
> Problem 3. Find a schedule with the most uniform distribution of the jobs among the machines.

Since these problems belong to NP class, a polynomial algorithm for solving them is unknown [1]. New approximate algorithms, however, appear on a regular basis. One way to improve them further is to determine the sufficient conditions of schedule optimality [2]. Approximate algorithms can use such conditions to define a termination condition as following: (1) for the current solution sufficient optimality condition is obtained (this means that the optimal solution is reached); (2) further improvement of the current best solution is impossible (in this case the optimality gap is known).

The goal of our research is to determine the sufficient conditions of schedule optimality for three aforementioned problems.

Our contribution is as follows. While conducting our analysis we have transformed optimality criteria for the analyzed problems into functions of makespan's lower boundary deviation. This makes possible to introduce auxiliary optimization problems of the mixed-integer programming problems class. The objective of these auxiliary problems is to determine a perfect schedule - the one that gives the perfect value of the corresponding source criterion for the given volume of jobs. Afterwards sufficient conditions of schedule optimality are determined for all three problems. The novelty of our research are conditions for Problems 2 and 3. Lastly, we demonstrate the proposed approach on a set of sample data.

2   The Related Works

A lot of research has been devoted to problems 1 and 2, especially to problem 1 and similar ones. Overview of results obtained for the different scheduling problems for uniform parallel machines is presented in the works [2, 3]. To solve such problems various methods were applied [4, 5]. The paper [6] proposes a meta-heuristic based scheduling, which minimizes execution time and cost. In the work [7] an efficient approximation algorithm for scheduling precedence-constrained jobs on machines with different speeds is proposed, while work [8] focuses on approximation schemes for scheduling on uniformly related and identical parallel machines. Research [9] demonstrates scheduling jobs on parallel machines while optimizing bi-criteria namely maximum tardiness and weighted flow time. Authors of [10] have developed heuristics for the problem 1 which is based on the optimal makespan of scheduling n independent single operation jobs on m uniform parallel machines. In the work [11] we can see an approach to solving of a makespan minimization problem for two uniform parallel machines, in [12] for uniform parallel machines with release times, and in [13] for multiple uniform machines. This paper applies an approach similar to one from the work [13] to solve a lower boundary of makespan for Problem 1. Problem 2 was also a subject of an extensive research. For instance, in [14] a new exact branch-and-bound algorithm solution for identical machine covering problem is proposed. The work [15] compares the performance relationship between two well-known longest-first heuristics. The work [16] researches a problem that is one of versions of the classical machine covering problem. In [17, 18] that precede the current work, an approach for solving these problems is described.

3   Analysis of the Problems

Let us assume that all p_j are integers, $p_1 \geq p_2 \geq \ldots \geq p_n$, $k_1 \leq k_2 \leq \ldots \leq k_m$ and let us introduce the following quantity: $P = \sum_{j=1}^{n} p_j$, $K = \sum_{i=1}^{m} \frac{1}{k_i}$. Let us define the perfect case as one where the amount of work P is performed in the shortest possible time. Let us define C^* to be the theoretical minimal time, in which all the machines are able to process all the jobs with amount of work P. Ideally, the schedule is *uniform*, and all the machines complete their jobs in time C^*. In this case $C^* = P/K$ [11]. On the other hand, the maximal jobs completion time can't be less than $p_1 k_1$. Taking this into account we get:

$$C_{\max} \geq \max(p_1 k_1; P/K). \tag{1}$$

Further we will assume that maximum in (1) corresponds to the second element.

3.1   Transforming Objectives of Problems to Functions of Deviations of C^*

Let us consider a schedule σ. In this schedule, let us denote the completion time of job j by $C_j(\sigma)$; the completion time of all the jobs on machine i by $S_i(\sigma) = \max_{j \in J_i(\sigma)} C_j(\sigma) = \sum_{j \in J_i(\sigma)} k_i p_j$; machine i *protrusion* by $\Delta_i(\sigma) = \max\{0; S_i(\sigma) - C^*\}$; machine i *reserve* by $R_i(\sigma) = \max\{0; C^* - S_i(\sigma)\}$. Let us define the following so called reduced values: reduced completion time of all the jobs on machine i:

$$S_i'(\sigma) = \frac{S_i(\sigma)}{k_i} = \sum_{j \in J_i(\sigma)} p_j; \quad \Delta_i'(\sigma) = \max\{0; S_i'(\sigma) - c_i^*\}; \quad R_i'(\sigma) = \max\{0; c_i^* - S_i'(\sigma)\}.$$

For the newly introduced values, the following equations hold: $S_i(\sigma) = C^* - R_i(\sigma) + \Delta_i(\sigma)$; $S_i'(\sigma) = c_i^* - R_i'(\sigma) + \Delta_i'(\sigma)$; $R_i(\sigma)\Delta_i(\sigma) = 0$, $i = 1, \ldots, m$.

Example 1. One of the possible schedules for n jobs on m machines with $m = 6$, $k_1 = 1$, $k_2 = 1, 3$, $k_3 = 1, 5$, $k_4 = 2$, $k_5 = 2, 3$, $k_6 = 2, 5$; $n = 20$, and vector of reference processing time $p = \{20, 9, 9, 8, 8, 7, 7, 7, 6, 6, 6, 6, 5, 5, 5, 4, 4, 3, 3, 2\}$ is shown in Fig. 1. Job number and processing duration of the corresponding machine are shown within the corresponding rectangles. C^*, $R_i(\sigma)$ and $\Delta_i(\sigma)$ are also depicted. Using the introduced notation let us state the optimality criterion for the analyzed problems:

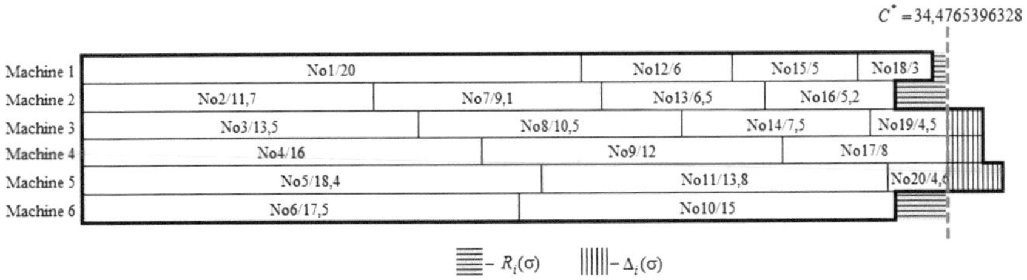

$C^* = 34,4765396328$

Machine 1	№ 1/20		№ 12/6	№ 15/5	№ 18/3
Machine 2	№ 2/11,7	№ 7/9,1	№ 13/6,5	№ 16/5,2	
Machine 3	№ 3/13,5	№ 8/10,5	№ 14/7,5	№ 19/4,5	
Machine 4	№ 4/16	№ 9/12	№ 17/8		
Machine 5	№ 5/18,4	№ 11/13,8	№ 20/4,6		
Machine 6	№ 6/17,5	№ 10/15			

$\equiv - R_i(\sigma)$ $||||| - \Delta_i(\sigma)$

Fig. 1. Graphical illustration of C^*, $R_i(\sigma)$ and $\Delta_i(\sigma)$ for *Example 1*: $m = 6$, $n = 20$.

Problem 1. The criterion of minimizing the maximal occupancy time (minimizing the makespan) is as follows:

$$C_{\max} = \max_i S_i(\sigma) = C^* + \max_i \Delta_i(\sigma) \rightarrow \min. \quad (2)$$

Since $C^* = const$, criterion $C_{\max} \rightarrow \min$ is reduced to minimizing the maximal protrusion:

$$\max_i \Delta_i(\sigma) \rightarrow \min. \quad (3)$$

Problem 2. The criterion of maximizing the minimum machine completion time is as follows: $\min_i S_i(\sigma) \rightarrow \max$. In fact, this criterion is symmetrical to the criterion (2): $\min_i S_i(\sigma) = C^* - \max R_i(\sigma) \rightarrow \max$. It is reduced to minimizing the maximal reserve: $\max_i R_i(\sigma) \rightarrow \min$.

Problem 3. Let us define the schedule with maximal uniform distribution the schedule, where maximal deviation of the completion time on all machines from the perfect completion time C^* of all jobs is minimized: $\max_i \{R_i(\sigma); \Delta_i(\sigma)\} \rightarrow \min$.

4 Sufficient Conditions of Optimality

4.1 Case When the Uniform Schedule Can Be Obtained (The First Sufficient Condition of Optimality)

As mentioned earlier, the uniform schedule (where $\sum_{i=1}^{m} \Delta_i(\sigma) = \sum_{i=1}^{m} R_i(\sigma) = 0$) is optimal. The uniform schedule can be obtained only when C^* and all $c_i^* = \frac{C^*}{k_i}$ are integers (otherwise equalities $\sum_{j \in J_i} p_j = c_i^*$, $i = 1, \ldots, m$, are not possible, because the left parts are assumed to be integers). Therefore, if the following holds for schedule σ:

$$S_i(\sigma) = k_i \sum_{j \subset J_i} p_j = k_i c_i^* = C^*, \, i = 1, \ldots, m \quad (4)$$

then it is optimal by all three criteria. Condition (4) is *the first sufficient condition of optimality for schedule* σ.

4.2   Case When the Uniform Schedule Cannot Be Obtained

Approach similar to one applied in [9] to research Problem 1 is also used in the work. This approach replaces set of jobs J with a set of P jobs of unit processing durations.

If some values c_i^* are not integers ($\exists i \,|\, c_i^* \notin Z$), it is not possible to obtain the uniform schedule. Let us determine sufficient conditions of optimality for three problems in this case. We will call a schedule, where some jobs are not assigned to any of the machines, an *incomplete* schedule. Let us suppose that a certain number of jobs is assigned to the machines (let them form a set $\bar{J} \subset J$); thus an incomplete schedule $\bar{\sigma}$ is obtained $\sum_{j \in \bar{J}_i} t_{ij} = k_i \lfloor c_i^* \rfloor$, $\sum_{j \in \bar{J}_i} p_j = \lfloor c_i^* \rfloor$, $i = 1, \ldots, m$, where $\bar{J}_i$ is a set of jobs processed by machine i in the obtained incomplete schedule $\bar{\sigma}$ (where $\lfloor a \rfloor$ is the largest integer less than or equal to a). Values c_i^*, $\lfloor c_i^* \rfloor$ and the set $\bar{J}$ for *Example 1* are shown in Fig. 2. Note that the cumulative duration of the jobs remaining unscheduled equals to $\delta = \sum_{j=1}^{n} p_j - \sum_{i=1}^{m} \sum_{j \in \bar{J}_i} p_j = \sum_{j=1}^{n} p_j - \sum_{i=1}^{m} \lfloor c_i^* \rfloor$.

Let us assume that δ jobs with durations of 1 are left unscheduled.

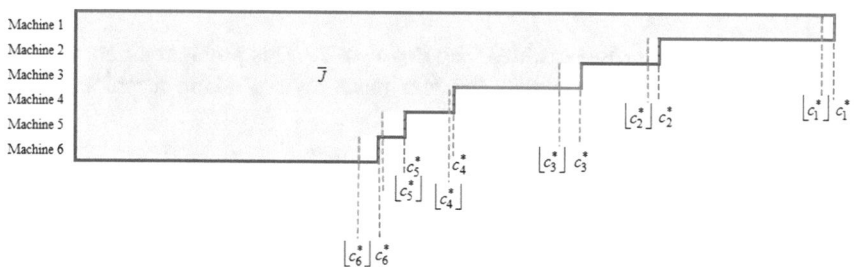

Fig. 2. Values c_i^*, $\lfloor c_i^* \rfloor$ and the set $\bar{J}$ for *Example 1*.

(a) Auxiliary Optimization Problem for Problem 1. Taking into account the criterion (3) for Problem 1, we are faced with the following problem: assign δ jobs to machines in such a way that the maximum protrusion of the complete schedule is minimal. As a result, we obtain a schedule, where jobs in volume of P are scheduled in the best way.

Let us introduce the following quantity: $e_i = c_i^* - \sum\limits_{j \in \bar{J}_i} p_j$, $i = 1, \ldots, m$, where e_i is the reserve of machine i in the incomplete schedule $\bar{\sigma}$, where δ jobs remain unscheduled. Taking into consideration previous definitions, we obtain $e_i = c_i^* - \lfloor c_i^* \rfloor$ ($e_i \geq 0$), $i = 1, \ldots, m$.

Mathematical Model of the Auxiliary Optimization Problem 1

Variables: x_i represent the number of «single» reference jobs to be assigned to machine i, $i = 1, \ldots, m$. *Constraints:* the total number of «single» reference jobs is δ: $\sum\limits_{i=1}^{m} x_i = \delta$; $x_i \geq 0$, x_i are integers, $i = 1, \ldots, m$. *Objective:* minimize the maximum protrusion (taking machines' performances into account):

$$\max_i \{k_i(x_i - e_i)\} \to \min. \tag{5}$$

The problem is reduced to the mixed-integer programming problem

$$y \to \min; \tag{6}$$

$$y \geq k_i(x_i - e_i), \quad i = 1, \ldots, m; \sum_{i=1}^{m} x_i = \delta; \ x_i \geq 0, \text{ are integers}, i = 1, \ldots, m. \tag{7}$$

Mathematical model proposed in [10] allows to obtain the optimal makespan of scheduling n independent single operation jobs on m uniform parallel machines. The number of variables in this model is $mn + 1$ and the number of constraints is $m + n$. In general case variable count in this model is $mP + 1$, number of constraints is $m + P$ while in model (6)–(7) the number of variables is m and the number of constraints is $m + 1$. Work [11] proposes an algorithm to determine a lower bound of makespan when analyzing Problem 1 for the case when the uniform schedule cannot be obtained. The result of applying this algorithm equals to the result that is based on the optimal solution of (6)–(7).

(b) Auxiliary Optimization Problem for Problem 2. This problem differs from (6)–(7) in objective function (minimize the maximum reserve taking machines' performances into account):

$$\min_i \{k_i(x_i - e_i)\} \to \max. \tag{8}$$

(c) Auxiliary Optimization Problem for Problem 3. This problem differs from (6)–(7) in objective function. In order to achieve maximal uniform loading of machines (taking machines' performances into account), it is required that:

$$\max_i \{|k_i(e_i - x_i)|\} \to \min. \tag{9}$$

(d) Second Sufficient Condition of Optimality. In the case when the uniform schedule cannot be obtained, the second sufficient condition of optimality can be determined for the problems.

Problem 1. Let x_i^*, $i = 1, \ldots, m$, be an optimal solution to the corresponding auxiliary optimization problem; $S_i^* = C^* + k_i(x_i^* - e_i)$ be the duration of processing the jobs on machine i in the «perfect» schedule, $S_{max}^* = \max_i S_i^*$. Hence, schedule σ, where $\max_i S_i(\sigma) = S_{max}^*$, is optimal.

Table 1. The optimal solution to the auxiliary Problem 1 for *Example 1*.

	Machine					
	1	2	3	4	5	6
c_i^*	34,4766	26,5204	22,9844	17,2382	14,9898	13,7906
$\lfloor c_i^* \rfloor$	34	26	22	17	14	13
x_i^*	1	0	1	0	1	1
S_i^*	35	33,8	34,5	34	34,5	35

Calculation results and optimal solution to the corresponding auxiliary optimization problem 1 by criterion (5) for *Example 1* are shown in Table 1. According to the results of Table 1 $S_{max}^* = 35$. The optimal solution to the auxiliary problem 1 for *Example 1* allows us to get the contour of the perfect schedule, which is shown in Fig. 3a and b. Since for schedule shown in Fig. 3b sufficient conditions of optimality are met, then it's the optimal solution of the Problem 1.

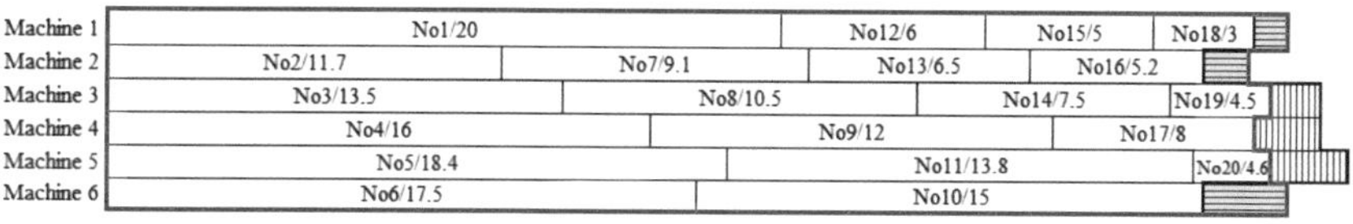

(a) The schedule from Fig. 1 and contour of the perfect schedule for the auxiliary Problem 1 (in red).

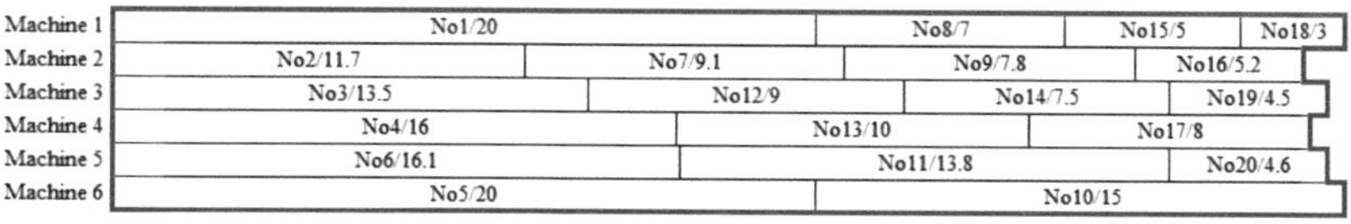

(b) The optimal solution to the optimization Problem 1 for *Example 1*.

Fig. 3. Solution to the optimization Problem 1 for *Example 1*.

Problem 2. Let x_i^*, $i = 1, \ldots, m$, be an optimal solution to the corresponding auxiliary optimization problem; $S_i^* = C^* + k_i(x_i^* - e_i)$ be the duration of processing the jobs on machine i in the «perfect» schedule, $S_{min}^* = \min_i S_i^*$. Hence, schedule σ, where $\min_i S_i(\sigma) = S_{min}^*$, is optimal.

Problem 3. Let x_i^*, $i = 1, \ldots, m$, be an optimal solution to the corresponding auxiliary optimization problem; $S_i^* = C^* + k_i(x_i^* - e_i)$ be the duration of processing the jobs on machine i in the perfect schedule, $H = \max_i\{|k_i(x_i^* - e_i)|\}$. Hence, schedule σ, where $\max_i\{|S_i(\sigma) - C^*|\} = H$, is optimal.

Optimal solutions for auxiliary problems 2 (by criterion (8)) and 3 (by criterion (9)) for *Example 1* are identical. This optimal solution and corresponding calculation results are shown in Table 2.

According to the results in Table 2 $S_{\min}^* = 34$ and $H = 0{,}623463671$. The optimal solution to the auxiliary problems 2 and 3 for *Example 1* allows us to get the contour of the perfect schedule, which is shown in Fig. 4a and b. Since for schedule shown in Fig. 4b sufficient conditions of optimality are met, then it is the optimal solution of the problems 2 and 3.

Table 2. The optimal solution to the auxiliary Problem 3 for *Example 1*.

	Machine					
	1	2	3	4	5	6
x_i^*	0	1	1	0	1	1
S_i^*	34	35,1	34,5	34	34,5	35

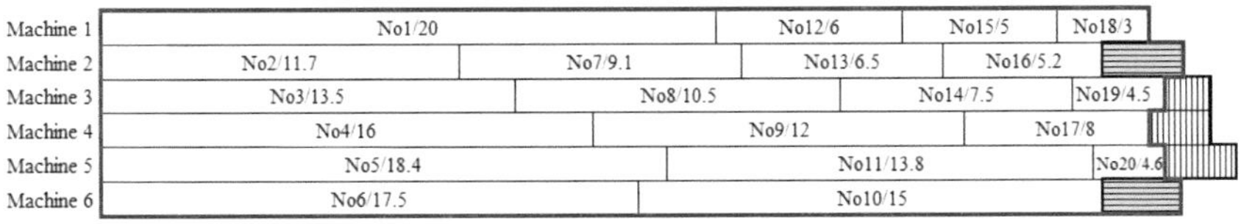

(a) The schedule from Fig. 1 and contour of the perfect schedule for the auxiliary Problems 2 and 3 (in red).

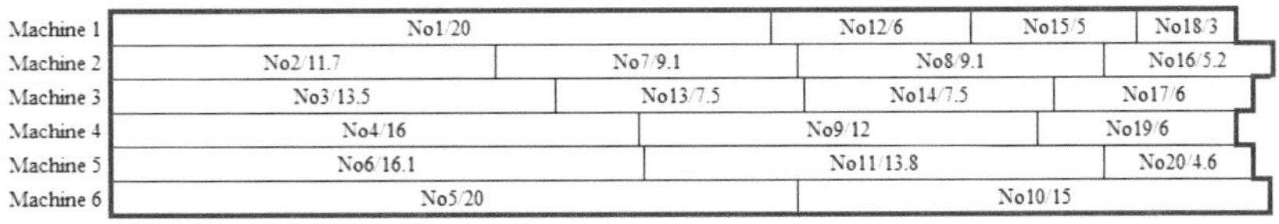

(b) The optimal solution to the optimization Problems 2 and 3 for *Example 1*.

Fig. 4. The solution to the optimization Problems 2 and 3 for *Example 1*.

5   Conclusions

In this work, an approach for solving three similar job scheduling problems on uniform machines was discussed. This approach is based on determining the sufficient conditions of schedule optimality according to the specified criteria.

To define these conditions, objective functions of analyzed problems were represented as functions of the theoretical minimal time, in which all the machines are able to process all the jobs. Assuming, that the original set of jobs is represented as a set of counterparts with unit processing durations, auxiliary optimization problems are introduced. Objective of these problems is construction of so called contour of the perfect schedule. The solutions of auxiliary problems were further used to determine the sufficient conditions of schedule optimality.

There are many algorithms for solving the problems in question. The sufficient conditions of schedule optimality defined in this paper can be incorporated in any of these algorithms, thereby increasing their effectiveness.

References

1. Pinedo, M.: Scheduling: Theory, Algorithms and Systems. Springer, New York (2008)
2. Zgurovsky, M., Pavlov, A.: Decision-Making in Network Systems with Limited Resources: Monograph. Naukova Dumka, Kyiv (2000)
3. Senthilkumar, P., Narayanan, S.: Literature review of single machine scheduling problem with uniform parallel machines. Intell. Inf. Manag. 2(8), 457–474 (2010)
4. Tawfeek, M.A., Elhady, G.F.: Hybrid algorithm based on swarm intelligence techniques for dynamic tasks scheduling in cloud computing. Int. J. Intell. Syst. Appl. (IJISA) 8(11), 61–69 (2016). https://doi.org/10.5815/ijisa.2016.11.07
5. Arya, L.K., Verma, A.: Child based level-wise list scheduling algorithm. Int. J. Mod. Educ. Comput. Sci. (IJMECS) 9(9), 24–31 (2017). https://doi.org/10.5815/ijmecs.2017.09.03
6. Kaur, S., Verma, A.: An efficient approach to genetic algorithm for task scheduling in cloud computing environment. Int. J. Inf. Technol. Comput. Sci. (IJITCS) 4(10), 74–79 (2012). https://doi.org/10.5815/ijitcs.2012.10.09
7. Chekuri, C., Bender, M.: An efficient approximation algorithm for minimizing makespan on uniformly related machines. J. Algorithms 41(2), 212–224 (2001)
8. Epstein, L., Sgall, J.: Approximation schemes for scheduling on uniformly related and identical parallel machines. In: Proceedings of 7th Annual European Symposium on Algorithms, pp. 151–162. Springer, Herzliya (1999)
9. Jeet, K., Dhir, R., Singh, P.: Hybrid black hole algorithm for bi-criteria job scheduling on parallel machines. Int. J. Intell. Syst. Appl. (IJISA) 8(4), 1–17 (2016). https://doi.org/10.5815/ijisa.2016.04.01
10. Sivasankaran, P., Ravi Kumar, M., Senthilkumar, P., Panneerselvam, R.: Heuristic to minimize makespan in uniform parallel machines scheduling problem. Udyog Pragati 33(3), 1–15 (2009)
11. Liao, C., Lin, C.: Makespan minimization for two uniform parallel machines. Int. J. Prod. Econ. 84(2), 205–213 (2003)
12. Koulamas, C., Kyparisis, G.: Makespan minimization on uniform parallel machines with release times. Eur. J. Oper. Res. 157(1), 262–266 (2004)

13. Liao, C., Lin, C.: Makespan minimization for multiple uniform machines. Comput. Ind. Eng. **54**(4), 983–992 (2008)
14. Walter, R., Wirth, M., Lawrinenko, F.: Improved approaches to the exact solution of the machine covering problem. J. Sched. **20**(2), 147–164 (2017)
15. Walter, R.: Comparing the minimum completion times of two longest-first scheduling-heuristics. CEJOR **21**, 125–139 (2013)
16. Azar, Y., Epstein, L.: On-line machine covering. In: ESA: European Symposium on Algorithms: Algorithms – ESA 1997, pp. 23–36. Springer, Austria (1997)
17. Sperkach, M.: The problem of defining the maximum start moment of execution of tasks with a common tough due date on parallel machines with different speeds. Sci. J. **63**, 12–18 (2015). NTUU "KPI", Informatics, Operation and Computer Science
18. Zhdanova, O., Sperkach, M.: Scheduling tasks on parallel devices of various productivity to maximize uniformity of devices' load. Sci. J. **1156**, 92–106 (2015). V. Karazin Kharkiv National University, series Mathematical Modelling. Information Technology, Automated Control Systems

MPLS Traffic Engineering Solution of Multipath Fast ReRoute with Local and Bandwidth Protection

Oleksandr Lemeshko, Oleksandra Yeremenko$^{(\boxtimes)}$, and Maryna Yevdokymenko

Kharkiv National University of Radio Electronics,
14 Nauky Avenue, Kharkiv, Ukraine
{oleksandr.lemeshko.ua,
oleksandra.yeremenko.ua}@ieee.org,
marina.ievdokymenko@nure.ua

Abstract. An MPLS Traffic Engineering solution of multipath Fast ReRoute with local and bandwidth protection is proposed. The novelty of the solution lies in the fact that the optimization problem of load balancing during fast rerouting is presented in the linear form provided the communication links bandwidth protection. This solution practically reduces the computational complexity of determining the routing variables responsible for the formation of the primary and backup paths and provides a balanced load of communication links that meet the requirements of the Traffic Engineering concept. The model provides implementation of local (link, node) and bandwidth protection schemes for fast rerouting with load balancing in telecommunication networks. The analysis of the proposed model has confirmed its adequacy and efficiency in terms of obtaining optimal solutions to ensure balanced load of network communication links and the implementation of necessary schemes for network elements (link, node, and bandwidth) protection.

Keywords: Routing · Flow-based model · MPLS · Traffic Engineering · Fast ReRoute · Protection schemes

1 Introduction

Despite the constantly increasing reliability of modern communication equipment, the problem of providing a certain level of resilience of telecommunication networks (TCN) is rather relevant. The main global causes of TCN failures include massive disasters, socio-political and economic factors, secondary failures, human factors (human-operator errors), network security threats, environmental problems, etc. In addition, among the main technological factors that lead to denial of service in the network we should point out failures of physical level, faults and overload of network equipment during its operation, errors in configuration and updating of terminal and network software [1–6]. Therefore, nowadays the task associated with the construction of the so-called resilient networks capable of providing a high level of Quality of Service (QoS) and Quality of Resilience (QoR) is urgent [5, 6].

© Springer Nature Switzerland AG 2020
Z. Hu et al. (Eds.): ICCSEEA 2019, AISC 938, pp. 113–125, 2020.
https://doi.org/10.1007/978-3-030-16621-2_11

Such a key process of the OSI (Open Systems Interconnection) network layer as routing is of great importance when increasing the TCN resilience. Within the frame work of standard routing solutions based on the periodic exchange of information on the state of TCN between routers, the Routing Information Protocol (RIP) can calculate a backup route that does not contain a failed network element within the period of 30 s. The Interior Gateway Routing Protocol (IGRP) can calculate a backup route within the period of 90 s. This does not take into account the fact that denial notifications in modern IP-networks can spread incorrectly often causing the problem of increasing the length of the route up to infinity (count-to-infinity problem), that is, to the looping of packets [1–3, 7]. Therefore, within the framework of fault-tolerant routing it is assumed that the backup route (routes) is simultaneously calculated with the primary one. Moreover, the backup route should not overlap with the primary route on the element of the network (router or communication link), which during fault-tolerant routing should be protected. Implementing the fault-tolerant routing allows real-time (tens of milliseconds) responding to a possible failure of network equipment, whereas in conventional routing protocols, the response time is measured in tens of seconds.

2 Classification of Fault-Tolerant Routing Means and Overview of Known Solutions

Currently the means of fault-tolerant routing can be classified according to the following criteria: the place of fault-tolerant routing implementation in the network; the type of reservation scheme used; the type of supported protection scheme and support of bandwidth protection (Fig. 1).

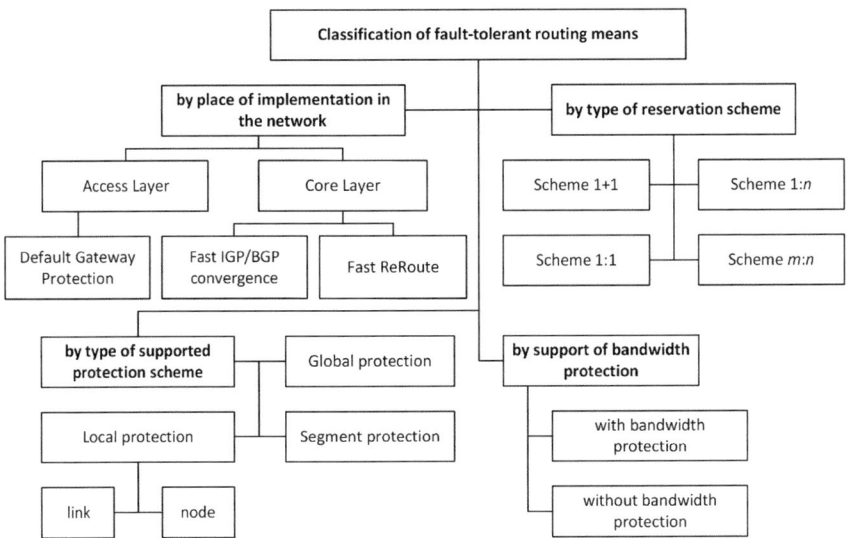

Fig. 1. Classification of fault-tolerant routing means.

According to the multilayer architecture of Next Generation Networks (NGN), fault-tolerant routing tasks can be solved both at the access layer and at the network core layer. At the access layer, this task is reduced to the default gateway protection, when the access networks are switched simultaneously to several edge routers, the interfaces of which are configured by the corresponding protocol as the virtual default gateway. In IP networks, such protocols include Hot Standby Router Protocol (HSRP); Virtual Router Redundancy Protocol (VRRP); Gateway Load Balancing Protocol (GLBP); Common Address Redundancy Protocol (CARP) [8, 9]. In order to increase the availability of edge routers in the case of the failure of the primary gateway, the protocol automatically switches traffic to the backup gateway. The results of the work [8] are devoted to the problems of such solutions. They are also oriented to support the load balancing by a set of interfaces of the virtual default gateway.

In the network core, the Fast IGP/BGP convergence and IP/MPLS Fast ReRoute [7] technologies are among the key solutions for improving its resilience. The Fast IGP/BGP convergence technology provides minimizing TCN response time for possible failures of its elements. This process is also referred to as network convergence or synchronization process of routing tables after the change of topology. The Fast ReRoute technology is used in IP/MPLS networks to protect the transport network elements – the link, node, path, and bandwidth of the network as a whole when the backup route is calculated together with the primary one in accordance with the implemented protection scheme.

It should also be noted that IP/MPLS networks, like most solutions related to increased network reliability, are based on the implementation of various reservation schemes [5, 10]:

- Scheme 1+1, in which the flow is transmitted both over the primary and over the backup route;
- Scheme 1:1 for each running route a backup path is created which should not contain a problem element of the network (link or node) that usually forms the primary path;
- Scheme 1:n, in which *one* backup path is created for n primary paths (facility backup);
- Scheme m:n, in which m backup paths are created for n primary (working) paths.

Generally, by the type of supported protection scheme in the implementation of fault-tolerant routing, for example, in the MPLS-FRR technology, local protection (link, node), global protection (path), and segment protection (sequence of elements of the network primary path) are distinguished [3, 5, 7]. The link protection scheme is the simplest solution and it involves creating a backup route bypassing the emergency link. When a failure occurs, the router virtually instantaneously switches the flow to the pre-computed backup route. The traffic is transmitted by the backup route until the calculation of the new primary route from the source to the destination is done. Solutions on this scheme have been obtained in [11–15]. Moreover, in [11] there has been proposed a nonlinear optimization link protection model for FRR while in [12], advanced linear solutions for both single path and multipath routing strategies are presented.

The node protection scheme is used in the case of shutdown of the router (node failure), caused, for example, by a power or overload failure. By analogy with the link protection scheme, a backup path is created that does not contain a protected node. In fact, the implementation of this scheme is to protect all links that are directly connected to the protected node. Generalized linear solutions to the node protection scheme are also presented in [12].

The path protection scheme in MPLS refers to global path recovery mechanisms. Moreover, along with the calculation of the primary Label Switched Path (LSP), a backup route, which does not include any element of the primary one, must also be calculated. In addition, in the course of implementing the protection of the path, the primary and backup routes contain shared source and destination nodes only. As soon as the signal protocol detects a failure, the flows are switched to the backup LSP used to calculate a new optimal path according to the control timer.

If there is a need to provide bandwidth protection, a necessary link resource is required for the successful transmission of packets both over the primary and over the backup route [8, 11–15]. The bandwidth protection scheme is implemented in the case when there is no enough available backup path, but it is necessary to ensure that there is a required bandwidth along this path. This is especially important for flows that are sensitive to bandwidth, delay, and jitter. In the works [5], the linear optimization model for the protection of the path is presented only for the case of the implementation of single path routing. However, in [13], a linear solution was proposed to protect both the route itself and its bandwidth for single path routing, while in [15] a bilinear path protection model for multipath routing was provided.

Thus, the considered protocol solutions provide an increase in the network resilience assuming the introduction of resource redundancy when, along with the determination of the primary path (PP) the backup path (BP) is simultaneously calculated in accordance with the implemented protection scheme. In this regard, when solving the tasks of the FRR, it is important to ensure the balanced use of the available network, first of all, the link resource, so that the protection of the network element does not lead to its overload and a significant decrease in QoS level. A large number of publications [16–24] are devoted to the scientific and practical direction of the implementation of the fast rerouting with load balancing in MPLS networks (MPLS Traffic Engineering Fast ReRoute, MPLS TE FRR). However, this concept can also be used in other types of networks and routing protocols [25–34].

Providing a consistent solution to the tasks of MPLS TE FRR usually leads to increased computing complexity and reduced scalability of protocol solutions. It is known that the efficiency of the protocol solution is largely determined by the adequacy and quality of the mathematical model of the calculation lying in its basis. As shown by the analysis [24], the order of FRR and TE is determined in the course of solving optimization problems of different complexity levels. In this case, implementation of the scheme of network bandwidth protection, as a rule, leads to the nonlinear formulation of the optimization problem and the corresponding increase in the computational complexity of the resulting solutions [8, 12, 14, 15]. In [13], an attempt was made to obtain the solution of the problem of TE FRR with the protection of bandwidth based on the introduction of a two-level hierarchy of calculations. However, optimization tasks, which are solved at different hierarchical levels, could be formulated in a linear form only for the case of implementing a single path routing strategy. Therefore, the scientific and practical task connected with the development of MPLS Traffic Engineering solution of multipath Fast ReRoute with local and bandwidth protection based on the adequate and effective linear optimization model supporting exactly the multipath routing strategy is relevant.

3 Flow-Based Fast ReRoute Model with Load Balancing in TCN

Suppose the structure of TCN is described using the graph $G = (R, E)$, in which $R = \{R_i; \ i = \overline{1, m}\}$ is a set of network routers, and $E = \{E_{i,j}; \ i, j = \overline{1, m}; \ i \neq j\}$ is a set of communication links. Let us denote the subset of routers as $R_i^* = \{R_j : \ E_{j,i} \neq 0; \ j = \overline{1, m}; \ i \neq j\}$, which are adjacent to the router R_i. The number of communication links in the network will be determined by $n = |E|$, and each of them will be matched to its bandwidth $\varphi_{i,j}$.

In the framework of this model, each k th unicast flow is associated with a number of functional characteristics: S_k is a source router; D_k is a destination router; λ^k is the average intensity of the k th packet flow measured in packets per second (1/s). Assume K is the set of packet flows that are transmitted over the network, then $k \in K$.

The result of the solution to the problem of fast rerouting with load balancing in the TCN is the calculation of two types of routing variables $x_{i,j}^k$ and $\bar{x}_{i,j}^k$, each of which characterizes the portion of the intensity of the k th flow in the communication link $E_{i,j} \in E$ included into the primary or backup path, respectively.

In the case when single path routing is used in the network, the two types of routing variables are superimposed with the constraints of the type

$$x_{i,j}^k \in \{0; 1\} \text{ and } \bar{x}_{i,j}^k \in \{0; 1\}, \tag{1}$$

but for a multipath routing strategy –

$$0 \leq x_{i,j}^k \leq 1 \text{ and } 0 \leq \bar{x}_{i,j}^k \leq 1. \tag{2}$$

To ensure the connectivity of the routes being calculated, the conditions for flow conservation separately for routing variables of the primary path are introduced:

$$\begin{cases} \sum\limits_{j:E_{i,j}\in E} x_{i,j}^k - \sum\limits_{j:E_{j,i}\in E} x_{j,i}^k = 0; \ k \in K, \ R_i \neq S_k, D_k; \\ \sum\limits_{j:E_{i,j}\in E} x_{i,j}^k - \sum\limits_{j:E_{j,i}\in E} x_{j,i}^k = 1; \ k \in K, \ R_i = S_k; \\ \sum\limits_{j:E_{i,j}\in E} x_{i,j}^k - \sum\limits_{j:E_{j,i}\in E} x_{j,i}^k = -1; \ k \in K, \ R_i = D_k; \end{cases} \tag{3}$$

and for variables of the backup path [12–15]:

$$\begin{cases} \sum_{j:E_{i,j}\in E} \bar{x}_{i,j}^k - \sum_{j:E_{j,i}\in E} \bar{x}_{j,i}^k = 0; \ k \in K, \ R_i \neq S_k, D_k; \\ \sum_{j:E_{i,j}\in E} \bar{x}_{i,j}^k - \sum_{j:E_{j,i}\in E} \bar{x}_{j,i}^k = 1; \ k \in K, \ R_i = S_k; \\ \sum_{j:E_{i,j}\in E} \bar{x}_{i,j}^k - \sum_{j:E_{j,i}\in E} \bar{x}_{j,i}^k = -1; \ k \in K, \ R_i = D_k. \end{cases} \quad (4)$$

4 Conditions for Local and Bandwidth Protection for Fast Rerouting with Load Balancing in TCN

As shown by the analysis [11, 12, 14, 15], in the process of fast rerouting, several basic schemes for protecting network elements: the node, link, path, and bandwidth can be supported. In [12, 14], the conditions have been obtained in an analytical form for supporting the mentioned protection schemes as elements of the corresponding mathematical models.

In [14], it is suggested that when implementing the link $E_{i,j} \in E$ protection scheme, the routing variables $\bar{x}_{i,j}^k$ that define the backup path are supposed to have additional constraints imposed, similar to (1). In this case, when implementing the single path routing strategy, the following constraint is applied:

$$\bar{x}_{i,j}^k \in \left\{0; \ \delta_{i,j}^k\right\}, \quad (5)$$

whereas under multipath routing

$$0 \leq \bar{x}_{i,j}^k \leq \delta_{i,j}^k, \quad (6)$$

where

$$\delta_{i,j}^k = \begin{cases} 0, \ \textit{under } E_{i,j} \textit{ link protection}; \\ 1, \ \textit{otherwise}. \end{cases} \quad (7)$$

Fulfilment of the conditions (5)–(7) ensures that the protected link $E_{i,j} \in E$ will not be used by the backup path in single path routing. The conditions (5)–(7) are linear in contrast to the nonlinear solutions proposed in [11]. This reduces the computational complexity of obtaining final protocol solutions.

In the realization of the node $R_i \in R$ protection scheme, the conditions (5)–(7) are generalized to protect the set of communication links incident to the protected node [12, 14]. Then, in the case of using the single path strategy, a system of conditions is introduced:

$$\bar{x}_{i,j}^k \in \left\{0; \ \delta_{i,j}^k\right\} \text{ at } R_j \in R_i^*, j = \overline{1,m}, \quad (8)$$

but for the multipath case –

$$0 \leq \bar{x}_{i,j}^k \leq \delta_{i,j}^k \text{ at } R_j \in R_i^*, j = \overline{1, m}, \tag{9}$$

where the choice of values $\delta_{i,j}^k$ is subject to the condition (7).

Thus, the fulfilment of the conditions (8) and (9) guarantees the protection of the node $R_i \in R$ by prohibiting the use of the backup path of all links that flow from the given node. Since only the transit routers are protected, the prohibition on using the outgoing links in accordance with the conditions (4) prevents the inclusion of a backup path and incoming links for the node R_i, which ultimately promotes the protection of the node as a whole. It should be noted that the conditions for the protection of predetermined nodes and network links are usually linear, and their inclusion does not critically affect the complexity of the calculation of routing variables $x_{i,j}^k$ and $\bar{x}_{i,j}^k$ that are responsible for the formation of a set of primary and backup paths.

Network bandwidth protection conditions, which are represented by the conditions for preventing overloading of communication links during the implementation of fast rerouting when some flows can switch to backup routes, have the form:

$$\frac{1}{2} \sum_{k \in K} \lambda^k \max \left[x_{i,j}^k, \bar{x}_{i,j}^k \right] \leq \varphi_{i,j}, E_{i,j} \in E. \tag{10}$$

The conditions (10) presented in [8] are proposed to be used in the following form

$$\frac{1}{2} \sum_{k \in K} \lambda^k \left[x_{i,j}^k + \bar{x}_{i,j}^k + \left| x_{i,j}^k - \bar{x}_{i,j}^k \right| \right] \leq \varphi_{i,j}, E_{i,j} \in E. \tag{11}$$

However, the conditions for network bandwidth protection (10) and (11) are nonlinear and this adversely affects the computational complexity of the corresponding protocol solutions. In [13], due to the introduction of the two-level hierarchy of calculations in accordance with the interaction prediction principle of the theory of hierarchical multilevel systems, these conditions were obtained in a linear form, but only for the case of the single path routing implementation. Therefore, in this paper, in order to provide a linear form of the conditions for the protection of network bandwidth in the implementation of both single path and multipath routing, it is proposed to introduce the following modified conditions for preventing overload in order to ensure load balancing in the network:

$$\sum_{k \in K} \lambda^k u_{i,j}^k \leq \alpha \varphi_{i,j}, E_{i,j} \in E \tag{12}$$

at

$$x_{i,j}^k \leq u_{i,j}^k \text{ and } \bar{x}_{i,j}^k \leq u_{i,j}^k, \tag{13}$$

where $u_{i,j}^k$ are also control variables

$$0 \leq u_{i,j}^k \leq 1 \tag{14}$$

and represent the upper bound (UB) of the values of the routing variables for the primary and backup paths, whereas α represents an additional control variable that determines the upper bound of the network communication links load and meets the following conditions

$$0 \leq \alpha \leq 1. \tag{15}$$

The optimality criterion for the solutions of the MPLS TE FRR problems, in analogy to the results obtained in the papers [18–20], will be the minimum of the bound introduced in (12), that is,

$$\min_{x,\bar{x},\alpha} \alpha. \tag{16}$$

Thus, the solution of the initial technological problem of fast rerouting with load balancing in telecommunication networks with the link, node, and bandwidth protection was reduced to solving the optimization problem of linear programming with criterion (16) under constraints (1)–(6), (8), (9), (12)–(15).

5   Investigation of the Proposed MPLS Traffic Engineering Solution of Multipath Fast ReRoute with Local and Bandwidth Protection

An analysis of the proposed model of fast rerouting with load balancing in the TCN was performed on a set of network configurations for a different number of flows and their characteristics. The features of the model TE FRR will be shown on the numerical example. The structure of the investigated network is shown in Fig. 2, and in the gaps of the network communication links, their bandwidth is indicated.

Let the network provide a solution to the problem of the fast rerouting of two flows. In this case, the packets of the first flow were transmitted from the node R_1 to the node R_{16}. Packets of the second flow were transmitted from R_5 to R_{12}. Assume that the intensity of these flows varied in the following limits: $\lambda^1 = 10 \div 400$ 1/s and $\lambda^2 = 10 \div 400$ 1/s. Let us consider how the upper bound of the network links utilization (15) behaves depending on the implemented scheme of protection of the link, node and bandwidth. Depending on the network utilization, determined by the values λ^1 and λ^2, the gain of the multipath solution with respect to the single path solution by criterion (16) varied. Table 1 shows the minimum and maximum values of this gain when protecting each of the network link separately. Thus, with the protection of communication links, the use of the model (1)–(16) makes it possible to improve the criterion (16) in general from 37.12% to 59.41%. For clarity, Fig. 3 shows the dependence of the upper bound of the communication links utilization on the values of the intensity of

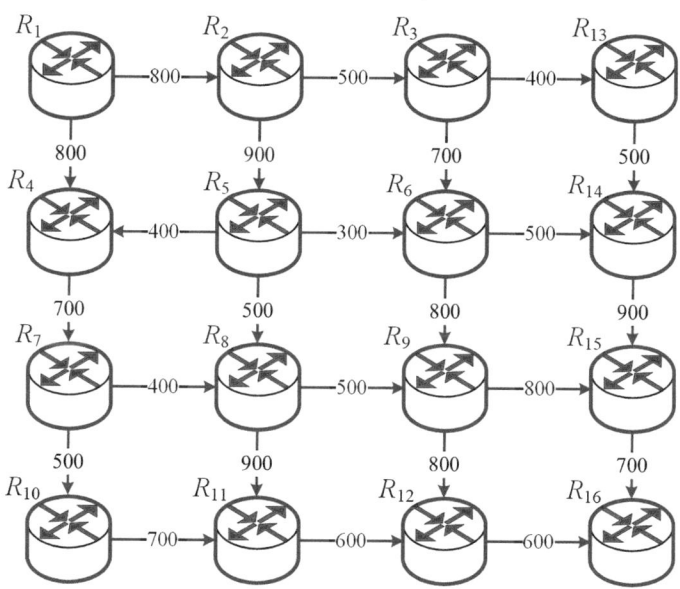

Fig. 2. The structure of the investigated model.

flows, if, for example, the link $E_{8,11}$ protection scheme for multipath (Fig. 3a) or single path (Fig. 3b) is implemented.

Table 2 shows the minimum and maximum values of the gain on the criterion (16) values when implementing multipath routing compared with the use of single path routing but for the protection of each network node separately in dependence on the varying intensities $\lambda^1 = 10 \div 400$ 1/s and $\lambda^2 = 10 \div 400$ 1/s.

Table 1. Minimum and maximum values of gain for the criterion (16) values for multipath routing compared to single path routing with protection of each of the network link separately.

Protected link	Gain, %		Protected link	Gain, %	
	Min	Max		Min	Max
$E_{1,2}$	28.57	58.33	$E_{7,10}$	41.18	58.33
$E_{2,3}$	28.57	61.54	$E_{8,11}$	44.44	61.54
$E_{1,4}$	37.5	58.33	$E_{9,12}$	16.67	60.55
$E_{2,5}$	47.37	58.33	$E_{10,11}$	41.18	58.33
$E_{3,6}$	44.44	61.54	$E_{11,12}$	23.08	58.33
$E_{5,4}$	37.5	61.54	$E_{3,13}$	47.37	61.54
$E_{5,6}$	37.5	61.54	$E_{13,14}$	47.37	61.54
$E_{4,7}$	23.08	37.05	$E_{6,14}$	47.37	61.54
$E_{5,8}$	40.17	61.54	$E_{14,15}$	47.37	61.54
$E_{6,9}$	44.44	61.54	$E_{9,15}$	47.37	61.54
$E_{7,8}$	44.44	61.54	$E_{15,16}$	16.67	58.33
$E_{8,9}$	28.57	61.54	$E_{12,16}$	28.57	58.33

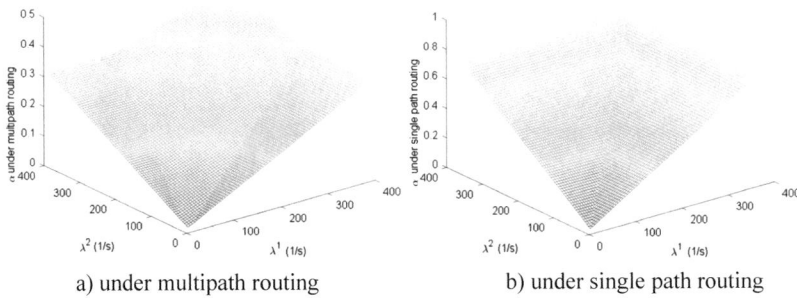

a) under multipath routing   b) under single path routing

Fig. 3. Dependence of the upper bound of network links utilization on the values of flows intensity under realization of the link $E_{8,11}$ protection scheme.

Table 2. Minimum and maximum values of the gain on the criterion (16) values when implementing multipath routing compared with the use of single path routing for the protection of each network node separately.

Protected node	Gain, %		Protected node	Gain, %	
	Min	Max		Min	Max
R_2	28.57	58.33	R_9	16.67	60.55
R_3	28.57	61.54	R_{10}	41.18	58.33
R_4	23.08	37.5	R_{11}	41.18	58.33
R_6	33.33	61.54	R_{13}	47.37	61.54
R_7	23.08	37.5	R_{14}	47.37	61.54
R_8	30.97	60.55	R_{15}	16.67	58.33

Thus, with the protection of network nodes, the use of model (1)–(16) allows improving the criterion (16) on average from 31.5% to 56.3%. Figure 4, for example, shows that the implementation of multipath routing with node R_9 protection allows improving the value of the criterion (16) from 16.67% to 60.55% compared to single path routing.

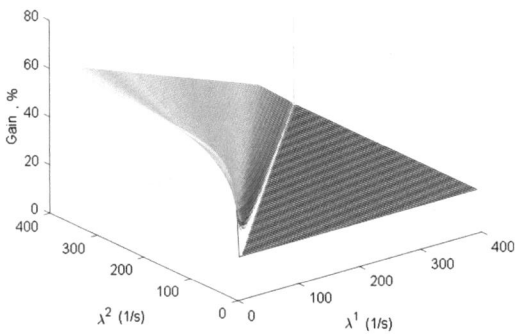

Fig. 4. Gain on the criterion (16) from the implementation of multipath routing in comparison with the use of single path routing (node protection R_9).

6   Conclusions

An MPLS Traffic Engineering solution and corresponding mathematical model of multipath Fast ReRoute with local and bandwidth protection in telecommunication networks presented by the expressions (1)–(9) and (12)–(16) has been proposed. The novelty of the proposed model is that a consistent solution to problems with TE and FRR with the protection of the link, node and bandwidth is provided during the solution of the linear optimization problem. The optimality criterion has been represented by the minimum of the upper bound for the utilization of the network communication links (15), (16) of the flows, which are transmitted both over the primary and over the backup paths. The transition from the nonlinear conditions of bandwidth protection (10), (11) to the linear analogue (12) has been achieved by some extension of the number of calculated variables (13) and (14), which define the upper bound for routing variables of the primary and backup paths. This approach focuses on reducing the computational complexity when calculating the routing variables responsible for the formation of the primary and backup paths, and ensures balanced loading of the communication links in accordance with the requirements of the Traffic Engineering concept.

The results of the analysis of the proposed model on a number of numerical examples have confirmed its adequacy and the possibility of obtaining optimal solutions to the problem of fast rerouting with load balancing in telecommunication networks in realizing various protection schemes of network elements (link, node) and bandwidth.

It has been shown that the implementation of the multipath routing strategy at TE FRR allowed decreasing the upper bound of communication links utilization on average from 37.12% to 59.41% for the link protection and from 31.5% to 56.3% for the node protection, which positively influenced the level of quality of service in the network as a whole.

It should be noted that the proposed solution can be used practically in deploying the so-called Hybrid WAN when using both MPLS and SD-WAN (Software-Defined Wide Area Networking) technologies together. At the same time, a reliable MPLS network is used to transmit traffic of the mission-critical and real-time applications that are sensitive to latency, jitter and packet loss. Whereas, SD-WAN facilities are used for the transmission of WAN applications traffic, the advantages of which include increased bandwidth and performance, allowing to utilize multiple, high-bandwidth, inexpensive Internet connections, simultaneously.

References

1. White, R., Banks, E.: Computer Networking Problems and Solutions: An Innovative Approach to Building Resilient, Modern Networks. Addison-Wesley Professional, Boston (2017)
2. White, R., Tantsura, J.E.: Navigating Network Complexity: Next-generation Routing with SDN, Service Virtualization, and Service Chaining. Addison-Wesley Professional, Boston (2015)

3. Monge, A.S., Szarkowicz, K.G.: MPLS in the SDN Era: Interoperable Scenarios to Make Networks Scale to New Services. O'Reilly Media, Inc., Sebastopol (2015)

4. Stallings, W.: Foundations of Modern Networking: SDN, NFV, QoE IoT, and Cloud. Addison-Wesley Professional, Boston (2015)

5. Rak, J.: Resilient Routing in Communication Networks. Springer, Switzerland (2015). https://doi.org/10.1007/978-3-319-22333-9

6. Alashaikh, A., Tipper, D., Gomes, T.: Exploring the logical layer to support differentiated resilience classes in multilayer networks. Ann. Telecommun. **73**, 1–17 (2017). https://doi.org/10.1007/s12243-017-0616-1

7. Cisco Networking Academy (Ed.) Routing Protocols Companion Guide. Pearson Education (2014)

8. Lemeshko, O., Yeremenko, O., Tariki, N.: Solution for the default gateway protection within fault-tolerant routing in an IP network. Int. J. Electr. Comput. Eng. Syst. **8**(1), 19–26 (2017)

9. Pavlik, J., Komarek, A., Sobeslav, V., Horalek, J.: Gateway redundancy protocols. In: 2014 IEEE 15th International Symposium on Computational Intelligence and Informatics (CINTI) Proceedings, pp. 459–464. IEEE (2014). https://doi.org/10.1109/CINTI.2014.7028719

10. Haider, A., Harris, R.J.: Recovery techniques in next generation networks. IEEE Commun. Surv. Tutor. **9**(1–4), 2–17 (2007). https://doi.org/10.1109/COMST.2007.4317617

11. Lemeshko, O., Arous, K., Tariki, N.: Effective solution for scalability and productivity improvement in fault-tolerant routing. In: 2015 Second International Scientific-Practical Conference Problems of Infocommunications Science and Technology (PIC S&T) Proceedings, pp. 76–78. IEEE (2015). https://doi.org/10.1109/INFOCOMMST.2015.7357274

12. Lemeshko, A.V., Yeremenko, O.S., Tariki, N.: Improvement of flow-oriented fast reroute model based on scalable protection solutions for telecommunication network elements. Telecommun. Radio Eng. **76**(6), 477–490 (2017). https://doi.org/10.1615/TelecomRadEng.v76.i6.30

13. Lemeshko, O., Yeremenko, O.: Enhanced method of fast re-routing with load balancing in software-defined networks. Journal of Electrical Engineering **68**(6), 444–454 (2017). https://doi.org/10.1515/jee-2017-0079

14. Yeremenko, O.S., Lemeshko, O.V., Tariki, N.: Fast ReRoute scalable solution with protection schemes of network elements. In: 2017 IEEE First Ukraine Conference on Electrical and Computer Engineering (UKRCON) Proceedings, pp. 783–788. IEEE (2017). https://doi.org/10.1109/UKRCON.2017.8100353

15. Yeremenko, O., Lemeshko, O., Tariki, N., Hailan, A.M.: Research of optimization model of fault-tolerant routing with bilinear path protection criterion. In: 2017 2nd International Conference on Advanced Information and Communication Technologies (AICT) Proceedings, pp. 219–222. IEEE (2017). https://doi.org/10.1109/AIACT.2017.8020105

16. Lin, S.C., Wang, P., Luo, M.: Control traffic balancing in software defined networks. Comput. Netw. **106**, 260–271 (2016). https://doi.org/10.1016/j.comnet.2015.08.004

17. Pan, P., Swallow, G., Atlas, A.: Fast reroute extensions to RSVP-TE for LSP tunnels, No. RFC 4090 (2005)

18. Seok, Y., Lee, Y., Choi, Y., Kim, C.: Dynamic constrained traffic engineering for multicast routing. In: International Conference on Information Networking, pp. 278–288. Springer, Heidelberg (2002). https://doi.org/10.1007/3-540-45803-4_26

19. Wang, Y., Wang, Z.: Explicit routing algorithms for internet traffic engineering. In: Proceedings Eight International Conference on Computer Communications and Networks (Cat. No.99EX370) Proceedings, pp. 582–588. IEEE (1999). https://doi.org/10.1109/ICCCN.1999.805577

20. Lemeshko, A.V., Evseeva, O.Y., Garkusha, S.V.: Research on tensor model of multipath routing in telecommunication network with support of service quality by greate number of indices. Telecommun. Radio Eng. **73**(15), 1339–1360 (2014). https://doi.org/10.1615/TelecomRadEng.v73.i15.30
21. King, D., Farrel, A.: The application of the path computation element architecture to the determination of a sequence of domains in MPLS and GMPLS, No. RFC 6805 (2012)
22. Paolucci, F., Cugini, F., Giorgetti, A., Sambo, N., Castoldi, P.: A survey on the path computation element (PCE) architecture. IEEE Commun. Surv. Tutor. **15**(4), 1819–1841 (2013). https://doi.org/10.1109/SURV.2013.011413.00087
23. Prabhavat, S., Nishiyama, H., Ansari, N., Kato, N.: On load distribution over multipath networks. IEEE Commun. Surv. Tutor. **14**(3), 662–680 (2012). https://doi.org/10.1109/SURV.2011.082511.00013
24. Wang, N., Ho, K., Pavlou, G., Howarth, M.: An overview of routing optimization for internet traffic engineering. IEEE Commun. Surv. Tutor. **10**(1), 36–56 (2008). https://doi.org/10.1109/COMST.2008.4483669
25. Koubàa, M., Amdouni, N., Aguili, T.: Efficient traffic engineering strategies for optimizing network throughput in WDM all-optical networks. Int. J. Comput. Netw. Inf. Secur. (IJCNIS) **7**(6), 39–49 (2015). https://doi.org/10.5815/ijcnis.2015.06.05
26. Mallapur, S.V., Patil, S.R., Agarkhed, J.V.: A stable backbone-based on demand multipath routing protocol for wireless mobile Ad Hoc networks. Int. J. Comput. Netw. Inf. Secur. (IJCNIS) **8**(3), 41–51 (2016). https://doi.org/10.5815/ijcnis.2016.03.06
27. Moza, M., Kumar, S.: Analyzing multiple routing configuration. Int. J. Comput. Netw. Inf. Secur. (IJCNIS) **8**(5), 48–54 (2016). https://doi.org/10.5815/ijcnis.2016.05.07
28. Krishna, S.R.M., Seeta Ramanath, M.N., Kamakshi Prasad, V.: Optimal reliable routing path selection in MANET through novel approach in GA. Int. J. Intell. Syst. Appl. (IJISA), **9**(2), 35–41 (2017). https://doi.org/10.5815/ijisa.2017.02.05
29. Smelyakov, K., Dmitry, P., Vitalii, M., Anastasiya, C.: Investigation of network infrastructure control parameters for effective intellectual analysis. In: 2018 14th International Conference on Advanced Trends in Radioelecrtronics, Telecommunications and Computer Engineering (TCSET) Proceedings, pp. 983–986. IEEE (2018). https://doi.org/10.1109/tcset.2018.8336359
30. Ruban, I.V., Churyumov, G.I., Tokarev, V.V., Tkachov, V.M.: Provision of survivability of reconfigurable mobile system on exposure to high-power electromagnetic radiation. In: Selected Papers of the XVII International Scientific and Practical Conference on Information Technologies and Security (ITS 2017), CEUR Workshop Processing, pp. 105–111 (2017)
31. Ageyev, D., Kirichenko, L., Radivilova, T., Tawalbeh, M., Baranovskyi, O.: Method of self-similar load balancing in network intrusion detection system. In: 2018 28th International Conference Radioelektronika (RADIOELEKTRONIKA) Proceedings, pp. 1–4. IEEE (2018). https://doi.org/10.1109/radioelek.2018.8376406
32. Yeremenko, O.S., Lemeshko, O.V., Nevzorova, O.S., Hailan A.M.: Method of hierarchical QoS routing based on the network resource reservation. In: 2017 IEEE First Ukraine Conference on Electrical and Computer Engineering (UKRCON) Proceedings, pp. 971–976. IEEE (2017). https://doi.org/10.1109/UKRCON.2017.8100393
33. Lakshman, N.L., Khan, R.U., Mishra, R.B.: MANETs: QoS and investigations on optimized link state routing protocol. Int. J. Comput. Netw. Inf. Secur. (IJCNIS) **10**(10), 26–37 (2018). https://doi.org/10.5815/ijcnis.2018.10.04
34. Najafi, G., Gudakahriz, S.J.: A stable routing protocol based on DSR protocol for mobile Ad Hoc networks. Int. J. Wirel. Microw. Technol. (IJWMT) **8**(3), 14–22 (2018). https://doi.org/10.5815/ijwmt.2018.03.02

Visualization of Expert Evaluations of the Smartness of Sociopolises with the Help of Radar Charts

Volodymyr Pasichnyk[✉], Danylo Tabachyshyn, Nataliia Kunanets, and Antonii Rzheuskyi

Lviv Polytechnic National University, Lviv, Ukraine
vpasichnyk@gmail.com, tabachyshyn.danylo@gmail.com,
nek.lviv@gmail.com, antonii.v.rzheuskyi@lpnu.ua

Abstract. The article covers the issue of visualization of the evaluation results by experts of various types of complex systems. The peculiarities of expert evaluation system "smartness sociopolis" are analyzed, namely, in such settlements are the main place of residence of the majority of humanity on the planet and the basic methods of visualization of expert assessments. The advantages of presenting the results of expert evaluation with the help of data visualization tools in comparison with text or tabular are noted. The radar charts have been selected as one of the tools for visualizing the evaluation results, which provide an opportunity to present peculiarities of the use of various criteria, to distinguish more weighty criteria and to take into account the opinion of each expert. The article proposes a mathematical model for evaluating of "smartness of sociopolis".

Keywords: Smart sociopolis · Smart cities · Visualization · Charts · Radar chart · Experts · Expert evaluations

1 Introduction

The city is the main place of residence of people on the planet. Cities every year increase their area, increasing the number of people in cities, less territory on the earth's surface remains without the influence of cities. All this calls for the updating of management tools for cities or territorial entities. The population and powers seek to receive a prestigious status for their settlement – "smart city". To achieve this status, settlements must achieve unprecedented success in various areas of the city's life, for example: economics, education, transport, public transport, medicine, ecology, the social sector and the involvement of information technology in the city's life. These areas become the criteria for estimation the smartness of sociopolises.

2 Related Work Section

In order to develop these industries, to create projects and develop sociopolises to the "smart" status, it is necessary to involve a large amount of data, money, time and people who are scientists, experts from different fields, residents, and, of course,

© Springer Nature Switzerland AG 2020
Z. Hu et al. (Eds.): ICCSEEA 2019, AISC 938, pp. 126–141, 2020.
https://doi.org/10.1007/978-3-030-16621-2_12

government officials. These people will form expert groups on each of the issues, on each project and direction of sociopolis development, they will hold a systematic analysis of the current state of affairs, investigate how sociopolis has historically developed, identify problems in different spheres, offer solutions to these problems, formulate recommendations and create projects for further development. It takes into account the importance of organically and harmoniously integration of different systems to create a "smart city" [1, 2].

To accomplish this series of tasks, experts need to submit information in a clear and easy-to-understand manner, and the latest information technology for presenting of information – visualization is well-applied for this purpose. Today, under visualization the interactive study of the visual representation of abstract data to enhance human knowledge is meant [3]. Such data contains text, numbers, graphic image. Data visualization provides information and data in the form of graphical presentation, thanks to which all the necessary information can be presented in a concise, easy-to-understand image. In this paper, the authors tried to develop their own method of visualization of data, obtained as a result of expert evaluation of the smartness of sociopolises not only by different criteria, but also taking into account the weight of each as a criterion and the weight of the opinion of each of the experts in the expert group. This will allow us to qualitatively and comprehensively evaluate the state of the city. Today, expert evaluation is one of the most common methods of obtaining and analyzing information, and plays an important role among the methods of statistical data analysis. Among the variety of methods of data statistical analysis for the evaluation of the obtained results, the expert evaluation methods take an important place. Nowadays, expert evaluation is the most widespread way of obtaining and analyzing qualitative information. The selection of experts among the possible candidates is an extremely important step, because they are responsible for evaluating the modern state of the city and they will create new projects for the development of the necessary industries in the city. Possible candidates for experts can be representatives of local authorities, academicians, local specialists in certain spheres, and the community as a whole [4–7], since the community of the city can best reflect today's problems, identify the most urgent ones that require urgent solutions both in short and long term. The expert is a highly skilled specialist in a certain field of activity who operates with the technologies of conducting expert evaluations and appropriate regulatory framework [8, 9]. Each expert who takes part in the evaluation of the smartness of the sociopolis, despite the weight of his opinion in comparison with other experts, must have the necessary knowledge about a certain sociopolis [10, 11], must be experienced, have knowledge in a particular subject area. If these conditions are not met, experts' valuation may be false, which will lead to significant losses both material and temporal.

3   Expert Evaluations of the Smartness of Sociopolises

In order to reduce the risk of making such decisions, it is necessary to use information processing technologies, namely, visualization of the results in such way that the expert or business analysts in the future could conveniently and quickly perceive information, effectively analyze it and predict the next steps, including possible negative consequences of these steps. Available methods of data processing automation allow business analysts to formalize qualitative and quantitative evaluations from experts and comprehensively evaluate the smartness of the sociopolis. During the survey, the expert is the main person who should have access to all available information related to the study.

For the role of the expert a person who is competent, has a high degree of analytical and broad thinking is elected. In general, there are two methods for selecting experts: subjective and objective. The objective includes documentary and experimental, while the subjective is self-estimation and mutual evaluation of candidates.

The expert evaluation is a versatile tool used in a variety of situations. This method is simple enough and does not require specific requirements from the initial information and its presentation. In this research, we partially use one of the methods of expert evaluation, namely the Delphi method, its purpose is to analyze the results of the expert survey. This method consists of the following steps:

- Experts evaluation;
- Summarizing expert estimations for each criterion;
- Determining the relative importance of each criterion by the importance coefficient;
- Calculation of the coefficient of unanimity of expert estimations according to expert rating estimations;
- An estimation of the experts unanimity using the Pearson criterion.

Taking into account the peculiarities of this task and based on the general Delphi method, the authors determined that the method of expert evaluation would be as follows:

- The expert estimation process begins with criteria of competence definition and selection of experts;
- The next step is to rank the experts and evaluate the coherence of the ranking results;
- After that, the "weight" of experts' opinions is calculated. The most experienced expert is the one whose sum of ranks will be the smallest;
- Next, a list of factors influencing the evaluation is compiled and their ranking is conducted.
- After that, the determination of total factor ranking, the choice of the most significant factor and evaluation of the coherence of the factor ranking using the Pearson criterion is conducted.
- And the last step is to visualize the result of expert evaluation of the smartness of sociopolis.

If the consistency score is unsatisfactory, then the process will have to be repeated again.

Therefore, the proposed algorithm is generally similar to the well-known Delphi method, or even its subspecies. But, it has features, such as: visualization of the results in the form of a radar chart. The steps to visualize the results of the evaluation of smartness start with the first evaluations of the criteria of experts. The visualization of evaluation is used for more convenience of the experts activity and more visual results [12–14]. One of these methods is a radar chart, also called a polar chart, a web chart, an irregular polygon, a spider chart or a stellar chart. Such presentation of information can be used to maximize the convenience of its understanding not only by experts but also by ordinary citizens, residents of sociopolis, for whom it is easier and more convenient to understand the graphic image than to delve into essence of dry figures. Also, the information in this form is more likely to be perceived, an accessible and understandable form is given to any object, subject or process in expert estimations.

Thus, visualization is a process of constructing a graphical data image. It helps in general data analysis to see the structure of the project, possible anomalies and relationships between objects in the structure [15].

The authors of this work have developed a methodology for selecting experts and methods for evaluating complex characteristics of sociopolis on the basis of the Delphi method and consider that the method of visualization quantitatively and qualitatively represents a complex set of values. In this case, the polygon chart, which is constructed in a polar coordinate system is used. Today, the theme of developing a methodology for visual representation of expert assessments of intelligent sociopolis remains relevant, especially in the form of radar charts. The subject of the research is the methods of visualization of the information obtained as a result of the processing of expert evaluations of the smartness of the sociopolis on various criteria in such a way that it is convenient and quick to perceive information and analyze it effectively.

The object of the paper is an expert evaluation of the smartness of the sociopolis.

The purpose of the paper is to develop a method of visualization of information obtained as a result of an expert evaluation of the smartness of sociopolis on various criteria using radar charts.

To achieve the goal you must accomplish the following tasks:

- To substantiate why expert assessments should be submitted in the form of a radar chart, which enables to display the result of the evaluation in the image and other necessary information;
- To propose the criteria for evaluation the wisdom of sociopolis their weighting factors for each expert, which will provide a reliable representation of the current situation, a correct understanding of the nature of the problem and the exact characteristics of its components;
- To determine the comprehensive characteristics of the wisdom of sociopolis both for each expert individually and for the general opinion of all experts, this evaluation will enable business analysts to calculate and analyze the results of the evaluation.
- To formulate conclusions derived from the evaluation results and provide recommendations on the use of the developed method of visualization of information.
- Applying expert estimates in the form of a polar chart.

The evaluation criteria are proposed in our previous studies. The city evaluation by a set of criteria and a method for obtaining quantitative estimates, based on the theory of fuzzy logic is presented in [16].

Various studies [17–19] show that a business analyst has a higher productivity of 17%, if using the information provided by the visual method.

Many scholars [20, 21] believe that due to the presentation of information, a person can remember such details, which in the text would not attract attention even by a careful reader who is thinking of reading and analyzing it. If the information comes from experts to the analyst or community of the sociopolis not only in the form of a text set about the state of things and further development, but with the corresponding drawings, schemes and other visual effects, then such information is undoubtedly perceptible by people much quicker and much easier and, as a result, a little more. In recent years, there has been tremendous change in the field of visual information, with such indicators as quality, volume and quantity.

The radar chart is a graphical representation of abstract data, in our case, this data is a result of the evaluation of the experts of the sociopolitical smartness, this chart is two-dimensional, with at least three variables (vectors) that are the criteria for which the evaluation is performed. These variables reflect the axes that have a common beginning. In the world, there are several options for naming radar charts: a web chart, a spider chart, a star chart, a web graph, an incorrect polygon, or a polar chart.

The result of the evaluation of experts will be presented in the form of vectors of the polar coordinate system, which form a polar chart. Each vector has indicators such as the angle of inclination and length. Figure 1 shows an example of a radar chart for calculating the smartness of a sociopolis for eighteen equally important criteria. In this case, the angle between the vectors is calculated by the following formula:

$$\beta = 2\pi/N$$

where N – number of criteria.

The length of each vector is equal to the quantitative score of the corresponding criterion. Ideally, the length of each vector should be equal to 100% (in this case, in Fig. 1, it should be equal to 1). Usually, the real length of each vector is less than 100% or 1, which corresponds to the real situation that has developed in the sociopoly for each of the criteria. The angle between the vectors characterizes the magnitude of the impact of the corresponding criterion on the outcome of experts' evaluation of the smartness of the sociopoly, since the larger the angle between the vectors, the greater the sector occupies this criterion, hence the overall share of the result of the evaluation depends on the indicators of this criterion. That is why it is important to first valuation the impact of each of the criteria for evaluation the wisdom of sociopolises and then evaluate the significance of these criteria. If certain criteria have a different effect on the general picture of the smartness of the sociopolis, the angles between the vectors will be different and can be defined as follows

$$\tilde{\beta} = \left\{ \beta_1 = 2\pi w_j / \sum\nolimits_{i=1}^{N} w_i, j = \overline{1,N} \right\} \tag{1}$$

Where $\bar{W} = \left\{ w_i, j = \overline{1,N} \right\}$ is weight coefficient of the j-th criterion for estimating the smartness of the sociopolis, which corresponds to the evaluation of a particular expert.

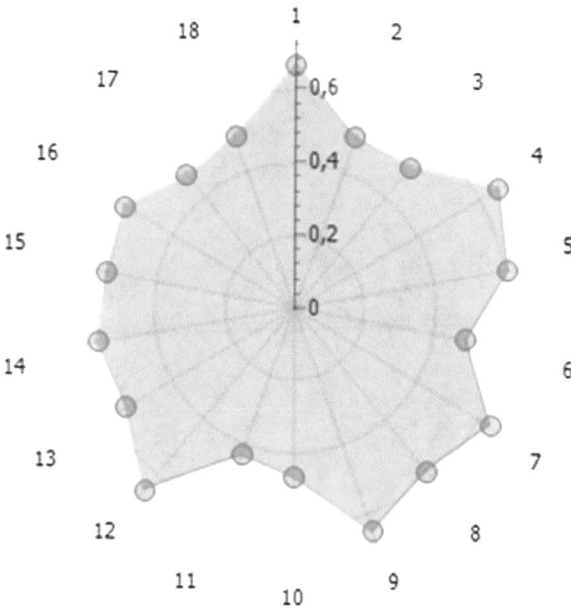

Fig. 1. Submission of the smartness of the sociopolis in the form of a radar chart.

If we depict each criterion in the form of a vector in a polar coordinate system and combine the points of their vertices, then we obtain the wrong polygon (in polygon $1, 2, 3, 4, \ldots, 17, 18$). The area of which is quantitatively and visually complex characterizes the evaluation of the smartness of the sociopolis for all characteristics simultaneously. The areas of individual polygons will characterize the wisdom of the sociopolis on certain particular characteristics. With the help of the form of this polygon we have the opportunity to visually obtain a qualitative characteristic of smartness for all the criteria at a time, and the form of the sector polygon - according to the corresponding criterion. The difference between the total area of the circle and the area of the polygon will be the proportion that still needs to be achieved at the moment to obtain the status of "reasonable sociopoly". If we divide the area of the polygon into sectors, then we will get evaluation of smartness only by a certain criterion, this estimate is more clearly depicted in Fig. 2.

In order that this method of visualization of the estimation of the smartness of the sociopolis is legitimate it is necessary that certain conditions be fulfilled, namely:

- The number of criteria must be at least three.
- The initial criterion must be on the positive axis of the ordinate.
- The area of the polar sector with its lateral rays, which are at an angle α must be divided in half by the vector-criterion.

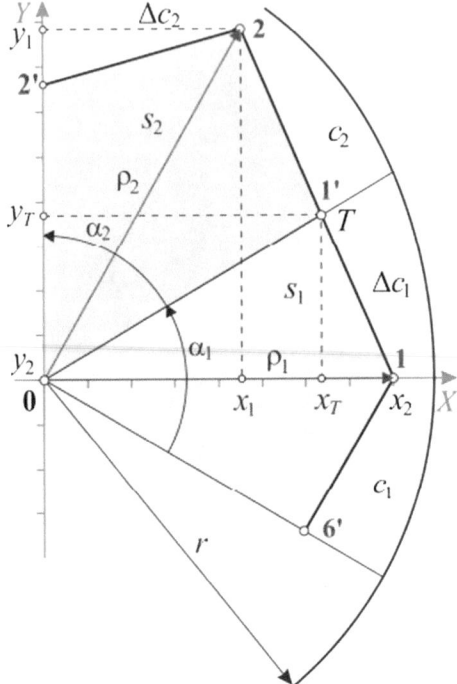

Fig. 2. Divide the polygon into sectors according to certain criteria

4   Algorithm for Calculating the Area of a Polygon

To find the area of the polygon, first of all you need to find the area of polygons for each criterion. To determine the area To find the area of each of the polygons we use an algorithm for calculation. First of all, we give the corresponding calculation scheme (Fig. 2), on which the coordinates of the points 1′ and 6′ remain unknown. A point T is formed due to the intersection of the segment [2, 1] with the lateral beam [0, 1′] of the polar sector. Here, the segment [2, 1] is a side of the irregular polygon $1, 2, \ldots, 6$ constructed in the Cartesian coordinate system, and the ray [0, 1′] is at an angle $\alpha_1/2$ to the ray [0, 6′] in polar coordinate system. Now we find the coordinates of the point T with the following algorithm.

So, from Fig. 2 it can be seen that the segment [2, 1] can be specified by the coordinates of two points - vertices of the vectors $\rho2$ and $\rho1$, namely 2 (x1, y1) and 1 (x2, y2). To find the analytic expression of this segment, we use linear interpolation. Expression of interpolation Its expression is described by an algebraic binomial $f(x) = ax + b$, which is given by two ends of the segment [2, 1]. Geometrically this means replacing the graph of the function f(x) with a straight line passing through points 2 (x1, y1) and 1 (x2, y2). The equation of this line will have the following form:

$$\frac{y - y_1}{y_2 - y_1} = \frac{x - x_1}{x_2 - x_1} \tag{2}$$

From here we obtain the formula of linear interpolation:

$$y = y_1 + \frac{y_2 - y_1}{x_2 - x_1} * (x - x_1), \tag{3}$$

With the aid of interpolation, we can find the values of the abscissas of the point T. To find the value of the abscissa of the point T, use the following formula:

$$y = y_0 + (x - x_0) \sin \alpha_1 / 2 \tag{4}$$

Where x_0 and y_0 these are the calculated coordinates of the center of the polar coordinate system. Equating the expressions (3) and (4) we obtain

$$y_1 + \frac{y_2 - y_1}{x_2 - x_1} * (x - x_1) = y_0 + (x - x_0) \sin \frac{\alpha_1}{2} \tag{5}$$

Find the value of this expression x

$$x = \frac{y_0 - y_1 + Ax_1 - x_0 \sin \frac{\alpha_1}{2}}{A - \sin \frac{\alpha_1}{2}}, \text{ where } A = \frac{y_2 - y_1}{x_2 - x_1} \tag{6}$$

Using formula (6), you can calculate the ordinates of the point T, after which the value of the abscissa of the point T can be calculated using formulas (3) or (4).

By this principle, we find the value of the point 6'. Now, having the coordinates of all points of a sector polygon, we can calculate the area of this polygon, for this we use the following formula:

$$s_1 = \frac{1}{2} \sum_{i=1}^{4} |y_i x_{i+1} - x_i y_{i+1}| \tag{7}$$

Where y_i, x_i are coordinates of vertices of a sector polygon. Also, we calculate the total area of the sector (c_1) by the formula

$$c_1 = \pi r^2 \alpha_1 / 2 \tag{8}$$

To find the coordinates of all points of a polygon, we need to generalize the above expressions (3), (4) and (6) and obtain the following system of mathematical equations:

$$\begin{cases} A_j = \frac{y_j - y_{j+1}}{x_j - x_{j+1}}, j = \overline{1, N} \\ x_j^T = \frac{y_0 - y_{j+1} + A_j x_{j+1} - x_0 \sin\frac{\alpha_j}{2}}{A_j - \sin\frac{\alpha_j}{2}}, \quad j = \overline{1, N} \\ y_j^T = \begin{cases} y_{j+1} + A_j \left(x_j^T - x_{j+1} \right) \\ y_0 + \left(x_j^T - x_0 \right) \sin\frac{\alpha_j}{2} \end{cases}, j = \overline{1, N} \end{cases} \tag{9}$$

Consequently, an algorithm for calculating the area of sector polygons is developed, by means of which one can calculate and visually estimate the smartness of the sociopolis according to the corresponding criteria. Also, this algorithm allows us to evaluate the state of each of the criteria we currently have for experts, as well as which sectors need to be further developed to achieve the "smart city" status.

In order to accurately and correctly estimate the smartness of the sociopolis, it is first necessary to select an expert group that will choose the criteria for which the evaluation will be conducted. Experts and criteria also need to give weight ratios, because some factors are more important and influential on the overall result, and these ranks will be made by the experts themselves. But experts may also have different levels of competence in different fields of activity, for example, in evaluation the smartness of the sociopolis, which may be representatives of government, academics, experts from certain industries and the community. And in each of these people qualifications are different, generalized by all the respondents who participate in the assessment of the smartness of the sociopolis, will be called experts. Taking into account the above, each expert is provided with certain weight coefficients, the values of which will indicate their awareness in a particular subject area [11].

Typically, the results of the survey from each expert are stored in the database. Obtaining of the results of the evaluation takes place in the form of a survey using a ranged scale for each of the criteria. Depending on the qualifications of the expert, each of them will also have different values of the weighting factors.

We introduce a set of weight coefficients for the criteria for assessing the smartness of a sociopolis, which is provided for each expert:

$$\tilde{W} = \left\{ \tilde{W}_i = \left\{ w_{i,k} = [0(1)M], k = \overline{1, K} \right\}, i = \overline{1, M} \right\}, \tag{10}$$

Where $w_{i,k}$ is weight coefficient i-th criterion for estimating the smartness of sociopolis, which is provided to the k-th expert, $0(1)M$ is range of values of grades from 0 to M, step - 1. K - number of experts and M - number of criteria. In this case there will be 10 experts and 6 criteria.

Table 1 shows the criteria for evaluating, selected 10 experts from different fields and their valuations.

In Table 1, the average value of weight coefficients for the i-th criterion for evaluation the smartness of the sociopolys can be defined as follows:

$$\tilde{W}^c = \left\{ w_i^c = \frac{1}{K} \sum_{k=1}^K w_{i,k}, i = \overline{1, M} \right\}, \tag{11}$$

Table 1. Evaluation criteria, weight factors and expert valuations.

Criteria for evaluation	Smart economics	Smart people	Smart leadership	Smart environment	Smart life	Smart mobility	Average score/total weight of the coefficient
Expert ratings/weight factors	4/8	5/7	8/10	8/6	5/8	3/9	5.5/48
	5/7	5/8	4/8	8/8	7/9	5/9	5.6/49
	5/9	4/10	7/9	7/6	6/6	4/10	5.5/50
	4/8	5/10	4/6	7/10	5/6	5/8	5/48
	4/9	6/9	6/10	8/10	7/5	3/7	5.6/50
	5/10	6/7	7/10	8/9	6/8	6/10	6.3/54
	3/8	5/10	5/7	8/7	4/6	7/9	5.3/47
	4/7	4/7	3/10	6/9	5/7	6/6	4.6/46
	5/8	5/9	7/8	5/9	3/8	5/5	5/47
	7/8	7/6	6/9	6/9	6/9	3/7	5.8/48
Average weight of coefficients	8.2	8.3	8.7	8.3	7.2	8.0	8.1

For each individual expert involved in the evaluation procedure, the set of quotes should be stored in the database. Also in the database are signs of the role of experts and the factors of their importance. The initial values of the weighting factors of experts, as a rule, are taken empirically, based on their qualifications. Table 2 shows the selection of 10 experts and calculates the weight of each expert's opinion. The criteria for competence were credibility and trust in experts among locals, position in local society, professionalism and education in the situation.

$\sum_{j=1}^{KR} y_{ij}$ is the sum of ranks obtained by the i-th expert in KR rankings (KR number of criteria). The smaller the number in the resulting rank, the more important the expert, and more must listen to his thoughts. The most experienced expert is 3 expert (in fact he has the smallest sum of ranks) and the "weakest" 6. δ is the weight of the opinion of each expert. Their thoughts are given weights $\delta_3 = 2$ and $\delta_6 = 1$ accordingly.

This is done to calculate the weight of each expert's opinion. To do this, we need to determine the coefficients a and c. We can identify them with the help δ_3 and δ_6. The calculation of a and c is performed by solving the following equation:

$$\delta_1 = a + c \sum_{j=1}^{KR} y_{ij} \tag{12}$$

To determine the coefficients a and c, we need to solve a system of equations

$$\begin{cases} 2 = a + c6 \\ 1 = a + c24, 5 \end{cases}$$

Thus, the method of selecting experts and evaluating their competence and weighing the opinions of each expert is proposed. The criteria for evaluation the smartness of sociopolis, their weighting factors for all experts, which will provide a

Table 2. Criteria for experts' competence and the weight of each expert's opinion

Competency criterion	Expert's number									
	1	2	3	4	5	6	7	8	9	10
Authoritativeness	7,5	10	1	6	5	7,5	2	3	9	4
Position	4	2	2	8	5	10	2	6	7	
Education	10	6	3	8	2	7	5	5	6	4
$\sum_{j=1}^{KR} y_{ij}$	21,5	18	6	22	12	24,5	9	15	20	14
Resulting rank	8	6	1	9	3	10	2	5	7	4
δ		1,16	1,35	2	1,14	1,68	1	1,84	1,51	1,24	1,57

reliable representation of the existing state of the sociopoly, have been selected. To determine the comprehensive indicator of the smartness of the sociopolis, we use the whole set of valuations of the experts of the evaluation participants, taking into account the weight of the thoughts of each of them. We introduce a set of estimates of the smartness of sociopolis, which can be put forward by any expert according to a certain criterion for its evaluation, namely.

$$\tilde{U} = \{u_i = [0(1)10], i = \overline{1,M}\} \tag{13}$$

Where is an estimation of the smartness of a sociopolis provided by an expert on the i-th criterion of its evaluation. Each individual evaluation by the relevant criterion provided by any expert belongs to this plurality:

$$\tilde{\tilde{X}} = \{\tilde{X}_i = \{x_{i,k} \in u_i, k = \overline{1,K}\}, i = \overline{1,M}\}, \tag{14}$$

Where $x_{i,k}$ is an estimation of the smartness of the sociopolis on the i-th criterion of its evaluation, which is provided by the k-th expert. For each expert we introduce such a concept as a comprehensive expert indicator of the smartness of a sociopolis on the relevant criterion for its evaluation, which can be calculated using the following formula:

$$\tilde{\tilde{G}} = \{\tilde{G}_i = \{g_{i,k} = x_{i,k}w_{i,k}\delta_k = \overline{1,K}\}, i = \overline{1,M}\}, \tag{15}$$

Where $g_{i,k}$ is a comprehensive expert intelligence indicator on the i-th criterion for its evaluation, which refers to the expert, taking into account the weight of the criterion and weight of the expert's opinion. For a generalized expert, the so-called complex expert indicator of the smartness of a sociopolis based on the relevant criterion for its estimation is calculated by the following formula

$$\tilde{\tilde{G}} = \left\{\tilde{G}_i = \left\{g_{i,k+1} = \frac{\sum_{j=1}^{K} x_{i,j}w_{i,j}\delta_j}{\sum_{j=1}^{K} \delta_j}, k = \overline{1,K}\right\}, i = \overline{1,M}\right\}, \tag{16}$$

Where $g_{i,k+1}$ is a complex expert indicator of the smartness of the sociopolis on the i-th criterion of its evaluation, which concerns the generalized $(K + 1)$ expert. If we consider that the estimation of expert's smartness $(x_{i,k})$ during the survey put on the 10-point scale, the weight factor of the evaluation criterion $w_{i,k}$ is also determined by the 10-point scale, and the weight factor of the expert δ_k is dimensionless value from 1 to 2, then a complex expert indicator of the smartness of the sociopolis $(g_{i,k})$ will have a value from 0 to 100. To calculate the complex outcome, we use the following formula for each expert:

$$\tilde{D} = \left\{ d_k = \delta_j \sum_{i=1}^{M} x_{i,k} w_{i,k} \middle/ \sum_{i=1}^{M} w_{i,k}, k = \overline{1,K} \right\}, \quad (17)$$

And an integrated generalized indicator for all experts in general will be calculated using the following formula

$$d^y = \sum_{k=1}^{K} d_k \middle/ \sum_{k=1}^{K} \delta_j \quad (18)$$

Table 3 presents the results of calculating the complex indicators of smartness of the sociopolys and their mean values taking into account the weighting factors of the evaluation criteria, as well as the importance of the opinion of each expert, both individually and in general.

Table 3. Results of calculation of complex indicators of smartness of sociopolis and their averaged values.

Criteria for evaluation	Experts' indicators										Average values	
	1	2	3	4	5	6	7	8	9	10	Rate	Evaluation
Smart economics	32	35	45	32	36	50	24	28	40	56	37,8	4,6
Smart people	35	40	40	50	54	42	50	28	45	42	42,6	5,2
Smart leadership	80	32	63	24	60	70	35	30	56	54	50,4	5,7
Smart environment	48	64	42	70	80	72	56	54	45	54	58,5	7,1
Smart life	40	63	36	30	35	48	24	35	24	54	38,9	5,4
Smart mobility	27	45	40	40	21	60	63	36	25	21	37,8	4,7
Average experts' evaluations	5,5	5,6	5,5	5	5,6	6,3	5,3	4,6	5	5,8	5,42	5,5
Estimates of experts taking into account their importance	6,38	7,56	11	5,7	9,41	6,3	9,75	6,95	6,2	9,11	7,85	7,97

5 Algorithm for Calculating the Area of the Wrong Polygon

Complex expert indicators of the smartness of the sociopolis will be presented in the form of vectors of the polar coordinate system, which should form a radar chart for each expert in particular and the generalized opinion of experts in general. Each such vector is characterized by the length and angle, respectively, of the preceding vector. As noted above, the length of the vector in any case should correspond to the quantitative meaning of the integrated index of smartness of the sociopolis by the corresponding criterion.

To find the coordinates of the vertices of the irregular polygon $1, 2, \ldots, 6$ (see Fig. 2) we use the following calculation algorithm. In the case of unequal influence of the criteria on the smartness of the sociopolys (see formula (1)), the angles between the corresponding vectors taking into account (10) can be determined by the following formula:

$$\widetilde{B_k} = \left\{ \beta_{i,k} = 2\pi w_{i,k} / \sum_{j=1}^{M} w_{j,k}, i = \overline{1,M} \right\}, k = \overline{1,K} \tag{19}$$

And for the mean value of the estimates of the smartness of the sociopolys (k = K + 1) taking into account (11), this formula will have the next form

$$\widetilde{B_k} = \left\{ \beta_{i,k} = 2\pi w_i^c / \sum_{j=1}^{M} w_j^c, i = \overline{1,M} \right\}, k = K+1 \tag{20}$$

Since the region of the polar sector with the angle β_j must be divided in half by the criterion vectors (formula (1)), then the first vector-criterion must be on the ordinate axis in the Cartesian coordinate system. Therefore, the starting angle of the angle $\beta_{1,k}$ ($\forall k \in K+1$), which corresponds to the 1st polar sector, we begin with the value of the angle $\alpha_{1,k} = -\beta_{1,k}/2$ ($\forall k \in K+1$), and all other angles will be calculated according to the following formula

$$\widetilde{A_k} = \left\{ \alpha_{i,k} = -\frac{\beta_{i,k}}{2}; \alpha_{i-1,k} + \beta_{i,k}, i = \overline{2,M} \right\}, k \epsilon K+1 \tag{21}$$

Taking into account (15), the value of the abscissa of the vertex of the vector-criterion in the Cartesian coordinate system of each vector-criterion can be determined by the following formula

$$\widetilde{A_k} = \left\{ a_{i,k} = g_{i,k} \sin(\alpha_{i,k}) i = \overline{1,M} \right\}, k \epsilon K+1 \tag{22}$$

and the value of its ordinate - by the formula

$$\widetilde{B_k} = \left\{ b_{i,k} = g_{i,k} \cos(\alpha_{i,k}) i = \overline{1,M} \right\}, k \epsilon K+1 \tag{23}$$

where $a_{i,k}$, $b_{i,k}$ – respectively, the values of the abscissa and ordinates of the vertex of the i-th criterion vector in the Cartesian coordinate system, which relates to the k-th expert. In order to make sure the calculations are correct, you need to perform such verification

$$\widetilde{C}_k = \left\{ c_{i,k} = \sqrt{a_{i,k}^2 + b_{i,k}^2}, i = \overline{1, M} \right\}, k \epsilon K + 1 \tag{24}$$

If $g_{i,k} = c_{i,k}$ ($\forall i, k$), then the calculations are done correctly. Otherwise, it is necessary to verify the correctness of the implementation of previous calculations.

By having the lengths of the values of the criterion vectors obtained by the formula (15) or (16), as well as the coordinates of their vertices, obtained by formulas (22) and (23), it is possible to construct polar charts for any of the experts in particular to obtain a generalized picture of the evaluation experts in general. As noted above, the form of an irregular polygon constructed at the vertices of the criterion vector for any expert gives a comprehensive evaluation of the smartness of the sociopolis on the chosen criteria for its evaluation. At the same time, the resulting area of the polygon will quantitatively characterize the smartness of the sociopolis according to all criteria. To find the area of an irregular polygon by the coordinates of its vertices, you can use the following formula.

$$S_k^{ip} = \frac{1}{2} \sum_{i=1}^{M} \left| a_{i,k} b_{i+1,k} - b_{i,k} a_{i+1,k} \right|, k \epsilon K + 1 \tag{25}$$

To establish the proportion of the smartness of a sociopolis, which we now have to divide the area of the polygon into the area of the circle in which this polygon is located:

$$z_k = \frac{S_k^{ip}}{\pi r^2}, k \epsilon K + 1 \tag{26}$$

where z_k – the share of smartness of sociopolis at the moment, which is determined by the data of k-th expert; r is the radius of the circle. As noted above, a comprehensive indicator of the wisdom of the sociopolis ($g_{i,k}$) will have a maximum value of 100, that is, a circle radius will be 100 units. The area of the circle is not filled, which is still to be improved and developed in the sociopoly to achieve the status of a "reasonable sociopoly".

6   Conclusions

It is clear that in this research the method for visualization of the results of a comprehensive expert's evaluation of sociopolis smartness was developed. The result of the evaluation is the set of irregular polygons constructed in the polar coordinate system, according to the estimates of individual experts, taking into account the importance of each of the criteria for evaluation and the value of the experts themselves. Such a mechanism of visualization of information provides the opportunity to qualitatively and quantitatively present set of values of complex indicators of sociopolis smartness for experts, public figures and community. Such results can be obtained through any surveys of different experts. The proposed method is suitable for presenting a plurality of experts' survey results with a division into an unlimited number of roles of participants in the evaluation of the sociopolis smartness, taking into account the importance of each of them. Also the algorithms for calculating the area of a polygon, sector polygons, system of criteria for evaluation of sociopolis smartness are developed. Knowledge of the polygonal area visually makes it possible to identify those areas that still require additional development to get the sociopolis the "smart" status.

References

1. Dirks, S.A., Keeling, M.: Vision of Smarter Cities: How Cities Can Lead the Way into a Prosperous and Sustainable Future. IBM Global Business Services, NY (2009)
2. Kanter, R.M., Litow, S.S.: Informed and Interconnected: A Manifesto for Smarter Cities. Business School General Management Unit, Harvard (2009)
3. Mazza, R.: Introduction to Information Visualization. Springer, London (2009)
4. Rzheuskyi, A., Kunanets, N., Kut, V.: Methodology of research the library information services: the case of USA university libraries. In: Advances in Intelligent Systems and Computing II, vol. 689, pp. 450–460 (2018)
5. Rzheuskyi, A., Kunanets, N., Kut, V.: The analysis of the United States of America universities library information services with benchmarking and pairwise comparisons methods. In: 12th International Scientific and Technical Conference on Computer Sciences and Information Technologies (CSIT), Lviv, pp. 417–420 (2017)
6. Rzheuskiy, A., Veretennikova, N., Kunanets, N., Kut, V.: The information support of virtual research teams by means of cloud managers. Int. J. Intell. Syst. Appl. (IJISA) 10(2), 37–46 (2018). https://doi.org/10.5815/ijisa.2018.02.04
7. Kunanets, N., Veretennikova, N., Pasichnyk, V., Gats, B.: E-Science: new paradigms, system integration and scientific research organization. In: Xth International Scientific and technical Conference CSIT 2015 on Computer Science and Information Technologies, pp. 76–81. Lviv Polytechnic Publishing House, Lviv (2015)
8. Litvak, B.G.: Development of management solutions. Delo (2000)
9. Veretennikova, N., Kunanets, N.: Recommendation systems as an information and technology tool for virtual research teams. In: Advances in Intelligent Systems and Computing II, vol. 689, pp. 577–587 (2018)
10. Kunanets, N., Lukasz, W., Pasichnyk, V., Duda, O., Matsiuk, O., Falat, P.: Cloud computing technologies in "smart city" projects. In: 9th International conference on Intelligent Data Acquisition and Advanced Computing Systems: Technology and Applications (IDAACS), Romania, pp. 339–342 (2017)
11. Bomba, A., Nazaruk, M., Pasichnyk, V., Veretennikova, N., Kunanets, N.: Information technologies of modeling processes for preparation of professionals in smart cities. In: Advances in Intelligent Systems and Computing, vol. 754, pp. 702–712 (2018)
12. Peleshko, D., Rak, T., Izonin, I.: Image superresolution via divergence matrix and automatic detection of crossover. Int. J. Intell. Syst. Appl. (IJISA) 8(12), 1–8 (2016). https://doi.org/10.5815/ijisa.2016.12.01
13. Izonin, I., Trostianchyn, A., Duriagina, Z., Tkachenko, R., Tepla, T., Lotoshynska, N.: The combined use of the wiener polynomial and SVM for material classification task in medical implants production. Int. J. Intell. Syst. Appl. (IJISA) 10(9), 40–47 (2018). https://doi.org/10.5815/ijisa.2018.09.05

14. Lytvyn, V., Vysotska, V., Peleshchak, I., Rishnyak, I., Peleshchak, R.: Time dependence of the output signal morphology for nonlinear oscillator neuron based on Van der Pol model. Int. J. Intell. Syst. Appl. (IJISA) **10**(4), 8–17 (2018). https://doi.org/10.5815/ijisa.2018.04.02
15. Kerren, A., Stasko, J.T., Fekete, J.-D., North, C.: Information visualization. In: LNCS State-of-the-Art Survey, vol. 4950, pp. 1–18. Springer, Heidelberg (2008)
16. Tabachishin, D.R., Lenko, V.S., Kunanets, N.E., Pasichnyk, V.V., Shcherbina, Yu.M.: Expert evaluation of "city smartness" with the use of fuzzy logic. Artif. Intell. **1**, 102–110 (2017)
17. Bederson, B., Shneiderman, B.: The Craft of Information Visualization: Readings and Reflections. Morgan Kaufmann Publishers Inc., San Francisco (2003)
18. Card, S.K., Mackinlay, J.D., Shneiderman, B.: Readings in Information Visualization: Using Vision to Think. Morgan Kaufmann Publishers, San Francisco (1999)
19. Heer, J., Card, S.K., Landay, J.: Prefuse: a toolkit for interactive information visualization. In: ACM Human Factors in Computing Systems CHI (2005)
20. Spence, R.: Information Visualization: Design for Interaction, 2nd edn. Prentice Hall, Englewood Cliffs (2007)
21. Ware, C.: Information Visualization: Perception for Design, 2nd edn. Morgan Kaufmann Publishers, San Francisco (2000)

Missing Data Imputation Through SGTM Neural-Like Structure for Environmental Monitoring Tasks

Oleksandra Mishchuk$^{(\boxtimes)}$ ⓘ, Roman Tkachenko ⓘ,
and Ivan Izonin ⓘ

Lviv Polytechnic National University, Lviv, Ukraine
oleksandra.myroniuk@gmail.com,
roman.tkachenko@gmail.com, ivanizonin@gmail.com

Abstract. The article describes a new missing data imputation method. It is based on the use of a high-speed neural-like structure of the Successive Geometric Transformations Model. The importance of the research is based on the analysis of disadvantages of the known methods for missing data processing. Various simple and complex algorithms are analyzed, among which are the arithmetic mean algorithm, regression modeling, etc. It is shown that the above-mentioned imputation methods in data monitoring of air pollution do not allow to obtain reliable results due to the low prediction accuracy. An effective method for processing data imputation through SGTM neural-like structure is proposed. An example of filling data by forecasting CO, NO and NO_2 missed parameters in data monitoring of air pollution is given. A comparison of the proposed method with the arithmetic mean algorithm is carried out. Accuracies of the data imputation by developed method and by arithmetic mean algorithm are based on calculated evaluation criteria: Root mean squared errors. Experimentally established that the data imputation method through SGTM neural-like structure has a three times higher accuracy of the data imputation than the arithmetic mean algorithm. The proposed approach can be used in various areas such as medicine, materials science, economics, science services, etc.

Keywords: Imputation methods · Missing data · Neural-like structure · Environmental monitoring · Successive Geometric Transformations Model

1 Introduction

Atmospheric air is a vital and integral part of the environment for the existence of people, plants and animals. With the continuous development of technologies, human activity is increasingly leading to changes in the environment, especially the air is exposed to harmful emissions. Environmental monitoring is carried out to control pollution of the environment and includes monitoring of atmospheric air pollution.

Monitoring of atmospheric air pollution is a system of the observation of changes in the state of the air environment [1]. The purpose of this observation system is not only passive statement of facts, but also their analysis, conducting of experiments, modeling of processes, forecasting changes [2]. To identify tendency to change the quality of the

© Springer Nature Switzerland AG 2020
Z. Hu et al. (Eds.): ICCSEEA 2019, AISC 938, pp. 142–151, 2020.
https://doi.org/10.1007/978-3-030-16621-2_13

atmospheric air it is necessary to form a monitoring system. The main procedures for monitoring atmospheric air pollution are reflected in the proposed flowchart in Fig. 1.

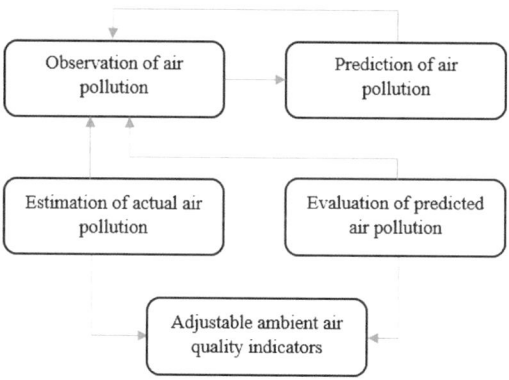

Fig. 1. The diagram of the monitoring of air pollution

In monitoring air pollution, it is important to determine the criteria for air pollution: the normalization of harmful emissions into the environment, calculation of their amount, maximum possible and maximum allowable concentrations (MPC/MAC) of harmful substances in the atmosphere, the degree of purification of atmospheric air [3]. In order to be able to determine whether the level of harmful impurities in the air exceeding established MPC or not, it is needed to have high-quality source information. Therefore, the data set, which consists of measured harmful impurities in the air, should be completed and representative for the investigated geographic area [4].

There are two approaches to the study of data sets. The first approach is to remove objects with missed values from the data set. This option is easy to implement and can be satisfactory with a small number of missing data [3]. However, even in cases where the percentage of deleted objects is relatively low, there is a high likelihood of deletion from a sample of important data and exceptions. This reduces the representativeness of the samples for training and testing. The second approach is to fill the missing data in one way or another and to analyze the received matrix. This option does not make a noticeable distortion of the impact on the results of the analysis, when the number of missing data is less than 20% [5].

2 Review and Analysis of Existing Methods

Since the use of any fill-in method can compile the sample structure, this may distort the real distribution of observations in the sample and reduce the actual significance of the results [6]. Therefore, when choosing a specific missing data imputation method, one must be taken into account that the possibility of its application depends essentially on the method of data analysis, which will be used in the future. The choice can be made from the imputation methods shown in Fig. 2.

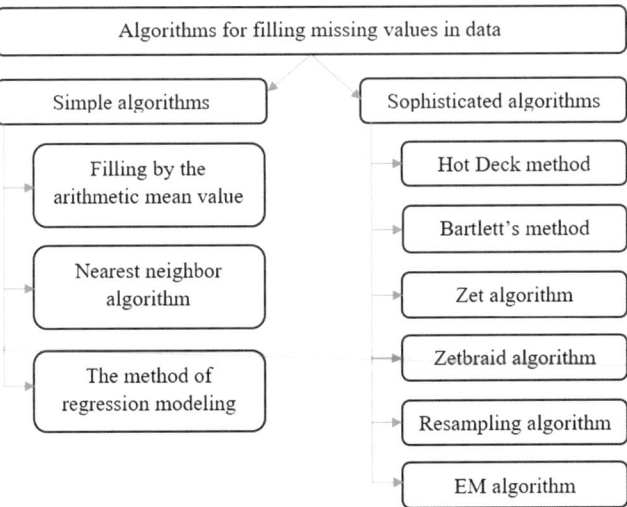

Fig. 2. Missing data imputation methods

Figure 2 reproduces the basic methods for processing missing data that are divided into two groups: simple and complex. Complex methods include:

- *Hot Deck* [7] is a method for handling missing data, when missing value is replaced with the nearest informational object;
- *Barlett's method* consists of two stages: substitutions instead of missing data the initial generated values in the first stage; conducting a covariance analysis of the target variable and constructing an indicator of completeness of observations for the target variable at the second stage [8];
- *Zet algorithm* consists of the selection of each value for filling missing data not from the whole set of observations [9], but from some of its part, that is called component matrix;
- *Zetbraid algorithm* is a consistent sequential selection of competent rows and columns, and each new competent matrix [10] is formed, then its performance is determined by the given criterion in prediction of missing data);
- *Resampling algorithm* is an iterative method, which involves changing the rows with missing data by randomly selected fully completed rows from the matrix, and then constructing a regression equation to predict the missing value [5];
- *EM algorithm* is also an iterative procedure for solving optimization problems of some functional through an analytic search for an extremum of a function) [3].

The peculiarity of these algorithms is the filling of missing data with values, which are selected by the algorithm [10]. But sophisticated methods are time-consuming and unwieldy in use [11]. At that time, such simple methods of filling missing data, as the arithmetic mean algorithm, k-nearest neighbor, and the regression [12] methods are less accurate than improved complex methods. Therefore, there was a need to find such a method that would be easy to use [13], while fast in time dependencies [14] and most accurate [15, 16].

The first step, in the research, was decided to investigate the first simple algorithm in the list of filling missing data methods, that is arithmetic mean algorithm [5]. This algorithm consists the replacing each missing value with the arithmetic mean of the observed values for that column by the formula (1):

$$A = \frac{1}{n} * \sum_{i=1}^{n} x_i \qquad (1)$$

where n is the number of items being averaged; x_i is the value of each individual item in the list of items being averaged.

All missing values for parameters CO, NO and NO_2 are calculated using arithmetic mean algorithm, but the mentioned algorithm can distort the distribution for the variable and leads to complications with summary measures including. So this algorithm was compared to new developed method of filling missing data.

3 Method of Filling the Missing Data on the Basis of the SGTM Neural-Like Structure

In the article, in contrast to the arithmetic mean algorithm, a method for processing missing data in the monitoring of atmospheric air pollution is provided, constructed using the neural network approach. According to this method, the appropriate training of the selected artificial neural network is initially carried out and in the application mode forecasts of missing data are obtained. Since, missing data may be present for various parameters in monitoring the state of air pollution, there is a need to study a large number of neural networks in different modes of use [17, 18].

The application of artificial neural networks of well-known architectures, such as single-layer and multilayer perceptrons, provides a significant amount of the time consuming to adjust their parameters [19]. So for the implementation of the method, a non-iterative linear neural-like structure of the Successive Geometric Transformations Model (SGTM) is selected [20], which is trained and function fully in automatic mode [21]. The investigated method for processing missing data through SGTM neural-like structure uses dependencies between parameters of the monitoring of air pollution.

The implementation of the proposed approach involves the creation of training and testing samples, where inputs include full data, and the output is the missing parameter that needs to be predicted. So, first of all, the parameters with missing elements in the dataset are determined. Then, separately, each defined parameter is filled by prediction. The main steps of the developed method are:

1. Delete from the sample all vectors that have missing data.
2. Randomly assign vectors to the new training set A (5000 vectors) and to the testing set B (1950 vectors). In both data sets enter to input vectors X_j all parameters except one containing missing data, and enter to output vector Y_j one of the parameters CO, NO, or NO_2.

3. Normalize the input vectors X_j. Thus, data samples A and B with $x_{i1}, \ldots, x_{i11}, y_i$ vectors are normalized as follows [20]: when the maximum value in each column. $max(X_j)$ is found, assign a value to one. The remaining values are seeking by the following formula (2):

$$x_{normed} = \frac{x_{ij}}{|max(X_j)|} \tag{2}$$

where x_{ij} is a determined value of the input vector; $max(X_j)$ stands for the maximum value of the input vector.

4. Choose the linear setting of the known SGTM neural-like structure (which includes several options for setting operating modes, determining the number of mass centers, normalizing inputs and outputs, defining activation function) through the implementation of experiments.

5. SGTM training on A sample for one of the parameters CO, NO, or NO_2, and further testing using B sample.

6. Determine the accuracy of the probable filling of missing data by calculating the prediction data root mean squared errors (RMSE_M) [21, 22] for parameters CO, NO, or NO_2, using the formula (3):

$$RMSE\_M = \frac{\sqrt{\sum_{i=1}^{n} (y_j' - y_j)^2}}{|max(Y_j)|} \tag{3}$$

where y_j' is a predicted output value; y_j is a benchmark output value for testing; $max(Y_j)$ is the maximum value of the output vector.

4   Simulation

The simulation of the proposed method for filling missing data took place on the real problem of CO, NO_2 and NOx urban pollution monitoring with on-field calibrated electronic nose by automatic Bayesian regularization [23]. The implementation of the developed fill-in method is performed on a data set, which contains 9,000 observations of hourly averaged responses from 5 metallic oxide chemical sensors embedded in an Air Quality Chemical Multi-Sensor Device [23]. The device was located on the field in a significantly polluted area, at road level, within an Italian city. Data were recorded from March 2004 to February 2005 (one year) representing the longest freely available recordings of on field deployed air quality chemical sensor devices responses. The data set consists of an hourly average concentration of carbon monoxide (CO), non-methane hydrocarbons (SnO_2), benzene (C_6H_6), titanium (Ti), total nitrogen oxides (NO) and nitrogen dioxide (NO_2), tungsten oxides (WO) and dioxide tungsten (WO_2), indium oxide (InO), air temperature (T), relative humidity (RH) and absolute (AH) humidity.

Parameters that need to be restored are indicated by the value "−200". Proofs of cross-sensitivity, as well as conceptual and sensor drifts, are present and described in [23], which ultimately affects the ability to evaluate the concentration of sensors. The part of the list of measurements provided by a commonly verified analyzer is given in Table 1.

Table 1. The list of hourly average measured concentration of air pollutants

CO	SnO_2	C_6H_6	Ti	NO	WO	NO_2	WO_2	InO	T	RH	AH
1,2	1185	3,6	690	62	1462	77	1333	733	11,3	56,8	0,7603
1	1136	3,3	672	62	1453	76	1333	730	10,7	60,0	0,7702
0,9	1094	2,3	609	45	1579	60	1276	620	10,7	59,7	0,7648
0,6	1010	1,7	561	-200	1705	-200	1235	501	10,3	60,2	0,7517
-200	1011	1,3	527	21	1818	34	1197	445	10,1	60,5	0,7465
0,7	1066	1,1	512	16	1918	28	1182	422	11,0	56,2	0,7366
0,7	1052	1,6	553	34	1738	48	1221	472	10,5	58,1	0,7353
1,1	1144	3,2	667	98	1490	82	1339	730	10,2	59,6	0,7417
2	1333	8,0	900	174	1136	112	1517	1102	10,8	57,4	0,7408
2,2	1351	9,5	960	129	1079	101	1583	1028	10,5	60,6	0,7691
1,7	1233	6,3	827	112	1218	98	1446	860	10,8	58,4	0,7552
1,5	1179	5,0	762	95	1328	92	1362	671	10,5	57,9	0,7352
1,6	1236	5,2	774	104	1301	95	1401	664	9,5	66,8	0,7951
1,9	1286	7,3	869	146	1162	112	1537	799	8,3	76,4	0,8393
...	...	...	...	...	...	...	...	...	...	...	...
2,2	1310	8,8	933	184	1082	126	1647	946	8,3	79,8	0,8778

Since, the SGTM neural-like structure includes several options for setting operating modes, for modeling, the following parameters were defined: linear type of structure, use of mass centers in the predicting missing data, normalization of inputs and outputs of training and testing samples. In addition, to understand the time expenditures, the parameters of the PC where the calculations took place should be described. So, in the research the laptop MI Pro is used, equipped with the eighth generation of Intel® Core™ i5-8250U CPU x-64 based processor and the NVIDIA GeForce MX15 discrete graphics accelerator with installed memory (RAM) 8 GB. It should be note, that proposed approach can be used in different Service Science area [24–26].

5 Results and Comparison

Visualization of simulation results of developed method for filling missing data through the SGTM neural-like structure for three different sets of data presented in Fig. 3. Scatter diagrams are performed using Orange software.

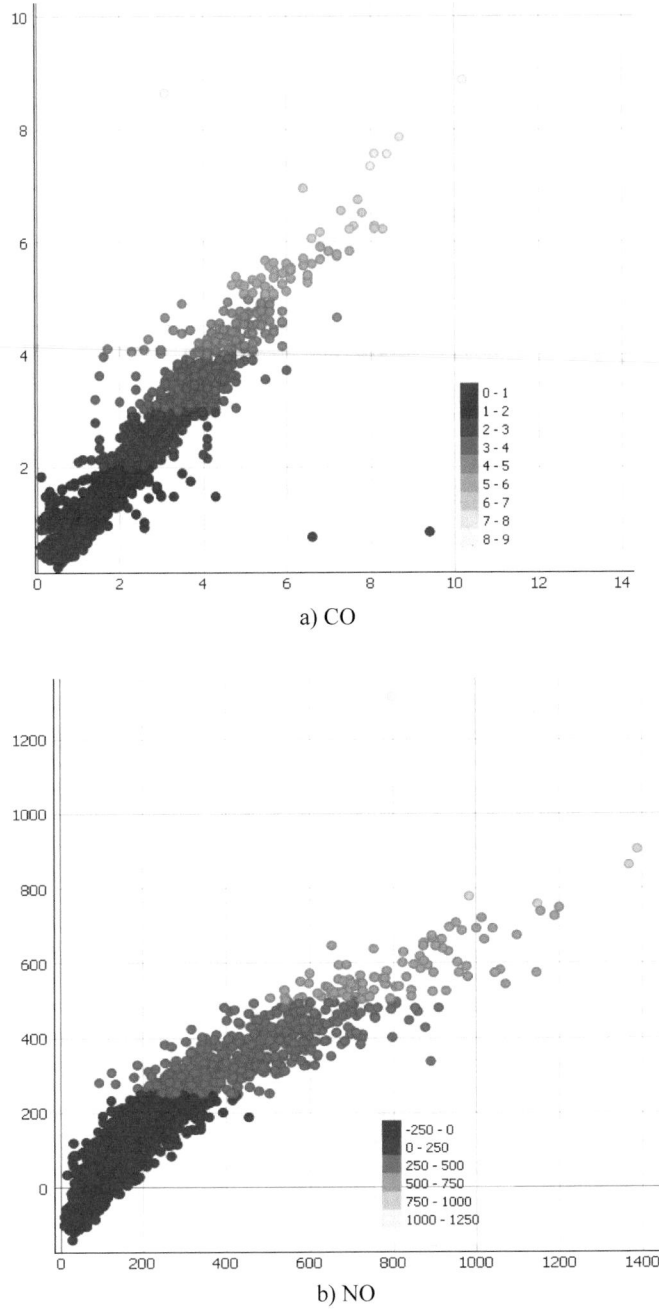

a) CO

b) NO

Fig. 3. Results of the proposed method of filling missing data in the form of diagrams scattering. On the *OX* axis is given real value, on the *OY* axis - predicted data: (a) carbon monoxide (CO); (b) nitrogen oxides (NO); (c) nitrogen dioxide (NO_2).

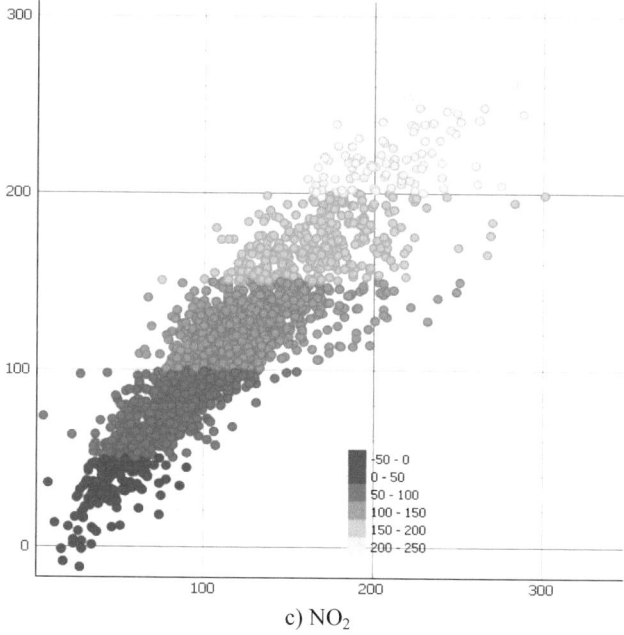

c) NO_2

Fig. 3. (*continued*)

Thus, after filling missing data in the environmental monitoring using the neural network method with non-iterative linear SGTM, errors were detected for each object with missing parameters. RMSE_M errors also found for filling missing data by replacing with the arithmetic mean value. Table 2 compares the RMSE_M errors found for the parameters CO, NO, or NO_2 by two investigated methods. In the table, it is observed that for example, the error of filling missing data using the neural network method is 5% for the CO parameter, which is almost three times less than the RMSE_M error (12%) for the same CO parameter, where the missing data are filled with using arithmetic mean algorithm.

Table 2. The comparison of investigated methods.

Method	RMSE_M for CO	RMSE_M for NO	RMSE_M for NO_2
Arithmetic mean algorithm	12,42347	15,21968	15,83967
SGTM neural-like structure	4,486251	8,159905	7,396727

6 Conclusions

In the research, the set of data on monitoring of atmospheric air pollution was investigated. The importance of effective solution of the task of filling the omissions in the data sets of environmental monitoring was substantiated. The main algorithms for filling missing data in arrays was analyzed. A new simple method of regressive filling of the missing data for environmental monitoring tasks by using the neural-like structure of the Successive Geometric Transformations Model was described. Also, it was determined that the developed method retains the filling rate at a large set of data. It was found that the method of filling missing data with SGTM is a three times more accurate than the arithmetic mean algorithm, while also easy to execute, since it does not require long-lasting debugging of structure and parameters.

References

1. Brauer, M., et al.: Exposure assessment for estimation of the global burden of disease attributable to outdoor air pollution. Environ. Sci. Technol. **42**(2), 652–660 (2012). https://doi.org/10.1021/es2025752
2. Shakovska, N., Shamuratov, O.: The structure of information systems for environmental monitoring. In: XIth International Scientific and Technical Conference "Computer Science and Information Technologies", Lviv, Ukraine, 6–10 September, pp. 102–107 (2016)
3. Zagnieva, I., Timonina, E.: Comparison of efficiency of missing data imputation algorithms, depending on the analysis method used. Public Opin. Monit. **1**(119), 41–55 (2014). (in Russian)
4. Skrynyk, O.: Recovery missing data in time series of meteorological parameters. Sci. Work. Ukr. Res. Hydro Meteorol. Inst. **260**, 46–53 (2011). (in Ukrainian)
5. Zloba, E., Yatskiv, I.: Statistical methods for recovering missing data. Comput. Model. New Technol. **6**(1), 51–61 (2002). (in Ukrainian)
6. Baraldi, A., Enders, C.: An introduction to modern missing data analyses. J. Sch. Psychol. **48**(1), 5–37 (2010). https://doi.org/10.1016/j.jsp.2009.10.001
7. Andridge, R., Little, R.: A review of hot deck imputation for survey non-response. Int. Stat. Rev. **78**(1), 40–64 (2010)
8. Engelberg, S.: Digital Signal Processing: An Experimental Approach, chap. 7, p. 56. Springer, New York (2008)
9. Snytiuk, V.: Evolutionary method of filling missing data. In: Collection of works VI MK "Intelligent Analysis of Information", Kyiv, pp. 262–271 (2006). (in Russian)
10. Kuznietsova, N.: Identification and processing of uncertainties in the form of incomplete data by methods of intellectual analysis. Syst. Res. Inf. Technol. **2**, 104–115 (2016). (in Ukrainian)
11. Kaminskyi, R., Kunanets, N., et al.: Recovery gaps in experimental data. In: Proceedings of the 2nd International Conference on Computational Linguistics and Intelligent Systems, Volume I: Main Conference Lviv, Ukraine, 25–27 June, pp. 110–118. CEUR-WS.org (2018)
12. Bodyanskiy, Ye., Tyshchenko, O., Kopaliani, D.: An evolving connectionist system for data stream fuzzy clustering and its online learning. Neurocomputing **262**, 41–56 (2017). https://doi.org/10.1016/j.neucom.2017.03.081
13. Babichev, S., Škvor, J., Fišer, J., Lytvynenko, V.: Technology of gene expression profiles filtering based on wavelet analysis. Int. J. Intell. Syst. Appl. (IJISA) **10**(4), 1–7 (2018). https://doi.org/10.5815/ijisa.2018.04.01
14. Lytvyn, V., Vysotska, V., Peleshchak, I., Rishnyak, I., Peleshchak, R.: Time dependence of the output signal morphology for nonlinear oscillator neuron based on Van der Pol model. Int. J. Intell. Syst. Appl. (IJISA) **10**(4), 8–17 (2018). https://doi.org/10.5815/ijisa.2018.04.02

15. Hu, Zh., Bodyanskiy, Ye., Tyshchenko, O.: A deep cascade neural network based on extended neo-fuzzy neurons and its adaptive learning algorithm. In: IEEE First Ukraine Conference on Electrical and Computer Engineering (UKRCON), Kyiv, Ukraine, 29 May–2 June, pp. 801–805 (2017)
16. Babichev, S., Korobchynskyi, M., Lahodynskyi, O., Basanets, V., Borynskyi, V.: Development of a technique for the reconstruction and validation of gene network models based on gene expression profiles. East.-Eur. J. Enterp. Technol. **114**(91), 19–32 (2018). https://doi.org/10.15587/1729-4061.2018.123634
17. Hu, Zh., Bodyanskiy, Ye., Tyshchenko, O.: A cascade deep neuro-fuzzy system for high-dimensional online possibilistic fuzzy clustering. In: XIth International Scientific and Technical Conference "Computer Science and Information Technologies" (CSIT 2016), Lviv, Ukraine, 6–10 September, pp. 119–122 (2016)
18. Hu, Zh., Bodyanskiy, Ye., Tyshchenko, O., Boiko, O.: A neuro-fuzzy Kohonen network for data stream possibilistic clustering and its online self learning procedure. Appl. Soft Comput. **68**, 710–718 (2018). https://doi.org/10.1016/j.asoc.2017.09.042
19. Bodyanskiy, Y., Vynokurova, O., Savvo, V., Tverdokhlib, T., Mulesa, P.: Hybrid clustering-classification neural network in the medical diagnostics of the reactive arthritis. Int. J. Intell. Syst. Appl. (IJISA) **8**(8), 1–9 (2016). https://doi.org/10.5815/ijisa.2016.08.01
20. Tkachenko, R., Yurchak, I., Polishchuk, U.: Neurolike networks on the basis of geometrical transformation machine. In: 2008 International Conference on Perspective Technologies and Methods in MEMS Design, Polyana, 21–24 May, pp. 77–80 (2008)
21. Tkachenko, R., Izonin, I.: Model and principles for the implementation of neural-like structures based on geometric data transformations. In: Hu, Zh., Petoukhov, S., Dychka, I., He, M. (eds.) Advances in Computer Science for Engineering and Education. Advances in Intelligent Systems and Computing, vol. 754, pp. 578–587. Springer, Cham, (2018)
22. Teslyuk, V., Beregovskyi, V., Denysyuk, P., Teslyuk, T., Lozynskyi, A.: Development and implementation of the technical accident prevention subsystem for the smart home system. Int. J. Intell. Syst. Appl. (IJISA) **10**(1), 1–8 (2018). https://doi.org/10.5815/ijisa.2018.01.01
23. De Vito, S., Piga, M., Martinotto, L., Di Francia, G.: CO, NO2 and NOx urban pollution monitoring with on-field calibrated electronic nose by automatic Bayesian regularization. Sens. Actuators B: Chem. **143**(1), 182–191 (2009)
24. Kaczor, S., Kryvinska, N.: It is all about services - fundamentals, drivers, and business models: the society of service science. J. Serv. Sci. Res. **5**(2), 125–154 (2013)
25. Rzheuskiy, A., Veretennikova, N., Kunanets, N., Kut, V.: The information support of virtual research teams by means of cloud managers. Int. J. Intell. Syst. Appl. (IJISA) **10**(2), 37–46 (2018). https://doi.org/10.5815/ijisa.2018.02.04
26. Dronyuk, I., Fedevych, O., Lipinski, P.: Ateb-prediction simulation of traffic using OMNeT++ modeling tools. In: 2016 XIth International Scientific and Technical Conference Computer Sciences and Information Technologies (CSIT), Lviv, pp. 96–98 (2016)

Study of Theoretical Properties of PSC-Algorithm for the Total Weighted Tardiness Minimization for Planning Processes Automation

Alexander Anatolievich Pavlov$^{(\boxtimes)}$ (ID), Elena Borisovna Misura (ID), Oleg Valentinovich Melnikov (ID), Iryna Pavlovna Mukha (ID), and Kateryna Igoryvna Lishchuk (ID)

National Technical University of Ukraine "Igor Sikorsky Kyiv Polytechnic Institute", 37, Prospekt Peremohy, Kyiv 03056, Ukraine
pavlov.fiot@gmail.com,
elena_misura@ukr.net, lishchuk_kpi@ukr.net,
oleg.v.melnikov72@gmail.com, mip.kpi@gmail.com

Abstract. We investigate and justify the efficiency of PSC-algorithm for solving the NP-hard in the strong sense scheduling problem of the total weighted tardiness of jobs minimization on one machine. The problem is used widely in the automation of planning processes in objects of different nature. As a result of the research, we identify and justify regions of the problem parameter values at which it is solved quickly and at which it requires a large computational effort. For each individual instance, a polynomial subalgorithm of $O(n^2)$ complexity constructs a sequence on which we determine the instance's structure and its characteristics designating the instance's complexity. We analyze the polynomial component of the PSC-algorithm which checks sufficient signs of optimality at all stages of the problem solving, in contrast to the existing methods. The optimal solution was achieved with the polynomial component for 27.3% of instances. We present experimental data on the solving time for problems with dimensions up to 300 jobs. We show that up to 68% of the generated benchmark instances are solved relatively quickly for any dimension. The PSC-algorithm is competitive by the computation time with the known method of Tanaka et al. and significantly exceeds it on instances from the determined regions of quick solving.

Keywords: Planning · Process automation · Scheduling ·
Combinatorial optimization · Exact algorithm · PSC-algorithm ·
Efficiency research · Total weighted tardiness

1 Introduction

Scheduling problems are used very widely in production process automation, in business process automation and service industry, for computer calculations optimization and in other areas. Almost all scheduling problems are NP-hard problems of

© Springer Nature Switzerland AG 2020
Z. Hu et al. (Eds.): ICCSEEA 2019, AISC 938, pp. 152–161, 2020.
https://doi.org/10.1007/978-3-030-16621-2_14

combinatorial optimization. Known exact methods for their solving generally have exponential complexity. Many state-of-the-art scheduling methods are described in the book [1] and in papers [2–4]. A review of efficient methods for exact exponential algorithms construction can be found in [5]. Some new scheduling methods, with an emphasis on cloud computing, are considered in a recent review [6]. Papers [7–9] research the methods for combinatorial problems solving using parallel computing systems. Parallel and cloud computing [10] and scheduling with restrictions on resources [11] are intensively developing fields of research.

In this paper, we research one of five most known NP-hard in the strong sense scheduling problems: the total weighted tardiness of jobs minimization on a single machine (TWT). It is a priority among the calendar and operational planning problems in networked systems with limited resources, project management systems, discrete-type productions (up to 80% of all production systems, including aircraft manufacturing and shipbuilding enterprises, as well as dual-use technologies). TWT problem has paramount importance as an independent problem as well as the basis for multistage planning systems. In particular, it is used at the fourth level of the four-level model of planning (including operative planning) and decision making presented in [12]. This model can be used for any objects with a network structure of technological processes and limited resources. Tardiness criteria optimization is crucial for just-in-time planning systems and other important planning models.

The TWT Problem Statement [13]. Given a set of n independent jobs $J = \{j_1, j_2, \ldots, j_n\}$, each job consists of a single operation. For each job j, we know its processing time $l_j > 0$, weight coefficient $\omega_j > 0$ and due date d_j. All jobs become available at the time point zero. Interruptions are not allowed. We need to build a schedule for one machine that minimizes the total weighted tardiness of the jobs:

$$f = \sum_{j=1}^{n} \omega_j \max\left(0, C_j - d_j\right) \qquad (1)$$

where C_j is the completion time of a job j.

This problem is studied from the 1960s. All known exact methods for its solving are based on enumeration methods (branch and bounds, dynamic programming, etc.) and are related to an exponential search. The best of known state-of-the-art exact dynamic programming algorithms for the TWT problem is limited by dimension of 300 jobs [14]. Extended dominance rules experimentally studied in [15] allow to derive 8 to 12% more arcs in the search tree, in average, compared to basic dominance rules. Kanet et al. [16] initiated consideration of precedence theorems involving more than two jobs. Tanaka et al. [14] also considered dominance rules including up to 4 jobs. Karakostas et al. [17] investigate particular cases of the TWT problem with restrictions on input data, they obtain exact polynomial time solving algorithms for such cases.

The theory of PSC-algorithms [18] is an alternative direction allowing to solve NP-hard problems in a general formulation exactly and efficiently. It allows for problems several times larger in size and also for the parallelization of computations. When sufficient signs of optimality of current solutions are fulfilled, PSC-algorithms, in contrast to existing methods, allow to obtain an exact solution by a polynomial subalgorithm. Otherwise, we obtain an efficient exact solution by an exponential subalgorithm or an approximate solution with an estimate of the deviation from the optimum for practical problem instances of large size (up to tens of thousands of jobs).

Exact PSC-algorithm for TWT problem solving presented in [13], in contrast to the existing methods, for the first time allows to determine the optimality of the obtained intermediate solution at the solving stage. Also it allows to classify the problem instances by their complexities and to attribute individual instances to a quick or hard solving region. We give a brief description of the investigated PSC-algorithm in Sect. 2.

This paper is devoted to the study of theoretical properties of the PSC-algorithm, justification of its efficiency, and its comparison with existing methods. We give the results of our study with the determination of the regions of quick solving in Sect. 3. We summarize the findings in Sect. 4.

2 Theoretical Foundations of the PSC-Algorithm for the Problem Solving

We now excerpt from [13] some definitions necessary to understand the paper. Let $j_{[g]}$ denote the number of a job occupying a position g in a schedule. Let $\overline{p, q}$ mean the interval of integer numbers from p to q: $\overline{p, q} = p, p+1, \ldots, q$.

Definition 1. We call $R_{j_{[g]}} = d_{j_{[g]}} - C_{j_{[g]}} > 0$ the time reserve of a job $j_{[g]}$.

Definition 2. We call a job $j_{[g]}$ tardy in some sequence if $d_{j_{[g]}} < C_{j_{[g]}}$.

Definition 3. A permutation of a job $j_{[g]}$ is its move into a later position $k > g$ with the shift of jobs in positions $\overline{g+1, k}$ to the left by one position.

Definition 4. A priority $p_{j_{[i]}}$ of a job $j_{[i]}$ is the value of $\omega_{j_{[i]}} \big/ l_{j_{[i]}}$.

Definition 5. The ordered sequence σ^{ord} is a sequence of all jobs $j \in J$ in non-increasing order of priorities p_j, i.e., $\forall j, i, j < i$: $p_j \geq p_i$, and if $p_j = p_i$, then $d_j \leq d_i$.

Definition 6. A free permutation of a non-tardy job $j_{[k]}$ in the sequence σ^{ord} is its permutation into such later position $q > k$ that $C_{j_{[q]}} \leq d_{j_{[k]}} < C_{j_{[q+1]}}$ if there is at least one tardy job in positions $\overline{k+1, q}$.

Definition 7. Sequence σ^{fp} is the sequence obtained after all free permutations in the ordered sequence σ^{ord}.

Definition 8. A tardy job $j_{[g]}$ in the sequence σ^{fp} is called a competing job if for at least one non-tardy preceding job $j_{[l]}$ in the sequence we have $d_{j_{[l]}} > d_{j_{[g]}} - l_{j_{[g]}}$.

Competing jobs in the sequence σ^{fp} are in non-increasing order of their priorities.

Definition 9. The time reserve of a job $j_{[i]}$ in a current sequence is called productive reserve if there is at least one such competing job $j_{[k]}$ in this sequence that satisfies the inequality $d_{j_{[i]}} > d_{j_{[k]}} - l_{j_{[k]}}$.

The PSC-algorithm [13] is based on jobs permutations and consists of a series of uniform iterations. The number of iterations is determined by the number of competing jobs in the sequence σ^{fp}. At each iteration, we check the possibility to use the time reserves of preceding jobs for the current competing job from the sequence σ^{fp}. This way, we construct an optimal schedule for jobs of the current subsequence which is bounded with the current competing job. We can eliminate some jobs from the set of competing jobs during the solving process. As a result of each iteration, the functional value decreases or remains unchanged. This allows to construct efficient approximation algorithms on the basis of the PSC-algorithm.

The PSC-algorithm contains two polynomial components and exact exponential subalgorithm. The complexity of the exact subalgorithm does not exceed $O(n^3)$ if the Sufficient Sign of Polynomial Solvability of the Current Iteration [13] is fulfilled at each iteration of the algorithm. The second polynomial component is the approximation algorithm to be used in case when the sufficient signs of optimality are not fulfilled or for solving complex instances of the problem.

3   Computational Studies of the PSC-Algorithm

Let us denote the number of tardy jobs in the sequence σ^{fp} by N_Z, the number of competing jobs by N_K, the number of jobs with non-zero time reserves by N_R, and the number of jobs with productive reserves by N_P.

We analyze the sequence σ^{fp}, the complexity of its construction is $O(n^2)$. By this we determine a structure of each individual instance we solve, i.e., values of N_Z, N_K, N_R, and N_P, the magnitude of reserves, the range and location of jobs with reserves, tardy and competing jobs. This allows to attribute the individual instance to a quick or hard solving region. The analysis of the problem structure also allows to classify special cases of the problem given in [13] by their complexity.

To obtain quantitative characteristics of efficiency of the PSC-algorithm, we used benchmark instances of dimensions from 40 to 300 jobs from [19]. There are five instances for each range of two parameters: the due date range R and the tardiness factor T of standard Fisher's generation scheme [20]. The complexity of the problem instances and, as a result, the time of their solving, depends on these parameters. According to the Fisher's scheme, processing times and weights of jobs are taken for each instance from a uniform distribution over interval [1, 100] and [1, 10], respectively. Then the due dates of jobs are generated randomly from a uniform distribution over the interval $[L \cdot (1 - T - R/2), L \cdot (1 - T + R/2)]$ where L is the sum of processing times of all jobs. The values of R and T were chosen from the set $\{0.2, 0.4, 0.6, 0.8, 1.0\}$. This gives 25 combinations of R and T parameters, 125 test instances in total for each dimension. The total number of benchmark instances is 875.

We solved all the instances with a PSC-algorithm on a personal computer with 2 GBytes of RAM and a Pentium IV processor that has 3.0 GHz frequency. We show in Table 1 the average solving time in seconds for each group of 5 instances depending on the parameters R and T for dimensions from 40 to 250 jobs. The average solving time for $n = 300$ is shown in Fig. 1.

Table 1. Average solving time (in seconds) for all test instances for each pair of R and T

T	0.2	0.4	0.6	0.8	1.0	0.2	0.4	0.6	0.8	1.0
R	$n = 40$					$n = 50$				
0.2	0.020	0.152	0.167	0.124	0.074	0.022	0.271	0.190	0.164	0.129
0.4	0.003	0.095	0.185	0.115	0.087	0.004	0.205	0.293	0.148	0.171
0.6	0.004	0.025	0.208	0.119	0.094	0.004	0.098	0.286	0.173	0.178
0.8	0.004	0.029	0.239	0.126	0.104	0.004	0.033	0.358	0.243	0.314
1.0	0.004	0.005	0.285	0.233	0.117	0.004	0.006	0.396	0.355	0.323
	$n = 100$					$n = 150$				
0.2	0.214	3.458	3.137	2.790	2.298	1.344	15.252	13.685	7.655	9.286
0.4	0.002	2.382	3.228	3.199	2.301	0.012	10.502	14.126	10.198	9.908
0.6	0.002	0.398	3.744	2.760	2.659	0.007	2.471	16.604	13.801	13.151
0.8	0.002	0.119	3.902	2.742	2.564	0.009	0.743	17.424	14.780	12.667
1.0	0.002	0.003	4.325	3.977	3.316	0.009	0.012	18.408	15.052	13.063
	$n = 200$					$n = 250$				
0.2	3.926	47.945	46.492	24.740	23.400	12.959	75.769	107.795	59.656	53.768
0.4	0.032	42.085	48.522	32.947	27.292	0.028	70.192	112.278	80.322	64.348
0.6	0.012	6.104	49.395	36.158	30.474	0.016	19.992	133.354	73.148	70.574
0.8	0.016	1.833	51.757	38.638	33.555	0.022	6.070	141.292	79.062	80.437
1.0	0.017	0.033	56.648	47.913	34.561	0.022	0.029	143.581	110.889	83.789

T R	0.2	0.4	0.6	0.8	1.0
0.2	B 20.703	224.383	212.315	C 108.588	109.600
0.4	0.047	208.650	216.029	144.681	119.809
0.6	0.030	31.605	266.458	D 145.509	E 128.567
0.8	A 0.031 B 9.490	279.319	155.604	138.392
1.0	0.032	0.047	295.623	327.207	145.601

Fig. 1. Regions of parameters R and T where the problem is solved relatively quickly (the average solving time in seconds for $n = 300$ is shown)

We also generated larger dimensions instances by the same generation scheme. The maximum time observed for solving a 1,000 jobs instance was about 56 h.

We analyzed the efficiency of the first polynomial component of the PSC-algorithm that checks the sufficient signs of optimality at all stages of the problem solving, in contrast to the existing methods. If at least one of the signs is fulfilled, we reach the optimal solution at the intermediate solving stage. The first polynomial component of the PSC-algorithm fulfilled for 27.3% of our benchmark instances (on average for all instances of all dimensions).

It is obvious from the Table 1 that the instances of any dimension are solved more quickly (often significantly) for some values of pairs of R and T parameters than in

Table 2. Values of N_Z, N_K, N_R, and N_P in different regions of the problem parameters, as a percentage of the problem dimension

n	N_Z	N_K	N_R	N_P	N_Z	N_K	N_R	N_P	N_Z	N_K	N_R	N_P
	Region A				Region B				Region C			
40	0.00	0.00	99.33	0.00	5.80	5.80	93.50	62.90	57.75	55.50	42.00	42.00
50	0.00	0.00	99.07	0.00	5.76	5.76	93.12	68.32	56.20	54.80	42.80	42.80
100	0.00	0.00	99.60	0.00	4.04	4.04	95.24	62.00	54.00	53.30	44.80	44.70
150	0.00	0.00	99.51	0.00	3.15	3.15	95.84	54.75	49.53	49.47	49.07	49.07
200	0.00	0.00	99.37	0.00	3.22	3.22	95.90	59.74	50.50	50.30	48.10	48.10
250	0.00	0.00	79.49	0.00	2.58	2.58	76.72	47.79	40.40	40.24	38.48	38.48
300	0.00	0.00	99.36	0.00	2.97	2.97	95.77	57.28	49.03	49.00	49.53	49.47
	Region D				Region E				The rest: "complex" instances			
40	61.00	50.75	37.50	37.50	84.10	67.10	15.50	15.50	32.94	30.31	66.19	64.31
50	57.40	46.80	41.80	41.80	82.88	64.72	16.48	16.48	31.45	28.00	67.40	65.75
100	56.10	47.90	43.10	43.10	84.60	71.48	14.76	14.76	31.43	29.70	67.43	67.18
150	39.20	38.47	59.07	59.07	57.71	57.17	40.99	40.99	24.12	23.87	75.03	74.72
200	37.65	36.85	61.05	61.05	57.84	57.34	40.58	40.54	24.01	23.78	74.98	74.55
250	30.12	29.48	48.84	48.84	46.27	45.87	32.46	32.43	19.21	19.02	59.98	59.64
300	35.83	35.70	62.93	62.80	56.85	56.44	41.99	41.99	22.67	22.52	76.28	75.79

other regions. Our analysis allows to distinguish 5 such regions of R and T values. In Fig. 1, we distinguish these regions by the average solving time in seconds for each pair of R and T values, on the example of 300 jobs instances.

Thus, in fact, the instances were solved relatively quickly by the PSC-algorithm for 17 out of 25 pairs of parameters R and T, i.e., in up to 68% of cases. Let us try to determine the cause of this.

Denote the found regions of parameter values by:

Region A: $R = 0.6...1.0$; $T = 0.2$.
Region B: $R = 0.2...0.4$; $T = 0.2$ and $R = 0.6...1.0$; $T = 0.4$.
Region C: $R = 0.2...0.4$; $T = 0.8$.
Region D: $R = 0.6...0.8$; $T = 0.8$.
Region E: $R = 0.2...1.0$; $T = 1.0$.

The parameters R and T affect the values of N_Z, N_K, N_R, and N_P determined in the sequence σ^{fp}. Table 2 shows these values as a percentage of the problem size.

We can draw the following conclusions from Table 2.

The parameters N_Z, N_K, N_R, and N_P differ insignificantly within each region of R and T values. A sufficiently large differences in the values of the parameters are only in the region E for dimensions of 40, 50, and 100 jobs. This difference is due to the use of different data generation schemes.

In the region A, free permutations almost always give the optimal solution: the number of competing jobs is zero or very small (which is true at large dimensions).

In the region B, there are few tardy and competing jobs, the number of iterations of the PSC-algorithm is generally small.

In the region C, about half of jobs have non-zero reserves and half of jobs are tardy. But the reserves of each job with a reserve are small in size, they are not enough to serve for all competing jobs. As the result, many competing jobs are excluded from the set of competing jobs.

In the region D, there are jobs with time reserves, but the reserves are small in size. So, many competing jobs are excluded from the set of competing jobs. This leads to a decrease in the number of iterations. The deviation of the functional value in the sequence σ^{fp} from the optimum is small.

In the region E, there are few jobs with reserves. This leads to a small deviation of the functional value in the sequence σ^{fp} from the optimum. The reserves are quickly exhausted at the current iterations of the PSC-algorithm.

In Table 3, we compare the average time to solve the same test instances (for all values of the parameters R and T) with known algorithm of Tanaka et al. [14]. For the method we compare, we know only the solving time for instances of up to 300 jobs dimension. We can state that the PSC-algorithm is competitive in terms of solving time with the method of Tanaka et al., taking into account the performances of the computer systems used. For the "simple" instances from the regions A...E, the PSC-algorithm is significantly faster. Also, we know from [21] that the algorithm of Tanaka et al. [14] requires a significant amount of RAM (up to 384 MBytes for 300 jobs instances). The PSC-algorithm from [13] requires only several complete or incomplete job sequences to be memorized, depending on the solving stage, i.e., it takes only a several kilobytes of memory. We also solved generated instances of dimension up to 2,000 jobs within the regions A...E.

Table 3. Average solving time (in seconds) by regions in comparison with known algorithm of Tanaka et al.

n	Region A		Region B		Region C		Region D		Region E		The rest		All instances	
	PSC	SSDP	PSC	SSDP	PSC	SSDP	PSC	SSDP	PSC	SSDP	PSC	SSDP	PSC	SSDP
40	0.00	0.00	0.02	0.05	0.12	0.13	0.12	0.13	0.09	0.11	0.20	0.12	0.11	0.09
50	0.00	0.00	0.03	0.12	0.16	0.21	0.21	0.21	0.22	0.23	0.29	0.26	0.17	0.18
100	0.00	0.01	0.15	0.99	2.99	3.63	2.75	3.57	2.63	2.92	3.52	4.35	2.14	2.75
150	0.01	0.05	0.92	3.17	8.93	13.83	14.29	16.50	11.62	12.45	15.13	17.84	9.21	11.26
200	0.02	0.13	2.39	8.85	28.84	34.47	37.40	43.41	29.86	34.98	48.84	52.54	27.38	31.82
250	0.02	0.27	7.82	35.06	69.99	76.75	76.10	97.37	70.58	70.81	111.89	115.45	63.18	72.08
300	0.03	0.48	12.38	51.80	126.63	157.10	150.56	172.92	128.39	136.72	253.75	245.74	131.53	142.80

PSC PSC-algorithm [13], Pentium IV, 3.0 GHz, 2 GBytes of RAM; *SSDP* algorithm of Tanaka et al. [14], Pentium IV, 3.4 GHz, 1 GByte of RAM

We also investigated the dependence between the deviation of the functional value in the sequence σ^{fp} from the optimum and the number of competing jobs. We can conclude that for any problem size the more is the number of competing jobs in the sequence σ^{fp}, the less is the average deviation of the functional value. If $N_K \geq 0.35n$, then the average percentage of deviation of the functional value in the sequence σ^{fp} from the optimum is no more than 5–6%. And if $N_K \geq n/2$, then this deviation is no more than 2.5% in average, which allows to use the sequence σ^{fp} as an approximate solution for the problem. We obtained a similar dependence in [22] for the particular case of TWT problem where all weights are equal.

Special cases of the problem given in [13] are not included in the commonly used benchmark instances. So, they expand the research domain of the algorithm's efficiency. Theoretically complex and simple cases of TWT problem were found in [13], i.e., the conditions verified by analyzing the sequence σ^{fp}. They show when the complexity of the problem solving will become near to the complexity of exhaustive search, and under which conditions the combinatorial search is significantly reduced. The applied data generation scheme does not allow to select all possible theoretically simple and complex cases of the problem given in [13]: such problems are generated in a special way. We expanded the scope of the efficiency research for the PSC-algorithm by including these special cases to analyze its complexity. We have confirmed the conclusions made in [13].

We leave for the future the study of the regions of R and T parameters in which the instances are hard-solving.

4   Conclusions

We have identified and justified the regions of values of the dataset generating parameters R and T in which the PSC-algorithm solves the problem instances of any dimension quickly, in comparison with other regions, due to: free permutations (in the region A, the solution is obtained by a polynomial subalgorithm); a small number of competing jobs (in the region B, we execute a small number of iterations of the PSC-algorithm); a small number of jobs with time reserves or small magnitude of reserves (in the regions C, D, and E reserves are quickly used up by higher priority jobs, competing jobs are excluded from the set of competing jobs, and the algorithm terminates).

We have shown that up to 68% of the generated test instances will be solved relatively quickly for any given dimension.

We have analyzed the efficiency of the first polynomial component of the PSC-algorithm. It fulfilled for 27.3% of our benchmark instances (on average for all instances of all dimensions).

We show that for any problem size the more is N_K, the less is the average deviation of the functional value in the sequence σ^{fp} from the optimum. If $N_K \geq 0.35n$, it is no more than 5–6%. And if $N_K \geq n/2$, then this deviation is no more than 2.5% in average, which allows to use the sequence σ^{fp} as an approximate solution to the problem.

We give the statistics of solving the test instances with dimensions up to 300 jobs and comparison with algorithm of Tanaka et al. [14] (for which we know the solving time for only up to 300 jobs instances). We show that the PSC-algorithm is competitive in terms of solving time with the method of Tanaka et al. and is significantly faster for the quick-solving instances. We also investigated and experimentally confirmed the complexity of solving special cases of the problem given in [13]. Thus, the results of statistical studies confirmed the high efficiency of the PSC-algorithm.

References

1. Pinedo, M.L.: Scheduling: Theory, Algorithms, and Systems. Springer, Cham (2016). https://doi.org/10.1007/978-3-319-26580-3
2. Heydari, M., Hosseini, S.M., Gholamian, S.A.: Optimal placement and sizing of capacitor and distributed generation with harmonic and resonance considerations using discrete particle swarm optimization. Int. J. Intell. Syst. Appl. (IJISA) 5(7), 42–49 (2013). https://doi.org/10.5815/ijisa.2013.07.06
3. Wang, F., Rao, Y., Wang, F., Hou, Y.: Design and application of a new hybrid heuristic algorithm for flow shop scheduling. Int. J. Comput. Netw. Inf. Secur. (IJCNIS) 3(2), 41–49 (2011). https://doi.org/10.5815/ijcnis.2011.02.06
4. Cai, Y.: Artificial fish school algorithm applied in a combinatorial optimization problem. Int. J. Intell. Syst. Appl. (IJISA) 2(1), 37–43 (2010). https://doi.org/10.5815/ijisa.2010.01.06
5. Fomin, F.V., Kratsch, D.: Exact Exponential Algorithms. Springer, Heidelberg (2010). https://doi.org/10.1007/978-3-642-16533-7
6. Soltani, N., Soleimani, B., Barekatain, B.: Heuristic algorithms for task scheduling in cloud computing: a survey. Int. J. Comput. Netw. Inf. Secur. (IJCNIS) 9(8), 16–22 (2017). https://doi.org/10.5815/ijcnis.2017.08.03
7. Garg, R., Singh, A.K.: Enhancing the discrete particle swarm optimization based workflow grid scheduling using hierarchical structure. Int. J. Comput. Netw. Inf. Secur. (IJCNIS) 5(6), 18–26 (2013). https://doi.org/10.5815/ijcnis.2013.06.03
8. Mishra, M.K., Patel, Y.S., Rout, Y., Mund, G.B.: A survey on scheduling heuristics in grid computing environment. Int. J. Mod. Educ. Comput. Sci. (IJMECS) 6(10), 57–83 (2014). https://doi.org/10.5815/ijmecs.2014.10.08
9. Sajedi, H., Rabiee, M.: A metaheuristic algorithm for job scheduling in grid computing. Int. J. Mod. Educ. Comput. Sci. (IJMECS) 6(5), 52–59 (2014). https://doi.org/10.5815/ijmecs.2014.05.07
10. Hwang, K., Dongarra, J., Fox, G.: Distributed and Cloud Computing: From Parallel Processing to the Internet of Things. Morgan Kaufmann, Burlington (2012)
11. Brucker, P., Knust, S.: Complex Scheduling, 2nd edn. GOR-Publications Series, Springer, Berlin, Heidelberg (2012). https://doi.org/10.1007/978-3-642-23929-8
12. Zgurovsky, M.Z., Pavlov, A.A.: Algorithms and software of the four-level model of planning and decision making. In: Combinatorial Optimization Problems in Planning and Decision Making: Theory and Applications, 1st edn. Studies in Systems, Decision and Control, vol. 173, pp. 407–518. Springer, Cham (2019). https://doi.org/10.1007/978-3-319-98977-8_9
13. Zgurovsky, M.Z., Pavlov, A.A.: The total weighted tardiness of tasks minimization on a single machine. In: Combinatorial Optimization Problems in Planning and Decision Making: Theory and Applications, 1st edn. Studies in Systems, Decision and Control, vol. 173, pp. 107–217. Springer, Cham (2019). https://doi.org/10.1007/978-3-319-98977-8_4
14. Tanaka, S., Fujikuma, S., Araki, M.: An exact algorithm for single-machine scheduling without machine idle time. J. Sched. 12(6), 575–593 (2009). https://doi.org/10.1007/s10951-008-0093-5
15. Kanet, J., Birkemeier, C.: Weighted tardiness for the single machine scheduling problem: an examination of precedence theorem productivity. Comput. Oper. Res. 40(1), 91–97 (2013). https://doi.org/10.1016/j.cor.2012.05.013
16. Kanet, J.: One-machine sequencing to minimize total tardiness: a fourth theorem for Emmons. Oper. Res. 62(2), 345–347 (2014). https://doi.org/10.1287/opre.2013.1253

17. Karakostas, G., Kolliopoulos, S., Wang, J.: An FPTAS for the minimum total weighted tardiness problem with a fixed number of distinct due dates. ACM Trans. Algorithms **8**(4), 1–16 (2012). https://doi.org/10.1145/2344422.2344430
18. Zgurovsky, M.Z., Pavlov, A.A.: Introduction. In: Combinatorial Optimization Problems in Planning and Decision Making: Theory and Applications, 1st edn. Studies in Systems, Decision and Control, vol. 173, pp. 1–14. Springer, Cham (2019). https://doi.org/10.1007/978-3-319-98977-8_1
19. Tanaka, S., Fujikuma, S., Araki, M.: OR-library: weighted tardiness (2013). https://sites.google.com/site/shunjitanaka/sips/benchmark-results-sips. Accessed 08 Nov 2018
20. Fisher, M.L.: A dual algorithm for the one machine scheduling problem. Math. Progr. **11**(1), 229–251 (1976). https://doi.org/10.1007/BF01580393
21. Ding, J., Lü, Z., Cheng, T.C.E., Xu, L.: Breakout dynasearch for the single-machine total weighted tardiness problem. Comput. Indust. Eng. **98**(C), 1–10 (2016). https://doi.org/10.1016/j.cie.2016.04.022
22. Pavlov, A.A., Misura, E.B., Melnikov, O.V., Mukha, I.P.: NP-hard scheduling problems in planning process automation in discrete systems of certain classes. In: Hu, Z., Petoukhov, S., Dychka, I., He, M. (eds.) Advances in Computer Science for Engineering and Education. ICCSEEA 2018. Advances in Intelligent Systems and Computing, vol. 754, pp. 429–436. Springer, Cham (2019). https://doi.org/10.1007/978-3-319-91008-6_43

Dosimetric Detector Hardware Simulation Model Based on Modified Additive Fibonacci Generator

V. Maksymovych[1], M. Mandrona[1,2(✉)], and O. Harasymchuk[1]

[1] Lviv Polytechnic National University, Lviv, Ukraine
volodymyr.maksymovych@gmail.com,
mandrona27@gmail.com, oleh.harasymchuk@gmail.com
[2] Lviv State University of Life Safety, Lviv, Ukraine

Abstract. In this paper, a hardware simulation model of a dosimetric detector is proposed, which is implemented on the basis of the Poisson pulse sequence generator, based on the modified Fibonacci source additive. The peculiarity of this structure is the possibility of an operational task of internal parameters and the type of dosimetric detector. The statistical characteristics, which show the conformity of the model's output signal with the real output signals of the dosimetric detectors in a wide range of exposure dose rates, are analyzed.

The purpose of the work is to develop and analyze the characteristics of the hardware simulation model of DD on the PPSG with the use of PNG based on the modified additive Fibonacci generator (MAFG).

Keywords: Modified additive Fibonacci generator · Dosimetric devices · Poisson pulse sequence generator · Statistical characteristics

1 Dosimetric Detector

1.1 Introduction

Throughout the period of development, humanity has always been exposed to ionizing radiation, because most of natural materials contain more or less radioactive elements. Contemporary technical progress, which led to the emergence of nuclear weapons, enterprises of nuclear power engineering, as well as violations of natural complexes, which causes radiation pollution of territories has led to the need of accurate, prompt and rapid detection of radiation emission, that is to improve dosimetric devices. Imitation models of dosimetric detectors (DD) allow to optimize the parameters of dosimetric devices when they are designed, adjusted and verified. They can be created on the basis of Poisson pulse sequence generator (PPSG). A lot of works are devoted to designing and analyzing the hardware characteristics of the PPSG, among which the following can be distinguished: [1–3], in which the Poisson sequence is formed by means of analog technology; [4–6], in which the Poisson sequence is formed on the basis of calculating the time interval to the next pulse; [7–9], in which the Poisson sequence is formed on the basis of the pseudorandom number generator (PNG). The

© Springer Nature Switzerland AG 2020
Z. Hu et al. (Eds.): ICCSEEA 2019, AISC 938, pp. 162–171, 2020.
https://doi.org/10.1007/978-3-030-16621-2_15

last way of PPSG realization, in our opinion, is the most promising, because, unlike the others, it provides a clear time-fixing, convenient controllability and reliability.

1.2 DD Structural Scheme

The statistical characteristics of the radiation particles flux, which fall into the DD, obey the Poisson distribution law [11–13]. The basis of the simulation model of radiation source (RS) may be the PPSG. The output signal DD can also be reproduced with the help of PPSG, taking into account the dead time of DD [14]. Based on this it is expedient to consider the simulation model of RS and DD as a whole.

Hardware simulation models of dosimetric detectors can be used at the stages of design, adjustment and testing of dosimetric devices. Their use simplifies these processes, and in many cases, avoids the use of physical radiation sources. Such simulation models can be implemented based on programmable logic integrated circuits (PLIC).

As the result of the research, authors proposed a PPSG scheme based on MAFG [10–12, 15] that showed in Fig. 1. MAFG itself consists of registers Rg1 - Rg5, combinational adders CA1 - CA3 and logical scheme (LS). Comparison scheme (CS) and logical element AND provide the formation of a pulsed flux with a Poisson distribution law. All structural elements of the MAFG, except of LS, operate in a binary-decimal code.

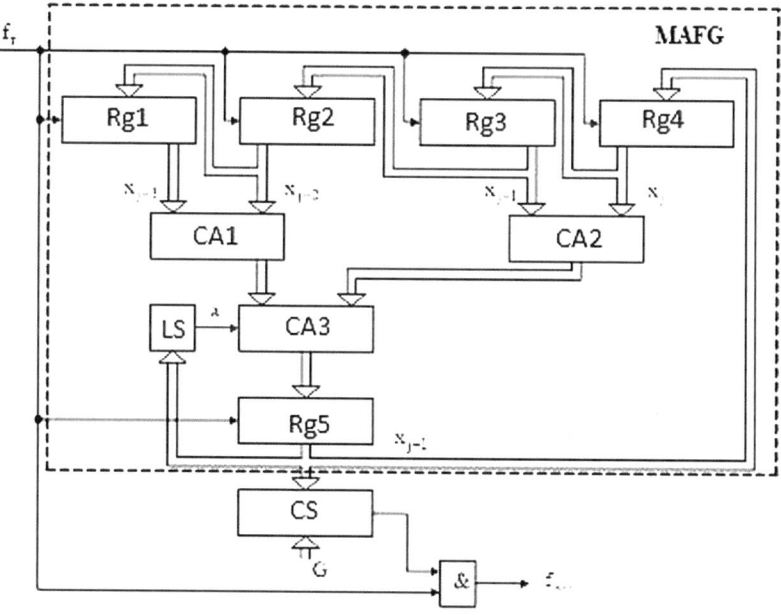

Fig. 1. Structural scheme of PPSG based on MAFG

At the output of the MAFG, that is, at the output of Rg5, a sequence of pseudo-random numbers is formed in accordance with the expression:

$$x_{j+1} = (x_j + x_{j-1} + x_{j-2} + x_{j-3} + a) mod \ m \tag{1}$$

where x_j, x_{j-1}, x_{j-2}, x_{j-3} – numbers in registers Rg4, Rg3, Rg2, Rg1 accordingly, $m = 10^q$, q – number of decades of scheme structural elements.

The value of the variable a is determined by the logical equation:

$$a = (a_{00} + a_{01} + a_{02} + a_{03)} \oplus \ldots \oplus (a_{q-10} \oplus a_{q-11} \oplus a_{q-12} \oplus a_{q-12}). \tag{2}$$

where $a_{ij}(i = 0, 1, 2, 3; j = 0, 1, 2, \ldots, q - 1)$ – the value of the bits of the binary-decimal number in Rg5.

The number of Eq. (2) members can be choose from the range – $0 \ldots 4*q$.

Since, in this case, the maximum value of the number at the MAFG output is equal to $x_{max} = 10^q$, then the theoretical average value of the pulse frequency at the PPSG output is determined by the equation:

$$f_{out} = \frac{G}{10^q} f_m, \tag{3}$$

where G – control code, f_m – clock pulse frequency.

Accordingly, the lower limit of the control code will be determined by the expression:

$$G_1 \geq \frac{10^q \cdot k_c \cdot n_{max}}{T_n}. \tag{4}$$

the upper – by expression:

$$G_2 = s \cdot X_{max}. \tag{5}$$

In this case, the value of the coefficient **s** is determined separately for the specific implementation of the PNG and, under certain conditions, is close to the value of 0,1.

The structural scheme of the DD model is presented in Fig. 2. It consists of the PNG on the basis of binary-decimals MAFG, as well as a frequency divider FD, comparison schemes CS1, CS2, counter Ct, trigger Tg, multiplexer MUX and logical elements AND1, AND2. The signal corresponding to the output of the DD is formed at the output of the logic element AND2, and the signal simulating the radiation source according to its characteristics - from the output of the logic element AND1.

The model can operate in two modes: with a not continuing type of DD dead time [16] and a continuing type of DD dead time [14–17].

Before starting the schemes work in the registers Rg1 - Rg5 the initial values of numbers are written – x(0) (Fig. 2), which allows for each new simulation experiment to receive a new sequence of the DD model output signal.

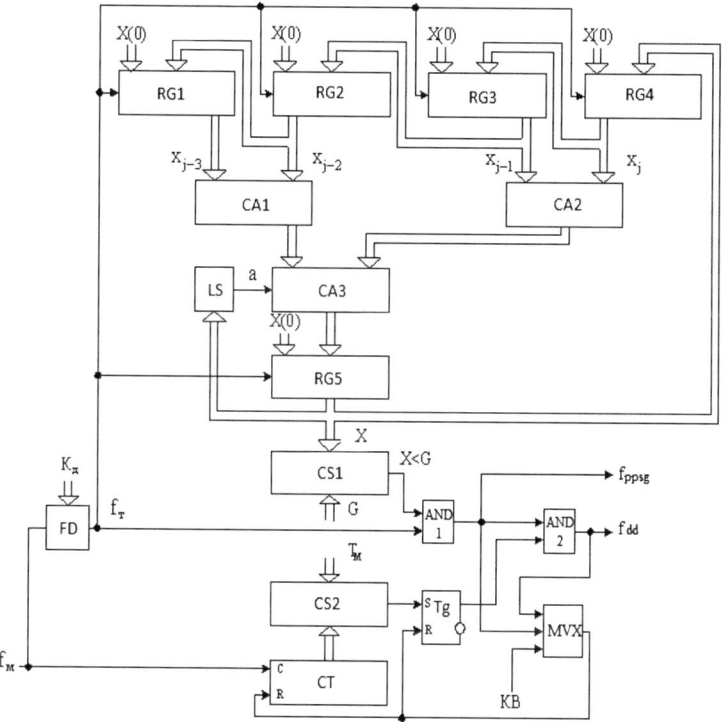

Fig. 2. Structural scheme of the DD model based on the MAFG

2 Statistical Characteristics

Table 1 shows the statistical characteristics of the DD model at such values $x_{max} = 10^6$, $f_m = 10^6$ Hz, $f_m = 2*10^5$ Hz $(K_{fd} = 5)$ i $T_v = 10$ s. Statistical characteristics are determined using the equations given below.

The average repetition period of the PPSG output pulses is determined by the expression:

$$f_{ppsg} = \frac{G}{X_{max}} f_m. \tag{6}$$

Value f_{ppsg} may also be found based on the values of exposure dose power (EDP) λ and sensitivity of the detector γ:

$$f_{ppsg} = \lambda \cdot \gamma. \tag{7}$$

Equations (6) and (7) determine the value of the control code G:

$$G = \frac{\lambda \cdot \gamma \cdot X_{max}}{f_m}. \tag{8}$$

The average number of pulses at the output of the PPSG for the time T_v is equal to:

$$k_{1avg} = T_v \cdot f_{ppsg}. \tag{9}$$

The average pulse frequency at the output of DD with not continuing type of its dead time τ_m is determined as follows [13]:

$$f_{out} = \frac{f_{ppsg}}{1 + f_{ppsg} \cdot \tau_m} = \frac{\lambda \cdot \gamma}{1 + \lambda \cdot \gamma \cdot \tau_m}. \tag{10}$$

The value of the control code T_m code is determined by the equation:

$$T_m = \tau_m \cdot f_m. \tag{11}$$

The average number of pulses at the output of the simulation model for the time T_v is equal to:

$$k_{2avg} = T_v \cdot f_{out} \tag{12}$$

The average pulse frequency at the output of the detector with a continuing type dead time is determined by the equation [18]:

$$f_{out} = f_{ppsg} e^{-f_{ppsg} \cdot \tau_m}. \tag{13}$$

It is known that for a PPSG output signal, the number of pulses k_1 of the Poisson pulse source, fixed in time T_v, with a reliable probability of 0.95, must be within range [16, 19]:

$$k_{1avg} - 2\sqrt{k_{1avg}} < k_1 < k_{1avg} + 2\sqrt{k_{1avg}}. \tag{14}$$

where k_{1avg} is determined by Eq. (9).

At $\tau_m \neq 0$ the output pulse flux of the DD model, as well as the output pulse flux of the dosimetric detector, does not correspond to the Poisson law of distribution. This is explained be the fact that during the dead time it is impossible to form output pulses. We have done and confirmed, as a result of the research, the assumption that the number of output pulses of the model k_2, fixed in time T_v, with a reliable probability of 0.95 is within range [14, 17]:

$$k_{2avg} - 2 \cdot \kappa_m \cdot \sqrt{k_{2avg}} < k_2 < k_{2avg} + 2 \cdot \kappa_m \cdot \sqrt{k_{2avg}}, \tag{15}$$

where k_{2avg} is determined by the Eq. (12), and k_m by equation:

$$\kappa_m = \frac{1}{1 + \lambda \cdot \gamma \cdot \tau_m}. \tag{16}$$

Table 1. Statistical characteristics of the DD model at $x_{max} = 10^6$, $f_m = 10^6$ Hz, $f_m = 2*10^5$ Hz ($K_{fd} = 5$) i $T_v = 10$ s

λ, μR/h	γ, Hz/ (μR/h)	τ_m, μs	G	T_m	fppsg, Hz	k_{1avg}	Type of DD dead time			
							Not continuing		Continuing	
							fdd, Hz	k2avg	fdd, Hz	k2avg
10	0,02	10	1	10	0,2	2	0,2	2	0,2	2
		100	1	100	0,2	2	0,2	2	0,2	2
	0,04	10	2	10	0,4	4	0,4	4	0,4	4
		100	2	100	0,4	4	0,4	4	0,4	4
20	0,02	10	2	10	0,4	4	0,4	4	0,4	4
		100	2	100	0,4	4	0,4	4	0,4	4
	0,04	10	4	10	0,8	8	0,8	8	0,8	8
		100	4	100	0,8	8	0,79994	7,9994	0,79994	7,9994
....										
10^3	0,02	10	100	10	20	200	19,9960	199,960	19,9960	199,960
		100	100	100	20	200	19,9601	199,601	19,9600	199,600
	0,04	10	200	10	40	400	39,9840	399,840	39,9840	399,840
		100	200	100	40	400	39,8406	398,406	39,8403	398,403
....										
10^4	0,02	10	1000	10	200	2000	199,601	1996,01	199,600	1996,00
		100	1000	100	200	2000	196,078	1960,78	196,040	1960,40
	0,04	10	2000	10	400	4000	398,406	3984,06	398,403	3984,03
		100	2000	100	400	4000	384,615	3846,15	384,316	3843,16
....										
10^5	0,02	10	10^4	10	2000	$2*10^4$	1960,78	19607,8	1960,39	19603,9
		100	10^4	100	2000	$2*10^4$	1666,66	16666,6	1637,46	16374,6
	0,04	10	$2*10^4$	10	4000	$4*10^4$	3846,15	38461,5	3843,15	38431,5
		100	$2*10^4$	100	4000	$4*10^4$	2857,14	28571,4	2681,28	26812,8

The number of pulses k_1 of the Poisson pulsed flux at the output of the PPSG, that is, at the output of the logical element AND1 (Fig. 2), fixed in time T_v, with a reliable probability 0,95 must be within the limits defined by the expression (14), where k_{1avg} is determined by the Eq. (9).

The number of output pulses of the DD model with the dead time of the not continuing type, that is, the output pulses of the logical element AND2, – k_2, fixed in time T_v, with a reliable probability 0,95 is within the limits defined by the expression (15), where k_{2avg} is determined by the Eq. (12), and k_m by Eq. (16).

In Figs. 3, 4, 5, 6, 7 and 8 the results of the study of statistical characteristics of the simulated DD model with the not continuing type of dead time are presented, for different values EDP – λ and different initial states $x(0)$ of MAFG registers.

At the same time, the following values were fixed: $q = 6$, $x_{max} = 10^6$, $f_m = 10^6$ Hz, $f_m = 2*10^5$ Hz (FD dividing coefficient - $K_{fd} = 5$) i $T_v = 10$ s, $\tau_m = 100$ fs, $\gamma = 0{,}04$ Hz/(μR/h).

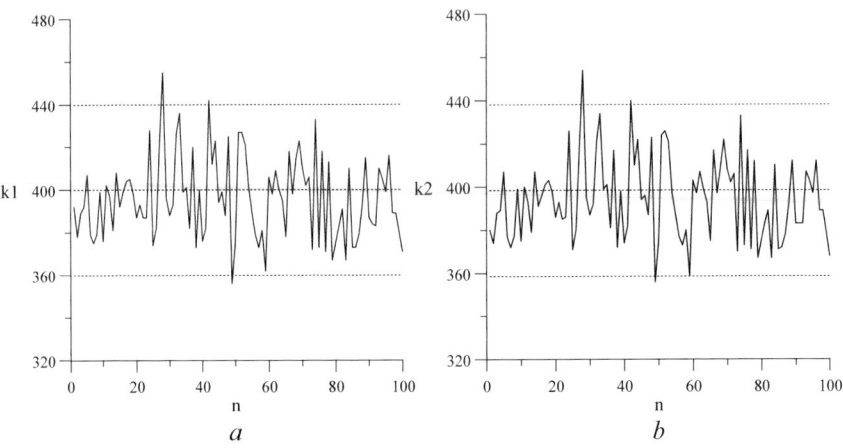

Fig. 3. Statistical characteristics of the simulation model DD at $\lambda = 1000$ $\mu R/h$ and initial states $x(0)$ of MAFG registers Rg1-Rg5 – 0, 0, 0, 0, 1 accordingly.

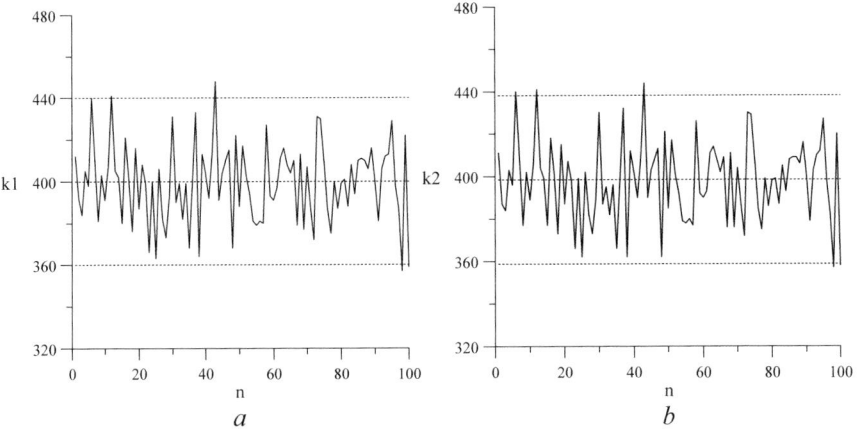

Fig. 4. Statistical characteristics of the simulation model DD at $\lambda = 1000$ $\mu R/h$ and initial states $x(0)$ of MAFG registers Rg1-Rg5 – 0, 0, 0, 0, 111111 accordingly.

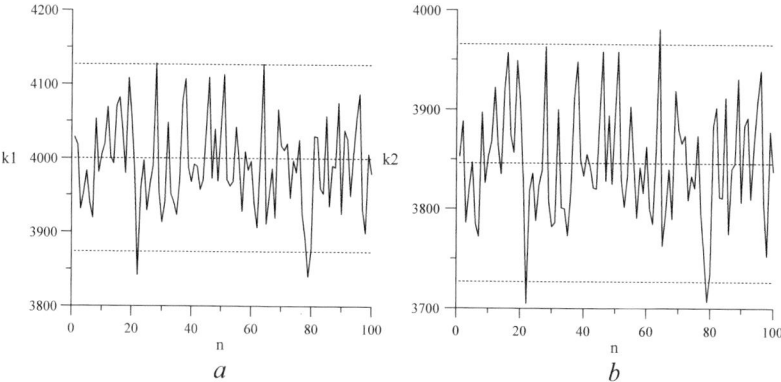

Fig. 5. Statistical characteristics of the simulation model of DD at $\lambda = 10000$ $\mu R/h$ and initial states $x(0)$ of MAFG registers Rg1-Rg5 – 0, 0, 0, 0, 1 accordingly.

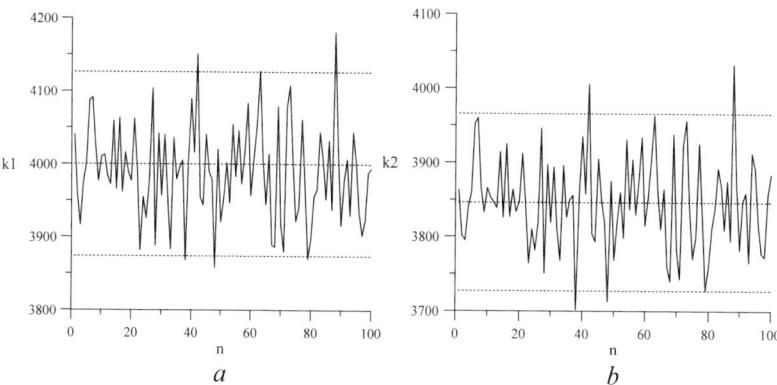

Fig. 6. Statistical characteristics of the simulation model of DD at $\lambda = 10000$ $\mu R/h$ and initial states $x(0)$ of MAFG registers Rg1-Rg5 – 0, 0, 0, 0, 111111 accordingly.

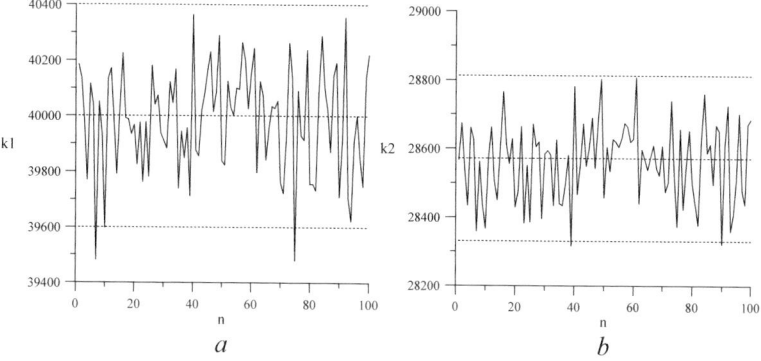

Fig. 7. Statistical characteristics of the simulation model of DD at $\lambda = 100000$ $\mu R/h$ and initial states $x(0)$ of MAFG registers Rg1-Rg5 – 0, 0, 0, 0, 1 accordingly.

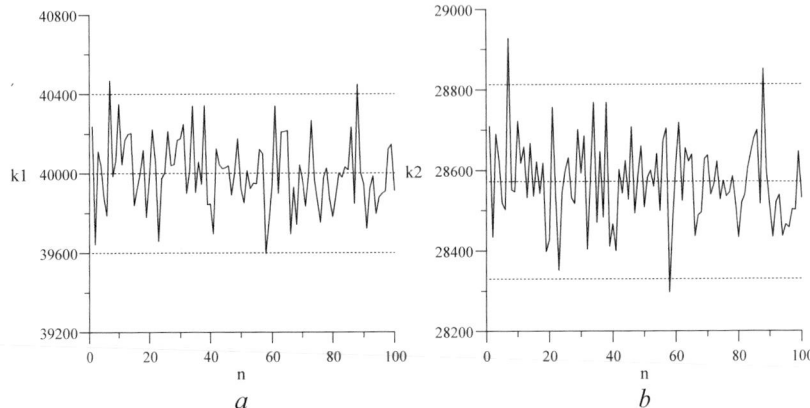

Fig. 8. Statistical characteristics of the simulation model of DD at $\lambda = 100000$ $\mu R/h$ and initial states $x(0)$ of MAFG registers Rg1-Rg5 – 0, 0, 0, 0, 111111 accordingly.

In Figs. 3, 4, 5, 6, 7 and 8 the following notations are accepted: n – sequence number of next interval T_y, the middle dashed lines correspond to the values k_{1avg} (9) and k_{2avg} (12), and the upper and lower – to the limits of inequalities (14) i (15).

3 Conclusions

Statistical characteristics which are shown in Figs. 3, 4, 5, 6, 7 and 8 confirm that the output pulse sequences of the DD model correspond to real DD output signals in a wide range of exposure dose power values λ. At the same time, it becomes possible to set the internal parameters of the DD model: value of dead time τ_m, sensitivity DD γ, as well as specify the type of DD (with a dead time of not continuing or continuing types).

The main advantage of the proposed hardware model of the dosimetric detector is that with its help, the processes of adjustment and testing of dosimetric devices can be carried out without the use of hazardous physical sources of radiation.

References

1. Lauch, J., Nachbar, H.U.: Random pulse generator with a uniformly distributed amplitude spectrum. Nucl. Instrum. Methods Phys. Res. Sect. A: Accel. Spectrometers Detect. Assoc. Equip. **267**(1), 177–182 (1988). https://doi.org/10.1016/0168-9002(88)90645-6
2. Divon, W.B., Rozen, B.: A random pulse generator. Nucl. Instrum. Methods **39**(1), 77–87 (1966). https://doi.org/10.1016/0029-554X(66)90046-2
3. Mashud, M.A.A., Hossain, M.S., Islam, M.N., Islam, S.: Design and development of PC based data acquisition system for radiation measurement. IJIGSP **5**(7), 34–40 (2013). https://doi.org/10.5815/ijigsp.2013.07.05
4. Khan, M.R.R., Kang, S.W.: Design and development of low-cost, highly stable regulated high-voltage power supply for radiation detector. IJEM **5**(2), 1–10 (2015). https://doi.org/10.5815/ijem.2015.02.01

5. Veiga, A., Spinelli, E.: A pulse generator with poisson-exponential distribution for emulation of radioactive decay events. Conference Paper, February 2016. https://doi.org/10.1109/lascas.2016.7451002
6. Heeger, D.: Poisson model of spike generation. Conference Paper, pp. 18–26 (2000)
7. Nyangaresi, V.O., Abeka, S., Rodrigues, A.: Multivariate probabilistic synthesis of cellular networks teletraffic blocking with poissonian distribution arrival rates. Int. J. Wirel. Microw. Technol. (IJWMT) 8(4), 14–39 (2018). https://doi.org/10.5815/ijwmt.2018.04.02
8. UsmanDilawar, M., Syed, F.A.: Mathematical modeling and analysis of network service failure in datacentre. Int. J. Mod. Educ. Comput. Sci. (IJMECS) 6(6), 30–36 (2014). https://doi.org/10.5815/ijmecs.2014.06.04
9. Harasymchuk, O.I., Dudykevych, V.B., Maksumovych, V.M., Smuk, R.T.: Generators of test pulse sequences for dosimetric devices. Heat Power Eng. Environ. Eng. Autom. 506, 186–192 (2004). Bulletin of the "Lviv Politechnic National University"
10. Maksymovych, V., Harasymchuk, O., Opirskyy, I.: The designing and research of generators of poisson pulse sequences on base of fibonacci modified additive generator. In: International Conference on Theory and Applications of Fuzzy Systems and Soft Computing, pp. 43–53 (2018)
11. Maksymovych, V., Harasymchuk, O, Mandrona, M.: Designing generators of poisson pulse sequences based on the additive fibonacci generators. J. Autom. Inf. Sci. 1–13. https://doi.org/10.1615/jautomatinfscien.v49.i12.10
12. Mandrona, M., Maksymovych, V., Harasymchuk, O., Kostiv, Y.: Implementation of modified additive lagged Fibonacci generator. Chall. Mod. Technol. 7(1), 3–6 (2016)
13. Ornatsky P.P.: Theoretical foundations of information-measuring equipment: studies for universities on spec. "Inform. - meas. technique" 2nd ed., overworked and added. Ornatsky. High school, Kiev, 455 p. (1983)
14. Bobalo, Y.Y., Dudykevych, V.B, Maksumovych, V.M., Horoshko, V.O., Bisuk, A.M., Smuk, R.T., Storonskiy, Y.B.: Methods and means of working out the output signals of the dosimetric detectors: Monograph, 200 p. Publishing of "Lviv Politechnic National University (2009)
15. Maksumovych, V.M., Smuk, R.T., Stronskiy, Y.B., Kostiv, Y.M.: Binary-decimal generator of Poisson pulsed flows. In: Computer Technologies of Printing: A Collection of Scientific Works, no 32, pp. 82–89. Publishing of the Ukrainian Academy of Printing, Lviv (2014)
16. Dudykevych, V.B., Maksumovych, V.M., Smuk, R.T.: Imitation models of a non-sustaining type of a dead-end dosimetric detector. Autom. Meas. Control. 530, 46–52 (2005). Bulletin of the "Lviv Politechnic National University"
17. Dudykevych V.B., Maksumovych, V.M., Kostiv, Y.M., Smuk, R.T.: Imitation model of dosimetric detector with dead times of non-extending and continuation. Scientific Bulletin of the Ukrainian National Forestry University: a collection of scientific and technical works, pp. 322–328 (2014). T.24.9
18. Nemets, O.F., Hofman, Y.V.: Handbook of Nuclear Physics: Handbook, 416 p. "Scientific thought", Kiev (1975)
19. Hurman V.E.: Probability theory and mathematical statistics. Study allowance for high schools. Pub. No. 5, 479 p. (1977). reworked and added by M., "High School"

Real-Time Flame Detection Using Hypotheses Generating Techniques

Dmytro Peleshko[1], Olena Vynokurova[1(✉)], Semen Oskerko[1],
Oleksii Maksymiv[2], and Orysia Voloshyn[3]

[1] IT Step University, 83a Zamarstynivs'ka Street, Lviv, Ukraine
dpeleshko@gmail.com, vynokurova@gmail.com,
semenosker@gmail.com
[2] Lviv State University of Life Safety, 35, Kleparivska Street, Lviv, Ukraine
aleks.maksymiv@gmail.com
[3] Lviv National Polytechnic University, 12 S.Bandery Street, Lviv, Ukraine
vop_ippt@ukr.net

Abstract. Nowadays the real-time object tracking in video streams is an important problem in smart city applications. One of the problem of smart city monitoring such as the flame detection task in real time is considered. The flame as the object of interest is one of the most complex objects tracking. This is due to the lack of time and space invariance in all features of the object of interest. In accordance with this fact, the computational processes are characterized by high computational complexity and latency of program tracking procedures. There are several basic approaches to minimizing computing costs and reducing time delays (shifts). Using the ROI is one of them. Thus, a new model of the hypothesis generator for the effective evaluation of ROI is proposed.

The task of the generator is to select the areas that contain a fire in the video stream. In this case, the main criterion for identifying the region is the index of complexitivity. Due to using the proposed model, the classification procedures quality is increased, and as a result, computing costs of the tracking methods is reduced in general. The many experiments based on both benchmarks and real data sets have confirmed the effectiveness of the proposed approach.

Keywords: Computer vision · Regions of Interest · Machine learning ·
Flame detection · Hypotheses Core Generator (HCG)

1 Introduction

Nowadays the real-time object tracking in video streams is an important task in computer vision applications [1–5, 23–29]. The flame detection is the special problem [6–9] in the tracking systems because the flame (as the region of interest) can be identified both an object and a process. If the flame is considered as an object than the main problem is the change of the geometrical shapes. In some cases, this change may have a rather high fluctuation character, which is determined both by internal and external factors. If the flame is considered as a process than the main problem is the fluctuating change in color characteristics at the time interval of the process. A typical up-to-date flame tracking scheme is shown in Fig. 1. Each of the stages, which is

© Springer Nature Switzerland AG 2020
Z. Hu et al. (Eds.): ICCSEEA 2019, AISC 938, pp. 172–182, 2020.
https://doi.org/10.1007/978-3-030-16621-2_16

shown in Fig. 1, is a set of mathematical procedures and methods that require significant computing resources. As a result, after implementation of each stage, the time delay can be very large. From a real-life point of view, the early detection of fire is a precondition for preventing the adverse effects. However, as a rule, the color parameters on the image of the fire, even in the short time interval from the moment of its appearance, can change very high. As a result, the identification of the flame becomes impossible if the tracking method does not take into account these changes.

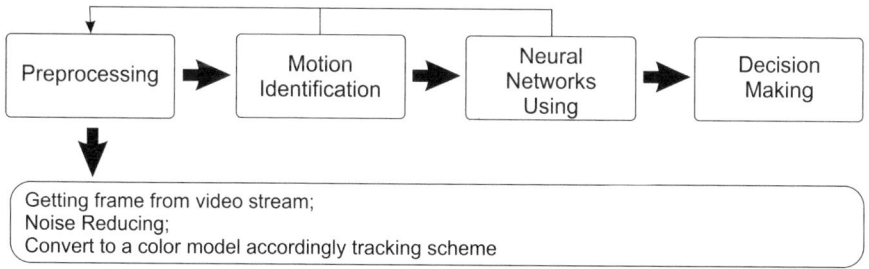

Fig. 1. A typical up-to-date flame tracking scheme

All these problems greatly complicate both tracking and possible flame analysis in real-time mode. Therefore, only those regions that only have an image with flame are tried to fed to the input of identifier. This is needed for reducing the number of analysis areas that have not a practical interest. As a result, the computing costs are reduced, which it allows reducing latency or adding a new application analysis. The methods, which is currently used to determine ROI, typically become highly specialized for tracking flame. For example, for cases when the fire is in the open air in the morning light or in the room during evening lighting, you need quite different techniques for determining the ROI. It is considered that the sources of fire are the same. If the sources of the flame, geometric distances, levels of smudging, atmospheric influences are different, then the task is complicated several times.

Therefore, we have proposed the new methodology of obtaining ROI, which is characterized by a high stability toward "open air"/"closed area" systems. The proposed method allows increasing the maximum distance to the fire, which can be identified and tracked. This makes it possible to organize control of large areas with a minimum number of video cameras. And, finally, it allows decreasing the tracking latency. The proposed evaluation of ROI is based on the use of the hypothesis generator. The task of the generator is to select the areas that contain a fire in the video stream. In this case, the main criterion for identifying the region is the index of complexivity. Due to this index, it is possible to minimize the geometric dimensions of the area with fire. The obtaining ROIs form a set of input data for tracking procedures. Usually, such procedures are based on methods of machine learning. In order to provide a visual representation of the effectiveness of the hypotheses generator, we have used the convolutional neural networks. The goal of the proposed approach is to increase the classification procedures quality and to reduce the computing costs of the tracking methods.

2 The Related Works

The most of approaches, which are used for detecting flames using the information about motion, color or both of these features together. The color models such as HSV [10], RGB [11, 12], YUV [13, 14], YCbCr [10, 15], or their combinations [16] are widely used for detecting the flame features in video streams. To track moving objects in the video sequences, a range of methodologies based on temporal or spatial features are used to identify the possible changes. Among the existed algorithms, the most usable and effective methods are the frame difference method [17], the method based on the optical flow [18], and background subtraction method [19]. In counterpart to the above-mentioned methods for in video streams, there are some methods which allow realizing the classification using the hidden Markov models [20], the local binary patterns [21], the wavelet analysis [22], etc.

3 The Main Problems of the Flame Detection in Video Streams

The complexity of the object of interest and the specificity of the burning process set the specific conditions for the solution of the tracking task with high quality.

The real time mode requires the minimal time delays in processing. In this case, the number of correctly identified objects (True positive, TP) should be maximized and the number of incorrectly identified objects (True negative, TN) should be minimized. Reducing the number of false identification of "true" objects (False positive, FP) and the number of objects, which are incorrectly belonged to the "false" category (False negative, FN) is natural in this case. The main problem is a combination of solving these tasks in one detector. Usually, when the quality of FP definition is high, the quality of FN is decreased. Achieving some optimal quality for both cases of FN and FN usually leads to a significant increase in computing resources. And this fact increases time delays and as result, we cannot solve the tracking task in real-time mode.

As a way to increase the quality of the tracking it is necessary to provide some degree of invariance to:

- The environmental influence – the closed or open places, a natural or artificial lighting, a time change of lighting, availability of periodic or aperiodic external noises.
- The fluctuations of parameters of the burning processes – changing the color in relation to changing the burning temperature or chemical properties of the source, the appearance of separation of the object of interest, non- separation in case of overlapping with other objects, the nonoccurrence of stable geometric forms of the tracking object.
- Using the neural networks greatly simplifies the solution of these problems. At the same time, an important factor is the possibility of additional training. The additional training provides the possibility of additional tuning in both boundary cases and in cases of retraining system when new information features or such images with objects of interest (which were not present in training datasets and have not the characteristic features) are fed.

4 The Hypotheses Generating Techniques for Video Streams Tracking Systems

The HCG of had been proposed taking into account the main features of the flame as an object of interest. Such generator allows obtaining the Regions of Interest (ROI). The general diagram for obtaining the ROI is shown in Fig. 2.

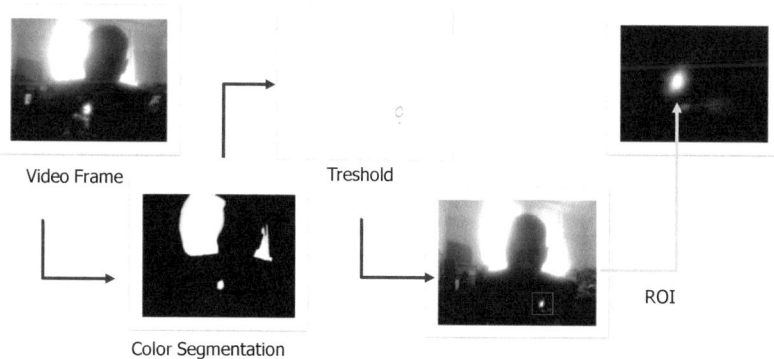

Fig. 2. The general diagram for obtaining the ROI

The generator consists of the next stages: pre-processing a video stream; color segmentation of a video frame; detecting the moving objects.

In comparison with the existed flame detection methods, the proposed core generator of hypotheses includes the advanced method of color segmentation. Also we propose a three-frame difference method instead of the classical algorithms of detecting moving objects. From the point of view of the color representation of the image, the flame has some characteristic singularities. In particular, the different sources of flame can give a flame of different colors in the same external conditions. Therefore, in the case of images the segmentation problem is sufficiently non-trivial. The first problem that was considered in the task of color segmentation of images with a flame was the choice of color model. As a way to solve this problem, the effectiveness of such color models as RGB, CIE L*a*b, HSV, and YCbCr was analyzed. For this purpose, a dataset was created from 150 randomly selected images with a flame. The example of the analyzed images is shown in Fig. 3. The randomness of choosing such images means that images have different DPI, noise levels and backgrounds. The image segmentation was performed using the k-means method. The choice of this method is dictated, on the one hand by the simplicity of implementation, and on the other hand by the minimization of the mean square deviation. The second characteristic was an important parameter in estimation of the color models effectiveness.

The k-means clustering method allowed to easily identify the boundaries of areas with the flame image. The number of clusters was set 4. The performed experiments have confirmed the adequacy of this number of clusters (Fig. 4).

Fig. 3. The example of an analyzed images

Although in some cases, it was necessary to increase the number of clusters to 8 (Fig. 5). Figure 4 shows the results of clustering in the case of 4 clusters, and Fig. 5 shows results for 8 clusters.

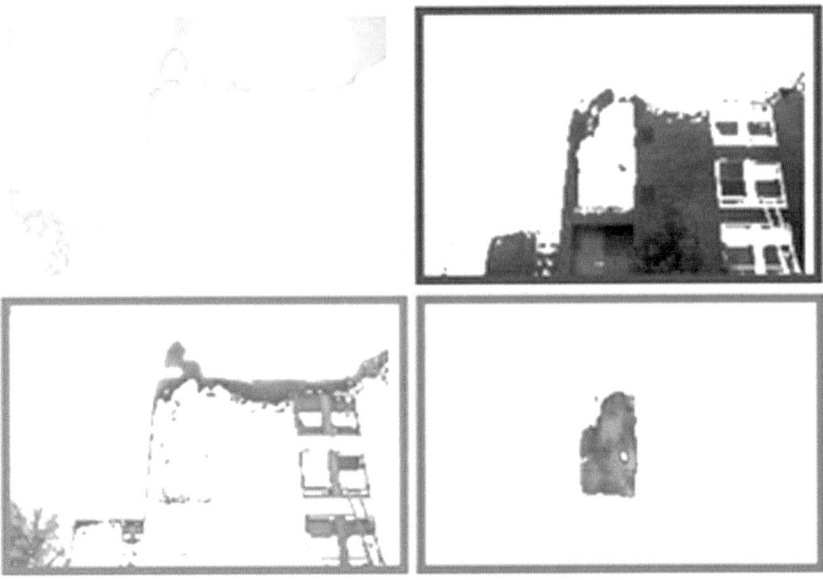

Fig. 4. The result of k-means clustering method for the image with 4 clusters

After identifying the boundaries of areas with flame image based on developed rules, the image pixels were checked for belonging or not belonging to the category of flame. After analyzing the obtained results, we can confirm that the HSV, YCbCr and L*a*b color models are approximately equally and well suited for segmentation of

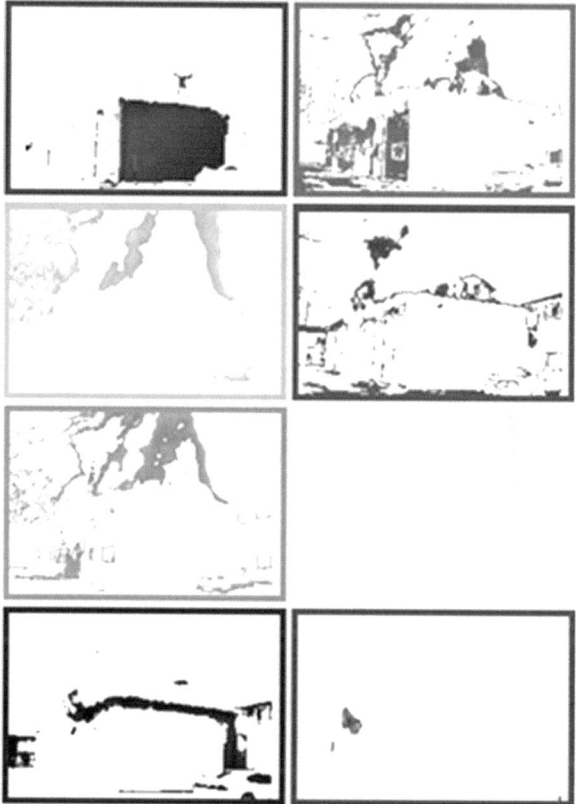

Fig. 5. The result of the k-means clustering method for the image with 8 clusters

images with a flame. The RGB color model is not suitable for such segmentation. It is clear that its fact is explained by color components with the same weight. For further research, the model L*a*b was chosen. This choice is determined by the higher stability of the model for cases of separation of areas with a fire image. Figure 6 shows the result of image segmentation based on the L*a*b color model. As mentioned above, there are fluctuation changes in the boundaries of the object of interest among the features of the fire image in the video stream. Therefore, the spatial methods of identification and tracking in the case of fire as an object are ineffective. In the other side, the fluctuations of the object boundaries can be used in the cases of the weak separation of objects of interest or the availability of periodic micro noises.

The developed method belongs to the class of optical flow methods. It is based on the classic method of frame difference, which allows determining the gradients of changes in color values for each pixel.

Fig. 6. The result of image segmentation based on the L*a*b color model

The discrete binary function $\triangle(x, y, t)$ of the color values differences (x, y, t) for the pixel with the coordinates (x, y) for the frame at the time t can be written in the form

$$\forall t > 0, \forall x \in [0, l] > 0, \forall y \in [0, h] :$$

$$\Delta(x, y, t+1) = \begin{cases} 1, \; |c(x, y, t) - c(x, y, t+1)| > T; \\ 0, \; |c(x, y, t) - c(x, y, t+1)| \leq T; \end{cases} \quad (1)$$

where T is some threshold value, which defines the threshold response of the frames difference method; l, h are the image width and height respectively.

The Eq. (1) is used for two frames difference. In practice, a three frames difference is used very often. In the case of a sufficient amount of memory, such approach does not significantly increase the computing resources but increases the accuracy of the estimation of belonging to individual classes.

However, analyzing the existed video data set, it has been experimentally established that information which is obtained by comparing only two frames (current and next) is not sufficient to accurately verify the presence or absence of a fire in a video stream. As a way to solve this problem, we have proposed to use the method with regard to three frames. Such method allows with respect to additional features of the motion of the object and providing a more accurate estimation of its belonging to a certain class. In particular, morphological operations can be used to identify borders, reduce noise, analyze shapes and textures, etc. There are 4 most general morphological operations: closing ($\diamond$), opening ($\circ$), dilation ($\ominus$), erosion ($\oplus$).

These operations can be written in the form:

$$X \Diamond H = (X \oplus H) \ominus H \tag{2}$$

$$X \circ H = (X \ominus H) \oplus H \tag{3}$$

$$X \ominus H = \left\{ (x, y) : H_{(x,y)} \subseteq X \right\} \tag{4}$$

$$X \oplus H = \left\{ (x, y) : H_{(x,y)} \cap X \neq \emptyset \right\} \tag{5}$$

where $H \subseteq \mathbf{R}^2$ is a structural element; X is an original image; $H_{(x, y)}$ is a set H to vector $(x, y) \in \mathbf{R}^2$.

During the progress of the development of the ROI generation approach, using the binarization and operation of opening to reduce very small areas and possible artifacts of the image (especially in the case of the video cameras with low-resolution) was proposed. The results of the noise reduction is shown in Fig. 7.

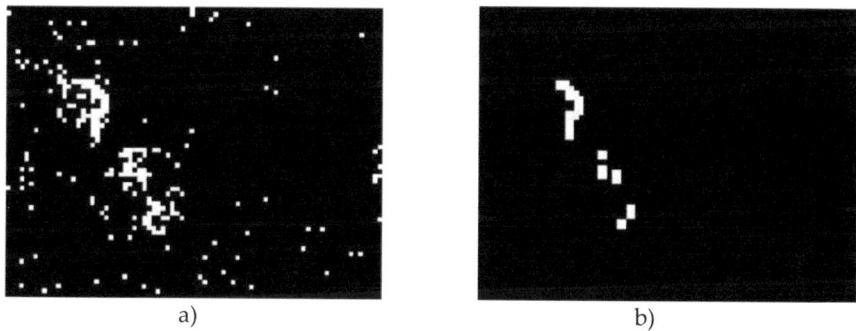

a)   b)

Fig. 7. The example of noise reduction based on the morphological operations: (a) the original binary image and (b) the image after applying the opening operation

5 Experiment Results

The experimental researches were performed based on the own dataset, which is consisted of 10 videos with a flame. Some of these videos (5 videos) were freely available on the Internet, some videos were filmed by authors in different conditions.

Total video duration was 16 min 46 s. We have obtained next results: the accuracy rate is 64.3% and the completeness rate is 99.9%. Figure 8 shows the results of the ROI generation. Figure 8(a) and (c) show the frames received from a video sequence, and Fig. 8(b) and (d) show the results of their processing by the proposed core generator of hypotheses.

a)   b)   c)   d)

Fig. 8. The results of proposed core generator of hypotheses based on video sequences in real time

The effectiveness of the new model of the hypothesis generator is determined by the decrease of the errors number in determining the ROI. Moreover, due to the principle of maximizing the index of complexitivity, it was possible to reduce the area of the ROI that includes the flame image. As a result, further ROI analysis we will already have smaller input data with smaller dimensions, which it can significantly reduce the computing complexity.

6   Conclusions

In the paper, the real-time flame detection approach using hypotheses generating techniques is proposed. And as a result, we have obtained reducing the time delay of operation of the common flame tracking procedures. It should be noted that reducing the area of ROI is also useful in cases of further analysis and prediction of such process as burning. For example, the estimation of temperature parameters will be more accurate, since the number of pixels that do not determine the flame will be significantly reduced. The computational experiments based on both benchmarks and real data sets have confirmed the effectiveness of the developed approach.

References

1. Chang, K., Liu, P., Yu, C.: Design of real-time video streaming and object tracking system for home care services. In: Proceedings of the 2016 IEEE International Conference on Consumer Electronics-Taiwan (ICCE-TW), Nantou, pp. 1–2 (2016). https://doi.org/10.1109/icce-tw.2016.7521004
2. Paul, G.V., Beach G.J., Cohen, C.J.: A realtime object tracking system using a color camera. In: Proceedings of the 30th Applied Imagery Pattern Recognition Workshop (AIPR 2001). Analysis and Understanding of Time Varying Imagery, Washington, DC, USA, pp. 137–142 (2001). https://doi.org/10.1109/aipr.2001.991216
3. Hung, M., Chang, C., Chen, J., Lin, J., Liao, T.: Design and implementation of real-time object tracking system using the Gaussian motion model and the otsu algorithm. In: Proceedings of the 2009 Second International Symposium on Knowledge Acquisition and Modeling, Wuhan, pp. 140–143 (2009). https://doi.org/10.1109/kam.2009.106

4. Lyu, C., Chen, H., Jiang, X., Li, P., Liu, Y.: Real-time object tracking system based on field-programmable gate array and convolution neural network. Int. J. Adv. Robot. Syst. **14**(1), 17–29 (2017). https://doi.org/10.1177/1729881416682705
5. Růžička, M., Mašek, P.: Real time object tracking based on computer vision. In: Březina T., Jabloński R. (eds.) Mechatronics 2013, pp. 591–598. Springer, Cham (2014). https://doi.org/10.1007/978-3-319-02294-9_75
6. Shen, D., Chen, X., Nguyen, M., Yan, W.Q.: Flame detection using deep learning. In: Proceedings of the 2018 4th International Conference on Control, Automation and Robotics (ICCAR), Auckland, pp. 416–420 (2018). https://doi.org/10.1109/iccar.2018.8384711
7. Muhammad, K., Ahmad, J., Mehmood, I., Rho, S., Baik, S.W.: Convolutional neural networks based fire detection in surveillance videos. IEEE Access **6**, 18174–18183 (2018). https://doi.org/10.1109/ACCESS.2018.2812835
8. Kim, O., Kang, D.-J.: Fire detection system using random forest classification for image sequences of complex background. Opt. Eng. **52**(6) (2013). https://doi.org/10.1117/1.oe.52.6.067202
9. Peleshko, D., Maksymiv, O., Rak, T., Voloshyn, O., Morklyanyk, B.: Core generator of hypotheses for real-time flame detecting. In: Proceedings of the 2018 IEEE Second International Conference on Data Stream Mining & Processing (DSMP), Lviv, Ukraine, pp. 455–458 (2018). https://doi.org/10.1109/dsmp.2018.8478446
10. binti Zaidi, N.I., binti Lokman, N.A.A., bin Daud, M.R., Achmad, H., Chia, K.A.: Fire recognition using RGB and YCBCR color space. ARPN J. Eng. Appl. Sci. **10**(21), 9786–9790 (2015)
11. Patel, P., Tiwari, S.: Flame detection using image processing techniques. Int. J. Comput. Appl. **58**(18), 13–16 (2012). https://doi.org/10.5120/9381-3817
12. Chen, T.-H., Kao, Ch.-L., Chang, S.-M.: An intelligent real-time fire-detection method based on video processing. In: Proceedings of the IEEE 37th Annual 2003 International Carnahan Conference on Security Technology, Taipei, Taiwan, pp. 104–111 (2003). https://doi.org/10.1109/ccst.2003.1297544
13. Rossi, L., Akhloufi, M., Tison, Y.: On the use of stereovision to develop a novel instrumentation system to extract geometric fire fronts characteristics. Fire Saf. J. **46**(1–2), 9–20 (2011). https://doi.org/10.1016/j.firesaf.2010.03.001
14. Prema, C.E., Vinsley, S.S., Suresh, S.: Multi feature analysis of smoke in YUV color space for early forest fire detection. Fire Technol. **52**(5), 1319–1342 (2016). https://doi.org/10.1007/s10694-016-0580-8
15. Çelik, T., Özkaramanlı, H., Demirel H.: Fire and smoke detection without sensors: image processing based approach. In: Proceedings of the 15th European Signal Processing Conference, Poznan, Poland, pp. 1794–1798 (2007). ISBN 978-839-2134-04-6
16. (Chao-Ho) Chen, T.-H., Wu, P.-H., Chiou, Y.-C.: An early fire-detection method based on image processing. In: Proceedings of the International Conference on Image Processing (ICIP), vol. 3, pp. 1707–1710 (2004). https://doi.org/10.1109/icip.2004.1421401
17. Celik, T.: Fast and efficient method for fire detection using image processing. ETRI J. **32**(6), 881–890 (2010). https://doi.org/10.4218/etrij.10.0109.0695
18. Rinsurongkawong, S., Ekpanyapong, M., Dailey, M.N.: Fire detection for early fire alarm based on optical flow video processing. In: Proceedings of the 9th International Conference on Electrical Engineering/Electronics, Computer, Telecommunications and Information Technology, Phetchaburi, Thailand, pp. 1–4 (2012). https://doi.org/10.1109/ecticon.2012.6254144

19. Guruh, F.S., Fajrian, N.A., Catur, S., Ricardus, A.P., Pulung, N.A.: Multi color feature, background subtraction and time frame selection for fire detection. In: Proceedings of the 2013 International Conference on Robotics, Biomimetics, Intelligent Computational Systems, Jogjakarta, Indonesia, pp. 115–120 (2013). https://doi.org/10.1109/robionetics. 2013.6743589

20. Toreyin, B., Dedeoglu, Y., Cetin, A.: Flame detection in video using hidden Markov models. In: Proceedings of the IEEE International Conference on Image Processing, Genova, Italy, vol. 2, pp. 213–216 (2005). https://doi.org/10.1109/icip.2005.1530284

21. Liu, Z.-G., Yang, Y., Ji, X.-H.: Flame detection algorithm based on a saliency detection technique and the uniform local binary pattern in the YCbCr color space. Signal, Image Video Process. **10**(2), 277–284 (2016). https://doi.org/10.1007/s11760-014-0738-0

22. Toreyin, B.U., Dedeoglu, Y., Cetin, A.E.: Wavelet based real-time smoke detection in video. In: Proceedings of the 13th European Signal Processing Conference, Antalya, Turkey, pp. 1–4 (2005). ISBN 978-160-4238-21-1

23. Araya, L.V., Espada, N., Tosini, M., Leiva, L.: Simple detection and classification of road lanes based on image processing. Int. J. Inf. Technol. Comput. Sci. (IJITCS) **10**(8), 38–45 (2018). https://doi.org/10.5815/ijitcs.2018.08.06

24. Hu, Z., Dychka, I., Sulema, Y., Valchuk, Y., Shkurat, O.: Method of medical images similarity estimation based on feature analysis. Int. J. Intell. Syst. Appl. (IJISA) **10**(5), 14–22 (2018). https://doi.org/10.5815/ijisa.2018.05.02

25. Chaudhary, A.S., Chaturvedi, D.K.: Analyzing defects of solar panels under natural atmospheric conditions with thermal image processing. Int. J. Image, Graph. Signal Process. (IJIGSP) **10**(6), 10–21 (2018). https://doi.org/10.5815/ijigsp.2018.06.02

26. Al-Ameen, Z., Zaman, A.H.: A low-complexity algorithm for contrast enhancement of digital images. Int. J. Image Graph. Signal Process. (IJIGSP) **10**(2), 60–67 (2018). https://doi.org/10.5815/ijigsp.2018.02.07

27. Gourav, Sharma, T., Singh, H.: Computational approach to image segmentation analysis. Int. J. Mod. Educ. Comput. Sci. (IJMECS), **9**(7), 30–37 (2017). https://doi.org/10.5815/ijmecs. 2017.07.04

28. Al-Othman, A.B., Al-Ameen, Z., Sulong, G.B.: Improving the MRI tumor segmentation process using appropriate image processing techniques. Int. J. Image, Graph. Signal Process. (IJIGSP) **6**(2), 23–29 (2014). https://doi.org/10.5815/ijigsp.2014.02.03

29. Youssef, K., Woo, P.-Y.: Difference of the absolute differences – A new method for motion detection. Int. J. Intell. Syst. Appl. (IJISA) **4**(9), 1–14 (2012). https://doi.org/10.5815/ijisa. 2012.09.01

Impact of Ground Truth Annotation Quality on Performance of Semantic Image Segmentation of Traffic Conditions

Vlad Taran[⊠], Yuri Gordienko, Alexandr Rokovyi, Oleg Alienin, and Sergii Stirenko

National Technical University of Ukraine "Igor Sikorsky Kyiv Polytechnic Institute", Kyiv, Ukraine
vladtkv@gmail.com

Abstract. Preparation of high-quality datasets for the urban scene under-standing is a labor-intensive task, especially, for datasets designed for the autonomous driving applications. The application of the coarse ground truth (GT) annotations of these datasets without detriment to the accuracy of semantic image segmentation (by the mean intersection over union—mIoU) could simplify and speedup the dataset preparation and model fine tuning before its practical application. Here the results of the comparative analysis for semantic segmentation accuracy obtained by PSPNet deep learning architecture are presented for fine and coarse annotated images from Cityscapes dataset. Two scenarios were investigated: scenario 1—the fine GT images for training and prediction, and scenario 2—the fine GT images for training and the coarse GT images for prediction. The obtained results demonstrated that for the most important classes the mean accuracy values of semantic image segmentation for coarse GT annotations are higher than for the fine GT ones, and the standard deviation values are vice versa. It means that for some applications some unimportant classes can be excluded and the model can be tuned further for some classes and specific regions on the coarse GT dataset without loss of the accuracy even. Moreover, this opens the perspectives to use deep neural networks for the preparation of such coarse GT datasets.

Keywords: Deep learning · Semantic segmentation · Accuracy · Cityscapes

1 Introduction

The current tendency in semantic image segmentation of traffic road conditions is making high quality images labeling to produce fine ground truth (GT) annotations for training and testing deep learning networks [1–3]. That stage of labeling is especially difficult, as it can take several hours to fine annotate single image. Considering the fact, that autonomous driving system can be used in different cities or regions, the model tuning is necessary to achieve good segmentation accuracy. Despite the availability of the proprietary (like Daimler AG [4], Tesla [5], etc.) and open-source datasets of traffic road conditions (like Cityscapes [6], KITTI [7], CamVid [8, 9], DUS [10], etc.), the lack of sample datasets for specific application regions hardens the model tuning stage.

© Springer Nature Switzerland AG 2020
Z. Hu et al. (Eds.): ICCSEA 2019, AISC 938, pp. 183–193, 2020.
https://doi.org/10.1007/978-3-030-16621-2_17

In the context of self-driving cars, for the some components like automatic braking and anti-collision systems, it may be not necessary to annotate all possible object classes for semantic segmentation tasks. It could be enough to select main classes such as a road, a car, a person/pedestrian, traffic lights, and traffic signs. It could also not necessary to make high quality annotations of the traffic condition while preparing training or testing datasets. It is assumed that coarse objects shapes may be suitable for the common object identification and localization on the road. Such a coarse annotation could speedup a datasets preparation for the model adaptation to the different use cases and model fine tuning before its practical application. But the problem is that the impact of decreasing ground truth annotation quality on performance of semantic image segmentation of traffic conditions was not investigated yet, especially for different classes and in the presence of the specific application regions, for example, for cities and countries with various architecture and urban life styles.

The main aim of this paper is to investigate the accuracy of semantic image segmentation by classes and the impact of the GT annotation quality on the performance of semantic image segmentation in the context of urban scene and traffic conditions understanding. The Sect. *2. Background and Related Work* gives the brief outline of the state of the art in urban scene datasets, networks used, and specific problems of dataset annotation. The Sect. *3. Experimental and Computational Details* contains the description of the experimental part related with the selected dataset, network, scenario, and metrics used. The Sect. *4. Results* reports about the experimental results obtained, the Sect. *5. Discussion* is dedicated to the discussion of these results, and Sect. *6. Conclusions* summarizes the lessons learned.

2   Background and Related Work

Currently, the street scenes are typically monitored by multiple input modalities and the obtained results are represented in the various relevant datasets:

- A stereo camera, traffic light camera, localization camera—Daimler AG - Research & Development proprietary solution and dataset [4];
- 8 surround cameras (in addition to 12 ultrasonic sensors and forward-facing radar; the radar can see vehicles through heavy rain, fog or dust)—Tesla Autopilot proprietary solution and dataset [5].
- A stereo camera rig that gives a diverse set of stereo video sequences recorded in street scenes from 50 cities, with high quality pixel-level annotations of 5 000 frames in addition to a larger set of 20 000 weakly annotated frames,—Cityscapes public dataset (https://www.cityscapes-dataset.com/) [6];
- Two stereo camera rigs (one for grayscale and one for color), 3D laser scanner, GPS measurements and IMU accelerations from a combined GPS/IMU system, timestamps—KITTI public dataset [7];
- Traffic cam videos—the open access image database Cambridge-driving Labeled Video Database (CamVid) (ftp://svr-ftp.eng.cam.ac.uk/pub/eccv/) [8, 9];
- Stereo image pairs—Daimler Urban Segmentation (DUS) public dataset [10].

Recently, several new networks appeared, like PSPNet [11], ICNet [12], DeepLab [13] and many others, and demonstrated the high performance with regard to the accuracy of semantic image segmentation (by the mean intersection over union—mIoU) and speed of prediction (by the inference time) of various object classes, for example, for some of these datasets.

In addition to this, there is the comparative analysis of ICNet and PSPNet performance with regard to several subsets of Cityscapes dataset including stereo-pair images taken by left and right cameras for different cities [14]. It was found that the distributions of the mIoU values for each city and channel are asymmetric, long-tailed, and have many extreme outliers, especially for PSPNet network in comparison to ICNet network. The results obtained demonstrated the different sensitivity of these networks to: (1) the local street view peculiarities (among different cities), (2) the change of viewing angle on the same street view image (right and left data channels). The differences with regard to the local street view peculiarities should be taken into account during the targeted fine tuning the models before their practical applications. For both networks, the information from the additional right data channel is radically different from the left channel, because it is out of the limits of statistical error in relation to the mIoU values. It means that the traffic stereo pairs can be effectively used not only for depth calculations (as it is usually used), but also as an additional data channel that can provide much more information about scene objects than a simple duplication of the same street view images.

In the view of the aforementioned background and the different quality of annotated images in various datasets, the aim of this work stated in Sect. 1. Introduction was formulated as impact study of ground truth annotation quality on performance of semantic image segmentation of traffic conditions. That is why we have tried to estimate the difference between accuracy values while applying fine and coarse annotated ground truth images.

3 Experimental and Computational Details

Here PSPNet network was used as semantic image segmentation model. It took the first place in ImageNet [15] scene parsing challenge in 2016, PASCAL VOC [16] and Cityscapes semantic segmentation benchmarks. The goal of this network was to extract complex scene context features. While solving the several common issues such as mismatched relationship, confusion categories and inconspicuous classes, it introduce the state-of-art pyramid pooling module which extend the pixel-level features to make the final prediction more reliable. This module separates the feature map into four different pyramid scales and form the pooled representation for different locations. It allows collecting levels of information, more representative than the global pooling.

For this investigation images from Cityscapes dataset were used that correspond to several German cities like Frankfurt (267 images), Munster (174 images), Lindau (59 images) with fine GT (Fig. 1), and Erlangen (265), Konigswinter (118), Troisdorf (138) with coarse GT (Fig. 2).

Because of the large amount of the recorded data, Cityscapes dataset provide fine and coarse annotated images. The fine image annotation means pixel-precision annotation for 30 classes for 3 cities, containing 5 000 images in total. The coarse image annotation means line-precision annotation for the remaining 23 cities, containing 20000 images in total, where a single image was selected every 20 s or 20 m driving distance and spending less than 7 min of the annotation time per image [17].

Fig. 1. Example of the original image from Cityscapes dataset (left) and its fine ground truth with pixel-wise precision (right) for Munster.

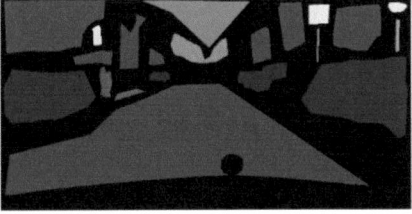

Fig. 2. Example of the original image from Cityscapes dataset (left) and its coarse ground truth with line-bounded precision (right) for Troisdorf.

PSPNet network, pre-trained on Cityscapes dataset, was applied to these data subsets to measure the image prediction accuracy (mIoU) and compare them for fine and coarse ground truth annotations for the corresponding cities. The whole workflow was carried out according to the scenarios 1 and 2 described in the Table 1, namely, for scenario 1 the fine GT images were used for training and prediction, and for scenario 2 the fine GT images were used for training and the coarse GT images were used for prediction. It should be noted that scenario 1 was already implemented in our previous work [14] and scenarios 3 4 are under work right now and will be published elsewhere [18]. In this, work the following main classes, which could be used in the autonomous driving tasks, were selected: a road, a car, a person/pedestrian, traffic lights and signs classes.

Table 1. Scenarios for comparison of impact of ground truth annotation quality.

Scenario	Training	Prediction	Implementation
1	Fine	Fine	[14], this work, [18]
2	Fine	Coarse	this work, [18]
3	Coarse	Fine	[18]
4	Coarse	Coarse	[18]

All experiments were conducted on the basis of TensorFlow framework [19] on the workstation with the single NVIDIA GTX 1080 Ti GPU card with CUDA 8.0 and CUDNN 7.

4 Results

The accuracy of semantic image segmentation was measured by the mean intersection over union (mIoU) parameter. It was used to compare the per-class prediction performance for images with the various segmentation quality (fine and coarse GT annotations) for the following classes: a road, a car, a person/pedestrian, traffic lights and signs. The results of accuracy measurements obtained by PSPNet network for different classes and quality of semantic image segmentation are shown for the corresponding cities in Fig. 3.

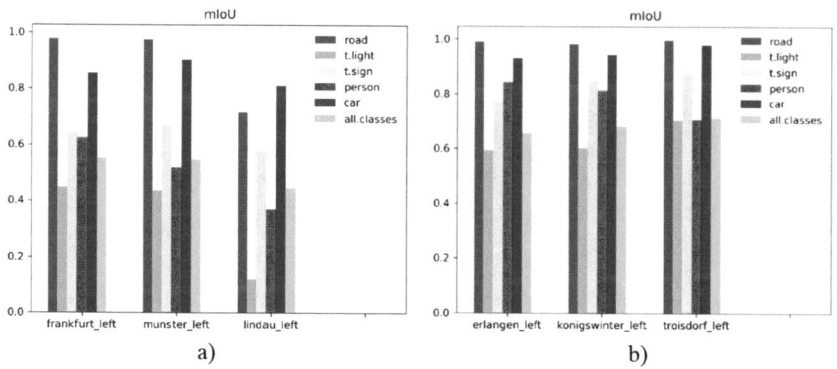

Fig. 3. Per-class accuracy (mIoU) values for each city with fine (a) and coarse (b) GT obtained by PSPNet network.

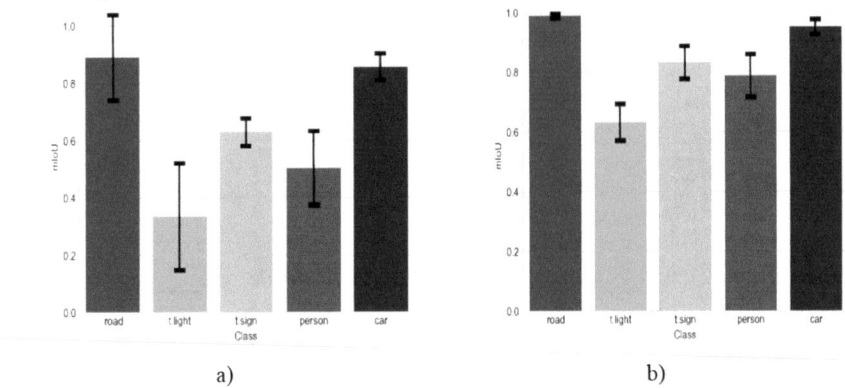

a)   b)

Fig. 4. Mean and standard deviations (shown by whiskers) of per-class accuracy (mIoU) values for all cities with fine (a) and coarse (b) GT.

The mean and standard deviations of per-class accuracy (mIoU) values for all cities are shown in Fig. 4 and listed in Table 2 for fine (Fig. 4a) and coarse (Fig. 4b) GT.

Table 2. The mean and standard deviations (shown by whiskers) of per-class accuracy (mIoU) values for all cities (Fig. 4).

Object class	Mean		Standard deviation	
	Fine	Coarse	Fine	Coarse
Road	0.89	0.99	0.14	0.007
Car	0.85	0.95	0.04	0.02
Traffic sign	0.62	0.83	0.05	0.05
Person	0.5	0.78	0.12	0.07
Traffic light	0.33	0.63	0.19	0.06

The common tendency is that the means of per-class accuracy (mIoU) for all cities with fine GT are lower than for the coarse GT, and it is vice versa for the standard deviations of mIoU.

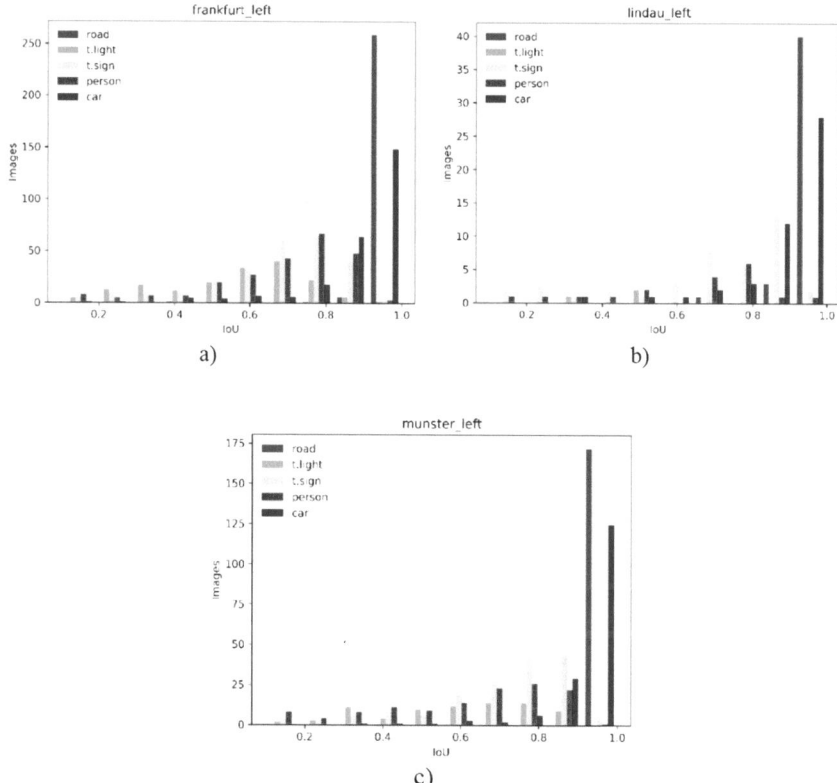

Fig. 5. Distributions of accuracy values (mIoU) for set of left images with fine ground truth from cities: (a) Frankfurt, (b) Lindau and (c) Munster.

The presented plots for fine and coarse GT allow us to make their comparison with regard to the accuracy (mIoU). Variations of mIoU (Fig. 3) among classes with fine and coarse GT can be very big, for example, for fine GT (Fig. 3a) the mIoU was equal to 0.5 ± 0.07, and the worst segmented was the traffic light class with the mIoU equal to 0.33 ± 0.19. The best segmented classes were the road and car ones, with the mIoU equal to 0.89 ± 0.14 and 0.85 ± 0.04, respectively (Fig. 4a, Table 2). The mIoU for the traffic sign class was equal to 0.62 ± 0.05, and for the person class it was equal to 0.5 ± 0.12, except for Lindau city, where it was equal to 0.38.

As to the coarse GT, the mIoU was greater than 0.6 for all classes (Fig. 3b). The road and car classes were the most accurately segmented ones with the mIoU equal to 0.99 ± 0.007 and 0.95 ± 0.02, respectively (Fig. 4b, Table 2). The person class accuracy was equal to 0.78 ± 0.07, except for Troisdorf city, where it was equal to 0.71 (Fig. 3b). The traffic light and sign classes have accuracy equal to 0.63 ± 0.06 and 0.83 ± 0.05 respectively.

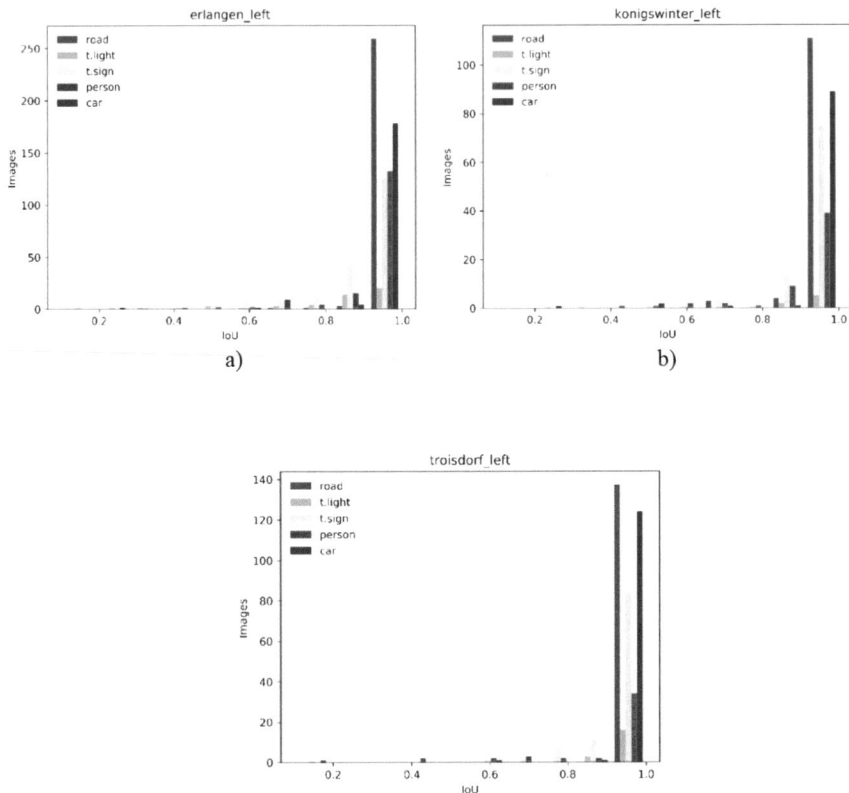

Fig. 6. Distributions of accuracy values (mIoU) for set of left images with coarse ground truth from cities: (a) Erlangen, (b) Konigswinter and (c) Troisdorf.

The distributions of the IoU values were determined for each type of GT annotation: fine—(Frankfurt, Munster, Lindau); coarse – (Erlangen, Konigswinter, Troisdorf), and presented in Figs. 5 and 6, respectively.

The qualitative analysis of distributions for the fine GT allows us to make the assumptions that they are asymmetric, and long-tailed (Fig. 5). In contrast, the distributions for coarse ground truth are short-tailed and have fewer outliers (Fig. 6).

5   Discussion

The results obtained allow us to compare the accuracy values of semantic image segmentation for images with different GT annotations: fine and coarse. While coarse annotated images are less detailed compared with fine annotated images, they still allowed us to identify, localize, and segment rough shapes of the objects from the preselected classes in the context of urban scenes. It is assumed that it can be suitable for autonomous car task where the object localization may be much more important than getting the exact shape of the object.

In general, the IoU values for coarse GT annotated images were greater than for the fine GT annotated ones. It can be explained by the assumption that objects in coarse GT images have more rough shapes (than in the fine GT images) that are more close to the correspondent pre-trained primitives inside the model itself and the accuracy is measured regarding their overlapping. In reverse, the objects in fine GT images have more detailed shapes (than in coarse GT images) that are far from to the correspondent pre-trained primitives inside the model itself.

But it should be realized that the presented results are based on the coarse segmentation analysis for only three cities from Cityscapes dataset. Therefore, the reliability of these results for other cities and countries should be verified by the wider range of images on other cities and countries that is under work right now. In addition, the current model was trained on Cityscapes dataset with fine GT annotations, namely for scenarios 1 and 2 (Table 1). And it is worth to estimate the segmentation accuracy for scenario 3 (Table 1), i.e. for the cases of training on coarse GT annotated images and prediction on fine GT images, and scenario 4 (Table 1), i.e. for the cases of training on coarse GT annotated images and prediction on coarse GT images. But these results are under work yet and will be reported in details elsewhere [18].

Considering the fact that creating quality datasets for urban scene understanding is a labor-intensive task, especially, for datasets designed for autonomous driving applications and model fine tuning before their practical applications, we proposed to limit object classes to only necessary ones. The further question can be raised about excluding some irrelevant classes (like a sky, a building, a vegetation and sidewalk) in some applications like emergency braking and anti-collision systems [20–22] and other applications [23–26].

The accuracy comparison for fine and coarse GT annotated images allow us to raise the question that deep neural networks may be used for creation of coarse GT annotated datasets which can be edited and used for fine tuning the pre-trained models for the specific application regions.

The additional promising development of this work could be related with the more rational segmentation where an object shape could be segmented not in the fine pixel-wise way, but by the means of only specific coarse features using lines, curves, arcs, etc. For example, such rational per-class segmentation for the road could be implemented by complex lines or arcs, for cars—by polygons, for traffic signs—by geometric primitives like ellipses and polygons, etc. But for the complex objects like pedestrians, some optimal methods should be investigated to distinguish moving people, as potentially more dangerous, and standing ones.

6   Conclusions

The comparative analysis of semantic segmentation accuracy was performed in this work. The results were obtained by PSPNet deep learning architecture for fine and coarse annotated images from Cityscapes dataset. The four possible scenarios of training/testing on fine/coarse GT images were proposed and two of them were considered: scenario

1—the fine GT images for training and prediction, and scenario 2—the fine GT images for training and the coarse GT images for prediction.

The results shown that for the most important classes (road, car, person/pedestrian, traffic light, traffic sign) mean accuracy values of semantic image segmentation for coarse GT annotations are higher than for fine GT ones, and the standard deviations are vice versa. Despite the coarse annotated images are less detailed compared with fine annotated ones, they still allowed us to identify, localize, and segment the rough shapes of the objects from the preselected classes in the context of urban scenes. Moreover, it was possible without detriment to the accuracy of semantic image segmentation (mIoU). It means that for some applications (like emergency braking and anti-collision systems) some unimportant classes (like sky, building, vegetation, sidewalk, etc.) can be excluded and the model can be tuned further for some classes and specific regions on the coarse GT dataset without loss of the accuracy even. Moreover, this opens the perspectives to use deep neural networks for the preparation of such coarse GT datasets, and the usage of the more rational segmentation where an object shape could be segmented not in a fine pixel-wise way, but by the means of only specific coarse features using lines, curves, arcs, etc.

References

1. Chouhan, S.S., Kaul, A., Singh, U.P.: Image segmentation using computational intelligence techniques: review. In: Archives of Computational Methods in Engineering, pp. 1–64 (2018)
2. Yu, H., Yang, Z., Tan, L., Wang, Y., Sun, W., Sun, M., Tang, Y.: Methods and datasets on semantic segmentation: a review. Neurocomputing **304**, 82–103 (2018)
3. Garciagarcia, A., Ortsescolano, S., Oprea, S., Villenamartinez, V., Martinezgonzalez, P., Garciarodriguez, J.: A survey on deep learning techniques for image and video semantic segmentation. Appl. Soft Comput. **70**, 41–65 (2018)
4. Franke, U., Pfeiffer, D., Rabe, C., Knöppel, C., Enzweiler, M., Stein, F., Herrtwich, R.G.: Making bertha see. In: ICCV Workshops, pp. 214–221 (2013)
5. Autopilot: Full Self-Driving Hardware on All Cars. Tesla Motors. https://www.tesla.com/autopilot. Accessed 26 Nov 2018
6. Cordts, M., Omran, M., Ramos, S., Rehfeld, T., Enzweiler, M., Benenson, R., Franke, U., Roth, S., Schiele, B.: The cityscapes dataset for semantic urban scene understanding. In: Proceedings of the IEEE Conference on Computer Vision and Pattern Recognition, pp. 3213–3223 (2016)
7. Geiger, A., Lenz, P., Stiller, C., Urtasun, R.: Vision meets robotics: the KITTI dataset. Int. J. Robot. Res. **32**(11), 1231–1237 (2013)
8. Brostow, G.J., Fauqueur, J., Cipolla, R.: Semantic object classes in video: a high-definition ground truth database. Pattern Recognit. Lett. **30**(2), 88–97 (2009)
9. Brostow, G.J., Shotton, J., Fauqueur, J., Cipolla, R.: Segmentation and recognition using structure from motion point clouds. In: European Conference on Computer Vision, pp. 44–57 (2008)
10. Scharwächter, T., Enzweiler, M., Franke, U., Roth, S.: Efficient multi-cue scene segmentation. In: German Conference on Pattern Recognition, pp. 435–445 (2013)
11. Zhao, H., Shi, J., Qi, X., Wang, X., Jia, J.: Pyramid scene parsing network. In: IEEE Conference on Computer Vision and Pattern Recognition (CVPR), pp. 2881–2890 (2017)

12. Zhao, H., Qi, X., Shen, X., Shi, J., Jia, J.: ICNet for real-time semantic segmentation on high-resolution images. arXiv preprint arXiv:1704.08545 (2017)
13. Chen, L.C., Papandreou, G., Schroff, F., Adam, H.: Rethinking atrous convolution for semantic image segmentation. arXiv preprint arXiv:1706.05587 (2017)
14. Taran, V., Gordienko, N., Kochura, Y., Gordienko, Y., Rokovyi, A., Alienin, O., Stirenko, S.: Performance evaluation of deep learning networks for semantic segmentation of traffic stereo-pair images. In: Proceedings of the 19th International Conference on Computer Systems and Technologies, pp. 73–80. ACM, September 2018
15. Deng, J., Dong, W., Socher, R., Li, L.J., Li, K., Fei-Fei, L.: Imagenet: a large-scale hierarchical image database. In: Computer Vision and Pattern Recognition 2009, pp. 248–255. IEEE (2009)
16. Everingham, M., Van Gool, L., Williams, C.K., Winn, J., Zisserman, A.: The pascal visual object classes (voc) challenge. Int. J. Comput. Vis. **88**(2), 303–338 (2010)
17. Cordts, M.: Understanding cityscapes: efficient urban semantic scene understanding. Doctoral dissertation, TechnischeUniversität (2017)
18. Taran, V., Gordienko, N., Kochura, Y., Gordienko, Y., Rokovyi, A., Alienin, O., Stirenko, S.: Optimization of ground truth annotation quality to simplify semantic image segmentation of traffic conditions (2018, submitted)
19. Abadi, M., et al.: Tensorflow: a system for large-scale machine learning. Oper. Syst. Des. Implement. **16**, 265–283 (2016)
20. Chae, H., Kang, C. M., Kim, B., Kim, J., Chung, C. C., and Choi, J. W.: Autonomous braking system via deep reinforcement learning. arXiv preprint arXiv:1702.02302, (2017)
21. Flores, C., Merdrignac, P., de Charette, R., Navas, F., Milanés, V., Nashashibi, F.: A cooperative car-following/emergency braking system with prediction-based pedestrian avoidance capabilities. IEEE Trans. Intell. Transp. Syst. **99**, 1–10 (2018)
22. Müller, M., Botsch, M., Böhmländer, D., Utschick, W.: Machine learning based prediction of crash severity distributions for mitigation strategies. J. Adv. Inf. Technol. **9**(1), 15–24 (2018)
23. Chaudhary, A.S., Chaturvedi, D.K.: Analyzing defects of solar panels under natural atmospheric conditions with thermal image processing. Int. J. Image, Graph. Signal Process. (IJIGSP) **10**(6), 10–21 (2018). https://doi.org/10.5815/ijigsp.2018.06.02
24. Bouzid-Daho, A., Boughazi, M.: Segmentation of abnormal blood cells for biomedical diagnostic aid. Int. J. Image, Graph. Signal Process. (IJIGSP) **10**(1), 30–35 (2018). https://doi.org/10.5815/ijigsp.2018.01.04
25. Memon, S., Bhatti, S., Thebo, L.A., Talpur, M.M.B., Memon, M.A.: A video based vehicle detection, counting and classification system. Int. J. Image, Graph. Signal Process. (IJIGSP) **10**(9), 34–41 (2018). https://doi.org/10.5815/ijigsp.2018.09.05
26. Sahoo, R.K., Panda, R., Barik, R.C., Panda, S.N.: Automatic Dead zone detection in 2-D leaf image using clustering and segmentation technique. Int. J. Image, Graph. Signal Process. (IJIGSP) **10**(10), 11–30 (2018). https://doi.org/10.5815/ijigsp.2018.10.02

Mechanical Experimental Platform Construction Based on BOPPPS Model

Mengya Zhang[1(✉)], Qingying Zhang[1], Ziye Wang[2],
and ChenSheng Huang[1]

[1] Wuhan University of Technology, Wuhan 430063, People's Republic of China
kmno40311@163.com
[2] Wuhan Institute of Physical Education,
Wuhan 430079, People's Republic of China

Abstract. With the continuous expansion of the demand for applied engineering talents in the society, in order to help students to develop comprehensive practical ability and innovation awareness, it is necessary to carry out high-quality experimental teaching work. Considering the status quo of mechanical experimental teaching and the requirements of talent development, mechanical experimental teaching reform based on BOPPPS model has been put forward in this paper, and a mechanical experimental teaching management platform has been built to improve the overall level of mechanical laboratory teaching.

Keywords: Mechanical experimental teaching · Teaching reform ·
Teaching management platform · BOPPPS model

1 Introduction

The BOPPPS teaching model originated in North America and is a model for teachers to design courses. This model has a good effect in strengthening the reflective function and interaction of teaching. In the mechanical experiment teaching, the BOPPPS model is introduced to integrate the teachers, students and administrators into the teaching management platform, so that the teaching and learning interaction, teaching and management interaction are combined.

In the teaching system, the scientific design of teaching content, organization form and assessment method make the BOPPPS model cover the three links before, during and after the class, so that the teacher can control every link of the experimental teaching, which can evaluate the student's learning effect and the teacher's teaching effect conveniently, and the quality and efficiency can be improved from the aspects of teaching, learning and management.

© Springer Nature Switzerland AG 2020
Z. Hu et al. (Eds.): ICCSEEA 2019, AISC 938, pp. 194–204, 2020.
https://doi.org/10.1007/978-3-030-16621-2_18

2 The Status Quo of Experimental Teaching of Mechanical Specialty and the Requirements for Talent Development

Since February 2017, the Ministry of Education has actively promoted the construction of new engineering, in order to cultivate diversified and innovative excellent engineering talents [1]. Therefore, all colleges and universities are more focused on the practicality of students' knowledge and skills. As the cradle of technical talents, the teaching practice of undergraduate colleges still differs greatly from the actual needs of the society, and it cannot meet the needs of continuous development of professional technology [2]. For this problem, we must pay close attention to the construction of the universities experimental teaching to enhance the overall competitiveness of students [3].

2.1 Analysis of Problems in Mechanical Experimental Teaching

At present, the problems in mechanical experimental teaching include the following aspects:

(1) The experimental content is obsolete

In the experimental teaching of universities, most of the experimental contents are set by the experimental instructors themselves. On the one hand, due to limitations of space and funds, the experimental equipment is not updated in time, so that the experimental content cannot be close to the development level of modern science and technology. On the other hand, the teacher's teaching content relies too much on experimental textbooks, but the mechanical engineering guides on the market are seriously lacking. Therefore, the teaching content has not been updated. For the reason that, teachers need to constantly understand the actual needs of the project, update the experimental equipment and design experimental content as much as possible to meet the needs of the enterprise development.

(2) The experimental arrangement is unreasonable

Even if the experiment time is short, many teachers are still "pay attention to theory and ignore practice". The most of teachers spend a lot of time teaching the relevant knowledge points and experimental principles. The left time for the students to experiment is limited, which leads to the students lacking enough time to explore. Students can only complete the task according to the steps of experimental guidance or teachers' demonstration, there is no time for them to think and create, greatly reducing the quality of teaching.

(3) Single experimental means

The experimental types can be divided into demonstration type, verification type, comprehensive design type and research innovation type. Most of the mechanical experiments belong to the demonstration type and the verification type [4]. The students passively accept the teacher's experimental demonstration and imitate the experiment steps to complete. Creativity and comprehensive ability are not improved.

Therefore, experimental teaching should establish a new teaching mode that take students as the main body and take teachers as the mainstay, heuristic teaching should be adopted to give full play to students' participation and enthusiasm, so that students can change from imitation to innovation.

2.2   Requirements for Talent Development in Experimental Teaching

Under the background of new engineering, mechanical students must grasp theoretical knowledge and have strong practical ability. In the aspect of knowledge learning, students must master the principle and methods of mechanical design. Furthermore, in the aspect of application of skills, students should have the abilities of innovative thinking, independent research and development, design, and manufacturing [5]. Only in this way can we satisfy the market demand for talents. Specific requirements include the followings:

(1)   self-learning ability [6];
(2)   hands-on ability;
(3)   comprehensive analysis ability;
(4)   ability to combine theory with practice;
(5)   innovative thinking.

3   Reform of Mechanical Experiment Teaching Based on BOPPPS Model

In order to break the traditional experimental teaching method of "teacher demonstration, student imitation", change the teaching subject, give full play to the student's main position, the BOPPPS teaching mode has been introduced to help teachers carry out effective teaching reform. Combined with the characteristics of mechanical professional practice, the experimental teaching reform based on BOPPPS needs to be carried out from three aspects: experimental content, experimental organization form and experimental evaluation mode.

3.1   Overview of the BOPPPS Model

The BOPPPS model consists of six elements: bridge-in, objective, pre-assessment, participatory, post-assessment, and summary [7]. Bridge-in is to let students generate learning motivation by inducing students' curiosity [8]; Objective has pointed out that the requirements and levels should be achieved through learning, which is mainly based on the cognitive, emotional and skill categories of Bloom's learning objectives [9]; Pre-assessment is mainly used to measure learners' knowledge, so as to guide teachers' follow-up teaching arrangements; Participatory is mainly based on active learning strategies to make the learner deeply involved in the classroom to achieve the teaching goal; Post-assessment is mainly to evaluate the level of the learners' knowledge that compare with the teaching goal [10]; Summary is mainly to give teachers and learners a chance to reflect. Students reflect on what they have learned. Teachers reflect on the problems in this class and play the foundation for the next course.

3.2   The Reform of Experimental Teaching Content

Judging from the current teaching situation of mechanical professional experiment in China, experimental teaching content has been neglect. But based on the characteristics of mechanical professional, students' demand for experimental teaching is increasingly strong. The theoretical knowledge of mechanical major is very complicated. It is very difficult for students to understand theoretical knowledge. In the teaching process, it is often necessary to combine experiments to learn, which can deepen students' understanding and memory of professional knowledge, and also help students to apply what they have learned to solve specific problems [11]. Therefore, it is necessary to update experimental content and add new comprehensive design and innovative experiments.

3.3   The Reform of Experimental Organizational Form

Under the guidance of the BOPPPS model, the organization of the mechanical experimental course is divided into three stages, which is shown in Fig. 1. Before the experiment, students should complete preview and pre-assessment, understand the purpose of the experiment; course introduction, teaching objectives and participatory learning are in the process of experiment; After the experiment, post- assessment and summary should been implemented.

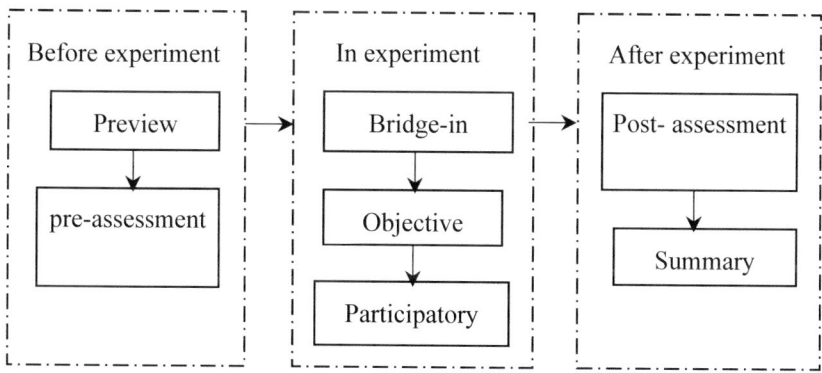

Fig. 1. Experimental teaching organization based on BOPPPS

(1) Before the experiment

Due to the tight time and heavy tasks of the experimental teaching, in order to give students more time to carry out the hands-on operation, the pre-assessment part of BOPPPS combined with preview is placed before the experiment. Before the experiment, students need to learn knowledge points and tutorials of mechanical experiment independently on the experimental teaching platform, and complete online pre-assessment, in order to enhance the students' self-learning ability while letting the teachers understand the level of knowledge of students. So that teacher can adjust teaching content to adapt the student's level.

(2) In the experiment

In the experiment, bridge-in of mechanical experiments can be combined with the actual project, which will lead students to think. At the same time, teacher should emphasize the teaching objectives of the experiment. On the one hand, combining with practice, students can understand theoretical knowledge better. On the other hand, students' ability to learn independently, analyze problems and solve problems can be improve and students' innovative thinking can be stimulated [12].

Students conduct experiments on a group basis, design experiments around the problem and complete the construction, commissioning and operation. This kind of experiment requires students to understand the operation method of experimental equipment while familiarizing with and mastering the basic theoretical knowledge, Moreover, It also requires students to integrate knowledge and practical experience, analyze and summarize the problems and determine the solution. Participatory experiments not only cultivate students' hands-on ability and collaborative spirit, but also stimulate students' creativity [13].

(3) After the experiment

Based on the BOPPPS model, after the experiment, students should carry out post-assessment and summary reflection. The post-test mainly includes the completion of the experimental report. The teacher can understand the learning effect of the students according to the post-test situation, find out the blind spots and the need to adjust and improve the teaching content and arrangement. Students can understand their own learning based on the results of the post-assessment, and do a good job review after class [14].

At the same time, for the professional characteristics of machinery, teachers can also provide follow-up learning resources for students to expand their knowledge, encourage students to participate in the national college student mechanical design competition, the national college students energy-saving emission reduction social practice and technology competition, the national university crane design competition, engineering training, independent innovation fund project, etc. So that students can continue to develop innovative thinking and exercise practical skills on the basis of experimental classes.

3.4 The Reform of Experimental Assessment Methods

Through the establishment of a comprehensive evaluation system for mechanical experiments, the whole process of students before, during and after the experiment was evaluated. The proportions were divided into:

(1) Pre-assessment accounted for 20%;
(2) Operation performance in the experiment accounted for 30%;
(3) The quality of the experimental report is 50%. The final weighted statistics give the results of the student's experimental course.

4   The Construction of Experimental Teaching Management Platform

The construction of teaching management platform is guided by deepening the reform of experimental teaching [15]. The reform of mechanical experiment teaching based on BOPPPS model needs to rely on the teaching management platform as the supporting system.

4.1   System Architecture

The mechanical professional experimental teaching management platform architecture mainly includes four aspects: support platform, basic resources layer, application services layer, and interactive layer, as shown in Fig. 2.

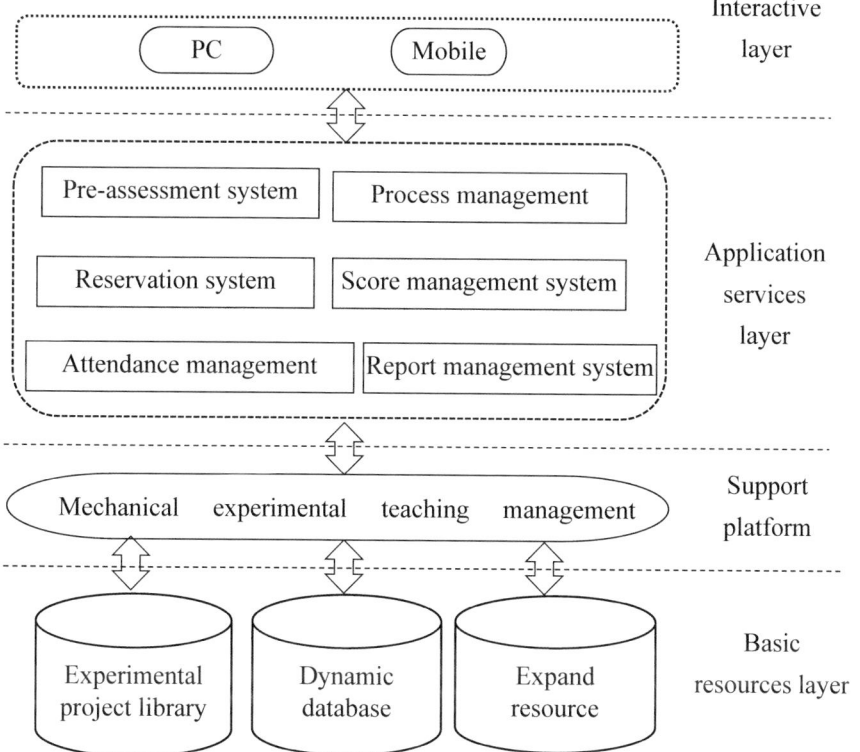

Fig. 2. Architecture design of mechanical experimental teaching platform

(1) Support platform construction

Supporting platform as a carrier, connect to the basic resource layer, for teachers to upload experimental-related learning materials, uplink application service layer, in order to support reservation system, pre-assessment system, attendance management system, process management system, score management system and report management system [16].

(2) Basic resource layer construction

The basic resources layer include three basic resource libraries: experimental project library, dynamic database, and expanded resource library [17]. The experimental project library contains all the compulsory experimental projects and open experimental projects of the mechanical experiment; the learning dynamic database can collect the dynamic data of the student learning process in real time to support the data analysis of the learning behavior; the expanded resource library contains various mechanical experimental software, competitions, exercises and papers.

(3) Application service development

Application services layer include reservation system, pre-assessment system, attendance management system, process management system, score management system and report management system, which can support online experiment release, configuration, management and real-time monitoring, etc.

(4) Interactive layer

The interactive layer is the support of the learning mode, which is used to support the change of the learning environment under the "Internet+" education method, and provides diverse learning platform support for students' self-learning and personalized learning [18].

4.2   The Function of Experimental Teaching Management Platform

The experimental teaching management includes experimental reservation system, experimental pre-assessment system, experimental attendance management, experimental process management system, experimental report management system, and experimental score management system, as shown in Fig. 3. The function realization of the experimental teaching management module can ensure the quality and order of the experimental teaching.

(1) Experimental reservation system

The teachers will carry out experimental arrangements according to the opening plan of the Academic Affairs Office before the experiment, open the teaching week, experimental projects and experimental workstations on the mechanical professional experimental teaching management platform. When the students pass the pre-assessment, they can freely choose the experimental time, the experimental workstation, and make an appointment for the relevant experiment.

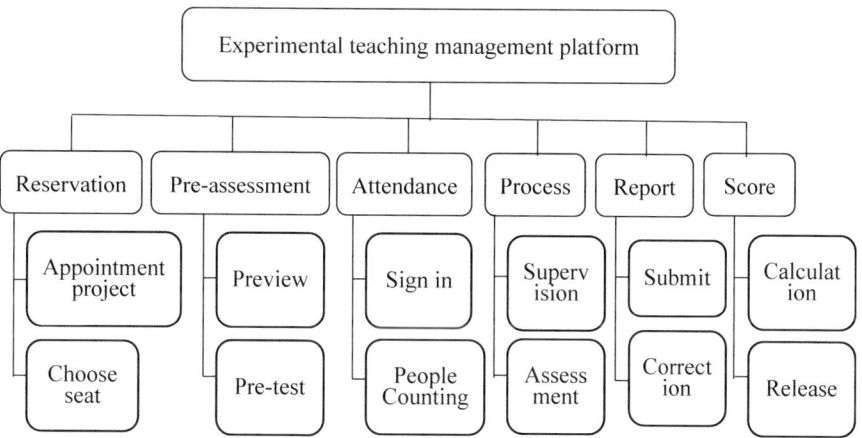

Fig. 3. Experimental teaching management platform functional architecture

(2) Experimental pre-assessment system

The students conduct self-learning according to the learning materials provided by the basic resource library (including mechanical related knowledge points, experimental instructions, experimental teaching videos, experimental equipment operation guides, etc.). And the teachers set pre-assessment for each experimental project on the platform, the students must conduct pre-set after they have finished preparation.

(3) Experimental attendance system

Students will go to the laboratory to make experiments according to the appointment time. The basic information of the students will be recorded by the student's credit card. After the card is swiped, the student's name, class, student number, experimental project name, experimental station, and experiment time are recorded in the system. The teacher can counts the number of students, the number of absentees, and the number of late arrivals according to the experimental attendance system.

(4) Process management system

By logging into the school's student experiment management system, students can effectively and timely view their experimental courses, including the time, location and content of the experimental course. Teachers can understand the usage information of experimental equipment at each workstation according to the experimental management system, directly score according to the performance of the student's experiments. Moreover, the teachers can interact with the students online and answer questions.

(5) Experimental report management system

After the experiment, the students upload the experimental report to the experimental report system. The teachers can view the student's experimental report online and make corrections. The students can also see the feedback result after the experiment, and can download the revised experimental report to analyze the problems. It is good for students' continuous improvement.

(6) Experimental score system

The experimental evaluation method based on the reform of BOPPPS model is divided into three parts, which are 20% of pre-assessment results, 30% of experimental operation results, and 50% of experimental report results. The pre-assessment results are directly imported through the pre-assessment system. The experimental operation results are directly imported through the experimental management system. The results of the experimental report are directly imported through the experimental report management system. After the three parts are obtained, the final results can be generated automatically in the experimental score system and can be released for student inquiry.

5 Effect Analysis of Experimental Teaching Reform and Platform Practice

According to the two-semester mechanical experiment teaching reform of our school, combined with the application of BOPPPS model and teaching management platform, the reform effect is remarkable, mainly reflected in the following aspects:

(1) The overall quality of students is improved

Comparing students' enthusiasm for learning, operation, ability to solve problems, nature of questions, and experimental reflection, it can be seen that students' abilities have been greatly improved, as shown in Table 1. They have changed their roles from imitating experiments passively to participating in experiments actively. They have developed the habit of independent thinking.

Table 1. Student ability comparison

Teaching method	Content				
	Learning motivation	Experimental operation	Ability to solve problems	Type of question	Experimental reflection
Traditional experimental teaching	Average	Imitation	Dependent on the teacher	Operational type	Shallow
Platform experiment teaching	Positive	Independence	Analysis independently	Thinking type	Deep

(2) The teacher's experimental teaching level has been improved

After the reform, the teacher no longer just teach knowledge based on the contents of the experimental guidebook, they design more innovative experimental projects, use heuristic, case-based, project-based teaching methods to stimulate students' interest in mechanical experiments, and use "coaching" communication skills to guide students to find ways to solve problems.

(3) The level of innovation competition has been improved

In recent years, there are more students of our school have participated in the National College Students' Mechanical Innovation Design Competition, the National University Student Energy Conservation and Emission Reduction Competition, and the "Challenge Cup" National College Students Extracurricular Academic Science and Technology Works Competition and other innovative competitions. The quality of the works has been improved, more and more students have won national awards.

6 Conclusions

The focus of the reform of experimental teaching methods is to cultivate students' innovative spirit and comprehensive quality according to market and enterprise needs. The BOPPPS model combined with the construction of the experimental teaching management platform embodies a student-centered teaching philosophy that can help teachers disassemble and analyze the teaching process, identify blind spots in teaching, and improve and enhance teaching effectiveness. Let students not only attend in class, but actively participate in wholeheartedly, thus effectively enhancing students' interest in learning, improving enthusiasm and initiative, practical ability, innovative ability and ability to analyze problems and solve problems, so as to achieve teaching goals.

References

1. Wen, H.: Exploration of experimental teaching in applied undergraduate universities under the background of "New Engineering". Educational Research, p. 163, October 2017
2. Keishing, V., Renukadevi, S.: A review of knowledge management based career exploration system in engineering education. Int J. Mod. Educ. Comput. Sci. (IJMECS) 8(1), 8–15 (2016). https://doi.org/10.5815/ijmecs.2016.01.02
3. Fetaji, B., Fetaji, M., Ebibi, M., Kera, S.: Analyses of impacting factors of ICT in education management: case study. Int. J. Mod. Educ. Comput. Sci. (IJMECS) 10(2), 26–34 (2018). https://doi.org/10.5815/ijmecs.2018.02.03
4. Liu, L., Zhang, D., Zhang, X., Ye, Y.: Practice on the construction of teaching team of mechanical basis. In: The 2nd Teaching Seminars on Higher Education Science and Engineering Courses, pp. 674–676, May 2017
5. Kotevski, Z., Tasevska, I.: Evaluating the potentials of educational systems to advance implementing multimedia technologies. Int. J. Mod. Educ. Comput. Sci. (IJMECS) 9(1), 26–35 (2017). https://doi.org/10.5815/ijmecs.2017.01.03

6. Duong, T.M.: EFL teachers' perceptions of learner autonomy and their classroom practices: a case study. Int. J. Educ. Manag. Eng. (IJEME) **4**(2), 9–17 (2014). https://doi.org/10.5815/ijeme.2014.02.02

7. Li, Y.: Effective teaching: application of BOPPPS model in ESL class. In: The 2016 2nd International Conference on Economy, Management, Law and Education, vol. 12, pp. 433–440 (2016)

8. Lou, S.-J., Dzan, W.-Y., Lee, C.-Y., Chung, C.-C.: Learning effectiveness of applying TRIZ-integrated BOPPPS. Int. J. Eng. Educ. **5**, 1303–1312 (2014)

9. Zhou, J., Fan, D., Wu, Z., et al.: Teaching practice of numerical control machining based on Boppps. In: The 11th International Conference on Modern Industrial Training, pp. 654–658, August 2017

10. Fu, Z., Lin, Z., Zhang, T.: Assessing the active learning in engineering education based on BOPPPS model. In: ASEE Annual Conference and Exposition, Conference Proceedings, June 2018

11. Li, Z., Huang, Y., Huang, Z.: Research on the reform of experimental teaching based on Fischertechnik model. In: Applied Mechanics and Materials, pp. 237–240 (2014)

12. Dominic, M., Francis, S.: An adaptable e-learning architecture based on learners' profiling. Int. J. Mod. Educ. Comput. Sci. (IJMECS) **7**(3), 26–31 (2015). https://doi.org/10.5815/ijmecs.2015.03.04

13. Alkhathlan, A.A., Al-Daraiseh, A.A.: An analytical study of the use of social networks for collaborative learning in higher education. Int. J. Mod. Educ. Comput. Sci. (IJMECS) **9**(2), 1–13 (2017). https://doi.org/10.5815/ijmecs.2017.02.01

14. Kaviyarasi, R., Balasubramanian, T.: Exploring the high potential factors that affects students' Academic Performance. Int. J. Educ. Manag. Eng. (IJEME) **8**(6), 15–23 (2018). https://doi.org/10.5815/ijeme.2018.06.02

15. Liu, M.-N.: A model of teaching resources management platform based on XML and web services. World Trans. Eng. Technol. Educ. **14**(1), 198–202 (2016)

16. Ye, W.: Design and realization of practical teaching management system on NET-based platform. In: Proceedings of 2012 International Conference on Information Management, Innovation Management and Industrial Engineering, ICIII 2012, no. 3, pp. 411–414 (2012)

17. Zhang, J., Pu, X., Zhang, Z.: Design and implement of teaching resources management network platform based on MVC. In: Applied Mechanics and Materials, pp. 999–1002 (2014)

18. Wang, S.: Innovation of interactive teaching mode of management communication based on online learning platform. Boletin Tecnico/Tech. Bull. **55**(6), 651–656 (2017)

Training Platform Construction of Omni-Media Live Broadcast of Sport Event

Ziye Wang[1(✉)], Qingying Zhang[2], and Mengya Zhang[2]

[1] Wuhan Sports University, Wuhan 430079, People's Republic of China
kathy8899@126.com
[2] Wuhan University of Technology, Wuhan 430063, People's Republic of China

Abstract. Aiming at the target of sport media professional talents cultivation, this paper discusses the construction, application, use effect and related problems of sports events all media live broadcast training platform. For sports communication specialty, it is an important task and sacred duty to cultivate high-quality professionals with solid theoretical basis and practical ability required in the workplace. The construction of all-media live training platform for sports events is an effective approach of personnel training for the students professionals, interns of provincial and municipal television stations, as well as some sports commentators from the we-media public plat. Complying with the principles of scientificity, subjectivity, coordination, professionalism and growth, the paper describes an integrated live broadcast training platform system and its construction. The design and evaluation of the system are carried out from the perspectives of hardware, software, equipment, personnel and management. Through a comprehensive analysis, the focus and direction of improvement are found, which is helpful for the training platform to increase its effectiveness and play a greater role in the training and cultivation of sports media professionals.

Keywords: All media · Sport event live broadcast · Training platform · Sport media professionals

1 Introduction

The development of science and technology carries the world into a new era of multi-media, even omni-media, or all media, as usually called, and brings the information communication some brand new ways and means [1]. Dissemination of sports events is currently accomplished not only by the traditional methods, such as television, radio and newspaper, but with the real time live broadcast [2], which provides the staff including hosts, commentators, and program directors updated requirements and higher pressure.

As a responsible unit of sport media professional (abbreviated as SMP) cultivating, colleges and universities should do their utmost to offer the students comprehensive equipment and condition to drill them systematically and provide them with specific abilities in live broadcasting of sports event [3], so as to enable them to be proficient and competent in the workplace in the future [4]. The construction of all-media live training platform for sports events is put forward under this background.

© Springer Nature Switzerland AG 2020
Z. Hu et al. (Eds.): ICCSEEA 2019, AISC 938, pp. 205–216, 2020.
https://doi.org/10.1007/978-3-030-16621-2_19

2   Research Background of All-Media Live Broadcasting of Sports Events

Sports game and activities are approaching into and becoming an essential content of public life nowadays [5]. The growing number of people is going to enjoy sports games and to participate in various bodybuilding and fitness activities. The propagating of sport events has getting increasingly greater demand and more channels.

In this era of all media, live broadcast of sports program is recognized as a kind of diversified modes of communication based on the backdrop of sport culture and social consumption. With the continuous promotion of "Internet+", and the combination, cross and fusion of sports science and journalism and communication, the live telecast of sports events has considerable popularity [6], and gets more abundant means and connotations. Sports communication has stepping into the age of "Micro Propagation" where multifarious instant messaging services, e.g. micro-blog, We-Chat, micro video, etc. go forward together, and shape an advanced pattern of integration and communication of different media [7].

Western Scholars proposed some research topics about media sports at the end of last century. Chinese researchers, by contrast, focus on the technique issues on the operational level, which belongs to a static study in essence, lacks new idea and originality, and is difficult or impossible to solve practical problems.

Researching all-media live broadcasting of sports events from the perspective of theory and practice is not only the need of international academic exchanges, but also the demand of encouraging people to get physical and mental health [8], and promoting our country to become a sports power.

3   Competence Requirements of Sports Media Professionals

Professional talents cultivated by sports media major need to possess high comprehensive quality and some specific skills and craftsmanship.

3.1   Being Jack of All Trades

SMPs have to be able to comprehensively understand and master knowledge of sports activities and events [9], to arm themselves with a wisdom mind and a keen perspective, to make quick and accurate judgments, a precise and reasonable expression, to accomplish enthusiastic while moderate interaction with the audience, and to deal with all kinds of unexpected situations on the field in time [10]. Besides, it is vital for them to acquire and know well some skill and the technical abilities to actualize the real-time live broadcasting of sports events and activities effectively, so as to help the audience understand the course of the competition, learn more about sports rules and knowledge and achieve advanced health concept [11]. With the continuous development of science and technology, new media forms, communication tools and forms are constantly emerging, while SMPs need to keep up with the pace of the times in order to make continuous progress and sustainable development.

Professional skills a SMP is asked to have mainly include:

(1) The ability of spot control;
(2) Interacting with audience;
(3) Accomplishing interview, producing and broadcasting concurrently,
(4) Using new media tools adroitly.

3.2 Strong Practical Ability

SMPs should be multi-disciplinary talents with interdisciplinary knowledge and versatility [12]. Under the all-media environment, the production process of news works is more complex, which can be brought to success by the persons who are proficient in specialist knowledge, and with a strong practical ability [13]. Only by fully grasping various developed and innovative technologies, can they, the SMP talents, produce excellent works and be competent for their job in the future [14].

In the process of professionals' cultivation, it is particularly momentous and necessary to take the practical ability strengthening as the top priority.

4 Basic Objectives and Principles of All-Media Live Training Platform Construction

Different from the experimental platform training of science and engineering, the live platform of sports events needs to pay more attention to the characteristics of sports communication, according to different sports events, grasp the rules, mobilize the mood of the audience, and guide them to watch the games enthusiastically and orderly.

4.1 Objectives of Training Platform Erecting

Sports events live media training platform, also called training system, center, or base, is installed for the sake of achieving the following objectives:

(1) Build it into a practical training system adapting to the changeability, complexity, instantaneity and scene of sports program production;
(2) Make it a fully digital experimental system with simple operation;
(3) Help it to become a teaching experiment system with low investment and high efficiency, which is connected with the industry.
(4) Turn the platform into an experimental system to realize multi-disciplinary accommodation, adapt to the learning characteristics of students majoring in sports media, and provide for all students of 4 majors of the college [15].

In a word, it is necessary to establish the training platform system into an excellent working base for cultivating compound and applied SMPs with innovative spirit and practical ability.

4.2 Principles of Training System Development

The development of all-media live training platform should follow the principles hereinafter:

(1) Scientific principle

To make a correct decision and design of the platform with a scientific thinking method, it is necessary to fully understand the demand of the system, evaluate the reliability of technical means and predict their development, list arguments and demonstration, and to put forward the plan and steps of implementation, so as to ensure the scientificity and rationality of the design and establishing of the training plat.

(2) Principle of subjectivity

The design of training platform should be focused on the cultivation of students and be developed on the basis of "ability-based" concept. According to the training objectives and requirements of SMPs, the function and structure of the platform are designed exquisitely to enable it serve the growth of professionals more effectively.

(3) Principle of harmony

Critical to the success of practical training platform fixing, some relationship must be treated appropriate, including the relationship between theoretical teaching and practical training, between teacher guidance and student operation, as well as the responsibility allocation and obligation definition of teaching managers, so that the platform can play a better role ideally.

(4) Principle of occupation

Focusing on the professional requirements and career demands of sports media professionals in the future, the establishing of all-media live training platform is carries out comprehensively for enhancing the practical ability of the student. Correct design of modules and contents of the platform is on account of deeply understand the actual requirements of employers.

(5) Principle of growth

The so-called "growth" here has multiple meanings. Firstly, it is necessary to help students grow up continuously, to enable them to have more and more practical abilities, and to effectively combine theoretical learning with practical training. The second is to make the platform itself has the ability to grow up, so that the procedure of training is called a process of continuous improvement and innovation. The third one is about the growth of teachers. In the progress of designing, constructing and experimenting of the training center, teachers' ability is also continuously increased and heightened.

5 Building of the All-Media Live Training Platform for Sports Events

The setting up of training platform involves lots of contents, including hardware, software, equipment, management, application and other aspects.

5.1   Composition of Integrated Platform System

The training center of all-media live broadcasting for sports events is a multi-station live broadcasting system [16], which mainly consists of three parts: the broadcasting station, the encoder, and the code stream broadcasting control system [17].

In the process of equipment selection, the latest wire-cast integrated live signal production system made by *TELESTREAM* company is adopted, which integrates signal acquisition, broadcasting switching, trans-coding and pushing into one device, and is capable to realize 8-channel signal pushing at the same time. The platform is also equipped with four Panasonic *AG-200 HD* cameras and two Black-magic *4K HD* cameras to meet the demands of *HD* live broadcasting and program production, while *AVID* system is used to package programs as required.

(1) Guide Station

There are many cameras in the studio. The broadcaster, like a program director, transmits the video signals to the video switchboard through the signal line. In the process of multi-signal broadcasting, the choice of different lenses and different locations definitely determines the expressive power of the picture presentation of film and video [18]. The director's responsibility is to instruct the "cameraman" to shoot, switch the scene and display special effects.

(2) Encoder

Soft compression is used to complete the coding process. A high performance server with a streaming media compression card supporting soft compression is in a position to complete the acquisition and coding of analog signals. The compression and broadcasting software supports a variety of streaming media formats and live broadcasting modes, as well as the support of adaptive streaming media technology.

(3) Broadcast control desk

Another important part of the network live broadcasting platform is the broadcast control of streaming media, which supports a large number of users to access the live broadcasting system at the same time, and realizes information service, live broadcasting, on demand and time-shift services. By connecting the encoder to the nearest network port through a network cable, the live video stream is retrieved by the broadcast server [19].

In fact, our school does not have a live broadcasting platform developed by ourselves, but generally relying on the existing public live broadcasting platform, such as Live Fish Fighting, Live Penguin, Live Dragon Beads, Always Broadcast and other available live telecasts to carry out live games and programs.

5.2   System Architecture of Live Broadcasting Platform

The system structure of live broadcast center is divided into four parts: user layer, business layer, service layer and front-end access layer. Their role is shown in Table 1. Figure 1 illustrates the architecture of the live training platform for sports events.

Table 1. System structure of live training platform

User layer	Provide a variety of systems, multi-terminal playback, third-party calls
business layer	Provide unified and centralized management and operation channels
Service layer	Including most of the business services in network TV scenarios
Front access layer	Supporting multiple types of signal access

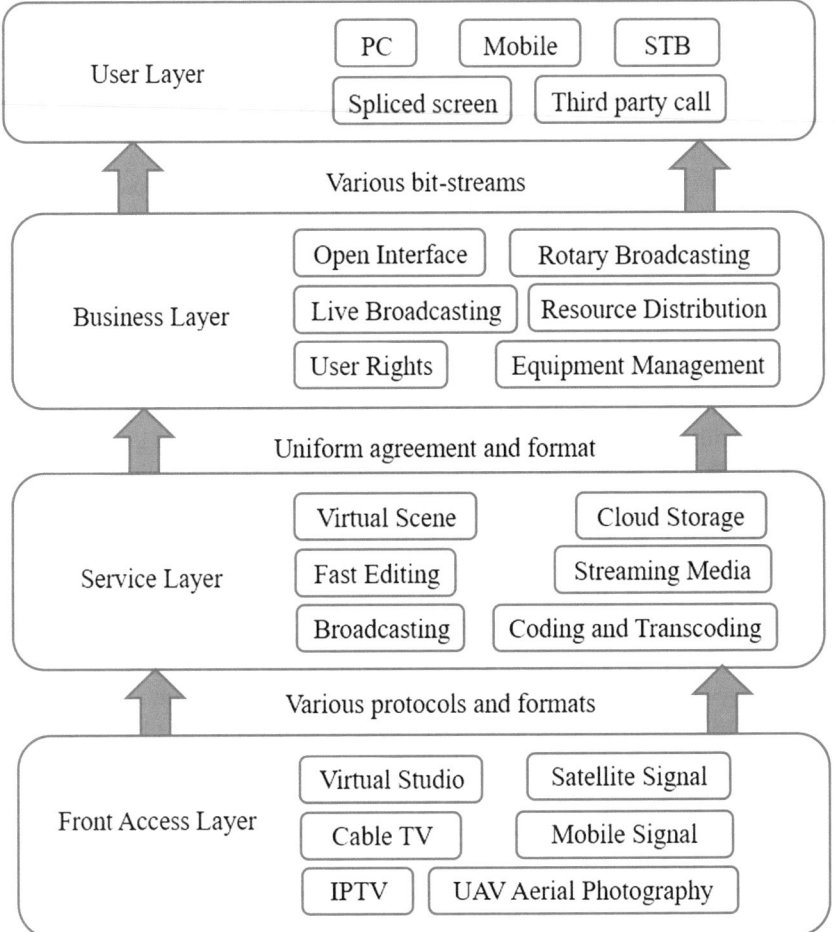

Fig. 1. Layered structure of live broadcast training platform

5.3   The Characteristics of Sport Competition Live Training System

As a typical one-stop solution, the training platform has complete functions and prominent features, including: high-definition three-dimensional scene library, importing three-dimensional scene by one key, virtual multi-screen video window opening, program diversification interacting, streaming media synchronous live broadcasting, multi-machine signal source inputting, professional-level broadcasting operation, two-dimensional and three-dimensional media signal arbitrary switching, high-quality signal source acquisiting, high-quality acquisition signal source *HDMI/SDI* broadcasting interface, and single/double/four channel(s) high-definition acquisition picture format selecting. At the same time, the platform primely supplies high-quality 3D graphics rendering to enhance the reality of the scene, is able to offer real physical light source projection, reflection/shadow simulation special effects, furnishes key frame movement in three-dimensional spectacle, and reflects the vivid effect and unlimited creativity.

The training platform truly embodies the concept and characteristics of "all media". Not only have many excellent video works been produced by the use of this live broadcasting system, but also other kinds of results in the form of radio broadcasting, newspaper manuscripts, ect. have gained remarkable achievements and fruits in the multi-channel communication of mainstream media and self-media.

5.4   Team Construction of Live Training Platform

Cooperating with the platform of all-media live training, a student practice team "WH Sports" has been set up, which had more than 30 members, was composed of graduate students and undergraduates, and divided into three groups: planning report, signal production and content operation.

The team joins lots of projects and participates in the on-site report and signal production of many sports competitions inside and outside school, including the whole process live report of *Wuhan Zhuo-er of the Chinese League A*, the signal production of *Hubei Men's Volleyball Home Event* of the *National Super League*, and the program pushing of *NBL Wuhan Contemporary Professional Basketball Club Home Event,* as well as the game event reporting of *Table Tennis Super Wuhan Club Home Event.* The program output from the platform is also selected and broadcast by the mainstream TV stations in the provinces and municipalities.

The team of live training platform not only plays an important role in specialty and discipline construction, but also has a good impact on the industry, and recognized as a unique and outstanding team in sports live broadcasting.

With this training platform, teachers and students are arranged to transcribe 64 matches of 2014 *Brazil World Cup* and dozens of matches in *Russia World Cup* in 2018 broadcasted or rebroadcasted by CCTV, to establish a corpus, and to carry out the research on CCTV World Cup hosting from the angles of broadcasting, hosting, commentary and commentary. Collating and summarizing the information of other sports and competitions are also the jobs of the team. Besides, they complete the live broadcasting of more than 100 sports competitions at all levels in the school every year, accumulate abundant video materials for all kinds of sports games and activities, promote the improvement of sports competitions, and to contribute to building a healthy nation, which is called the Chinese dream.

6   The Effect Analysis of All-Media Live Training Platform for Sports Events

6.1   Functional Evaluation of Live Broadcast Training Platform

The establishment of training platform provides a more effective tool for teachers and students to enhance their abilities and career development. It is necessary to comprehensively evaluate the implementation effect of the platform [20] and make it play a better role through continuous improvement and optimization [21].

After extensive investigation, research and demonstration, 12 evaluation criteria are determined to estimate the application effect of the live training platform, which are:

(1)   Live broadcasting quantity, quality and level;
(2)   Radio broadcasting and newspaper publication (be rebroadcasted, reprinted);
(3)   Intercommunication and sharing with mainstream media;
(4)   Student Competition Award-quantity and level;
(5)   Platform utilization rate;
(6)   Discipline and professional construction;
(7)   Cooperation between industry, university and research;
(8)   Scientific research achievements;
(9)   Teaching research projects;
(10)   Academic exchanges;
(11)   Richness of the database; and
(12)   Operation effect.

Using *Fuzzy Comprehensive Evaluation Method*, several experts are invited to grade the various indicators according to the above 12 aspects, and rank the training platforms according to the scoring situation of each university, so as to judge the effect of the training platforms.

Assuming that n experts participate in the evaluation, they are scored according to the consideration of 12 indicators separately.

If the score given by the expert *No.j* for the index i is X_{ij},

In the N scores of the index i, the relative score X_i of the index i can be determined by eliminating the highest score X_{max} and the lowest score X_{min} and calculating the average value as below:

$$X_i = \frac{\sum_{j=1}^{n} X_{ij} - X_{max} - X_{min}}{n - 2}$$

Among all 12 indicators, if the weight of the indicator *No.i* is W_i, the total score of the training platform is obtained as follows:

$$S = \sum X_i W_i$$

The comparison of the use effects of the different all-media live training platforms for sports events among M universities participated in the evaluation is made by ranking the scores in sequence, and then the situation and the using effect of the platforms are easy-to-read.

In addition, the shortcomings are able to be found according to the scores of different indicators in each training platform, which points out the direction for the improvement of the system.

For the purpose of effect analysis, 4 similar live broadcasting platforms in different universities are selected, while 15 experts are invited to score on 12 indicators above. The final results show that our school's live platform ranks first, with a score of *94.5*. Among the indicators, our scientific research achievements (No.8) and teaching research projects (No.9) are relatively weak, with *74.8* and *76.1* points respectively. This is the focus and key points of our work next period of time.

6.2   The Use Effect of All-Media Live Training Platform for Sports Events

The construction of all-media live training base has achieved good results, mainly reflected in four aspects, discussed below.

(1) By the application of the platform building, the teaching and practical condition of the subjects has been improved.

The purpose of the project is to build the sports event live training platform into an excellent base of teaching, research and practice with high efficiency and great effect. In line with "Internet + sports" discipline groups, competition organization, commentary and commentary, and the construction of related courses, some effective ways and means of all media live broadcast of sports events are explored, which gives full play to the advantages of the school's rich resources about sports and athletic activities, match organization, and talents. The development of the training base offers a strong guarantee foe deeper study of "Internet + event organizing", provides a greater academic impact, and has formed a prominent disciplinary superiority.

(2) With the construction of training platform of all-media live broadcast, the overall quality of teachers and subject teams has been elevated.

Utilizing the conditions of all-media live training platform as fully as possible, the research and practice of live broadcasting with omnimedia are continuously promoted. The tasks of the practical system erecting also include deeply exploring the use value of the experimental platform, to make it acting on the international convention and in line with domestic advanced level in all media live broadcasting, so that our voice could be heard by the world. It is extremely meaningful to comprehensively put forward the mounting of an innovative talent training and development system of "Talents + Projects + Platforms", and to promote the joint training and sharing of innovative talents, the cooperative training of industry, universities and research institutes, and the "order-based" talent training model.

(3) Improving the scientific research ability and innovation through the construction of the platform.

The all media live broadcasting training platform of sport event may play a great part in scientific research and innovative practice by means of the full combination of theory and practice. The targets and contents are to study the problems of system planning, system control, coordination of various factors and their links, decision-making, improvement of live broadcasting schemes, and application of various technical means needed to solved problems in the live broadcasting of sports events in the era of all-media, and then to carry out scientific research deeply and improve the level of innovation practice constantly.

(4) Raise the quality of education by the utilization of the training system.
    The application of the training platform has increased the teaching practice of cultivating practical ability, aroused the students' interest in learning, brought into play the students' subjective consciousness, realized the comprehensive assessment of knowledge and ability, improved the teaching concept, methods and means, and improved the teaching effect.

By the structuring of this practice platform, the course teaching and curriculum building achieve fruitful results, provide students with real scene of live training places and opportunities, exercise students' practical operation ability, so that they can get better exercise and more opportunities for growth. At the same time, it provides comprehensive support for curriculum reform and specialty development, and expands professional influence constantly.

Supported by the all-media live training platform for sports events, outstanding professional talents stand out and emerge in an endless stream. A number of students actively participate in various international, national, regional and inter-school competitions which stimulate their vitality nicely. They actively participated in the national and provincial competitions, such as golden microphone contests, sports events commentary contests and some other competitions, achieved many honors and medals, and have got highly praised by the industry.

7   Conclusions

It is the call of the times and the need of industry development and talent cultivation to build a full-media live training platform for sports events. It is significant for SMPs in the era of all-media to possess high comprehensive quality. They are expected not only to be proficient in theoretical knowledge, but also be good at using knowledge to solve practical problems. Only by having a strong operational ability, can talents be competent at their job and performance remarkable in performance in the future.

The establishing of the platform involves many links such as hardware, software and equipment. It also depends on the cooperation of managers, users and coordinators to maximize its effectiveness. The research includes but is not limited to the operation and application, while the more important thing is to seek countermeasures through multi-disciplinary media research by means of the study of the training system to improve the media literacy and sports literacy of the people, stimulate the enthusiasm of the public to participate in sports, promote the physical and mental health of the nation, and enhance their social interaction and social adaptability.

Research on the establishment of all-media live training platform for sports events is to give full play to the theoretical study and practical training, and their effective combination, of sports media, so that the all-media live broadcast of sports events is hopeful to become an important tool for the construction of sports power and the improvement of national quality.

References

1. Centieiro, P., Cardoso, B., Eduardo Dias, A.: If you can feel it, you can share it!: a system for sharing emotions during live sports broadcasts. In: Proceedings of the 11th Conference on Advances in Computer Entertainment Technology Held at Funchal, Portugal on 11–14 November, article No. 15 (2014)
2. Li, T.: The impact of new media on the spread of sports events. Public Commun. Sci. Technol. (05), 118–119 (2015). (in Chinese)
3. Zhong, P., Li, M.: Analysis on the development of sports event communication under the new media environment. J. Chongqing Univ. (Soc. Sci. Ed.) **19**(06), 143–147 (2013). (in Chinese)
4. Li, L.: Sports news communication in the age of fusion media. News Front. (04), 137–138 (2016). (in Chinese)
5. Lu, L., Wu, W.: Analysis of the coexistence between sport and the media in the modern society. In: Proceedings of 4th International Conference on Education, Management and Computing Technology (ICEMCT 2017), pp. 1289–1294 (2017)
6. Zang, W.: The concept of "Internet +" and news dissemination of major sports events. News Res. Guide, (20), 34 (2015). (in Chinese)
7. Xu, Z., Qiu, X., He, J.: Construction of all-media resource service platform under the background of triple play. J. Comput. Theor. Nanosci. **13**(12), 9852–9856 (2016)
8. Alkhathlan, A.A., Al-Daraiseh, A.A.: An analytical study of the use of social networks for collaborative learning in higher education. Int. J. Mod. Educ. Comput. Sci. (IJMECS) **9**(2), 1–13 (2017). https://doi.org/10.5815/ijmecs.2017.02.01
9. Cao, Y., Liu, Y.: Attainment of press-type host and cultivating under the background of convergence Media. Sci. Technol. Commun. **6**(22), 208–209 (2014)
10. Khan, A.A., Madden, J.: Speed learning: maximizing student learning and engagement in a limited amount of time. Int. J. Mod. Educ. Comput. Sci. (IJMECS) **8**(7), 22–30 (2016). https://doi.org/10.5815/ijmecs.2016.07.03
11. Kun, Z.: Assessment of media professionals with four dimensions. Youth Report. (25), 68–69 (2015). (in Chinese)
12. Yin, L.: Cultivation of inter-disciplinary media talent under the background of convergence media. Beijing Educ. (02), 17–18 (2015). (in Chinese)
13. Kotevski, Z., Tasevska, I.: Evaluating the potentials of educational systems to advance implementing multimedia technologies. Int. J. Mod. Educ. Comput. Sci. (IJMECS) **9**(1), 26–35 (2017). https://doi.org/10.5815/ijmecs.2017.01.03
14. Robles, A.C.M.O.: Blended learning for lifelong learning: an innovation for college education students. Int. J. Mod. Educ. Comput. Sci. (IJMECS) **4**(6), 1–8 (2012). https://doi.org/10.5815/ijmecs.2012.06.01
15. Papadimitriou, A., Gyftodimos, G.: The role of learner characteristics in the adaptive educational hypermedia systems: the case of the MATHEMA. Int. J. Mod. Educ. Comput. Sci. **9**(10), 55–68 (2017)

16. Rick, C., Maria, H., Marvin, W., et al.: Augmenting live broadcast sports with 3D tracking information. IEEE Multimedia **18**(4), 38–47 (2011)
17. Lonchamp, J.: A platform for CSCL practice and dissemination. In: Sixth IEEE International Conference on Advanced Learning Technologies (ICALT 2006) held at Kerkrade, Netherlands on 5–7 July, pp. 66–70 (2006)
18. Sergio, I., Eduardo, M., Arantza, I., et al.: A friendly location-aware system to facilitate the work of technical directors when broadcasting sport events. Mob. Inf. Syst. **8**(1), 17–43 (2012)
19. Ma L., Su Z., Cao S.: A study of fragmentation and reorganization mechanism in video production and distribution process. In: Applied Mechanics and Materials, pp. 974–977 (2013)
20. El Haji, E., Azmani, A., El Harzli, M.: Using AHP method for educational and vocational guidance. Int. J. Inf. Technol. Comput. Sci. (IJITCS) **9**(1), 9–17 (2017). https://doi.org/10.5815/ijitcs.2017.01.02
21. Liu, S., Chen, P.: Research on fuzzy comprehensive evaluation in practice teaching assessment of computer majors. Int. J. Mod. Educ. Comput. Sci. (IJMECS) **7**(11), 12–19 (2015). https://doi.org/10.5815/ijmecs.2015.11.02

A Multi-object Tracking Method Based on Bounding Box and Features

Feng Liu$^{(\boxtimes)}$, Wei Jia, and Zhong Yang

College of Informatics, Huazhong Agricultural University, Wuhan, China
liufeng@mail.hzau.edu.cn

Abstract. Multi-object tracking is a key research problem in computer vision area, and with the fast development of the deep learning based image and video processing algorithms, the performance and accuracy of multi-object tracking methods are dramatically improved. However, current multi-object tracking methods mostly focus on human and seldom animals, and usually there are too many parameters, which make these methods very complicated and very hard to use in practical scenarios. In order to solve these problems, we proposed an easy-use multi-object tracking method based on bounding boxes and object appearance features. In our method, we take animals as the tracking object. In order to count the number of them in a closed area we firstly tracking and identify them based on the fact that two different objects cannot appears in a same video frame and the trajectory of an animal is continuous. Then, we store the appearance features of each individual animal and when it cannot be identified by tracking, we use appearance feature to re-identify it. Thirdly, we combined these two methods and when the whole system converged to stable state, we can get the total mumble of these animals. The results show that our method can tracking the multi-object accurately and can be easily used in practice.

Keywords: Multiple object tracking · Bounding box · Appearance features · Tracking by detection

1 Introduction

With the development of deep learning, object-tracking problem, including single object tracking and multiple object tracking, attracts more and more attentions in the computer vision area. Among all the methods, the tracking-by-detection paradigm is most commonly used [1, 2]. In this kind of paradigm, an object detector is firstly used in identifying the targets and then these identified objects across several video frames associate the trajectory [3]. Some typical factors are usually taken into consideration, such as the object appearance model, motion model and interaction model [4]. There also exist some object tracking methods using traditional image processing algorithms, such as [5–7]. However, the mainstream of object tracking methods are mostly based on deep leaning.

To track the objects, usually the first step is to detect them and locate them by the bounding box. A lot of deep learning object detecting method have been proposed, such as SSD [8], MTCNN [9], Fast R-CRNN [10] and Yolo [11]. After get the

© Springer Nature Switzerland AG 2020
Z. Hu et al. (Eds.): ICCSEEA 2019, AISC 938, pp. 217–227, 2020.
https://doi.org/10.1007/978-3-030-16621-2_20

bounding box of the moving objects, the feature maps of the corresponding targets can be generated by such algorithms as FaceNet [12], DeepFace [13], SphereFace [14] and Baidu Face recognition model [15].

In some scenarios, the real time tracking results are necessary, therefore a lot of paper focus on online multiple object tracking. In [16], a CNN-based framework for online multiple object tracking is proposed by utilizing single object tracker and in addition, the spatial-temporal attention mechanism is used to handle the drift caused by occlusion and interaction. A multiple object tracking named SORT (Simple Online and Realtime Tracking) is designed in [17]. SORT integrate appearance information to solve the longer periods of occlusions. In [18], Xian etc. use the markov decision processes to formulate the online multiple object tracking problem as a decision making model. In order to achieve real-time tracking goal, some researchers utilize hardware to accelerate algorithms [19, 20]. In [20], Singh etc. using the hardware of Xilinx ML510 FPGA board to accelerate the tracking method and achieve real-time performance. In order to handle occlusion problem, a real-time object tracking algorithm is proposed in [21], which can adaptively estimate the object scale and trains the radial basis function neural network to solve the occlusion problem. Because of multiple object tracking problem is a time series process, Deep RNN network are also widely used. In [4], based on RNN, the appearance cues, motion priors and interactive forces are combined to solve the multi-target tracking problem.

Although most researchers currently focus on human tracking, there are other scenarios where the multiple object tracking can play an import role. In [22], in order to track generic objects, a real-time deep object tracker named Re3 is proposed, which utilize the object appearance, motion and the changing features over time.

From the discussion above, we can see that the current methods usually uses so many parameters and make the application very complicate. To simply the tracking approach and improve the performance, in [3], a high speed method is proposed without using image information. This method only bases on the bounding box and only uses the IOU (intersection over union) value to track objects over time.

By analyzing the current research work, we can find that their work most focus on human and seldom other types of objects. Furthermore, we also want to simplify the method and improve the performance and accuracy of multiply object tracking method. Motivated by [3], we also use the bounding box to track object. But different with [3], in order to solve the long period occlusion problem, we also combine the appearance features. In our method, we take animals as the tracking object. In order to count the number of them in a closed area we firstly tracking and identify them based on the fact that two different objects cannot appears in a same frame and the trajectory of animal is continuous. Then, we store the appearance features of each individual animal and when it cannot be identified by tracking, we use appearance feature to re-identify it. Thirdly, we combined these two methods and when the whole system converged to stable state, we can get the total mumble of these animals. Therefore, we can achieve the goal of tracking the animals based on bounding boxes and features and get their total number.

The rest of this paper is organized as follows. In Sect. 2, we will describe the research problem in detail. In Sect. 3, we will propose our method of multiple-object tracking by bounding box and features. We will present the experiment results in Sect. 4 and draw the conclusion in Sect. 5.

2 Problem Description

In order to count the number of the animals in a closed area, we use the multiple object tracking method to track and identify the distinct objects. In this scenario, the target objects are unchanged during a period. Because of the similarity between animals, the appearance features are not sufficient to identify each individual. Therefore, we combined the appearance features and bounding box to tracking animals.

Suppose there are a certain number of animals and we denote the total number is N. This group of animals is denoted as $O = \{o_i | i \text{ in } (1, 2, \ldots N)\}$. We can capture the video by a camera for a certain period. Suppose the number of frames in this video is T and we denote these frames as $F = \{f_j | j \text{ in } (1, 2, \ldots T)\}$. In a certain frame f_j, o_{ij} represent that the object o_i appears in this frame, and all the objects appear in frame f_j are recorded as $O_j = \{o_{ij} | o_i \text{ appears in } f_j\}$. Therefore, in this video, the objects series of appearing can be recorded as $\{O_1, O_2, \ldots, O_T\}$. After we get the bounding box of each object, we can calculate the appearance feature map by the deep learning based model such as FaceNet [12], which needs to be trained ahead. We denote the feature calculating function as $\Omega(\cdot)$ and $m_i^j = \Omega(o_i^j)$ represents the feature map of o_i^j. Therefore, the corresponding feature map set of O_j is represented by (1), and the time series of feature map of a certain object o_i is denoted as (2).

$$\Psi^j = \{\Omega(o_i^j) | \text{where } o_i \text{ appears in the frame } f_j\} \tag{1}$$

$$\Phi_i = \{\Omega(o_i^j) | \text{where } f_j \text{ contains the object } o_i\} \tag{2}$$

For a certain frame, there will be multiple objects. The relationship of these different objects can be described by intersection over union (IoU) as shown in (3). In (3), $A(\cdot)$ represents the function that calculates the pixels of the certain area. $o_p \cap o_k$ refers to the intersection area of the corresponding bounding boxes of object o_p and o_k, meanwhile $o_p \cup o_k$ refers to the union area of these two bounding boxes.

$$IoU(o_p, o_k) = \frac{A(o_p \cap o_k)}{A(o_p \cup o_k)} \tag{3}$$

When $IoU(o_p, o_k)$ is zero, we also calculate the distance between two bounding box, as shown in (4).

$$D(o_p, o_k) = \frac{distance(center(o_p), center(o_k))}{\max(width(o_p), height(o_p), width(o_k), height(o_k))} \tag{4}$$

So, given the frame set F of a video record, we want to tracking the objects O both by bounding box and appearance features and then get the total object number N.

3   Method

To solve the problem describe in Sect. 2, there are two facts we need to utilize. Firstly, the two objects in the same frame cannot be identical. Secondly, the trajectory of an object is continuous. Based on these we can identify the object in and between frames.

3.1   Checking Identical Objects

For a certain video frame f_j, we check if any two bounding boxes belong to a same object by (5). In (5) $I(o_p^j, o_k^j) = 1$ represents that object o_p^j and o_k^j are identical and otherwise they are two different objects. We can set an IoU threshold and according to the IoU values of these two objects to check if these two objects are identical or not.

$$I(o_p^j, o_k^j) = \begin{cases} 1, & \text{if } IoU(o_p^j, o_k^j) > IoU_{threshold} \\ 0, & else \end{cases} \qquad (5)$$

3.2   Identify Objects in Continuous Frames Based on IoU

For two continuous frames, f_{j-1} and f_j, we also check any two objects, which separately located in these two frames, are identical or not by IoU value. However, things become more complicated in this scenario and we cannot get the result only by a simple threshold value. Because of occlusion usually happens, some object may appears in frame f_{j-1}, but does not appears in frame f_j and vice versa. Therefore, we need to use both the IoU value and appearance features.

O^{j-1} and O^j respectively denote the object set in frame f_{j-1} and f_j. $o_p^{j-1} \in O^{j-1}$ and $o_k^j \in O^j$ respectively represent the two objects in this two frames. The problem is two check whether any o_p^{j-1} and o_k^j pair is identical or not.

Firstly, we calculate the IoU matrix on each pair of objects. This matrix is as shown in Table 1.

Table 1. The IoU matrix of all the object pairs.

	o_1^j	o_2^j	o_3^j	...	o_N^j
o_1^{j-1}	$IoU(o_1^{j-1}, o_1^j)$	$IoU(o_1^{j-1}, o_2^j)$	$IoU(o_1^{j-1}, o_3^j)$		$IoU(o_1^{j-1}, o_N^j)$
o_2^{j-1}	$IoU(o_2^{j-1}, o_1^j)$	$IoU(o_2^{j-1}, o_2^j)$	$IoU(o_2^{j-1}, o_3^j)$		
...					
o_M^{j-1}	$IoU(o_M^{j-1}, o_1^j)$	$IoU(o_M^{j-1}, o_2^j)$	$IoU(o_M^{j-1}, o_3^j)$		$IoU(o_M^{j-1}, o_N^j)$

Secondly, we set a *IoU* threshold between the two continuous frames, which is referred to as $IoU^{j-1,j}_{Threshold\_tracking}$. Any pair of objects (o^{j-1}_p, o^j_k) whose *IoU* value is greater than $IoU^{j-1,j}_{Threshold\_tracking}$ is considered to be potential identical objects. Therefore, we can construct an *IoU* matrix which constructed by the potential identical objects as shown in Table 2. For simplicity, we suppose that there are three objects and four objects in frame f_{j-1} and f_j respectively.

Thirdly, we select the maximal value in the potential identical object sets, and the corresponding pair of objects are considered identical. Then, we remove these two objects from the object sets and delete the corresponding row and column of Table 2.

Table 2. The *IoU* matrix of potential identical objects.

	o^j_1	o^j_2	o^j_3	o^j_4
o^{j-1}_1	$IoU(o^{j-1}_1, o^j_1)$ > *Threshold*	$IoU(o^{j-1}_1, o^j_2)$ > *Threshold*	$IoU(o^{j-1}_1, o^j_3)$ > *Threshold*	$IoU(o^{j-1}_1, o^j_4)$ > *Threshold*
o^{j-1}_2	$IoU(o^{j-1}_2, o^j_1)$ > *Threshold*	$IoU(o^{j-1}_2, o^j_2)$ > *Threshold*	$IoU(o^{j-1}_2, o^j_3)$ > *Threshold*	$IoU(o^{j-1}_2, o^j_4)$ > *Threshold*
o^{j-1}_3	$IoU(o^{j-1}_3, o^j_1)$ > *Threshold*	$IoU(o^{j-1}_3, o^j_2)$ > *Threshold*	$IoU(o^{j-1}_3, o^j_3)$ > *Threshold*	$IoU(o^{j-1}_3, o^j_4)$ > *Threshold*

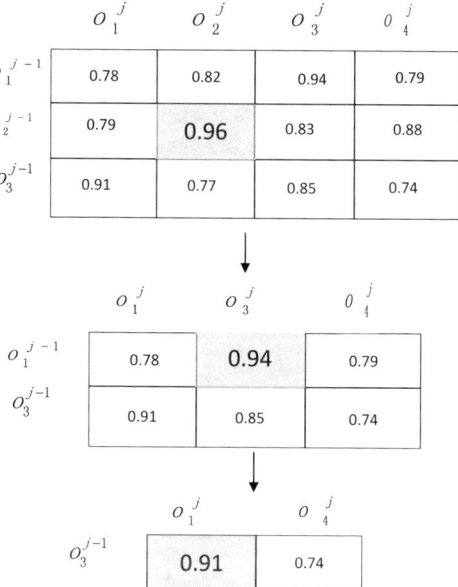

Fig. 1. The process of searching identical objects based on the *IoU* values.

We repeat this process until one of the object set o^{j-1} and O^j is empty. This process is shown in Fig. 1. For clearly describe the process, we assign some specific IoU values to the matrix in Fig. 1. From Fig. 1, we can see that, in each round, we find the maximum of all the values and then remove the corresponding identical objects from the object sets o^{j-1} and O^j. In this way, we can find the optimum pair from all possible object pairs. For example, in the first round in Fig. 1, two object pairs (o_2^{j-1}, o_2^j) and (o_2^{j-1}, o_4^j) are all larger than the threshold value, but we choose the pair (o_2^{j-1}, o_2^j) as the identical object, because its value is the largest.

3.3   Identify the Objects in Continuous Frames Based on Appearance Features

In Sect. 3.2 we discussed the IoU based method to search the identical object pairs in two continuous frames. However, for the occlusion phenomena usually happen during the tracking period, we may lost some objects if we only use the IoU value. Some method such as [4] utilize the motion model and the interaction model to handle the occlusion issue. However, I think these models are so complicated. Therefore, we use the feature map to identify objects.

Firstly, we believe that two objects in the same frame cannot be identical. In the first frame of this video, the objects we detect are different and we use the pre-trained feature map's generating model to calculate the feature map of every object and store them for further utilization. As shown in Fig. 2, for example, the objects o_1, o_2 and o_3 are recognized as three different objects.

Secondly, if we can track the object by the bounding box and IoU values. We add their feature maps for these identified objects to the corresponding feature-map record list. For example, in the second frame f_2, there are four objects, among which object o_2 and o_3 are identified by the method proposed in 3.2. Then we add the feature maps of object o_2 and o_3 to the corresponding feature-map record list. If occlusion happens to o_1 and it cannot be tracked, no new feature map is added. For object o_4, it is a new comer and we add its feature map to the feature-map record list.

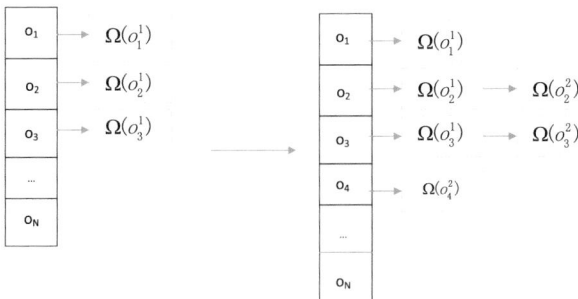

Fig. 2. The feature-map record lists

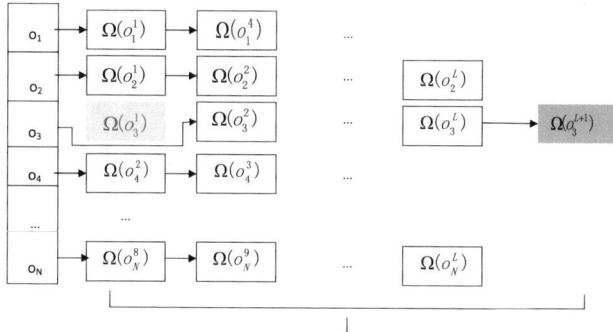

Fig. 3. The update process of feature-map record lists

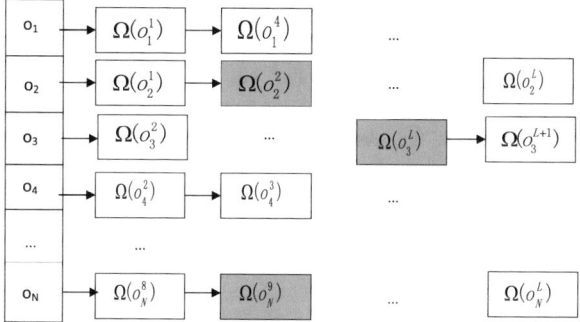

Fig. 4. Analyzes the feature maps for every object and remove the outliers

Thirdly, after some frames, object o_1 appears again. Now, we cannot judge it is a new comer or a re-appearance object. In this occasion, the *IoU* based tracking method cannot work. Therefore, we calculate the distance between feature map of o_1 and all the object feature map records as shown in Fig. 2, and try to find the minimum value of the distance. If the minimum distance value is less than a pre-defined threshold, we believe that this object is the object that has been identified before. Otherwise, if the minimum distance value is greater than the threshold, we treat this object as a new comer and add its feature map to the feature-map record list.

For online tracking, when more video frames are being analyzed, the number of feature-map records for each object increases. According to the principle of locality, we keep the latest L feature-map records for each object. When new feature map comes, as shown in Fig. 3, it will add to the end of the feature-map list of the corresponding tracking object and remove the earliest one.

Furthermore, the identification of the object only by the feature maps may also not be accurate. We may add the feature map to the wrong list. To handle this issue, we use the strategy as show in Fig. 4. From Fig. 4, we can see that, we apply the update mechanism, which periodically analyzes the feature maps for every tracking objects and remove the outliers. To find the outliers, we first calculate the distance of each pair of feature maps for every object. The feature map whose distance from the center of feature maps is larger than a certain threshold is regarded as the outlier. Then we remove it from the feature-map record list.

3.4　Multiple Object Tracking by Both Bounding Box and Appearance Features

In this section, we will discuss how we can track the multiple objects by both bounding box and appearance features.

For the objects in two continuous frames as shown in Fig. 5, we categorize the object pairs into three different sets. The first type of set contains pairs whose IoU value is great than $IoU_{Threshold\_tracking}^{j-1,j}$, which means these objects can be tracked only by IoU value. The second type of set include the pairs whose IoU value is less than $IoU_{Threshold\_tracking}^{j-1,j}$ but greater than zero. For this type of objects, we consider both IoU value and appearance features. We assign weights to these two factors. For example, the IoU factor weight increases while the IoU value goes up. The third type of object pairs include the objects whose IoU value is zero, which means they can not be tracked by bounding box. In this case, we mainly use the appearance feature maps to identify the object. In this condition, we also take the distance of two bounding boxes into consideration, which is calculated by (4).

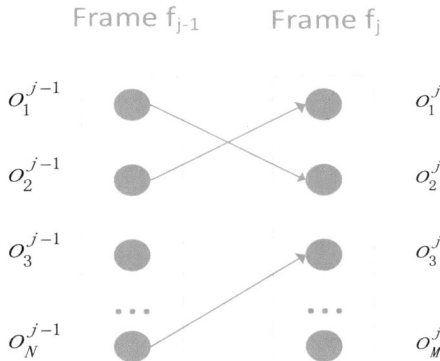

Fig. 5. There different types of object pairs in two continuous frames

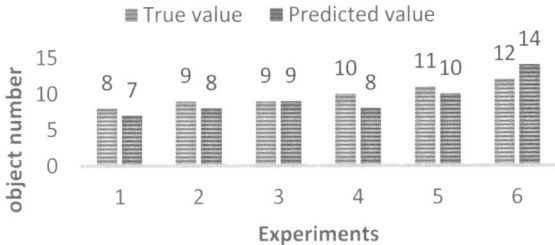

Fig. 6. Experiments results

By this method, after several rounds, the feature maps list is stable. Then we can get the number of the feature-map record list and the total number of the objects.

4   Experiment Results

We test our methods by the tracking videos. We take a lot of videos in the farms. Firstly, we based on MobileNet SSD [8] object detection model and transfer learning method to training animal detection model. By doing this, we can generate the bounding box in each frame of the video and based on these bounding boxes we can calculate the IoU values of any pair of bounding box. Secondly, we training the FaceNet model to generate animal appearance feature maps. These two steps for training object-detection model and feature-map generating model are offline and takes a lot of time and GPU computing resources. After doing this, we apply our method proposed in this paper to tracking the objects and count the number.

From Fig. 6, we can see that the experiment results are very close to the true number. In practical, some animals remain staying in the corner and are occluded by others. They will never show before the camera in a certain period. That is why the experiment results are usually less than the true value. But for the other conditions, if we capture the video for a longer period, the experiment result is larger than the true value. This is because that the same animal's feature map changes dramatically with different pose and in different position and this may cause that the system takes it as different objects.

5   Conclusions

In this paper, we propose a new multiple object tracking method that comprehensively utilize the bounding boxes generated by object detection algorithm and the appearance features generated by feature map calculating algorithm. By the proposed method we can tracking animals and count their total number. The results show that although the deep learning based model training process is time consuming, the running process is effective and this method can real-time track the multi-object accurately and can be used in practical scenarios.

Acknowledgment. This project was financially supported by the Fundamental Research Funds for the Central Universities (Grant No. 2662017JC028) and Hubei Provincial Natural Science Foundation of China (Grant No. 2015CFB437).

References

1. Luo, W., Xing, J., Zhang, X., Zhao, X., Kim, T.-K.: Multiple object tracking: a literature review. Comput. Sci. (2014)
2. Elbahri, M., Kpalma, K., Taleb, N., Chikr El-Mezouar, M.: A novel object position coding for multi-object tracking using sparse representation. Int. J. Image, Graph. Signal Process. (IJIGSP) **7**(8), 1–12 (2015). https://doi.org/10.5815/ijigsp.2015.08.01

3. Bochinski, E., Eiselein, V., Sikora, T.: High-speed tracking-by-detection without using image information. In: IEEE AVSS (2017)
4. Sadeghian, A., Alahi, A., Savarese, S.: Tracking the untrackable: learning to track multiple cues with long-term dependencies. In: 16th IEEE International Conference on Computer Vision, pp. 300–311. IEEE, Venice (2017)
5. Jatoth, R.K., Gopisetty, S., Hussain, M.: Performance analysis of alpha beta filter, kalman filter and meanshift for object tracking in video sequences. IJIGSP 7(3), 24–30 (2015). https://doi.org/10.5815/ijigsp.2015.03.04
6. Tirandaz, H., Azadi, S.: Utilizing GVF active contours for real-time object tracking. Int. J. Image, Graph. Signal Process. (IJIGSP) 7(6), 59–65 (2015). https://doi.org/10.5815/ijigsp. 2015.06.08
7. Rao, G.M., Satyanarayana, C.: Object tracking system using approximate median filter, kalman filter and dynamic template matching. Int. J. Intell. Syst. Appl. (IJISA) 6(5), 83–89 (2014). https://doi.org/10.5815/ijisa.2014.05.09
8. Liu, W., Anguelov, D., Erhan, D., Szegedy, C., Reed, S., Fu, C.-Y., Berg, A.C.: SSD: single shot multibox detector. In: 14th European Conference on Computer Vision, pp. 21–37. Springer, Amsterdam (2016)
9. Zhang, K., Zhang, Z., Li, Z., Qiao, Y.: Joint face detection and alignment using multitask cascaded convolutional networks. IEEE Signal Process. Lett. 23(10), 1499–1503 (2016)
10. Girshick, R.: Fast R-CNN. In: 15th IEEE International Conference on Computer Vision, pp. 1440–1448. IEEE, Santiago (2015)
11. Redmon, J., Divvala, S., Girshick, R., Farhadi, A.: You only look once: unified, real-time object detection. In: 29th IEEE Computer Society Conference on Computer Vision and Pattern Recognition, pp. 779–788. IEEE, Las Vegas (2016)
12. Schroff, F., Kalenichenko, D., Philbin, J.: FaceNet: a unified embedding for face recognition and clustering. In: IEEE Conference on Computer Vision and Pattern Recognition, pp. 815–823. IEEE, Boston (2015)
13. Taigman, Y., Yang, M., Ranzato, M.A., Wolf, L.: DeepFace: closing the gap to human-level performance in face verification. In: 27th IEEE Conference on Computer Vision and Pattern Recognition, pp. 1701–1708. IEEE, Columbus (2014)
14. Liu, W., Wen, Y., Yu, Z., Li, M., Raj, B., Song, L.: SphereFace: deep hypersphere embedding for face recognition. In: 30th IEEE Conference on Computer Vision and Pattern Recognition, pp. 6738–6746. IEEE, Honolulu (2017)
15. Liu, J., Deng, Y., Bai, T., Wei, Z., Huang, C.: Targeting ultimate accuracy: face recognition via deep embedding. arXiv:1506.07310v4 (2015)
16. Chu, Q., Ouyang, W., Li, H., Wang, X., Liu, B., Yu, N.: Online multi-object tracking using CNN-based single object tracker with spatial-temporal attention mechanism. In: 16th IEEE International Conference on Computer Vision, pp. 4846–4855. IEEE, Venice (2017)
17. Wojke, N., Bewley, A., Paulus, D.: Simple online and realtime tracking with a deep association metric. In 24th IEEE International Conference on Image Processing, pp. 3645–3649, IEEE, Beijing (2018)
18. Xiang, Y., Alahi, A., Savarese, S.: Learning to track: online multi-object tracking by decision making. In: 15th IEEE International Conference on Computer Vision, pp. 4705–4713. IEEE, Santiago (2015)
19. Tayyab, M., Qadri, M.T., Ahmed, R., Dhool, M.A.: Real time object tracking using FPGA development kit. Int. J. Inf. Technol. Comput. Sci. (IJITCS) 6(11), 54–58 (2014). https://doi.org/10.5815/ijitcs.2014.11.08
20. Singh, S., Saini, R., Saurav, S., Saini, A.K.: Real-time object tracking with active PTZ camera using hardware acceleration approach. Int. J. Image, Graph. Signal Process. (IJIGSP) 9(2), 55–62 (2017). https://doi.org/10.5815/ijigsp.2017.02.07

21. Ramaravind, K.M., Shravan, T.R., Omkar, S.N.: Scale adaptive object tracker with occlusion handling. Int. J. Image Graph. Signal Process. (IJIGSP) **8**(1), 27–35 (2016). https://doi.org/10.5815/ijigsp.2016.01.03
22. Gordon, D., Farhadi, A., Fox, D: Re3: real-time recurrent regression networks for visual tracking of generic objects. arXiv:1705.06368v3 (2018)

Traffic Orchestration in Data Center Network Based on Software-Defined Networking Technology

Yurii Kulakov$^{(\boxtimes)}$, Alla Kohan, and Sergii Kopychko

National Technical University of Ukraine "Igor Sikorsky Kyiv Polytechnic Institute", 37 Peremohy Avenue, Kyiv 03056, Ukraine
ya.kulakov@gmail.com, a.v.kohan433@gmail.com,
kopychko.sn@gmail.com

Abstract. This paper addresses the issues of traffic orchestration in Data Center Networks (DCN) using Software-Defined Networking (SDN). A generalized structure of a Traffic Engineering (TE) system in DCN using SDN is given. The features analysis of SDN organization and operation which allows increasing the efficiency of DCN is provided. The expediency of traffic design in DCN based on multi-path routing is given.

A method of traffic orchestrating is proposed, which simplifies the traffic reconfiguration procedure and ensures the network load balance due to the centralized approach of generating routing information in SDN controller and using multi-path routing. The routing information organization performed by the wave routing algorithm, in this case, the paths are being created from all intermediate nodes to the end node.

A method for centralized organization of the route information is proposed, which allows avoiding re-organization of route information for previously created sections of the route. The procedure and an example of multi-path routing tables organization are given in this paper.

The results of traffic orchestration modeling process with network load change are presented. It is noted from simulation results, that the traffic reconfiguration of the route does not affect its metric.

Keywords: Data Center Network · Software-Defined Networking · Multi-path routing · Streaming algorithm · Traffic orchestration

1 Introduction

At the present time due to the growing needs for computing power and data volumes, one of the crucial tasks is to increase the efficiency of Data Center Networks (DCN). A modern DCN differ by large dimensionality and diverse equipment including mobile communication tools and mobile hotspots. Therefore, the process of managing this type of networks, in particular traffic orchestration, becomes more complicated. Software-Defined Networking (SDN) [1–3] is currently being used to solve these problems.

Figure 1 shows the general structure of Traffic Engineering (TE) system consisting of a Data Center Network (DCN), SDN controller and TE manager. The DCN collects

© Springer Nature Switzerland AG 2020
Z. Hu et al. (Eds.): ICCSEEA 2019, AISC 938, pp. 228–237, 2020.
https://doi.org/10.1007/978-3-030-16621-2_21

and transmits to the SDN information about network load, network flows, and channel status through the data control plane interface. SDN controller provides the TE manager with the necessary information to orchestrate traffic. After controller modifies the routing information in SDN switches by updating their routing tables to select the best route considering minimizing power consumption and channel congestion.

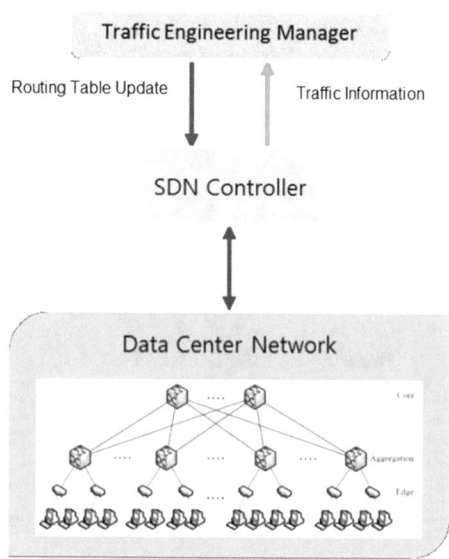

Fig. 1. General structure of Traffic Engineering (TE) system

This leads to reduce energy consumption about 41% and increase the maximum utilization of communication channels by 60% compared with static TE methods [4].

Centralized generation of multiple paths based on multi-path routing in SDN allows reducing TE time and improving traffic quality of service (QoS) [5], however, the known methods of multiple paths organization have a high time complexity [6]. In paper [7], the modified method for a multiple paths organization is proposed, which has less time complexity compared with the known methods of multiple paths organization. An improved multi-path routing method was also proposed in [8], which takes into account the characteristics of SDN organization, in particular, due to the presence of a central controller in the network, hence, reduces the time required to generate multiple access routes to network resources.

The main disadvantage of methods above is that routes are created by each network node and not considering the paths already created by other nodes. This leads to the reorganization of individual sections of already created paths.

Therefore, the development of multi-path routing methods and traffic orchestration in DCN is important, taking into account the features and advantages of SDN [9–11].

2 SDN-Based DCN Traffic Orchestration Method

2.1 Organization of Route Information

Based on the analysis of the known methods of multi-path traffic routing and taking into account the organization features of DCN-based SDN operation, the modified route generation algorithm is proposed. This algorithm is based on the advantages of centralized and decentralized routing methods. In SDN controller based on modified channel state routing algorithm an entire set of paths between different nodes is created, taking into account the DCN topology. While a path between two remote nodes is created, paths between their internal nodes are generated simultaneously. This allows to reduce the time complexity of the paths organization by eliminating the paths re-organization between nodes of a previously created path.

The routing information is generated in the reverse direction from the end node V_n to the initial node V_1 on the path P_i. For each following adjacent node V_j of the path P_i the routing table is created (Table 1).

Table 1. The route table of node V_j

Table of node V_i			
Nodes		Route metric	Route load to V_l
Addressee	Adjacent		
V_n	V_l	$M_{i,n}$	D_l

2.2 Procedure of the Multi-path Routing Tables Organization

1. Define a set of nodes $W_1 = \{V_n\}$;
2. $D_i = 0$;
3. $J = 0$;
4. for $j = j + 1$ step 1, create a set of nodes $W_j + 1 = \{V_i | i = 1, \ldots k\}$ adjacent to nodes of set $Z_i\{V_n, V_l, M_i, n, d_i\}$, where k is the sum of the powers of the node set $W_{j1} = \{V_i | i = 1, \ldots k\}$;
5. if $W_{j+1} = \varnothing$ then go to 10 do
6. for i = 1 step 1 to k calculate $Z_i\{V_n, V_l, M_i, n, d_i\}$
7. if $d_j > D_i$ then $D_i = d_j$
8. end;
9. go to 4
10. end.

An Example of the Multi-path Routing Tables Organization

Consider an example of the routing tables organization for transferring information from the node V_7 to the node V_{14} of DCN using the Fat Tree topology (Fig. 2).

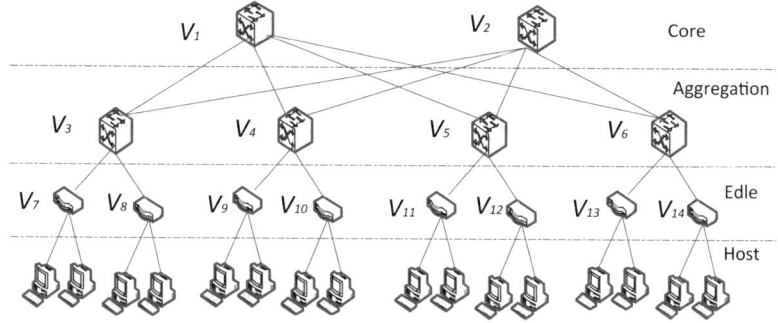

Fig. 2. Fat Tree topology example

Table 2 shows initial values of coefficients d_i of channel load values $L_i = (V_j, V_m)$.

Table 2. Channel load values

L_1	L_2	L_3	L_4	L_5	L_6	L_7	L_8	L_9	L_{10}	L_{11}	L_{11}
V_{14}	V_6	V_6	V_5	V_5	V_2	V_2	V_2	V_1	V_1	V_1	V_3
V_6	V_2	V_1	V_2	V_1	V_3	V_4	V_5	V_3	V_4	V_5	V_7
0.1	0.2	0.7	0.5	0.6	0.6	0.2	0.4	0.3	0.2	0.4	0.1

The Sequence of the Route Tables Generation is as Follows

1. A set of nodes $a = \{V_6\}$, adjacent to node V_{14} is being formed.
2. *For node V_6, generates a route table as follows* (Table 3):

Table 3. The route table of node V_6

Table of node V_6			
Nodes		Route metric	Route load to V_{14}
Addressee	Adjacent		
V_{14}	V_{14}	$M_{i,14}$	0.1

3. $j = 1$;
4. A set of nodes $W_2 = \{V_1, V_2\}$ adjacent to the node V_6 is being formed.
5. Channel load factor $L_2 = (V_6, V_2)$ is $d_2 = 0.2$.
6. $d_2 = D_6$ then $D_i = d_2 = 0.2$.
7. *For node V_2, generates a route table as follows* (Table 4):

Table 4. The route table of node V_2

Table of node V_2			
Nodes		Route metric	Route load *to* V_6
Addressee	Adjacent		
V_{14}	V_6	$M_{i,14}$	0.2

8. Channel load factor $L_3 = (V_6, V_1)$ is $d_3 = 0.7$.
9. $d_3 = D_i$ then $D_i = d_3 = 0.7$.
10. *For node V_1, generate a route table as follows* (Table 5):

Table 5. The route table of node V_1

Table of node V_1			
Nodes		Route metric	Route load *to* V_6
Addressee	Adjacent		
V_{14}	V_6	$M_{i,14}$	0.7

11. $j = 2$;
12. A set of nodes $W_3 = \{V_3, V_4, V_5\}$ adjacent to the node V_2 is being formed.
13. $V_3 = L_9 = (V_3, V_1)$; $V_3 \subset L_6 = (V_3, V_2)$.
14. Channel load factor $L_9 = (V_3, V_1)$ is $d_9 = 0.3$.
15. $d_9 = D_6$ then $D_9 = 0.7$.
16. Channel load factor $L_6 = (V_3, V_2)$ is $d_9 = 0.5$.
17. $d_2 = D_6$ then $D_6 = 0.5$.
18. *For node V_3, generate a route table as follows* (Table 6)

Table 6. The route table of node V_3

Table of node V_3			
Nodes		Route metric	Route load to V_3
Addressee	Adjacent		
V_{14}	V_1	$M_{i,14}$	0.7
V_{14}	V_2	$M_{i,14}$	0.5

$$V_4 \subset L_9 = (V_4, V_1); V_4 \subset L_6 = (V_4, V_2).$$

19. Channel load factor $L_{10} = (V_4, V_1)$ is $d_{10} = 0.3$.
20. $d_{10} = D_2$ then $D_4 = 0.2$.
21. Channel load factor $L_6 = (V_4, V_2)$ is $d_9 = 0.6$.
22. $d_2 = D_6$ then $D_6 = 0.5$.
23. *For node V_4, generate a route table as follows* (Table 7)

Table 7. The route table of node V_4

Table of node V_4			
Nodes		Route metric	Route load to V_3
Addressee	Adjacent		
V_{14}	V_1	$M_{i,14}$	0.2
V_{14}	V_2	$M_{i,14}$	0.2

24. $V_5 = L_9 = (V_5, V_2); V_5 \subset L_{11} = (V_5, V_1)$.
25. Channel load factor $L_{11} = (V_5, V_1)$ is $d_{11} = 0.2$.
26. $d_{11} = D_6$ then $D_5 = 0.4$.
27. Channel load factor $L_6 = (V_5, V_2)$ is $d_2 = 0.5$.
28. $d_2 = D_6$ then $D_6 = 0.5$.
29. *For node V_5, generate a route table as follows* (Table 8)

Table 8. The route table of node V_5

Table of node V_5			
Nodes		Route metric	Route load to V_5
Addressee	Adjacent		
V_{14}	V_1	$M_{i,14}$	0.7
V_{14}	V_2	$M_{i,14}$	0.5

30. Then, the route tables for the nodes are updated as follows (Tables 9, 10 and 11):

Table 9. The route table of node V_1

Table of node V_1			
Nodes		Route metric	Route load to V_1
Addressee	Adjacent		
V_{14}	V_4	$M_{i,14}$	0.2
V_{14}	V_5	$M_{i,14}$	0.6
V_{14}	V_6	$M_{i,14}$	0.7

Table 10. The route table of node V_2

Table of node V_2			
Nodes		Route metric	Route load to V_2
Addressee	Adjacent		
V_{14}	V_4	$M_{i,14}$	0.2
V_{14}	V_5	$M_{i,14}$	0.4
V_{14}	V_6	$M_{i,14}$	0.2

Table 11. The route table of node V_3

Table of node V_3			
Nodes		Route metric	Route load to V_1
Addressee	Adjacent		
V_{14}	V_1	$M_{i,14}$	0.3
V_{14}	V_2	$M_{i,14}$	0.5

Based on the above routing tables, data is transmitted from node V_7 to node V_{14} node with the minimum value $D_1 = 0.3$ in the following order: $V_7 \to V_3 \to V_1 \to V_4 \to V_2 \to V_6 \to V_{14}$.

In case of link load increasing of the selected route, changes have to be made to its routing table. This leads to a reconfiguration of this path. For example, increasing the load on channel $L_{10}(V_1, V_4)$ to $d_{10} \geq 0.5$ will change the contents of the routing table for node V_1 (Table 12):

Table 12. The route table of node V_1

Table of node V_1			
Nodes		Route metric	Route load to V_1
Addressee	Adjacent		
V_{14}	V_4	$M_{i,14}$	**≥ 0.5**
V_{14}	V_5	$M_{i,14}$	0.4
V_{14}	V_6	$M_{i,14}$	0.7

This will change the route accordingly: $V_7 \to V_3 \to \underline{V_1 \to V_5} \to V_2 \to V_6 \to V_{14}$.

2.3   Traffic Orchestration Procedure

1. The transmitting node, based on own route table, determines the path existence with a valid QoS value to the receiving node.
2. If the valid path exists, proceed to step 4.
3. If a valid path is absent, the transmitting node sends a request to SDN controller to create new or reconfigure existing paths.
4. Among the existing paths, the transmitting node select a path with a valid metric and a minimum value of Di (refer below).
5. The sending node informs SDN controller about the selected path and begins the process of data transmission.
6. SDN controller revises the metrics and adjacent paths.
7. SDN controller updates the metrics in the nodes of the selected path and in the adjacent paths.
8. The process of data transmission is complete. The transmitting node informs the SDN controller.

9. SDN controller revises the metrics and adjacent paths.
10. SDN controller updates the metrics in the nodes of the selected path and in the adjacent paths.
11. Finish of the traffic orchestration procedure.

The reconfiguration operation is performed dynamically. SDN controller receives information about the state of the channels from the network switches, and corrects routing tables of the corresponding network switches. This enables to quickly reconfigure the paths and optimize the loading of data transmission channels.

Figure 3 presents the custom modeling application for path reconfiguring procedure considering the changes in the channels load.

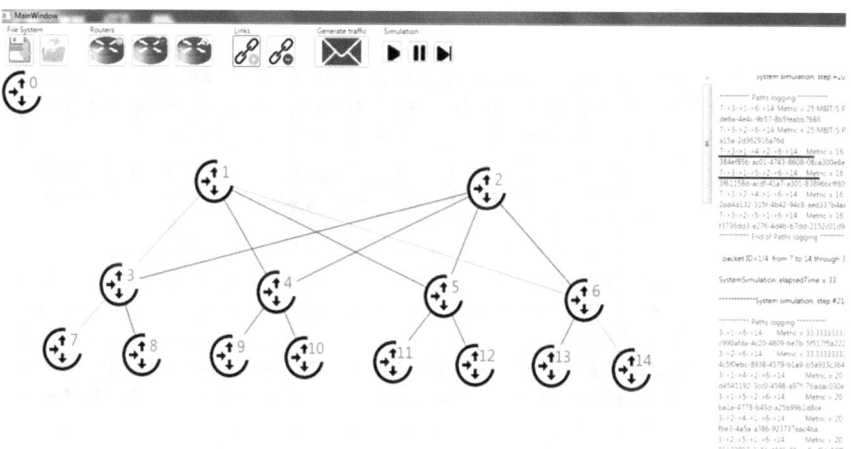

Fig. 3. Modeling application for path reconfiguring procedure

Figure 4 presents the simulation results of the path reconfiguration procedure with increasing load on the channel.

```
********** Paths logging **********
7->3->1->6->14 Metric = 25 MBIT/S PathId = d09840c6-
de6a-4e4c-9b57-8b5feabb7686
7->3->2->6->14 Metric = 25 MBIT/S PathId = d335903c-b381-47f4-
a15a-2d362916a76d
7->3->1->4->2->6->14    Metric = 16,6666666666667 MBIT/S PathId =
384ef85b-ac01-4743-8608-08ca300e8ec4
7->3->1->5->2->6->14    Metric = 16,6666666666667 MBIT/S PathId =
3f61158d-acdf-41a7-a301-83896bcff80f
```

Fig. 4. Path reconfiguration procedure simulation results

In this case, traffic is quickly redirected to node V5 instead of node V4, without additional delay and packet loss in node V1. The process route reconfiguring has no delay. Based on the simulation results, we declare that the proposed method of traffic orchestration makes possible to almost eliminate the delay or packets loss in the process of re-routing. Hence, the more paths are generated in SDN controller, the less chance of delay or packet loss.

3   Conclusion

The paper proposes a method of traffic orchestration, considering the features of the SDN organization, in particular due to the presence of a central controller in the network, reduces the time required to generate multiple routes of access to network resources and simplify the rerouting procedure.

Existence of multiple routes allows to almost eliminate the delay or loss of packets in the process of traffic rerouting. In this case, the more paths are generated in SDN controller, the less chance of delay or packet loss.

Further improvement of traffic orchestration methods could be connected with traffic load forecasting and the nature of the load change.

References

1. Isong, B., Kgogo, T., Lugayizi, F.: Trust establishment in SDN: controller and applications. Int. J. Comput. Netw. Inf. Secur. (IJCNIS) **9**(7), 20–28 (2017). https://doi.org/10.5815/ijcnis. 2017.07.03
2. Kumar, P., Dutta, R., Dagdi, R., Sooda, K., Naik, A.: A programmable and managed software defined network. Int. J. Comput. Netw. Inf. Secur. (IJCNIS) **9**(12), 11–17 (2017). https://doi.org/10.5815/ijcnis.2017.12.02
3. Sahoo, K.S., Mishra, S.K., Sahoo, S., Sahoo, B.: Software defined network: the next generation internet technology. Int. J. Wirel. Microwave Technol. (IJWMT) **7**(2), 13–24 (2017). https://doi.org/10.5815/ijwmt.2017.02.02
4. Han, Y., Seo, S., Li, J., Hyun, J., Yoo, J.H., Hong, J.W.: Software defined networking-based traffic engineering for data center networks. In: Asia-Pacific Network Operations and Management Symposium (2014). https://ieeexplore.ieee.org/document/6996601
5. Chemerinsky, E., Smeliansky, R.: On QoS management in SDN by multipath routing. In: Proceedings International Science and Technology Conference (Modern Networking Technologies) (MoNeTeC) (2014). https://ieeexplore.ieee.org/document/6995581
6. Moza, M., Kumar, S.: Analyzing multiple routing configuration. Int. J. Comput. Netw. Inf. Secur. (IJCNIS) **8**(5), 48–54 (2016). https://doi.org/10.5815/ijcnis.2016.05.07
7. Kulakov, Y., Kogan, A.: The method of plurality generation of disjoint paths using horizontal exclusive scheduling. Adv. Sci. J. **10**, 16–18 (2014). https://doi.org/10.15550/ASJ.2014.10
8. Kulakov, Y., Kopychko, S., Gromova, V.: Organization of network data centres based on software-defined networking. In: Proceedings International Conference on Computer Science, Engineering and Education Applications ICCSEEA 2018, pp. 447–455 (2018). https://link.springer.com/book/10.1007/978-3-319-91008-6

9. Basit, A., Qaisar, S., Syed, H., Ali, M.: SDN orchestration for next generation inter-networking: a multipath forwarding approach. IEEE Access **5**, 13077–13089 (2017). https://ieeexplore.ieee.org/document/7879870

10. Shu, Z., Wan, J., Lin, J., Wang, S., Li, D., Rho, S., Yang, C.: Traffic engineering in software defined networking: measurement and management. IEEE Access **4**, 3246–3256 (2016). http://www.ieee.org/publications_standards/publications/rights/index.html

11. Abbasi, M.R., Guleria, A., Devi, M.S.: Traffic engineering in software defined networks: a survey. J. Telecommun. Inf. Technol. **4**, 3–13 (2016)

Software/Hardware Co-design of the Microprocessor for the Serial Port Communications

Oleksii Molchanov$^{(\boxtimes)}$, Maria Orlova, and Anatoliy Sergiyenko

Igor Sikorsky Kyiv Polytechnic Institute, Kyiv 03056, Ukraine
oleksii.molchanov@gmail.com, aser@comsys.kpi.ua

Abstract. The eight-bit stack processor architecture is proposed, which is designed for the FPGA implementation. The microprocessor with this architecture has small hardware costs, reduced software amount, and ability to add up to hundred new user instructions to its instruction set. The microprocessor architecture is adapted for programming the serial port communications and is able to perform the data stream parsing.

Keywords: Stack processor · Forth · FPGA · VHDL

1 Introduction

In recent years, field programmable gate arrays (FPGAs) became a reasonable alternative to von Neumann architecture-based approaches in computer systems design. Increasing demand of performance and energy-efficiency is neatly fulfilled by capabilities that FPGAs propose.

A need to organize the data exchange through the interfaces such as I2C, SPI, Ethernet and others often occurs when a system on an FPGA is developed. At the same time, it is more rational to use the microprocessor core, which has both small hardware costs, and simple programming and debugging procedures. In addition, such a microprocessor can replace the finite state machines, which are needed for control of a designed system. RISC processors can be considered as those, which match described characteristics.

There is a small selection of universal RISC processors offered by FPGA manufacturers such as Xilinx Picoblaze, Microblaze, Altera Nios, or clones of common microcontrollers, such as i8051 [1–3]. But in many cases, the data exchange is performed using a simple protocol and at a relatively small speed, such as in the case of the I2C interface. In this situation, the capabilities of the RISC microprocessors are used inefficiently.

For the implementation of many application-specific systems in FPGA, it is important to have a configurable microcontroller with both minimized hardware and software. This is dictated by the fact that the memory blocks, which are embedded in FPGA, have significantly limited volume. It is desirable to have such a microcontroller which instruction set can be manually adjusted by the programmer to the needs of the project, to simplify programming, allow program subroutines reuse and, as a result, to

© Springer Nature Switzerland AG 2020
Z. Hu et al. (Eds.): ICCSEEA 2019, AISC 938, pp. 238–246, 2020.
https://doi.org/10.1007/978-3-030-16621-2_22

minimize the program length. Its instruction set has to be adapted for scheduling the data transfer through the interfaces. Implementation of the architecture of such a microprocessor is the goal of this work.

This paper is organized as follows: Sect. 2 gives overview of related works; Sect. 3 describes architecture of the developed microprocessor; Sect. 4 contains modeling results; conclusion of this work is presented in Sect. 5.

2 Related Works

The are several related works.

In [4] authors work on 8-bit input/output processor for performing USB operations. It has the stack architecture and its instruction set consists of seventeen 14-bit instructions. This processor was tested on Altera Cyclone II FPGA and proven to become a good substitution for the big USB IP Cores.

The parameterizable bitstream concept and its hardware implementation using the small processor core is introduced in [5]. This concept introduces the fast generation of parameterizable configurations in the commercial FPGAs and its implementation states significantly reduces the resources (up to 80%) in comparison to the MicroBlaze soft processor when it is used as a configuration generation engine.

The novel soft processor core that executes the native Forth language is presented in [6]. The core is designed to be a replacement of an embedded controller running Forth in a VM. The branch prediction architecture, which is part of the designed core, was introduced in order to the execution speed up.

In [7] a RISC 16-bit microcontroller Little16, is proposed. This novel microcontroller provides the small amount of silicon utilization with highly improved performance for the efficient data flow control. It was tested on Xilinx Zynq7000 FPGA platform and yielded a clock speed of 311 MHz at the cost of 366 LUT-6 blocks and 310 Flip-Flops.

The work [8] presents the 8-bit RISC processor with the reduced instruction set (29 instructions) and pipelining. It has 8-bit ALU, two 8-bit I/O ports, eight 8-bit general purpose registers and 4-bit flag register. The proposed processor is verified in the Xilinx Spartan-6 SP605 Evaluation Platform.

The high performance and low power MIPS microprocessor and its implementation in FPGA are proposed in [9]. The authors use different methods to achieve the high performance and low power consumption. They are unfolding transformation, C-slow retiming technique, and double edge registering. The design was tested in Quartus II 9.1 and Stratix II FPGAs and has demonstrated the high performance of the proposed microprocessor.

Other related works are [10–12]. All mentioned above projects show the high interest to the small processor IP cores which are utilized for the simple control tasks. Many of them have the minimized hardware volume, but a few processors are well fitted for the serial communications.

3 Microprocessor Architecture for Serial Port Communications

3.1 Stack Processors

The stack processor architecture is distinguished among all microprocessor architectures. Its instruction set differs in that the operands have implicit addressing because they are usually placed in a few fixed stack registers. Therefore, such instructions have a short instruction length because they have the implicit register addressing. Since these instructions support algorithms that actively use the stack addressing, the programs that are composed for this processor occupy very small memory volume [13].

Various authors have developed several projects of stack processors, which are implemented in the FPGA and which are available for reproduction [14–16]. All of them have 16-bit instructions and process 16-bit data. It is shown in [16], that the stack processor has approximately 2.5 times less program length than the program for the Xilinx MicroBlaze processor when the data exchange protocols through the serial interfaces are implemented. In addition, all stack processors allow the designer to extend the instruction set. To do so, the appropriate changes should be made to the description of the processor at the register transfer level.

Consequently, the architecture of the stack processor provides both firmware amount and hardware costs minimization. In addition, it is easy to develop compilers for such architecture, because, as a rule, its instruction set is a subset of the Forth language commands [17]. It is known that this language is convenient for both grammatical parsing of lines and for the interpretation of high-level language operators. The stack processor assembly language has the same syntax as the Forth language [13]. Therefore, it is attractive to develop the stack processor architecture, which gives not only minimized hardware costs but also simplified implementation of user instructions, which are adapted to the serial port communications.

3.2 SM8 Microprocessor

The structure of the developed SM8 microprocessor is shown in Fig. 1. This processor has the well-known two-stack architecture. It consists of a program counter PC, Data RAM, Program ROM, instruction register IR, user instruction encoder UIE, return address stack RS, data stack DS, ALU and peripheral registers R0, ..., Rn, n < 32. The registers T, N, P are the top registers of the DS-stack and are designed to store the operands and the ALU results. The Program ROM has the volume up to 7936 bytes, and the Data RAM has up to 256 bytes, and both of them have a common address space.

The SM8 microprocessor instructions are sampled in Table 1. All instructions, except CALL, LIT, and IF, have the 8-bit length. The branch and input-output instructions are executed in two cycles, and the rest of the instructions are single cycle instructions. Due to the frequent use of the CALL, LIT, and IF instructions, the average duration of one instruction execution is 1.5 clock cycles.

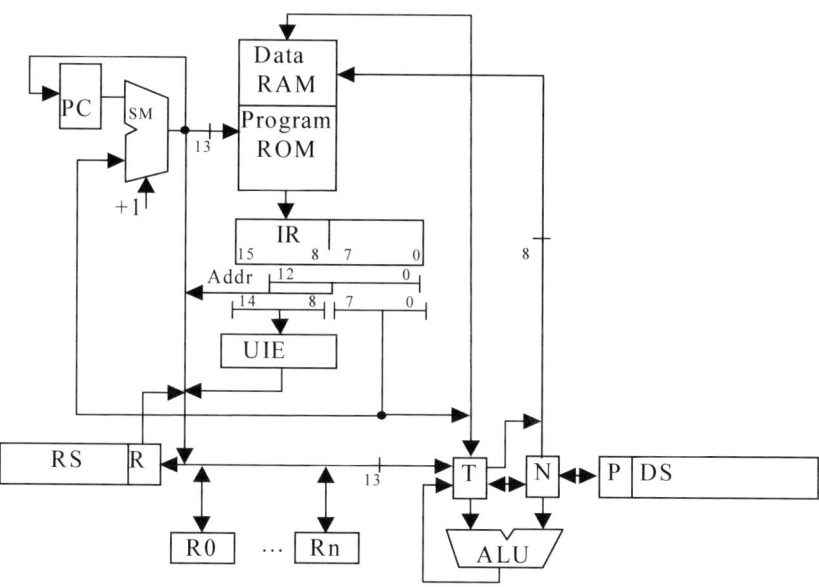

Fig. 1. Structure of the SM8 microprocessor

3.3 User-Defined Instructions

The user-defined instructions are implemented as follows. First, the instruction code is associated with the specified address in the user subroutine library, where the user-defined subroutine is located. Second, when the control-flow approaches this instruction, it writes the instruction code to the IR register. Then the code is encoded by UIE to the address of the subroutine, associated with it. The return address (address of the next instruction) is saved in the RS stack. After that, the control-flow is passed to the first action (sub-instruction) of the subroutine (subroutine is 'called') and all actions which it contains are executed. The return of the control-flow from the subroutine to the next instruction is performed by the RET instruction. This subroutine can also read and process the operand fields that follow the byte of the opcode. But the return address in the R register must be properly corrected.

This instruction is coded by a single byte comparing to the two-byte CALL instruction. Therefore, the user instructions can save the software memory volume comparing to the equivalent instruction implementation using the CALL instruction.

The user-defined instructions can be stored in both Program ROM and Data RAM. Thus, a microprocessor can store a certain dynamic data processing script, which is formed by the user instructions and respective data bytes for them. It can perform a line parsing as well. For example, this line can be a string of decimal calculator operations and digits.

Table 1. Instruction set of the SM8 microprocessor

Name	Instruction	Description
CALL	001 Addr	PC + 1 → R, PC = Addr, subroutine call
INR	010 n	Rn → T, data receiving
OUTR	011 n	Rn = T, data sending
NOP	0000 0000	No operation
LIT	0000 0001 B	B → T, constant input
IF	0000 0010 D	PC = PC + D by T = 0, else PC = PC + 1
DUP	0000 0110	N = T
SWAP	0000 1001	X = T, T = N, N = X
@	0000 1010	T = RAM[T], memory reading
!	0000 1011	RAM[T] = N, memory writing, T →
R>	0000 1100	R → T
>R	0000 1110	T → R
RET	0000 1101	R → PC, return from the subroutine
DROP	0000 1111	T →, stack purge
NOT	0001 0000	T = not T
OR	0001 0001	T = T or N
AND	0001 0010	T = T and N
XOR	0001 0011	T = T xor N
ADD	0001 1000	T = T + N
INC	0001 1001	T = T + 1
SUB	0001 1010	T = T − N
DEC	0001 1011	T = T − 1
	1xxx xxxx	User instruction

3.4   Dynamic Reconfiguration

A common problem for many FPGA-based architectures is the reconfiguration process. Usually, the need in reconfiguration leads to recompilation of a hardware circuit, which is a very CPU-intensive and time-consuming task (it can last from minutes to hours). The solution to this problem was presented in several works [18, 19] as an implementation of task-specific architectures that can be reconfigured 'on-the-fly'. For example, in [19] the authors propose the segment-based architecture for the XML filtering. The sequence of configured segments implements the XPath pattern of the interesting part of the whole XML. This pattern can be changed 'on-the-fly', and the hardware reconfiguration takes from nano- to microseconds.

Another approach is implemented in SM8. The processing script, which is saved in RAM, can be rewritten or loaded from other memory, for example, from ROM. Such simple rewriting allows the system changing the behavior of a chip in terms of microseconds. As far as such action is performed in a synchronization point, neither data loss nor wrong behavior happens. In such a way, the segments in [19] are reconfigured without stop of the input data processing with preserving all currently processed XML node tree parts.

3.5   Assembler of the SM8 Microprocessor

An assembler was developed for programming the SM8 microprocessor. The assembler is written in Java and is called from the command line. Below, an example of a program in the SM8 assembly language is shown, which performs a single-byte transfer to the I2C port.

```
DEFINE nap 9        \ memory address width
DEFINE WAITRDY 82h \ user instruction - wait for port is ready
DEFINE DELAYN  83h \ user instruction - delay for N cycles
EQU START  2
EQU A_SEND 4
EQU D_SEND 5
EQU STOP   12
EQU PAUSE  15
ORG 256             \ program segment begin
\Write byte to I2C
: WR2I2C (r1 - I2C address, r2 - inner address, r3 - byte,
                        r8 - I2C data, r9 - I2C control)
   lit START outr r9
   inr r1 outr r8 lit A_SEND outr r9 WAITRDY
   inr r2 outr r8 lit D_SEND outr r9 WAITRDY
   inr r3 outr r8 lit D_SEND outr r9 WAITRDY
   lit STOP   outr r9
   lit PAUSE  outr r9
   lit 100    DELAYN xor if END
;
: DELAYN           (N -- - N cycles)
   dec dup ifn DELAYN
;
: WAITRDY          (do while rdy=1)
   inr r10         \0-th bit = rdy
   lit 1 and if WAITRDY
;
: END
```

The assembly language of the SM8 core uses the syntax of the Forth language. Therefore, the comments here are enclosed in parentheses or followed after a backslash. The label follows a colon. Operators and literals are separated by spaces. A semicolon indicates the subroutine return instruction.

Some special operators (called the pragmas) are used in the script for the special purpose. Table 2 contains the description of all pragmas.

As it is seen from the example above, none of the subroutines contain the RET instruction. It is explained by the fact that the semicolon sign represents the RET in struction in the Forth language. The user also can specify its own RETs in his sub-routine if needed.

The assembler generates two VHDL files, which contain the data and programs in the memory and the user instruction encoder content. As a result, this assembly language by its user properties occupies an intermediate position between the usual assembly language and the high-level language. Thanks to this, writing and debugging of programs is significantly accelerated. Besides, the VHDL model of the SM8 microprocessor core is equipped with a disassembler. Such a feature significantly simplifies the program debugging in the VHDL simulator.

Table 2. Pragmas set for SM8 assembler

Name	Arguments	Description
DEFINE	<name> <value>	Associates <name> with <value> for assembler. <name> can be a common setting attribute (like nap – memory address width) or name of a user-defined instruction. <value> is a value for a setting or code of instruction
EQU	<name> <value>	Defines constant with name <name>. Each occurrence of <name> in script is replace with <value>
ORG	<shift>	Defines a shift of next and all the following instructions in memory to address <shift>

4   Experimental Results

The SM8 microprocessor core is described in VHDL and has synthesized for different FPGA circuits. Table 3 presents the results of the SM8 microcontroller synthesis in the Xilinx Spartan-6 FPGA while setting the optimization parameters for hardware costs. Also, the parameters of the microprocessors, which were synthesized in the same conditions, are presented in this table for a comparison. The analysis of the table shows that the SM8 microprocessor has the lowest hardware costs in the look-up tables (LUTs), and in configured logical blocks (CLBs), and the highest speed in millions of instructions per second (MIPS) among stack processors. This is explained by the fact that the reduction of the data bit width up to eight digits reduces both hardware costs and delay in ALU.

Table 3. Parameters of the microprocessor core configured in Xilinx FPGA

Microprocessor	Instruction bit width	Hardware costs		Max. clock frequency, MHz	Speed, MIPS
		LUTs	CLBs		
FS8051 [20]	8, 16, 24	1293	470	89	30
KCPSM6 [2]	18	87	26	140	70
MSL16 [14]	16	235	61	100	67
b16-small [15]	16	280	73	100	50
J1 [16]	16	342	93	106	70
SM8	8, 16	181	50	140	94

Other synthesis results comparison is presented in Table 4. These results show core parameters for some Intel FPGAs.

The waveforms from the ActiveHDL simulator showing the first clock cycles of the processor operation after its reset are presented in Fig. 2. The cop signal shows the opcode as a result of the disassembler function.

Table 4. Microprocessor core parameters in Intel FPGA

Microprocessor	FPGA	Hardware volume	Max. clock frequency, MHz	Speed, MIPS
Nios II/f [21]	MAX10	2268 LE	150	135
Nios II/f [21]	Cyclone 5	867 ALM	170	150
SM8	MAX10	1164 LE	150	100
SM8	Cyclone 5	210 ALM	205	140

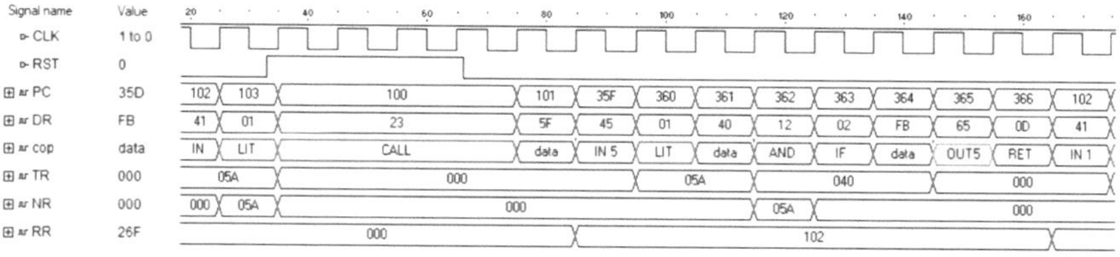

Fig. 2. First cycles of simulation in ActiveHDL

5 Conclusions

The proposed SM8 microprocessor core has small hardware costs at high performance and reduced hardware volume. It is designed to implement simple control algorithms, for example, to control the data exchanges through the I2C interface. The core is described in VHDL and can be implemented in an FPGA of any series. The programmer has the ability to add his own instructions to the instruction set without changing the core description. The developed assembler provides to write and compile the programs written in the Forth language style. This simplifies the design of devices that implement the protocols for the serial port communications through interfaces such as RS232, I2C, SPI, Ethernet.

References

1. Nurmi, J. (ed.): Processor Design. System-on-Chip Computing for ASICs and FPGAs. Springer, Netherlands (2007)
2. Chapman, K.: PicoBlaze for Spartan-6, Virtex-6, and 7-Series (KCPSM6). Xilinx, Inc., San Jose (2012)
3. Meyer-Baese, U.: Digital Signal Processing with Field Programmable Gate Arrays, 4th edn. Springer, Heidelberg (2014)

4. Al-Dujaili, A., Hiung, L.H., Tan, S.: ASH1: a stack-based input/output processor for USB operations. In: Proceedings of 2015 IEEE International Circuits and Systems Symposium (ICSyS), pp. 76–79. IEEE (2015)
5. Abouelella, F., Bruneel, K., Stroobandt, D.: Efficiently generating FPGA configurations through a stack machine. In: Proceedings of 2010 International Conference on Field Programmable Logic and Applications, pp. 35–39. IEEE (2010)
6. Hanna, D.M., Jones, B., Lorenz, L., Porthun, S.: An embedded Forth core with floating point and branch prediction. In: Proceedings of 2013 IEEE 56th International Midwest Symposium on Circuits and Systems (MWSCAS'2013), pp. 1055–1058. IEEE (2013)
7. Hin, W.K., Chiu-Sing, C., Oliver: Little16 - small scale 16-bit controller architecture for FPGA systems flow control. In: Proceedings of TENCON 2015 - 2015 IEEE Region 10 Conference, pp. 1926–1929. IEEE (2015)
8. Jeemon, J.: Low power pipelined 8-bit RISC processor design and implementation on FPGA. In: Proceedings of 2015 International Conference on Control, Instrumentation, Communication and Computational Technologies (ICCICCT), pp. 476–481. IEEE (2015)
9. Daghooghi, T.: Design and development MIPS processor based on a high performance and low power architecture on FPGA. Int. J. Mod. Educ. Comput. Sci. (IJMECS) 5(5), 49–59 (2013). https://doi.org/10.5815/ijmecs.2013.05.06
10. Afzal, S., Hafeez, F., Akhter, M.O.: Single chip embedded system solution: efficient resource utilization by interfacing LCD through softcore processor in Xilinx FPGA. IJIEEB 7(6), 23–27 (2015). https://doi.org/10.5815/ijieeb.2015.06.04
11. Oyetoke, O.O.: A practical application of ARM Cortex-M3 processor core in embedded system engineering. Int. J. Intell. Syst. Appl. (IJISA) 9(7), 70–88 (2017). https://doi.org/10.5815/ijisa.2017.07.08
12. Rani, A., Grover, N.: Novel design of 32-bit asynchronous (RISC) microprocessor & its implementation on FPGA. Int. J. Inf. Eng. Electron. Bus. (IJIEEB) 10(1), 39–47 (2018). https://doi.org/10.5815/ijieeb.2018.01.06
13. Koopman, P.: Stack Computers: The New Wave. Ellis Horwood, Mountain View (1989)
14. Leong, P.H.W., Tsang, P.K., Lee, T.K.: A FPGA based forth microprocessor. In: Proceedings of the IEEE Symposium on Field-Programmable Custom Computing Machines (FCCM). IEEE, Napa Valley (1998)
15. Paysan, B.: b16-small—Less is More. In: Proceedings EuroForth 2004, 9 July 2006
16. Bowman, J., Garage, W.: J1: a small Forth CPU Core for FPGAs. In: Proceedings EuroForth'2010, pp. 1–4, January 2010
17. Kelly, M., Spies, N.: Forth: A Text and Reference. Prentice Hall, Englewood Cliffs (1986)
18. Najafi, M., Sadoghi, M., Jacobsen, H.-A.: Configurable hardware-based streaming architecture using online programmable-block. In: ICDE 2015, Seoul, South Korea, 13–17 April, pp. 819–830 (2015)
19. Teubner, J., Woods, L., Nie, C.: XLynx—an FPGA-based XML filter for hybrid XQuery processing. ACM Trans. Database Syst. (TODS) 38(4), Article 23 (2013)
20. Maslennikov, O., Shevtshenko, J., Sergyienko, A.: Configurable microcontroller array. In: Proceedings Parallel Computing in Electrical Engineering, PARELEC'02, 25 September 2002, Warsaw, Poland, pp. 47–49. IEEE (2003)
21. Nios II Performance Benchmarks. DS-N28162004. Intel, pp. 1–7 (2018)

Hardware Implementation Neural Network Controller on FPGA for Stability Ball on the Platform

Peter Kravets and Volodymyr Shymkovych[✉]

National Technical University of Ukraine
"Igor Sikorsky Kyiv Polytechnic Institute", 37 Prosp. Peremohy, Kyiv, Ukraine
shymkovych.volodymyr@gmail.com

Abstract. This study is about a development and investigation of the neural network controller of the stabilization system of a moving object on a plane with its hardware implementation on the FPGA. It consists of a designing balloon balancing model, hardware and software for this layout. The platform ball balancing system consists of a black plastic plate with a white table tennis ball, a drive mechanism for tilting a plate around two axes, a digital video camera, tracking the position of the ball, hardware and software that processes the information and manages the system in real time mode. The physical modeling of the system is carried out and the equation of motion of the ball in the plane is deduced. The equations of motion are nonlinear and unsuitable for the synthesis of control systems based on linear control theory. A generalized block diagram of a neurocontroller based on the FPGA, which implements the basic components of neural network control systems, is developed. To control the position of the ball on the platform a neural network control system with feedback was developed. The scheme is a classical scheme of specialized inverse training.

Keywords: Neural networks · FPGA · Non-linear systems · Real-time control

1 Introduction

Neural network control systems represent a new high-tech direction in the control theory and relate to a class of nonlinear dynamical systems [1, 2]. High speed through the parallelization of incoming information coupled with the ability to teach neural networks makes this technology very attractive for the creation of control devices in automatic systems [2]. Neural networks can be used to construct regulatory and adjustment devices, reference, adaptive, nominal and inverse-dynamic models of the object. On the basis of these models, observation and evaluation of the parameters of the control object (OC), observation and estimation of the magnitude of the perturbing systems in operation, the search or computation of the optimal control change control program, the identification of the OC, the prediction of the state of the OC, etc. is performed [2, 3]. Ability to study for any given principle of operation allows you to create automatic control systems, optimal in speed, energy consumption, etc., while, of course, the implementation of several principles of operation and transition from one to another. Trained neural networks do not require significant time-consuming

© Springer Nature Switzerland AG 2020
Z. Hu et al. (Eds.): ICCSEEA 2019, AISC 938, pp. 247–256, 2020.
https://doi.org/10.1007/978-3-030-16621-2_23

calculations, and therefore systems with neural networks have much better dynamics. They are a universal means for modeling complex nonlinear OCs and finding solutions in improper problems [2–4].

Ways of realization of neural network control systems should be guided by the wide application in industrial conditions. Being versatile and flexible. To study and adapt in real time, to be simple and cheap, therefore the most promising means can be considered FPGA [5–7]. With the advent of FPGA, the design of digital chips has ceased to be the fate of exclusively large enterprises with volumes of production in tens and hundreds of thousands of crystals. The design and production of a small batch of unique digital devices became possible in the conditions of design and development units of industrial enterprises, in research and educational laboratories and even in the conditions of home radio amateur sites. Industrially produced "billets" of programmable chips with electrical programming and automated translation process of the user circuit into a sequence of programming pulses make the design of new digital devices comparable to software development [8–15].

To construct the layout of the neural network controller as a control object, a system for balancing the ball is selected on the platform, capable of setting the ball to a given point. The platform leaning on each of the two horizontal axes controls the position of the ball on the platform. For each axis, the tilt angle of the platform is carried out under the action of an electric motor. The position of the ball on the platform is traced with the camcorder. Such an object of control is a complex multi-dimensional, multifunctional object and designed to demonstrate the work of neural network control systems synthesized on previously developed components [16–18].

The purpose of the work is to develop a model of the neural network controller of the ball stabilization system on a platform with FPGA hardware implementation.

2 Layout Design

The layout will consist of two subsystems, a subsystem for determining the position of the ball on the plane and the subsystem controlling the tilt angle of the platform. The primary task is to determine the position of the ball, precisely, reliably and in a non-bulky and inexpensive way. The various options that were considered are listed below. The relative advantages and disadvantages are also indicated:

1. With touch screen, when the touch screen is pasted onto the platform: lack of information, it can be difficult to implement.
2. Tracking the position of the ball with a digital camera: not expensive, requires the use of additional software, requires the use of additional structure to install the camera.
3. Using a resistive grid on a platform (2D potentiometer): the method is excessive and cumbersome.
4. Grid of infrared sensors: the method is cumbersome and excessive.
5. Tracking 3D motion with a ball using an infrared ultrasonic transponder attached to a ball that exchanges signals with three remotely located towers: very precise measurements, the need for an additional device as a whole is very expensive.

Based on the above positive aspects and the disadvantages associated with each choice, it was decided to track the position of the ball with a digital video camera.

The next task was to design the mechanism of the slope of the plate. The plate should lean over its two axes in order to be able to balance the ball. For this design, the following options were considered:

1. Two actuators are attached to two corners of the plate, which is supported by a ball joint in the center, as two necessary degrees of motion.
2. Installing the plate on the suspension rings. One engine converts a pivot suspension, which provides one degree of rotation; the second engine rotates the plate relative to the ring thus providing a second angle of rotation.
3. Using the device – a belt and a pulley, turn the plate with two motors.

In this case, the first option is selected. Figure 1 shows the appearance of the entire system in the assembly, including the mechanism of the slope of the plate and reading the position of the ball with the camcorder.

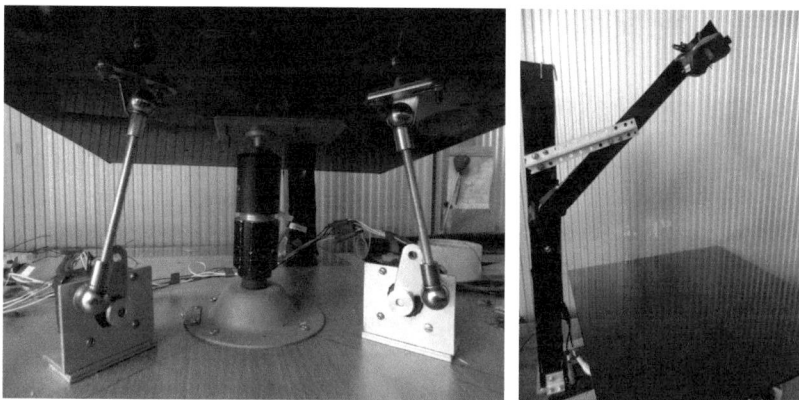

Fig. 1. Platform control and camcorder

The ball balancing system on the platform consists of a black plastic acrylic-lined plastic plate with a white table tennis ball, a drive mechanism for tilting the plate around two axes, a digital video camera, balloon tracking, hardware and software that manages information and controls the system in real time.

Each engine interacts with one of the axes of the plate angle and is connected to the plate, using the spatial coupling mechanism, as shown in Fig. 1. Each of the sides of the mechanism of spatial coupling consists of a parallelogram. This ensures that, for small movements around the equilibrium, the angles of the plate are equal to the corresponding angles of the engine. The plate is connected to the stand with the help of a central movable support.

So the stand consists of: a sensor for finding the position of the ball – a digital video camera; two drives, which are mounted perpendicularly to each other and are responsible for turning on X and Y plane; as equipment for the implementation of the regulator – board with a FPGA processor; As a recognition object, a platform with a label in the center acts; control object – the ball.

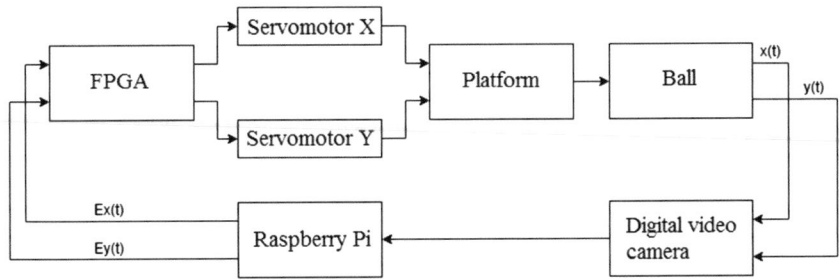

Fig. 2. Block diagram control system of stabilization object

The block diagram control system of stabilization object is shown in Fig. 2. The block diagram consists of the following components: a board with an FPGA processor – implements a neural network controller wish adaptation algorithm based on the language of the description of the VHDL architecture: an error is encountered at the input; at the output we obtain the value of the PWM signal – the turning angle of the servomotor, which controls the axes of the plane with the ball; Raspberry Pi's single-board computer – is in the back of the ASU, is responsible for processing the video stream – the input stream video from the digital video camera is received, at the output we get an error, which is submitted to the regulator; As a sensor for capturing the position of an object, a digital video camera is in place, it is positioned above the plane with a ball; There are two actuators in the ASU: servo responsible for the inclination of the plane at different angles – they regulate the position of the ball; as a management object serves as a ball for table tennis.

3   Physical Modeling of the System

The following assumptions are used in the simulation of the above-described physical system:

1. It is assumed that the friction-slip between the ball and the plate is high enough to prevent the ball from slipping onto the plate. This limits the degree of system freedom, and makes the equation of motion simpler.
2. The rotation of the ball around its vertical axis is expected to be negligible.
3. Friction-rolling between a ball and a plate - to neglect.

4. It is assumed that there will be a small movement of the plate with equilibrium configuration. This ensures that the angles of the plate will be approximately equal to the angles of the engine.
5. The plate is supposed to have maximum symmetry.
6. The physical model of the controlled object is shown in Fig. 3, where the X-Y-Z is the base of the frame. The platform has two degrees of freedom, and its orientation is determined by two angles (q_1 and q_2), which represent the rotation points (1–2). The frame x''-y''-z'' is a plate fixed to the reference frame, while x'-y'-z' is intermediate.

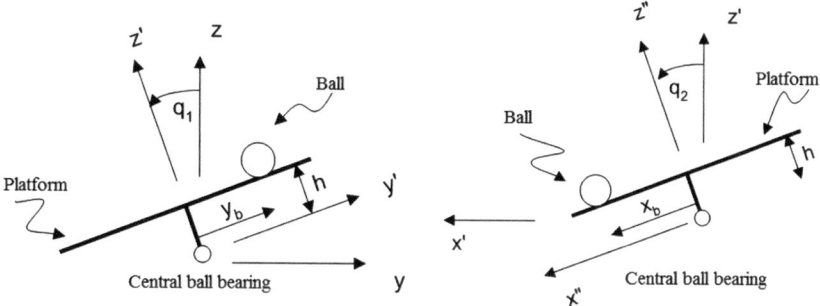

Fig. 3. Physical simulation of the ball movement on the platform

The scheme of the spatial relationship is shown in Fig. 4. The connection is rigidly attached to the base of the plate and connected to the ground at the point O. The two motors are connected to the traction by fixed couplings. The rest are on ball joints.

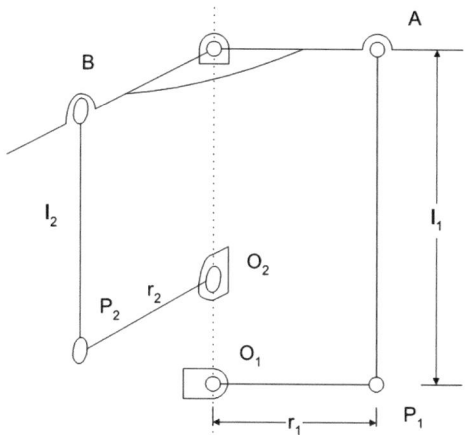

Fig. 4. Platform control mechanism

Calculation of the degree of freedom: the number of solids $(N) = 5$; number of axes $(P) = 2$; number of U-shaped compounds $(U) = 1$; number of ball joints $= 4$;

The number of excess degrees of freedom is two (The rotations of two vertical ties in their axes constitute two reserve degrees of freedom, since they do not affect the state of the system in any case).

Thus, the plate and the mechanism of spatial communication have two degrees of freedom, as expected. It also equals the number of inputs to the system, controlling the signal by two electric motors.

Based on the foregoing, it is obvious that of the four variables – θ_{m1}, θ_{m2}, q_1, q_2, but only two are independent, since the mechanism has two degrees of freedom. Thus, there are two following kinematic equations of communication, which transform the angle of rotation of the motor shafts into the angle of the platform

$$(r_1 \cos q_1 - r_1 \cos \theta_{m1})^2 + (r_1 \sin q_1 - r_1 \sin \theta_{m1} + l_1)^2 = l_1^2,$$
$$(r_2 \cos \theta_{m2} - r_2 \sin q_2 \cos q_1 + l_2)^2 + (r_2 \cos q_2 - r_2 \cos \theta_{m2})^2 + r_2 \sin q_2 \sin q_1 = l_2^2. \tag{1}$$

It is noticed that, in the general case, the angle of rotation of the plate q_2 is related to the angles of rotation of the engines θ_{m1} and θ_{m2} by the complex nonlinear equations given above. However, for small movements on the balance of balls on the platform, it is possible to reduce the expressions to the following linear expressions:

$$q_1 = \theta_{m1},$$
$$q_2 \cong \theta_{m2}.$$

The validity of this assumption is also verified experimentally. It has been established that for the corresponding range of work, the correspondence between the angle of rotation of the engine and the angle of inclination is satisfactory.

Equations of motion for this system can be obtained using Newton's laws or the Lagrange equations. For this case, these methods have been used to test the final results. A complete nonlinear mathematical model is described by a system of equations:

x – coordinate:

$$m_b g r_b \sin q_2 \cos q_1 - m_b r_b \left[\begin{array}{l} (h+r)\ddot{q}_2 - y_b \ddot{q}_1 \sin q_2 - x_b \dot{q}_2^2 - x_b \dot{q}_1^2 \sin^2 q_2 + \\ + (h+r_b)\dot{q}_1^2 \sin q_2 \cos q_2 - 2\dot{y}_b \dot{q}_1 \sin q_2 + \ddot{x}_b \end{array} \right] \tag{2}$$
$$- I_b((\ddot{x}_b/r_b) + \ddot{q}_2) = 0$$

y – coordinate:

$$m_b g r_b \sin q_1 - m_b r_b \left[\begin{array}{l} x_b(\ddot{q}_1 \sin q_2 + \dot{q}_2 \dot{q}_1 \cos q_2) - (h+r_b)(\ddot{q}_1 \cos q_2 + \dot{q}_2 \dot{q}_1 \sin q_2) + \\ + \dot{q}_2 \dot{q}_1 (h+r_b) \sin q_2 - \dot{y}_b \dot{q}_1^2 + x_b \dot{q}_2 \dot{q}_1 \cos q_2 + 2\dot{x}_b \dot{q}_1 \sin q_2 + \ddot{y}_b \end{array} \right]$$
$$- I_b(\ddot{y}_b/r_b + \ddot{q}_1 \cos q_2 \dot{q}_2 \dot{q}_1 \sin q_2) = 0$$

$$\tag{3}$$

The equations of motion are nonlinear and unsuitable for the synthesis of control systems based on linear control theory. Such a control object as a ball on a platform is a complicated nonlinear object, as can be seen from Eqs. (2) and (3), so it requires a neural network control system that belongs to a class of nonlinear dynamical systems. Next, consider the general structural diagram of the neurocontroller.

4 General Structure Diagram of the Neurocontroller

Modern facilities for managing technical and technological objects include a computer kernel, the means for entering analogue and digital information, means for outputting analogue and digital information, data exchange facilities with other computing devices, visualization elements and operational management. Such controls are based on industrial computers, freely programmable controllers, specialized controllers and microcomputers. Next, the controller is considered for controlling dynamic objects, in which as a computing core of the FPGA.

Figure 5 presents a generalized structure diagram of a neurocontroller, where a computer based neural network based on artificial neural networks is used as a computing core.

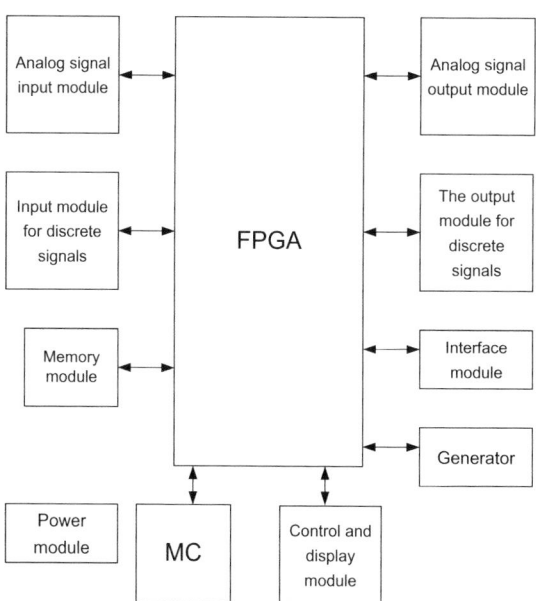

Fig. 5. General structure diagram of the neurocontroller

The structure of the neurocontroller includes: FPGA – a programmable logic integrated circuit; MC – microcontroller; analog signal input module; analog output signal output module; the module of input of discrete signals; the module of output of discrete signals; memory module; module of interfaces; control and display module; generator; FPGA is the central element of this structure, which implements neural network control algorithms, including the algorithm of training the network in "on line" mode.

The microcontroller (MC) performs auxiliary functions, such as loading a configuration sequence in a FPGA chip, implements a surveillance function.

Input/output modules of analog and digital signals for input/output information from/to the system or control object.

The memory module provides data storage and the FPGA configuration sequence.

In this block diagram, there are two computing cores: a microcontroller (MIC) and an FPGA. Depending on the management tasks, the microcontroller can be: the main element of the neural network controller, and the FPGA is assigned the role of a "fast" calculator fragments of the general algorithm of the controller connected with the neural computing; to supplement on an equal program of work of the FPGA, or to be a purely technical element of the neural network controller, which provides the operating modes of the FPGA. However, in all cases, the FPGA is intended for the implementation of the neural network elements of the control system, therefore, in the future, we will focus mainly on the FPGA.

5   Development of Management System and Software

To control the position of the ball on the platform a neural network feedback control system is presented, shown in Fig. 6. The scheme is a classical scheme of specialized inverse training. By the method described in [17], we synthesize a neural network component such as the inverse (inverse) model of the control object. In Fig. 6, the neurocontroller includes an inverse model of the control object, a component of its debugging, developed by the method described in [18] and the inverse relationship.

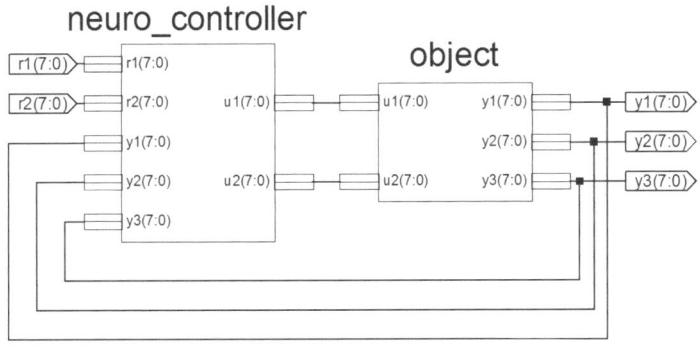

Fig. 6. Control system model

The scheme shown in Fig. 6 is a model of the entire control system, including the object on the basis of previously developed neural network components, in the FPGA loaded the neurocontroller itself.

6 Conclusions

A generalized block diagram of a neurocontroller based on the FPGA, which implements the basic components of neural network control systems, is developed.

The model of the neurocontroller for the system of stabilization of the moving object on a limited plane was implemented and its research was carried out, which confirmed the high efficiency of work.

Such a control object as a ball on a platform is a complicated nonlinear object, therefore, for managing it, a neural network control system, which belongs to a class of nonlinear dynamic systems, was developed. The neural network control system, developed in this work, functions and adapts in real time and generates rather complicated management functions. When the control system is on, the ball will be balanced at any given point on the plate. It can also be aimed at moving from one point to another and staying there. Using this control system you can set the trajectory of the ball movement on the platform, for example, circular motion.

The built and tested system can be additionally used as an excellent test bench for testing various other control schemes. In the future, more efficient controllers can be developed on the basis of FPGAs to achieve even higher performance.

References

1. Kravets, P.I., Shymkovych, V.M., Samotyy, V.: Method and technology of synthesis of neural network models of object control with their hardware implementation on FPGA. Paper Presented at the Proceedings of the 2017 IEEE 9th International Conference on Intelligent Data Acquisition and Advanced Computing Systems: Technology and Applications, IDAACS 2017, vol. 2, pp. 947–951 (2017). https://doi.org/10.1109/idaacs.2017.8095226
2. Samotyy, V., Telenyk, S., Kravets, P., Shymkovych, V., Posvistak, T.: A real time control system for balancing a ball on a platform with FPGA parallel implementation. Tech. Trans. **5**, 109–118 (2018)
3. Kravets, P.I., Lukina, T.I., Zherebko, V.A., Shimkovich, V.N.: Methods of hardware and software realization of adaptive neural network PID controller on FPGA-Chip. J. Autom. Inf. Sci. **43**(4), 70–77 (2011)
4. Doroshenko, A., Shymkovych, V., Fedorenko, V.: Software means of modeling of the vector type of reactive engine control system. In: CEUR Workshop Proceedings 2139, pp. 296–305 (2018)
5. Artem, V., Volodymyr, S., Ivan, V., Vladyslav, V.: Research and development of a stereo encoder of a FM-transmitter based on FPGA. In: Hu, Z., Petoukhov, S., Dychka, I., He, M. (eds.) Advances in Computer Science for Engineering and Education. ICCSEEA 2018. Advances in Intelligent Systems and Computing, vol. 754. Springer, Cham (2019). https://doi.org/10.1007/978–3-319-91008-6_10

6. Kravets, P.I., Shymkovych, V.M., Zubenko, G.A.: Technology of hardware and software implementation of artificial neurons and artificial neural networks by means of FPGA. Visnyk NTUU "KPI" Informatics, Operation and Computer Science, № 55, pp. 174–180 (2012)

7. Mukhin, V., Volokyta, A., Heriatovych, Y., Rehida, P.: Method for efficiency increasing of distributed classification of the images based on the proactive parallel computing approach. Adv. Electr. Comput. Eng. **18**(2), 117–122 (2018). https://doi.org/10.4316/AECE.2018. 02015

8. Hu, Z., Mukhin, V., Kornaga, Y., Volokyta, A., Herasymenko, O.: The scheduler for distributed computer systems based on the network centric approach to resources control. Paper Presented at the Proceedings of the 2017 IEEE 9th International Conference on Intelligent Data Acquisition and Advanced Computing Systems: Technology and Applications, IDAACS 2017, vol. 1, pp. 518–523 (2017). https://doi.org/10.1109/idaacs.2017. 8095135

9. Chu, P.P.: RTL Hardware Design Using VHDL. Coding for Efficiency, Portability, and Scalability, p. 669. Wiley, Hoboken (2006)

10. Mohammed, R.H., Elnaghi, B.E., Bendary, F.A., Elserfi, K.: Trajectory tracking control and robustness analysis of a robotic manipulator using advanced control techniques. Int. J. Eng. Manuf. (IJEM) **8**(6), 42–54 (2018). https://doi.org/10.5815/ijem.2018.06.04

11. Karam, E., Mjeed, N.: Modified integral sliding mode controller design based neural network and optimization algorithms for two wheeled self balancing robot. Int. J. Mod. Educ. Comput. Sci. (IJMECS) **10**(8), 11–21 (2018). https://doi.org/10.5815/ijmecs.2018.08.02

12. Teslyuk, V., Beregovskyi, V., Denysyuk, P., Teslyuk, T., Lozynskyi, A.: Development and implementation of the technical accident prevention subsystem for the smart home system. Int. J. Intell. Syst. Appl. (IJISA) **10**(1), 1–8 (2018). https://doi.org/10.5815/ijisa.2018.01.01

13. Geraldine Bessie Amali, D., Dinakaran, M.: A new quantum tunneling particle swarm optimization algorithm for training feedforward neural networks. Int. J. Intell. Syst. Appl. (IJISA) **10**(11), 64–75 (2018). https://doi.org/10.5815/ijisa.2018.11.07

14. Sharma, A., Kulshrestha, S., Daniel, S.B.: Machine learning approaches for cancer detection. Int. J. Eng. Manuf. (IJEM) **8**(2), 45–55 (2018). https://doi.org/10.5815/ijem.2018.02.05

15. Mirzadeh, M., Haghighi, M., Khezri, S., Mahmoodi, J., Karbasi, H.: Design adaptive fuzzy inference controller for robot arm. Int. J. Inf. Technol. Comput. Sci. (IJITCS) **6**(9), 66–73 (2014). https://doi.org/10.5815/ijitcs.2014.09.09

16. Kravets, P.I., Shimkovich, V.N., Ferens, D.A.: Method and algorithms of implementation on PLIS the activation function for artificial neuron chains. Elektronnoe Modelirovanie **37**(4), 63–74 (2015)

17. Kravets, P.I., Shimkovich, V.N., Omelchenko, P.: Neural network components of the systems of control of dynamic objects and their hardware-software implementation on FPGA. Visnyk NTUU "KPI" Informatics, operation and computer systems, № 59, pp. 78–85 (2013)

18. Kravets, P.I., Shimkovich, V.N.: Method of optimization of weight coefficients of neuron networks by means of genetic algorithm under implementation on programmed logical integral circuits. Elektronnoe Modelirovanie **35**(3), 65–75 (2013)

Perfection of Computer Algorithms and Methods

Convolution Polycategories and Categorical Splices for Modeling Neural Networks

Georgy K. Tolokonnikov[(⊠)]

Federal Scientific Agro-Engineering Center VIM of the Russian
Academy of Sciences, 5, 1-st Institutsky pr.,
109428 Moscow, Russian Federation
admcit@mail.ru

Abstract. The work relates to the categorical approach to general theory of systems, different from well-known Goguen category direction in Computer Science. Developed by the author generalized polycategories (convolution polycategories and categorical splices) are used. The composition of arrows in generalized polycategories is replaced by convolutions. Models for artificial neural networks of various architectures, including those for quantum neuron networks, are found in the categorical approach to the theory of systems. Neural networks are modeled by associative composite convolutional polycategories of the corona type. Previously unknown nature of branching connections of individual neurons with many others is revealed within this model. Categorical splices, as a separate categorical formation, simulate dynamic systems, including classical and quantum mechanics. It is shown how functor and splices, modeling system-forming P.K. Anokhin's factor, collect an integral system from individual classical or quantum particles. Categorical splices as well simulate simplicial connections in R. Atkin's q-analysis, which are used for physiological and mental traffic in a cognitom by K.V. Anokhin. As a result, the neuro-graph model of "brain-mind" is detailed in relation to simplicial connections in neural network as a model of "brain-mind", and in possible models of artificial neural networks.

Keywords: Computer Science · Artificial Intelligence ·
Category theory of systems · Polycategories · Duality · Category splices ·
Atkin's q-analysis · Combinatorial homotopy groups · Mental traffic ·
Artificial neural networks

1 Introduction

The categorical theory of systems, which has arisen recently on the basis of new mathematical objects called convolutional polycategories [1–4], made it possible to advance in a number of unsolved problems of functional and biomachsystems. It has become a categorical language for numerous approaches to the notion of a system from Mesarovic systems and Vilems systems to ergatic systems (human factors), functional systems and biomachsystems [1]. In the second section of the article, this categorical approach is applied to the theory of artificial neural networks of various architectures. It

© Springer Nature Switzerland AG 2020
Z. Hu et al. (Eds.): ICCSEEA 2019, AISC 938, pp. 259–267, 2020.
https://doi.org/10.1007/978-3-030-16621-2_24

is shown that neural networks are modeled by associative composite convolutional polycategories of the corona type, within the framework of this model, the nature of branching compounds of individual neurons with many others is revealed. In the third section of the article category splices, as a separate categorical formation, simulates simplicial connections in the Atkin's q-analysis [5, 6], which are used for physiological and mental traffic in the "cognitom" by Anokhin [7, 8]. As a result, the neurograph model of "brain-mind" is detailed in relation to simplicial connections in a neural network, as a model of "brain-mind" and in possible models of artificial neural networks. The results of the work are given in the conclusion.

2 Simulation of Neural Networks by Convolution Polycategories

Computer Science has a fundamental theoretical part (Abramsky categorical quantum mechanics, functional programming, etc.) and an applied part, which, in particular, includes engineering work on the design of artificial neural networks for robots, for pattern recognition and for solving other engineering problems [9–14]. This report refers to the fundamental theoretical part of Computer Science. Engineers offer specific artificial neural networks and, as a rule, do not think about the mathematical nature of the objects they use. Before the author's work, even the main object "artificial neural network" was understood intuitively and did not have an exact mathematical model. One of the still uncorrected errors in the definition of an artificial neural network arose when describing a neural network proposed by a group of PDP scientists back in the 1980s and results from a misunderstanding that the neurons shown in the Fig. 1 are mathematically different objects.

Fig. 1. Images of a neuron (a neuron soma is depicted as a rectangle) and its axon.

The right shows the vertex of the graph (soma of the neuron) and the three edges of the graph emerging from it, and the image on the left is not a graph, since the axon branch point by definition PDP is not included in the set of vertices and edges defining the graph. However, it is precisely from neurons of the type depicted on the left that all neural networks are clearly built, and everywhere the networks are referred to as graphs. But these networks are not a graphs.

From the point of view of the theoretical part of Computer Science, the determination of the mathematical nature of artificial neural networks is a very important task. However, the solution of this problem before the introduction of convolutional poly categories by the author was impossible, since the neural network is a convolutional polycategory of a special type, which is claimed by the theorem below. A similar situation occurred in the case of the notion of an algorithm. It was used everywhere on an intuitive level until the 1930s. But the appeared precise definition of the algorithm allowed solving a number of difficult problems, in particular, connected with the proof of the non-existence of some algorithms. Ascertaining the nature of artificial neural networks, in addition to developing their theory in Computer Science, also leads to useful results for engineers. As an example, we have corrected errors in the construction of the "conjugate Osovsky graph" for the beautiful formula for calculating derivatives found by him for the backpropagation method [9]

$$\frac{\partial E}{\partial w_{ij}^{(m)}} = u_i'^{(m)} v_j^{(m-1)}.$$

Osovsky's proposed change in the direction of the arrows is not enough; the exact wording uses the application of two types of duality (polyarrows duality and convolution-polyarrows duality) of a convolutional polycategory that represents a neural network. It became clear what Osovsky was doing on an intuitive level when building a "conjugate graph" for his formula.

A convolution polycategory is a collection of polygraphs and convolutions (analogous to the usual composition of arrows, multi- and polymaps in category theory (for details, see [1–4])).

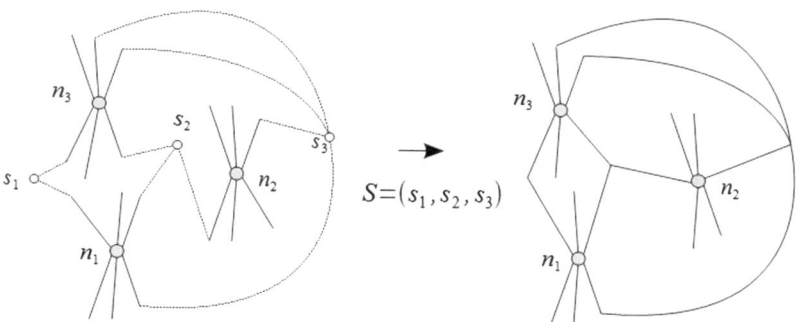

Fig. 2. Convolution of three polymaps (left) and the result of convolution in the form of a new polymaps (right).

An example of the convolution S of three polymaps is shown in Fig. 2. The neuron, as an adder, is represented as a function of many variables, which could be considered [15] as a polymap of the multicategory Set, if the neurons were interconnected as in Set (where one input is connected to one output). For a neural network, neuron compounds can be modeled by convolution in an associative composite convolutional polycategory according to the following theorem, proved in [2]. Before the formulation, we note that a convolution, branching one output into several inputs into polymaps of a convolutional polygraph, is called a corona convolution or simply a corona (see Fig. 3).

Theorem. Let we have an artificial neural network with neurons having several inputs and one output with its activation function for each neuron. The neuronal connections are performed by the available output which branches into a finite number of lines connected to the inputs of other neurons. Then neural networks, constructed from these neurons, form an associative composite convolutional polycategory with a set of corona type convolutions.

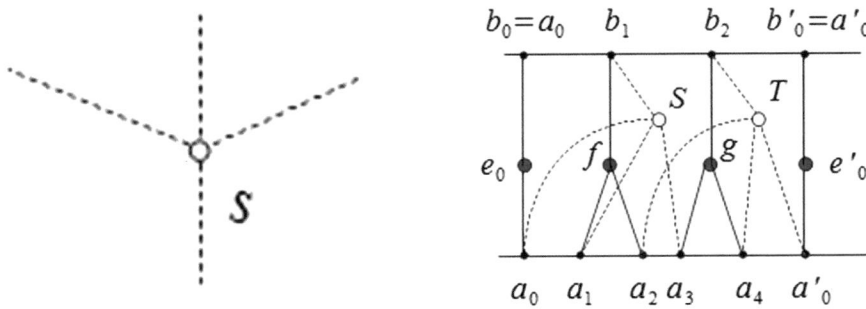

Fig. 3. Example of corona convolution (left) and an example of the Hopfield network (right).

The Hopfield network, shown in Fig. 3, consists of two neurons f and g, e with indices - single neurons, a, b with indices - respectively, inputs and outputs of the network, S and T are two corona convolutions.

Let a multilayer neural network of the type [9] p. 55 (see the diagram of this network below) be given, as a polygraph of a convolution polycategory [2]. Apply to it the duality mapping by polyarrows duality (corresponds to the "change of arrow directions" in categories) and further the convolution-polyarrows duality (corresponds to the replacement of soma neurons with the names of suitable convolutions). As a result, we obtain a neural network as a convolution polygraph of a convolutional polycategory, which we call conjugate to a given source network, if we select its parameters as shown below (see the network diagram and the conjugate network).

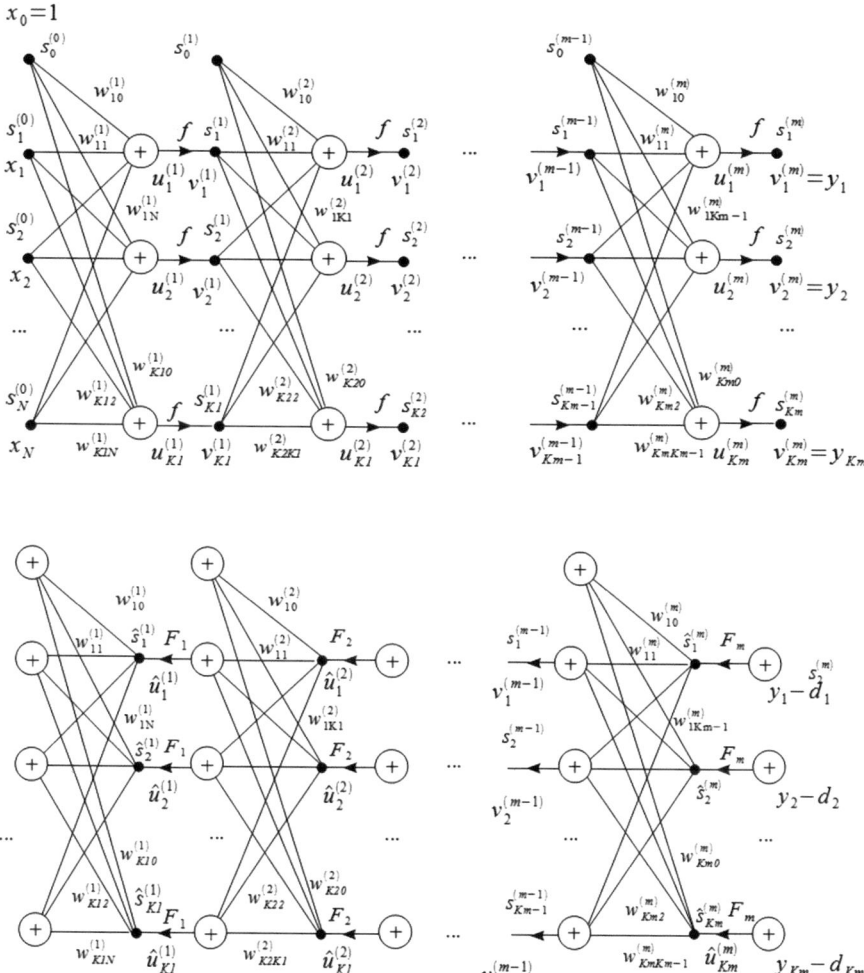

Fig. 4. Multi-layer neural network of type [9] p. 55 as a polygraph of convolutional polycategory (above) and the result of applying duality, which we call paired with this source network, if we select its parameters, as shown below.

In the above Osovsky formula, the corresponding values from the lower polygraph (Fig. 4) are substituted into the first factor, and the corresponding values from the upper polygraph are substituted into the second factor. Additional elements of the polygraph as compared to [9] are also neurons, into which convolutions are translated according to duality transformations.

3 Category Splices, Particle Mechanics and the Simplex Connection by R. Atkin

It is possible to go from convolution polycategies to categorical splices, if you omit orientation in polygraphs. It is more convenient, however, to consider category splices as a separate categorical essence with its axiomatics and models. Dynamic systems, classical and quantum particle mechanics are modeled with the help of category splices. Graphical representation of categorical splices formal language formulas is similar to polycategory case. For example, two simple splices modeling two material points of classical mechanics and their convolution are presented in Fig. 5.

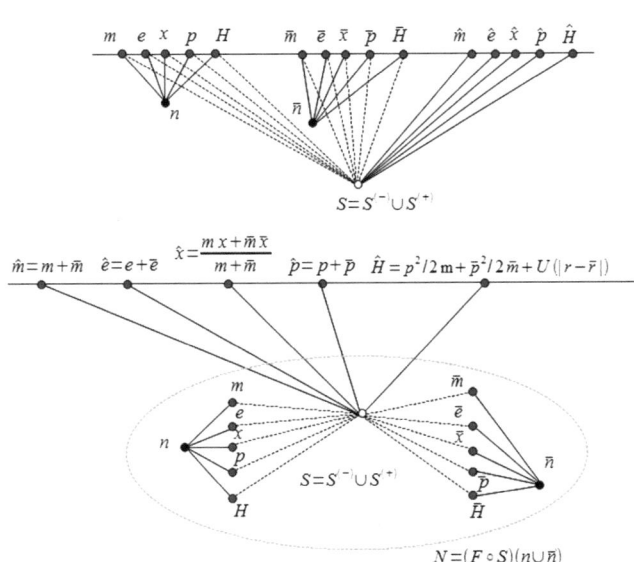

Fig. 5. The upper part of the figure shows two splices simulating two particles and convolution, the lower part represents the result of the action of convolution, which is splice, modeling a holistic system of two particles.

A similar convolution (Fig. 6) collects an integral system of N particles in quantum mechanics.

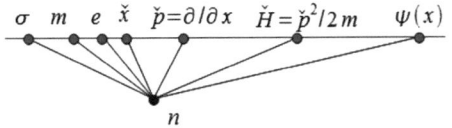

Fig. 6. The splice of a quantum particle. There is a wave function (spinor) of the particle state, spin and other observables.

Quantum-mechanical systems are modeled by categorical splices in the categorical theory of systems. There is a representation theorem analogous to Theorem for quantum neural networks. By the given arrows of the directed graph one can construct by means of a path composition from one vertex to another. According to the available directed graph, a free category is constructed based on these paths. The arrows, if get abstracted from orientation, are 1-simplexes and the graph is 1-simplicial set. If there is a simplicial set consisting of simplexes of different dimension intersecting along the faces, one can speak of simplicial paths connecting various simplexes in the same way as the path of the graph joins its boundary points.

One can talk about networks loading paths with traffic. Such simplicial paths proved to be very useful in information networks of social phenomena. R. Atkin's q-analysis lies at the core of applications here [6, 7]. The study of simplicial paths led R. Atkin to the concept of pseudo-homotopy groups, the rigorous theory of which, was constructed by R. Lubenbacher and his colleagues [16]. Pseudo-homotopy groups obtained a more precise name for combinatorial homotopy groups, which, in fact, are purely combinatorial objects and have little in common with homotopy theory in algebraic topology.

The example of "people-interests" from Johnson's book [7] illustrates the idea of simplicial connections (as seen from the table) and the corresponding images of Pete and Tim are connected in a simplicial way in the network of interests (Fig. 7).

	g	p	c	s	f	pt	h	l	gr	ck	n	sc
Pete	1	1	1	1	0	0	0	0	0	0	0	0
Sam	0	1	1	1	1	0	0	0	0	0	0	0
Sue	0	0	0	0	1	1	1	1	0	0	0	0
Jane	0	0	0	0	0	0	1	1	1	1	0	0
Tim	0	0	0	0	0	0	0	0	1	1	1	1

(g – gaming, p – pabs, c – cars, s – sport, f – fashion, pt – painting, h – history, l – literature, gr – gardening, ck – cooking, n – nature, sc – science)

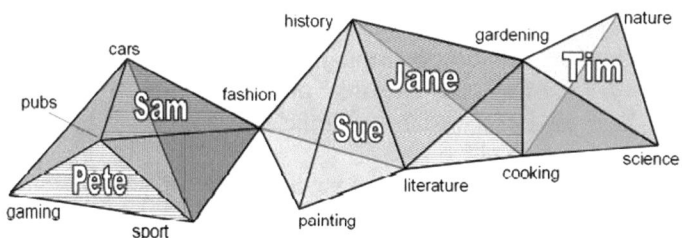

Fig. 7. Simplicial network "people-interests".

Simplicial paths of this kind can be used to describe physiological and mental traffic in the cognitom model proposed by Anokhin [7, 8] as network-networks (the Johnson hypernetwork [6]). In [1], a formalization was given for this intuitive cognitom model.

Such simplicial paths are modeled by a very special case of categorical splices with composite convolutions. The splices diagram of Fig. 8 for the example is obtained for splice with a composite convolution (corresponds to common faces for a pair of simplexes-cones for which gluing can be performed).

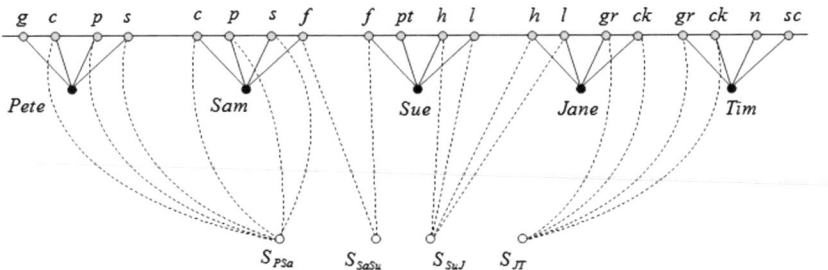

Fig. 8. A splice scheme for the simplicial "people-interests" network.

Just as in the indicated case of constructing a free category by a graph, there is a procedure for constructing a category splice along a given set of cones. R. Atkin and R. Lubenbacher do not consider similar category operations on simplicial paths.

Since category splices are embedded in convolutional polycategories, the categorical platform proposed by the author in [1, 2] for the brain and its both mental and physiological traffic in the form of a neurograph or neurocategory contains R. Atkin's simpicial networks as special cases.

4   Conclusion

Convolutional polycategories and categorical splices have numerous applications, described in [2]. In this paper we propose a polycategorical model for artificial neural networks, including quantum neural networks, as a convolutional associative composite polycategory with convolutions of the corona type, which made it possible, in particular, to reveal the nature of branching neuron connections in neural networks. Thus, we gave for the Computer Science theory a solution to the problem of the mathematical description of the nature of artificial neural networks, which were previously considered at an intuitive level.

The modeling of classical and quantum mechanics of particles in the form of categorical splices is described. This allows us to include dynamic systems of quantum and classical mechanics in the author's categorical theory of systems, as well as artificial networks of quantum neurons.

Based on the proposed modeling of simplicial connections of Atkin's q-analysis with the help of category splices, a more detailed consideration of K.V. Anokhin's approach to the cognitom previously included in the categorical platform for the brain with mental and physiological traffic in the form of neurocategories, was carried out.

The listed results are new and original, having applications in Computer Science.

References

1. Biomachsystems: Theory and Applications, 2016. Volume 1, Rosinformagrotekh, Moscow, 230 p., Volume 2, Rosinformagrotekh, Moscow, 228 p. (2016) (in Russian)
2. Tolokonnikov, G.K.: Manifesto: neurographs, neurocategories, category splices. Biomash-sistemi [Biomachsystems] **1**(1), 59–114 (2017). (in Russian)
3. Tolokonnikov, G.K.: Classification of functional and other types of systems when modeling with convolution polycategories. Neurokomputeri: razrabotka, primenenie [Neurocomput-ers: Development, Application], no. 6, pp. 8–18 (2018). (in Russian)
4. Chernoivanov, V.I., Sudakov, S.K., Tolokonnikov, G.K.: Category theory of systems, functional systems and biomachsystems. In: Collection of Proceedings of the International Scientific and Technical Conference "Neuroinformatics 2017", part 1, vol. 2, pp. 131–138, part 2, vol. 2, pp. 139–147 (2017). (in Russian)
5. Atkin, R.H.: Mathematical Physics. Arima Publishing, Bury St Edmunds (2010)
6. Johnson, J.: Hypernetworks in the Science of Complex Systems. Series on Complexity Science, vol. 3, p. 348. Imperial College Press, London (2014)
7. Anokhin, K.V.: Cognitom: in search of a general theory of cognitive science. In: Sixth International Conference on Cognitive Science Kaliningrad, pp. 26–28 (2014). (in Russian)
8. Vityaev, E.E.: Formalization of cognitom. Neiroinformatika [Neuroinformatics] **9**(1), 26–36 (2016). (in Russian)
9. Osowski, S.: Neural networks for information processing, Moscow, 448 p. (2017). (in Russian)
10. Geron, A.: Hands-On Machine Learning with Scikit-Learn and TensorFlow, p. 581. O'Reilly, Sebastopol (2017). ISBN 978-1-491-96229-9
11. Karande, A.M., Kalbande, D.R.: Weight assignment algorithms for designing fully connected neural network. Int. J. Intell. Syst. Appl. (IJISA) **10**(6), 68–76 (2018). https://doi.org/10.5815/ijisa.2018.06.08
12. Dharmajee Rao, D.T.V., Ramana, K.V.: Winograd's inequality: effectiveness for efficient training of deep neural networks. Int. J. Intell. Syst. Appl. (IJISA) **10**(6), 49–58 (2018). https://doi.org/10.5815/ijisa.2018.06.06
13. Hu, Z., Tereykovskiy, I.A., Tereykovska, L.O., Pogorelov, V.V.: Determination of structural parameters of multilayer perceptron designed to estimate parameters of technical systems. Int. J. Intell. Syst. Appl. (IJISA) **9**(10), 57–62 (2017). https://doi.org/10.5815/ijisa.2017.10.07
14. Leinster, T.: Higher Operads, Higher Categories. London Mathematical Society Lecture Note Series, Cambridge University Press, 410 p. (2004)
15. Barcelo, H., Kramer, X., Laubenbacher, R., Weaver, C.: Foundations of a connectivity theory for simplicial complexes. Adv. Appl. Math. **26**, 97–128 (2001)

Algorithms for Forming a Knowledge Base for Decision Support Systems in Cybersecurity Tasks

V. A. Lakhno[(✉)] [ID]

Department of Computer Systems and Networks,
National University of Life and Environmental Sciences of Ukraine,
Kyiv, Ukraine
valss21@ukr.net

Abstract. The article proposes a general structure of the modular decision support system (DSS) in cybersecurity tasks. A model is described for the subsystem of fuzzy inference (FI). Based on the FI rules for input values that can be obtained from sensors, multi-agent systems, SIEM sensors that detect the presence of threats, cyberattacks, anomalies, it is suggested to determine the output values for estimating the protection degree of critical computer systems (CCS) using DSS. The model assumes that the input values for the FI subsystem were obtained as a result of the fuzzification procedure in the corresponding module. Each element of the output values characterizes the presence or absence of a sign of unforeseen situations associated with anomalies, attacks or other attempts to interfere with the work of the CCS without authorization. An algorithm is proposed for forming a knowledge base of unforeseen (emergency) and typical situations in CCS. The algorithm differs from the known ones in that it made it possible to form a set of cases of typical responses to threats, anomalies and attacks in CCS, as well as rules for the output for authentication of unforeseen situations which are primarily associated with a targeted destructive impact on CCS. The use of the "fuzzy logic output" module allows one to display the state of the most vulnerable components of CCS as a multiparameter "image". The obtained multiparameter "image" can be applied in DSS for a qualitative assessment of the security of CCS.

Keywords: Critical computer systems · Cybersecurity ·
Decision support system · Multiparameter "image" · Security evaluation

1 Introduction

In the process of operation of critical computer systems (CCS) one of the priority tasks of processing data coming from various devices that are part of the complex information security systems (hereinafter referred to as CISS, for which we primarily mean hardware and software components) is to obtain information about the status of protection components. Efficiency and accuracy during an operational assessment of the degree of protection of CCS can be complicated by the influence of the factors listed below. The first - incoming data (from SIEM sensors, multi-agent systems, sensors

© Springer Nature Switzerland AG 2020
Z. Hu et al. (Eds.): ICCSEEA 2019, AISC 938, pp. 268–278, 2020.
https://doi.org/10.1007/978-3-030-16621-2_25

detecting the presence of threats, cyberattacks, anomalies, then the abbreviation sensor subsystems – SenS) can be different in their parameters. Second, in the process of obtaining data, the influence of external influences influencing the authenticity of the monitored characteristics is possible. The third is that the response to destructive interventions is limited to a time frame, with minimal time left for the results of the analysis. Fourth, there may be situations when a combination of the estimated parameters in the CCS leads to "vagueness" during making decisions on the assessment of current security conditions for CCS (unlike typical ones). For these reasons, operational and effective solutions in the analysis of complex targeted cyberattacks on CCS and the corresponding decision-making procedures require the use of special analytical systems [1, 2]. It is obvious that such systems should be based on modern authentication methods for CCS and CISS states. It is also advisable to use the potential of intellectualized adaptive decision support systems (DSS) in cybersecurity (CS) tasks, as well as the recognition of threats, anomalies and cyber-attacks [3]. In the period of avalanche-like growth of the complexity and quantity of cyber-attacks on CCS [4], and the increase in the number of parameters coming from sensory SenS CISS, it became necessary to introduce adaptive expert (AE) and DSS into CISS structures. This is necessary for complex multi-criteria analysis of data from SenS CISS, which generate data for assessing the security of CCS. Note that one of the most effective methods for solving this class of problems is the method that assumes the construction of AE and DSS based on the theory of fuzzy sets (TFS). It is also possible to use a fuzzy logic device (FL) [3, 5, 6].

The use of AE and DSS can minimize the impact of the "human factor" on the quality of decisions made. In addition, the speed of decision-making increases. Possible situations related to the diversion of information security personnel to routine work are being reduced. Also, in the end, the cost of ownership of similar complexes is reduced.

2   Literature Review and Problem Posing

In recent years, the scenarios for the realization of cyber-attacks have become more complex [4, 7]. The increase in the number of anomalies recorded in CCS and other attempts of unauthorized interference in the operation of complex digital systems is recorded [4, 8]. In these conditions, a line of research has emerged on the intellectualization of decision support procedures in the process of recognition of threats, cyberattacks and anomalies. Analysis of existing world experience [4, 6–10] confirms that the extensive approach to solving the tasks of cybersecurity of CCS due to the increase of means and measures for information protection (IP), often does not lead to the expected result. A promising area of research has been the work devoted to the creation of intelligent decision support systems (DSS) [3, 6] and expert systems (ES) [5, 11] in problems of assessing the security of information objects. These studies have not yet been completed.

In [3, 5, 6, 11], the experience of implementing commercial DSS and ES for analysis of threats, attacks and anomalies is analyzed. It is noted that commercial systems have a closed nature, and their acquisition by individual companies or organizations is associated with significant financial costs.

Thus, considering the polemics in [1, 2, 5, 10, 11], it seems relevant to develop new and improve the existing models and algorithms for adaptive DSS, which are involved in data processing processes from various SenS subsystems of cybersecurity and (CS) protection of information (PoI) in CCS.

3   Purpose and Objectives

The aim of the research is to develop new or improve the existing models and algorithms for adaptive DSS, which are involved in the processes of data analysis from SenS subsystems of cybersecurity and information security in CCS.

The objectives of the article are:

To develop a module of "fuzzy logical inference" for expert examination of data from SenS CCS;

To develop an algorithm that forms the knowledge base (KB) about typical and unforeseen situations (US) in CCS (which provides expert analysis of the degree of security of CCS).

4   Methods and Models

The general structure of the developed modular decision support system in cybersecurity tasks is shown in Fig. 1. The Fuzzy Inference Module and the algorithms that generate KB typical (reference) and unforeseen situations for DSS for CCS security analysis are described below.

The module "FLB" is intended for realization of the system of fuzzy output (FI). In Fig. 1 are designated as 6–6k. Based on the rules of fuzzy inference (FI) on the input values of SenS $\left\{ Rd\left(R_{i,n_i}(t)\right), m\left(R_{i,n_i}(t)\right)\right\}$ $\left(i = \overline{1,k}; \ n_i = \overline{1,N_i}\right)$, the output values are determined $\left\{ \left(Rdc_{i,n_i}^j(t)\right), m\left(Rdc_{i,n_i}^j(t)\right)\right\}$ $\left(i = \overline{1,k}; \ n_i = \overline{1,N_i}; j = \overline{1,J}\right)$. In this case, we assume that the input values were obtained as a result of the fuzzification procedure in the corresponding module (module 5, see Fig. 1). Each element of the output values, in turn, characterizes the presence (absence) of the US flag (hereinafter, the notation - is introduced *EmS*). Then a specific indication of an unforeseen situation (j), for example, which arose because of a cyber-attack on CCS, can be described using the variables described below. The following variables are entered: $Md_{i,n_i}^j(t)$ – discrete state characteristic $n_i Param$, for example, the number of bytes from the source to the destination, the number of bytes to the client, the connection flags, the number of root accesses, the number of file creation operations, the number of shell requests, the number of operations to access file control, etc. (we accept by KDD 99) [12, 13]. This variable can take one of the following values, see Fig. 2, where "1" is deviation from the norm from the top; "0" is a typical situation (or norm); "−1" is deviation from the norm below; "2" is deviation from the norm below or above; $m\left(Md_{i,n_i}^j(t)\right)$ – experimental evaluation of the degree of explication of influence n_i value (parameter *Param*) (i) subsystem CCS for the emergence $j - EmS$ (for example, anomalies in the CCS network), at each time point (t).

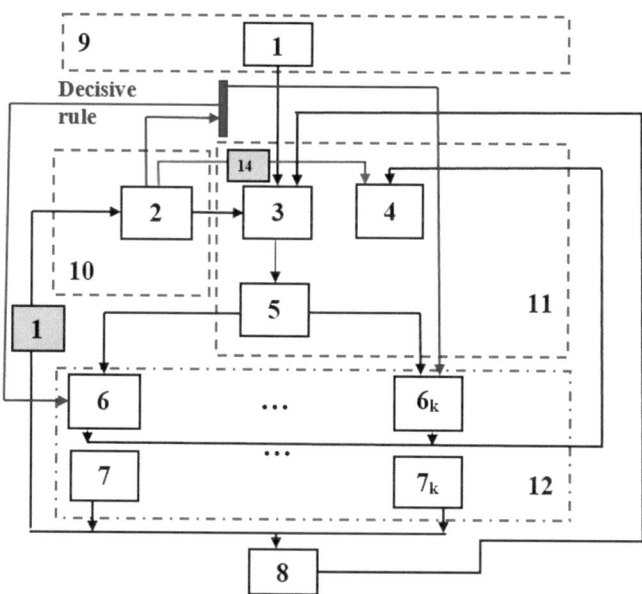

Designations adopted on the scheme: 1 - input device; 2 - server; 3 - visualization module; 4 - defuzzification module; 5 - modulus of fuzzification; 6 (6k) - modules of the subsystem of fuzzy inference; 7 (7k) - output devices; 8 - the module for outputting the results of the analysis and recommendations on how to overcome unforeseen situations; 9 - module for primary processing of information coming from sensors, multi-agent systems, sensors that determine the presence of threats, cyber-attacks, anomalies; 10 - server module with knowledge bases (KB); 11 - the module of the analysis of the basic parameters of functioning CCS with the integrated estimation of security; 12 - modules of employees of information security and cyber security services CCS (by the number of CCS subsystems); 13 - new rules (recommendations) added to the KB; 14 - recommendations on how to overcome unforeseen situations related to CS CCS.

Fig. 1. General structure of the modular decision support system in cybersecurity tasks.

The rules for identifying $j - EmS$ can be described in the following way:

$$If \quad Rd\big(R_{i,n_i}(t)\big) \vee Md^j_{i,n_i}(t) = 1 \ that \ is \ (j) \tag{1}$$

with a degree of explication $m\big(Md^j_{i,n_i}(t)\big)$;

$$If \quad Rd\big(R_{i,n_i}(t)\big) \vee Md^j_{i,n_i}(t) = -1 \ that \ is \ (j) \tag{2}$$

with a degree of explication $m\left(Md_{i,n_i}^j(t)\right)$;

$$If \quad \left(Rd\left(R_{i,n_i}(t)\right) = 1 \; or \; Md_{i,n_i}^j(t) = -1\right) \vee \left(Md_{i,n_i}^j(t) = 2\right) \quad that \; is \; (j) \qquad (3)$$

with a degree of explication $m\left(Md_{i,n_i}^j(t)\right)$.

Figure 2 shows an example of the formation of variables for the module of the fuzzy output subsystem during the filling of the KB DSS with TCP traffic data (for a specific CCS, for example, the TCP-flood attack). The graph shown in the green line shows the actual average traffic threshold. If a deviation from the norm (red line) is recorded for 30 s in the KB DSS, a corresponding one that describes the anomaly or threat is recorded. If within a given interval the mean values of the metrics exceed the threshold values, the analyst will be informed of a deviation from the norm from above or below ("1" or "−1"). Information about the rejection is also available for the administrator of the CCS network to view.

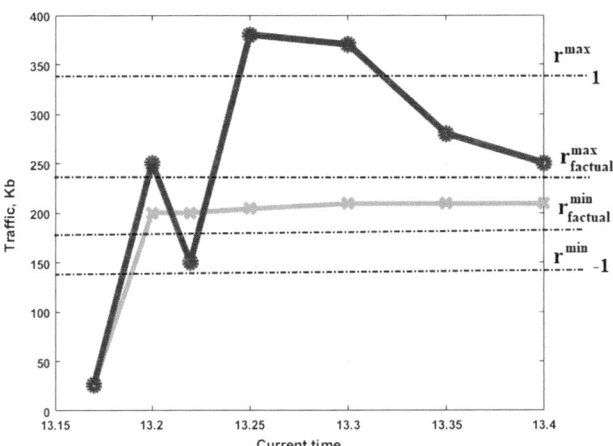

"0" – a typical situation (or norm); "1" - deviation from the norm from above; "-1" - deviation from the norm below; "2" - deviation from the norm below or above

Fig. 2. Formation of a variable for the module of the subsystem of fuzzy output.

At each moment (t) for detection *EmS* on the basis of (j) only one of the three rules may change (1)–(3).

We believe that each rule described in KB DSS for the recognition of threats, anomalies and cyber-attacks in CCS, i.e. $j - EmS$ can match the value of deviations $R_{i,n_n}(t)$ in the moment (t) for n_i *EmS* i subsystem CCS with the degree of explication of influence n_i parameter on *EmS* which is $m\left(Md_{i,n_i}^j(t)\right)$.

As a result of the use of rules (1)–(3) for n_i *Param i* subsystems CCS can form (J) output values $\left\{ \left(Rdc^j_{i,n_i}(t) \right), m\left(Rdc^j_{i,n_i}(t) \right) \right\}$ $\left(i = \overline{1,k}; \ n_i = \overline{1,N_i}; j = \overline{1,J} \right)$. Note that for fixed (for example, critical numbers $j - EmS$), we assume:

If there is an active rule from (1)–(3) then it is possible $j - EmS$,

$$Rdc^j_{i,n_i}(t) = m\left(R_{i,n_i}(t)\right) \vee m\left(Rdc^j_{i,n_i}(t) = m\left(Md^j_{i,n_i}(t) \right) \right);$$

If there are no active rules from (1)–(3) then it is not possible $j - EmS$

$$Rdc^j_{i,n_i}(t) = 0 \vee m\left(Rdc^j_{i,n_i}(t) = m\left(Md^j_{i,n_i}(t) \right) \right).$$

In the defuzzification module (J) output values $\left\{ \left(Rdc^j_{i,n_i}(t) \right), m\left(Rdc^j_{i,n_i}(t) \right) \right\}$ $\left(i = \overline{1,k}; \ n_i = \overline{1,N_i}; j = \overline{1,J} \right)$ is a numerical value (fuzzy). For this value, in the future, using DSS, we define a set of recommendations on how to overcome unforeseen situations (*EmS*).

Weight coefficients, for example, for j–th *EmS*, are defined as follows:

$$\xi^j_i(t) = \frac{\sum\limits_{n_i=1}^{N_i} m\left(Rdc^j_{i,n_i}(t) \right) \cdot Rdc^j_{i,n_i}(t)}{\sum\limits_{n_i=1}^{N_i} m\left(Rdc^j_{i,n_i}(t) \right)}. \tag{4}$$

We believe that *EmS* (j^*) occurred in i subsystem CCS if:

$$\xi^{j^*}_i(t) = \max_{j=\overline{1,J}}\left(\xi^j_i(t)\right) \vee \xi^{j^*}_i(t) \geq thv_i, \tag{5}$$

where thv_i – the threshold value of the degree of detection (for example, once detected, partially detected, not detected) *EmS*. We believe that $thv_i \in [0,1]$.

If $\xi^{j^*}_i(t) < thv_i$, then *EmS* in i subsystem in moment (t) not identified. In this situation, you should begin an interactive analysis of the expert on CS and DSS. With this, you can use the algorithm for generating the KB for the FI system, see Fig. 1.

The algorithm that forms KB for typical (reference) and unforeseen situations for DSS is described below.

The input data for the fuzzy inference subsystem include: allowed boundaries $\left[r^{min}_{i,n_i}(t), r^{max}_{i,n_i}(t) \right]$ which determine the typical operating modes of CCS; the function of belonging to the unforeseen modes in the CCS for the monitored parameters for situations that deviate from the acceptable limits.

Personal KB formation for typical CCS operating modes is filled based on the instructions for CCS staffing behaviour.

Based on the results of the software and hardware components of the CISS, the KB DSS is filled with actual data on the regular operating modes of the system.

We assume that the function $m\left(R_{i,n_i}(t)\right)$ for actual tolerances is a piecewise linear function by which the degree of explication (belonging) is established: If "0", then *Param* it is in the actual zone of the allowed boundaries; If "1", then *Param* it is outside the calculated range of allowed boundaries.

Thresholds $tv_{q_{n_i}}(t)$, $\left(q_{n_i} = \overline{1, Q_{n_i}}\right)$ were calculated upon request to KB using the following formulas:

$$tv_{q_{n_i}}(t) = tv_{Q_{n_i}} \cdot \left((q_{n_i} - 1)^{st_{n_i}} / (Q_{n_i} - 1)^{st_{n_i}}\right) \,\&\, st_{n_i} = u \cdot (1/su), \tag{6}$$

where st – the degree of filling the KB for a particular parameter; number of tests that are taken into account in statistics; su – the number of DSS tests for which the KB is supposed to be full (KB fragment of the DSS prototype "Threat Analyzer" [14]).

The procedure for forming a KB including rules for the fuzzy output module in the process of automated processing of indications from SIEM sensors, multi-agent systems, sensors detecting the presence of threats, cyber-attacks, anomalies is described below.

A new or existing rule for detecting an unforeseen situation in CCS by value n_i – parameter $\left(n_i = \overline{1, N_i}\right)$ for the subsystem $\left(i = \overline{1, k}\right)$ CCS is formed on the basis of $Rd\left(R_{i,n_i}^j(t), Md_{i,n_i}^j(t)\right)$. Next, we should determine the degree of explication of influence $m\left(Md_{i,n_i}^j(t)\right)$ n_i – the occurrence $\left(j = \overline{1, J}\right)$ *EmS*. The rules described by expressions (1)–(3) are used.

Baseline values for detection $\left(j = \overline{1, J}\right)$ *EmS* : $Md_{i,n_i}^j(t)$ – parameters that characterize the discrete states of n_i – parameter $\left(n_i = \overline{1, N_i}\right)$ for subsystems $\left(i = \overline{1, k}\right)$ CCS, which influences the occurrence $\left(j = \overline{1, J}\right)$ *EmS*; $m\left(Md_{i,n_i}^j(t)\right)$ – parameters that characterize the expert evaluation of the degree of explication of the influence of n_i parameter on the appearance $\left(j = \overline{1, J}\right)$ *EmS*.

The received values were recorded in the KB DSS "Threat Analyzer" [14], with the initial description by the experts. It is also possible to enter data in the KB with automatic/automated collection of indications from SIEM sensors, multi-agent systems, sensors detecting the presence of threats, cyberattacks, anomalies. If an expert has discovered a new unforeseen (non-standard) situation with CCS security, he can respond to messages from the DSS window interface. After that, the expert records in KB. The new record characterizes the current state of the subsystems that identified abnormal situations. The following describes some of the results of testing the prototype of the modular decision support system in cybersecurity tasks "Threat Analyzer" [14]. These results continue the research, the results of which were previously presented in [14, 15].

5   Simulation Experiment

In order to test the model for the "fuzzy inference" module and the algorithm for generating the KBs of unforeseen and typical cybersecurity situations in CCS, test experiments were conducted for the DSS Threat Analyzer prototype.

Figure 3 shows the histogram of the time comparison (in minutes) spent by the experts independently (red columns) and using DSS (blue columns) to assess the characteristics of unauthorized access to the CCS server.

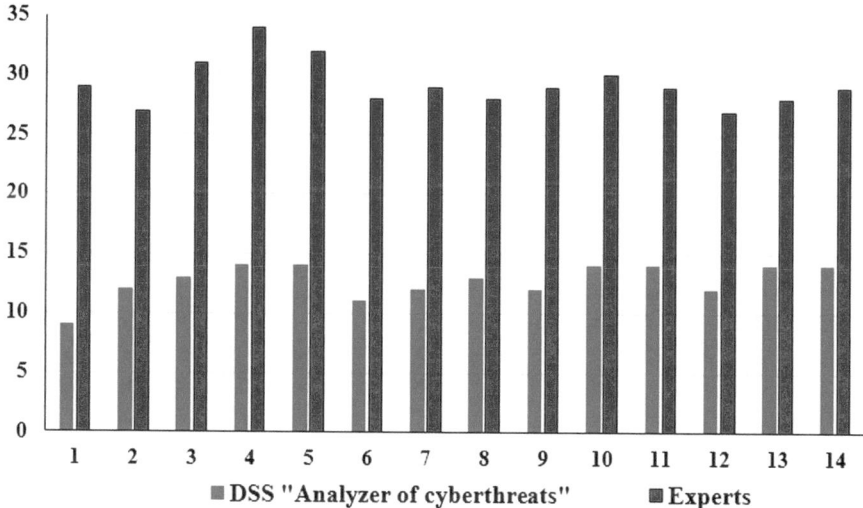

Fig. 3. Time spent for assessing the characteristics of unauthorized access to CCS Information Resources by experts independently and with the help of DSS "Threat Analyzer".

The results show that without using the DSS prototype "Threat Analyzer" experts were more optimistic about the overall state of protection for CCS and its servers. But at the same time, the time spent without automating the collection and processing of indications from SIEM sensors, multi-agent systems, sensors detecting the presence of threats, cyber-attacks, anomalies, was 1.2–2.5 times higher.

During the testing of the DSS prototype, the mechanisms of interaction between experts and DSS "Threat Analyzer" were also tested in the process of synthesizing the guiding rules in the tasks of assessing the degree of protection of CCS.

6 Discussion

The developed algorithms for expert study of data from SenS and formation of KB of unforeseen and typical situations in CCS allow: accumulating knowledge concerning unforeseen situations in CCS; reducing the time spent on taking decisions in the process of analyzing threats, anomalies and cyberattacks in CCS. This allows one to promptly assess the situation and prevent its development into an emergency one in real time, in unforeseen situations, caused by attempts to unauthorized influence on the system; increasing the objectivity of decisions taken during the analysis of data from CISS sensors and sensors in CCS; expanding the functional capabilities and efficiency of the personnel of the information security and cybersecurity units of CCS. This is achieved by the ability to automate the decision support process if the deviations from the typical (regular) operating modes of the CCS are recorded.

Certain drawbacks of the work can be situations caused by a lack of data from SenS CISS. However, we note that the process of forming the KB for tests has passed quite quickly and with sufficient personnel qualification, this drawback will not lead to a decrease in the degree of security of the CCS.

It is established that the use of the DSS prototype "Threat Analyzer", equipped with modules from automatic and automated collection of information from SIEM sensors, multi-agent systems, sensors detecting threats, cyber-attacks, anomalies, reduced the costs of organizing complex GIS by 12–15% alternative methods [5, 6, 10]. The described solutions complement existing studies [6, 14], in the context of solving the management tasks of CCS protection based on the implementation of the integrated systems of cybersecurity of intelligent DSS and expert systems. The investigations carried out are a continuation of the works [6, 14, 15].

A promising direction of the development of this work is the filling of KB and logical rules of the prototype DSS taking into account the expansion of test information and the results of approbation "Threat Analyzer". It is also planned to expand the functionality of the proposed DSS prototype "Threat Analyzer" and to test it on a larger number of CCS enterprises.

7 Conclusion

For the first time in the work:

A model for the module "fuzzy inference", which is intended for the implementation of the subsystem of fuzzy inference (FI) is proposed. Based on the fuzzy inference (FI) rules, the input values of SenS determine the output values for the DSS security rating of the CCS. The model assumes that the input values for the FI subsystem were obtained as a result of the fuzzification procedure in the corresponding module. Each element of output values characterizes the presence (absence) of a sign of an unforeseen situation related to anomalies, attacks or other attempts at unauthorized interference with the work of CCS;

An algorithm for forming the knowledge base of unforeseen and typical situations in CCS is suggested. The algorithm differs from the known ones in that it made it possible to form a set of cases of typical responses to threats, anomalies and attacks in the CCS, as well as rules for the withdrawal for authentication of unforeseen situations, which are primarily associated with a targeted destructive impact on the CCS.

It is shown that the proposed models and algorithms for the expert study of SenS data differ from the known ones, they use elements of theories of expert systems, fuzzy sets, as well as the fuzzy logic apparatus. The use of the "fuzzy inference" module allows displaying the state of the most vulnerable CCS components as a multiparameter "image" (MPI). The received MPI is used in DSS for a qualitative assessment of the processes taking place in CCS. Possible "fuzziness" was considered in assessing the situation with the security of CCS.

References

1. Cherdantseva, Y., Burnap, P., Blyth, A., et al.: A review of cyber security risk assessment methods for SCADA systems. Comput. Secur. **56**, 1–27 (2016)
2. Abu Samra, A.A., Qunoo, H.N., Al Salehi, A.M.: Distributed malware detection algorithm (DMDA). Int. J. Comput. Netw. Inf. Secur. (IJCNIS) **9**(8), 48–53 (2017). https://doi.org/10.5815/ijcnis.2017.08.07
3. Lakhno, V., Boiko, Y., Mishchenko, A., Kozlovskii, V., Pupchenko, O.: Development of the intelligent decision-making support system to manage cyber protection at the object of informatization. Eastern-Eur. J. Enterp. Technol. **2**(9), 53–61 (2017). https://doi.org/10.15587/1729-4061.2017.96662
4. Hu, Z., Khokhlachova, Y., Sydorenko, V., Opirskyy, I.: Method for optimization of information security systems behavior under conditions of influences. Int. J. Intell. Syst. Appl. (IJISA) **9**(12), 46–58 (2017). https://doi.org/10.5815/ijisa.2017.12.05
5. Hu, Z., Gnatyuk, S., Koval, O., Gnatyuk, V., Bondarovets, S.: Anomaly detection system in secure cloud computing environment. Int. J. Comput. Netw. Inf. Secur. (IJCNIS) **9**(4), 10–21 (2017). https://doi.org/10.5815/ijcnis.2017.04.02
6. Akhmetov, B., Lakhno, V., Boiko, Y., et al.: Designing a decision support system for the weakly formalized problems in the provision of cybersecurity. Eastern-Eur. J. Enterp. Technol. **1**(2(85)), 4–15 (2017). https://doi.org/10.15587/1729-4061.2017.90506
7. Hu, X., Xu, M., Xu, S., Zhao, P.: Multiple cyber attacks against a target with observation errors and dependent outcomes: characterization and optimization. Reliab. Eng. Syst. Saf. **159**, 119–133 (2017)
8. Yang, Y., Xu, H.Q., Gao, L., Yuan, Y.B., McLaughlin, K., Sezer, S.: Multidimensional intrusion detection system for IEC 61850-based SCADA networks. IEEE Trans. Power Delivery **32**(2), 1068–1078 (2017)
9. Wong, K., Dillabaugh, C., Seddigh, N., Nandy, B.: Enhancing Suricata intrusion detection system for cyber security in SCADA networks. In: IEEE 30th Canadian Conference on Electrical and Computer Engineering (CCECE), pp. 1–5 (2017)
10. Akhmetov, B., Lakhno, V., Akhmetov, B., Alimseitova, Z.: Development of sectoral intellectualized expert systems and decision making support systems in cybersecurity. In: Silhavy, R., Silhavy, P., Prokopova, Z. (eds.) Intelligent Systems in Cybernetics and Automation Control Theory. CoMeSySo 2018. Advances in Intelligent Systems and Computing, vol. 860, pp. 162–171 (2019). https://doi.org/10.1007/978-3-030-00184-1_15
11. Elhag, S., Fernández, A., Bawakid, A., Alshomrani, S., Herrera, F.: On the combination of genetic fuzzy systems and pairwise learning for improving detection rates on intrusion detection systems. Expert Syst. Appl. **42**(1), 193–202 (2015)
12. Moustafa, N., Slay, J.: The evaluation of network anomaly detection systems: statistical analysis of the UNSW-NB15 data set and the comparison with the KDD99 data set. Inf. Secur. J.: Glob. Perspect. **25**(1–3), 18–31 (2016)
13. Villaluna, J.A., Cruz, F.R.G.: Information security technology for computer networks through classification of cyber-attacks using soft computing algorithms. In: IEEE 9th International Conference on Humanoid, Nanotechnology, Information Technology, Communication and Control, Environment and Management (HNICEM), pp. 1–6 (2017)

14. Lakhno, V., Kazmirchuk, S., Kovalenko, Y., Myrutenko, L., Zhmurko, T.: Design of adaptive system of detection of cyber-attacks, based on the model of logical procedures and the coverage matrices of features. Eastern-Eur. J. Enterp. Technol. **3**(9), 30–38 (2016). https://doi.org/10.15587/1729-4061.2016.71769
15. Lakhno, V., Tkach, Y., Petrenko, T., Zaitsev, S., Bazylevych, V.: Development of adaptive expert system of information security using a procedure of clustering the attributes of anomalies and cyber attacks. Eastern-Eur. J. Enterp. Technol. **6**(9), 32–44 (2016). https://doi.org/10.15587/1729-4061.2016.85600

Digital Signature Authentication Scheme with Message Recovery Based on the Use of Elliptic Curves

Svitlana Kazmirchuk, Ilyenko Anna$^{(\boxtimes)}$ (ID), and Ilyenko Sergii (ID)

National Aviation University, Kyiv 03058, Ukraine
{sv.kazmirchuk,ilyenko.a.v,ilyenko.s.s}@nau.edu.ua
http://www.nau.edu.ua

Abstract. Digital signature authentication scheme provides secure communication between users. Digital signatures guarantee message integrity and authentication information about the origin of a message. In the present paper we describe existing algorithms for the formation and verification of the digital signature. The conducted studies made it possible to determine that the schemes with message recovery differ from the schemes with the addition that they do not completely hash messages, but instead use masking functions and redundancy of the message. It was determined that the most effective and optimal for further use is the Nyrberg-Rueppel scheme, which is based on elliptical curves (ECNR). In this paper, we present a new digital signature scheme with message recovery based on elliptic curve cryptograph on the base of the State standart 4145-2002. Elliptic curve cryptosystem provides greater security compared to integer factorization system and discrete logarithm system, for any given key size and bandwidth. The main difference between the proposed scheme was the replacement of the hash function with the hash token function, which makes the signature and verification procedure reversed and allows you to recove messages from the signature r-component.

Keywords: Digital signature · Verification · Confidentiality ·
Elliptical curves · Hash token · Discrete logarithm

1 Introduction

Digital signature authentication schemes provide secure communication with minimum computational cost for real time applications, such as electronic commerce, electronic voting, etc. Under the notion of electronic digital signature we will understand the type of electronic signature obtained by the cryptographic transformation of the set of electronic data, which is added to this set or logically combines with it and allows to confirm its integrity and identify the signer [4, 5]. An electronic digital signature is a tool that allows you to create the legal basis for electronic document flow (including the Internet) and which can be a means of protecting the document flow system. In Ukraine, the main document governing the procedure for the creation and verification of an electronic-digital signature is the State standard of Ukraine 4145-2002. «Information Technology. Cryptographic protection of information. Digital signature based on elliptical curves. Formation and verification» [3].

© Springer Nature Switzerland AG 2020
Z. Hu et al. (Eds.): ICCSEEA 2019, AISC 938, pp. 279–288, 2020.
https://doi.org/10.1007/978-3-030-16621-2_26

Digital signature are becoming increasingly popular, which make it possible not only to verify signatures, but also to recove the original digital signature message. This means that the signing document will not be transmitted by the open communication channel, but will be a signature component, and the checking subscriber will receive access to the message only in case of confirmation of authorship and integrity. Thus, such a structure of digital signature provides another security service – confidentiality.

Digital signature authenticated schemes, have the following properties: confidentiality; authentication; non-repudiation; message recovery. The State standard of Ukraine 4145-2002 does not have the ability to recove the message. Therefore, the purpose of the work was to create a system for ensuring the integrity, reliability and confidentiality of information on the basis of the use of an electronic-digital signature on the basis of elliptic curves using the State standard of Ukraine 4145-2002 with the recovering of an information message [3].

The purpose of this article is to analyze and consider the practical bases of the formation and verification of electronic digital signature scheme. In this paper, we propose a new digital signature scheme with message recovery using public keys and State standard of Ukraine 4145-2002 based on Elliptic Curve Discrete Logarithm Problem. The public key and the private key of the proposed scheme are agreed upon between the user and server through secure channel. The main difference between the proposed scheme was the replacement of the hash function with the hash token function, which makes the signature and verification procedure reversed and allows you to recove messages from the signature r-component. It is computationally infeasible for the adversary to find the private key from the publicly known information. This approach allows solving the problem of information protection, including stored information, processed and transmitted in modern information networks on the basis of confidentiality and integrity.

2 Related Work

There are different classifications of modern electronic digital signature schemes. They can be classified according to the mechanism of construction (symmetric and asymmetric), with the recovery of a message or without, one-time and multiple, deterministic and probabilistic, on the problem that underlies them [12–14].

All schemes of electronic-digital signatures can also be divided into two major classes: conventional digital signatures (with the addition) and electronic digital signatures with message recovery. In the digital signature, the recovery part or the full message can be recovery from the digital signature, that is, to verify the digital signature, you only need to know the digital signature and possibly the public key certificate. In the digital signature with the addition - the digital signature joins the message and is stored and transmitted with it, and to verify the digital signature must necessarily have a public key certificate [1, 7, 18].

In Ukraine, the main document governing the procedure for the creation and verification of the digital signature is the «State standard of Ukraine 4145-2002. Information Technology. Cryptographic protection of information. Digital signature based on elliptical curves. Formation and verification» [3]. The State standard of Ukraine

4145-2002 establishes a digital signature mechanism based on the properties of groups of points of elliptic curves over $GF(2^m)$, and the rules for applying this mechanism to messages sent by communication channels and/or processed in computerized general-purpose systems. The application of this standard guarantees the integrity of the signed message, the authenticity of its author, and the untruthfulness of authorship [3, p. 6]. The scheme of the formation of an electronic-digital signature under the State standard of Ukraine 4145-2002 refers to schemes with message completion, but there are schemes with message recovery. The signature of the recovery of the message, in comparison with the supplement signature, provides an additional security service – confidentiality [6, 8, 21].

Structurally, the schemes with message recovery differ from the schemes with the addition that they do not gage completely the message (but in different algorithms can partly use the hex functions), and instead use the functions of masking and finding the losses of the message. Such an algorithm has its advantages: a user who verifies a signed message will only be able to access it if the signature is verified; in addition to ensuring the integrity of the message, it can also ensure its confidentiality; electronic-digital signatures with recovery can provide a smaller amount of signature in small messages; using the functions of generating the redundancy of the message (the stage of creating a pre-signature), which provides a flexible function (for example, choosing the type of redundancy, you can determine whether the part of the message will be recovered or completely all the message).

In 2003, the international standard ISO/IEC 15946-4 was adopted. It included 5 independent algorithms for e-mail retrieval (ECNR, ECMR, ECAO, ECPV, ECKNR), cryptographic transformations based on elliptic curves. Subsequently, this standard was refined and adopted in 2006 as ISO/IEC 9796-3 [1, 2, 8] to replace existing one. The standard ISO/IEC 9796-3 distributes and refines the algorithms specified in ISO/IEC 15946-4 and since 2008 is the main standard for signatures with message recovering. Signatures have a common Nyberg-Rueppel (NR) scheme, but they are used to optimally use the r-components of the modified algorithm before the signature [11].

The RSA with recovery feature has the advantage that it is the most common, has high speeds during signing/verification, and is also versatile, as it is also suitable for encryption/decryption. NR schemes, unlike RSA, are usually effective for a short message, because in this case all messages will be recovered. However, these algorithms have an advantage in the cryptostability and complexity of the mathematical problems on which they are based. The use of redundancy functions can guarantee lower volumes of signatures on short messages. In addition, due to the fact that the ECNR scheme operates on the use of elliptic curves, it allows the use of smaller parameters and keys [6, 15, 17, 20]. Comparing schemas without recovery messages, it was found that RSA and El-Gamal algorithms with the same key size would have the same cryptosecurity (approximately 2.7 • 1028 for a key of 1024 bits). But the El-Gamal algorithm is nicely faster than RSA when signing a document, but inferior in speed at verification. The advantage of the El-Gamal scheme is that at the given level of stability of the digital signature algorithm, the integer numbers involved in the calculations have a record 25% shorter (in RSA, the multiplier should be 1024 bits, and El-Gamal 512 bits), which reduces the complexity of the calculations by almost twice and allows you to noticeably reduce the amount of memory used. In addition, the

El-Gamal signing procedure does not allow you to calculate digital signatures under new messages without the knowledge of a secret key. However, the scheme El-Gamal inferior to the scheme of RSA in the impossibility of recovery the message and that the length of the digital signature is 1.5 times greater, which increases the time of its calculation [9, 16]. The advantage of the DSA and ECDSA algorithms is a smaller signature size than RSA and El-Gamal (on average 320 bits), because the main system parameters are 160 bits by default. Also, when checking the signature, most operations with numbers are also carried out by a module of a length of 160 bits, which reduces the amount of memory and computing time [10]. However, it is believed that ECDSA is more crypt proof due to the complexity of the problem of discrete algorithmicity along points of an elliptic curve. In addition, the secret key in ECDSA is unique, and not just random, as in DSA, which improves the reliability of the algorithm. The DSA algorithm also has the drawback that it is intended only for the signature/verification of electronic documents, and not for their encryption/decryption, in contrast to all other considered algorithms [15–21]. By the criterion of the problem underlying, one can conclude that the most crypto-resistant algorithms are based on the problem of discrete logarithm in the group of points of the elliptic curve.

That is why the most efficient and optimal for the further simulation is the ECNR algorithm, and its circuit, shown in the international standard ISO/IEC 9796-3 [1], will be used. In detail, we can see that both schemes, in accordance with the State standard of Ukraine 4145-2002 and ECNR, have almost the same structure. Therefore, the ECNR algorithm is ideally suited for the improvement of the State standard of Ukraine 4145 and this procedure will require minimal changes [3].

3   Proposed Digital Signature Scheme with Message Recovery

The previous sections provided the advantages and reasonable choice of algorithms of the State standard of Ukraine 4145-2002 and ECNR. Both algorithms are based on the mathematical problem of discrete logarithm in the group of points of an elliptic curve. And in a detailed review, one can see that both schemes have an almost identical structure. Therefore, the ECNR algorithm is ideally suited for modifying the State standard of Ukraine 4145-2002 and this procedure will require minimal changes [8, 9]. In this paper, we propose a new Digital Signature scheme with message recovery based on elliptic curve. The proposed scheme is divided into three phases: Finding the general parameters, Signature formation, and Signature verification.

3.1   Finding the General Parameters of a Digital Signature and Generating Keys

General options are selected as follows:

1. Choosing the main Galois field $GF(2^m)$. The text of the standard gives the recommended values of the power of the field m under polynominal and normal bases.
2. Choosing an elliptical curve of the form

$$y^3 + xy = x^3 + Ax^2 + B; A, B \in GF(2^m), B \neq 0, A \in \{0, 1\} \qquad (1)$$

3. Calculation of the base point of the elliptic curve P.
4. Calculate the random element of the main field using a random sequence generator;
5. Calculate the element of the main field $w = u^3 + Au^2 + B$;
6. Solve the equation $z^2 + uz = w$ (need to get a number of solutions to equation k and one of the solutions z);
7. Coordinates (x_p, y_p) describe the base point of the elliptic curve P $x_p = u, y_p = z$.
8. Finding the order of the base point P: n should be an odd integer, $n \geq \max(2^{160}, 4([2^{m/2}] + 1))$.
9. The condition must be fulfilled $2^{mk} \neq 1 \mod n, k = 1, \ldots, 32$.

Cryptographic keys are calculated by the standard for the elliptical curves method: personal key d is located using the random sequence generator $d \in [1, n-1]$; public key calculated as $Q = -dP$. Point coordinates $Q(x_Q, y_Q)$ should belong $GF(2^m)$ and be the solution of the Eq. (1), $Q \neq O$ (O – infinitely distant point of the elliptic curve), $nQ = O$.

3.2   Formation of the Signature

The scheme is not specified, but if the field is a polynomial basis, it is necessary to check the polynomial $f(t)$ for primitiveness (before checking the coefficients A and B). Upon termination of this stage we have the following parameters: the degree of expansion m; elliptic curve of the form $y^3 + xy = x^3 + Ax^2 + B$; the base point of the elliptic curve $P(x_p, y_p)$; the order of the base point n; secret key d; open key Q.

Note that unlike the usual algorithm according to the State standard of Ukraine 4145-2002, we will use not the transfer of parameters to the binary string, but their conversion into octets. Thus, the following functions will be used: I2OSP – primitive transformation of integers into octet lines, OS2IP – primitive conversion of octet lines into integers, EC2OSP – primitive transformation of the elliptic curve into octet lines, OS2ECP – primates the conversion of octet lines into an elliptic curve. Random parameter and pre-signature are computed unchanged. That is, there is a random number $e \in [1, n-1]$, the point of the elliptic curve $R(x_R, y_R) = eP$ (be sure to check that $x_R \neq 0$) and the prefix $F = x_R$. And after that moment changes will be extended to the ECNR algorithm. The previous chapters indicated that the Nyberg-Rueppel schemas lacked the fact that they could not always recovery all messages. Recovery is complete if the message is relatively small. And although the message will not be used in the work of a large volume, but the option still need to be described for them.

To do this, the initial message M needs to be split into 2 parts: M_{rec} is the recovered part and M_{clr} is the part that is being sent (this means that M_{clr} will be added to the signature for data transmission). That is, we can say that $M = M_{rec}||M_{clr}$, and henceforth they will be used for their length len_M, len_rec and len_clr respectively (obviously $len\_M = len\_rec + len\_clr$). When working with an electronic-digital signature with the recovery of the message it is advisable to use redundancy. In this case, it is convenient to choose a short and long redundancy and accept their values as

$len\_1 = 64\,bits$ and $len\_2 = 136\,bits$ [1]. Now you can check the condition and how it is executed, then you can be sure that the message will be recovered completely:

$$len\_M \leq len\_n - len\_1 - 1 \tag{2}$$

That is, it means that $M_{rec} = M$, $len\_rec = len\_M$, $len\_clr = 0$ (M_{clr} does not exist) and $len\_h = len\_1$ (the length of the result of the hash token). If condition 2 is not satisfied then it is assumed that $len\_rec = len\_n - len\_2 - 1$, $len\_clr = len\_M - len\_rec$ and $len\_h = len\_2$. In this case, we have a part that is not recovered and will be sent together with the signature of M_{clr}. According to ISO/IEC 9796-3 for ECNR and NR algorithm the hex token of the recovered part is calculated:

$$\delta(M) = Hash(I2OSP(len\_rec, 4) \,\|\, I2OSP(len\_clr) \,\|\, Mrec \,\|\, Mclr) \,\|\, Mrec, \tag{3}$$

As a function of hashing, used GOST 34.311, because it is recommended by the State standard of Ukraine 4145. In this case, only the first 64 bits of the received digest are used in case of full recovery of the message, and in case of partial recovery, the first 136 bits are used.

Then you need to convert the result of the hash token and pre-signature to integers:

$$\delta(M) = OS2IP(\delta(M)), \ F1 = OS2IP(F) \bmod n. \tag{4}$$

Now you can go directly to the signature calculation. First, the reciprocal component $r : r_1 = (\delta_1 + F_1) \bmod n$ and $r = I2OSP(r_1, L(n))$ are calculated, the irreversible component is found as $s = (e + dr_1) \bmod n$. Thus, a pair of numbers (r, s) (if desired, they can be converted to a numeric string in accordance with State standard of Ukraine 4145-2002) and a part of the M_{clr} message in the event of partial recovering of the message will be transmitted. Despite the fact that structurally the algorithms according to the State standard of Ukraine 4145-2002 and ECNR are very similar, however, in the algorithm according to the State standard of Ukraine 4145-2002 necessarily use the hash function to find the message digest. However, it does not make it possible to further recover the message, so this part uses the function of the scheme Nyrberg-Rueppel. In order to be able to recove the message, it was also added the opportunity to split it into two parts, and the condition for checking the message length according to the base point of the elliptical curve and the short loss margin is added (default value). Instead the hash functions, the masking function is calculated using the hash token. Also, to find the inverse signature component, scalar assembly is used instead of scalar multiplication (to provide signature reciprocity and subsequent recovering). Different and received signatures – in the usual algorithm for the formation of electronic-digital signature in accordance with the State standard 4145-2002 this (iH, M, D), although the first and not mandatory. But when updating with a recove, we get (M_{clr}, r, s) where M_{clr} will be used only in isolated cases, and the pair (r, s) and represents the signature D. In addition, after creating the signature, the proposed method adds and ensures the service of data privacy, as well as increases cryptographic stability.

3.3   Verification of the Signature

At first, it is recommended to perform a standard check of the general basic parameters and the public key:

1. The degree of expansion of the field m should be selected from the tables of the State standard of Ukraine 4145-2002 (and in the case of a polynominal basis, it is also necessary to perform a primitive test) [3].
2. Check the equation of the elliptic curve and the order of the base point. The coefficients of the elliptic curve must satisfy the requirements: $A, B \in GF(2^m)$, $B \neq 0$, $A \in \{0, 1\}$. The order of the base point n should be an odd integer, $n \geq \max(2^{160}, 4([2^{m/2}] + 1))$. In addition, the condition must be fulfilled $2^{mk} \neq 1 \bmod n$, $k = 1, \ldots, 32$.
3. Check the base point. Coordinates $(x_p, y_p) \in GF(2^m)$ and is the solution of the Eq. (1), $P \neq O$ (O – infinitely distant point of the elliptic curve), $nP = O$.
4. Check the public key. Coordinates $(x_Q, y_Q) \in GF(2^m)$ and is the solution of the Eq. (1), $Q \neq O$ (O – infinitely distant point of the elliptic curve), $nQ = O$.

You must also check the signature components: $0 < r < n$ and $0 < s < n$. If at least one of these requirements is not met, the signature is considered invalid. Next is the octet line r converts to an integer: $r' = OS2IP(r)$. It is recoved before the pre-signature $F_1 = sP + r'Q$, turns into octet lines $F = EC2OSP(F_1)$, and then an integer number $F' = OS2IP(F) \bmod n$. After that, the masked part of the signature is recovered $\delta' = (r' - F') \bmod n$ and turns into octet lines $\delta_1(M) = I2OSP(\delta')$. So we get a string consisting of a hash token and an initial message.

The following analysis follows: if there were three components in the part of the received signature, that is, the M_{clr} parameter, then it is concluded that the message is not fully recovered, and the redundancy value will be 136 bits. If M_{clr} is not present then the message is fully recovered, and its redundancy is 64 bits.

Obviously, the result of the hash token len_h is the first lossy bits $\delta_1(M)$, which we denote as H'. This means that all subsequent bits - and is a M_{rec} message transmitted. Now you need to re-mask. In our case, we already have M_{rec} and M_{clr} and know their length. And we make a check:

$$Hash(I2OSP(len\_rec, 4) \,||\, I2OSP(len\_clr) \,||\, Mrec \,||\, Mclr)_{len\_h} = H'. \qquad (5)$$

If equality is true, the signature is verified, and the recovered message is $M = M_{rec} \,||\, M_{clr}$. If the equality is not confirmed, the signature is considered invalid and the message is not recovered.

4   Results

As a result of the proposed scheme, a software implementation of the procedure for the formation and verification of a digital signature in C# was obtained using the CryptoLib library. This library fully supports the State standard of Ukraine 4145-2002 and is certified for use in Ukraine at the state level. The algorithms were tested in the

Crypto ++5.6.0 software environment on a dual-core Intel Core 1.83 GHz processor running Windows 8 32 bit x86 (Table 1). So you can draw a conclusion about the changes in the signature verification procedure. Stages of checking general parameters and public key remained unchanged. The verification procedure in the algorithm according to the State standard of Ukraine 4145-2002 with message recovery differs first of all in the type of signature. In version with message recovery, a pair (r, s) is octet and an integer, and in the standard form it is one line. Then the pre-signature is recovered completely the same way. However, there is a change in the very principle of final verification. If in the standard version of the algorithm is r' and tested with the signature obtained from the r-component, then the recovery algorithm requires more action. After finding the disguised message, it is necessary to separate it from the «mask» (hash token), to calculate the hash token with the values found, and only then can you make a comparison. Another important difference is that, in the usual version of the algorithm, integers are compared, and in the case of recovery, the results of masking, that is, the semantic value is analyzed.

Table 1. Comparison of modern algorithms for the formation and verification of digital signature

Algorithm	Open key, bit	Minimum length of signature, bit	Time of formation the signature, ms	Time of verification the signature, ms	Ability to recove a message
RSA	512–4096	1024	1,48	0,07	No
DSA	1024–3072	1024	0,42	0,52	No
El Gamal	1024–4096	1024	0,83	3,84	No
NR	1024–4096	1024	3,78	7,78	Yes
ECDSA	112–570	256	2,88	8,53	No
ECNR	112–570	256	10,61	21,54	Yes
State standard of Ukraine 4145-2002	163–768	256	1,24	1,67	No
Proposed scheme (modification of 4145-2002)	112–768	162	1,98	2,83	Yes

5 Conclusions

Thus, in this article we give a full description of the modification of the State standard for the creation and verification of electronic-digital signature using elliptic curves. Given that the schemes of the algoritm of the State standard of Ukraine 4145 and ECNR are similar, the modification chosen did not require major changes. The main difference was the replacement of the hesh function with the masked use of a hash token, which makes the signature and verification procedure turned upside down and allows you to restore messages from the signature r-components.

As a result of the modification, the size of the public key has decreased by 32%, that is, the minimum size of the public key is 116 bits, and the length of the electronic-digital signature is 37%, that is, the minimum size of an electronic-digital signature is 162 bits. Thus, in this way, the modification of the algorithm for the formation and verification of an electronic-digital signature in accordance with the State standard of Ukraine 4145-2002 is carried out using the ECNR algorithm with the ability to provide it with the function of message recovery. This adds a privacy service, reduces the amount of signature, increases cryptographic stability, but slows down the signature/verification procedure with the same initial parameters.

References

1. ISO/IEC: ISO/IEC 9796-3:2006, Information technology – Security techniques – Digital signature schemes giving message recovery – Part 3: Discrete logarithm based mechanisms (2006)
2. ISO/IEC: ISO/IEC 15946-4, Information technology – Security techniques – Cryptographic techniques based on elliptic curves – Part 4: Digital signatures giving message recovery (2001)
3. The state standard of Ukraine 4145-2002, Information Technology – Cryptographic protection of information – Digital signature based on elliptical curves. Formation and verification (2002). (in Ukrainian)
4. Law of Ukraine: About electronic documents and electronic document circulation (2003). (in Ukrainian)
5. Law of Ukraine: On electronic digital signature (2003). (in Ukrainian)
6. Koblitz, N.: Elliptic curve cryptosystems. Math. Comput. **48**, 203–209 (1987)
7. Miyaji, A.: Another countermeasure to forgeries over message recovery signature. IEICE Trans. Fundam. **E80-A**(11), 2192–2200 (1997)
8. Shevchuk, O.A.: Particulars of digital signatures with message recovery. Appl. Radio Electron. **9**(3), 489–492 (2010). (in Ukrainian)
9. Alguliev, R.M., Imamverdiev, Ya.N.: Study of international and national standards for digital signatures on elliptic curves. Secur. Inf. **2**(69), 2–7 (2005). (in Russian)
10. Moldovyan, D.N.: New signature generation mechanism in EDS schemes based on the complexity of discrete logarithmization and factorization. Secur. Inf. **4**(71), 81–93 (2005). (in Russian)
11. Gorbenko, Yu.I, Shevchuk, A.A.: Analysis of authorities and areas of digital signatures to the standard ISO. Appl. Radio Electron. **8**(3), 304–314 (2009). (in Ukrainian)
12. Ilyenko, A.V.: Evaluating the effectiveness of the optimized cryptographic system Gentry of conditions for ensure the confidentiality of information. Sci.-Based Technol. **1**(33), 41–45 (2017). (in Ukrainian)
13. Ilyenko, A.V.: Modern ways of improving the procedure for the formation and verification of digital signature. Sci.-Based Technol. **1**(37), 61–66 (2018). (in Ukrainian)
14. Ilyenko, A.V., Ilyenko, S.S.: Program module using the procedure for the formation and verification of electronic digital signature. Sci.-Based Technol. **3**(39), 345–354 (2018). (in Ukrainian)
15. Alimoradi, R.: A study of hyperelliptic curves in cryptography. Int. J. Comput. Netw. Inf. Secur. (IJCNIS) **8**(8), 67–72 (2016). https://doi.org/10.5815/ijcnis.2016.08.08

16. Goyal, R., Khurana, M.: Cryptographic security using various encryption and decryption method. Int. J. Math. Sci. Comput. (IJMSC) **3**(3), 1–11 (2017). https://doi.org/10.5815/ijmsc.2017.03.01

17. Kamboj, D., Gupta, D.K., Kumar, A.: Efficient scalar multiplication over elliptic curve. Int. J. Comput. Netw. Inf. Secur. (IJCNIS) **8**(4), 56–61 (2016). https://doi.org/10.5815/ijcnis.2016.04.07

18. Agarkhed, J., Ashalatha, R.: Security and privacy for data storage service scheme in cloud computing. Int. J. Inf. Eng. Electron. Bus. (IJIEEB) **9**(4), 7–12 (2017). https://doi.org/10.5815/ijieeb.2017.04.02

19. Nyberg, K., Rueppel, R.A.: Message recovery for signature schemes based on the discrete logarithm problem. In: De Santis, A. (ed.) Advances in Cryptology-EUROCRYPT'94, LNCS, vol. 950, pp. 175–190. Springer, Verlag (1995)

20. Menezes, A.J., Vanstone, S.A., Van Oorschot, P.C.: Handbook of Applied Cryptography. CRC Press, London (1996)

21. Schneier, B.: Applied Cryptography, 2nd edn. Wiley, Hoboken (2015)

Decomposition Method for Synthesizing the Computer System Architecture

Vadym Mukhin[1(✉)], Nina Kuchuk[2], Nataliia Kosenko[3],
Roman Artiukh[4], Alina Yelizyeva[5], Olga Maleyeva[5],
Heorhii Kuchuk[6], and Viktor Kosenko[4]

[1] National Technical University of Ukraine "Kiev Polytechnic Institute",
Kiev, Ukraine
v_mukhin@i.ua
[2] V. N. Karazin Kharkiv National University, Kharkiv, Ukraine
n.kuchuk@karazin.ua
[3] O. M. Beketov Kharkiv National University of Urban Economy,
Kharkiv, Ukraine
natalija.kosenko@kname.edu.ua
[4] Southern National Design and Research Institute of Aerospace Industries,
Kharkiv, Ukraine
{roman.artyuh, viktor.kosenko}@nure.ua
[5] National Aerospace University "Kharkiv Aviation Institute", Kharkiv, Ukraine
{a.elizeva, o.maleyeva}@khai.edu
[6] National Technical University "Kharkiv Polytechnic Institute",
Kharkiv, Ukraine
kuchuk@kpi.kharkov.ua

Abstract. The paper considers the problem of synthesizing the architecture of a modern computer system (CS) for managing a complex distributed object. The main stages of the architecture synthesis are analyzed. The necessity of preliminary decomposition, structurization and formalization of the system is substantiated. The main stages of the hierarchical sequence of the decomposition process are given. The decomposition method for synthesizing the architecture of a computer system is developed on the basis of the stratification of the directed graph of CS implementation options. The simulation model of this process was developed. The modelling results showed the feasibility of using the developed method for synthesizing the architecture of a computer system with a large number of nodes. It is shown that the use of the decomposition method is unpractical when there are less than 60 nodes. Using more than 100 nodes is practically advantageous. This method, unlike the existing ones, enables getting results in a reasonable time when using more than 1000 nodes. These results are particularly significant in the context of operational reconfiguration of a system with a large number of nodes. Moreover, the more the number of nodes, the greater the advantage of the method of decomposition. Thus, while synthesizing the computer system architecture to control a complex distributed object at the stage of developing the relationship among the components, the method of decomposition should be used to reduce the time of the process.

Keywords: Computer system · Synthesis · Control system decomposition · Stratification

© Springer Nature Switzerland AG 2020
Z. Hu et al. (Eds.): ICCSEEA 2019, AISC 938, pp. 289–300, 2020.
https://doi.org/10.1007/978-3-030-16621-2_27

1 Introduction

A large number of control systems have been developed, implemented and operated within many modern industries, which enables analyzing the results of their operation and specifying certain advantages and disadvantages. There is an inevitable tradeoff among a number of functions of corresponding control systems, their universalization and actual implementation of specific computer networks (CN). This tradeoff can be implemented while synthesizing the computer system architecture.

The subject matter of the study is a computer system that supports the control complex of a compound distributed object. The object of the study is the process of synthesizing the architecture of a computer system.

The problem of synthesizing the architecture of a modern computer system (CS) to control a complex distributed object (CDO) is a complex problem. It cannot be solved without preliminary decomposition, structurization and formalization. Many scientists have described a number of probable approaches to solving this problem so far [1–3]. According to these papers, the following stages of CS architecture synthesis can be singled out as:

- Synthesizing the organizational structure of the object management system (MS);
- Synthesizing the CS architecture which provides the requirements for system operation;
- Synthesizing the architecture of the basic computer network (BCN).

The main task of synthesizing the organizational structure of the object MS is the development of the logical structure of the object MS. Such a structure should include the main components of the object and the interrelations among them. This structure will determine the input data for synthesizing the CS architecture using which a CDO can be controlled to solve certain problems. Such an architecture is a logical set of necessary components stratified by the corresponding levels. In addition, it should take into account the relationship between the components and levels that assure exchanging both service information and user information.

The synthesis of the basic computer network involves developing the physical structure for the CS architecture obtained at the previous stage.

In addition, in the process of synthesis, at each stage, the distribution of the set of tasks to be solved in the set of the components used by the CS should tend to the extremum according to the given quality criterion considering all applicable restrictions.

2 Problem Analysis and Setting the Problem

Approaches to synthesizing the architecture of the CDO computer system are considered in many scientific papers [4–7]. According to these papers, the synthesis involves the following basic stages:

- Formalizing the structure and components, including determining the required number of components and the required number of levels over which they should be distributed;
- Formalizing interrelations between components, including reproducing necessary links in the general hierarchy of levels;
- Formalizing tasks implemented by such CSs, including their optimal distribution by components.

The organizational structure of the CDO MS should give an unambiguous answer to the question of the distribution of processes among the subsystems at different levels as well as the distribution of the entire set of possible functions and methods of their implementation along the necessary subsystems.

While solving the task of the synthesis, it is necessary to determine:

- Sets of principles and methods that should be implemented;
- A set of interrelated functions implemented by the CS;
- Sets of components of the CDO CS that are interrelated and stratified by levels;
- Mapping of the elements of the set of interrelated functions implemented by the CDO CS to the set of components of the basic computer network.

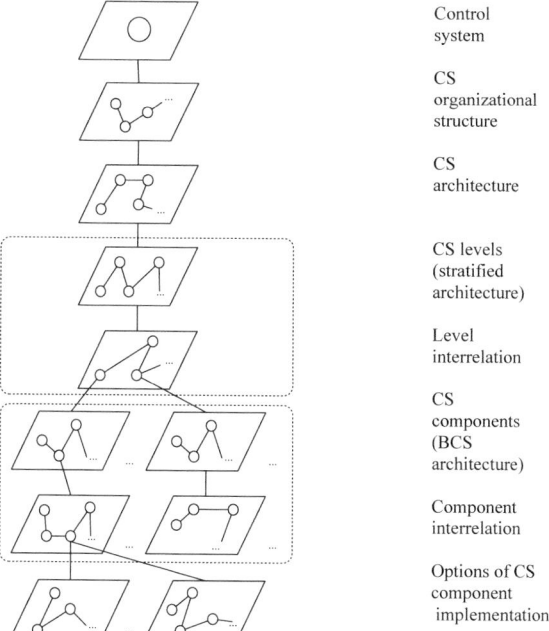

Control system

CS organizational structure

CS architecture

CS levels (stratified architecture)

Level interrelation

CS components (BCS architecture)

Component interrelation

Options of CS component implementation

Fig. 1. The hierarchy diagram of the stages of CDO CS synthesis

Papers [8, 9] suggest the decomposition of the synthesis task. Paper [10] suggests the hierarchical sequence of the processes of decomposition (Fig. 1).

An important task of the initial stage of the synthesis is determining and formalizing the requirements for the CDO CS [11, 12]. CS requirements can be implemented in many ways but they are formalized in the most convenient way with the use of the mathematical apparatus of graph theory [13]. Thus, the relationships for the synthesized CS can be formalized.

When mapping a set of interrelated functions and tasks of the CDO CS, which are specified in the form of corresponding graphs, a multigraph component can be obtained for a set of CS components, some parts of which correspond to probable variants of distribution of functions in BCS components and the arcs reflect the interrelations between them.

The optimal option for implementing the CS will be the subgraph of the directed multigraph, which is simultaneously an acceptable option for implementing the organizational structure of the CDO MS.

The corresponding model can be represented by the directed graph

$$G_A = (V_D, R_F, \Theta), \tag{1}$$

where $V_D = \{v_d, d = \overline{1, d_0}\}$ is a set of probable options for data organization; $R_F = \{r_f, f = \overline{1, f_0}\}$ is a set of probable options of function implementation; $\Theta = (\theta_{df})$, $d = \overline{1, d_0}$; $f = \overline{1, f_0}$; dim $\Theta = d_0 \times f_0$ is the matrix reflecting the interrelation of data and functions. In this case, $\theta_{df} = 1$ if R_F function is used for developing a set of v_d data or $\theta_{df} = 0$ in other cases.

The options of data implementation and the ways of their development can be set in graph G. It enables mapping the variants of input data transformation into output data to select the optimal option under given conditions of optimality and available constraints [10, 14]. However, a high dimension of the graph (1) does not usually enable finding an optimal variant over the feasible amount of time [15, 16]. That is why approximate methods of finding an option from a set of applicable ones are used in practice. Papers [17–19] shows that while stratifying the variants if CS architecture implementation, the search time decreases but the solution accuracy increases.

So, the goal of the work is to develop the decomposition method of synthesizing the architecture of a computer system on the basis of the stratification of the directed graph of the options of CS implementation.

3   Problem Solving

In the directed graph (1) two subsets of terminal vertices the can be singled out: S is source vertices; $\Im$ is sink vertices. Then, if N is a number of nonterminal graph (1) vertices

$$V_D = \bigcup_{k=1}^{N_s} s_k \bigcup_{j=1}^{N_t} t_j \bigcup_{i=1}^{N} x_i, \tag{2}$$

where x_i is a nonterminal of graph (1) vertex, $s_k \in S$; $t_j \in \Im$; $N_s = \text{card } S$; $N_t = \text{card } \Im$;

Let us also introduce $E = V_D \bigcup R_F$.

Let us introduce the relation of continuity in G_A net in the following way. Let j element be a descendant of i element, designate $j \in \sigma(i)$, if

- i is an arc incoming into j vertex;
- i is a vertex with the outgoing arc $- j$.

In other words, is i is the net vertex, $\sigma(i)$ is a set of outgoing arcs and if i is an arc, то $\sigma(i)$ is a vertex this arc come into.

Let $i \in \sigma^{-1}(j)$, if $j \in \sigma(i)$. In other words, $\sigma^{-1}(j)$ is a set of elements whose descendant is j element. It should be noted that for the relation of continuity of σ elements can be introduced in any directed circuit free graph. G_A net with a set of E elements where the relation σ is given will be designated through $G_A(E, \sigma)$.

Let us call (a, b)-path the sequence of elements $l_0 = a, l_1, l_2, \ldots l_k = b$ where $\forall i = 1, 2, \ldots, k$ l_i element is a descendant of l_{i-1}, and the path that connects the source of the net with the sink of the net is called the terminal path.

Let us consider that all net elements are given by their numbers. Proper element numbering (similarly to proper vertex numbering) will be the one where if $j \in \sigma(i)$, $i < j$, where i, j are the numbers of net elements.

It should be noted that the elements of final circuit-free digraph $G_A(E, \sigma)$ can always be numbered properly. Actually, one-to-one mapping ψ of graph $G_A(E, \sigma)$ in graph $G'_A(E, R_F)$ whose set of vertices corresponds to the set of vertices of graph $G_A(E, \sigma)$ and the relation of vertex adjacency R_F corresponds to the relation of continuity of σ elements in G graph. It is known that in graphs of $F_A(E, R_F)$ type, proper numbering can be given [17]. Consequently, it can be given on the set of elements E of $G_A(E, \sigma)$ graph, which follows from the mutual one-oneness of ψ mapping.

Mixed (s, t)-cut set or mixed terminal cut set will be called such a set of P network elements in which there is not a single terminal path if all the elements of the net are withdrawn from it.

Element ℓ will be called redundant in P cut set if a set of $\{P \backslash \ell\}$ elements is also a cut set. A cut set that does not contain redundant elements is a minimal one and a cut set that contains redundant elements is a redundant one. In a redundant cut set there always is an element that can be withdrawn without violating a certain property of a cut set. It should be noted that several minimal cut sets can be made from one redundant cut set excluding various redundant vertices.

Thereafter, mixed terminal cut sets will be called merely cut sets.

The algorithm of cut set enumeration should include the way of making an initial cut set, the method of making new cut sets from the ones that have already been built, the procedure of excluding the redundant elements and the order of enumeration.

Let us select a set of elements $P_0 = \sigma(s)$ as an initial cut set. It is obvious that there cannot be redundant elements in it.

In the built initial cut set all the elements are set in ascending order.

From every built cut set $P = \{\ell_1, \ell_2, \ldots, \ell_n\}$, $\ell \leq n$ of new cut sets can be built. If $\tilde{r} \in \sigma(\ell_i)$,

$$P_i = ((P \backslash \ell_i) \cup \sigma(\ell_i)) \backslash I_i, \qquad (3)$$

where I_i is a set of redundant elements in P_i.

Let us set the elements in each newly built cut set in ascending order and the elements with identical numbers will be replaced by one and the same. All built cut-sets will be pit in the general list of cut sets lexicographically and by ascending order. It should be noted. That cut sets are generated from P by changing various elements ℓ_i for $\sigma(\ell_i)$, one and the same minimal cut set can be built which should be put in the list only the first time round.

To decrease a number of exchanges in the context of lexicographical ordering, elements from P can be changed for their σ-images in the order that is the reverse order to P, that is in the order $\ell_n, \ell_{n-1}, \ldots, \ell_1$. Moreover, every built minimal cut set should be immediately put in the list so that the lexicographical ordering of the whole list is not broken.

Due to the proper enumeration of the net elements, each newly built cut set is placed in the lexicographically ordered list after the cut set according to which it was built.

Let us consider the way of excluding redundant elements from each cut set generated from P

$$\tilde{P} = (P \backslash l_i) \cup \sigma(l_i). \qquad (4)$$

In $\tilde{P}$ cut set, r_j element is redundant if all (s, r_j)-paths or all (r_j, t)-paths pass through one more element $r_k \in P$. But r_k in $\tilde{P}$ cut set cannot exist for the accepted way of generating cut sets in (s, r_j)-paths of similar elements. That is why redundant element should be found taking into consideration only the segments of (r_j, t)-paths.

Let us set the elements in the generated cut set $\tilde{P}$ in ascending order. Then, to check if r_j elements are redundant, only such segments of (r_j, t)-paths, where the number of the last element is not greater than the number of the last element in $\tilde{P}$ cut set can be chosen.

Let use designate $\tilde{P} = \{r_1, r_2, \ldots, r_m\}$ – cut set, generated from $P = \{\ell_1, \ell_2, \ldots, \ell_n\}$ cut set by changing ℓ_i for $\sigma(\ell_i)$. The procedure of excluding redundant elements will look like as follows.

For all j from 1 to m the following steps should be performed:

- Let $\Sigma = \sigma(r_j) = \{y_1, \ldots, y_p\}$;
- If there is $y \in \Sigma$ where $y > r_m$, r_j is the nonredundant element in $\tilde{P}$;
- If $\Sigma \in \tilde{P}$, r_j is the redundant element in $\tilde{P}$;

- If for all $y \in \Sigma$ there is $y \leq r_m$, $\phi = \Sigma \backslash (\Sigma \cap \tilde{P})$ is created and $\Sigma = \sigma(\phi)$ is obtained, after that let us pass on to 2.

The considered cut sets are selected from the list of cut sets built in the lexicographical order according to the numbers of their constituent elements. When the cut set is viewed, new cut sets are generated from it by replacing those ℓ_i elements for which $\sigma(\ell_i)$ does not have t sink.

The whole algorithm of enumeration of minimal cut sets has $O(N^2)$ intensity.

Let us consider the way of converting the list of cut sets into a net. In the formalized view, this task can be presented as follows.

There is μ list that contains M of $\xi_1, \ldots, \xi_M$ lines and has the following properties:

- $\xi_i = (a_{i1}, \ldots, a_{im})$ line contains m_i of the symbols of $A = \{a_i\}_{i=1}^N$ alphabet that are not repeated.
- Symbols in the line can be interchanged in any way.
- There are no $\xi_i, \xi_j \in \mu$, where $\xi_i \leq \xi_j$.
- There is even one pair of symbols a_i, a_j where there is no $\xi_k \in \mu$ line where $a_i \in \xi_k$, $a_j \in \xi_k$. In this case, a_i and a_j symbols are said to be incompatible.

Digraph $G'_A(Y, R_F)$ with minimum vertices should be built with the following properties.

- One symbol $a \in A$ matches each $y \in Y$ vertex; this vertex is designated as y_a;
- For each line $\xi_i \in \mu$, $i = 1, 2, \ldots, M$, there is a way in $G'_A(Y, R_F)$ graph that passes through the vertices with symbols $a_{i1}, a_{i2}, \ldots, a_{im}$; this path is called the one that matches ξ_i line and is designated as μ_{ξ_i};
- There is no such a way in $G'_A(Y, R_F)$ graph where $x_{a_i} \in \mu$ and $x_{a_j} \in \mu$ if symbols a_i and a_j are incompatible.

This actually means that symbols in the lines of the list should be ordered so that the most compatible symbols in lines are joined and there are no incompatible symbols in one line.

Let us select the three symbols a_i, a_j and a_k where a_i symbol belongs simultaneously to two lines – ℓ_p and ℓ_q which contains incompatible symbols $a_j \in S_p$ and $a_k \in S_q$. Moreover, the symbols should be placed in lines ℓ_p and ℓ_q so that

$$a_i < a_j, a_i < a_k \tag{5}$$

or

$$a_i > a_j, a_i > a_k. \tag{6}$$

Relations (5) are written as

$$f = a_i(a_j, a_k), \tag{7}$$

and relation (6) is

$$\bar{f} = (a_j, a_k)a_i. \tag{8}$$

Let (7) be called direct and (8) reverse.

Forms f_1 and f_2 are called equivalent if $f_1 = a_i(a_j, a_k)$, a $f_2 = a_i(a_k, a_j)$.

Forms f_1 и f_2 are alternative to one another if $f_1 = \bar{f}_2$ or $f_2 = \bar{f}_1$.

Let us consider that f_i contradicts f_j is the general system of inequalities created by f_i and f_j is incompatible. It is obvious that $f = a_i(a_j, a_k)$ can contradict only the forms of the following type:

$$f_1 = a_j(a_i, a_\ell), f_4 = (a_\ell, a_i)a_k$$

$$f_2 = a_k(a_i, a_\ell), f_3 = (a_j, a_\ell)a_i, \tag{9}$$

where $\ell \in \left(\overline{1, N}\right)$.

Let forms (9) be designated as:

$$f \vdash f_1; f \vdash f_2; f \vdash f_3; f \vdash f_4. \tag{10}$$

For $\bar{f} = (a_j, a_k)a_i$, it should be noted that

$$\bar{f} \vdash \bar{f}_1, \bar{f} \vdash \bar{f}_2, \bar{f} \vdash \bar{f}_3, \bar{f} \vdash \bar{f}_4. \tag{11}$$

Let us construct the initial system of forms $\Sigma = \Sigma_0$ that contain direct forms for all similar triples of symbols. Let us enumerate the forms of this system (for example, Q). Let construct the matrix of form contradictions using relations (10) and (11) as well as the rule:

$$v_{ij} = \begin{cases} 1, & \text{if } f_i \vdash f_j; \\ 0, & \text{if } f_i \text{ does not contradict } f_j; \\ -1, & \text{if } \bar{f}_i \vdash f_j. \end{cases} \tag{12}$$

where $f_i, f_j \in \Sigma$.

Let us introduce additional vectors D^+, D^- and Δ, where

$$d_i^+ = \sum_{\substack{j=1 \\ v_{ij} > 0}}^{Q} v_{ij}, d_i^- = \sum_{\substack{j=1 \\ v_{ij} < 0}}^{Q} v_{ij}, \delta_i = \sum_{j=1}^{Q} v_{ij} = d_i^+ + d_i^-. \tag{13}$$

Let us change the system of Σ forms by changing its constituent forms for the alternative ones where necessary so that resultant system $\Sigma = \Sigma^*$ be obtained that contains the minimum of forms contradicting one another. Value $\delta_i > 0$ points out that f_i form should be changed for f_i^*. While changing f_i form for $\bar{f}_i$, the i-th line and i-th column of V matrix are changed as well as some components of vectors D^+, D^- and Δ according to the following rule:

- Change v_{ij} for $-v_{ij}$, in the i-th line of V matrix $j = 1, \ldots, Q$, and

$$d_i^+ (new) = d_i^- (old), \quad d_i^- (new) = d_i^+ (old),$$
$$\delta_i(new) = -\delta_i(old); \tag{14}$$

- Change v_{ki} for $-v_{ki}$, $k = 1, \ldots, Q$ in the i-th column of V matrix, and

$$d_k^- (new) = d_k^- (old) - v_{ki}(old),$$
$$d_k^+ (new) = d_k^+ (old) - v_{ki}(old), \tag{15}$$
$$\delta_k(new) = \delta_k(old) - 2v_{ki}(old).$$

Acting similarly for those f_i, in which $\delta_i > 0$, the general number of contradictions in Σ system is gradually being decreased until all $\delta_i \leq 0$, $i = 1, \ldots, Q$. In addition, the system of forms $\Sigma = \Sigma^*$ with the minimum number of contradictions will be obtained. It gives partial (or complete) order on a set of a symbols. This will enable collecting the maximum number of symbols in lines to build the desired graph.

4   Discussion

To analyze the efficiency of the suggested method of synthesizing the computer system architecture, the simulating model of this process was developed. All input data of the model were recorded while modelling except for a number of computer system nodes (N). The time for finding the optimal option of the computer system synthesis was considered as initial data. The results of modelling are presented in Fig. 2.

The analysis of modelling results showed that the suggested method is feasible for synthesizing the architecture of the computer system with many nodes. Thus, the graphs in Fig. 2a showed that the suggested method lies a little bit behind in time. The obvious reason for this is time spending on G_A graph stratification. However, the method of decomposition has more advantages when a number of nodes increases (Fig. 2b, c). It should be noted that when there is a small number of nodes, the application of the decomposition method is not necessary. However, if there are more than 100 nodes, there will be an advantage in the synthesis time. And when there are more than 1000 nodes, this method, unlike the existing ones, enables getting a result over a reasonable amount of time. These results are particularly significant in the context of the operational reconfiguration of a system with a large number of nodes.

Thus, while synthesizing the computer system architecture to control a complex distributed object at the stage of developing the relationship among the components, the method of decomposition should be used to reduce the time of the process.

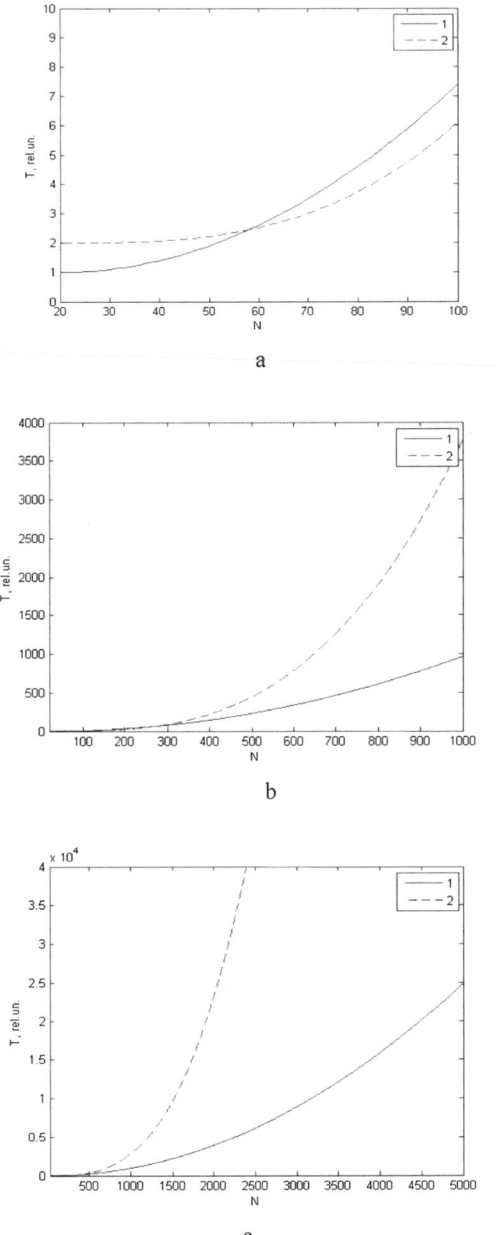

Fig. 2. Dependence of the time for finding the optimal solution on the net dimension: 1 – net decomposition along the levels of stratification; 2 – without net decomposition; a – small systems ($N < 100$); b – middle systems ($100 < N < 1000$); c – large systems ($N > 1000$)

5 Conclusions

The paper dealt with the task of synthesizing the architecture of a modern computer system to control a complex distributed object. The basic stages of the architecture synthesis were analyzed. The necessity of preliminary decomposition, structurization and formalization of the system was substantiated. The basic stages of the hierarchical sequence of the process of decomposition are presented. The decomposition method of synthesizing the computer system architecture was developed on the basis of the stratification of the directed graph of the options of CS implementation. The simulating model of this process was developed. The modelling results proved the feasibility of the suggested method while synthesizing the architecture of the computer system with many nodes. Thus, when $N < 60$, this method is not feasible. When a number of nodes of the computer system lie within $60 < N < 100$, there is no practical advantage in the synthesis time. However, if $N > 100$, the advantage becomes evident. Moreover, the more the number of nodes, the greater the advantage of the method of decomposition.

Further studies are related to the enhancement of the proposed method. In particular, it is planned to consider a two-stage synthesis, which enables determining the minimum possible set of components for the realization of their location, the number and nature of the relationships among them. In the process of synthesizing, the requirements for accessibility of network components as well as the environmental impact should be taken into consideration.

References

1. Dinari, H.: A survey on graph queries processing: techniques and methods. Int. J. Comput. Netw. Inf. Secur. (IJCNIS) **9**(4), 48–56 (2017). https://doi.org/10.5815/ijcnis.2017.04.06
2. Hu, Z., Mukhin, V., Kornaga, Y., Lavrenko, Y., Barabash, O., Herasymenko, O.: Analytical assessment of security level of distributed and scalable computer systems. Int. J. Intell. Syst. Appl. (IJISA) **8**(12), 57–64 (2016). https://doi.org/10.5815/ijisa.2016.12.07
3. Mukhin, V.Ye.: Adaptive approach to safety control and security system modification in computer systems and networks. In: Proceedings of the 5-th IEEE Workshop on Intelligent Data Acquisition and Advanced Computing Systems: Technology and Applications (IDAACS'2009), Rende (Cosenza), Italy, 21–23 September 2009, pp. 212–217 (2009)
4. Hu, Z., Mukhin, V., Kornaga, Y., Lavrenko, Y., Herasymenko, O.: Distributed computer system resources control mechanism based on network-centric approach. Int. J. Intell. Syst. Appl. (IJISA) **9**(7), 41–51 (2017). https://doi.org/10.5815/ijisa.2017.07.05
5. Kuchuk, G.A., Akimova, Yu.A., Klimenko, L.A.: Method of optimal allocation of relational tables. Eng. Simul. **17**, 681–689 (2000)
6. Gelenbe, E., Pujolle, G.: Analysis and Synthesis of Computer Systems, 2nd edn. Advances in Computer Science and Engineering, Texts, vol. 4, 309 p. (2010)
7. Ageev, D.V., Ignatenko, A.A., Kopylev, A.N.: Method for determination of flow parameters in different parts of multiservice telecommunications network, taking into account the effect of self-similarity. Electron. Sci. Spec. Ed.- J. "Probl. Telecommun." **3**(5), 8–37 (2011). http://pt.journal.kh.ua/2011/3/1/113_ageyev_method.pdf. Accessed 23 Mar 2017

8. Kuchuk, N., Mozhaiev, O., Mozhaiev, M., Kuchuk, H.: Method for calculating of R-learning traffic peakedness. In: 2017 4th International Scientific-Practical Conference Problems of Infocommunications Science and Technology, PIC S and T 2017 – Proceedings, pp. 359–362 (2017). https://doi.org/10.1109/infocommst.2017.8246416

9. Mukhin, V., et al.: The method of variant synthesis of information and communication network structures on the basis of the graph and set-theoretical models. Int. J. Intell. Syst. Appl. (IJISA) **9**(11), 42–51 (2017). https://doi.org/10.5815/ijisa.2017.11.06

10. Akande, A.J., Fidge, C., Foo, E.: Limitations of passively mapping logical network topologies. Int. J. Comput. Netw. Inf. Secur. (IJCNIS) **9**(2), 1–11 (2017). https://doi.org/10.5815/ijcnis.2017.02.01

11. Mozhaev, O., Kuchuk, H., Kuchuk, N., Mykhailo, M., Lohvynenko, M.: Multiservice network security metric. In: 2nd International Conference on Advanced Information and Communication Technologies, AICT 2017 – Proceedings, pp. 133–136 (2017)

12. Mukhin, V.Ye., Bidkov, A.Ye., Thinh, V.D.: The forming of trust level to the nodes in the distributed computer systems. In: Proceedings of XIth International Conference "Modern Problems of Radio Engineering, Telecommunications and Computer Science TCSET'2012", Lvov, Slavsko, 21–24 February 2012, p. 362 (2012)

13. Mukhin, V.Ye.: The security mechanisms for grid computers. In: Proceedings of the 4-th IEEE Workshop on Intelligent Data Acquisition and Advanced Computing Systems: Technology and Applications (IDAACS'2007), Dortmund, Germany, 6–8 September 2007, pp. 584–589 (2007)

14. Zhang, S., Hu, M., Yang, J.: TreePi: a novel graph indexing method. In: IEEE 23rd International Conference on Data Engineering, Istanbul, pp. 966–975 (2007)

15. Kuchuk, G., Kharchenko, V., Kovalenko, A., Ruchkov, E.: Approaches to selection of combinatorial algorithm for optimization in network traffic control of safety-critical systems. In: Proceeding of IEEE East-West Design and Test Symposium (EWDTS'2016), pp. 384–389 (2016). https://doi.org/10.1109/ewdts.2016.7807655

16. Kuchuk, G., Nechausov, S., Kharchenko, V.: Two-stage optimization of resource allocation for hybrid cloud data store. In: 2015 International Conference on Information and Digital Technologies, IDT 2015, Zilina, Slovakia (2015). https://doi.org/10.1109/dt.2015.7222982

17. Yan, X., Yu, S.P., Han, J.: Graph indexing: a frequent structure-based approach. In: ACM SIGMOD International Conference on Management of Data 2004, pp. 335–346 (2004)

18. Khan, Z.A., Siddiqui, J., Samad, A.: Linear crossed cube (LCQ): a new interconnection network topology for massively parallel system. IJCNIS **7**(3), 18–25 (2015). https://doi.org/10.5815/ijcnis.2015.03.03

19. Meligy, A.M., Ibrahem, H.M., Othman, E.A.: Communication centrality in dynamic networks using time-ordered weighted graph. IJCNIS **6**(12), 21–27 (2014). https://doi.org/10.5815/ijcnis.2014.12.03

Method of Distributed Two-Level Storage System Management in a Data Center

Eduard Zharikov[1](✉) , Sergii Telenyk[1,2] ,
and Oleksandr Rolik[1]

[1] National Technical University of Ukraine
"Igor Sikorsky Kyiv Polytechnic Institute", Kyiv, Ukraine
zharikov.eduard@acts.kpi.ua, stelenyk@pk.edu.pl,
o.rolik@kpi.ua
[2] Cracow University of Technology, Cracow, Poland

Abstract. Modern applications in cloud computing, internet of things and machine learning are I/O intensive. They use data storage systems as the main resource of a data center. The advent of new storage technologies makes it possible to increase the performance of I/O operations by integrating devices with different performance within the data storage system by utilizing the storage-tiering approach. To prevent data loss and service downtime, the data storage systems must ensure fault tolerance using data replication management. As modern hybrid IT infrastructures are based on hyperconverged systems, the development of new methods and models for storage management in order to ensure high performance of I/O operations, high availability and fault tolerance becomes an urgent need. The authors propose the management method based on the model of a distributed two-level data storage system. The proposed method uses the algorithms for data migration between fast and slow levels of the data storage system and the algorithms for replication of data between the nodes of distributed data storage. The simulation results indicate that the proposed management method allows increasing performance of I/O operations with files and evenly placing replicas of data blocks on the data center nodes.

Keywords: Data center · Storage tiering · Data replication · High availability · Hyperconverged system

1 Introduction

Effective digital transformation demands the implementation of new technologies and methods for data intensive computing and data storage system management [1] in order to make further improvements in productivity and decision-making processes. Recent trends in cloud computing, internet of things and machine learning are based on data storage systems as the main resource of the data center. Modern cloud applications are I/O intensive. They are usually deployed in virtualized environments such as virtual machines (VMs) and containers.

The file system plays an important role in ensuring I/O operations to modern cloud applications while providing functions such as snapshots, integrity checking,

© Springer Nature Switzerland AG 2020
Z. Hu et al. (Eds.): ICCSEEA 2019, AISC 938, pp. 301–315, 2020.
https://doi.org/10.1007/978-3-030-16621-2_28

replication, deduplication, compression, and erasure coding. These functions are usually implemented at the levels of file system, operating system, hypervisor or storage system. The file systems that support the distributed storage of objects, files, and blocks [2–6] have built in functions to provide a given performance and fault tolerance. Distributed storage in most file systems is based on a replication mechanism [7].

As storage systems grow in size and complexity, they increasingly encounter disk failures. That may lead to data loss and service downtime. In order to ensure fault tolerance, it is necessary to provide data replication management using storage nodes depending on the number of I/O requests and the amount of free space. In that case, data replicas should be placed on the storage nodes in such a way as to evenly distribute data and I/O requests across all storage nodes to achieve increased data access performance, high availability and fault tolerance [8–10].

To solve the problem caused by I/O bottlenecks without significant loss of capacity while maintaining low cost, data migration between the fast tier storage devices and the slow tier storage devices becomes a common approach [11–14]. Tiered storage is also a promising trend for next generation of distributed storage and processing systems based on hyperconverged infrastructure [15]. When partitioning a storage system by two levels, a problem arises to manage data migrations between levels. In that case, VMs and containers serve the data at the fastest storage media. Therefore, the management of data storage systems in virtualized and consolidated environments is an important scientific and practical problem. The ultimate goal is to efficiently utilize data storage resources by providing a given quality of I/O operations and fault tolerance while minimizing capital and operating costs for the maintenance of data storage system.

In order to solve these problems, this paper proposes a dynamic model of distributed two-level data storage system and a data management method that allow ensuring fault tolerance of the data storage, as well as increased data access performance for virtual machines and containers in cloud data centers.

The rest of this paper is organized as follows. Section 2 presents the related work. The model of distributed two-level data storage with replication is proposed in Sect. 3. The method of two-level data storage management is described in Sect. 4. The method of data replication management is described in Sect. 5. Section 6 describes the modeling of the two-level storage system and replication of data blocks. Section 7 demonstrates the simulation results and the relevant discussions of data migration and replication. Conclusions and future work are summarized in Sect. 8.

2   Related Work

Data storage systems have attracted more attention recently due to the fact that modern cloud applications impose new requirements to capacity and performance of a storage system.

Development of new methods and models for storage management has been for some time an area of interest for many years in literature [12–14, 16–19]. In [16], the authors propose an automatic data placement manager to handle virtual machine disk files allocation and migration in an all-flash multi-tier data center to best utilize the storage resource, optimize the performance, and reduce the migration overhead. Study

[17] presents mathematical models for storage system resource management in order to minimize the cost of using existing storage, to improve the quality of data access, and to evenly distribute files between levels of data storage system. The scheme for storing data using a two-level data storage is proposed in [18] with the aim of making applications faster to launch. At the level of the file system, logical hybrid disks are created with corresponding block pointers in the file-system that belong to the applications when they are started. This allows eliminating delays when searching for data blocks in the corresponding tables and place frequently used blocks on high-performance devices. As a result, the launch of applications is accelerated, and the performance of the data storage increased. Distributed file system, which takes into account the performance of different levels of data storage system, is presented in [19]. The proposed file system is based on HDFS [4] and allows to apply the placement policies (network-aware and tier-aware) for data management both within the levels of the data storage on one node and in the distributed setting. Using multi-criteria optimization, the authors proposed schemes for ensuring fault-tolerance, load balancing and I/O performance.

The replication mechanism is studied, improved and adapted to different use cases in literature [20–23]. The main directions of improvement of replication techniques are ensuring the reliability of data storage, the even distribution of data over replication nodes, the uniform loading of storage nodes and the high performance of I/O operations. In [21], the algorithm of replica placement using HDFS file system is proposed. The method of intelligent data placement selects the least loaded rack and the least loaded node of the cluster in order to evenly distribute data blocks on storage nodes. An improved strategy for dynamically locating a replica node is proposed in [22]. The strategy aims to even load storage nodes and takes into account the file heat and the node load as well as hybrid cloud environment characteristics. The average of the file heat and the average of the node load is also used to determine the number of file replicas. It is supposed that the file located on the replication node completely. The focus of the proposed strategy is on determining the number of file replicas. The authors in [23] proposed an adaptive file replication strategy in the data center that takes into account the availability of the file, the last time the replica was requested, number of access, and size of replica when selecting a replication node. The main criterion for choosing a storage node for replica placement is to achieve a balanced load of all storage nodes in the data center.

3   The Model of Distributed Two-Level Data Storage with Replication

The model of a distributed two-level data storage system is composed of data (files and blocks), fast-level devices, slow-level devices, and physical machines (PMs). Each PM has a local storage directly attached to it. Thereby, each server also represents a storage node. The direct attached storage (DAS) system includes several data storage devices, which are arranged at several levels, depending on their performance. In the proposed model the two-level data storage is used. Data storage of the server node is composed of fast level disks and slow level disks.

Table 1. Variables of the model of distributed two-level data storage with replication

f_{im}	The size of the file i on the PM m, $i = \overline{1, n_m}$, $m = \overline{1, M}$
c	The size of data block
q_{jm}	The number of access times to the j-th data block of the PM m at the fast level for a given time, $j = \overline{1, t_m}$, $m = \overline{1, M}$
p_{jm}	The number of access times to the j-th data block of the PM m at the slow level for a given time, $j = \overline{1, t_m}$, $m = \overline{1, M}$
B	The matrix of file splitting into data blocks ($b_{ij} = 1$ if j-th block is a part of the file j, $b_{ij} = 0$ otherwise), $i = \overline{1, n_m}$, $j = \overline{1, t_m}$
R	The matrix of data block placement on physical servers (replication matrix), $r_{jm} = 1$ if j-th data block is placed on the PM m, $r_{jm} = 0$ otherwise)
s_{km}	The size of the k-th fast level device in the PM m, $k = \overline{1, u_m}$, $m = \overline{1, M}$
h_{lm}	The size of the l-th slow level device on the PM m, $l = \overline{1, v_m}$, $m = \overline{1, M}$
X	The matrix of data block placement at fast level devices, x_{kj} - logical variable of placing j-th block at k-th fast level device, $k = \overline{1, u_m}$, $j = \overline{1, t_m}$
Y	The matrix of data block placement at slow level devices, y_{lj} - logical variable of placing j-th block at l-th slow level device, $l = \overline{1, v_m}$, $j = \overline{1, t_m}$
Z	The vector of latencies of data transmission from the current PM m to other PMs of the data center, which store replicas of the PM m, $z_m = 0$. The vector index corresponds to the PM number
E	The vector of performance indicators of each data storage node, which stores the value of free space at a slow level for each PM of the data center. The vector index corresponds to the PM number
W	The vector of performance indicators of each data storage node, which stores the value of the number of access times to data blocks at both levels for each PM of the data center. The vector index corresponds to the PM number
C_k	The cost of using the k-th fast level device, $k = \overline{1, u_m}$
G_k	The cost of storing one data block at the k-th fast level device $k = \overline{1, u_m}$
D_l	The cost of using the l-th slow level device, $l = \overline{1, v_m}$
g_l	The cost of storing one data block at the l-th slow level device, $l = \overline{1, v_m}$
d_{vk}	The cost of migration of the data block from the l-th slow level device to the k-th fast level device, $k = \overline{1, u_m}$, $l = \overline{1, v_m}$

The proposed model consists of $M \in \mathbb{N}$ PMs, each of which stores n_m files of local virtual machines or containers. In the general case, the number of files in each local data storage is variable. Each file is represented by several blocks of the same size. The block size depends on a file system and an operation system. Denote by t_m the number of blocks that store files of the m-th PM. The number of blocks may vary during data processing. The number of fast level devices on the m-th PM is denoted by u_m, and the number of slow level devices is denoted by v_m. Table 1 describes the notations used in this paper.

3.1 The Model of Direct and Reverse Migration of Data Blocks

The data block access occurs at the fast level. If the required data block is absent at the fast level, this data block is processed as follows: (i) reading data block from the slow level, (ii) sending data block to the appropriate memory exchange channels, (iii) writing data block to the fast level and (iv) deleting that data block from the slow level. Further I/O operations with this data block is performed at the fast level. If to write a data block to the fast level is impossible then the I/O operations with this data block are performed at the slow level until the enough space appears at the fast level. In the case of reverse data block migration, this block is initially written to the slow level, and then removed from the fast level. Accordingly, when migrating a data block from the slow level to the fast level, this block is initially written to the fast level, and then removed from the slow level. Thus, there is only one instance of the data block in a local storage system.

The reason for the data block migration between levels is the presence/absence of I/O requests to the data block for some time. Thus, it is necessary to maximize the average number of I/O operations with data blocks that are placed at the fast level (1).

$$\max \sum_{k=1}^{u_m} \frac{\sum_{j=1}^{t_m} q_{jm} x_{kj}}{\sum_{j=1}^{t_m} x_{kj}}. \tag{1}$$

The number of data blocks on the current PM is defined as follows: $t_m = \frac{\sum_{i=1}^{n_m} f_i}{c}$.

The following are some constraints to be satisfied:

1. The presence of free space at the fast level devices and at the slow level devices can be expressed as follows:

$$\sum_{k=1}^{u_m} \sum_{j=1}^{t_m} x_{kj} c < \sum_{k=1}^{u_m} s_k, \sum_{l=1}^{v_m} \sum_{j=1}^{t_m} y_{lj} c < \sum_{l=1}^{v_m} h_l. \tag{2}$$

2. Only one instance of the j-th data block can be at the fast level and at the slow level in the local storage system:

$$\sum_{k=1}^{u_m} x_{kj} = 1, \sum_{l=1}^{v_m} y_{lj} = 1 \tag{3}$$

3. Each data block j belongs to only one file:

$$\sum_{i=1}^{n_m} b_{ij} = 1, j = \overline{1, t_m}. \tag{4}$$

The number of storing devices at each storage level is determined, on the one hand, by the capabilities of the interfaces, and on the other hand, by the cost of using these devices. The cost of storing data at all fast level devices and at all slow level devices are determined as follows:

$$\sum_{k=1}^{u_m} C_k, \sum_{l=1}^{v_m} D_l. \tag{5}$$

The cost of storing data blocks at the k-th fast level device and at the l-th slow level device are determined as follows:

$$\sum_{j=1}^{t_m} G_k x_{kj}, \sum_{j=1}^{t_m} g_l y_{lj}. \tag{6}$$

When storing data at the fast level devices becomes unreasonable (expensive), then it is necessary to migrate this data to the level with lower performance where data storage will be cheaper. The cost of data migration between two devices of different levels is determined as follows:

$$\sum_{l=1}^{v_m} \sum_{k=1}^{u_m} d_{lk}. \tag{7}$$

Thus, given the expressions (5)–(7), the objective function of minimizing the cost of storing data in the local data storage with the data block migrations between devices of the fast and slow levels is as follows:

$$\min \left[\sum_{k=1}^{u_m} (C_k + \sum_{j=1}^{t_m} G_k x_{jk}) + \sum_{l=1}^{v_m} (D_l + \sum_{j=1}^{t_m} g_l y_{lj}) + \sum_{l=1}^{v_m} \sum_{k=1}^{u_m} d_{lk} \right], \tag{8}$$

subject to (2)–(4).

3.2   Replication Model

To ensure data protection and availability in a distributed environment, the proposed model leverages a replication factor (RF). The replication factor shows how many copies of data block must be stored at separate nodes in the data center. If the replication factor is equal to three then this ensures the data exists in three independent locations and are fault tolerant. However, this comes at the cost of storage resources as full copies are required.

The meta data management server stores the matrix $\mathbf{R}$ and the vectors $\mathbf{E}$, $\mathbf{W}$, whose elements are calculated by the meta data management server as the data arrives from the data storage nodes. The vector $\mathbf{E}$ is used by the replication method on each storage node m in order to determine the replication nodes with the maximum amount of free space at the slow level devices. The elements e_m of the vector $\mathbf{E}$ are determined as follows:

$$e_m = \sum_{l=1}^{v_m} h_{lm} + \sum_{k=1}^{u_m} s_{km} - \sum_{j=1}^{t_m} cr_{jm}. \tag{9}$$

The vector $\mathbf{W}$ takes into account the number of access times to data blocks at the fast and the slow levels of the replication nodes. The elements w_m of the vector $\mathbf{W}$ are determined as follows:

$$w_m = \sum_{j=1}^{t_m} (p_{jm} + q_{jm}) r_{jm}. \tag{10}$$

The size of storage space that currently used up on the storage node including replicas from other storage nodes should not exceed the total size of the disk drives of this node:

$$\exists m = \overline{1, M} : \sum_{j=1}^{t} cr_{jm} \leq \sum_{l=1}^{v_m} h_{lm}. \tag{11}$$

The number of copies of data blocks at all data storage nodes should satisfy the following equation:

$$\exists j = \overline{1, t} : \sum_{m=1}^{M} r_{jm} = RF. \tag{12}$$

Thus, taking into account expressions (9)–(12), to evenly distribute data blocks and I/O requests across all storage nodes it is needed to achieve the minimum of the standard deviation of the values of free space at the slow level and to achieve the minimum of the standard deviation of the values of the access times to data blocks on each storage node:

$$\min[\sigma_E + \sigma_W], \tag{13}$$

subject to (2)–(4).

4   The Method of Two-Level Data Storage Management

The idea of the two-level storage management method is based on a data migration between a fast storage level and a slow storage level as follows. The local pool of disks is divided into two levels named fast level and slow level. Applications running in a VM at any given time do not use all data on logical volumes (disks). If there are requests for certain files for some time, then it is necessary to migrate the appropriate data blocks from the slow level to the fast level. If I/O requests for those files completed and the data blocks are no longer needed, then these data blocks migrate to disks at the slow level, resulting in the release of disk space at the fast level. Therefore, in order to have enough free space at the fast level, it is necessary to constantly maintain a portion of the free space that can be used to store a "hot" data. The new data are always placed at the fast level disks.

To ensure the preparation of the list of files for migration to the slow level it is necessary to sort files according to two criteria: by file size and by the number of access times to each file. Each of the criteria may have some weight that should be determined experimentally. The Algorithm 1 composes the list of files to be migrated from the fast level disks to the slow level disks. The Algorithm 1 is based on the file sorting by increasing the number of access times to files and by decreasing the file size.

Denote by L^F the list of fast-level files and denote by $migrL^F$ the list of fast-level files defined for migration to the slow level. The function $size(L_i^F)$ returns the size of the i-th file.

Algorithm 1. Preparing the list of files for migration to the slow level
Input: L^F
Output: $migrL^F$
1.   The list initialization, $migrL^F \leftarrow$ NULL
2.   **if** $\sum_{k=1}^{u_m} s_k < Th^{SSD}$ **then**
3. $L^F \leftarrow$ sorting L^F by increasing the number of access times to files and by decreasing the file size
4. $wsize \leftarrow size(L_0^F)$, $i \leftarrow 1$, $migrL^F \leftarrow L_0^F$
5.   **while** $wsize < Th^{SSD}$ **do**
6. $migrL^F \leftarrow L_i^F$, $wsize \leftarrow wsize + size(L_i^F)$, $i \leftarrow i+1$
7.   **end while**
8.   **end if**
9.   **return** $migrL^F$

If possible, new files are always stored at the fast level of storage system. Since the amount of free space at the fast level should not be less than a specified threshold, the files with the least number of I/O requests and with the largest size should migrate from the fast level. The migration process runs in parallel with the regular I/O operations. Data blocks migrate to the slow level until the free space at the fast level exceeds a threshold Th^{SSD}. The complexity of the Algorithm 1 is $O(2a \times \log 2a)$ where a is the number of files in the list L^F.

The value of Th^{SSD} depends on the I/O operation intensity with large files and is selected experimentally. The algorithm takes into account that there is always free space at the slow level for file migration. The algorithm for the migration of data blocks to the slow level (Algorithm 2) works as follows:

Algorithm 2. Migration of the file blocks from the fast level to the slow level
Input: $migrL^F$ is the list of files for migration to the slow level
Output: *MRes* is a result of file migration
1. $i \leftarrow 1$, $MRes \leftarrow$ false
2.   **while** $\sum_{k=1}^{u_m} s_k - size(migrL^F) < Th^{SSD}$ **then**
3.   Migrate blocks of the i-th file to the slow level
4.   Delete blocks of the i-th file at the fast level
5.   Delete the i-th file from L^F and $migrL^F$
6. $i \leftarrow i+1$, $MRes \leftarrow$ true
7.   **end while**
8. **return** *MRes*

The complexity of the Algorithm 2 is O(a) where a is the number of files in the list $migrL^F$. The algorithm of data migration from the slow level to the fast level (Algorithm 3) works as follows:

Algorithm 3. Migration of the file blocks from the slow level to the fast level
Input: F is a file requested for I/O operation
Output: a result of file migration
1. $k \leftarrow 1$, $Mg \leftarrow$ true
2. **repeat**
3.   **if** (the file F located at the fast level) **then**
4.   Perform I/O operation with the file blocks at the fast level
5.   **else**
6.   Configure I/O operation with the blocks of the file F from the slow level
7.   Perform I/O operation with the file blocks at the slow level
8.   **while** ($k <= u_m$) and Mg **do**
9.   **if** $s_{km} - size(F) > 0$ **then**
10.   Write the blocks of the file F to the device k at the fast level
11.   Configure I/O operation with blocks of the file F using the fast level
12.   Delete the blocks of the file F from the device at the slow level
13. $Mg \leftarrow$ false
14.   **end if**
15. $k \leftarrow k+1$
16.   **end while**
17.   **if** $k=u_m$ and Mg **then**
18.   Call the Algorithm 2
19.   **end if**
20.   **end if**
21. **until** (I/O operations with the file continue)
22. **return** true

The complexity of the Algorithm 3 is O(u_m) where u_m is the number of the fast level devices on the m-th PM. Algorithm 3 works every time when I/O operations with a certain file start. In this case, Algorithm 2 can be launched in response to a lack of free space at the fast level. It can be also launched at specific intervals of time. Defining such intervals is beyond the scope of this paper.

5   The Method of Data Replication Management

To build and rebuild the cross-node replica structure the following three approaches are designed based on the indicators of the current performance of node operation such as the amount of free RAM, the current state of storage capacity at the fast and the slow levels, the network utilization, etc.

The method of data replication management is carried out in three stages: (i) determining the indicators of current performance of node operation, (ii) selecting storage nodes with the best indicators with respect to the objectives (13), (iii) replicating data blocks to selected storage nodes.

The definition of replication nodes is performed by the Algorithm 4. When searching for replication nodes, for each i-th node, except of the current one (from which the replication will take place), the replication coefficient p^r is calculated as follows:

$$p_i^r = p^e(1/e_i) + p^w w_i + p^z z_i,$$
$$p^e + p^w + p^z = 1, \tag{14}$$

where p^w is the coefficient taking into account the importance of the number of I/O operations with data blocks, p^e is the coefficient taking into account the importance of the free space at storage node, p^z is the coefficient taking into account the importance of the latency of the data transmission indicator. The complexity of the Algorithm 4 is O(M) where M is the number of PMs.

Algorithm 4. Replication of the data block from the storage node (server) m
Input: W, E, Z
Output: rL is a list of server IDs of servers that store the data block replica
1. $rL \leftarrow$ NULL, $sL \leftarrow$ NULL
2.   **for** i=1 **to** M **do**
3.   **if** $(e_i \diamond$ NULL) and $(w_i \diamond$ NULL) and $(z_i \diamond$ NULL) and $(i \diamond m)$ and $(e_i > Th^{HDD})$ **then**
4. $sL \leftarrow (i, p^e(1/e_i) + p^w w_i + p^z z_i)$
5.   **else** $sL \leftarrow (i, $NULL$)$
6. $sL.sortIncreasingReplCoef()$
7.   **end for**
8. $i \leftarrow 1, j \leftarrow 0$
9.   **while** i<RF **do**
10.   **if** $sL_j \diamond$ NULL **then**
11. $rL.add(sL_j), i \leftarrow i+1$
12.   **end if**
13. $j \leftarrow j+1$
14.   **end while**
15.   **return** rL

After determining the replication nodes, the data block replica is written on each storage node in the list rL at the fast level, and then, if the data block is not changed, the migration of this block to the slow level is performed (Algorithm 2). Also, the correspondent values in the matrix $\mathbf{R}$ are updated.

The proposed method makes it possible to adjust the value of the coefficients in the Eq. (14) by estimating the standard deviation of the parameters on the current management step by using the [24] method. The correspondent coefficient can be increased for the parameter whose standard deviation value increases. Thus, the minimization of objective (13) can be achieved by selecting the coefficients of the Eq. (14).

6 Modeling of the Two-Level Storage System and Replication of Data Blocks

To study the model of a distributed two-level data storage and the proposed management method, a simulation application has been implemented using Java programming language. The simulations were conducted on the computer with an Intel i7-3632QM processor and 8 GB of RAM running Windows 10 Pro 64bit.

Process Monitor tool [25] was used to create the input data to the simulation application. Process Monitor allows to record a list of processes running in the operating system and accessing files on disks. Input data collection for simulation includes several stages. At the first stage, the data about OS files accessed by Windows applications during 8 h was collected using the Process Monitor.

The second stage is to save the monitoring results represented by the Process Monitor report in CSV file. Third stage is to form an input data using CSV file for studying the model of a distributed two-level data storage and assessing the quality of algorithms presented in Sect. 4. The following information about the activity of the disk subsystem is provided on the input of each VM during the simulation: an absolute file path; file size; whether its size changed during the Process Monitor run; and the number of requests to the file. To complete the third stage of input data preparation, a separate application was developed to convert the Process Monitor data to the format of the simulation application.

To simulate a distributed two-level data storage with replication and to study the proposed management method, the following initial data must be specified: the number of servers that are required for simulation; the number of devices that are at the fast level on each of the server; the number of devices that are at the slow level. The characteristics of data storage devices that are placed at the fast level: the latency is 1 ms; the average file access time is 1 ms; the bandwidth of the disk device is 300 MBps; disk space is 64 GB. The characteristics of data storage devices at the slow level: the latency is 6 ms; the average file access time is 9 ms; the bandwidth of the disk device is 120 MBps; disk space is 500 GB.

During the experiment, the I/O operations with files of five virtual machines is modeled on each server. Thus, five CSV files have been used to work with the simulation application. Each file contains input data about I/O operations with OS files obtained at the first stage (8742 MB in size). Data migration between the levels of the data storage occurs if there is a certain number of I/O operations to a specific OS file. The data migration helps to reduce latency of data access by storing frequently requested files on fast level of data storage when server experiences high I/O workload generated by virtual machines. During simulation, the replication factor equals to 3 ($RF = 3$). However, the current version of the simulation application does not support block-level I/O operations. Thus, when accessing a file, it is completely written to the fast level.

7 Evaluation of Simulation Results

During the simulations it has been observed that the I/O operations generated by modeled virtual machines did not lead to exhaust disk space at the fast level of the repository. The threshold for maintaining free space at each fast level drive is set to 100 MB based on the size of the largest OS file in the input data. Data transmission latency between servers was generated randomly in the range of 2–8 ms.

7.1 Simulation of Migrations

The simulation results demonstrated the dependencies of the file access time on its size. The simulations were performed for two configurations of the data storage: with a two-level data storage and with a one-level data storage. The one-level data storage was modeled by slow level devices. Ten simulations have been performed for each data storage configuration. The average results were obtained and analyzed.

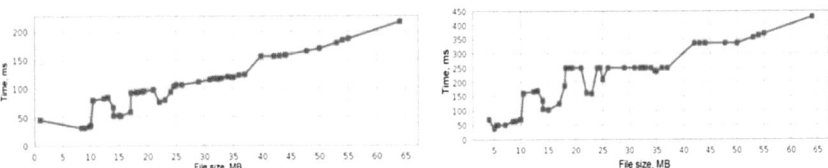

Fig. 1. The time of write operations against file size using two-level data storage (on the left) and using one-level data storage (on the right)

Since write operations in the two-level data storage are usually performed at the fast level the recording time is determined by the performance of the fast level devices as shown in Fig. 1. Thus, the average recording time of such files is slightly increasing.

As also shown in Fig. 1, write operations using the one-level data storage take more time because the performance of the slow level devices is lower than the performance of the fast level devices.

First read operation of the file is performed using the slow level device with the simultaneous writing of this file to the fast level device. Further I/O operations with the file will take place using fast level devices. Thus, as shown in Fig. 2, read operations

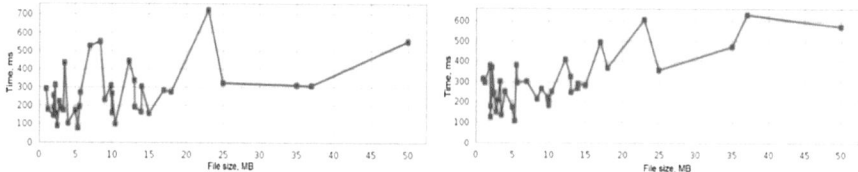

Fig. 2. The time of read operations against file size using two-level data storage (on the left) and using one-level data storage (on the right)

using two-level data storage take less time because hot data (i.e. frequently accessed data) is stored at the fast level. As also shown in Fig. 2, read operations using one-level data storage take more time because the performance of the slow level devices is lower than the performance of the fast level devices.

Thus, considering the average values of reading and writing operations with a specific number of files using the two-level data storage and one-level data storage it can be seen that the use of two-level data storage can reduce the file access latency for virtual machines performing simultaneous I/O operations.

7.2   Simulation of Replication

The simulation of replication is performed considering the additional limitation on file replication that is all blocks of one file can be replicated only to one server. Ten simulation runs were conducted to obtain an average standard deviation of data size on each server before and after replication process. At the initial stage each storage node (server) contains on average the amount of data as follows: Server 1 – 12375 MB, Server 2 – 10517 MB, Server 3 – 11613 MB, Server 4 – 9219 MB, Server 5 – 11427 MB. Average standard deviation of data size in all servers is 1.18 GB.

All servers starting from Server 1 to Server 5 are considered to find out which one may potentially be used for replica placement during simulation. First, the replica location was fined out for data of Server 1, then the replica location was fined out for data of Server 2 and so on.

As a result, the replica locations for all servers were obtained. Each storage node after replication process contains on average the amount of data as follows: Server 1 – 34415 MB, Server 2 – 34319 MB, Server 3 – 32349 MB, Server 4 – 33207 MB, Server 5 – 32163 MB. Average standard deviation is 1.03 GB.

Thus, considering the standard deviation of data size in all servers before replication and after replication, it can be concluded that the proposed replication method allows to evenly place replicas of data on the servers of the data center without bottlenecks when creating new and modifying existing data blocks.

8   Conclusions

In this paper, the management method based on the model of a distributed two-level data storage system has been proposed. In order to manage I/O operations in a distributed storage based on a DAS system, the proposed method uses the algorithms for data migration between fast and slow levels of the data storage and the algorithms for replication of data between the nodes of distributed data storage.

The proposed method uses the algorithms for data migration based on the criterion of minimizing the cost of data storage. Release of disk space at the fast level by migrating unused files to the slow level is performed using two criteria: (i) files with biggest size migrate first and (ii) files on which I/O operations occur less frequently migrate first. In order to ensure fault tolerance for a data storage system, the method of data replication with a given replication factor has been developed considering: (i) the number of I/O operations with data blocks, (ii) the amount of free space at the storage node, and (iii) the latency of data transmission between the storage nodes.

To study the proposed model and the management method, the simulation application has been developed. As a result of simulations, it has been revealed that the use of two-level data storage with the proposed management method allows to reduce the required volume of storage devices at the fast level namely to reduce the cost of data storage. The proposed management method also allows to reduce the waiting time for completing I/O operations with files for virtual machines performing simultaneous I/O operations. It also allows to evenly place replicas of data blocks on the nodes of the data center without bottlenecks when creating new and modifying existing data blocks.

To address current limitations of the simulation application, future work should refine it in considering the data block exchange. Another area for future work is to conduct experiments to empirically detect weights and replication coefficients using real world storage workloads.

References

1. Newman, D.: Top 10 Digital Transformation Trends For 2019 Framework (2018). https://www.forbes.com/sites/danielnewman/2018/09/11/top-10-digital-transformation-trends-for-2019/#17d55d1e3c30. Accessed 22 Dec 2018
2. DeCandia, G., Hastorun, D., Jampani, M., Kakulapati, G., Lakshman, A., Pilchin, A., Sivasubramanian, S., Vosshall, P., Vogels, W.: Dynamo: amazon's highly available key-value store. In: ACM SIGOPS Operating Systems Review, vol. 41, no. 6, pp. 205–220 (2007)
3. Ghemawat, S., Gobioff, H., Leung, S.T.: The Google file system. In: SOSP'03 Proceedings of the Nineteenth ACM Symposium on Operating Systems Principles, vol. 37, no. 5, pp. 29–43 (2003)
4. Shvachko, K., Kuang, H., Radia, S., Chansler, R.: The hadoop distributed file system. In: 2010 IEEE 26th Symposium on Mass Storage Systems and Technologies (MSST), vol. 11, pp. 1–21 (2010)
5. Calder, B., Wang, J., Ogus, A., Nilakantan, N., Skjolsvold, A., McKelvie, S., Haridas, J.: Windows Azure Storage: a highly available cloud storage service with strong consistency. In: Proceedings of the Twenty-Third ACM Symposium on Operating Systems Principles, pp. 143–157 (2011)
6. Weil, S.A., Brandt, S.A., Miller, E.L., Long, D.D., Maltzahn, C.: Ceph: a scalable, high-performance distributed file system. In: Proceedings of the 7th Symposium on Operating Systems Design and Implementation, pp. 307–320. USENIX Association (2006)
7. Tanenbaum, A.S., Van Steen, M.: Distributed Systems: Principles and Paradigms. Prentice-Hall, Upper Saddle River (2007)

8. Shieh, F., Arani, M.G., Shamsi, M.: An extended approach for efficient data storage in cloud computing environment. Int. J. Comput. Netw. Inf. Secur. (IJCNIS) **7**(8), 30–38 (2015). https://doi.org/10.5815/ijcnis.2015.08.04

9. Kaur, P., Mahajan, M.: Integration of heterogeneous cloud storages through an intermediate WCF service. Int. J. Inf. Eng. Electron. Bus. (IJIEEB) **7**(3), 45–51 (2015). https://doi.org/10.5815/ijieeb.2015.03.07

10. Seera, N.K., Jain, V.: Perspective of database services for managing large-scale data on the cloud: a comparative study. Int. J. Mod. Educ. Comput. Sci. (IJMECS) **7**(6), 50–58 (2015). https://doi.org/10.5815/ijmecs.2015.06.08

11. Gregg, B.: ZFS L2ARC (2008). http://www.brendangregg.com/blog/2008-07-22/zfs-l2arc.html. Accessed 22 Dec 2018

12. Chen, F., Koufaty, D.A., Zhang, X.: Hystor: making the best use of solid state drives in high performance storage systems. In: Proceedings of the International Conference on Supercomputing, pp. 22–32 (2011)

13. Guerra, J., Pucha, H., Glider, J.S., Belluomini, W., Rangaswami, R.: Cost effective storage using extent based dynamic tiering. In: FAST, vol. 11, pp. 1–14 (2011)

14. Arteaga, D., Zhao, M.: Client-side flash caching for cloud systems. In: Proceedings of International Conference on Systems and Storage, pp. 1–11 (2014)

15. Rolik, O., Telenyk, S., Zharikov, E.: Management of services of a hyperconverged infrastructure using the coordinator. In: Proceedings of International Conference on Theory and Applications of Fuzzy Systems and Soft Computing, pp. 456–467. Springer, Cham (2018)

16. Yang, Z., Hoseinzadeh, M., Andrews, A., Mayers, C., Evans, D.T., Bolt, R.T., Swanson, S.: AutoTiering: automatic data placement manager in multi-tier all-flash datacenter. In: 2017 IEEE 36th International Proceedings of Performance Computing and Communications Conference (IPCCC), pp. 1–8 (2017)

17. Bodanyuk, M.E., Karnaukhov, O.K., Rolik, O.I., Telenyk, S.F.: Management of data storage systems. Electron. Commun. **5**(76), 81–90 (2013). (in Ukrainian)

18. Ryu, J., Lee, D., Han, C., Shin, H., Kang, K.: File-system-level storage tiering for faster application launches on logical hybrid disks. IEEE Access **4**, 3688–3696 (2016)

19. Kakoulli, E., Herodotou, H.: OctopusFS: a distributed file system with tiered storage management. In: Proceedings of the 2017 ACM International Conference on Management of Data, pp. 65–78 (2017)

20. Milani, B.A., Navimipour, N.J.: A comprehensive review of the data replication techniques in the cloud environments: major trends and future directions. J. Netw. Comput. Appl. **64**, 229–238 (2016)

21. Ibrahim, I.A., Dai, W., Bassiouni, M.: Intelligent data placement mechanism for replicas distribution in cloud storage systems. In: Proceedings of IEEE International Conference on Smart Cloud (SmartCloud), pp. 134–139 (2016)

22. Zhao, Y., Li, C., Li, L., Zhang, P.: Dynamic replica creation strategy based on file heat and node load in hybrid cloud. In: Proceedings of 19th International Conference on Advanced Communication Technology (ICACT), pp. 213–220 (2017)

23. Mansouri, N.: Adaptive data replication strategy in cloud computing for performance improvement. Front. Comput. Sci. **10**(5), 925–935 (2016)

24. Welford, B.P.: Note on a method for calculating corrected sums of squares and products. Technometrics **4**(3), 419–420 (1962)

25. Russinovich, M.: Process Monitor (2018). https://docs.microsoft.com/en-us/sysinternals/downloads/procmon. Accessed 22 Dec 2018

Probability Models for Validity Evaluation

Drobiazko Iryna, Sapsai Tetiana[(✉)], Tarasenko Volodymyr,
and Teslenko Olexandr

National Technical University of Ukraine "Igor Sikorsky Kyiv
Polytechnic Institute", Kiev, Ukraine
{drobyazko, vtarasen, teslenko}@scs.kpi.ua,
stgkin77@gmail.com

Abstract. Testing is widely used for detection of malfunctions in digital systems. However besides malfunctions in digital systems, there is a possibility of faults that cannot be defected by testing tools. As a result of faults, a digital system may not go to a wrong state. A false command may cause the object of control to go to the wrong state, which is confirmed by practical cases. Modern elements of digital systems are created by using a nanometer range of design limits and they operate at a high clock frequency. High energy particles (SEE – Single Event Effects) affect the operation of microcircuits that results in an increase in probability of faults. This fact necessitates a more critical evaluation of a quality of computer device designs according to reliability of their functioning. This article represents further development of methods for optimization of signal probability model with the aim to decrease the volume of calculations. The obtained results can be used at the stage of computer systems design for the preliminary evaluation of reliability of their functioning under fault conditions.

Keywords: Data distortions · Probability models · Validity ·
Probability of faults · Functioning under fault conditions

1 Introduction

Principles of dependable computer systems design are being actively researched over the last years. Validity is one of the important dependability indices as an estimation of correctness of device functioning under conditions of faults and malfunctions caused by natural or technogenic factors [1, 2]. For example, as a result of SEE, electromigration, interference, etc. [3–5]. Asynchronous processors, which are promising with regard to speed and energy costs, are especially sensitive to faults [6, 7]. At the same time computer devices can have implicit correction of input data distortions and malfunctions – called self-correction [8]. Additionally, such properties can be artificially (explicitly) added by using certain high-reliability device design methods [9–11]. The idea of self-correction is in the following. Name a set of input codes of a device as X. Set the equality relation of device output values for any elements from X on the $X \times X$ set (Cartesian product). This relation is an equivalence relation, which splits the X set into equivalence classes (insensitivity classes). If insensitivity classes contain more than one element from X, then input data corruption within bounds of such classes will not cause an error. As the result, device reliability increases. Calculation of device functioning is

© Springer Nature Switzerland AG 2020
Z. Hu et al. (Eds.): ICCSEEA 2019, AISC 938, pp. 316–324, 2020.
https://doi.org/10.1007/978-3-030-16621-2_29

important on the design stage to make a choice of implementation according to the particular project requirements. The problem implies that in the general case the volume of calculations of reliability is proportional to an exponential function on bits of input data. Authors [8] proposed methods of decreasing the volume of calculations by taking into account decompositions. This allowed a considerable decrease in the volume of calculations for the one, separately taken, output. The methods proposed in [8] have a sub exponential volume of calculations in the case of complex calculation for all outputs of the device. In this connection, there appears the problem of further development of probability models that consider physical nature of distortions, device design and provide a possibility of polynomial dependence of the volume of calculations on the number of bits of input data.

2 Main Part

It is suggested to examine probability models of output signal values (hereinafter referred to as a "probability model") of logical information transformers (LIT) – combination circuits (CC, digital stateless automatons). LIT output signal distortions (hereinafter referred to as "codes") can be caused by malfunctions (failures) inside of device and/or input data distortions.

Represent LIT output probability model caused by input data distortions in the form of a JF table (JF array, $\|JF\|$ matrix), with table rows named $w \in W$, where W is an ordered set of possible binary r-digit LIT output codes. According to the project requirements, columns are also named e binary codes, $e \in W$. Real CC produce codes from W set with any input codes, i.e. it is assumed that any input code from W set, according to project requirements, while distorted, can be transformed into binary code from the same set with some probability. This probability value depends on a particular LIT and equals the sum of input code probabilities which result in e code output instead of w code output. Such sum is written into $JF(w, e)$ table cell (JF matrix or as $JF[\hat{w}, e]$ array item, where e, $\hat{w}$ are sequential numbers of e, w codes in the W set). Validity of LIT functioning equals the sum of element values with $e = w$.

Represent LIT output code probability model, caused by malfunctions (failures) inside device in the form of a failure table $JZ(w, e)$ (failure matrix $\|JF\|$, failure array JZ $[\hat{w}, e]$), where rows are codes from W set, and columns are binary codes e, which belong to a set of all possible r-digit binary codes. Hence, it is assumed that, unlike input data distortions, a failure in LIT itself with some probability can cause transformation of any code from W set into arbitrary binary code of the same number of digits. This probability is written in the $JZ(w, e)$ table cell ($\|JF\|$ matrix or $JZ[\hat{w}, e]$ array item). LIT functioning validity equals the sum of values with $e = w$. It is obvious that the sum of values in each row of JZ table equals 1.

It can be assumed that internal LIT failures and input data distortions are independent random events. Additionally, LIT output codes given input data distortions can also be distorted due to internal LIT failures. Thus, multiplication of appropriate probabilities can be used to obtain a $JFZ(w, e)$ probability table ($\|JFZ\|$ matrix, $JFZ[\hat{w}, e]$ array) including both input data distortions and malfunctions.

Name all r-digit codes from the W set as $a_1, a_2, ..., a_{|W|}$, and, similarly, name $b_1, b_2, ..., b_{|W|}$. Given $c \in \{0, 1, ..., S\}$, where $S = 2^r - 1$, name all possible r-digit codes as $c_0, c_1, ... c_S$. Then

$$JFZ(a_i, c_j) = JF(a_i, b_1) * JZ(b_1, c_j) + JF(a_i, b_2) * JZ(b_2, c_j) + ... + JF(a_i, b_{|W|}) * JZ(b_{|W|}, c_j)$$
$$(i = 1, 2, ..., |W|, j = 1, 2, ..., S)$$

This shows that $\|JFZ\|$ matrix is formed via multiplication of $\|JF\|$ and $\|JZ\|$ matrices.

$$\|JFZ\| = \|JF\| \times \|JZ\| \qquad (1)$$

A $JFZ(w, e)$ table and a $JFZ[\hat{w}, e]$ array can be created from $\|JFZ\|$ matrix when needed.

Such probability model can be used for primary device inputs by using simulated information transformer (primary input data distortion model). Input of such transformer can receive non-distorted data, while output can produce data, distorted by the effect of external factors. Represent the probability model of simulated transformer output codes caused by external factors in the form of a JX table (JX array, $\|JX\|$ matrix), with table rows being w codes, $w \in \{0, 1, ..., S\}$, where $S = 2^r - 1$, and columns also being binary codes $e \in \{0, 1, ..., S\}$. Transformation probability of w code into e code is written into $JX(w, e)$ table cell ($JX[w, e]$ array or $\|JX\|$ matrix item). The sum of matrix diagonal elements defines the input data validity. Consider forming JX given primary input values are independent random values. In computer engineering possible signal distortions are typically classified into 3 types: "constant 0" distortion, "constant 1" distortion, "inversion" distortion. Let x_i be a value of digit i of code, for which create a $Jx_i(w, e)$ table, ($Jx_i[w, e]$ array). Then:

$$Jx_i[0, 0] = p_0 p_w + p_0 p_{k0}$$
$$Jx_i[0, 1] = p_0 p_{k1} + p_0 p_{inv}$$
$$Jx_i[1, 0] = p_1 p_w + p_1 p_{k0}$$
$$Jx_i[1, 1] = p_1 p_{k1} + p_1 p_{inv}$$

Where

– p_0 is a probability of 0 input signal value
– p_w is a probability of no distortions
– p_{k0} is a probability of "constant 0" distortion
– p_{k1} is a probability of "constant 1" distortion
– p_{inv} is a probability of "inversion" distortion

Note that Jx_i probabilities are the sum of appropriate conditional events because distortion event (or lack of it) depends on the probability of receiving 0 or 1 on the input of simulated information transformer. Validity of digit i is $Jx_i[0, 0] + Jx_i[1, 1]$.

Name code $w_j = <w_{1j}, w_{2j}, ..., w_{rj}>$, $w_j \in \mathbf{W}, j \in \{0, 1, ..., 2^r - 1\}$, code $e_i = <w_{1j},$ $e_{2,i}, ..., e_{r,i}>$ $i = 0, 1,..., 2^r - 1$, and determine JX[e, w] assuming input code digit values are independent random values:

$$JX\left[w_j, e_i\right] = Jx_1\left[w_{1j}, e_{1i}\right] \times Jx_2\left[w_{2j}, e_{2i}\right] \times ... \times Jx_r\left[w_{rj}, e_{ri}\right] \tag{2}$$

3 Model Use Examples

Duplication and the use of majority function is one of the methods of increasing device reliability. Calculate validity of majority function operation using the proposed probability model. Assume that majority function receives some value being tripled. Majority function forms output value by using a voting principle. The idea of using majority function is that single distortions are the most probable. Create a probability table for single distortions (Tables 1 and 2):

Table 1. JX table for single distortions

In\out	000	001	010	011	100	101	110	111
000	0,364	0,015	0,015	0	0,015	0	0	0
111	0	0	0	0,015	0	0,015	0,015	0,546

Table 2. JF table for single distortions

w/e	0	1
0	0,409	0,000
1	0,000	0,591

Validity is 1.

Consider validity of majority function operation given double distortions (Tables 3, 4 and 5).

Table 3. JX table for single and double distortions

In\out	000	001	010	011	100	101	110	111
000	0,364	0,014	0,014	0,001	0,014	0,001	0,001	0
111	0	0,001	0,001	0,014	0,001	0,014	0,014	0,546

Table 4. JF table for single and double distortions

w/e	0	1
0	0,406	0,003
1	0,003	0,588

Validity is 0,994

Table 5. *JZ* failure table

w\e	0	1
0	0,99	0,01
1	0,001	0,999

Multiplying *JF* matrix on *JZ* matrix we'll get probability values on majority function output with input distortions and failures (Tables 6 and 7).

Table 6. *JFZ* table

w\e	0	1
0	0,40491	0,004090
1	0,00059	0,590409

Validity is 0,995319

Table 7. *JF1Z* table

w\e	0	1
0	0,401943	0,007057
1	0,003558	0,587442

Validity is 0,989385

In general, direct practical use of the provided probability model is limited by the amount of calculations. Amount of floating point multiplication operations for LIT with m inputs is $2^{2m}(m-1)$, which limits practical use with a value m $\leq$ 16 (without use of supercomputers). Because the model works with device output signal probabilities, it is possible to significantly reduce required amount of calculations. We shall examine validity of calculations for an adder, as a typical computer engineering device.

Note the feature of binary adder – values of less significant digits of sum do not depend on values of more significant digits of arguments, while the most significant digits of sum depend on less significant digits of arguments via the carry-over signal. Table 8 shows the data distortion model for primary inputs for the first digit of adder (*JX* table). Rows are denoted by non-distorted input data values in the c_{in}, x_0, y_0 order. Columns are denoted by data values that can be obtained as a result of distortions. Cells contain product values according to (2). For example,

$$JX[011, 101] = J_{cin}[0, 1] \times J_{x0}[1, 0] \times J_{y0}[1, 1]$$

X = {000, 001, …, 111} set is being split into the following four insensitivity classes – {000}, {100, 010, 001}, {011, 101, 110}, and {111}. Carry-over and sum values do not change within bounds of the specified classes. Table 8 is split to 16 parts by the intersections insensitivity classes. As the result, the JCS_0 probability model for

Table 8. *JX* model for the first digit of adder

w\e (result)	000	001	010	100	011	101	110	111
000 (00)	1 (00)	2 (01)			3 (10)			4 (11)
001	5 (00)	6 (01)			7 (10)			8 (11)
010 (01)								
100								
011	9 (00)	10 (01)			11 (10)			12 (11)
101 (10)								
110								
111 (11)	13 (00)	14 (01)			15 (10)			16 (11)

carry-over (c_{out0}) and sum (s_0) for the first adder digit (see Table 9) is created. Cells of Table 9 show the appropriate intersections of insensitivity classes from the Table 8. Table 9 rows are denoted by non-distorted output data values in the c_{out0} and s_0 order. Columns are denoted, with the same order, by output values of the first adder digit with distortions applied to the input data.

Table 9. JCS0 model

w\e	00	01	10	11
00	1 (true)	2 (false)	3 (true)	4 (false)
01	5 (false)	6 (true)	7 (false)	8 (true)
10	9 (true)	10 (false)	11 (true)	12 (false)
11	13 (false)	14 (true)	15 (false)	16 (true)

Cells of Table 9 contain sums of distortion probabilities, calculated from appropriate products in intersections of insensitivity classes from the Table 8. Products are calculated based on probability arrays $Jx_0[0, 0]$, $Jx_0[0, 1]$, $Jx_0[1, 0]$, $Jx_0[1, 1]$, $Jy_0[0, 0]$, $Jy_0[0, 1]$, $Jy_0[1, 0]$, $Jy_0[1, 1]$, $Jc_{in0}[0, 0]$, $Jc_{in0}[0, 1]$, $Jc_{in0}[1, 0]$, $Jc_{in0}[1, 1]$, being defined via research. For example, the $JCS_0[01, 00]$ value is written to the cell labeled with a number 5:

$$JCS_0[01, 00] = JX[001, 000] + JX[010, 000] + JX[100, 000],$$

where

$$JX[001, 000] = Jc_{in0}[0, 0] \times Jx_0[0, 0] \times Jy_0[1, 0]$$
$$JX[010, 000] = Jc_{in0}[0, 0] \times Jx_0[1, 0] \times Jy_0[0, 0]$$
$$JX[100, 000] = Jc_{in0}[1, 0] \times Jx_0[0, 0] \times Jy_0[0, 0]$$

Hereafter we are interested in non-distorted sum values in the first digit. Cells of Table 9 are marked with the presence of match (false, true) of the correct sum value and a sum value received after input data distortions.

Note the following. Firstly, the value of sum in the first digit doesn't affect values of sums in the next digits. Secondly, when calculating validity of the adder as a whole, need to consider only probability of the correct sum in the first digit. This allows using JCout model for calculating adder validity – the probability model of carry-over signal values assuming non-distorted sum value (see Table 10). The cells in Table 10 are marked with appropriate numbers from the Table 9.

Table 10. JCout model

w\e	0 (true)		1 (true)	
0 (true)	1	6	3	8
1 (true)	9	14	11	16

The considered models can be built for arbitrary i-th digit of the adder given defined probabilities $Jx_i[0,0], Jx_i[0,1], Jx_i[1,0], Jx_i[1,1], Jy_i[0,0], Jy_i[0,1], Jy_i[1,0], Jy_i[1,1]$, which do not depend on values of probabilities in other digits, and probabilities $Jc_{ini}[0,0] = JCout_{(i-1)}[0,0]$, $Jc_{ini}[0,1] = JCout_{(i-1)}[0,1]$, $Jc_{ini}[1,0] = JCout_{(i-1)}[1,0]$, $Jc_{ini}[1,1] = JCout_{(i-1)}[1,1]$, calculated from the previous digits using Table 10.

Proposed models allow calculating validity values with a linear dependency of amount of calculations on the number n of adder digits given input data distortions. The complex of applications was developed for experimental research, which helped confirming the match of calculation results using this method and without it for numbers $n \leq 8$. Calculations for $n > 8$ without using the proposed method are problematic because of significant amount of calculations.

Table 11. An example of input data distortions

Type\digit	Carry-over	x_0	x_1	...	x_{63}	y_0	y_1	...	y_{63}
No distortions	0,99986	0,99979	0,99997	...	0,99982	0,99989	0,99982	...	0,99984
Constant 0	0,00006	0,00009	0	...	0,00008	0,00006	0,00005	...	0,00008
Constant 1	0,00005	0,00005	0,00002	...	0,00002	0,00004	0,00007	...	0,00001
Inversion	0,00003	0,00007	0,00001	...	0,00008	0,00001	0,00006	...	0,00007
Jx_i\digit	Carry-over	x_0	x_1	...	x_{63}	y_0	y_1	...	y_{63}
[0, 0]	0,49996	0,49994	0,499985	...	0,49995	0,499975	0,499935	...	0,49996
[0, 1]	0,00004	0,00006	0,000015	...	0,00005	0,000025	0,000065	...	0,00004
[1, 0]	0,000045	0,00008	0,000005	...	0,00008	0,000035	0,000055	...	0,000075
[1, 1]	0,499955	0,49992	0,499995	...	0,49992	0,499965	0,499945	...	0,499925

Table 11 shows adder input data distortion probabilities, set with help of a random number generator.

The experimental research results for adders with simultaneous distortions on all inputs are shown in Table 12. Columns are denoted with a number of decimal digit after the separator, which start non-zero decimal digit values of distortion probability. For example, the column named 5 shows distortion probability values from Table 11.

Table 12. Results of experimental research

Number of digits in adder	Number of the first non-zero digit after the radix point in the probability of distortions			
	5	4	3	2
8	0,998491	0,985033	0,861543	0,264539
16	0,996975	0,970192	0,741436	0,068708
32	0,994148	0,943058	0,559996	0,005471
64	0,988716	0,892827	0,325823	0,000041

The obtained results correspond to the known ones for the case of high radiation power, for example in a space, where the computer device validity is nearly zero. The following result is also expected: validity decreases with an increase of number of digits in the adder given the same external conditions.

Single distortions are the most probable under normal climate conditions. The experimental research was conducted under conditions that only one column of Table 11 has non-zero distortion probability. The results have shown that adder validity does not depend on the place of distortion. That is, do not depend on column numbers in Table 11 with non-zero distortion probabilities. This result is also expected, as bool functions of sum and carry-over are symmetric.

4   Conclusions

Provided examples confirm the adequacy of calculation results using the proposed models to both expected and practically known results. Possibility of taking into account both input data distortions and internal device malfunctions (failures) is a property of the proposed models. And all possible single and group distortions can also be taken into account.

The models are effective to create the methods for a further decrease in the volume of reliability calculations. Using an adder, one of the main computer devices, as an example, it has been shown analytically and experimentally that the employment of the proposed models provide a linear dependence of the volume of reliability calculations on bits of input data.

References

1. Avizienis, A., Laprie, J.C., Randell, B., Landwehr, C.: Basic concepts and taxonomy of dependable and secure computing. IEEE Trans. Dependable Secur. Comput. 1(1), 11–33 (2004)
2. Kharchenko, V.S.: Dependability and dependable systems: elements of methodology. Radioelectron. Comput. Syst. 5(17), 7–19 (2006)
3. Bilal, B., Ahmed, S., Kakkar, V.: An insight into beyond CMOS next generation computing using quantum-dot cellular automata nanotechnology. Int. J. Eng. Manuf. (IJEM) 8(1), 25–37 (2018). https://doi.org/10.5815/ijem.2018.01.03
4. Wang, L., Wang, G.: Big data in cyber-physical systems, digital manufacturing and industry 4.0. Int. J. Eng. Manuf. (IJEM) 6(4), 1–8 (2016). https://doi.org/10.5815/ijem.2016.04.01
5. Normand, E.: Single event upset at ground level. IEEE Trans. Nuclear Sci. 43, 2742–2750 (1996). http://web.archive.org/web/20131021190327/http:/pdf.yuri.se/files/art/2.pdf
6. Rani, A.: Grover, N,: Design and implementation of control Unit-ALU of 32 bit asynchronous microprocessor based on FPGA. Int. J. Eng. Manuf. (IJEM) 8(3), 12–22 (2018). https://doi.org/10.5815/ijem.2018.03.02
7. Rani, A., Grover, N.: Novel design of 32-bit asynchronous (RISC) microprocessor & its implementation on FPGA. Int. J. Inf. Eng. Electron. Bus. (IJIEEB) 10(1), 39–47 (2018). https://doi.org/10.5815/ijieeb.2018.01.06
8. Klyatchenko, Y., Tarasenko, G., Tarasenko-Klyatchenko, O., Tarasenko, V., Teslenko, O.: Optimization of processor devices based on the maximum indicators of self-correction. In: Advances in Computer Science for Engineering and Education, vol. 754, pp. 380–390. Springer (2018)
9. Wang, X., Li, S., Liu, F., Fan, X.: Reliability analysis of combat architecture model based on complex network. Int. J. Eng. Manuf. (IJEM) 2(2), 15–22 (2012). https://doi.org/10.5815/ijem.2012.02.03
10. KalaiKaviya, K., Balasubramanian, D.P., Tamilselvan, S.: Design of a optimized parallel array multiplier using parallel prefix adder. Int. J. Eng. Manuf. (IJEM) 3(2), 40–50 (2013). https://doi.org/10.5815/ijem.2013.02.03
11. Rajaei, R.: Design of a radiation hardened register file for highly reliable microprocessors. International Int. J. Eng. Manuf. (IJEM) 6(5), 11–21 (2016). https://doi.org/10.5815/ijem.2016.05.02

Data Warhouses of Hybrid Type: Features of Construction

Valentyn Tomashevskyi[1], Andrii Yatsyshyn[1],
Volodymyr Pasichnyk[2(✉)], Nataliia Kunanets[2],
and Antonii Rzheuskyi[2]

[1] National Technical University of Ukraine "Igor Sikorsky
Kyiv Polytechnic Institute", Kiev, Ukraine
`simtom@i.ua`, `yatsyshyn@hotmail.com`
[2] Lviv Polytechnic National University, Lviv, Ukraine
`vpasichnyk@gmail.com`, `nek.lviv@gmail.com`,
`antonii.v.rzheuskyi@lpnu.ua`

Abstract. Actual scientific and practical task of creating information technology for construction of distributed data warehouses of hybrid type taking into account the properties of data and statistics of queries to the storage is considered in the article. The analysis of the problem of data warehouses construction taking into account data properties and executable queries is carried out. The conceptual, logical and physical models of distributed storages and inter-level transitions procedures are proposed. Location of data on the nodes, data replication routes are determined by criterion of the minimum total cost of data storage and processing using a modified genetic algorithm.

Keywords: Hybrid data warehouses · Data models ·
Distributed data warehouses · Data replication routes

1 Introduction

The introduction of information technologies and computer systems into all areas of humanity has led to the need to accumulate and process large volumes of data of various types and formats such as: texts, images, geospatial data, media files. These data are stored in a structured, weakly structured and unstructured form in repositories and databases. The effective use of such data is based on their consolidation and depends on technologies of work with heterogeneous data. These technologies are located and processed on various spatially dispersed technical means, primarily with the help of Big Data technologies. Thus, there are problems of designing hybrid and virtual databases, parallel data processing, data structuring, etc. An overview of the current state of problems is related to data consolidation, their physical distribution by methods of databases and data warehouses structuring and data warehouses content management.

© Springer Nature Switzerland AG 2020
Z. Hu et al. (Eds.): ICCSEEA 2019, AISC 938, pp. 325–334, 2020.
https://doi.org/10.1007/978-3-030-16621-2_30

2   Related Works Section

The issues of data consolidation were considered in the works [1, 2]. For physical data distribution, the "Virtual Database" architecture, Big Data technology [3–5] (platform Hortonworks based on Hadoop and Apache Foundation solutions) are mainly used. The data warehouses structuring on the basis of common platforms, for example, Hadoop, is investigated in the works [6, 7], methods of structuring databases and data warehouses, data warehouses processing are considered in the works [8–11].

However, the existing approaches do not take into account the peculiarities of construction of distributed data warehouses of hybrid type for storing data from different types of sources on geographically distributed computing facilities and ensuring efficient storage operation. In particular, there are no comprehensive studies of the availability of different types of data sources, different levels of data abstraction, interlevel transitions between them, and the possibility of increasing the efficiency of the data warehouse facility during its operation is not considered. In existing models of data warehouse construction, the efficiency of the logical and physical data distribution is not investigated. Thus, despite the presence of significant results of scientific researches of various aspects of data warehouses construction and operation, the problem of construction of effective distributed data warehouses of hybrid type remains unsolved. **The aim** of the research is to reduce the total cost of data storage and processing in distributed data warehouses of hybrid type due to the new information technology created on the basis of a complex of mathematical models and methods of analysis and evaluation of data processing in the territorial bodies of Ukraine.

3   Construction of Distributed Data Warehouses of the Hybrid Type

The reorganization processes of the territorial structure of Ukraine require new approaches to the processing of unstructured data, generated at institutions of different levels. Information was being uploaded to the central data warehouse in manual mode for a long time. Establishing a decentralized system has determined the need for distributed data storage. Data must be uploaded both to the central data warehouse and to the storage of appropriate level. Thus, there was a problem of efficient processing of operational, primary and analytical data of large volumes, taking into account the relationships between them both at the central and at the local levels. The analysis shows that the following types of data are stored and processed in the bodies of state and municipal administration. *Operational data*, which include all indicators affecting the budget process and are used for decision making. The data is obtained in real time and does not require data integrity checks, for example, information about bank transactions, exchange rates or stock indicators. These data can have a different structure, stored distributed (throughout Ukraine) and have large volumes – hundreds of gigabytes per year. *Primary data* is the result of decision making based on operational data and have a standard well-defined structure, stored distributed (throughout Ukraine), their volumes - gigabytes per year. *Data for analysis* are obtained from the primary data by presenting in some form, in particular aggregation, their volumes - gigabytes per year. *Metadata* - information on numerical indicators used in the budget process and have tree structure (clearly specified or with additional attributes, inherent to separate data elements), their volumes - tens of MB per year. *Accompanying data* (documents, educational materials, scientific publications, etc.) in unstructured form, in separate applications. In some cases, the data are analyzed and fall under one of the previous categories of data, their volumes - to gigabytes per year.

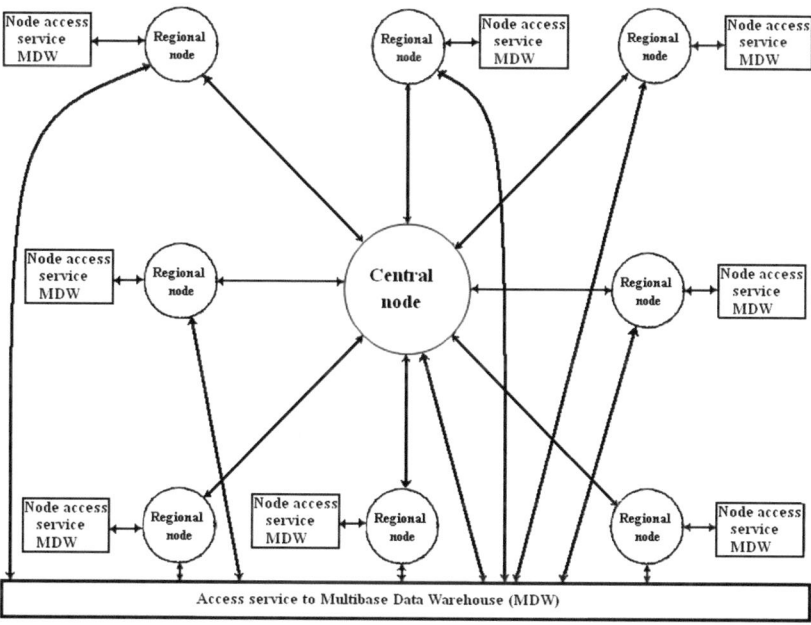

Fig. 1. The structure of the data storage levels of a typical public finance management system.

To solve the problem of storage, processing and exchange of these volumes of data, information technology for the construction of distributed hybrid data warehouses is proposed. It is convenient for creation of a number of information systems, in particular for the construction of a distributed information management system for budget funds. The compilation and control over the implementation of state and local budgets takes place at the following levels: center level; region level and district level (territorial communities, department). The central level provides the collection, processing, storage of data applied at all levels of management. At this level, server equipment, central switching equipment, data storage devices are concentrated. At the regional and district level, data are collected, stored and processed only for a given region, territorial community or district. In addition, at the regional level, the network for data transmission, aggregation of these channels for the district level and the interface ensuring for connecting regional financial institutions to the departmental information network are provided. At the district level, a point of access of local level institutions to the departmental network and routing of user connections to the web portal of the data management system is provided. Information technology provides storage and efficient processing of various properties and types of data. Different databases are used, including relational, multidimensional, NoSQL databases, and also a file storage to perform these processes at logical level. For each of the nodes, it is determined which databases will be used to ensure the efficiency of the system to store data at the physical level. That is, to ensure the required level of speed and not exceed the limits set for the total cost of the system. A multibase data warehouse, which operates on the basis of the developed information technology for constructing distributed data warehouses of the hybrid type is proposed to use to implement an information system that solves this practical problem. It allows distributed storage and efficient processing of different types of data. A typical data management system which consists of the following levels was developed on the basis of information technology of construction of the distributed data warehouses of hybrid type. The level of data storage of a typical system of public finance management (Fig. 1) is hosted on database servers and contains multibase data storage (relational database management system, multidimensional database, NoSQL database, file storage), which receives requests through access services and according to them provides data or modifies data of the storage. It also optimizes its work based on the method of construction of distributed data warehouses of the hybrid type.

4   Use Distributed Hybrid Data Warehouses to Manage the Data of the Territorial Communities

The developed information technology provides an opportunity to create a data management system at each level of its use, where a typical information management system for budget funds (Fig. 2) with modules that are executed on servers of various levels of applications is formed:

- Data provision module designed to interact with data in the repository, including export and import modules, data analysis and reporting. These modules are located on servers that have a network latency (network latency) to servers that have deployed multibase data warehousing services for a multi-database storage.
- Application modules that process data in accordance with business finance management processes and are located on servers that are centrally located, in the case of using services remotely and locally, in the case where the application operates with the data of a particular region.
- Support modules for related processes that have a low level of criticality and can be located on both central and local servers.
- Multibase Data Warehouse Management Systems, a multi-database data warehouse that processes data warehousing.

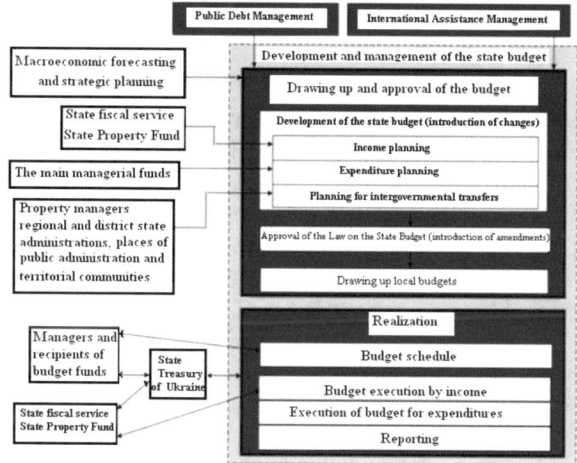

Fig. 2. Information flows to manage budget funds.

5 Interface Level of the Information System

Interaction with users is provided at the interface level of the developed information technology [12–14], which includes web servers, CMS servers, file servers, secure access servers (Fig. 3). These modules allow clients (other applications and end users) to work with applications for data management. Service modules are hosted on both central and local servers, while applications clients are mobile in relation to these services and can only be used if there is a network connection to access services.

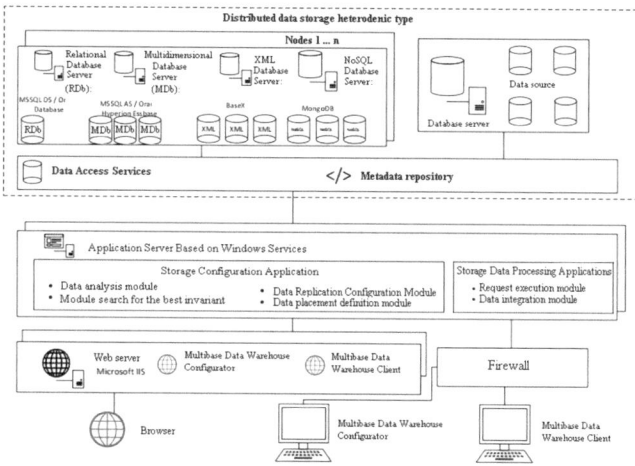

Fig. 3. The architecture of IT construction based on a mixed three-tier web-based architecture with database servers.

6   Advantages of Distributed Data Warehouses of the Hybrid Type

One of the main factors to consider when constructing distributed data warehouses of the hybrid type is the ability to effectively execute different query classes for data of different levels of structuring. That is why the concept of building distributed data warehouses of the hybrid type as a multilevel system of data storage and processing is proposed. The peculiarity of the concept lies in the innovative approach to building data warehouses, formed by information resources from existing and predefined data sources. It is proposed to transform the levels of description of the repository, starting with known (by metadata) logical models of data sources and establishing interconnections between the elements of the metadata and the elements of the conceptual model of the repository. Based on this mutual compatibility, a logical model of the repository is constructed. The logical model, in turn, relates the elements of the database of the repository with the essences of the conceptual model. The transition between these relationships occurs by choosing models to represent these sources of data, and based on the logical model, a physical repository model is developed that includes the deployment and replication data configurations. The information technology developed for the construction of distributed data warehouses of the hybrid type provides: analysis of data sources, including data structuring; placement of data in the storage media according to their structuring; execution of user requests for storage of storage with the necessary additional operations; optimization of the data warehouse by the collected query execution statistics. For interaction of users with information systems built on the basis of this information technology, its implementation ensures: the convenience of obtaining data for the end user; providing information on user requests, regardless of the media where the data is located and the functions that it supports; providing data in various submissions - tabular, "Pivot", hierarchical, textual, etc. Extensibility of data, metadata and storage functions; the possibility of distributed work of users with the delimitation of access to data. Based on the above requirements, the developed IT is based on a mixed three-tier Web-oriented architecture with database servers, applications and clients (browsers, thick clients), which is shown in Fig. 3. The peculiarity of this architecture is the ability to use the functional application of information technology both in public networks using browser and protocols (HTTP/S), and in secure networks with the help of thick clients and their protocols. If necessary, clients from different locations of database servers, applications, and web servers are also deployed distributed. According to this architecture, the data is stored in a distributed hybrid data warehouse (used by the MBSD) consisting of databases (relational, multidimensional, XML, NoSQL with their corresponding management systems) consisting of nodes and data sources hosted in storage systems under managing the appropriate database servers and data access services to provide external interaction with the data store that stores and uses the appropriate metadata repository. Using the data from these services, modules on application servers build a distributed hybrid data warehouse, scheduling and data collection to fulfill queries. To build a distributed hybrid data store, an application is used to configure the data repository containing the following modules: data analysis, which analyzes the data structuring and definition of the repository areas; search for an invariant that defines a better model for data representation; placement of data, which places the data areas of the repository on its nodes; data replication setup that configures replication of storage areas between nodes. To operate a distributed hybrid data store, an application is used to process the data warehouse, consisting of the following modules: execution of queries that receives requests from users and sends datasets according to these queries; data integration, which allows import and export of data from external systems and in the opposite direction. User access to software module resources is via the Multibase Data Warehouse Multi-Database Data Warehouse and the Multibase Data Warehouse Multi-Database Data Warehouse located on web servers. The user interacts with these applications through a thin client (browser or GUI client) that is installed on end-user computers. The Usage Diagram describes the interaction between software usage and actors, and reflects the functional requirements for information technology software for constructing distributed hybrid-type data stores from the perspective of the end-user. The Multibase Data Warehouse Management System consists of the following applications: MDW Manager, a graphical utility for designing a repository; MDW Manager Web, Web-based design for storage; MDW Client, a "thick" client, through which users receive query results; MDW Client Web, Web-based applications for processing data storage; MDW Service Build, a service that builds storage and analysis of data; MDW Service Query, a service that provides queries and interaction with sources and data carriers; MDW Test App, research application. The purpose of the creation of the information and analytical system is to ensure transparency and publicity of work with data both at the level of territorial communities, and at the national level through the dissemination of public information on the results of the activities of state and municipal bodies. The main tasks of the developed information technology are: to increase of awareness of citizens and business about the work of state authorities and the budget process; enabling every member of society to see which decisions are made and for what purposes the taxes they are paid are spent; creating conditions for a transparent dialogue between the authorities and the public in the course of decision-making; ensuring open government policy for the world community, which increases the investment attractiveness of the country and its regions and contributes to the growth of investment in the Ukrainian economy; increase transparency of payments in the field of social security, formation, deduction and distribution of social benefits and benefits; provision of the possibility for the public to be monitored for the decisions taken and target budget expenditures as an instrument for combating corruption; reduction of the time of service of applications and questions of citizens; modeling and forecasting the development of territorial communities; verification of the effectiveness of the decisions taken; automation of the choice of the optimal solution; providing flexible visualization capabilities. The efficiency of using information technology for designing distributed data warehouses is determined by: aggregate cost of storing and processing data in a built-in repository; time of building a data warehouse; speed of queries (there are no requests that are late); number of parallel replicated data. To evaluate these indicators, a set of experiments with test data was conducted. During the experiments, a series of requests for test data was executed. The sample consisted of 320 requests, of which 165 were late. Late requests are those whose performance is less than 15,000 1/ms. The minimum, average, and maximum performance rates for queries in each data region were evaluated (Fig. 4). The minimum performance at the repository is 0.96 1/ms, the average is 554543.8278 1/ms, 19533753.64 1/ms.

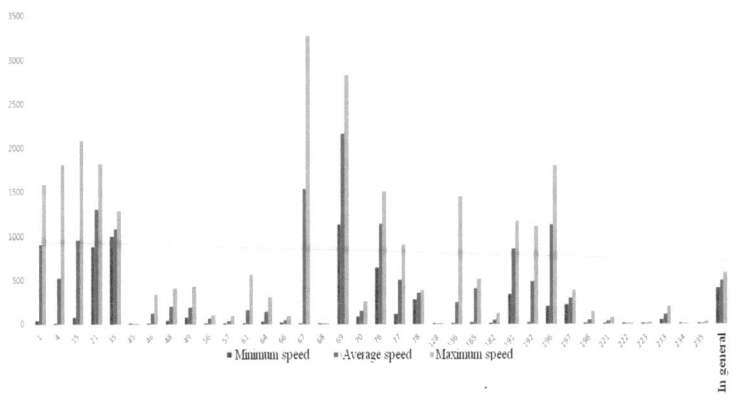

Fig. 4. The performance of querying the data sources by regions (x-axis – numbers of regions, the axis of ordinates the execution speed of queries, data elements for a millisecond, 1/ms).

To select the configuration of the location and replication of the data in the repository, optimization tasks are solved using the genetic algorithm using the phase of the initial formation of the initial population with the bee-keeping algorithm. In order to check the quality of the proposed algorithm, a series of tests of the execution time of the algorithm was performed at a different dimension of the problem. A comparison of the proposed (modified genetic) algorithm performed at certain values of the parameters was carried out with a bee-based algorithm, a genetic algorithm and a full-fledged method to find optimal values (all algorithms coincided with the optimum). The results are shown in Figs. 5, 6.

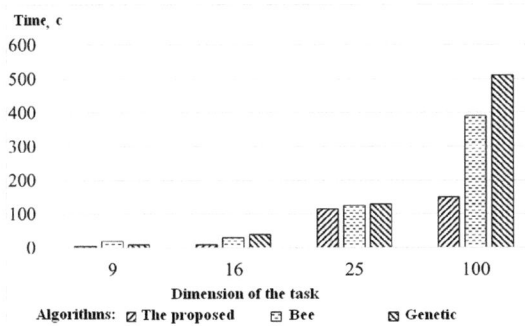

Fig. 5. Diagram of comparison of the time of execution of algorithms depending on the dimension of the problem.

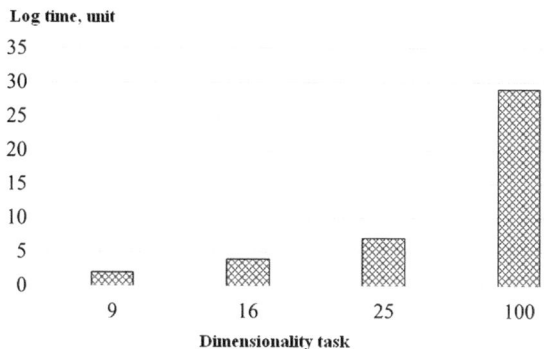

Fig. 6. The graph of the logarithm of the runtime of the algorithm a complete search depending on the dimension of the problem.

The execution time of the brute force method is shown in Fig. 5 using a logarithmic scale because they have a large difference in order (from 102 to 1029).

Based on the results obtained, we can conclude that the proposed algorithm can reduce the execution time of the algorithm by 10% compared to the time of genetic and up to 50% bee, and orders of magnitude compared with the time of the full search method. This result was obtained due to the initial population closer to the optimum.

7   Conclusions

The information technology of construction of the distributed data storages of hybrid type on the basis of a complex of mathematical models and methods of the analysis and evaluation of data processing in data storages of type on optimization of their structure is developed. This allows us to create information systems that effectively process heterogeneous data that is stored distributed. Information technology of construction of distributed data warehouses of hybrid type was used in applications in the creation of public finance management system of the Ministry of Finance of Ukraine, which allowed to reduce to 25–30% the cost of data storage and processing.

References

1. Prokhorov, I., Kolesnik, N.: Development of a master data consolidation system model (on the example of the banking sector). Procedia Comput. Sci. **145**, 412–417 (2018)
2. Jovanovic, P., Romero, O., Simitsis, A., Abell, A.: Incremental consolidation of data-intensive multi-flows. Trans. Knowl. Data Eng. **28**(5), 1203–1216 (2016)
3. Merendino, A., Dibb, S., Meadows, M., Quinn, L., Wilson, D., Simkin, L., Canhoto, A.: Big data, big decisions: the impact of big data on board level decision-making. J. Bus. Res. **93**, 67–78 (2018)

4. Bomba, A., Nazaruk, M., Pasichnyk, V., Veretennikova, N., Kunanets, N.: Information technologies of modeling processes for preparation of professionals in smart cities. In: Advances in Intelligent Systems and Computing, vol. 754, pp. 702–712 (2018)
5. Kunanets, N., Lukasz, W., Pasichnyk, V., Duda, O., Matsiuk, O., Falat, P.: Cloud computing technologies in "smart city" projects. In: 9th International Conference on Intelligent Data Acquisition and Advanced Computing Systems: Technology and Applications (IDAACS), Romania, pp. 339–342 (2017)
6. Mavridis, I., Karatza, H.: Performance evaluation of cloud-based log file analysis with Apache Hadoop and Apache Spark. J. Syst. Softw. **125**, 133–151 (2017)
7. Glushkova, D., Jovanovic, P., Abelló, A.: Mapreduce performance model for Hadoop 2.x. Inf. Syst. **79**, 32–43 (2019)
8. Rzheuskyi, A., Kunanets, N., Kut, V.: Methodology of research the library information services: the case of USA university libraries. In: Advances in Intelligent Systems and Computing II, vol. 689, pp. 450–460 (2018)
9. Rzheuskiy, A., Veretennikova, N., Kunanets, N., Kut, V.: The information support of virtual research teams by means of cloud managers. Int. J. Intell. Syst. Appl. (IJISA) **10**(2), 37–46 (2018). https://doi.org/10.5815/ijisa.2018.02.04
10. Ansari, M., Smith, J.S.: Warehouse Operations Data Structure (WODS): a data structure developed for warehouse operations modeling. Comput. Ind. Eng. **112**, 11–19 (2017)
11. Bani, F.C.D., Suharjito, Diana, Girsang, A.S.: Implementation of database massively parallel processing system to build scalability on process data warehouse. Procedia Comput. Sci. **135**, 68–79 (2018)
12. Peleshko, D., Rak, T., Izonin, I.: Image superresolution via divergence matrix and automatic detection of crossover. Int. J. Intell. Syst. Appl. (IJISA) **8**(12), 1–8 (2016). https://doi.org/10.5815/ijisa.2016.12.01
13. Izonin, I., Trostianchyn, A., Duriagina, Z., Tkachenko, R., Tepla, T., Lotoshynska, N.: The combined use of the Wiener polynomial and SVM for material classification task in medical implants production. Int. J. Intell. Syst. Appl. (IJISA) **10**(9), 40–47 (2018). https://doi.org/10.5815/ijisa.2018.09.05
14. Lytvyn, V., Vysotska, V., Peleshchak, I., Rishnyak, I., Peleshchak, R.: Time dependence of the output signal morphology for nonlinear oscillator neuron based on Van der Pol model. Int. J. Intell. Syst. Appl. (IJISA) **10**(4), 8–17 (2018). https://doi.org/10.5815/ijisa.2018.04.02

Routing Method Based on the Excess Code for Fault Tolerant Clusters with InfiniBand

Goncharenko Olexandr[1], Pavlo Rehida[1], Artem Volokyta[1(✉)], Heorhii Loutskii[1], and Vu Duc Thinh[2]

[1] National Technical University of Ukraine
"Igor Sikorsky Kyiv Polytechnic Institute", Kyiv, Ukraine
alexandr.ik97@ukr.net, pavel.regida@gmail.com,
artem.volokita@kpi.ua, georgijluckij80@gmail.com
[2] Ho Chi Minh City University of Food Industry, Ho Chi Minh, Vietnam
thinhvd@cntp.edu.vn

Abstract. The main problem of modern computing systems is a low bandwidth of message switching channels. The reason for this is rapid growth of processor element's bandwidth relatively to link's bandwidth. Therefore, a lot of attention is now being given to the task of a significant increase in network bandwidth. In designing of high performance parallel systems this task is the most actual. This article will look at the InfiniBand technology as the solution to this problem and how to transfer information in it. Among the methods considered, the most attention is paid to the wormhole transmission method, because it allows you to quickly transmit information, but has significant disadvantages. To eliminate these disadvantages a switching method is proposed based on the topology de Bruyn and its improvement by using the excess code with the numbers 0, 1, −1. Using this method allows not only to eliminate the main disadvantages of the wormhole, but also significantly increase fault tolerance and have a simple routing algorithm.

Keywords: de Bruyn topology · Wormhole routing · Excess code ·
Fault tolerance · Clusters · InfiniBand

1 Introduction

The main problem of modern computing systems is a low bandwidth of message switching channels. The reason for this is rapid growth of processor element's bandwidth relatively to link's bandwidth. Therefore, a lot of attention is now being given to the task of a significant increase in network bandwidth. In designing of high performance parallel systems this task is the most actual.

The main approach of solving these problems is a rejection of shared architectures of type PCI-X and using opposite approaches that use methods of sequential data transfer and peer-to-peer connections. This is due to the fact that PCI-X with a bus frequency of 133 MHz and a width of 64 bits allows to achieve a speed of 1 Gbit/s, which is not enough. The systems that provide the opportunity to increase the speed of data transfer should, first of all, include the standard technology, called InfiniBand [1–

© Springer Nature Switzerland AG 2020
Z. Hu et al. (Eds.): ICCSEEA 2019, AISC 938, pp. 335–345, 2020.
https://doi.org/10.1007/978-3-030-16621-2_31

3]. This technology implements a network approach to solving the problem of the effectiveness of data exchanges mechanisms. High-performance architecture is based on components, by which it is logical to create complex distributed computing systems [4–11]. These are the main machine connection channel adapter (HCA), target machine connection adapter (TCA), multistage switch and router.

As a rule, InfiniBand is not implemented in a single device, but involves the use of many small and inexpensive building blocks that can be connected in various combinations. The set of such blocks forms a certain topological organization, with the help of which the bandwidth of the communication system can be varied. For example, one physical channel can be replaced by four, which will be perceived by the system as one channel. Therefore, the bandwidth of InfiniBand is not fixed, because on the basis of this technology it is possible to form any excess count of channels between two points. Routing is performed dynamically, i.e. the network itself chooses the best route to data transfer.

In Sect. 1 of this paper, we consider the main issue of high performance parallel systems and clusters and main approaches to solving them with these advantages and disadvantages. In Sect. 2 described a topology, that built on shuffles and routing trees, that can be used with wormhole switching method and removes its disadvantages. In Sect. 3 this method will be improved with using excess code.

2   Problem Statement

The exchange protocols can be solved on the principles of packet switching or on mechanisms of preliminary connection. This allows using this technology for block and stream types of traffic.

Such well-known methods of information transfer, such as circuit switching, packet switching, virtual pipeline packet switching, wormhole, can be implemented on the basis of InfiniBand technology.

Switching of channels has two phases: establishing a connection and sending a message. During the connection, the physical channels are reserved. Upon receiving the header, the receiver sends an acknowledgment. After receiving the acknowledgment header, the sender starts the data transfer. The main disadvantage of this commutation is the ability to block other transmission channels.

When switching packets, the messages that are transmitted are broken into packets, and each packet is transmitted to the receiver in a pipelined manner. Packet transmission implies the need to buffer the entire packet completely, which, accordingly, increases the data transfer time [12–14].

Virtual pipeline packet switching (VCT) differs from the packet switching in that it transmits data immediately after a decision is made about a further recipient. In case the header cannot be transferred to the next node, it is stored in the input buffer until the corresponding output buffer is freed.

The main difference between a wormhole and a VCT is that transmission control issues are solved at the level of fleets. This allows you to significantly reduce the size of the buffers and, accordingly, increase the data transfer rate [15–18]. Wormhole method is the most efficient, simple, cheap and fast data transfer approach. The main disadvantage of this method is the ability to block transmissions. A deadlock situation is also possible for the entire data transmission system.

For example, consider four nodes N1–N4 and four packets P0–P3, each of which is intended for nodes N3, N4, N1, N2, respectively. Each packet has taken three of the four buffers it needs in different nodes and makes an attempt to occupy the fourth. As a result, we get a deadlock situation (see Fig. 1).

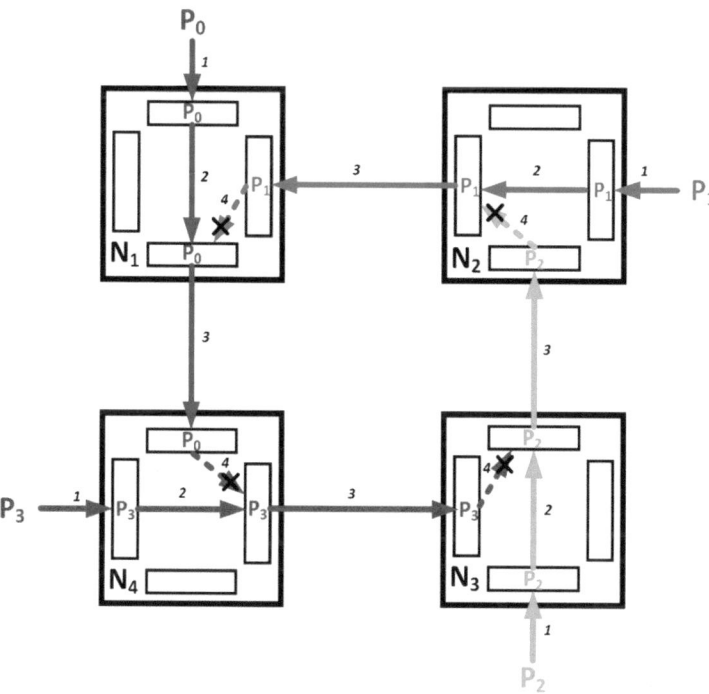

Fig. 1. Deadlock situation in the system using wormhole

To solve the deadlock problem in wormhole has been proposed methods such as the use of additional memory in each node, the implementation of virtual channels, combining the wormhole with the circuit switching method.

When using additional memory, in case of blocking the further transfer, it is necessary to copy the entire package into memory and store until the release of the corresponding channel. This greatly complicates and slows down the message transfer process.

In the case of virtual channels, the physical channel is divided in time between several logical channels. This allows you to use each of the logical channels regardless of the status of the others (see Fig. 2), which leads to a decrease in bandwidth and an increase in transmission time.

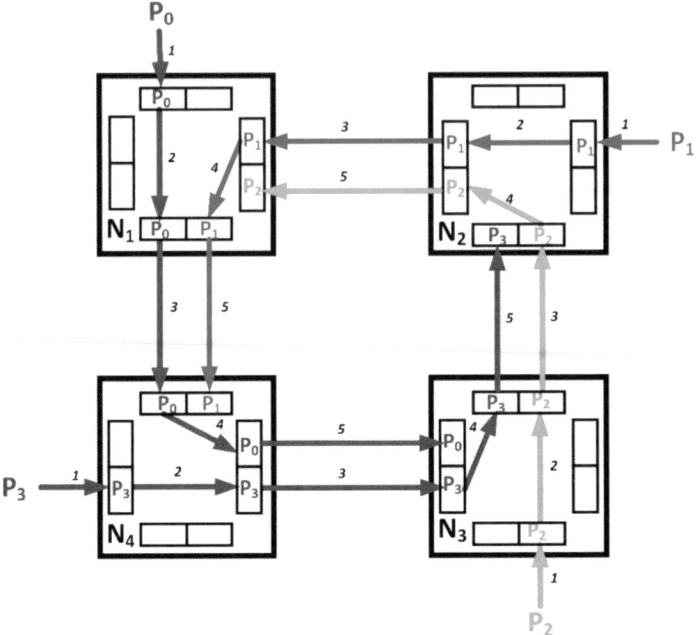

Fig. 2. System using wormhole with virtual channels

Combining the wormhole with the circuit switching method consist of two stages. At the first stage, a connection is established between the nodes, which allows you to reserve a channel for data transmission and, accordingly, eliminate the possibility of blocking. The second step is to send a message use wormhole method. This method increases the total transmission time.

3   The Essence of the Proposed Method Without Excess Code

The proposed switching method is based on the de Bruyn topology and wormhole. The connections between the nodes in the topology are determined by the shifts to the left and right with a fill of 0 and 1 in the bit that has been released. Consider the formation of connections on the example of the node number $7_{10} = 0111_2$ in the system of 16 nodes. On shifting to the left, we obtain the following neighboring nodes:

$$1110_2 = 14_{10}, \tag{1}$$

$$1111_2 = 15_{10}, \tag{2}$$

and on shifting to the right –

$$0011_2 = 3_{10}, \tag{3}$$

$$1011_2 = 11_{10} \tag{4}$$

As a result, two alternative trees can be distinguished (see Figs. 3 and 4).

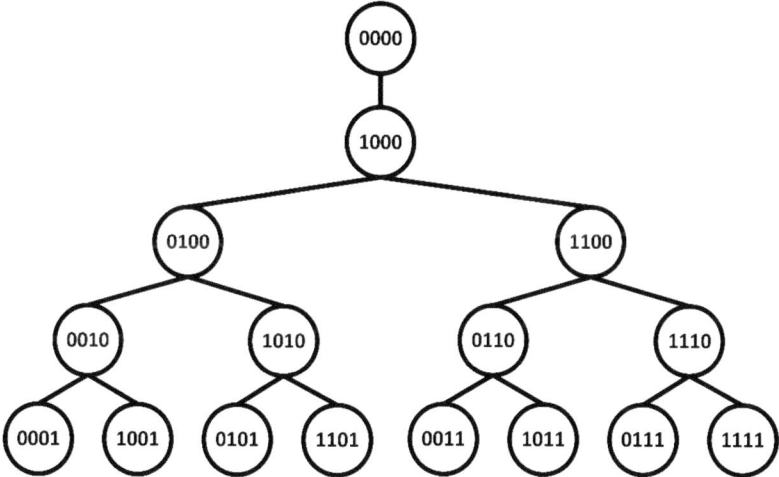

Fig. 3. Trees of routing, built from nodes 0

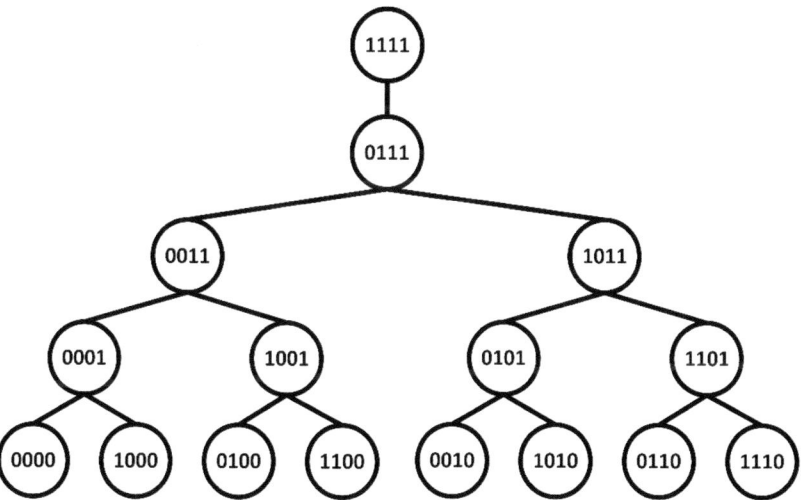

Fig. 4. Trees of routing, built from nodes 15

In the first tree in the lower tier there are nodes with odd numbers, and in the second tree - nodes with even numbers. This gives two alternative routes between any nodes of the system, which means: if a blockage occurs when moving along one of the trees, you can switch to an alternative tree and continue to transfer data via another route. Consider an example of simultaneous data transfer from 1 to 10 and from 9 to 11 (Figs. 5 and 6).

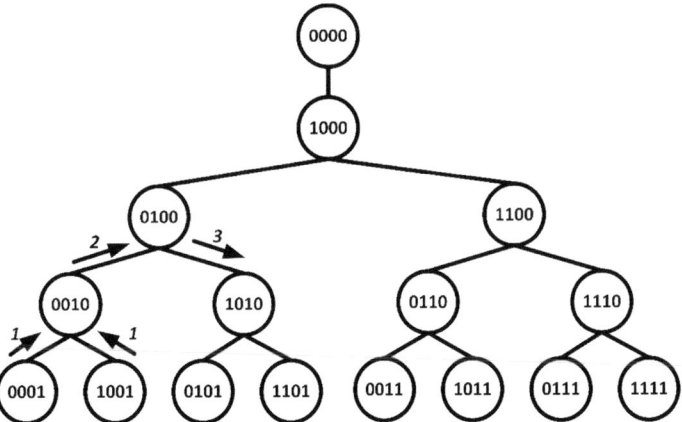

Fig. 5. Using trees to solve a conflict

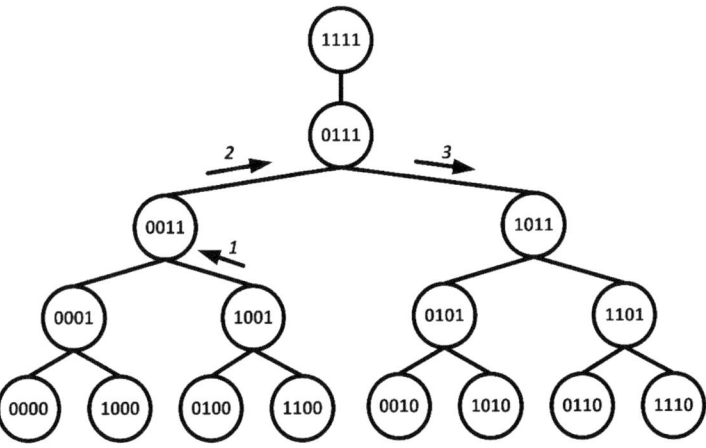

Fig. 6. Using trees to solve a conflict

Steps of data transfer:

1. Node 1 transfers data to node 2, node 9 transfers data to node 2. node 2 is still receive the data from node 1. Use the alternative route. Node 9 transfers data to node 3.
2. Node 2 transfers data to node 4, node 3 transfers data to node 7.
3. Node 4 transfers data to node 10, node 7 transfers data to node 11.

The example above, when the topology de Bruyn allows you to get away from the locks. Also, the use of the proposed graph also increases the fault tolerance of the system, since the transition to the alternative route can be performed not only when the node is busy, but also when it is faulty.

4   Using the Excess Code to Improve the Proposed Method

4.1   What Is the Excess Code 0/1/−1?

The excess code 0/1/−1 is based on the conventional binary code, it has the same weights for the bits, but contains, in addition to the digits 0 and 1, also the digit −1, which is written with the symbol T. This code has the following features:

- Initially contains negative numbers. For example, the number −3 in a 3-bit code is represented as 0TT. The smallest number that can be represented in the N-bit excess code is $-2^N + 1$.
- Provides several ways to represent the same number (see Fig. 7). For example, the number −3 can be represented as 0TT, T01 and T1T.

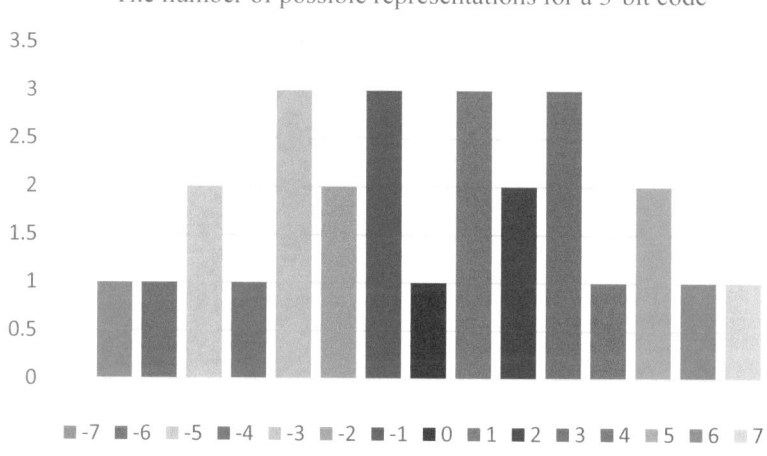

Fig. 7.   The number of possible representations of numbers for a 3-bit excess code

4.2   How It Used in Topology Synthesis

Improved topology based on shifts like previous, but in released bit can be inputted not even 0 and 1 but also −1 (T). For example, in previous topology node $3_{10} = 011_2$ has neighbors 001, 101, 110 and 111. In topology based on excess code (Fig. 8) it has neighbors $11T = 5_{10}$ and $T01 = -3_{10}$ in addition. But node 3 has alternative representations, as result, there are other nodes with number 3 and they have own set of neighbors.

Count of nodes in topology according to count of bits in address shown in Table 1.

Table 1. Topology parameters

Parameter	Base topology			Redundant topology		
	N = 2	N = 5	N = 8	N = 2	N = 3	N = 5
Count of nodes	8	32	256	9	27	243
Diameter	3	5	8	2	3	5
Power without links between nodes with same number	4	4	4	6	6	6
Power	4	4	4	7	8	10
Count of routing trees	2	2	2	3	3	3

It is possible to choose approximately the same number of nodes for the binary and ternary systems (32, 27 and 256, 243) and build fault-tolerant topologies with additional route trees.

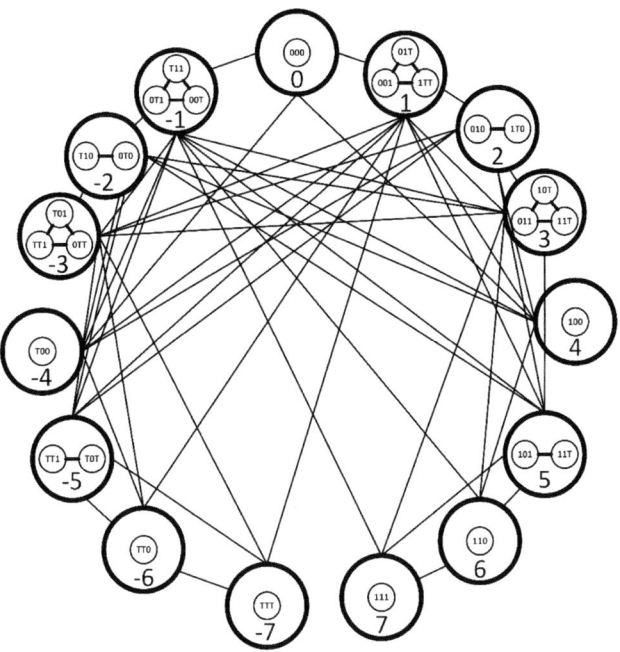

Fig. 8. Topology based on excess code 0/1/−1 (without explicitly excess connections)

For example, node 10T (3_{10}) has followed neighbors: 010 (2_{10}), 110 (6_{10}), T10 (-2_{10}), 0T0 (-2_{10}), 0T1 (-1_{10}) and 0TT (-3_{10}). And we can see the next feature of this topology: excess connections between nodes. It can additionally increase fault tolerance of the system and been used to parallel data transfer between 2 nodes.

4.3   Main Advantages (and Disadvantages) of Topology

Since the number in the excess code can sometimes be represented in more than one way, the consequence of this is that there can be more than one node with the same number. As a result, when the node with a number, for example, 011 fails, the node T01, theoretically, may assume the role of the node 011, thereby restoring the topology in terms of a user. However, the main problem for this is that the node 011 and T01 have different communication due to different codes, as a result, if the node T01 want to take on the role of 011, it needs to be done so that it can interact with the same connections as the node 011, intercepting messages addressed to 011 in case of failure.

The second advantage of using excess code is 3 possible routing trees instead of 2 (see Fig. 9). As a result, in case of failure of even 2 nodes in 2 trees, it is possible to use an alternative tree, which, given the possibility of restoring the topology, can significantly increase the system fault tolerance even in comparison with the basic switching method.

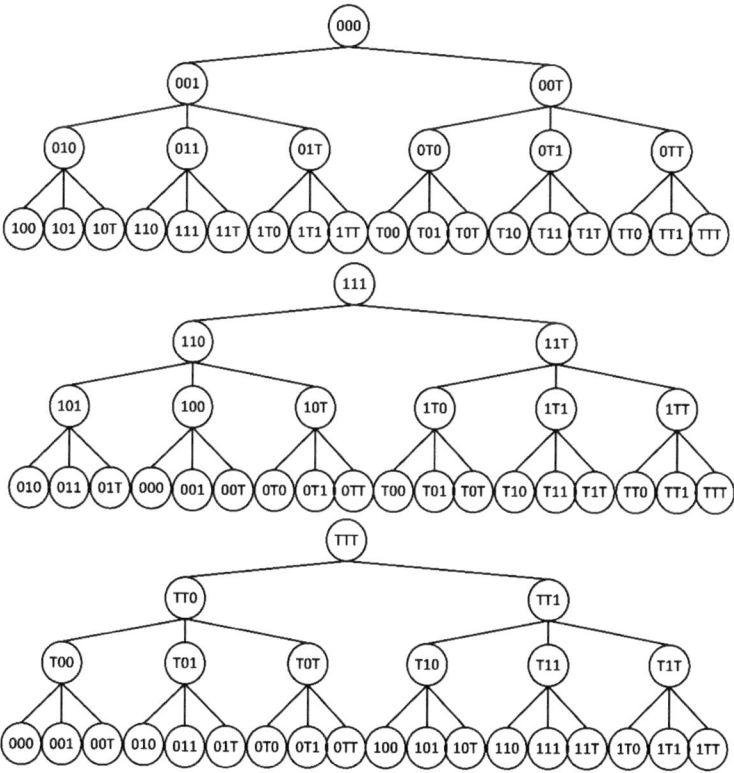

Fig. 9. Routing trees of topology, based on 3-bit excess code

The third advantage of topology is the excess connections between nodes, that can be used for the particular solution of main problem of using excess nodes, because if some nodes with same numbers have same connections, there are no necessary to duplicate it explicitly.

In addition, using the InfiniBand feature to provide multiple physical channels as one, you can increase bandwidth by using all alternative routes to simultaneously transmit information and using all connections between nodes with the same number but different code.

5 Conclusion

In this work, research of the effectiveness of the proposed method for increasing the fault tolerance of routing in clusters based on InfiniBand is done. We made a description of the alternative route trees and showed it in the corresponding figures. The results of the measurements show that the developed method of constructing a "quasi-quantum" topological organization can improve the efficiency of routing for distributed computer systems. Usage of excess code in the encoding of the node by additional digits "-1" and forming an alternative routing tree don't changes time of routing and allows you to reduce the diameter of the system, but leads to an increase in the power of the nodes. In addition, fault tolerance of distributed systems potentially may be increased.

References

1. InfiniBand Trade Association. InfiniBand architecture specification, volume 1, release 1.2.1. http://www.infinibandta.com. Accessed 12 Dec 2018
2. InfiniBand Trade Association. InfiniBand architecture specification, volume 2, release 1.2.1. http://www.infinibandta.com. Accessed 12 Dec 2018
3. Gamess, E., Ortiz-Zuazaga, H.: Low level performance evaluation of InfiniBand with benchmarking tools. Int. J. Comput. Netw. Inf. Secur. (IJCNIS) 8(10), 12–22 (2016). https://doi.org/10.5815/ijcnis.2016.10.02
4. Dupas, R., Grebennik, I., Lytvynenko, O., Baranov, O.: An heuristic approach to solving the one-to-one pickup and delivery problem with three-dimensional loading constraints. Int. J. Inf. Technol. Comput. Sci. (IJITCS) 9(10), 1–12 (2017). https://doi.org/10.5815/ijitcs.2017.10.01
5. Russoniello, A., Gamess, E.: Evaluation of different routing protocols for mobile ad-hoc networks in scenarios with high-speed mobility. Int. J. Comput. Netw. Inf. Secur. (IJCNIS) 10(10), 46–52 (2018). https://doi.org/10.5815/ijcnis.2018.10.06
6. Kravets, P.I., Shymkovych, V.M., Samotyy, V.: Method and technology of synthesis of neural network models of object control with their hardware implementation on FPGA. In: Proceedings of the 2017 IEEE 9th International Conference on Intelligent Data Acquisition and Advanced Computing Systems: Technology and Applications, IDAACS 2017, pp. 947–951. IEEE, Bucharest (2017)
7. Volokyta, A., Shymkovych, V., Volokyta, I., Vasyliev, V.: Research and development of a stereo encoder of a FM-transmitter based on FPGA. In: Advances in Computer Science for Engineering and Education, ICCSEEA 2018. Advances in Intelligent Systems and Computing, vol. 754. Springer, Cham (2018)

8. Mukhin, V., Volokyta, A., Heriatovych, Y., Rehida, P.: Method for efficiency increasing of distributed classification of the images based on the proactive parallel computing approach. Adv. Electr. Comput. Eng. **18**(2), 117–122 (2018)
9. Hu, Z., Mukhin, V., Kornaga, Y., Volokyta, A., Herasymenko, O.: The scheduler for distributed computer systems based on the network centric approach to resources control. In: Proceedings of the 2017 IEEE 9th International Conference on Intelligent Data Acquisition and Advanced Computing Systems: Technology and Applications, IDAACS 2017, pp. 518–523 (2017)
10. Hu, Z., Mashtalir, S.V., Tyshchenko, O.K., Stolbovyi, M.I.: Clustering matrix sequences based on the iterative dynamic time deformation procedure. Int. J. Intell. Syst. Appl. (IJISA) **10**(7), 66–73 (2018). https://doi.org/10.5815/ijisa.2018.07.07
11. Babichev, S., Korobchynskyi, M., Mieshkov, S., Korchomnyi, O.: An effectiveness evaluation of information technology of gene expression profiles processing for gene networks reconstruction. Int. J. Intell. Syst. Appl. (IJISA) **10**(7), 1–10 (2018). https://doi.org/10.5815/ijisa.2018.07.01
12. Cole, R.J., Maggs, B.M., Sitaraman, R.K.: On the benefit of supporting virtual channels in wormhole routers. J. Comput. Syst. Sci. **62**(1), 152–177 (2001)
13. Ni, L.M., McKinley, P.K.: A survey of wormhole routing techniques in direct networks. Computer **26**(2), 62–76 (1993)
14. Linder, D., Harden, J.: An adaptive and fault tolerant wormhole routing strategy for k-ary n-cubes. IEEE Trans. Comput. **40**(1), 2–12 (1991)
15. Park, H., Agrawal, D.P.: Efficient deadlock-free wormhole routing and virtual-channel reduction in shuffle-based networks. J. Comput. Syst. Sci. **46**(2), 165–179 (1997). https://www.sciencedirect.com/science/article/pii/S0743731597913800
16. Doverspike, N., Jha, V.: Comparison of routing methods for DCS-switched networks. Interfaces **23**(2), 21–34 (1993)
17. Mohapatra, P.: Wormhole routing techniques for directly connected multicomputer systems. ACM Comput. Surv. **30**(3), 374–410 (1998)
18. Washington, N., Perros, H.: Performance analysis of traffic-groomed optical networks employing alternate routing techniques. Lecture Notes in Computer Science, vol. 4516, pp. 1048–1059. Springer, Heidelberg (2007)

The Methods to Improve Quality of Service by Accounting Secure Parameters

Tamara Radivilova$^{(\boxtimes)}$ ⓘ, Lyudmyla Kirichenko ⓘ,
Dmytro Ageiev ⓘ, and Vitalii Bulakh

Kharkiv National University of Radio Electronics,
Avenue Nauki, 14, Kharkiv, Ukraine
tamara.radivilova@gmail.com

Abstract. A solution to the problem of ensuring quality of service, providing a greater number of services with higher efficiency taking into account network security is proposed. In this paper, experiments were conducted to analyze the effect of self-similarity and attacks on the quality of service parameters. Method of buffering and control of channel capacity and calculating of routing cost method in the network, which take into account the parameters of traffic multifractality and the probability of detecting attacks in telecommunications networks were proposed. The both proposed methods accounting the given restrictions on the delay time and the number of lost packets for every type quality of service traffic. During simulation the parameters of transmitted traffic (self-similarity, intensity) and the parameters of network (current channel load, node buffer size) were changed and the maximum allowable load of network was determined. The results of analysis show that occurrence of overload when transmitting traffic over a switched channel associated with multifractal traffic characteristics and presence of attack. It was shown that proposed methods can reduce the lost data and improve the efficiency of network resources.

Keywords: Self-similarity · Attack detection · Quality of service · Buffering · Routing · Fractality · Traffic management

1 Introduction

A computer network is a complex and expensive system that solves critical tasks and serves many users. Characteristics of service quality reflect critical network properties: performance, reliability and security [1]. Qualities of service (QoS) methods ensure the stable operation of modern services: IP-telephony, video and radio broadcasting, interactive distance learning, etc. QoS methods are aimed at improving the performance characteristics and network reliability and reduce variations in delays and packet loss during periods of overload network [2, 3].

Quality of service characteristics reflect the negative impact of the queues mechanism on traffic transmission. This effect can be expressed in a temporary decrease in the rate of traffic transmission, in the delivery of packets with variable delays and in the loss of packets due to a buffer nodes overload. QoS methods are aimed at compensating the negative effects of temporary overloads occurring in packet-switched networks.

© Springer Nature Switzerland AG 2020
Z. Hu et al. (Eds.): ICCSEEA 2019, AISC 938, pp. 346–355, 2020.
https://doi.org/10.1007/978-3-030-16621-2_32

These methods use various queue management, reservation and feedback algorithms to reduce the negative impact of queues to acceptable to user's level [3–6].

Experimental and numerical studies conducted in recent decades indicate that many multiservice networks traffic has self-similar properties [2, 7–10]. Self-similar traffic causes significant delays and packet losses, even if the total intensity of all flows is far from the maximum allowable values [7, 10, 11].

A big problem for service providers is to ensure QoS in terms of self-similar traffic, avalanche traffic of intruders, the sources of which are various temporal nodes. This type of behavior is associated with threats such as distributed denial of service (DDoS) attacks, Internet worms, phishing, viruses, email spam and others [13, 14]. The amount of traffic that is generated due to infection, and subsequent traffic bursts can disrupt the normal work of network and create an additional risk to network devices (routers, switches). Security is becoming a critical characteristic of all services and plays a crucial role in the profitability of service providers [15, 16].

When the transfer rate reaches several gigabits, to suppress emerging threats providers must supply protection that ensure reliability and does not affect network performance as a whole [16–18]. The task of functioning to ensure network security determines how a provider can effectively offer a larger volume of services with higher efficiency with a greater degree of manageability [19–21].

The aim of this work is to develop a method of buffering and control channel capacity and routing method in the network, which are based on the multifractal properties of traffic and parameters of network security.

2 Self-similar and Multifractal Traffic's Properties

The self-similarity of random processes lies in preservation of probabilistic characteristics when changing the time scale. A stochastic process $X(t)$ is self-similar with a parameter H if the process $a^{-H}X(at)$ is described by the same laws of finite-dimensional distributions as $X(t)$ $Law\{a^{-H}X(at)\} = Law\{X(t)\}, \forall a > 0$, where the parameter H, $0 < H < 1$ represents the degree of self-similarity of process and calls the Hurst parameter. The parameter $H > 0.5$ characterizes the measure of the long-term dependence of the stochastic process. The initial moments of self-similar random process can be expressed as $M[|X(t)|^q] = C(q) \cdot t^{qH}$, where the value $C(q) = M[|X(1)|^q]$ [5].

Multifractal traffic can be defined as the expansion of self-similar traffic by taking into account the scalable properties of statistical characteristics of second and higher orders. For the moments of multifractal processes, the relation $M[|X(t)|^q] = c(q) \cdot t^{qh(q)}$ is executed, where $c(q)$ is some deterministic function; $h(q)$ is the generalized Hurst index which is a nonlinear function in the general case. Value $h(q)$ at $q = 2$ coincides with the value of the degree of self-similarity H.

Multifractal traffic has a special structure that persists on many scales: there is always some number of very large bursts with a relatively small average traffic level. As a characteristic of multifractal traffic, it was proposed to use the following parameters: traffic intensity λ, Hurst parameter H and the range of generalized Hurst index $\triangle h = h(q_{min}) - h(q_{max})$. For monofractal processes, the generalized Hurst index does not depend on the parameter q and is a straight line: $h(q) = H$, $\triangle h = 0$. The more process heterogeneity, i.e. the higher emissions are in traffic, leads to the greater range $\triangle h$. The degree of bursts corresponds to the gravity tails distribution. The coefficient of variation σ_{var} can be considered as the simplest quantitative characteristic of the tail distribution $\sigma_{var}(T) = \sigma(T)/M(T)$, where T is a random variable the values of which are the number of events in a given time interval [5].

A node can be represented as a queuing system, at each unit of time it receives input data that arrives during this system working period, processes and sends them. The network traffic processing system in this paper is a node with adjustable performance or the data processing speed (the number of packets per unit of time that it can process) and the buffer memory (buffer) with the specified volume (the amount of data it can hold) in which the node puts traffic that it did not manage to process considered step (unit of time) of this work [22–24].

Intrusion detection systems are used to detect network attacks. Their performance and efficiency are evaluated using the parameters of cost, resource utilization, detection rate. Moreover, if it is possible to classify attacks and normal traffic (events), then these are observable parameters. Depending on the nature of the attack and the probability of its detection, four possible outcomes are used [13, 15, 16]:

- True Positive (TP): process that is actually an attack and are successfully classified and called attack.
- False Positive (FP): A normal and legitimate process is classified as an attack.
- True Negative (TN): a process that is actually normal and legitimate, and successfully marked and detected as normal.
- False Negative (FN): attack is incorrectly classified as normal or legitimate action.

The main difficulty lies in the fact that the number of false positive detections is very large. It is clear that a high FP value will lead to less effective detection, and a high FN value will make the system vulnerable to intruders.

3   Method of Buffering and Control of Channel Capacity

The simulation is a main tool for research networks with self-similar flows. The simulations of channel load and the queues formation in the buffer for the realization of fractal traffic were studied in [12, 15, 20]. The simulation results allow to calculate the dependence of the values $Net_i^{k_i}(T)$ network bandwidth of k channel of i-th node and parameters of incoming data flows $\{\lambda, H, \sigma_{var}, P_{sec}\}$ on the size of the buffer memory $Q_w^{new} = f(Net_i^{k_i}, \lambda, H, \sigma_{var}, P_{sec})$, where $P_{sec} = (P_{TP}, P_{FP}, P_{TN}, P_{FN})$ is the probability of detection attacks on system resources (i.e. buffer overload, DDoS attacks).

The $P_{sec} = (P_{TP}, P_{FP}, P_{TN}, P_{FN})$ calculates by using the Machine Learning algorithm that shown in [16, 21, 25–27]. Each input flow is sent to a queue Q_w of limited size. Queuing time is dependent on qs. The normal traffic transmitted through the communication node is provided by the calculated values of the buffer size Q_w^{new} (the loss does not exceed the specified percentage). Similarly, a functional dependence of a specified buffer size Q_w and traffic parameters on a channel capacity $Net_i^{newk_i} = \varphi(Q_w, \lambda, H, \sigma_{var}, P_{sec})$ were obtained. It is possible to avoid network overload by managing of buffers size and/or data flows based on the calculation of the value of the maximum permissible load in accordance with the obtained results. To determine maximum allowable load of the channel for the given size of channel capacity and the buffer memory are used the functional dependencies $Net_i^{newk_i} = \varphi(Q_w, \lambda, H, \sigma_{var}, P_{sec})$ and $Q_w^{new} = f(Net_i^{k_i}, \lambda, H, \sigma_{var}, P_{sec})$. By predicting the start of overload, the required size of the channel capacity $Net_i^{newk_i}$ and/or buffer memory Q_w^{new} can be allocated. If the calculated buffer size Q_w^{new} is larger than the existing Q_w and the probability of detection attacks on system resources $P_{sec} = (P_{TP}, P_{FP}, P_{TN}, P_{FN})$ is high the system allocated the required size of buffer memory. But if probability of detection attacks on system resources $P_{sec} = (P_{TP}, P_{FP}, P_{TN}, P_{FN})$ is low the system creates the alert and doesn't allocate the required size of buffer memory. The required size of channel capacity can be determined from the received data. If the requested capacity size $Net_i^{newk_i}$ is larger than available $Net_i^{k_i}$ and the probability of detection attacks on system resources P_{sec} is high, the system provides the requested resource, thus distributing the rest to the second channel capacity. But if P_{sec} is low the system created the alert and doesn't provide the requested resource.

This method can be used in nodes (switch, router and i.e.) for preventing of network overload. It allows to reduce a loss packets and increase channel utilization and network performance.

4 Calculating of Routing Cost Method in the Network

To provide QoS is required routes selection based on separate safety flows, at the same time, different flows that are sent to one recipient may be directed by various pathes. In addition, the paved pathes can be changed in case of overload. The shortest paths (Low Cost Routing) between the incoming edge router and others are calculated by the routing protocol. Lets consider the calculating of routing cost method based on security parameters and fractal structure of traffic [4, 23, 25, 26].

The communication links between nodes with the maximum capacity $Link_{lk} = \{L_{lk}\}$, $lk = 1, 2, \ldots$, that are divided into $k \in K$ channels with bandwidth $Net_{lk}^k(t) = \{Net_{lk}^k\}$ at time t [15, 26, 27]. Assume that at each time t traffic of intensity $\lambda_{Net_i}^{qs}(t)$ relating to one of the classes of service qs-th with requirements QoS. Each qs corresponds to the maximum percentage of loss l_{qs} and the maximum delay value τ_{qs}. The variable $X_i^{qs}(t) \in X$ is the coefficient of loss qs traffic that transmitted by the path $Net_{lk}^k(t)$ to the i-th node at the moment t. It is assumed that the probability of package error in the path can be neglected and losses occur only in node because buffer overloads. The coefficient of losses for all nodes in the network:

$$0 \leq X_i^{qs}(t), \ \sum_{i=1}^{N} X_i^{qs}(t) \leq l_{qs}. \tag{1}$$

Thus, the restriction (1) show that the total loss for the traffic $\lambda^{qs}_{Net_i}(t)$ routed at time t, should not exceed the maximum permissible values for the class of service l_{qs}. Loss is defined as the ratio of the discarded data to received data. Value $X^{qs}_i(t) \to \min$ is subject to minimization.

Restrictions imposed by the delay time are similar: $0 \leq T^{qs}_i(t), \sum_{i=1}^{N} T^{qs}_i(t) \leq \tau_{qs}$, where $T^{qs}_i(t)$ is the average waiting time of package qs-th class of service in the queue at the i-th node [21, 28, 29]. Performing this restriction helps to ensure that the packets delivery does not exceed the maximum permissible values for a given class of service τ_{qs}.

All input traffic is divided at the qs flows so that to ensure the transmission requirements of all classes $QS(t)$ in full. The channels set of traffic QoS $K = K(Net^k_{lk}, P_{lk}, L_k)$, where $P_{lk} = \{p^1_{lk}, \ldots, p^k_{lk}\}$ is allowable set of pathes to the path L_k, that is defined for each traffic channel.

The value of routing cost c_{lk} is assigned to the communication link lk and may depend on several parameters, particularly the reliability, speed, and length. The cost of path p^k_{lk} is equal to the sum of communication lines cost: $C^k_{lk} = \sum_{lk \in p^k_{lk}} c_{lk}$. If $Netx^k_{lk}(t)$ is bandwidth that is forwarded to allowable path p^k_{lk} of transmission traffic channel $\lambda^{qs}_{Net_i}(t)$, then following relation holds $\sum_{t \in T; lk=1}^{L_k} Netx^k_{lk}(t) = Net^k_{lk}, \forall k \in K, \forall lk \in \{1, \ldots, L_k\}$.

In [10, 13, 15] it was shown that at values $H \geq 0,9$ or at $H > 0,5$ and simultaneously values $\sigma_{var} \geq 3$ (which roughly corresponds to the values $\Delta h > 1$) the amount of loss is greater than 5–10%. When passing traffic with strong fractal properties it needs to timely increase bandwidth of communication lines. To reflect the changes in the multifractal properties of flows, cost of paths C^k_{lk} are updated in regular intervals and recalculated by the formula

$$Cnew^k_{lk} = \begin{cases} C^k_{lk}, & H \leq 0,5; \quad P_{sec} < 0.6 \\ C^k_{lk} + (H - 0.5)C_0, & 0.5 < H < 0.9, \ \sigma_{var} \leq 1, P_{sec} > 0.6; \\ C^k_{lk} + (H - 0.5)(\sigma_{var} - 1)C_0, & 0.5 < H < 0.9, \ 1 < \sigma_{var} < 3, P_{sec} > 0.6; \\ C^k_{lk} + C_0, & H \geq 0.9 \text{ or } H > 0.5, \ \sigma_{var} \geq 3, P_{sec} > 0.6. \end{cases}$$

where $C^k_{lk} = \sum_{lk \in p^k_{lk}} c_{lk}$ is determined in accordance with the objective function, value C_0 is selected by the network administrator considering network topology and probability of detection attacks on system resources $P_{sec} = (P_{TP}, P_{FP}, P_{TN}, P_{FN})$. The routing algorithm is not changed (path cost $Cnew^k_{lk} = C^k_{lk}$) if the traffic has independent values $(H = 0.5)$ or has antipersistent properties $(H < 0.5)$ and $P_{sec} < 0.6$. If the value $0.5 < H < 0.9$ and the dispersion of data is small $(\sigma_{var} \leq 1)$ the value C^k_{lk} increases in proportion to the Hurst exponent value. If the Hurst exponent value $0.5 < H < 0.9$, $P_{sec} > 0.6$ and dispersion is large $(1 < \sigma_{var} < 3)$ the value C^k_{lk} increases in proportion to three characteristics. The cost with a maximum value $C^k_{lk} + C_0$ is obtained at $H \geq 0.9$, $P_{sec} > 0.6$ or persistent traffic $(H > 0.5)$ with a coefficient of variation $\sigma_{var} \geq 3$. After recalculating the value of all paths, the announcement of the state of paths are sent between routers.

This method can be used to increase utilization channel by rerouting most important data flows to alternative low load channels.

5 Simulation of the Developed Methods

To carry out simulation work of the proposed methods program modules were developed by using Python. The input system receives the generated information flows having predetermined fractal properties with parameters similar to the real traffic [22, 25–27, 29]. This traffic was sent from the sender to receiver through 14 nodes: routers, nodes and firewall by using various parthes. By using model realizations traffic with predetermined properties in experiments a various network parameters can be determined in a variety of loading and operation modes. The network load was changed from 20% to 90%. All traffic parameters were changed and some of them have attacks.

The dynamic control of channels capacity and sizes of nodes buffer memory were performed during simulation with method of buffering and control channel capacity (MBCCC).

In the dynamic channel capacity control mode, a forecast of flows requirements in the necessary channel capacity $Net_i^{newk_i}$ were made in the subsequent time interval $t + \Delta$ (ms). Based on these data the necessary resource of capacity $Net_i^{newk_i}$ was allocated for self-similar traffic critical to the time of transmission and the remaining traffic by the remaining resource for a time Δ.

Similarly, in the buffer size control mode of nodes the buffer memory requirements of the node Q_w^{new} were calculated for a time $t + \Delta$. Based on this data required resources of buffer memory Q_w^{new} (MB) node were allocated for a while.

During the simulation with Routing method (RM) calculation of routing costs $Cnew_{lk}^k$ of fractal traffic in channels at the time t were calculated taking into account the required throughput $Net_{lk}^k(t)$, using the minimum cost criterion and probability of detection attacks P_{sec}, with limiting the QoS for the uniform use of traffic channels of different QoS classes in the case of multifractal flows. The fractal traffic of high qs class of service was routed along most free, secure, lowest cost paths taking into account traffic properties. It is possible to multiplex low-priority traffic into separate flows so that its life time does not expire.

The effectiveness of proposed methods we can evaluate by analysis of the simulation results: changing a buffer memory size and the dynamic distribution of a channel capacity.

During the experiment, the amount of channels utilization, value of lost data, value of jitter, probability of detection attacks on system resources P_{sec} were measured. The QoS parameters that were obtained during the experiments by using proposed methods are shown on Fig. 1 and in Table 1. The channel utilizations are shown in Fig. 1(a) and lost packets are shown on Fig. 1(b).

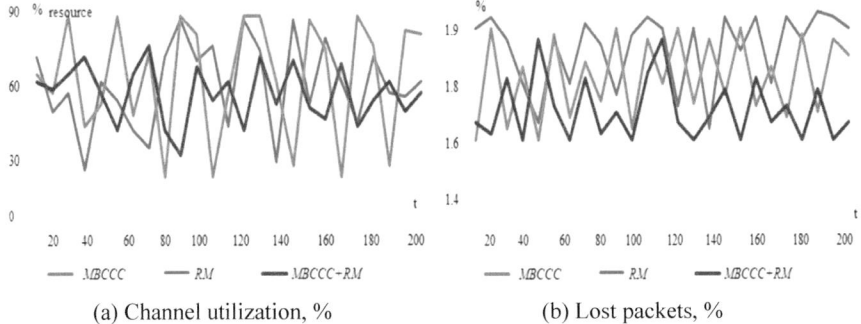

(a) Channel utilization, %   (b) Lost packets, %

Fig. 1. QoS parameters by using the proposed methods.

Table 1. The parameters of quality of service.

Methods	Channel utilization, %	Value of lost data, %	Jitter, ms	P_{sec} (TP), %	P_{sec} (FP), %
Method of buffering and control of channel capacity	0.69	1.88	54	0.62	0.31
Routing method in the network	0.64	1.9	38	0.65	0.33
Using both methods	0.52	1.73	33	0.71	0.28

With the same volume of transmitted information in network the lost of self-similar traffic during transmission is noticeably lower with simultaneous use of proposed methods. It is similarly for parameters of channel utilization and jitter. The probability of attack detection is improved when both proposed methods are used.

The results of the experiments were realized and they brought the following effect: increasing the utilization of data transmission channels, by redirecting the most critical information flows to less loaded alternative channels, more efficient usage of system resources, improving the quality of service, and reduction of data loss.

6   Discussion and Conclusion

In this paper by collecting and analyzing traffic in real-time the network QoS parameters were improve. Methods for ensuring QoS are proposed. They are taking into account the parameters of the probability of attacks detection in telecommunications networks during the transmission of fractal traffic over the network. The occurrence of congestion during the transmit of traffic through a switched communication channel is associated with the multifractal characteristics of traffic. Depending on the multifractality parameters, the current channel load, the node buffer size it is possible to determine the maximum permissible load of this network.

A method for predicting network congestion based on the calculation of the degree of communication channel load when traffic monitoring considering the probability of detection attacks. A method for estimating the cost of routing is described and based on accounting the fractal properties of network traffic, security parameters and specified restrictions on delay time and the number of lost packets.

The proposed methods can reduce network data loss and increase the efficiency of network resources, provide higher performance and throughput, reduce costs by redirecting the most critical information flows to less loaded alternative channels, and reduce response time and the amount of lost data. In future work it is planned to investigate the influence and dependence of various types and levels of attacks on the network QoS parameters.

References

1. Yu-Xing, Y.: Research for network security and reliability and performance assess. In: Proceedings of the Fourth International Conference on Intelligent Systems Design and Engineering Applications, Zhangjiajie, pp. 444–447 (2013). https://doi.org/10.1109/isdea. 2013.506
2. Daradkeh, Y.I., Kirichenko, L., Radivilova, T.: Development of QoS methods in the information networks with fractal traffic. Int. J. Electron. Telecommun. **64**(1), 27–32 (2018). https://doi.org/10.24425/118142
3. Acharya, H.S., Dutta, S.R., Bhoi, R.: The impact of self-similarity network traffic on quality of services (QoS) of telecommunication network. Int. J. IT Eng. Appl. Sci. Res. (IJIEASR) **2**, 54–60 (2013)
4. Czarkowski, M., Kaczmarek, S., Wolff, M.: Influence of self-similar traffic type on performance of QoS routing algorithms. INTL J. Electron. Telecommun. **62**(1), 81–87 (2016). https://doi.org/10.1515/eletel-2016-0011
5. Ivanisenko, I., Kirichenko, L., Radivilova, T.: Investigation of self-similar properties of additive data traffic. In: Proceedings of 2015 Xth International Scientific and Technical Conference "Computer Sciences and Information Technologies" (CSIT), pp. 169–171. IEEE (2015). https://doi.org/10.1109/stc-csit.2015.7325459
6. Pramanik, S., Datta, R., Chatterjee, P.: Self-similarity of data traffic in a Delay Tolerant Network. In: 2017 Wireless Days, pp. 39–42 (2017). https://doi.org/10.1109/wd.2017. 7918112
7. Bhat, U.N.: An Introduction to Queueing Theory: Modeling and Analysis in Applications. Birkhauser/Springer, Boston (2015)
8. Hu, Z., Mukhin, V., Kornaga, Y., Lavrenko, Y., Herasymenko, O.: Distributed computer system resources control mechanism based on network-centric approach. Int. J. Intell. Syst. Appl. IJISA. **9**(7), 41–51 (2017)
9. Ivanisenko, I., Radivilova, T.: The multifractal load balancing method. In: Proceedings of Second International Scientific-Practical Conference Problems of Infocommunications Science and Technology (PIC S&T), pp. 122–123. IEEE (2015). https://doi.org/10.1109/ infocommst.2015.7357289
10. Aggarwal, A., Verma, R., Singh, A.: An efficient approach for resource allocations using hybrid scheduling and optimization in distributed system. Int. J. Educ. Manag. Eng. (IJEME) **8**(3), 33–42 (2018). https://doi.org/10.5815/ijeme.2018.03.04

11. Radivilova, T., Kirichenko, L., Ivanisenko, I.: Calculation of distributed system imbalance in condition of multifractal load. In: Proceedings of Third International Scientific-Practical Conference Problems of Infocommunications Science and Technology (PIC S&T), pp. 156–158. IEEE (2016). https://doi.org/10.1109/infocommst.2016.7905366

12. Kirichenko, L., Ivanisenko, I., Radivilova, T.: Dynamic load balancing algorithm of distributed systems. In: Proceedings of the 13th International Conference on Modern Problems of Radio Engineering, Telecommunications and Computer Science (TCSET), pp. 515–518. IEEE (2016). https://doi.org/10.1109/tcset.2016.7452102

13. Gupta, N., Srivastava, K., Sharma, A.: Reducing false positive in intrusion detection system: a survey. Int. J. Comput. Sci. Inf. Technol. 7(3), 1600–1603 (2016)

14. Deka, R., Bhattacharyya, D.: Self-similarity based DDoS attack detection using Hurst parameter. Secur. Commun. Netw. 9(17), 4468–4481 (2016)

15. Ageyev, D., Kirichenko, L., Radivilova, T., Tawalbeh, M., Baranovskyi, O.: Method of self-similar load balancing in network intrusion detection system. In: Proceedings of 28th International Conference Radioelektronika (RADIOELEKTRONIKA), pp. 1–4. IEEE (2018). https://doi.org/10.1109/radioelek.2018.8376406

16. Bulakh, V., Kirichenko, L., Radivilova, T., Ageiev, D.: Intrusion detection of traffic realizations based on machine learning using fractal properties. In: Proceedings of 2018 International Conference on Information and Telecommunication Technologies and Radio Electronics (UkrMiCo), pp. 1–4. IEEE (2018)

17. Sharma, R., Chaurasia, S.: An integrated perceptron Kernel classifier for intrusion detection system. Int. J. Comput. Netw. Inf. Secur. (IJCNIS) 10(12), 11–20 (2018). https://doi.org/10.5815/ijcnis.2018.12.02

18. Chahal, J.K., Kaur, A.: A hybrid approach based on classification and clustering for intrusion detection system. Int. J. Math. Sci. Comput. (IJMSC) 2(4), 34–40 (2016). https://doi.org/10.5815/ijmsc.2016.04.04

19. Nasr, A.A., Ezz, M.M., Abdulmaged, M.Z.: An intrusion detection and prevention system based on automatic learning of traffic anomalies. Int. J. Comput. Netw. Inf. Secur. (IJCNIS) 8(1), 53–60 (2016). https://doi.org/10.5815/ijcnis.2016.01.07

20. Kirichenko, L., Radivilova, T.: Analyzes of the distributed system load with multifractal input data flows. In: 14th International Conference The Experience of Designing and Application of CAD Systems in Microelectronics (CADSM) Proceedings, pp. 260–264. IEEE (2017). https://doi.org/10.1109/cadsm.2017.7916130

21. Popa, S.M., Manea, G.M.: Using traffic self-similarity for network anomalies detection. In: Proceedings of 20th International Conference on Control Systems and Computer Science, pp. 639–644. IEEE (2015)

22. Betker, A., Gamrath, I., Kosiankowski, D., Lange, C., Lehmann, H., Pfeuffer, F., Simon, F., Werner, A.: Comprehensive topology and traffic model of a nationwide telecommunication network. IEEE/OSA J. Opt. Commun. Netw. 6(11), 1038–1047 (2014). https://doi.org/10.1364/JOCN.6.001038

23. Han, D., Chung, J.-M.: Self-similar traffic end-to-end delay minimization multipath routing algorithm. IEEE Commun. Lett. 18(12), 2121–2124 (2014). https://doi.org/10.1109/LCOMM.2014.2362747

24. Lemeshko, A.V., Evseeva, O.Y., Garkusha, S.V.: Research on tensor model of multi-path routing in telecommunication network with support of service quality by greate number of indices. Telecommun. Radio Eng. 73(15), 1339–1360 (2014). https://doi.org/10.1615/TelecomRadEng.v73.i15.30

25. Lakshman, N.L., Khan, R.U., Mishra, R.B.: MANETs: QoS and investigations on optimized link state routing protocol. Int. J. Comput. Netw. Inf. Secur. (IJCNIS) 10(10), 26–37 (2018). https://doi.org/10.5815/ijcnis.2018.10.04

26. Lemeshko, O., Yeremenko, O.: Enhanced method of fast re-routing with load balancing in software-defined networks. J. Electr. Eng. **68**(6), 444–454 (2017). https://doi.org/10.1515/jee-2017-0079
27. Yeremenko, O., Lemeshko, O., Persikov, A.: Secure routing in reliable networks: pro-active and reactive approach. In: Advances in Intelligent Systems and Computing II, CSIT 2017. Advances in Intelligent Systems and Computing, vol. 689, pp. 631–655 (2018). https://doi.org/10.1007/978-3-319-70581-1_44
28. Fan, C., Zhang, T., Zeng, Z., Chen, Y.: Optimal base station density in cellular networks with self-similar traffic characteristics. In: Proceedings of 2017 Wireless Communications and Networking Conference (WCNC), pp. 1–6. (2017). https://doi.org/10.1109/wcnc.2017.7925881
29. Bulakh, V., Kirichenko, L., Radivilova, T.: Time series classification based on fractal properties. In: Proceedings of Second International Conference on Data Stream Mining & Processing (DSMP), pp. 198–201. IEEE (2018). https://doi.org/10.1109/dsmp.2018.8478532

Keystroke Pattern Authentication of Computer Systems Users as One of the Steps of Multifactor Authentication

Olena Vysotska[1(✉)] and Anatolii Davydenko[2]

[1] National Aviation University, Kiev, Ukraine
Lek_Vys@ukr.net
[2] Pukhov Institute for Modeling in Energy Engineering of NAS of Ukraine,
Kiev, Ukraine
davidenkoan@gmail.com

Abstract. In this paper, an authentication program was created for Ukrainian-speaking users of computer systems based on their keyboard style. To develop the algorithm of this program, a series of experiments were conducted. Based on the results of the experiments, the optimal handwriting characteristics were selected, which were subsequently analyzed for the implementation of recognition, also the requirements for educational samples and the stages of their selection and preliminary processing are determined. Besides considered the most critical parameters, setting which significantly increases the likelihood of correct recognition. This system is proposed to use as one of the stages of multifactor authentication.

Keywords: Multifactor authentication · Identification · Biometrics · Keystroke pattern · Computer systems

1 Introduction

From year to year biometric systems are used increasingly in various branches. In the estimation of J'son & Partners Consulting predictable market volume will increase from $20 billion in 2018 to $40 billion in 2022, herewith from 2016 till 2022 compound annual growth rate (CAGR) will make up 18,6% for biometric market [1].

One of the reference directions, in which it is advisable to use biometric systems, is the information security, namely the biometric authentication. According to the data of Spiceworks IT-network analytics, biometric authentication will be implemented into 86% of companies in North America and Europe till 2020. Such conclusions were based on the Spiceworks' poll of 500 staff members in these regions. According to the results, 62% of companies have already implemented such way of authentication, and additional 24% will do it in next two years [2]. The poll participants consider the usage of biometrics more reliable, than the usage of pin-codes or login-password pair. However, herewith only 10% of polled consider this method is enough. The rest tend towards that it is advisable to use multifactor authentication, where biometric system is one of the components of authentication system.

© Springer Nature Switzerland AG 2020
Z. Hu et al. (Eds.): ICCSEEA 2019, AISC 938, pp. 356–368, 2020.
https://doi.org/10.1007/978-3-030-16621-2_33

IBM Security subdivision of IBM Corporation released a global research [3] at the beginning of 2018, where consumer opinion about digital identity and authentication is analyzed. According to the results, when login any application and device, for users the security level is more important, than the usability feature. An additional point is that, according to the data of this research, younger generation trust traditional identification with password less. In order to increase personal informational security level when login, they prefer to use biometric systems, multifactor authentication and password manager.

From the above it is possible to make a conclusion, that developing of biometric systems as one of the components of multifactor authentication is one of the relevant objectives now.

An objective of this paper was to develop biometric authentication system for computer systems (CS) users as one of the components of multifactor authentication. This system can be used for solving other tasks, for example, for accessing any objects.

To solve the problem, the following actions were performed. First, the existing types of biometric authentication systems were analyzed [4–9, 11–15]. Based on the analysis and taking into account the fact that biometric authentication is more often used when login any CS or when accessing an information resource, it is proposed to use the keystroke pattern as an human's analyzed feature in this paper. Then the features of a human's keystroke pattern were determined for further usage in the authentication process. After that, there was made an analysis of the selected features of the keystroke pattern in order to determine their usability for further identification. Further, there was made a selection of training samples of CS users' patterns for increasing probability of correct identification. Then the program was written for implementation one of the steps of multifactor authentication, namely for performing the keystroke pattern authentication of CS users. The neural network was used as a recognition mechanism in this work [10]. As a result, on the basis of the analysis, with the help of the written program, it was concluded that it is advisable to use recognition systems by the human's keystroke pattern to implement one of the stages of the authentication systems of CS users.

2   Problem Solving

Recognition biometric systems are based on the analysis of any human feature. All of them are divided into statics (by fingerprint, face, iris, etc.) and dynamic (by keystroke pattern, handwriting, voice). The keystroke pattern is a dynamic human feature, in other words, this parameter isn't always stable, and therefore probability of correct recognition of a person by this feature isn't perfect. This fact is the reason of that it is recommended to use the keystroke pattern authentication systems in combination with other authentication methods in order to provide a high level of security.

The work of any biometric recognition system is composed of two stages:

1. Accumulation of training samples for all recognized objects.
2. Recognition of objects based on previously accumulated samples.

Depending on the implementation of each specific biometric system, there may be some additional actions. For example, if necessary, you can continue to accumulate a database of training samples (DBTS) at the recognition stage yet. Depending on the biometric authentication method used, the minimum allowable DBTS size will vary. For example, if the recognition method uses a fingerprint, then there should be several samples for each user in the database (for several fingers and taking into account that the finger can be placed at different angles on the scanner).

If the method for recognition uses keystroke pattern, as in this paper, then the size of the database will be many times as large. It is due to the fact that the fingerprint, as mentioned earlier, is a static parameter and doesn't change during a human's life, and the keystroke pattern is a dynamic parameter and, depending on various factors, can change, though not significantly.

The need for a large volume of the DBTS is also explained by the specifics of the neural network functioning, which was used as a recognition mechanism in this paper. Usually the DBTS reaches a few hundred and sometimes thousands of samples for each user.

During the recognition by keystroke pattern, the dynamics of typing any key word (phrases, combinations of letters) is usually analyzed, that means the time intervals between keystrokes when typing this word. You can also analyze other parameters, such as the quantity (percentage) of incorrect characters entered, the speed change when typing the beginning and the end of a word (slowdown or acceleration), etc.

Depending on the type of key phrase, there are several types of keystroke pattern recognition technologies:

1. By the dynamics of typing any constant key phrase.
2. By the dynamics of typing free text – the ever-changing phrase.
3. By the dynamics of typing one word from a pre-selected set of words.
4. By the dynamics of typing letter combinations, that repeat in all the words of a pre-selected set of words.

Each of the considered options has its advantages and disadvantages and you can choose the best one only empirically in each particular case.

The quality of recognition also depends on the quality of the selected controlled parameters. That means, the more high-quality parameters are in the sample, the more efficient is the use of this authentication method. Therefore, there should always be a function of determining the quality of the selected controlled parameters in the biometric authentication system, because these parameters may vary for different groups of users. For example, in the case of keystroke pattern authentication, the time of typing symbols of one word can be typical for one group of people, and the time of typing symbols of another word – for another group. As a rule, it is due to the specifics of the area in which the company operates. If the organization is associated with aviation, then typing the words from the field of aviation is typical, if it is a medical institution, then it is appropriate to analyze the medical terms input. In addition to that, if authentication must be performed not only at login, but also in the process in order to check whether there was an unauthorized change of user, then the analysis of the characteristics of input of the words related to the enterprise, also increases the possibility of authentication, since such words are entered more often than others. Also an important factor when choosing words (phrases) to analyze the dynamics of their input, is the language used. This is explained by the fact that each language has its own specificity in the construction of phrases, its most frequently used characteristic words.

In Fig. 1 there are shown two examples of the values distribution of one feature of three people. The first example (Fig. 1a) illustrates the case when a feature is suitable for performing by its authentication value, because the value areas of the feature of three people are quite clearly delimited and the spread of the values of the analyzed feature for each user is minimal.

The second example (Fig. 1b), on the contrary, illustrates the case when this characteristic isn't suitable for the classification of a computer system users, because the values of these characteristics are not separated for each user and, in addition, they have a sufficiently large spread. In order to translate these visual conclusions into a quantitative assessment of suitability of the characteristic for usage in its classification, such mathematical concepts as expectation and variance can be used.

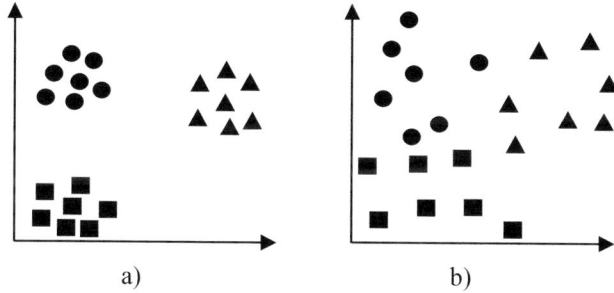

a)   b)

Fig. 1. Examples of the values distribution of one feature of three people where: ▲ denotes the first user's characteristic; ● denotes the second user's characteristic; ■ denotes the third user's characteristic.

The quality of the controlled parameter can be calculated as follows. Assuming that the set X_i is the set of values of the i-th parameter of the legal user's samples, and Y_i is the set of values of the i-th parameter of the offender's samples, we can say that the quality of the i-th parameter depends on the ratio of variances and the distance between the centers of the sets. The smaller the dispersion values and the greater the distance between the centers of the shared sets, the better the quality of this parameter. Numerically, the quality of the controlled parameter can be calculated as follows:

$$q = \sqrt{\frac{(m_{X_i} - m_{Y_i})^2}{\sigma_{X_i}^2 + \sigma_{Y_i}^2}} \tag{1}$$

where m_{X_i} denotes the expectation of the values of the monitored i-th parameter of the set X_i;

m_{Y_i} denotes the expectation of the values of the monitored i-th parameter of the set Y_i;

σ_{X_i} denotes the variance of the values of the monitored i-th parameter of the set X_i;

σ_{Y_i} denotes the variance of the values of the monitored i-th parameter of the set Y_i.

In the case, if the computational resources are limited, then the following formula can be used to determine the quality of the monitored parameter:

$$q = \frac{\left|m_{X_i} - m_{Y_i}\right|}{\sigma_{X_i} + \sigma_{Y_i}} \qquad (2)$$

The higher is q, the higher is the quality of the monitored parameter.

In addition, the more analyzed parameters there are, the higher is the quality of biometric authentication. But when choosing the number of analyzed parameters, it is necessary to proceed from the required security level, because if you use an unreasonably large number of analyzed parameters, then unnecessarily large resources will be spent during authentication. In other words, if the samples are too large, correspondingly, there will be a large database of these samples and there will be needed a lot of time and memory to process and store all this information. However, if you select too few parameters to analyze, then the authentication quality will be too low. According to the results of the experiments, for example, when the number of analyzed parameters changes from 3 to 6, the probability of correct recognition increases by about 50%, and when changing from 6 to 11 signs, the probability increases only by about 7–10%.

The same is the case with the number of accumulated samples in the database – the higher its' amount is, the higher is the authentication quality, but the more resources are spent.

The optimal number of samples and the number of features in them depends not only on the authentication technology used, but also on the particular users, that will be recognized by this system. Because different people have different degrees of uniqueness and stability of certain biometric characteristics.

Thus, it is impossible to specify in advance the particular values of these system parameters (the number of samples and features in them). Each protection system must have a subsystem for their determining them in each case, and focus on the average (recommended) values of these system parameters can lead to inefficiency of this protection system.

The choice of words (phrases, combinations of letters), the input parameters of which will be analyzed, the determination of the number and list of analyzed parameters is appropriate to be performed not on the basis of the analysis of a large (full) DBTS, but analyzing small (trial) database. That means that it makes sense to divide the stage of DBTS accumulation into several steps:

1. Arbitrary choice of several variants of the key phrase (words, letter combinations).
2. Accumulation of a small (trial) DBTS for each of the selected options.
3. Analysis of the characteristics of the accumulated trial DBTS (according to the formula considered earlier), in order to determine the key phrase with the best characteristics for recognition by this phrase input.
4. Check whether the selected keyword can provide the required recognition accuracy. If not, then return to the first stage. If so, pass to the next step.
5. Accumulation of the required DBTS volume for the selected key phrase.

After selecting the most optimal pattern characteristics for recognition, it is necessary to pre-process the accumulated data. When typing a key phrase or word during the accumulation DBTS, for a number of reasons, a user may accidentally press the wrong key, and make so a mistake or just think unusual long after it. All this leads to the appearance of damaged data in the accumulated database, and, consequently, to a decrease in the probability of correct recognition by this authentication system. Therefore, if during key phrase or a word input, during the accumulation of training data, an error was made or between the input of two near-by symbols of the word there was an unacceptably long pause, then this input attempt is better to stop, not to save this sample in the database, and to propose the test user to enter this word again. If the quality of the educational data isn't checked at the moment of DBTS accumulation, but later, then such a sample must be removed from the accumulated database.

However, all people have a different degree of care, that's why the number of mistakes made, the frequency of their repetition and words, when entering which there was a failure – it is enough individual features of a person, which can then be used further to identify users, so it is also desirable to save these data in the DBTS.

A number of experiments were made to verify the assumptions made. After analyzing their results, we can say that on the percentage of errors made by the user (Och), the probability of correct recognition of the user by the system quite strongly depends.

The percentage of errors made by the user is calculated using the following formula:

$$Och = \frac{100 * k_o}{kol + k_o} \tag{3}$$

where k_o denotes the number of wrong keys pressed;
kol denotes the number of correct keys pressed.

In addition to that, the probability of correct recognition of all users by the system is strongly influenced by the amplitude (spread) of the percentage of errors among users. This parameter is calculated using the following formula:

$$A_{Och} = Och_{max} - Och_{min} \tag{4}$$

where Och_{max} denotes the maximum error rate among all users;
Och_{min} denotes the minimum error rate among all users.

These facts should be taken into account when creating and configuring the authentication system in each particular case. So, according to the results of the experiments, the probability of recognition of an attentive CS user (that means the minimum error rate) is about 50% higher than not very careful.

An important step in the pre-processing of training samples is the exclusion of DBTS notes with gross value deviations of at least one of the features from the values of the same feature in other samples of his pattern (gross errors) [11]. Indeed, in order to improve the recognition quality it is necessary to make the selection of educational data. This need is explained by the fact that the keystroke pattern, as mentioned earlier, is a dynamic characteristic, in other words, this parameter doesn't have one hundred percent stability.

Because unlike static characteristics (fingerprint, retina, iris, etc.), keystroke pattern is subject to various factors (mood, distress, environment, etc.). Any person can be distracted by some external circumstances when typing (he may think, he may have something to get sick, he may have something or someone to prevent) and because of this make an unusually large (or vice-versa small) pause. And if the characteristics of his input were recorded in the DBTS, then the further subsequent analysis of these data will show significant instability of the analyzed parameters, especially if the time deviations of this user were quite a lot. But if a small instability is typical for keystroke pattern and the neural network [11–16] copes with this problem quite well, namely this mechanism is used in this paper to perform authentication, then a large values spread of any analyzed parameter leads to a decrease in the accuracy of recognition and, consequently, to a decrease in the effectiveness of this protection system. Therefore, we can say that the exclusion of educational data with a gross error of at least one of the analyzed features will lead to a certain increase in the quality of recognition.

Do some comparison with other methods. There are different algorithms for determining which training samples should be excluded. For comparison there was analyzed the algorithm AD described in [17], which solves this problem by sorting and excluding extreme values (using mathematical expectation, variance and Student's table) and proposed in this paper an algorithm AP that solves this problem by comparing the characteristic with its average value (excludes the analyzed sample in a case of a significant value deviation of at least one feature from its average value). Based on the results of the experiments, it can be concluded that the developed algorithm in most cases is a more effective mean to improve the recognition quality of CS users, namely it:

1. Is simple.
2. Consumes less time and computing resources.
3. In most cases is more effective.

The first two advantages of the proposed algorithm are explained by the fact that:

1. In this algorithm, a selection is performed on all analyzed features at once, in contrast to the algorithm described in [17], in which many actions are performed several times.
2. In contrast to the proposed algorithm, the algorithm described in [17] performs sorting for each feature, which takes a lot of time and memory, especially if the sorted array is large.

Great efficiency of the developed algorithm, in most cases, is expressed in the following:

1. Analysis of the analyzed features quality after the elimination from the DBTS with gross errors according to the proposed algorithm showed that all parameters have approximately the same, quite good quality level, in contrast to the case when the eliminations were performed according to another considered algorithm [17], in which the quality level for different analyzed features varies greatly among themselves (some very good, and some very bad).

2. With a relatively large number of analyzed features, the proposed algorithm provides better recognition quality, and with a small number (insufficient for qualitative recognition) of the analyzed features, the other considered algorithm [17] is more effective.

All this confirms that the use of the algorithm developed in this work was preferable in this case. According to the results of the experiments we can say that the quality of the analyzed characteristics, after the elimination of samples with gross errors, improved by about 3–4 times.

Such a selection is preferably to perform not only in the processing of the resulting (accumulated in sufficient volume for recognition) DBTS, but also when processing a trial DBTS at the stage of selecting the optimal key phrases. However, at this stage, you may be able to perform a more coarse selection of the educational data, that is, you may be able to allow a larger value of maximum deviation percentage than at the recognition stage. The impact of this coefficient on the accuracy of the authentication is determined through experimentation.

And only after all the considered preparatory actions directly recognition of users by keystroke pattern can be performed. In this paper, this problem was solved using a neural network [10].

Thus, the algorithm of the CS users recognition system based on the keystroke pattern, using a neural network, is the following:

1. Random selection of several options of the key phrase.
2. The accumulation of trial DBTS for each option.
3. The elimination of the records with gross deviations of at least one of the characteristics from its average value (coarse selection) from each trial DBTS.
4. Analysis of the characteristics of the trial DBTS in order to determine the key phrase with the best characteristics for recognition the input of this phrase.
5. Check whether the selected keyword can provide the required recognition accuracy. If not, then return to the first stage. If so, pass to the next step.
6. Accumulation of the required amount of DBTS for the selected key phrase.
7. Elimination of records with gross deviations of the value of at least one of the characteristics from its average value from the resulting DBTS (the permissible deviation of the value of the analyzed characteristic from its average value depends on the required recognition accuracy).
8. Choosing the best characteristics in the resulting DBTS, that will be used for the further recognition.
9. Setting the parameters of the authentication system, on which the quality of recognition depends.
10. Perform authentication of the CS users.

All stages, except the last one, are performed once when installing this authentication system and then, if necessary, for adjustment.

3 Experimental Part

To determine the influence of the parameters considered earlier, on the probability of correct recognition of Ukrainian-speaking users of computer systems by their keyboard handwriting, a database of training samples of the required volume was accumulated, after which a series of experiments were conducted. For this purpose, a specially created program was used. All experiments can be divided into three stages:

1. At the first stage, the data for all users were analyzed, while all training samples were considered, that is, the time of the character set of each word. At this stage, the influence of the following parameters on the effectiveness of the neural network to solve the problem was determined:

   1.1. The presence of averaging training samples, in which the characteristics of each of the ten records were averaged. At the same time, the accuracy of the data increased, but their number decreased.

   1.2. The presence of selection by words, that is, "unknown" instances were compared with all educational data or only with the same words. If the selection was not carried out, then only the first sign was inaccurate, and if the selection was carried out, then there were no inaccuracies, but the amount of training data decreased 10 times, because there were 10 words in the set.

   1.3. The number of feature.

   1.4. The amount of training data for each user.

2. At the second stage, the effect on the probability of correct recognition of Ukrainian users of a computer system such parameters as the percentage of errors in typing (*Och*) (user care when working on the keyboard), amplitude (*AOch*) of the percentage of errors among users was investigated. In addition, the influence of these factors on the dependence of correct recognition on the parameters considered earlier was determined.

3. At the third stage, the effect on the probability of correct recognition of Ukrainian-speaking users of a computer system of such a parameter as the number of gross errors in the values of attributes in the training samples was investigated. It was also investigated which of the algorithms for the exclusion of training samples with gross errors (considered [17] or proposed) is the best. After choosing the best algorithm, the maximum permissible percentage of each characteristic deviation from its average value was determined. In addition, the influence of this factor on the dependence of correct recognition on the parameters considered earlier was determined.

According to the results of the experiments, the following conclusions can be drawn:

1. When selecting by words the probability of a correct answer is about 2% lower than in the first one, hence it follows that it is better to learn more than the selection of words, because without the selection, the words "suffer" only the first feature.

2. In cases where averaging is not performed, the probability of correct recognition increased by about 3–4%, indicating that the amount of training data has a greater impact on the outcome than a slight increase in the accuracy of the training data.

3. With an increase in educational samples from 1000 to 1500, the probability of correct recognition increased by 2%, and with a decrease from 1000 to 500 the probability of correct recognition decreased by 2%.
4. The change in the number of attributes from 3 to 6 increased the probability of the correct answer by about 12–22%, and the more users in the system, the better the result.
5. The greatest influence is the number of errors (Och) when typing, and the smaller the error rate amplitude (A_{Och}) is among users, the greater the likelihood of correct recognition. That is, the more careful the users are when typing the text, the greater the likelihood of correct recognition, in addition, it is desirable that the level of user's attention is approximately the same.
6. After the elimination of gross sample errors with any of the algorithms (discussed [17] or proposed), the quality of the signs and, accordingly, the probability of correct recognition increases significantly.
7. After the exclusion of educational samples with gross errors in the proposed algorithm, the quality of the signs is much better and the probability of correct recognition is much greater than after the exclusion for the considered [17] algorithm.

The results of some experiments are shown in the graphs (Figs. 2, 3 and 4).

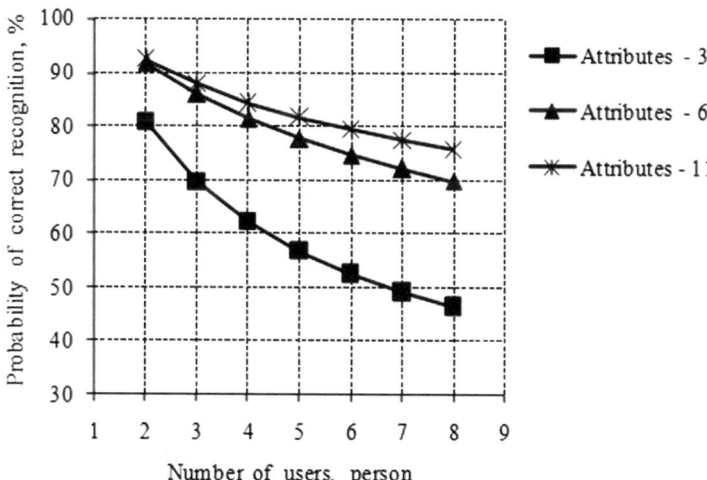

Fig. 2. Influence of the number of attributes on the probability of correct recognition of users of the computer system, provided that the exclusion of training samples with gross errors is performed by the algorithm AP, selection by words – there is, training samples – 700

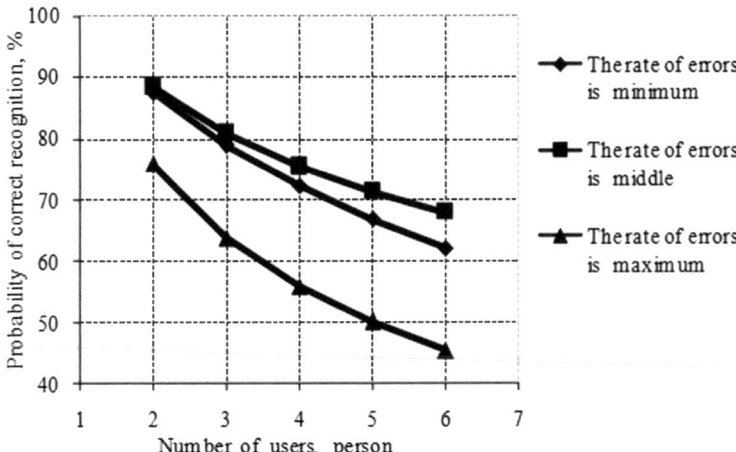

Fig. 3. Influence of the number of errors during typing on the probability of correct recognition of users of the computer system, provided that attributes – 6, training samples – 1500, selection by words – no, averaging – no

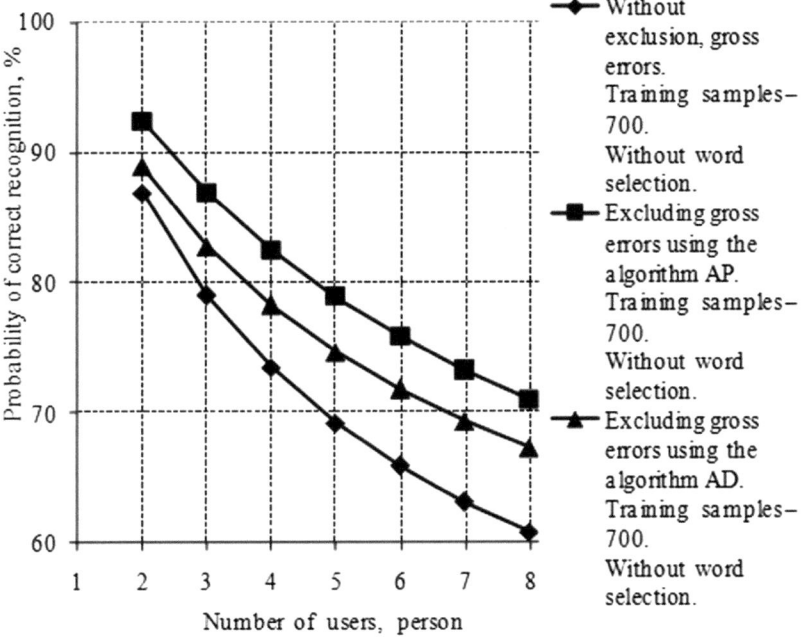

Fig. 4. Impact of the exclusion of training samples with gross errors on the probability of correct recognition of users of the computer system, provided that attributes - 6

4 Conclusions

In this work, on the basis of the analysis, in order to create an authentication system, a biometric dynamic method of recognition was chosen, namely the method based on the analysis of human keystroke pattern. Then there were selected the characteristics of pattern, which were analyzed further for the implementation of recognition; there were determined the requirements for training samples and the stages of their selection. On the basis of the created algorithm, a program was written to perform authentication of CS users, with help of which a series of experiments were conducted. Based on the results of the experiments, recommendations (requirements) for pre-processing data for recognition and adjustment of the most critical parameters of the system were given.

Thus, it can be concluded that the considered authentication method is quite effective as one of the stages of multi-factor authentication.

References

1. World market of the biometric systems, 2015–2022, 19 January 2017. http://json.tv/ict_telecom_analytics_view/mirovoy-rynok-biometricheskih-sistem-2015-2022-gg-20170119025618
2. Arthur Galeev.: Almost all companies in the U.S. and Europe will use biometrics in two years, 30 March 2018. http://safe.cnews.ru/news/top/2018-03-26_pochti_vse_kompanii_v_ssha_i_evrope_budut_ispolzovat
3. IBM Future of Identity Study: Millennials Poised to Disrupt Authentication Landscape, 29 January 2018. https://www-03.ibm.com/press/us/en/pressrelease/53646.wss
4. Vysotska, O., Davydenko, A.: Classification of biometric authentication systems. Collection of Scientific Works of the Pukhov Institute for Modeling in Energy Engineering of NAS of Ukraine, vol. 27, pp. 108–114 (2004). (in Russian)
5. Vysotska, O., Davydenko, A.: Determination of critical parameters when choosing a biometric authentication system. Modeling and information technologies. Collection of Scientific Works of the Pukhov Institute for Modeling in Energy Engineering of NAS of Ukraine, vol. 27, pp. 80–86 (2004). (in Russian)
6. Benafia, A., Mazouzi, S., Sara, B.: Handwritten character recognition on focused on the segmentation of character prototypes in small strips. Int. J. Intell. Syst. Appl. (IJISA) 9(12), 29–45 (2017). https://doi.org/10.5815/ijisa.2017.12.04
7. Hamd, M.H., Ahmed, S.K.: Biometric system design for Iris recognition using intelligent algorithms. Int. J. Mod. Educ. Comput. Sci. (IJMECS) 10(3), 9–16 (2018). https://doi.org/10.5815/ijmecs.2018.03.02
8. Zoubida, L., Adjoudj, R.: Integrating face and the both Irises for personal authentication. Int. J. Intell. Syst. Appl. (IJISA) 9(3), 8–17 (2017). https://doi.org/10.5815/ijisa.2017.03.02
9. Angadi, S.A., Hatture, S.M.: Biometric person identification system: a multimodal approach employing spectral graph characteristics of hand geometry and palmprint. Int. J. Intell. Syst. Appl. (IJISA) 8(3), 48–58 (2016). https://doi.org/10.5815/ijisa.2016.03.06
10. Kallan, R.: Basic concepts of neural networks. Translate from English. Publishing House "Williams" (2001). (in Russian)

11. Vysotska, O.: The influence of the elimination of educational data with gross errors on the dependence of the efficiency of probabilistic neural networks for authentication of computer systems users by keystroke pattern on various parameters. Modeling and information technologies. Collection of Scientific Works of the Pukhov Institute for Modeling in Energy Engineering of NAS of Ukraine, vol. 28, pp. 3–10 (2005). (in Russian)

12. Rao, G.A., Kishore, P.V.V., Kumar, D.A., Sastry, A.S.C.S.: Neural network classifier for continuous sign language recognition with selfie video. Far East J. Electron. Commun. **17** (1), 49 (2017)

13. Kishore, P.V.V., Rao, G.A., Kumar, E.K., Kumar, M.T.K., Kumar, D.A.: Selfie sign language recognition with convolutional neural networks. Int. J. Intell. Syst. Appl. (IJISA) **10**(10), 63–71 (2018). https://doi.org/10.5815/ijisa.2018.10.07

14. Hu, Z., Bodyanskiy, Y.V., Kulishova, N.Y., Tyshchenko, O.K.: A multidimensional extended neo-fuzzy neuron for facial expression recognition. Int. J. Intell. Syst. Appl. (IJISA) **9**(9), 29–36 (2017). https://doi.org/10.5815/ijisa.2017.09.04

15. Shimada, M., Iwasaki, S., Asakura, T.: Finger spelling recognition using neural network with pattern recognition model. In: SICE 2003 Annual Conference, vol. 3, pp. 2458–2463. IEEE (2003)

16. Hu, Z., Tereykovskiy, I.A., Tereykovska, L.O., Pogorelov, V.V.: Determination of structural parameters of multilayer perceptron designed to estimate parameters of technical systems. Int. J. Intell. Syst. Appl. (IJISA) **9**(10), 57–62 (2017). https://doi.org/10.5815/ijisa.2017.10.07

17. Rastorguev, S.P.: Program methods of information security. Teaching guide. Penza State University. Publishing House of Penza State University (2000). (in Russian)

Complexity Estimation of GL-Models for Calculation FTMS Reliability

Alexei Romankevich[ID], Ivan Maidaniuk[ID], Andrii Feseniuk[(✉)][ID], and Vitaliy Romankevich[ID]

National Technical University of Ukraine
"Igor Sikorsky Kyiv Polytechnic Institute", Kyiv, Ukraine
romankev@scs.kpi.ua, andrew_fesenyuk@ukr.net

Abstract. The work is devoted to determining the complexity of the models used in the calculation of the reliability characteristics of fault-tolerant multiprocessor systems. In particular, we are talking about GL-models, which reflect the behavior of systems in the flow of failures. The focus is on determining the complexity of GL-models when converting them by introducing additional edges into the model graph. The upper and lower bounds of the number of such edges were obtained during the transformation of the model, which is associated with the solution of the problem of increasing the system reliability.

Keywords: Reliability · Fault-tolerance · Multiprocessor systems · GL-models

1 Introduction

Currently, fault-tolerant reconfigurable multiprocessor systems (FTMS) are becoming more widespread, the main advantage of which is high reliability. Such systems are able to reconfigure when a certain set of modules fail and to continue to function fully [1]. In particular, the article [2] considers *m-out-of-n* systems (of *n* components) that are stable (that is, remain operable) to any failures, the multiplicity of which does not exceed a fixed value *m*. We call such systems basic and denote as K(m, n). Systems the behavior of which in the flow of failures is different from the basic ones are called non-basic.

Reliability becomes the most important requirement when talking about control system of critical application such as nuclear power plant, aviation, health monitoring system for critical patients etc. because the operational failure of such system could have catastrophic consequences. The critical application control systems are usually designed as multiprocessor systems with embedded testing, diagnostics, fault-tolerance and reconfiguration.

When designing such systems, one of the main tasks faced by the developer is to achieve the required level of reliability or, more precisely, the estimation of the probability of its failure-free operation (PFO) [3–5]. Known methods for calculating PFO FTMS mainly focus on basic systems [6–9]. The article [10] describes a universal calculation method based on conducting statistical experiments with models that adequately reflect the system's response to failures of its components. GL-models can be used as such models.

© Springer Nature Switzerland AG 2020
Z. Hu et al. (Eds.): ICCSEEA 2019, AISC 938, pp. 369–377, 2020.
https://doi.org/10.1007/978-3-030-16621-2_34

2   GL-Model

According to [11], in accordance with the efficiency of the modeled FTMS as a whole, the connection of a non-oriented graph R is put, the total number of edges of which is denoted as r.

A = $\{a_1, a_2, \ldots, a_r\}$ is the set of edges of the graph R. Each element of the set is denoted (marked) in a certain way by the selected Boolean function: $f_1(x_1, \ldots, x_n), \ldots, f_r(x_1, \ldots, x_n)$ where x_i is indicator variable ($i = 1\ldots n$), which reflects the state of the i-th element of the FTMS, i.e. $x_i = 1$, if the i-th element of the system is efficient, and $x_i = 0$ otherwise. It should be noted that not every edge function depends on all variables. $X = (x_1, \ldots, x_n)$ is the system state vector, characterizing the state of all elements of the system. The edge a_i, denoted by the function $f_i(x_1, \ldots, x_n)$, remains in the graph G if $f_i(x_1, \ldots, x_n) = 1$, otherwise it is removed from it.

There are many ways to form the functions of the GL model depending on the selected graph R, the degree of fault tolerance m, the requirements for the model itself and the features of the system.

3   Formulation of the Problem

One of the most universal is the GL-model proposed in [13], the so-called MFE (minimum of falling edges) GL-model, a feature of which is the loss of only two edges when a vector with ($m + 1$) failure appears. An example of such a model is given in [10].

There are two ways to transform the MFR model for non-basic FTMS s: by drawing additional edges and by modifying the edge functions. We are using the first method below.

The complexity of the model directly affects the time of the experiment with it, and, therefore, the total time and the calculation accuracy of the PFO. In this regard, the calculation of model parameters in the first stages of the design of the FTMS, as well as determining the transformation complexity of an already created model when transforming the system itself is of practical interest.

In [13], a basic algorithm for constructing the MFE of the GL-model of the basic FTMS K(m, n) for arbitrary values of m and n is described, based on minimizing the so-called canonical GL-model. The number of edges of the MFE model constructed according to [11] is easy to determine by the formula:

$$r = n - m + 1.$$

The question of estimating the number of additional edges in the non-basic model is more interesting.

4   Complexity Estimation of the GL-Model of Non-basic FTMS

We should remind that the basic FTMS K(m, n) continues to function if the number of failed modules does not exceed the value of m. We define W = $\{w_1, w_2, w_3, \ldots, w_{|W|}\}$ as the set of specified system state vectors with $m + 1$ faulty processor, with the appearance of which the system retains its functionality. It is clear that the appearance of a vector from W leads to the disappearance of a pair of main edges. Let's denote the set of such pairs as L, and its power by l.

In accordance with the algorithm described in [12], the edges of the graph of the original GL-model are split into two or more non-empty, non-intersecting S-subsets: in one S-subset there are only such edges, a sample of two of which does not match any of the elements from L-set. One of the important criteria for splitting a graph into S-subsets is the minimality of their number. Further, additional edges are drawn in accordance with the division of the graph into S-subsets. In fact, dividing the edges set of a graph into S-subsets determines the number and drawing of additional edges, which serve to preserve the connectivity of the graph when certain pairs of edges disappear. Since the drawing of additional edges is actually determined by dividing the graph into S-subsets, we will assume that S-subsets block pairs of edges from the L-set.

Further, to reduce the calculations, preserving the connectivity of a graph with the exclusion of edge pairs (from the set) will be called a block of the L-set, or simply a block. We will also say that S-subsets block pairs of edges. Let the number of S-subsets be μ, and their powers are $s_1, s_2, \ldots, s_\mu$, respectively.

At the design stage of the model and making changes to it, the developer should know the limits of the number of introduced additional edges (to ensure the adequacy of the model). On the other hand, the limits in which there may be a number of vectors blocked by a specified number of additional edges are of great interest. The task can be formulated as following:

- To calculate the bounds of μ, based on the given r (the number of the model main edges) and l (the power of the L-set);
- To determine the bounds for l for given μ and r.

It is clear that the value $l \in [1..C_r^2]$, and μ is in the range from 1, if the L-set is empty, to r, when each edge defines one S-subset (in this case, $l = C_r^2$).

Let $\mu_{max}(l)$ be such a number of S-subsets that can block any set of elements of L-set with a power of l, meeting the condition that the number of S-subsets is minimal. $\mu_{min}(l)$ is the minimum number of S-subsets that ensures the possibility of blocking at least one set of l pairs of edges, and adding at least one element to the L-set definitely would require dividing into bigger number of S-subsets than $\mu_{min}(l)$.

Let $l_{min}(\mu)$ be such a minimal value of the power of the L-set, for blocking of which the division of the graph edges into μ S-subsets is required, it is such a combination of l pairs that removing at least one element from the L-set would require less than μ S-subsets (taking into account the requirement of minimality to their number).

Let $l_{max}(\mu)$ be the maximum number of pairs of edges for blocking of which it is enough to divide the graph into μ S-subsets.

It is possible to prove the mutual invertibility of pairs of functions $\mu_{min}(l) - l_{max}(\mu)$ and $l_{min}(\mu) - \mu_{max}(l)$, which allows to find one of each pair for determining the sought boundary functions.

Firstly, let's search $l_{min}(\mu)$. To do this, we use the concept of the graph V, [12]. Figure 1 shows an example of a V-graph, its coloring, and the formation of a graph S for a model with six edges (r = 6).

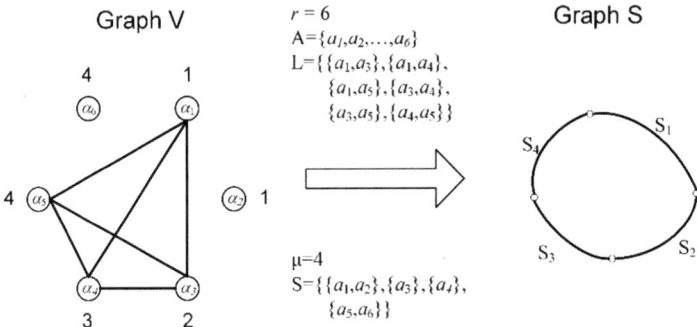

Fig. 1. The formation of the graph S based on the coloring of the V-graph

Let μ be the chromatic number of the graph V being formed, based on the GL-model of some FTMS and the L-set. Since l is the number of edges of the V-graph, it is sufficient to calculate the minimum number of edges forming the V-graph requiring μ colors for the correct coloring of their vertices to determine $l_{min}(\mu)$. It is known that the chromatic number is not less than the number of vertices in the clique of the graph. Consequently, it suffices to choose such a number of edges, which would determine a complete subgraph consisting of μ vertices when it is optimally conducted. Thus, the task is reduced to determining the number of edges of a complete graph, which is known that the sought bound is:

$$l_{\min}(\mu) = \mu(\mu - 1)/2. \tag{1}$$

Based on the fact that $l_{min}(\mu)$ and $\mu_{max}(l)$ are mutually inverse functions, it is easy to determine the upper bound of the number of S-subsets from (1):

$$\mu_{max}(l) = \frac{1 + \sqrt{1 + 8l}}{2}. \tag{2}$$

Now let's search for $l_{max}(\mu)$. As before, μ is the number of S-subsets, the power of each of them is $s_1, s_2, \ldots, s_\mu$, and r is the number of edges in the original graph. To search for the border turn to the V-graph again.

We set the problem: what should be the structure of a V-graph with the maximum number of edges, so that it would remain μ-colored. We divide the vertices of the sought V-graph (let's remember that the vertices of the V-graph correspond to the edges of the GL-model) into μ groups and draw all possible edges between the vertices from different groups. It is clear that such a graph is μ-colorable, since the vertices belonging to one group define an internally stable set. However, depending on the division method, the number of edges that can be drawn while maintaining the μ-colorability property will be different.

To solve the problem, it suffices to choose the power of the groups such that the number of edges of the V-graph would be maximum. Since all vertices belonging to the same group will have the same color, the edges corresponding to them in the GL-model will belong to the same S-subset. Based on the above, the L-set should be formed as a set of pairs $\left\{ a_{s_i}^t, a_{s_j}^z \right\}$ where $t = 1 \ldots s_i, z = 1 \ldots s_j, i = 1 \ldots \mu - 1, j = (i+1) \ldots \mu$, and $a_{s_i}^t$ is an edge of S_i-subsets. Let us find such $s_1, s_2, \ldots, s_\mu$, that the power of L will be maximal. It is obvious that: $1 \leq s_i \leq r - \mu + 1$ where $i = 1 \ldots \mu$. We also remind that $\sum_{i=1}^p s_i = r$.

Let we have $S_1, S_2, \ldots, S_\mu$ - subsets, and the corresponding powers of which: $s_1, s_2, \ldots, s_\mu$. Based on the above, we write down:

$$l_{max}(s_1, s_2, \ldots, s_\mu) = \sum_{i,j=1, j>i}^{\mu} (s_i s_j). \tag{3}$$

The solution of the problem is reduced to the search for a conditional extremum. We use the method of Lagrange multipliers.

Let's introduce some additional variables that were adopted in solving such problems, in addition, we will assume that all functions and variables introduced below will get real values, which does not limit generality.

Let $x_i = s_i (i = 1 \ldots \mu)$, in the context of this solution, x_i will represent not the boolean indicator state variable of the system module, but the real one.

$$U = l_{max}(s_1, s_2, \ldots, s_\mu) = f(x_1, x_2, \ldots, x_\mu) = \sum_{\substack{j > i \\ i,j = 1}}^{\mu} x_i x_j.$$

We need to find the maximum of the function of n variables.

$$U = f(x_1, x_2, \ldots, x_\mu) = \sum_{\substack{j > i \\ i,j = 1}}^{\mu} x_i x_j. \tag{4}$$

With one connection condition:

$$F(x_1, x_2, \ldots, x_\mu) \sum_{i=1}^{p} x_i - r = 0. \tag{5}$$

Since the function is symmetric and linear, we can express any variable through others. From (5) we express x_1, then we get the following function $x_1(x_2, \ldots, x_\mu) = r - \sum_{i=2}^{\mu} x_i$.

Let's substitute the left part of (4) into (5), as a result of which we get:

$$U = f(x_1(x_2, \ldots, x_\mu), x_2, \ldots, x_\mu) = \Phi((x_2, \ldots, x_\mu)). \tag{6}$$

Having the function (6) and the condition of connection, we write the Lagrange function: $\Psi = U - \lambda F$ where λ is an indefinite factor. Thus, the search for extremum is reduced to solving the following system:

$$\begin{cases} \frac{\partial \Psi}{\partial x_1} = 0 \\ \frac{\partial \Psi}{\partial x_2} = 0 \\ \dots \\ \frac{\partial \Psi}{\partial x_n} = 0 \\ F = 0 \end{cases} \Leftrightarrow \begin{cases} \sum_{i \neq 1} x_i - \lambda = 0 \\ \sum_{i \neq 2} x_i - \lambda = 0 \\ \dots \\ \sum_{i \neq n} x_i - \lambda = 0 \\ \sum_{i=1}^{n} x_i = r \end{cases} \qquad (7)$$

If we subtract the first equation of system (7) from the rest, we get the following:

$$\left\{ \sum_{i \neq 1} x_i - \lambda = 0; x_2 = x_1; \dots; x_n = x_1; \sum_{i=1}^{\mu} x_i = r. \right.$$

As a result, we obtained: $x_1 = x_2 = \dots = x_\mu$, then the final solution of the system: $x_i = \frac{r}{\mu}$ $(i = 1 \dots \mu)$, and $\lambda = \frac{r(\mu-1)}{\mu}$.

We have found a point of possible, unconditional extremum, and now we need to check whether this point is an extremum. For this, it is necessary to check the condition of a fixed sign $d^2\Psi$ at this point.

$d^2\Psi = (dx_1 \partial/\partial x_1 + dx_2 \partial/\partial x_2 + \dots + dx_p \partial/\partial x_p)^2 \Psi$, taking into account that $dx_1 = -\sum_{i=2}^{\mu} dx_i$:

$$d^2\Psi = \left[\left(dx_2 \frac{\partial}{\partial x_2} + \dots + dx_p \frac{\partial}{\partial x_p} \right) - \left(dx_2 + dx_3 + \dots + dx_p \right) \frac{\partial}{\partial x_1} \right]^2 \Psi =$$

$$\left(\sum_{i=2}^{\mu} dx_i \frac{\partial}{\partial x_i} \left[\sum_{j=2}^{\mu} dx_j \frac{\partial}{\partial x_j} - \sum_{j=2}^{\mu} dx_j \frac{\partial}{\partial x_1} \right] - \right.$$
$$\left. \sum_{i=2}^{\mu} dx_i \frac{\partial}{\partial x_1} \left[\sum_{j=2}^{\mu} dx_j \frac{\partial}{\partial x_j} - \sum_{j=2}^{\mu} dx_j \frac{\partial}{\partial x_1} \right] \right) \Psi =$$

$$\sum_{i=2}^{\mu} \sum_{j=2}^{\mu} \frac{\partial^2 \Psi}{\partial x_i \partial x_j} dx_i dx_j - \sum_{i=2}^{\mu} \sum_{j=2}^{\mu} \frac{\partial^2 \Psi}{\partial x_i \partial x_1} dx_i dx_j - \sum_{i=2}^{\mu} \sum_{j=2}^{\mu} \frac{\partial^2 \Psi}{\partial x_1 \partial x_j} dx_i dx_j$$
$$+ \sum_{i=2}^{\mu} \sum_{j=2}^{\mu} \frac{\partial^2 \Psi}{\partial x_1 \partial x_1} \cdot dx_i dx_j = \sum_{i=2}^{\mu} \sum_{j=3}^{\mu} 1 - 2 \sum_{i=2}^{\mu} \sum_{j=2}^{\mu} 1 + \sum_{i=2}^{\mu} \sum_{j=2}^{\mu} 0$$
$$= (\mu - 1)(\mu - 2) - 2(\mu - 1)(\mu - 1) = -\mu(\mu - 1).$$

Thus, $d^2\Psi < 0$ if $\mu > 1$. Since μ (the number of S-subsets) cannot be less than 2 for a non-empty L-set, so the found point is a maximum.

Substitute in the formula (2.6) the value for $s_i = \frac{r}{\mu}$ $(i = 1, 2, \dots, \mu)$ and get:

$$l_{max}(r, \mu) = \frac{r^2}{2} \frac{(\mu - 1)}{\mu}. \qquad (8)$$

Function (8) is real, but $l_{max}(r, \mu)$ must take natural values, like $s_1, s_2, \dots, s_\mu$, then $s_i = [r/\mu]$ or $s_i = [r/\mu] + 1$. Let $s_1 = s_1 = [r/\mu]$, and the other $s_2, \dots, s_\mu$ are equal to either $[r/\mu] + 1$, or $[r/\mu]$, which is defined by the value $\{r/\mu\}$, which we denote by v. Based on the fact that $\sum_{i=1}^{\mu} s_i = r$ and that the mutual position of S-subsets to each other does not matter, we can determine $s_1, s_2, \ldots, s_{\mu-\nu} = [r/\mu]$, and $s_{\mu-\nu+1} \ldots s_\mu = [r/\mu] + 1$. Now we substitute the new values of the arguments in the formula (3):

$$
\begin{aligned}
l'_{max}(r,\mu) &= C^2_{\mu-\nu}[r/\mu]^2 + C^2_\nu([r/\mu]+1)^2 + C^1_\nu C^1_{\mu-\nu}[r/\mu]([r/\mu]+1), \\
l'_{max}(r,\mu) &= \frac{\mu(\mu-1)}{2}[r/\mu]^2 + \frac{\nu-1+2[r/\mu](\mu-1)}{2}\nu.
\end{aligned}
\tag{9}
$$

Formula (9) gives exact values. Based on a large enough number of experiments, it was found that the average value of the absolute accuracy, that is the difference between (8) and (9) depends on the value of r and is equal to:

$$
\left| l'_{max}(r,\mu) - l_{max}(r,\mu) \right| \cong 0.0386\,r.
$$

Further, in the calculations we will use the function $l_{max}(r,\mu)$, because it does not contain integer division and, therefore, is easier to handle.

We should remind that $l_{max}(\mu)$ и $\mu_{min}(l)$ are mutually inverse functions, therefore, by representing l as the argument of the function μ, we easily determine the lower bound of the number of S-subsets:

$$
\mu_{min}(l,r) = \frac{r^2}{r^2 - 2l}.
\tag{10}
$$

Thus, the results obtained above can be represented as the following inequalities:

$$
\frac{r^2}{r^2 - 2l} \le \mu(r,l) \le \frac{1+\sqrt{1+8l}}{2}.
$$

$$
\frac{\mu(\mu-1)}{2} \le l(\mu,r) \le \frac{\mu(\mu-1)}{2}\left[\frac{r}{\mu}\right]^2 + \frac{\nu-1+2\left[\frac{r}{\mu}\right](\mu-1)}{2}\nu.
$$

As an example (Fig. 2), we find the exact values of the boundaries for the quantity μ and approximations by formulas (2.5), (10) for the GL model with 8 edges.

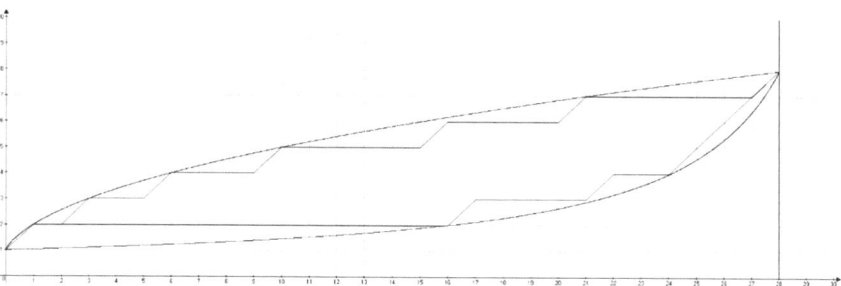

Fig. 2. Accurate values and approximations of $\mu_{min}(l,r)$ and $\mu_{max}(l)$

We should note that the formula obtained above for the lower bound of the number of S-subsets coincides with the results given in [90] for the lower bound of the chromatic number of an arbitrary graph, which confirms the correctness of the presented reasonings.

The results allow us to estimate the complexity of the GL model of a non-basic FTMS constructed on the base of K (m, n) model already in the early stages of system design. This allows the developer to determine the resources in advance (computational and time) which are necessary to calculate the reliability of the FTMS with the required accuracy.

5   Conclusions

The article considers the problem of determining the complexity of the GL-model, reflecting the behaviour of a fault-tolerant multiprocessor system in the flow of failures of its processors. The task is of practical importance, because this indicator determines the time for performing one statistical experiment with a model and, therefore, the accuracy of calculating the reliability parameters of the system.

By now, the authors have completed a large amount of work in this area. In particular, effective algorithms have been developed for generating the edge functions of the GL model and their further minimization. Programmes developed in accordance with these algorithms make it easy to solve the problem of creating a GL model for any real values of the number of processors and the number of allowed failures. The use of these programmes to calculate the reliability parameters of a fault-tolerant multiprocessor (for example, 25–30 processors) complex object control system is performed in real time on one standard computer.

In order to modify the GL-model when transforming the system to increase the degree of fault tolerance on a certain set of state vectors of a system of higher multiplicity, we chose the method of additional edges with its functions, which actually block the loss connectivity of the graph of the model in these cases. Boundary estimates are obtained for the number of such additional edges depending on the power of the set of vectors defining the deviation of the system from the base one.

References

1. Avižienis, A., Laprie, J., Randell, B.: Dependability and its threats: a taxonomy. In: Building the Information Society, pp. 91–120 (2004)
2. Kuo, W., Zuo, M.J.: Optimal Reliability Modeling: Principles and Applications. Wiley, Hoboken (2003)
3. Suo, B., Cheng, Y.-S., Zeng, C., Li, J.: Calculation of failure probability of series and parallel systems for imprecise probability. Int. J. Eng. Manuf. (IJEM) 2(2), 79–85 (2012). https://doi.org/10.5815/ijem.2012.02.12
4. Rahdari, D., Rahmani, A.M., Aboutaleby, N., Karambasti, A.S.: A distributed fault tolerance global coordinator election algorithm in unreliable high traffic distributed systems. Int. J. Inf. Technol. Comput. Sci. (IJITCS) 7(3), 1–11 (2015). https://doi.org/10.5815/ijitcs.2015.03.01

5. Wang, X., Li, S., Liu, F., Fan, X.: Reliability analysis of combat architecture model based on complex network. Int. J. Eng. Manuf. (IJEM) **2**(2), 15–22 (2012). https://doi.org/10.5815/ijem.2012.02.03
6. Keshtgar, S.A., Arasteh, B.B.: Enhancing software reliability against soft-error using minimum redundancy on critical data. Int. J. Comput. Netw. Inf. Secur. (IJCNIS) **9**(5), 21–30 (2017). https://doi.org/10.5815/ijcnis.2017.05.03
7. Kaswan, K.S., Choudhary, S., Sharma, K.: Software reliability modeling using soft computing techniques: critical review. Int. J. Inf. Technol. Comput. Sci. (IJITCS) **7**(7), 90–101 (2015). https://doi.org/10.5815/ijitcs.2015.07.10
8. Wason, R., Soni, A.K., Rafiq, M.Q.: Estimating software reliability by monitoring software execution through opcode. Int. J. Inf. Technol. Comput. Sci. (IJITCS) **7**(9), 23–30 (2015). https://doi.org/10.5815/ijitcs.2015.09.04
9. Thomas, M.O., Rad, B.B.: Reliability evaluation metrics for internet of things, car tracking system: a review. Int. J. Inf. Technol. Comput. Sci. (IJITCS) **9**(2), 1–10 (2017). https://doi.org/10.5815/ijitcs.2017.02.01
10. Romankevich, A., Feseniuk, A., Maidaniuk, I., Romankevich, V.: Fault-tolerant multiprocessor systems reliability estimation using statistical experiments with GL-models. In: Advances in Intelligent Systems and Computing, vol. 754, pp. 186–193 (2019)
11. Romankevich, A.M., Karachun, L.F., Romankevich, V.A.: Graph-logic models for analysis of complex fault-tolerant computing systems. Elektronnoe modelirovanie **23**(1), 102–111 (2001). (in Russian)
12. Romankevich, A.M., Ivanov, V.V., Romankevich, V.A.: Analysis of fault-tolerant multiprocessor systems with complex fault distribution based on cyclic GL-models. Elektronnoe modelirovanie **26**(5), 67–81 (2004). (in Russian)
13. Romankevich, V., Potapova, K., Bakhtari, H.: The some properties of model's of k-out-of-n system's behavior in the stream of faults. In: Proceedings of IEEE East-West Design & Test Symposium, Armenia, Yerevan, p. 763, September 2007

Game Method of Event Synchronization in Multi-agent Systems

Petro Kravets, Volodymyr Pasichnyk, Nataliia Kunanets,
and Nataliia Veretennikova[✉]

Information Systems and Networks Department,
Lviv Polytechnic National University, Lviv, Ukraine
krpo@i.ua, vpasichnyk@gmail.com,
nek.lviv@gmail.com, nataver19@gmail.com

Abstract. The adaptive game method of event synchronization in multiagent systems in the conditions of uncertainty is developed. The essence of a method consists in alignment of delays of event approach based on action supervision of the next players. The formulation of stochastic game is executed and the game algorithm for its solving is developed. The parameter influences on convergence of a game method are investigated by means of a computer experiment that allows to study the dependence of the training time on the stochastic game of agents from the basic parameters of the algorithm and permits to assert that partial compensation of uncertainty is ensured by the agent ability to self-learning and adaptive decision-making strategies. The obtained work results are used in the construction of multi-agent systems of various purposes, ensuring the work coordination of the components, message transmission between agents, construction of communication protocols, promoting self-organization of multi-agent systems.

Keywords: Multiagent systems · Event synchronization · Stochastic game · Self-learning recurrent method · Uncertainty conditions

1 Introduction

The tasks and challenges faced by people in the modern information society are characterized by considerable complexity, elements of uncertainty, and large needs in material and intellectual resources. To solve them, it is necessary to cooperate with the efforts of many people and involve powerful means of computing. Such cooperation can lead to uncontrolled or chaotic operating modes of systems critically related to the human factor. In this regard, it is important to develop and use intelligent software agents that can work in a computer network with the minimum necessary interaction with an operator, make their own decisions and act on behalf of their owner or developer.

A software agent is an autonomous software system with elements of artificial intelligence that can interact with other agents and people using the resources of the information network during the task solution.

© Springer Nature Switzerland AG 2020
Z. Hu et al. (Eds.): ICCSEEA 2019, AISC 938, pp. 378–387, 2020.
https://doi.org/10.1007/978-3-030-16621-2_35

Agents have the following characteristics:

(1) Autonomy: the agents are completely or partially independent;
(2) Intelligence: the ability to make decisions on their actions in a networked information environment based on analysis of current information;
(3) Specialization: as a rule, agents carry out highly specialized functions;
(4) Decentralization: each agent has no information about the entire system and, in this regard, there are no agents that run the system as a whole;
(5) Location: the decision making by each agent is carried out on the basis of available local information and collaboration between agents within the multi-agent system.

Multi Agent System (MAS) is a system formed by several intelligent agents that interact with each other while solving a task [1–3].

MAS operation, as a rule, is implemented under a priori uncertainty conditioned by internal or external factors [4–6]. Uncertainty of the system can be:

(1) Structural when there is an unknown exact structure of the system and the connection between its elements;
(2) Algorithmic when there is an unknown algorithm for the system functioning;
(3) Informational when there is an obscurity, a lack of complete information necessary for decision-making;
(4) Linguistic when there is an ambiguity of the statement while exchanging messages between agents;
(5) Target when there is an unknown global purpose of the system's operation;
(6) Social which is due to the collective interaction of agents, when the actions of one of the agents affect the solution choice by other agents;
(7) Stochastic when there is an influence on the system of uncontrolled external factors.

Partial compensation of uncertainty is ensured by the ability of agents to self-learning and adaptive decision-making strategies.

MAS is used for distributed task solving of logistics, e-commerce, military affairs, geoinformation systems, emergency response, social event modeling, human resource management, distance learning, scheduling, security and investment management, network management, power system management, for building strategic computer games, business games, professional simulators, network technologies and distributed computing, information search on the Internet, mobile communication organizations and others.

The success of the distributed task solving depends on the level of agent coordination. Coordination is a process of coordinated and arranged operation of all sections of the MAS. Coordination can be centralized or decentralized. The methods of decentralized agent coordination will be more effective with the growth of the structural and functional complexity of the distributed system under conditions of uncertainty.

Coordination is needed to adjust the individual goals and behavior of agents, in which each agent improves or does not worsen the value of its usefulness function, and the system improves the quality of solving a common task. Methods for solving the coordination problem are based on the results of the classical theory of management, the study of operations, the theory of games, planning and the results of other areas of mathematics and cybernetics.

A partial manifestation of coordination is a synchronization of agents. Synchronization is an acquisition of a single work rhythm by the objects of a distributed system. The synchronization may be caused by a weak interaction between the objects or the action of external force. The effect of synchronization was discovered by Christian Huygens in the XVII century as a result of observing the agreed course of two pendulum clocks caused by the weak influence of own oscillations of their joint support.

In multi-agent systems, synchronization is necessary to ensure the coordinated work of their constituents, the message transmission between agents, and the support of self-organization conditions, when the distributed system behaves as a holistic, artificially formed organism. Self-organization is a purposeful process of creation, reproduction, organization or improvement of the organization (structure and functions) of a complex dynamic system through the internal factors without the corresponding external influence [7, 8].

In the theory of synchronization two main parts are distinguished:

(1) The classical theory of synchronization, which studies phenomena in connected periodic self-oscillation systems;
(2) The theory of chaotic synchronization, which studies the cooperative behavior of chaotic systems.

Among chaotic synchronization systems there are three main types:

(1) Full (or identical) synchronization, where the states of linked objects are completely identical;
(2) Generalized synchronization, where object outputs are linked through a function;
(3) Phase synchronization, when the establishment of some relations between the phases of interacting objects, the result of which is a coincidence of their characteristic frequencies or characteristic time scales.

There are the following types of object synchronization:

(1) Forced synchronization of objects using an external source of signals, for example, clock rate generator;
(2) Free spatially-distributed synchronization of objects among themselves.

Event synchronization in MAS is necessary for the formation of a co-ordinated communication between agents in the course of solving their common task for the search of global coherent solutions. Under uncertainty, global event synchronization is achieved using adaptive protocols of local interaction between elements of MAS.

Taking into account the factors of competition, interaction, cooperation, training for the investigation of processes of coordination and synchronization of MAS in conditions of uncertainty, it is advisable to use the model of stochastic game [9].

Game synchronization of events in the MAS is an actual scientific and practical problem that has not been sufficiently studied at the present time. Unlike the synchronization networks of oscillators described by systems of differential equations, the stochastic games of MAS investigate the complex behavior of networks of intellectual agents with various decision-making models in conditions of uncertainty based on artificial intelligence methods. In the stochastic game, self-training agents go to select the optimal strategies (or actions), reconstructing their own vectors of dynamic mixed strategies (or conditional probabilities of action options). Under certain conditions, which are determined by the parameters of the environment, the parameters of the game method and the criteria for choosing solutions, the self-learning of the stochastic game leads to the synchronization of the strategies of the agents.

The **aim** of this work is to construct a stochastic game method of identical pro-partially distributed synchronization of multi-agent systems. In order to achieve the goal, the formulation of the stochastic game task is executed, the method is proposed and an algorithm is developed for its solution, as well as the results of computer simulation of the stochastic game are analyzed.

2   Statement of the Game Task of Event Synchronization

We will consider a set of agents D that can observe the states of each other. Let each agent $i \in D$ signals the implementation of some event at random moments of time $t_1^i, t_2^i, \ldots$ This event is available for observation by neighboring agents from a subset $D_i \in D \, \forall i \in D \, (D = \bigcup_{i \in D} D_i)$ at a time interval $[t_{n-1}^i, t_n^i)$, where $n = 1, 2 \ldots$ indicates the serial number of the time point. In the extreme case $D_i = D \, \forall i \in D$ each agent observes the events of all other agents. By adjusting the amount of time interval $\Delta t_n^i = t_n^i - t_{n-1}^i$ between their own events, agents tend to synchronize events within local subsets D_i. The synchronization essence is to align the time of event occurrence of each agent with the time of the events of its neighbors. For this purpose, agents randomly and independently select one of the pure strategies $U^i = (u^i(1), u^i(2), \ldots, u^i(N))$, which is the current amount of time interval from the moment of the event's implementation t_{n-1}^i:

$$u^i(k) = k\Delta t_{\max}/N, \, k = 1..N,$$

where $\Delta t_{\max}$ is maximum time interval between events.

The method of forming time moments t_n^i defines the kind of stochastic game with synchronous or asynchronous choice of pure strategies. In the synchronous game, all agents simultaneously and independently choose a pure strategy for each iteration $n = 1, 2 \ldots$ In the asynchronous game, a pure strategy is chosen by only agents for whom $t_n^i = \min_{j \in D}(t_n^j)$.

At the time t_n^i, the agent with the number i calculates the average deviation of the time fronts of the events in the subset of neighboring agents D_i from the implementation time of their own event:

$$\delta_n^i = |D_i|^{-1} \sum_{j \in D_i} |t_n^i - t_n^j| + \mu_n^i \, \forall i \in D, \tag{1}$$

where μ_n^i is white noise simulating the inaccuracy of measuring the time of event occurrence.

The calculated deviation is used to form the current game loss $\xi_n^i \in R^1$:

$$\xi_n^i = \delta_n^i. \tag{2}$$

The binary loss $\xi_n^i \in \{0, 1\}$ is determined by the sign of the differential of the current deviation δ_n^i:

$$\xi_n^i = \begin{cases} 0, & if \quad \delta_n^i - \delta_{n-1}^i \leq 0 \\ 1, & if \quad \delta_n^i - \delta_{n-1}^i > 0 \end{cases}. \tag{3}$$

The current losses $\xi_n^i = \xi_n^i(u_n^{D_i})$ are functions of common strategies $u^{D_i} \in U^{D_i} = \underset{j \in D_i}{\times} U^j$ of agents from local subsets $D_i \subseteq D$, $D_i \neq \varnothing \; \forall i \in D$.

Effectiveness of event synchronization is determined by the functions of average losses:

$$\Xi_n^i = \frac{1}{n} \sum_{\tau=1}^{n} \xi_\tau^i \; \forall i \in D. \tag{4}$$

The aim of the game is to minimize the functions of average losses:

$$\overline{\lim_{n \to \infty}} \, \Xi_n^i \to \min \; \forall i \in D. \tag{5}$$

Consequently, based on calculations of random current losses $\{\xi_n^i\}$ players have to choose pure strategies $\{u_n^i\}$ in such way as to ensure that the goal set (5) is achieved in the course of time $n = 1, 2, \ldots$. Depending on the method of sequence formation $\{u_n^i | \forall i \in D, \; n = 1, 2, \ldots\}$ the multi-criteria problem (5) has solutions that satisfy one of the conditions of collective optimality by Nash, Pareto, etc. [10].

To solve the game, it's necessary to build a method for generating pure strategy sequences $\{u_n^i\} \forall i \in D$, so that the synchronization of agent events within local subsets $D_i \in D \; \forall i \in D$ leads to the global synchronization of the events of all agents in the set D.

3   Game Method for Solving the Event Synchronization Task

Generating time intervals between events $\{u_n^i | \forall i \in D, \; n = 1, 2, \ldots\}$ will be done on the basis of discrete dynamic distributions $p_n^i = (p_n^i(u_1^i), p_n^i(u_2^i), \ldots, p_n^i(u_N^i)) \; \forall i \in D$, which in terms of game theory are called mixed strategies. Vectors $p_n^i \in S_\varepsilon^N$ take values on unit ε- simplexes:

$$S_\varepsilon^N = \left\{ p^i \in R^N \middle| p^i(j) \geq \varepsilon \, (j = 1..N), \sum_{j=1}^{N} p^i(j) = 1 \right\}, \; \varepsilon \in (0, \min_i N^{-1}).$$

Let random losses $\{\xi_n^i\}$ are independent $\forall u_n \in U$, $\forall i \in D$, $n = 1, 2, \ldots$, have a constant mathematical expectation $M\{\xi_n^i(u^{D_i})\} = v(u^{D_i}) = const$ and a limited second moment $\sup_n M\{[\xi_n^i(u^{D_i})]^2\} = \sigma^2(u^{D_i}) < \infty$. Stochastic characteristics of random losses are not known to apriori agents.

We will define the function of average losses for the matrix game:

$$V^i(p^{D_i}) = \sum_{u^{D_i} \in U^{D_i}} v^i(u^{D_i}) \prod_{j \in D_i; u^j \in u^{D_i}} p^j(u^j),$$

where $p^{D_i} \in S_{\varepsilon}^{D_i}$, $S_{\varepsilon}^{D_i} = \prod_{j \in D_i} S_{\varepsilon}^N$, $u^{D_i} \in U^{D_i}$.

Taking into account the value of the gradient $\nabla_{p^i} V^i = M\left\{\frac{\xi^i}{e^T(u_n^i)p_n^i} e(u_n^i) \middle| p_n^i = p^i\right\}$, on the basis of the method of stochastic approximation [11, 12] we obtain such self-learning recursive method of changing the vectors of mixed strategies:

$$p_{n+1}^i = \pi_{\varepsilon_{n+1}}^N\left\{p_n^i - \gamma_n \frac{\xi_n^i e(u_n^i)}{e^T(u_n^i)p_n^i}\right\}, \tag{6}$$

where $\pi_{\varepsilon_{n+1}}^{N_i}$ is a projection operator for a single ε- simplex; $\varepsilon_n > 0$ is an expansion parameter of ε- simplex; $\gamma_n > 0$ is a parameter of the learning step; $e(u_n^i)$ is a unit vector indicator of the choice of a pure strategy $u_n^i = u^i$; $e^T(u_n^i)$ is a vector column.

Parameters γ_t and ε_t in (6) determine the rate of agent training. To ensure the convergence of the stochastic game learning process, these parameters are given as positive, monotonically decreasing values:

$$\gamma_t = \gamma_0/t^\alpha, \quad \varepsilon_t = \varepsilon_0/t^\beta, \tag{7}$$

where γ_0, $\alpha > 0$; ε_0, $\beta > 0$.

The conditions for the convergence of the stochastic game gradient method (6) to the equilibrium point by Nash with probability 1 and in the mean square are defined in [13].

Vectors of mixed strategies are used for random selection of solutions:

$$u_n^i = \left\{u_i[k] \middle| k = \arg\left(\min_k \sum_{j=1}^k p_n^i(u_n^i[j]) > \omega\right), k = 1..N\right\} \forall i \in D, \tag{8}$$

where $\omega \in [0, 1]$ is a real random number with a uniform distribution.

Effectiveness of event synchronization is estimated by:

(1) A function of average losses:

$$\Xi_n = |D|^{-1} \sum_{i \in D} \Xi_n^i, \tag{9}$$

where Ξ_n^i is calculated according to (4);

(2) A synchronization level, that is the average number of agents with synchronized events:

$$K_n = \frac{1}{n}\sum_{\tau=1}^{n}\sum_{i\in D}\chi\left(t_\tau^i = t_\tau^{\max}\right),\qquad(10)$$

where $\chi()\in\{0,1\}$ is an indicator event function, $t_\tau^{\max} = \max\limits_{i\in D}(t_\tau^i)$ is a maximum value of the time fronts of the agents.

The Steps of the Method for Solving the Task of Event Synchronization

Step 1. Set the initial parameter values, namely $n = 0$ is an initial time; $L = |D|$ is a number of players; $D_i\ \forall i\in D$ is a set of neighboring players that determine the interaction structure of agents; N is a number of pure strategies of players; $\Delta t_{\max}$ is a maximum value of the time interval between events; $U^i = (u^i(1), u^i(2), \ldots, u^i(N))$, $u^i(k) = k\Delta t_{\max}/N$, $k = 1..N$, $i = 1..L$ are vectors of pure player strategies; $p_0^i(j) = 1/N$, $j = 1..N$, $i = 1..L$ are vectors of mixed player strategies; $\gamma > 0$ is a parameter of the learning step; $\alpha\in(0,1]$ is an order of the learning step; ε is a parameter of ε- simplex; $\beta > 0$ is an order of an ε- simplex expansion rate; $d > 0$ is a noise dispersion; $\mu_n\sim Normal(0,d)$ is a random variable with normal distribution; $n_{\max}$ is a maximum number of the method steps.

Step 2. Perform a random selection of time intervals $u_n^i\in U^i$, $i = 1..L$ between events according to (8).

Step 3. Calculate the following moments of time $t_n^i = t_{n-1}^i + u_n^i$, $i = 1..L$ for beginning of agent events.

Step 4. Get the value of current losses $\xi_n^{\varepsilon i}$, $i = 1..L$ according to (1)–(3).

Step 5. Calculate parameter values γ_n, ε_n according to (7).

Step 6. Calculate the elements of mixed strategy vectors p_n^i, $i = 1..L$ according to (6).

Step 7. Calculate the quality characteristics of decision making Ξ_n (9) and K_n (10).

Step 8. Set the next time moment $n := n + 1$.

Step 9. If there is $n < n_{\max}$, then go to step 2, otherwise it is an end.

4   Results of Computer Simulation

For research, a multiagent model of event synchronization in biological systems was built, for example, the spontaneous emergence of rhythmic light by a group of fireflies from the family Lampyridae. In this model, agents synchronize the change of their states (lighting and damping of fireflies).

Let L agents be placed in a straight line. The distance between the agents may be arbitrary within the limits of the availability of observing signals from neighboring agents. Each agent $i\in D$ records the time moment $t_n^j\ \forall j\in D_i$ of receiving signals from neighboring agents and calculates their current deviation δ_n^i from the generation moment of their own signal t_n^i. For an example, choose a discrete way of forming the current deviation of the time moments:

$$
\delta_n^i = \begin{cases} \left| t_n^{i-1} + t_n^{i+1} - 2t_n^i \right|, & \text{if } i > 1 \text{ and } i < L; \\ \left| t_n^{i+1} - t_n^i \right|, & \text{if } i = 1; \\ \left| t_n^{i-1} - t_n^i \right|, & \text{if } i = L. \end{cases}
$$

The current deviations of the time moments are subjected to white noise μ_n^i, which simulates the measurement error, and is used as current players' losses:

$$
\xi_n^i = \delta_n^i + \mu_n^i.
$$

The value of white noise is calculated as a Gaussian random value with zero mathematical expectation and dispersion $d > 0$:

$$
\mu_n^i = \sqrt{d} \left(\sum_{j=1}^{12} \omega_{j,n} - 6 \right),
$$

where $\omega \in [0, 1]$ is a real random number with a uniform distribution.

The initial time moments of the event occurrence take random values $t_0^i \leq 50$, $i = 1..L$. The game starts with untrained mixed strategies: $p_0^i[j] = 1/N$, $j = 1..N$, $i = 1..L$. The maximum time interval between events in all experiments is equal to $\Delta t_{max} = N$. The simulation of a stochastic game is carried out during $n_{max} = 10^4$ iterations or until a predetermined level $K_n = K$ of agent synchronization is reached.

In Fig. 1 there is a logarithmic scale showing the function graphs of average losses and the average number of synchronized players. Here and further results are obtained for such parameters of the game method, namely $L = 100$, $N = 2$, $\gamma_0 = 1$, $\varepsilon_0 = 0.999/N$, $\alpha = 0.5$, $\beta = 1$, $d = 0.01$ (except for cases when the effect of these parameters on the convergence of the stochastic game is studied).

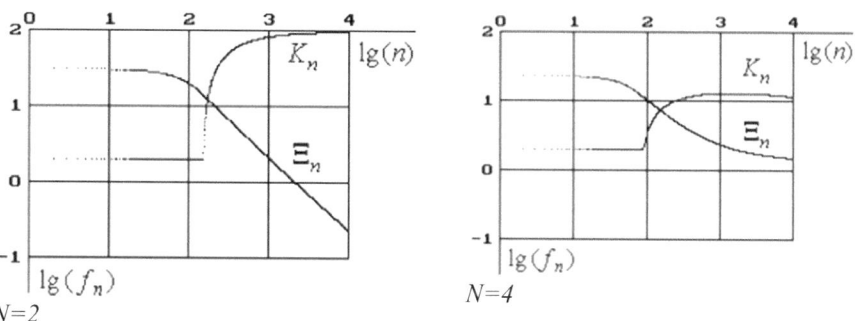

Fig. 1. Stochastic game synchronization characteristics

Reducing the average losses Ξ_n and increasing the number of synchronized agents K_n confirm the convergence of the game method through the implementation of the criteria system (5). As the number of strategies increases N, the number of synchronized agents reduces and the value of the current loss function increases.

In Fig. 2 the time fronts of an event in the MAS for a trained stochastic game of agents $L = 50$ are shown. Agents start the game with different delays t_0^i, $i = 1..L$. Nevertheless, they enter the event synchronization mode with a small amount (~ 100) of steps.

As can be seen in Fig. 2, the result of studying a stochastic game is to equalize the time fronts of the onset of agent events at $N = 2$. The increase in the number of pure strategies (for example, at $N = 4$) leads to the emergence of frontal noise caused by the random choice of time intervals between events.

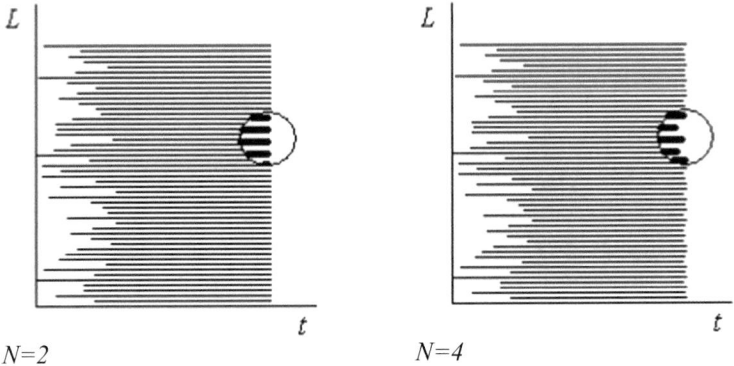

$N=2$ $N=4$

Fig. 2. Time chart for event synchronization

5 Conclusion

Considered game method with proper configuration of its parameters provides synchronization of events in MAS. Theoretically, the values of such parameters should satisfy the basic conditions of stochastic approximation. In practice, the values of parameters that ensure the convergence of the game method to one of the points of collective optimality, can be specified as a result of computer simulation. The conducted researches allow to assert that the game method (6)–(8) provides event synchronization for variants of synchronous and asynchronous choices of pure strategies, variants of non-binary and binary losses.

The results obtained in this paper allow us to determine the optimal values of the parameters of the game method for solving the task of synchronizing MAS events in the shortest time.

The results of this research should be used to coordinate the work of component parts of multi-agent systems of various purposes, transmission of messages between agents, construction of communication protocols, support of conditions for the self-organization of multi-agent systems.

In the proposed MAS synchronization model, the intellectual level of the agents is limited to the possibilities of the theory of stochastic automata with variable structure. Increasing the agent intelligence is possible by attracting artificial intelligence methods, for example, artificial neural networks, Bayesian decision-making networks, fuzzy logic, reinforcement learning, etc.

References

1. Byrski, A., Kisiel-Dorohinicki, M.: Evolutionary Multi-Agent Systems: From Inspirations to Applications, 224 p. Springer (2017)
2. Radley, N.: Multi-Agent Systems – Modeling, Control, Programming, Simulations and Applications, 284 p. Scitus Academics LLC (2017)
3. Rachid, B., Hafid, H.: Distributed monitoring for wireless sensor networks: a multi-agent approach. Int. J. Comput. Netw. Inf. Secur. (IJCNIS) **6**(10), 13–23 (2014). https://doi.org/10.5815/ijcnis.2014.10.02
4. Barabash, O., Shevchenko, G., Dakhno, N., Neshcheret, O., Musienko, A.: Information technology of targeting: optimization of decision making process in a competitive environment. Int. J. Intell. Syst. Appl. (IJISA) **9**(12), 1–9 (2017). https://doi.org/10.5815/ijisa.2017.12.01
5. Amato, C.: Decision-making under uncertainty in multi-agent and multi-robot systems: planning and learning. In: Proceedings of the Twenty-Seventh International Joint Conference on Artificial Intelligence (IJCAI 2018), pp. 5662–5666 (2018)
6. Sahai, A., Sankat, C.K., Khan, K.: Decision-making using efficient confidence-intervals with meta-analysis of spatial panel data for socioeconomic development project-managers. Int. J. Intell. Syst. Appl. (IJISA) **4**(9), 92–103 (2012). https://doi.org/10.5815/ijisa.2012.09.12
7. Demir, O., Lunze, J.: Event-based synchronization of multi-agent systems. In: IFAC Proceeding Volumes, vol. 45, Issue 9, pp. 1–6. Elsevier (2012)
8. Veretennikova, N., Kunanets, N.: Recommendation systems as an information and technology tool for virtual research teams. In: Advances in Intelligent Systems and Computing II, vol. 689, pp. 577–587 (2018)
9. Zhang, W. (ed.): Self-organization: Theories and Methods, 255 p. Nova Science Publishers, USA (2013)
10. Roy, S., Biswas, S., Chaudhuri, S.S.: Nature-inspired swarm intelligence and its applications. Int. J. Mod. Educ. Comput. Sci. (IJMECS) **6**(12), 55–65 (2014). https://doi.org/10.5815/ijmecs.2014.12.08
11. Yin, G., Wang, L.Y., Zhang, H.: Stochastic approximation methods – powerful tools for simulation and optimization: a survey of some recent work on multi-agent systems and cyber-physical systems. In: AIP Conference Proceedings, vol. 1637, p. 1263 (2014)
12. Chakroborty, S., Hasan, M.B.: A proposed technique for solving scenario based multi-period stochastic optimization problems with computer application. Int. J. Math. Sci. Comput. (IJMSC) **2**(4), 12–23 (2016). https://doi.org/10.5815/ijmsc.2016.04.02
13. Ummels, M.: Stochastic Multiplayer Games: Theory and Algorithms, 174 p. Amsterdam University Press (2014)

Optimization of the Method of Technical Analysis of Cryptocurrency Price Differences Movements

Lesya Lyushenko$^{(\boxtimes)}$ ⓘ and Anastasiia Holiachenko

National Technical University of Ukraine
"Igor Sikorsky Kyiv Polytechnic Institute", Pr. Peremohy, 37, Kyiv, Ukraine
lyushenkol@gmail.com,
anastasiia.shapovalova75@gmail.com

Abstract. Cryptocurrencies increasingly cover the world market and grow rapidly. Cryptotrading, namely cryptocurrency arbitrage is one of the most reliable methods of obtaining benefits from this market. This article describes the approaches to processing data on the movements of the cryptocurrency exchange rate in the market, their features and disadvantages. The two main methods of cryptocurrency analysis (fundamental and technical) are described in more details. Based on the stated theses and mathematical statistics, a method of technical analysis of cryptocurrency price differences on the same exchange has been developed to achieve the maximum number of transactions with minimal losses. The method is suitable for both monotonous currency price leaps and trend movements of the currency price, because it adapts automatically and consistently with given parameters to each move in any direction. Based on the created method, some software algorithms have been developed that allow visualization of all related processes. The thorough work was carried out on the optimization of the developed algorithms by the selection of the most relevant parameters. In the article, graphs of the results and pseudo-code of the algorithms and its optimization are presented. The algorithms of modified cryptocurrency arbitrage ensure the maximum profit for a trader.

Keywords: Cryptocurrency · Cryptobot ·
Methods of cryptocurrency analysis · Cryptocurrency volatility ·
Cryptocurrency arbitrage · Mathematical statistics

1 Introduction

As the cryptocurrencies market grows, many people decided to invest their money in Bitcoin [1]. The investors have received a threefold profit (it has gained wide publicity at a rate of about 7,000 per a dollar and in only six months it rose up to 22,000 dollars). A lot of other cryptocurrencies have appeared such as Ethereum, XRP, Litecoin and many others [2, 3]. As the cryptocurrency market expanded rapidly, it became more difficult to work with it. There are two methods of the market analysis: fundamental and technical. Fundamental analysis for cryptocurrencies makes estimates of the rate of its cryptocurrency price to the market indicators at the current moment [4]. The market indicators include:

© Springer Nature Switzerland AG 2020
Z. Hu et al. (Eds.): ICCSEEA 2019, AISC 938, pp. 388–397, 2020.
https://doi.org/10.1007/978-3-030-16621-2_36

- Price volatility
- Trading volumes
- Ongoing transactions
- Project team, its development plan
- Urgency of the problem it solves [3].

Technical analysis of the cryptocurrencies evaluates the cryptocurrency price leaps in the market. It allows to determine general trends or forecast future currency price movements [5].

In this article, the proposed method applied only to the technical analysis. It uses the main principles of cryptocurrency arbitrage. The last one is a type of cryptotrading. Cryptocurrency arbitrage based on getting the profit from cryptocurrency price differences on different exchanges.

The technical analysis makes the assessment of the provided data spreads as accurately and quickly as possible and predicts the profit with a certain set of rules. In this regard, the better the algorithm of crypto-oriented software is grounded, the less risks of the profit losses are. Therefore, the actual task is to improve methods of the cryptocurrency price movements analysis [5, 6].

The novelty of the developed method is the optimization of the "buy-sell" model in order to increase the profits from cryptocurrency price leaps analysis. The optimization comes from dynamic criteria adjustment when deciding on cryptocurrency arbitrage.

2   Guidelines for Working with Cryptocurrency Arbitrage

Cryptocurrency arbitrage is based on getting the profit from cryptocurrency exchange rate differences on different exchanges [7]. In addition, there are more complex financial instruments that use the differences on the same exchange. The main features of cryptocurrency arbitrage are:

- High volatility
- Possible big leaps of cryptocurrency prices on different exchanges
- Temporary delays in transactions (they can change the outcome of the situation in the opposite direction
- Unpredictability of currency price rises or fall.

The cryptocurrency arbitrage is based on Dow's theory. Its postulates were announced in a series of articles published by an American financial journalist, Charles Dow. Considering Dow's theory and technical analysis methods, the three basic principles of cryptocurrency arbitrage operation were obtained:

1. Accidents are excluded. There is always the cause of an incident, otherwise it would not have happened at all.
2. Patterns exist. Each trend showing active growth or fall in a currency price can be affected only by a force comparable to it, thereby changing the direction of movement. Weaker factors can only lead to temporary fluctuations which will not change the general trend indicators.

3. Events are repeated. Most likely, what has already happened in the history of the market can happen again and it is likely that the consequences will be identical to the previous ones.

Optimization of the method of technical analysis of cryptocurrency price differences movements was developed to reflect the guidelines outlined above. The method is suitable for both monotonous currency price leaps and trend movements.

3 Optimization of the Method of Technical Analysis

3.1 The Optimization of "Buy - Sell" Model

In the virtual world, a bot is a program that is configured to perform repetitive actions. The bots for cryptocurrency trading are set to mechanical trading on the exchanges according to specified parameters. The simplest bots buy cryptocurrencies when the course falls and sell when it grows. That is the "buy-sell" model [8]. There are two types of crypto-oriented bots - trading and arbitration.

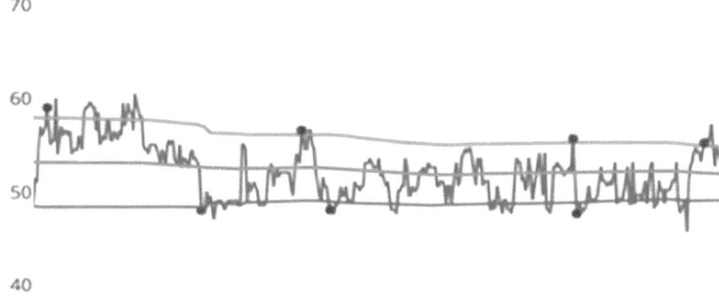

Fig. 1. This chart gives the overview of volatile price difference flows of two different cryptocurrencies in terms of the rate to US dollars on the same exchange. The created method, based on the "buy-sell" concept, should make optimum decisions on making the transactions.

The cryptocurrency trading bots automatically execute trades within a single cryptocurrency exchange, earning income as a result of buying currencies at a low price and selling at a higher one ("Cryptohopper", BTC Robot, etc.).

The cryptocurrency arbitrage bots only trade within several exchanges. Income is earned as a result of buying currency on the exchange where the rate is lower and selling on another exchange where the rate is higher (Gekko, Blackbird).

The method is created for cryptocurrency arbitrage bots. It is based on the "buy-sell" model too, but it earns profit from the movements of the cryptocurrency price differences between two different currencies on the same exchange. It allows a trader to gain the maximum profit. In this case, the profit is made not by the currency price movements, but by the fluctuations of the currencies price differences (Fig. 1).

This chart gives the overview of volatile price difference flows of two different cryptocurrencies in terms of the rate to US dollars on the same exchange. The created method, based on the "buy-sell" concept, should make optimum decisions on making the transactions:

– When crossing the upper dispersion limit in case the last transaction was made after crossing the lower dispersion limit.
– When crossing the lower dispersion limit in case the last transaction was made after crossing the upper dispersion limit.

The red dots on the graph show the places where a certain set of the transaction was performed (4 transactions):

– The first two "buy-sell" transactions close the previous set of transactions;
– The second two "buy-sell" transactions open the new set of transactions.

While crossing the lower dispersion limit the first currency is always bought, and the second one is always sold. While crossing the upper dispersion limit, everything goes another way: the second currency is always bought, and the first currency is always sold. At the completion of each set of transactions (two opening transactions and two closing transactions are implied), the bot makes a profit.

3.2 Data Processing and Analysis

The model uses standard parameters for analysis such as the median, lower and upper dispersion limits [9]. These parameters directly depend on the current currency price movement and will automatically change when a trend line is determined over the last n hours of historical price data (n is a number of hours needed for optimal detection of trends and exclusion of erroneous leaps).

Income data is a set of prices differences during a certain period. This data should be analyzed. Based on the data analysis, optimal additional data sets are generated: median line, and two dispersion lines.

Current median point search.

It is necessary to find a current point of the median line of the differences price movement of two different cryptocurrencies on the exchange.

The following pseudo-code for *current median point:*

```
Input data: data, power level, number of hours (n)
Output:current median point
temp_data: = data for the last n hours from data
sum: = 0
for each temp_el element from temp_data:
temp_el: = temp_el ^ power
sum: = sum + temp_el
length: = number of elements in the temp_data array
result: = sum / length
```

Current dispersion point search.

It is necessary to find a current point of the dispersion line of the differences price movement of two different cryptocurrencies on the exchange.

The following pseudo-code for *current dispersion point*:

```
Input data: data, dispersion rate (rate), number of
hours (n)
 Output: current dispersion point.
 temp_data: = data for the last n hours from data
 sum: = 0
 for each temp_el element from temp_data:
 temp_el: = temp_el ^ 2
 sum: = sum + temp_el
 ength: = number of elements in the temp_data array
 last_meadiana_square: = sum / length
 sum: = 0
 for each temp_el element from temp_data:
 sum: = sum + temp_el
 last_mediana: = (sum / length) ^ 2
 result: = (last_mediana_square - last_mediana) ^ (1/2)
* rate
```

The last pseudo-code can be simplified using the previously mentioned search algorithm of current median point search:

```
 last_mediana_square: = result of the active median
search algorithm (input parameters: data data, power lev-
el: = 2, number of hours n)
 last_mediana: = result of the current median point
search algorithm (input parameters: data data, power lev-
el: = 1, number of hours n)
 last_mediana: = (sum / length) ^ 2
 result: = (last_mediana_square - last_mediana) ^ (1/2)
*rate
```

The following pseudo-code is created to generate the full median set:

```
Input data: data, number of hours (n)
 Output: full median set
 mean_list: = empty array
 k: = 0 for each el element from data:
 k: = k + 1
 temp_data: = data for the last k hours from data
 add an element to the mean_list (the result of the cur-
rent median point search algorithm (input parameters: da-
ta temp_data, power: = 1, number of hours n)
 result_list: = mean_list
```

The algorithm created to generate the full dispersion set works similarly:

```
Input data: data, dispersion rate (rate), number of
hours (n)
Output: full dispersion set
disp_list: = empty array
k: = 0
for each el element from data:
k: = k + 1
temp_data: = data for the last k hours from data
add an element to the mean_list (the result of the ac-
tive scatter search algorithm (input parameters: data
temp_data, coefficient koef : = rate, number of hours n)
result_list: = disp_list
last_mediana: = (sum / length) ^ 2
result: = (last_mediana_square - last_mediana) ^ (1/2)
* rate
```

Data processing and analysis have to be performed according to the following algorithm:

- Generate the full median set from given data;
- Generate the full dispersion set from given data;
- Divide the full dispersion set into two sets (upper dispersion line and lower dispersion line);
- Each time a new data point is added, the last points of each set are recalculated online.

3.3 Optimization of Algorithm Parameters

In subsect. 3.2 the algorithms for automatic finding and updating the median line set and dispersion lines set (at each point for the last n hours) are described. To start them, two additional parameters are required:

- The dispersion rate;
- The number of hours which determines the difference price movement trend. It was decided to create an algorithm which optimizes these parameters to obtain the maximum profit.

The dispersion rate defines how close the upper and lower dispersion lines are to the median line. It directly affects the number of transactions. The lower it is, the more often they will occur. There are commissions that exchanges get from each transaction. If the number of transactions is too big, and at the same time the profit brought for each transaction is too small, then there may not be made a profit at all [9, 10]. The number of hours defines the latest trend in the chart. It is a decisive factor, in case the trend is determined incorrectly, the transaction is going to be unprofitable at all.

The range of dispersion rates and the number of hours that could be relevant in this algorithm were selected and combined into two data sets.

The concept of this algorithm is to run each possible combination of parameters from two data sets mentioned above and get the new profit set. When a cycle is started, the dispersion rate is fixed it runs with each value from the selected range of the number of hours with a certain step. After receiving the profit result, the algorithm works with the next value from dispersion rate set with a certain step and again runs with each value from number of hours set with a certain step. This algorithm works until the dispersion rate set is proceed.

3.4   Visualization of the Results

There were generated the following charts to visualize the algorithm results:

- The accumulative income chart (Fig. 2);
- The drawdown chart (Fig. 3);
- The statistical chart with all data sets and transaction marks (Fig. 4).

The figures show the examples of results after analyzing difference price data between two currencies XBTUSD and XBTU18 on the same exchange (Bitmex) for 1 day. Also, an example of the optimization algorithm graph (from Subsect. 3.3) for the same period of time is given (Fig. 5).

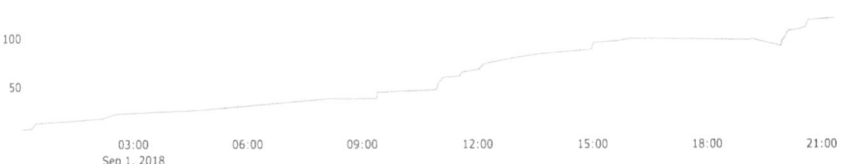

Fig. 2. The example of accumulative graph of profits for 1 day of differences between XBTUSD and XBTU18 with such parameters as a coefficient of 1 and 6 h of capture.

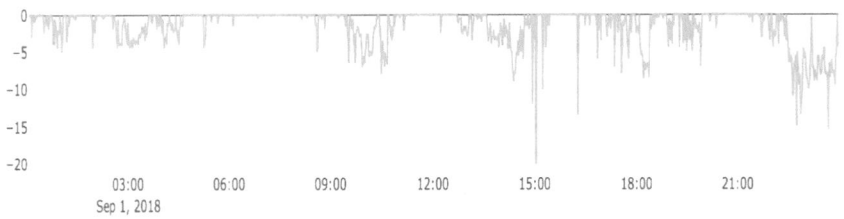

Fig. 3. The example of drawdown chart for 1 day of differences between XBTUSD and XBTU18 with such parameters as a coefficient of 1 and 6 h of capture.

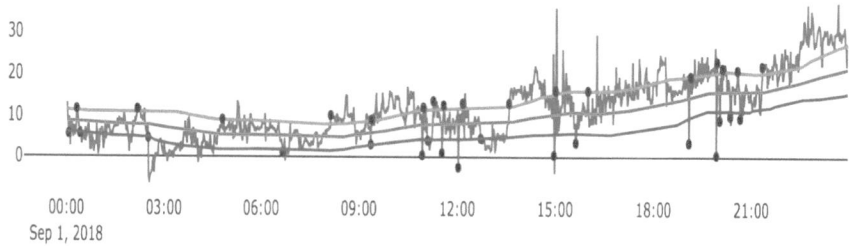

Fig. 4. The example of drawdown graph for 1 day of differences between XBTUSD and XBTU18 with such parameters as a coefficient of 1 and 6 h of capture. The blue line is the actual value of the differences, the orange line is the median, the green line is the upper modified dispersion, the purple line is the lower modified dispersion, the red dot is the completion of a set of operations.

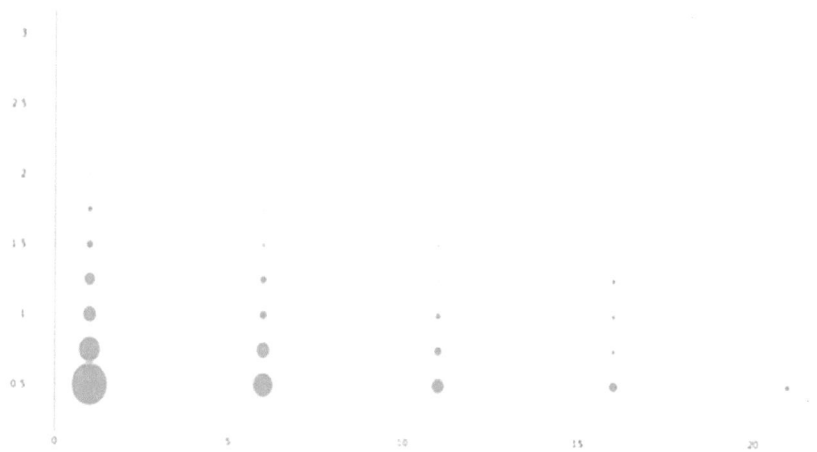

Fig. 5. The result of optimization algorithm run on 1-day difference prices data between XBTUSD and XBTU18 with such parameters as a dispersion rate range from 0.5 to 3, step 0.25, and a range of hours from 1 to 21, step 5 h.

Optimization algorithm (from Subsect. 3.3) generates the new profit data sets running each possible combination of parameters from. Table 1 is providing an example for the display: the profit data sets obtained from the optimization algorithm on 1-day (difference prices data between XBTUSD and XBTU18 with such parameters as a dispersion rate range from 0.5 to 3, step 0.25, and a range of hours from 1 to 21, step 5 h). The result table shows that the biggest profit can be earned by dispersion rate 0,5- and 1-h time period (Table 1). The visualized version of this table is also provided (Fig. 5).

Table 1. The generated profit set of optimization algorithm on 1-day.

Dispersion rate\hours	1	6	11	16	21
0,5	492,82	286,58	184,68	133,04	84,6
0,75	291,53	192,43	96,61	21,87	32,55
1	186,2	122,65	75,98	27,24	5,64
1,25	147,97	125,54	59,65	42,65	0
1,5	105,49	86,97	42,22	14,59	0
1,75	89,88	54,53	39,54	6,44	0
2	32,36	16,76	15,8	0	0
2,25	25,79	0	7,22	0	0
2,5	17,2	0	0	0	0
3	0	0	0	0	0

Consequently, for maximum profit with minimum risks, the operations should be implemented fully:

- Run the optimization algorithm from Subsect. 3.3 for each currency pair and set optimal parameters;
- Run the algorithm from Subsect. 3.2.

4   Conclusions

As described in the article, the following results were achieved:

- An optimized "buy-sell" model of a cryptocurrency bot was created to obtain the maximum profit with minimum financial investments;
- A method of technical analysis was developed and implemented as an algorithm for cryptocurrency arbitrage bots;
- The algorithm was optimized for a number of parameters to obtain the maximum profit. The most important parameters are the dispersion coefficient and time range.
- The developed method gets the profit from the movements of differences between two currencies on the same exchange. This allows the trader to extract the maximum profit with minimum money spent because in this case, the profit is obtained not from the currency movement but from the leaps of currencies differences.

References

1. Umair, A., Ashfaq, U., Khan, M.G.: Recent trends, applications, and challenges of brain-computer interfacing (BCI). Int. J. Intell. Syst. Appl. (IJISA) **9**(2), 58–65 (2017). https://doi.org/10.5815/ijisa.2017.02.08
2. Waves: Waves Platform - Blockchain for the people. https://wavesplatform.com/. Accessed 06 Nov 2018

3. Coleman, L.: Bitfinex's Lesson: Has the Time for Regulation Arrived?. Cryptocoins News. https://www.cryptocoinsnews.com/bitfinexs-lesson-has-the-time-for-regulation-arrived/. Accessed 07 Nov 2018
4. Investors Underground: Intoduction to Technical Analysis. https://www. investorsunderground.com/technical-analysis/. Accessed 08 Nov 2018
5. Thu, T.N.T., Xuan, V.D.: Supervised support vector machine in predicting foreign exchange trading. Int. J. Intell. Syst. Appl. (IJISA) **10**(9), 48–56 (2018). https://doi.org/10.5815/ijisa.2018.09.06
6. Hansen, J.D.: Virtual Currencies: International Actions and Regulations. Perkins Coie. https://www.perkinscoie.com/en/news-insights/virtual-currencies-international-actions-and-regulations.html. Accessed 05 Nov 2018
7. Kaur, J., Kaur, K.: Internet of Things: a review on technologies, architecture, challenges, applications, future trends. Int. J. Comput. Netw. Inf. Secur. (IJCNIS) **9**(4), 57–70 (2017). https://doi.org/10.5815/ijcnis.2017.04.07
8. Voorhees, E.: The True Cost of Bitcoin Transactions. http://bitcoin.xyz/erik-voorhees-true-cost-bitcoin-transactions-2/. Accessed 08 Nov 2018
9. Al-Maqaleh, B.M., Al-Mansoub, A.A., Al-Badani, F.N.: Forecasting using artificial neural network and statistics models. Int. J. Educ. Manag. Eng. (IJEME) **6**(3), 20–32 (2016). https://doi.org/10.5815/ijeme.2016.03.03
10. Blockchain.info: Bitcoin Wallet – Blockchain. https://blockchain.info/wallet/. Accessed 08 Nov 2018

Statistical Research of Efficiency of Approximation Algorithms for Planning Processes Automation Problems

Alexander Anatolievich Pavlov$^{(\boxtimes)}$ ⓘ, Elena Borisovna Misura ⓘ,
Oleg Valentinovich Melnikov ⓘ, Iryna Pavlovna Mukha ⓘ,
and Kateryna Igoryvna Lishchuk ⓘ

National Technical University of Ukraine "Igor Sikorsky Kyiv Polytechnic
Institute", 37, Prospekt Peremohy, Kyiv 03056, Ukraine
pavlov.fiot@gmail.com, elena_misura@ukr.net,
lishchuk_kpi@ukr.net, oleg.v.melnikov72@gmail.com,
mip.kpi@gmail.com

Abstract. We study the efficiency of polynomial approximation algorithms for solving NP-hard scheduling problems to minimize total tardiness and total weighted tardiness on one machine. The algorithms are based on PSC-algorithms for the considered problems given in [Zgurovsky & Pavlov, Combinatorial Optimization Problems in Planning and Decision Making: Theory and Applications, Springer, 2019]. They were developed by excluding procedures related to exponential enumeration. The approximation algorithms have a polynomial complexity and can solve problems of any practical dimension. We give a method for determining the maximum possible deviation of the approximate solution of the total tardiness problem from the optimum for each individual problem instance. We present experimental data on the solving time and the actual percentage of the deviation of the functional value from the optimum. The deviation was less than 5% after the execution of the approximate algorithm for the weighted tardiness problem and less than 2.4% for the total tardiness problem. This confirms the high efficiency of the approximation algorithms.

Keywords: Planning · Process automation · Scheduling ·
Combinatorial optimization · Approximation algorithm · PSC-algorithm ·
Efficiency research · Total tardiness · Total weighted tardiness

1 Introduction

Scheduling problems are found in very wide applications during planning process automation in objects of various nature in different spheres of human activity: in production processes planning, aircraft manufacturing and shipbuilding, project management etc. Almost all of them are NP-hard problems of combinatorial optimization. This complicates the development of not only exact, but also efficient approximation algorithms.

State-of-the-art scheduling methods are described in the book [1] and in papers [2–4]. A review of efficient methods for exact exponential algorithms construction can be

© Springer Nature Switzerland AG 2020
Z. Hu et al. (Eds.): ICCSEEA 2019, AISC 938, pp. 398–408, 2020.
https://doi.org/10.1007/978-3-030-16621-2_37

found in [5]. Some new scheduling methods are considered in [6–9]. Parallel and cloud computing [10] and scheduling with restrictions on resources [11] are intensively developing fields of research.

The purpose of this paper is the efficiency research of approximation algorithms for solving two scheduling problems:

- NP-hard in the strong sense problem of the total weighted tardiness of jobs minimization on a single machine (TWT) [12];
- Its particular case: NP-hard problem of the total tardiness of jobs minimization on a single machine (TT) [13].

The two problems are important for discrete-type production planning. In particular, the TWT problem is used at the fourth level of the four-level model of planning (including operative planning) and decision making presented in [14]. This model can be used for any objects with a network structure of technological processes and limited resources. Optimization by tardiness criteria is crucial for just-in-time planning systems and other important planning models. The TT problem is used, for example, for planning process automation in innovative software development and to determine an efficient portfolio of orders, in terms of the expected profit, and automation of its execution planning process [15].

The TWT Problem Statement [12]. Given a set of independent jobs $J = \{j_1, j_2, \ldots, j_n\}$, each job consists of a single operation. For each job j, we know its processing time $l_j > 0$, weight coefficient $\omega_j > 0$ and due date d_j. All jobs become available at the time point zero. Interruptions are not allowed. We need to build a schedule for one machine that minimizes the total weighted tardiness of the jobs:

$$f = \sum_{j=1}^{n} \omega_j \max\left(0, C_j - d_j\right) \tag{1}$$

where C_j is the completion time of a job j.

This problem is studied from 1960s. The best of available state-of-the-art exact dynamic programming algorithms for TWT problem is limited by dimension of 300 jobs [16]. However, exact methods are inefficient for large dimensions due to the exponential increase in complexity. Therefore, various metaheuristic approaches have been developed, e.g. [17]: simulated annealing [18], tabu-search [19], variable neighborhood descent [20]. The best approximation algorithm we know for TWT problem solving is given by Ding et al. [17]. However, all of these algorithms also require huge computation time on large dimensions. In particular, the algorithm from [20] solves 1000 jobs problems in several weeks or months [21]. The theory of PSC-algorithms [22] is an alternative direction allowing to solve NP-hard problems in a general formulation exactly and efficiently. It allows for problems several times larger in size and also for the parallelization of computations. When sufficient signs of optimality of current solutions are fulfilled, PSC-algorithms allow to obtain an exact solution by a polynomial subalgorithm. Otherwise, we obtain an efficient exact solution by an exponential subalgorithm or an approximate solution with an estimate of the deviation from the optimum for practical problem instances of large size (up to tens of thousands of jobs).

In this paper, we study the efficiency of approximation algorithms for TWT [12] and TT [13] problems solving. The algorithms are based on PSC-algorithms and were developed by excluding procedures related to exponential enumeration. The approximation algorithms have a polynomial complexity and can solve problems of any practical dimension.

The rest of the paper is organized the following way. We give a brief description of the PSC-algorithms in Sect. 2. The structure of the approximate algorithms we study and the heuristics they are constructed on are in Sect. 3. In Sect. 4, we derive a theoretical estimate of the maximum possible decrease in the functional value after executing a specified number of iterations of the exact PSC-algorithm for TT problem. This allowed us to obtain an estimate of the maximum possible deviation from the optimum for the approximation algorithm of TT problem solving. In Sect. 5, we experimentally research the approximation algorithms showing the solving time and the percentage of deviation of the obtained functional value from the optimum. We summarize the findings in Sect. 6.

2 Basic Provisions for the PSC-Algorithms for TWT and TT Problems Solving

We now excerpt from [12] some definitions necessary to understand the paper. Let $j_{[g]}$ denote the number of a job occupying a position g in a schedule. Let $\overline{p,q}$ mean the interval of integer numbers from p to q: $\overline{p,q} = p, p+1, \ldots, q$.

Definition 1. We call $R_{j_{[g]}} = d_{j_{[g]}} - C_{j_{[g]}} > 0$ the time reserve of a job $j_{[g]}$.

Definition 2. A priority $p_{j_{[i]}}$ of a job $j_{[i]}$ is the value of $\omega_{j_{[i]}} / l_{j_{[i]}}$.

Definition 3. We call a job $j_{[g]}$ tardy in some sequence if $d_{j_{[g]}} < C_{j_{[g]}}$.

Definition 4. A permutation of a job $j_{[g]}$ is its move into a later position $k > g$ with the shift of jobs in positions $\overline{g+1,k}$ to the left by one position.

Definition 5. Interval of permutation of a job $j_{[g]}$ into a position k is defined by the sum of the processing times of jobs in positions $\overline{g+1,k}$.

Definition 6. An insertion of a job $j_{[g]}$ is its move into an earlier position $p < g$ with the shift of jobs in positions $\overline{p,g-1}$ to the right by one position.

Definition 7. Interval of insertion $I_{j_{[g]}}$ of a job $j_{[g]}$ into a position $p < g$ is defined by the sum of the processing times of jobs in positions $\overline{p,g-1}$. The position p is defined by condition

$$\sum_{i=p-1}^{g-1} l_{j_{[i]}} < C_{j_{[g]}} - d_{j_{[g]}} \le \sum_{i=p}^{g-1} l_{j_{[i]}}, \tag{2}$$

and if (2) is false for every position, then $p = 1$. Thus, the tardiness of the job in the position p must be zero or minimal.

Definition 8. The ordered sequence σ^{ord} is a sequence of all jobs $j \in J$ in non-increasing order of priorities p_j, i.e., $\forall j, i, j < i : p_j \geq p_i$, and if $p_j = p_i$, then $d_j \leq d_i$.

Definition 9. A free permutation of a non-tardy job $j_{[k]}$ in the sequence σ^{ord} is its permutation into such later position $q > k$ that $C_{j_{[q]}} \leq d_{j_{[k]}} < C_{j_{[q+1]}}$ if there is at least one tardy job in positions $\overline{k+1, q}$.

Definition 10. Sequence σ^{fp} is the sequence obtained after all free permutations in the ordered sequence σ^{ord}.

Definition 11. A tardy job $j_{[g]}$ in the sequence σ^{fp} is called a competing job if we have $d_{j_{[l]}} > d_{j_{[g]}} - l_{j_{[g]}}$ for at least one non-tardy preceding job $j_{[l]}$ in this sequence.

Competing jobs in the sequence σ^{fp} are in non-increasing order of their priorities.

Definition 12. The time reserve of a job $j_{[l]}$ in a current sequence is called productive reserve if at least one competing job $j_{[k]}$ in this sequence satisfies the inequality $d_{j_{[l]}} > d_{j_{[k]}} - l_{j_{[k]}}$.

A special case of the TWT problem is the TT problem. In it, the weights of all jobs are equal to one, so priority of job j is the value of $1/l_j$.

PSC-algorithms for TWT and TT problems are based on jobs permutations and consists of a series of uniform iterations. The number of iterations is determined by the number of competing jobs in the sequence σ^{fp}. At each iteration, we check the possibility to use the time reserves of preceding jobs for the current competing job from the sequence σ^{fp}. Such way, we construct an optimal schedule for jobs of the current subsequence which is bounded with the current competing job. We can eliminate some jobs from the set of competing jobs during the solving process.

We perform the following at each current iteration of the PSC-algorithms:

1. Tardy jobs use the existing reserves of non-tardy jobs. We do it by:

   - Determination and extension of the interval of insertion;
   - Performing the insertion;
   - Ordering jobs by priority at the interval of insertion;
   - Step-by-step optimization for each tardy job at the interval of insertion.

2. Tardy jobs use the reserves released by jobs which previously used them. We do it:

   - Permutating these jobs into a later position;
   - By recursive step-by-step optimization for each tardy job at the interval of permutation.

As a result of each iteration of optimization (that has an exponential complexity), the functional value decreases or remains unchanged.

3 Structure of Approximation Algorithms for the Problems

Approximation algorithms for TWT [12] and TT [13] problems solving were constructed on the basis of the PSC-algorithms of their solving taking into account the following heuristics [12].

Heuristic 1. We perform the insertion procedures only for competing jobs. We eliminate the procedure of priority ordering at the intervals of insertion. We only shift the jobs within the interval of insertion into later positions. We perform for each tardy job within the interval of insertion only permutations that decrease the functional value and have polynomial complexity.

Heuristic 2. We exclude the recursive step-by-step optimization. It is replaced with permutations that decrease the functional value and have polynomial complexity.

The structure of Approximation Algorithm AA [12] is as the following:

1. Specify a desired restriction on the execution time of the algorithm.
2. Build the sequence σ^{fp} and determine the set of competing jobs.
3. Execute the maximum number of iterations of the exact PSC-algorithm satisfying the imposed restriction on the execution time. If we have obtained a solution in this time, the algorithm terminates. Otherwise, we continue solving process with the approximation algorithm, that is, with the simplified PSC-algorithm taking into account Heuristics 1 and 2.

The structure of the approximation algorithm for TT problem solving is similar, with the following main difference [13]. According to Definition 8, competing jobs in the sequence σ^{fp} are arranged in non-decreasing order of their processing times. This has allowed to simplify the rules for eliminating unpromising permutations and insertions and to reduce their number. Also, the rules for eliminating some jobs from the set of competing jobs have been formulated and justified [13].

4 Estimation of the Deviation of the Approximate Solution from the Optimum for TT Problem

4.1 Construction of an Estimate of the Maximum Possible Decrease in the Functional Value in the Current Sequence After Executing a Specified Number of Iterations of the Exact PSC-Algorithm

Suppose that we have obtained a sequence σ^{cur} as a result of executing a specified number of iterations of the exact algorithm. We need to estimate the maximum possible decrease in the functional value (1), during the optimal schedule construction by the exact PSC-algorithm, for each unconsidered competing job in this sequence.

Algorithm for Obtaining the Estimate of Deviation has the following 6 steps.

1. Partition the set of competing jobs from the sequence σ^{fp} into such subsequences σ^k, $k = \overline{1, r}$, that $\forall \sigma^k \ \forall j_{[i]} \in \sigma^k \ C_{j_{[i]}}\left(\sigma^{fp}\right) > d_{j_{[i]}}$.
2. Select from σ^{cur} the next competing job j_s for which we have not completed the iteration of optimization. If all competing jobs were considered, go to step 6.

3. Determine a subsequence σ^k, $k = \overline{1,r}$, of the sequence σ^{fp}, which the job j_s belongs to. Let it be a subsequence σ^t.
4. Analyze the decrease in the functional value for job j_s based on a comparison with the decrease in the functional value for the previous competing job j_g from the subsequence σ^t for which we performed the last iteration of optimization.

 - If such job j_g is not found (j_s is the first competing job in the subsequence σ^t), then the maximum possible decrease in the functional value for job j_s during the optimal schedule construction is equal to its tardiness in the current sequence σ^{cur}:

$$\Delta f(j_s) = C_{j_s} - d_{j_s}.$$

 - If a job j_g is found, then
$$\Delta f(j_s) = \begin{cases} \Delta f(j_g) & \text{if } d_{j_g} \leq d_{j_s}; \\ \Delta f(j_g) + \left(d_{j_g} - d_{j_s}\right) & \text{if } d_{j_g} > d_{j_s}. \end{cases}$$

5. Determine the actual decrease in the functional value for job j_s as the minimum $\min\left(\Delta f(j_s), C_{j_s} - d_{j_s}\right)$ where the tardiness of job j_s is defined in the sequence σ^{cur}. Go to step 2.
6. The maximum possible decrease in the functional value for the current sequence σ^{cur} is

$$\Delta f_{\max}(\sigma^{cur}) = \sum_{i \in K'} \Delta f(j_i)$$

where K' is the set of competing jobs from the subsequence σ^{cur} for which we did not perform an iteration of optimization.

The described algorithm for constructing an estimate of deviation is based on the fact that competing jobs in the sequence σ^{fp} are arranged in non-decreasing order of their processing times.

A theoretical estimate of the maximum possible deviation of the functional value in the current sequence after the sequence σ^{fp} construction is considered in [23].

4.2 Estimation of the Maximum Possible Deviation of the Functional Value from the Optimum After Executing the Approximation Algorithm for TT Problem Solving

The estimate of the maximum possible deviation from the optimum of the functional value obtained by the Approximation Algorithm is equal to

$$\Delta_{\max}^{AA} = f^{AA} - \left(f(\sigma^{cur}) - \Delta f_{\max}(\sigma^{cur})\right)$$

where f^{AA} is the functional value in the schedule obtained by the Approximation Algorithm AA and $f(\sigma^{cur})$ is the functional value in the sequence σ^{cur} obtained after executing the iterations of the exact PSC-algorithm.

5   Computational Studies of the Approximation Algorithms

To carry out the computational studies of efficiency of approximation algorithms for TWT [12] and TT [13] problems, we used:

- Benchmark instances of 40...300 jobs dimensions from [24] for TWT problem;
- Benchmark instances of 100...300 jobs dimensions from [25] for TT problem.

In both cases, the set of benchmark instances was generated using standard Fisher's generation scheme [26]. The complexity of the problem instances and the time of their solving depends on two parameters: the due date range R and the tardiness factor T. For each instance, processing times and weights of jobs were taken from a uniform distribution over the interval [1, 100] and [1, 10], respectively. Then the due dates of jobs were generated randomly from a uniform distribution over the interval $[L \cdot (1 - T - R/2), L \cdot (1 - T + R/2)]$ where L is the sum of processing times of all jobs. The values of R and T were chosen from the set {0.2, 0.4, 0.6, 0.8, 1.0} (case $T = 1.0$ was not considered for TT problem). This gives 25 combinations of R and T parameters for the TWT problem and 20 combinations for TT problem. For each pair of R and T, five instances were generated for the TWT problem (125 test instances for each dimension) and ten instances for TT problem (200 instances for each dimension). The total number of test instances is 1675.

We solved all the instances by a corresponding approximation algorithm on a personal computer with 2 GBytes of RAM and a Pentium IV processor that has 3.0 GHz frequency.

We show in Table 1 the average solving time in seconds by the Approximation Algorithm AA for both problems, depending on the dimension, without use of the iterations of exact PSC-algorithms (without specifying a time limit). We also give the comparison with the approximation algorithm of Ding et al. [17].

Table 1. Average solving time (in seconds) by approximation algorithm without iterations of the exact PSC-algorithm and comparison with the approximation algorithm of Ding et al.

n	TWT problem: AA	TWT problem: BDS	TT problem: AA
40	0.003	0.003	$-^*$
50	0.005	0.008	–
100	0.021	0.06	0.019
150	0.057	8.77	0.049
200	0.098	54.39	0.083
250	0.181	128.80	–
300	0.286	251.10	0.257

AA Approximation algorithm AA [12, 13], Pentium IV, 3.0 GHz, 2 GBytes of RAM; *BDS* approximation algorithm of Ding et al. [17], Intel Xeon E5440, 4 cores, 2.83 GHz, 2 GBytes of RAM
* The OR-library [25] does not contain instances of the TT problem for this dimension

We can state that the Approximation Algorithm AA works significantly faster than the approximation algorithm of Ding et al. [17]. We also investigated it on generated instances of 10,000 jobs dimension. The average computation time was approximately 1 h 17 min for the TWT problem and 1 h 9 min for TT problem.

We also calculated the percentage of deviation Δ_{dev} of the functional value in the resulting sequence σ from the known optimal solution f_{opt} for each test instance in the following way:

$$\Delta_{dev} = \frac{f(\sigma) - f_{opt}}{f(\sigma)} \times 100\%.$$

Instances of any dimension are solved more quickly (often significantly) for some values of R and T parameters than in other regions. Therefore, now we show the percentage of deviation Δ_{dev} for TWT and TT problems broken down by the values of R and T parameters. Table 2 (for TWT problem) and Table 3 (for TT problem) show average percentage of deviation from the optimum depending on R and T after executing iterations of exact algorithm with a two minutes time limit. Table 4 (for TWT problem) and Table 5 (for TT problem) show average percentage of deviation from the optimum depending on R and T after continuing the approximation algorithm after the iterations of exact algorithm. For both the problems, the percentage of deviation is zero for $n \leq 200$ because all instances were solved by the iterations of exact algorithm.

Table 2. Average percentage of deviation from the optimum depending on R and T after executing iterations of exact algorithm with a two minutes time limit: TWT problem

T	0.2	0.4	0.6	0.8	1.0	0.2	0.4	0.6	0.8	1.0
R	$n = 250$					$n = 300$				
0.2	0	0	0.64	0	0	0	3.39	2.15	0.19	0.01
0.4	0	0	2.35	0	0	0	4.56	4.11	1.03	0.18
0.6	0	0	6.74	0	0	0	0	3.32	1.94	0.36
0.8	0	0	7.01	0	0	0	0	4.39	1.80	0.64
1.0	0	0	7.85	1.37	0	0	0	7.98	2.48	0.71

Table 3. Average percentage of deviation from the optimum depending on R and T after executing iterations of exact algorithm with a two minutes time limit: TT problem

T	0.2	0.4	0.6	0.8
R	$n = 300$			
0.2	0	1.84	0.59	0
0.4	0	0.38	1.23	0.08
0.6	0	0	1.73	0.04
0.8	0	0	0.72	0.03
1.0	0	0	3.88	0.05

Table 4. Average percentage of deviation from the optimum depending on R and T after executing iterations of exact and approximation algorithm: TWT problem

T	0.2	0.4	0.6	0.8	1.0	0.2	0.4	0.6	0.8	1.0
R	$n = 250$					$n = 300$				
0.2	0	0	0.37	0	0	0	1.94	1.25	0	0
0.4	0	0	1.39	0	0	0	2.81	2.42	0.62	0
0.6	0	0	4.05	0	0	0	0	1.99	1.18	0.20
0.8	0	0	4.27	0	0	0	0	2.67	1.11	0.36
1.0	0	0	4.86	0.86	0	0	0	4.63	1.55	0.41

Table 5. Average percentage of deviation from the optimum depending on R and T after executing iterations of exact and approximation algorithm: TT problem

T	0.2	0.4	0.6	0.8
R	$n = 300$			
0.2	0	1.11	0.36	0
0.4	0	0.25	0.77	0.05
0.6	0	0	1.10	0
0.8	0	0	0.46	0
1.0	0	0	2.39	0

We see from Tables 2, 3, 4, and 5 that the exact algorithm solved to optimality about 80% of test instances in 2 min for 250 jobs dimension (TWT problem) and about 45% of 300 jobs instances for both TWT and TT problems, depending on the values of the R and T parameters. In the region of "more complex" instances, after executing iterations of exact algorithm, we obtained approximate solutions with an actual percentage of deviation from the known optimum of less than 8% for TWT problem and less than 4% for TT problem. In the same region, after continuing the execution of the approximation algorithm, we have got a solution with less than 5% deviation for TWT problem and less than 2.4% for TT problem. At the same time, we have solved to optimality about 9% of the remaining instances by the approximation algorithm.

We also generated "very difficult" instances of the problems according to the theoretical conditions from [12] satisfying of which approaches the solution time to that of an exhaustive search. The deviation of the functional value from the optimum for these problems, obtained by the Approximation Algorithm AA, was very small (less than 1%).

Thus, the statistical studies confirm the high efficiency of the Approximation Algorithm AA for both TWT [12] and TT [13] problems.

6 Conclusions

We give basic principles and theoretical foundations of PSC-algorithm and approximation algorithms for TWT [12] and TT [13] problems solving and show differences between the algorithms. We derive an estimate for the maximum possible deviation from the optimum of the solution obtained by the approximation algorithm for TT problem solving. We carry out a computational analysis of the solving time for the problems instances for the approximation algorithm (without execution of iterations of the exact algorithm). Also, we analyze the actual percentage of the deviation of the functional value from the optimum.

We show that the exact algorithm from [12] optimally solves about 80% of the benchmark instances in 2 min for 250 jobs dimension (TWT problem) and about 45% of 300 jobs instances for both TWT and TT problems. In the region of "more complex" instances, after the approximation algorithm execution, the actual deviation was less than 5% for TWT problem and less than 2.4% for TT problem. The average computation time for 10,000 jobs instances was approximately 1 h 17 min for the TWT problem and 1 h 9 min for TT problem.

Our findings confirm the high efficiency of the Approximation Algorithm AA [12] both for TWT and TT problems. We recommend to use the Approximation Algorithm AA to solve instances of a real practical dimension (up to tens of thousands of jobs), as well as in solving "difficult" cases of the problems of a smaller dimension.

References

1. Pinedo, M.L.: Scheduling: Theory, Algorithms, and Systems. Springer, Cham (2016). https://doi.org/10.1007/978-3-319-26580-3
2. Heydari, M., Hosseini, S.M., Gholamian, S.A.: Optimal placement and sizing of capacitor and distributed generation with harmonic and resonance considerations using discrete particle swarm optimization. Int. J. Intell. Syst. Appl. (IJISA) 5(7), 42–49 (2013). https://doi.org/10.5815/ijisa.2013.07.06
3. Wang, F., Rao, Y., Wang, F., Hou, Y.: Design and application of a new hybrid heuristic algorithm for flow shop scheduling. Int. J. Comput. Netw. Inf. Secur. (IJCNIS) 3(2), 41–49 (2011). https://doi.org/10.5815/ijcnis.2011.02.06
4. Cai, Y.: Artificial fish school algorithm applied in a combinatorial optimization problem. Int. J. Intell. Syst. Appl. (IJISA) 2(1), 37–43 (2010). https://doi.org/10.5815/ijisa.2010.01.06
5. Fomin, F.V., Kratsch, D.: Exact Exponential Algorithms. Springer, Heidelberg (2010). https://doi.org/10.1007/978-3-642-16533-7
6. Soltani, N., Soleimani, B., Barekatain, B.: Heuristic algorithms for task scheduling in cloud computing: a survey. Int. J. Comput. Netw. Inf. Secur. (IJCNIS) 9(8), 16–22 (2017). https://doi.org/10.5815/ijcnis.2017.08.03
7. Garg, R., Singh, A.K.: Enhancing the discrete particle swarm optimization based workflow grid scheduling using hierarchical structure. Int. J. Comput. Netw. Inf. Secur. (IJCNIS) 5(6), 18–26 (2013). https://doi.org/10.5815/ijcnis.2013.06.03
8. Mishra, M.K., Patel, Y.S., Rout, Y., Mund, G.B.: A survey on scheduling heuristics in grid computing environment. Int. J. Mod. Educ. Comput. Sci. (IJMECS) 6(10), 57–83 (2014). https://doi.org/10.5815/ijmecs.2014.10.08
9. Sajedi, H., Rabiee, M.: A metaheuristic algorithm for job scheduling in grid computing. Int. J. Mod. Educ. Comput. Sci. (IJMECS) 6(5), 52–59 (2014). https://doi.org/10.5815/ijmecs.2014.05.07
10. Hwang, K., Dongarra, J., Fox, G.: Distributed and Cloud Computing: From Parallel Processing to the Internet of Things. Morgan Kaufmann, Burlington (2012)
11. Brucker, P., Knust, S.: Complex Scheduling. GOR-Publications Series, 2nd edn. Springer, Heidelberg (2012). https://doi.org/10.1007/978-3-642-23929-8

12. Zgurovsky, M.Z., Pavlov, A.A.: The total weighted tardiness of tasks minimization on a single machine. In: Combinatorial Optimization Problems in Planning and Decision Making: Theory and Applications, 1st edn. Studies in Systems, Decision and Control, vol. 173, pp. 107–217. Springer, Cham (2019). https://doi.org/10.1007/978-3-319-98977-8_4

13. Zgurovsky, M.Z., Pavlov, A.A.: The total earliness/tardiness minimization on a single machine with arbitrary due dates. In: Combinatorial Optimization Problems in Planning and Decision Making: Theory and Applications, 1st edn. Studies in Systems, Decision and Control, vol. 173, pp. 219–263. Springer, Cham (2019). https://doi.org/10.1007/978-3-319-98977-8_5

14. Zgurovsky, M.Z., Pavlov, A.A.: Algorithms and software of the four-level model of planning and decision making. In: Combinatorial Optimization Problems in Planning and Decision Making: Theory and Applications, 1st edn. Studies in Systems, Decision and Control, vol. 173, pp. 407–518. Springer, Cham (2019). https://doi.org/10.1007/978-3-319-98977-8_9

15. Pavlov, A.A., Khalus, E.A., Borysenko, I.V.: Planning automation in discrete systems with a given structure of technological processes. In: Hu, Z., Petoukhov, S., Dychka, I., He, M. (eds.) Advances in Computer Science for Engineering and Education, ICCSEEA 2018, Advances in Intelligent Systems and Computing, vol. 754, pp. 177–185. Springer, Cham (2019). https://doi.org/10.1007/978-3-319-91008-6_18

16. Tanaka, S., Fujikuma, S., Araki, M.: An exact algorithm for single-machine scheduling without machine idle time. J. of Sched. **12**(6), 575–593 (2009). https://doi.org/10.1007/s10951-008-0093-5

17. Ding, J., Lü, Z., Edwin Cheng, T.C., Xu, L.: Breakout dynasearch for the single-machine total weighted tardiness problem. Comput. Ind. Eng. **98**(C), 1–10 (2016). https://doi.org/10.1016/j.cie.2016.04.022

18. Potts, C.N., Van Wassenhove, L.N.: Single machine tardiness sequencing heuristics. IIE Trans. **23**(4), 346–354 (1991). https://doi.org/10.1080/07408179108963868

19. Bilge, Ü., Kurtulan, M., Kıraç, F.: A tabu search algorithm for the single machine total weighted tardiness problem. Eur. J. of Oper. Res. **176**(3), 1423–1435 (2007). https://doi.org/10.1016/j.ejor.2005.10.030

20. Geiger, M.J.: On heuristic search for the single machine total weighted tardiness problem – Some theoretical insights and their empirical verification. Eur. J. of Oper. Res. **207**(3), 1235–1243 (2010). https://doi.org/10.1016/j.ejor.2010.06.031

21. Geiger, M.J.: New instances for the single machine total weighted tardiness problem. Research report RR-10-03-01. Helmut Schmidt Universität, Hamburg (2010)

22. Zgurovsky, M.Z., Pavlov, A.A.: Introduction. In: Combinatorial Optimization Problems in Planning and Decision Making: Theory and Applications. Studies in Systems, Decision and Control, vol. 173, pp. 1–14. Springer, Cham (2019). https://doi.org/10.1007/978-3-319-98977-8_1

23. Pavlov, A.A., Misura, E.B., Melnikov, O.V., Mukha, I.P.: NP-hard scheduling problems in planning process automation in discrete systems of certain classes. In: Hu, Z., Petoukhov, S., Dychka, I., He, M. (eds) Advances in Computer Science for Engineering and Education. ICCSEEA 2018. Advances in Intelligent Systems and Computing, vol. 754, pp. 429–436. Springer, Cham (2019). https://doi.org/10.1007/978-3-319-91008-6_43

24. Tanaka, S., Fujikuma, S., Araki, M.: OR-library: weighted tardiness (2013). https://sites.google.com/site/shunjitanaka/sips/benchmark-results-sips. Accessed 08 Nov 2018

25. Kara, B.Y.: OR-library: total tardiness (2001). http://bkara.bilkent.edu.tr/start.html. Accessed 08 Nov 2018

26. Fisher, M.L.: A dual algorithm for the one machine scheduling problem. Math. Program. **11**(1), 229–251 (1976). https://doi.org/10.1007/BF01580393

Data Mining and Nonlinear Non-stationary Processes Forecasting by Using Linguistic Modeling Method

Tatiana Shulkevich, Yurii Selin$^{(\boxtimes)}$ (ID), and Vilen Savchenko (ID)

National Technical University of Ukraine "Igor Sikorsky Kyiv Polytechnic Institute", 37, Prosp. Peremogy, Kyiv 03056, Ukraine
`selinyurij@online.ua`

Abstract. The report focuses on the actual task of development of mathematical tool for data mining and nonlinear non-stationary processes forecasting. The proposed mathematical tool can use in solving the problems of data analysis of various nature for nonlinear non-stationary processes forecasting. The results of its application for prediction of such processes are present.

Keywords: Nonlinear non-stationary processes · Forecasting · Mathematical models · Linguistic modeling

1 Introduction

Intensive science development has led to the appearance of a large number of methods for forecasting the behaviour of processes of various nature obtained in the form of time series (Bidyuk et al. 2013), (Montgomery et al. 2015), (Mishra et al. 2018) (Hu et al. 2016). Note that almost all of them are nonlinear and non-stationary (we can say exclusively about piecewise linearity and piecewise non-stationarity).

That is why the objects of the research are non-linear non-stationary processes in ecology, economics and finance. The necessity of developing a mathematical tool for the creation of data mining methods and forecasting of nonlinear non-stationary physical processes of different nature is substantiated in order to increase the adequacy of mathematical models of nonlinear non-stationary processes and improve the quality of forecasts estimates which are calculated by using created models. The linguistic modeling method is described in details. The informative and mathematical formulation of the task of finding linguistic patterns of time series is formulated. The concept of linguistic modeling was introduced, and the method of applying this approach was gradually described to solve the problem of forecasting of nonlinear non-stationary physical processes of different nature. The results of practical testing of the proposed mathematical tool for nonlinear non-stationary processes forecasting by using linguistic modeling method are presented. The results of the comparative analysis of the adequacy of the developed method with several existing ones are given and the quality of the obtained forecasting values is analyzed. The analysis shows high adequacy of the models developed by using linguistic modeling method and high quality of received forecasts.

Z. Hu et al. (Eds.): ICCSEEA 2019, AISC 938, pp. 409–418, 2020.
https://doi.org/10.1007/978-3-030-16621-2_38

2 The Linguistic Modeling Method

The Linguistic Model Construction. To achieve the purpose, the task of finding the linguistic pattern for the time series should be solved, which includes:

(a) Calculation of the subtractions series of the output time series;
(b) Choice of intervaling criterion of subtractions series;
(c) Intervaling of certain subtractions series according to the chosen criterion;
(d) Determining the linguistic chain for the certain subtractions series;
(e) Determining the transition matrix for each possible symbols pair in the linguistic chain of the certain subtractions series.

The input data for this task is the value of the time series.

The initial data for this task is the linguistic pattern of time series (dynamic process), that is:

– The set of intervals obtained as a result of intervaling of subtractions series of certain order from the time series;
– The transition matrix, constructed on the intervals set (described above) and by time series.

The specified linguistic pattern is constructed separately for subtractions series of input time series (Baklan and Selin 2005), (Baklan and Selin 2006), (Baklan et al. 2017), (Gurung et al. 2017). In this way the set of linguistic patterns is obtained, which is an intermediate result of forecasting problem by using linguistic modeling method.

Applying of linguistic modeling method for constructing linguistic patterns of input time series will be considered below.

Usage of pattern recognition methods for implementation of forecasting procedures is one of the approaches. So let's focus just at the pattern recognition. Most of variety of mathematical methods for solving pattern recognition problems can be divided into two main types.

The first type may be positioned in relation with the decision theory, or as it is also called as discriminatory approach. In this case, objects are characterized by the sets of numbers - the results of some measurements set, which are called signs. Pattern recognition by using this approach is usually carried out by partitioning the signs space into the areas (Fu 1968), (Mezzi and Benblidia 2017), (Ali et al. 2016).

The second type is developed within the syntactic (or structural) approach. The feature of this approach is the pattern recognition. The information about patterns structure is important. And it is required from the recognition procedure that it enables not only to refer the object to the certain type (that is, to determine its classification), but also to give the description of those object sides that exclude the possibility of its classification into another type.

When the objects are complex and the number of possible descriptions is big, it is inconvenient to consider that each description defines a class. We have such a situation in the problems of identifying images, fingerprints. The same situation is observed during the dynamic pattern recognition. In these cases, recognition can be carried out by using the description of each object and not simply by the classification methods.

In order to represent the hierarchical structural information existing in each pattern, we use simple partial patterns. This approach is based on an analogy between the patterns structure and the languages syntax. Within the syntactic approach it is assumed that patterns are composed of partial patterns connected in a different variety of ways in the same way as phrases and sentences are formed by connections of words and the words consist of symbols. They are also called non-derived elements.

It is easy to see that such approach is useful in the cases where it is easier to recognize the chosen partial patterns than the patterns themselves. The rules of com-position of non-derived elements are traditionally given by so-called pattern description language grammar. The pattern recognition process is carried out after non-derived elements identification in the object and making object description. Directly the recognition consists of the syntactic parsing (or grammatical parsing) of the "sentence" that describes the object. This procedure detects whether this sentence is syntactically correct according to the given grammar. In parallel syntactic parsing gives certain structural sentence description.

The syntactic approach to the pattern recognition makes it possible to describe a sufficiently big set of complex objects by using the small enough non-derived elements set and grammatical rules. And in such case we make use of recursive grammar tools.

The grammatical rule (or substitution rule) can be applied any number of times, so it is possible to provide some structural characteristics of the infinite set of sentences in a compact enough way. The practical benefit of this approach depends on the ability of recognition of non-derived elements of patterns and their composition operations.

Traditionally various relationships that is defined between partial patterns can be represented by logical and mathematical operations.

Symbolic sequences processing puts some problems. Symbols are grouped into words, words form sentences, and not in a free way, but in accordance with certain rules. To identify regularities in the sentence location, it is necessary to define the representation, within which these laws could not only be described, but also be found in symbolic sequences.

These questions are fundamental for studying linguistic characteristics of symbolic sequences. Just as for the linguistic requirements, theory of formal grammar proposed by Noam Chomsky in the middle of the last century became one of the main branches of mathematical linguistics.

According to the stages of constructing linguistic model, the initial task consists of the following subtasks: obtaining subtractions series subtask; intervaling subtask; lin-guistization subtask; determining transition matrix subtask.

Obtaining Subtractions Series Subtask. The purpose of this subtask is obtaining series characterizing the dynamics of changing of the mouse cursor movement: the speed (the subtractions series of the 1st order), the acceleration (the subtractions series of the 2nd order), etc. Thus, the subtractions series are the derivatives of the initial series.

Given: Integers vector $\overline{X}$ with power $n = |\overline{X}|$.

Results: Integers vector $\overline{D}$ with power $k = |\overline{D}|$.

Limitation: $\forall d_i \in \overline{D} : d_i = x_{i+1} - x_i$

where

$$i \in [0; n-1); x_i, x_{i+1} \in \overline{X}; \tag{1.1}$$

$$k = n - 1 \tag{1.2}$$

Intervaling Subtask. The purpose of this subtask is the construction of the user's alphabet by splitting sorted subtractions series into the set of intervals, where each set element characterizes certain alphabet letter.

Given: hypothetical alphabet power a;

- Integers vector $\overline{D}$ with power $k = |\overline{D}|$.

Results:
Vector of integers pairs $\overline{I}$ with power $n = |\overline{I}|$.
Limitations:

$$\forall x \in \overline{I} : x^1 \le x^2 \tag{1.3}$$

$$\forall x_i, x_{i+1} \in \overline{I} : x_i^2 < x_{i+1}^1 \tag{1.4}$$

where $i \in [0; n-1)$;

$$n \le a \tag{1.5}$$

$$a << k \tag{1.6}$$

$$\exists x \in \overline{I} :: \forall d \in \overline{D}, d \in [x^1; x^2] \tag{1.7}$$

$$\forall d_i, d_{i+1} \in \overline{D} : d_i \le d_{i+1} \tag{1.8}$$

where $i \in [0; k-1)$

$$x_0 \in \overline{I} : x_0 = \left(-\infty; x_1^1\right) \tag{1.9}$$

$$x_n \in \overline{I} : x_n = \left(x_{n-1}^2; +\infty\right) \tag{1.10}$$

Linguistization Subtask. The purpose of this subtask is to obtain a linguistic chain by determining the corresponding alphabet letter for each value of the subtractions series. The alphabet letter uniquely corresponds to a certain interval from the set of intervals obtained as the result of the previous subtask solution.

Given: Integers vector $\overline{D}$ with power $k = |\overline{D}|$, that corresponds to the limitation presented in formula 1.7;

Vector of integers pairs $\overline{I}$ with power $n = |\overline{I}|$ with limitations presented in formulas 1.3, 1.4, 1.9 and 1.10.

Results:

Integers vector $\overline{A}$ with power $k = |\overline{A}|$.

Limitation:

$$\forall x_i \in \overline{A} : \exists d_i \in D, \exists y_j \in \overline{I}, d_i \in \left[y_j^1 ; y_j^2 \right], x_i = j, \tag{1.11}$$

where $i \in [0; k)$, $j \in [0; n)$.

The general scheme of transition from the time series in numerical form to the linguistic chain is illustrated in the Fig. 1.

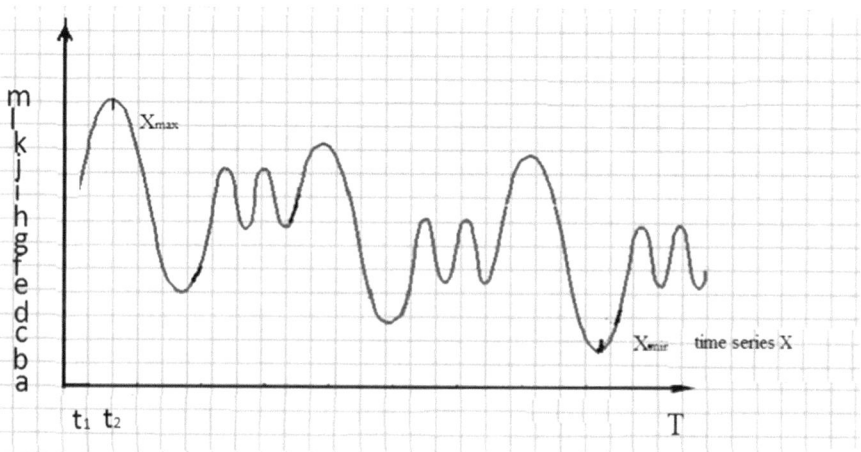

Fig. 1. The general scheme of transition into the linguistic chain.

Determining Transition Matrix Subtask. The purpose of this subtask is determining transition matrix between two alphabet letters in sentence. The alphabet and its letters are defined in the intervaling subtask, and sentences are defined in the linguistization subtask.

Given: Integers vector $\overline{A}$, that corresponds to the limitation 1.11 with power $k = |\overline{A}|$;

The power of set of intervals n, obtained as the result of solving intervaling subtask.

Results: square matrix with rational numbers $\overline{P}$ with dimensionality n.

Limitation:

$$\forall x_{ij} \in \overline{P} : \quad x_{ij} \in [0.0; 1.0] \tag{1.12}$$

where $i, j \in [0; n)$.

The obtained sequence is analyzed for the presence of grammatical constructions. At the output the list of grammatical constructions with the probabilities of their presence in the process as well as the probability matrix of transitions from one symbol to another symbol is obtained. This stage is closely correlated with hidden Markov processes modelling, as well as with the method of "similar" trajectories.

3   Practical Implementation

By means of the above describing mathematical tool and developed algorithms several time series (Swiss International Air Lines stock, Dow Jones Industrial Average, value of gold) were analyzed.

4000 values of timeslots with gradual reduction of the series size up to 200 with step 200 were taken for the maximal file size.

The calculation results are illustrated in Figs. 2 and 3.

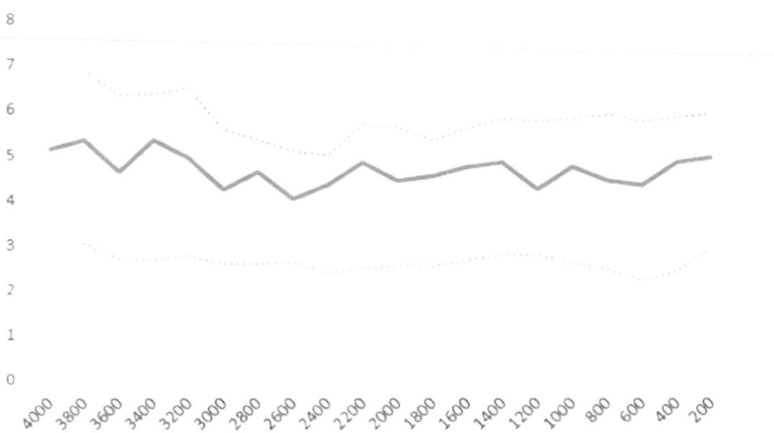

Fig. 2. Changing number of successful trend forecasts for the different dimensionality of input series.

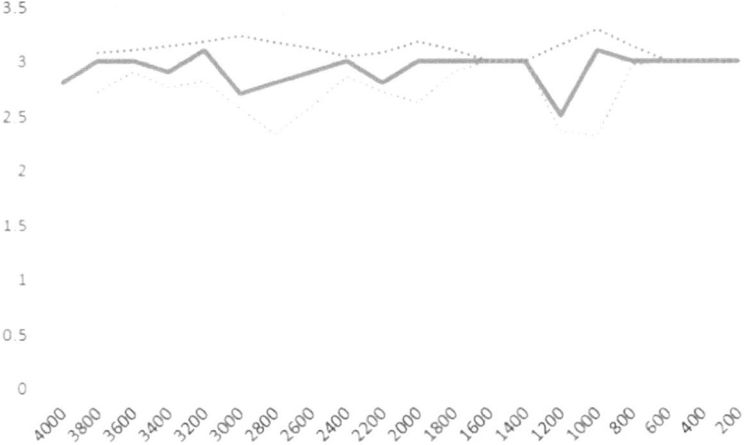

Fig. 3. Changing number of successful forecasts of time series values for the different dimensionality of input series.

The calculations were repeated at different dimensionality values of the input series for the obtained series without trend (subtractions series). The obtained results are illustrated in Figs. 4 and 5.

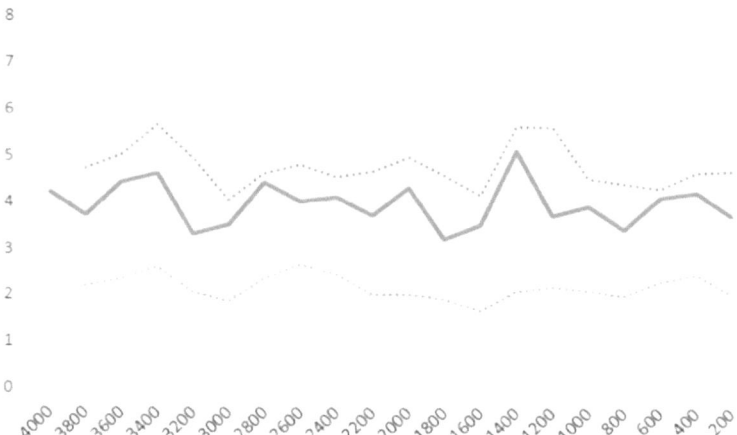

Fig. 4. Changing number of successful trend forecasts for the initial series without trend (subtractions series).

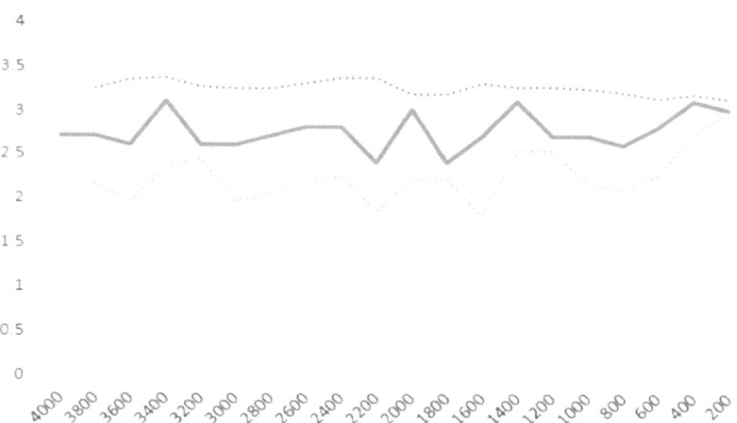

Fig. 5. Changing number of successful forecasts of time series values for the initial series without trend (subtractions series).

Corresponding symbol encodes one or another range of values depending on the alphabet power, thus research of the dependence of the number of alphabet symbols on forecasts quality was carried out. Dimensionality of input series was 400 values. Experiments results are illustrated in Figs. 6 and 7.

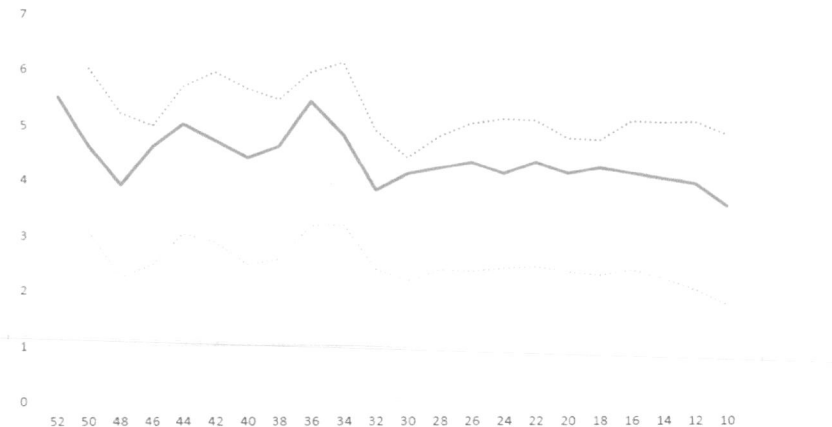

Fig. 6. Changing number of successful trend forecasts for different alphabet dimensionality.

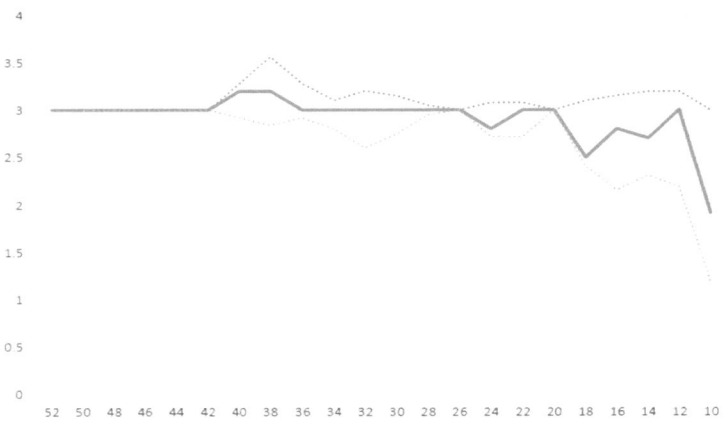

Fig. 7. Changing number of successful forecasts for the different alphabet dimensionality.

Efficiency of the developed linguistic modeling method was compared with the autoregressive method of order 10, autoregressive method of order 10 with the trend of order 3 and the group method of data handling (GMDH).

Adequacy of the model was calculated according to the criterion of determination R^2, the Durbin–Watson statistical criterion. Also sum of forecast errors squares $\sum e^2(x_i)$ was calculated (Bidyuk et al. 2013), (Khashei et al. 2015).

Mean absolute (percentage) error of forecast (MAPE) and mean square error of forecast (MSE) were calculated for the forecast quality estimation. The following table shows study's summary (Table 1).

The comparison results determine following. The described linguistic model is absolutely adequate. The forecast quality obtained by using linguistic modeling method

Table 1. Comparison of methods effectiveness

Model type	Model adequacy			Forecast quality	
	R^2	DW	$\sum e^2$	MAPE/MSE	MAPE/MSE
AR(10 + t_3)	0,98	2,04	57,28	1,25	5,28
AR(10)	0,98	2,02	61,22	2,81	13,15
GMDH	0,99	2,02	55,7	2,75	4,95
Linguistic model method	0,99	2,01	53,6	2,65	4,86

at the available data is better than by autoregressive method of order 10, autoregressive method of order 10 with the trend of order 3 and the group method of data handling.

4 Conclusions

Thus the mathematical tool was developed for calculation of quantitative and qualitative indicators of time series of economical and ecological nature by using linguistic modelling. It is characterised by the persistence of obtained results and provides the improvement of forecasts quality of corresponding indicators. The structures of new mathematical models of nonlinear non-stationary processes are formed. They differ by simplification of the model constructing procedure and provide high-level adequacy description of researched processes.

The presented method is universal both for the type of obtained information and for the presence of nonlinearities and non-stationarities in it, but it has the general lack of all statistical methods – the lack of historical information.

References

Baklan, I.V., Selin, Y.: Analysis of the behavior of economic time series using structural approaches. Bull. Kherson Natl. Tech. Univ. **2**, 27–31 (2005). (in Russian)

Baklan, I.V., Selin, Y.: Structural approach to the analysis and prediction of behavior of time series. Bull. Kherson Natl. Tech. Univ. **5**, 29–34 (2006). (in Russian)

Baklan, I.V., Savchenko, V.V., Selin, Y., Shulkevych, T.V.: Some aspects of nonlinear non-stationary processes forecasting. Syst. Terminol. **6**(113), 31–42 (2017)

Bidyuk, P.I., Romanenko, V.D., Tymoshchuk, O.L.: Time series analysis. Tutorial, NTUU "KPI", Kyiv (2013). (in Ukrainian)

Gurung, D., Chakraborty, U.K., Sharma, P.: An analysis of the intelligent predictive string search algorithm: a probabilistic approach. Int. J. Inf. Technol. Comput. Sci. (IJITCS) **9**(2), 66–75 (2017). https://doi.org/10.5815/ijitcs.2017.02.08

Fu, K.S.: Sequential Methods in Pattern Recognition and Machine Learning. Academic Press, New York (1968)

Hu, Z., Bodyanskiy, Y.V., Tyshchenko, O.K., Boiko, O.O.: Adaptive forecasting of non-stationary nonlinear time series based on the evolving weighted neuro-neo-fuzzy-ANARX-model. Int. J. Inf. Technol. Comput. Sci. (IJITCS) **8**(10), 1–10 (2016). https://doi.org/10.5815/ijitcs.2016.10.01

Khashei, M., Montazeri, M.A., Bijari, M.: Comparison of four interval ARIMA-base time series methods for exchange rate forecasting. Int. J. Math. Sci. Comput. (IJMSC) **1**(1), 21–34 (2015). https://doi.org/10.5815/ijmsc.2015.01.03

Mezzi, M., Benblidia, N.: Study of context modelling criteria in information retrieval. Int. J. Inf. Technol. Comput. Sci. (IJITCS) **9**(3), 28–39 (2017). https://doi.org/10.5815/ijitcs.2017.03.04

Mishra, N., Soni, H.K., Sharma, S., Upadhyay, A.K.: Development and analysis of artificial neural network models for rainfall prediction by using time-series data. Int. J. Intell. Syst. Appl. (IJISA) **10**(1), 16–23 (2018). https://doi.org/10.5815/ijisa.2018.01.03

Montgomery, D.C., Jennings, C.L., Kulahci, M.: Introduction to Time Series Analysis and Forecasting, 2nd edn (2015)

Ali, S.K., Azeez, Z.N., Ouda, A.A.H.: A new clustering algorithm for face classification. Int. J. Inf. Technol. Comput. Sci. (IJITCS) **8**(6), 1–8 (2016). https://doi.org/10.5815/ijitcs.2016.06.01

Multicriteria Choice of Software Architecture Using Dynamic Correction of Quality Attributes

Bodnarchuk Ihor[1](✉)🆔, Duda Oleksii[1]🆔,
Kharchenko Aleksandr[2]🆔, Kunanets Nataliia[3]🆔,
Matsiuk Oleksandr[1]🆔, and Pasichnyk Volodymyr[3]🆔

[1] Ternopil Ivan Puluj National Technical University,
Ruska Street, 56, Ternopil 46000, Ukraine
bodnarchuk.io@gmail.com, oleksij.duda@gmail.com,
oleksandr.matsiuk@gmail.com
[2] National Aviation University,
Kosmonavta Komarova Avenue 1, Kyiv 03058, Ukraine
kharchenko.nau@gmail.com
[3] Lviv Polytechnic National University, St. Bandera Street, 12, Lviv, Ukraine
nek.lviv@gmail.com, vpasichnyk@gmail.com

Abstract. The optimization problem of the procedure for the software system architecture selection is considered taking into account changes of requirements in iterative technologies of design and reengineering. The survey of this problem was made in order to find available formal methods or techniques for architecture decisions operative correction in modern agile approaches. The conclusion is made that there is the significant lack of such methods that allow to assess dynamically final quality of the product during architecture design without repetitive labor consuming expert methods. The solution is offered on the base of software characteristics correction which have been taken into account on previous iteration or modification of existing software system during reengineering. This method prevents recalculations for the evaluation and the selection process. The method of pairwise substitution is applied for alternatives characteristics correction. Its general concept is based on compensation on supremacy of a criterion or an attribute change. Multicriteria optimization of substitution is being carried out using nonlinear scalar convolution what improves validity of selected software architecture decision.

Keywords: Software architecture · Software architecture quality ·
Agile software design · Flexible techniques · Software quality ·
Quality requirements

1 Introduction

Modern technologies of software systems design, such as extreme programming, SCRUM techniques and some other are iterative actually [1–4]. Changes into requirements or restrictions can be made during each iteration what cause the changes in corresponding parts of the project [5].

© Springer Nature Switzerland AG 2020
Z. Hu et al. (Eds.): ICCSEEA 2019, AISC 938, pp. 419–427, 2020.
https://doi.org/10.1007/978-3-030-16621-2_39

The series of methods was worked out for quality assessment of the software architecture (SA) alternatives and selection of the best one among them. The most known are scenario-based methods ATAM, SAAM, CBAM and others [6, 7] are labor-intensive considerably.

The survey of the method for architecture assessment and quality assurance in agile techniques of software design shown significant lack of mathematical formalism here. For example, standard Saaty's Analytic Hierarchy Process (AHP) is used to formalize the problem in [8]. But there are not represented solutions for cases where big quantity of architecture alternatives and/or quality attributes are available. Traditional mentioned above scenario-based methods are still proposed for architecture assessment in agile techniques [9, 10]. Thus, they become a bottle neck for fast responding of new change requests. In the articles [11, 12] the intervals of values changes for quality criteria weights were obtained. These intervals allow do not to rearrangement of alternatives for estimating of compromise decision on the SA choosing.

Appearance of articles where AHP is applied for assessment of alternatives allowed to avoid some disadvantages proper to scenario methods, to formalize and to automate the procedure of assessment [13, 14]. The significant disadvantage of AHP is limited quantity of alternatives for simultaneous assessment ($n \leq 7 \pm 2$) what is caused by inconsistency of the pairwise comparisons matrix (PCM) [15]. Pavlov offered the modification of AHP in [16] to solve this problem, where weight indices are estimated from the condition of inconsistency minimization of the PCM. This transforms initial problem to the mathematical programming problem. Applying of modified AHP to the problem of SA alternative assessment for case of large number of alternatives are discussed in [17].

Comparative assessments of the alternatives are determined with using of expert information in AHP method, and in case of requirements change to the SWS on the current iteration of development it is necessary to execute repetitively expert assessment and calculation of estimations for quality criteria of alternatives.

Because of simultaneous processes running on some stages of software development lifecycle (SDLC) in iteration techniques with using of some base architecture the significant changes in this variant will require corrections of the majority parts of the project that increase time of the project ant its cost. Therefore, partial replacement of the base architecture will be more effective, i.e. correction of certain its characteristics and corresponding amends of the SA. Actually, the technics for this process have been worked out and discussed in this paper as a new improvement of quality assurance process in project management and software development.

The procedure of correction on the base of substitution of quality criteria of base architecture and its following optimization of this correction is offered for this. It will allow to make minimum changes in all parts of the project and do not exceed the project budget and time.

The mathematical model of correction is developed using axiomatic theory of criteria importance [18] and the results of the research are shown in [19]. The problem of optimization of global quality criterion for the architecture under restrictions defined by the model of correction is determined for selection of the proper values of criteria corrections. Since application of linear scalar convolution as a global criterion has essential restrictions [20] we suggest to discuss the usage of nonlinear scalar convolution. The iteration procedure of simplex planning is used for formalizing of the process of criteria weights estimation [21]. Obtained optimal values of the criteria corrections are used for modification of the base SWS architecture with purpose to take into account changes of requirements to the SWS under design or reengineering.

Thus, we will formulate the problem of multicriteria selection for the field of software architecture under reengineering what is especially important for flexible methods like SCRUM. Then the main concepts of comparability by substitution for quality criteria of software architecture is described with supplementary artifacts like criteria comparability, their co-replaceability, importance and restrictions for criteria substitution. Further the problem of software architecture selection is formulated as a multicriteria optimization problem that can be solved with nonlinear convolution. Finally use-case for described method is represented.

2   The Model of the Quality Criteria Correction for SA Alternatives

The best architecture variant among the set of the alternatives

$$\{A_i\}, \ (i = \overline{1,n}) \tag{1}$$

was selected relatively to the vector quality criterion (2)

$$\{K_s\}, \ (s = \overline{1,m}). \tag{2}$$

Their assessments are calculated for the quality criteria and then a multicriteria selection of the architecture is executed on the base of obtained assessments. As it was discussed earlier the problem of the alternative's assessments for a separate quality criterion can be solved most effectively with using AHP. The set of alternative architectures (1) is examined in the process of SWS design or reengineering. The relative assessments of quality criteria values for each alternative architecture have been calculated

$$\{K_{is}\}, (i = \overline{1,n}; \ s = \overline{1,m}). \tag{3}$$

The alternative $\{A_i\}$ is chosen as an alternative for correction. It is selected with taking into account additional factors, although it is not the best among alternatives for some criteria. An existing architecture was considered during reengineering. The problem of correcting its characteristics is formulated in order to make it better for some criteria and not worse for other criteria in the rest of alternatives. Concept of criteria comparability by substitution [17] was used during development of the criteria correction model.

The comparability concept by substitution of criteria K_r and K_s means that for any alternative A_i the compensation on superiority of some changes of criterion K_r is possible by some changes of criterion K_s. The ratio between possible changes in the K_r and K_s is defined by the essence of these criteria and the trade-off is achieved relatively to their importance. That is, if A_i^p is an alternative that substitutes the alternative A_i by correction of K_r and compensation of K_s then their corrected values will be equal to:

$$\overline{K}_r^{ip} = \overline{K}_r^i - \delta_r, \overline{K}_s^{ip} = \overline{K}_s^i + \delta_{si}, \delta_{si} = f(r, s, \overline{K}, \delta_r) \tag{4}$$

where $\overline{K}$ is a vector of criteria values.

The compensation on the substitution ratio for the set of the vector $\overline{K}^i$ components for an alternative A_i that we want to make better than an alternative A_j can be written as:

$$\delta \overline{K}_r^{ir_z} = C_r^{ir_z} \cdot \delta K_r^i, \ r_z \in R_i^2(r), \ r \in R_i^1 \tag{5}$$

where the right side of this expression is possible decreasing of the component $\overline{K}_r^i$ value for increasing of the component $\overline{K}_{r_z}^i$ value; R_i^1 is a set of indices r for which $\overline{K}_r^{iz} > \overline{K}_r^j, j = \overline{1, n}; i \neq j; R_i^2(r)$ is a set of indices given for R_i^1 such that the components $\overline{K}_r^i, r \in R_i^1$ may participate in substitution of the components $\overline{K}_s^i, s \in R_i^2(r); C_r^{ir_z}$ are chosen coefficients of proportionality.

The components of the vector $\overline{K}^i$ after substitution are estimated as follows:

$$\overline{K}_r^{ip} = \overline{K}_r^i - \sum_{r_z \in R_i^2(r)} C_r^{ir_z} \cdot \delta \overline{K}_{r_z}^i, \quad r \in R_i^1;$$

$$\overline{K}_r^{ip} = \overline{K}_{r_z}^i + \sum_{r \in R_i^1} \sum_{r_z \in R_i^2(r)} \delta \overline{K}_{r_z}, \quad r_z \in s, \ s \overline{\in} R_i^1, \ r_z \in R_i^2(r). \tag{6}$$

The values of the criteria correction (5) are estimated under the condition of global architecture quality criterion optimization:

$$Y(A_i) = \sum_{j=1}^{n} \alpha_j \cdot K_j(A_i). \tag{7}$$

Mathematical models of the optimization problems for estimation of (5) depend on the decision-making strategy. For a Pareto-optimal strategy the optimization model with taking into consideration (6) will look as follows:

$$\max_{i=\overline{1,m}} F(K(A_i)), \tag{8}$$

under restrictions:

$$d_r, d_s \geq 0, \ r \in L_j^1, \ s \overline{\in} R_j^1; \ \overline{K}_r^j - \sum_{r_z \in R_j^2(r)} \delta \overline{K}_r^{jr_z} \geq \max_i \left(\delta \overline{K}_r^i \right) + d_r,$$

$$i \in \overline{1, n}, \ i \neq j, \ r \in R_j^1; \ \overline{K}_s^j + \sum_{r \in R_j^1} \sum_{r_z \in R_j^2(r)} \frac{1}{d_r^{jr_z}} \delta \overline{K}_r^{jr_z} \geq \max_i \left(\overline{K}_s^i \right) + d_s;$$

$$i = \overline{1, n}, \ i \neq j, \ r \in R_j^1, \ r_z = s, \ \exists r, s \in R_j^2(r);$$

$$\sum_{r_z \in L_j^2(r)} \delta \overline{K}_r^{jr_z} \leq b_{K_j^i}, \ i \neq j, \ r \in R_j^1, \ r_z = s. \tag{9}$$

The unknown variables for optimization of the criteria (8) are left part of (5) and s. d_i, $i = \overline{1,n}$ are given coefficients that are defined from the essence of each criterion, and with that $d_i \geq 0$, $i = \overline{1,n}$. A linear scalar convolution of partial criteria is selected as rule for a global criterion

$$Y(A_i) = F(K(A_i)) = \sum_{j=1}^{m} \alpha_j K_j(A_i), \quad i = \overline{1,n}. \tag{10}$$

where α_j are weight indices of criteria that meet conditions $\sum_{j=1}^{m} \alpha_j = 1$, $\alpha_j \geq 0$.

The problem of linear programming (8) and (9) will be obtained when (10) is substituted into (8). The optimal values of the corrections (5) will be found upon solution of this equation. However, as it was previously described by Pavlov et al. [16], applying a linear scalar convolution can assure the estimating of optimum for problems of multicriteria selection only for the condition when the set of solutions is convex and the functions $K_j(A_i)$ are concave. The set of solutions for the problem (8), (9) is discrete and finite in our case, i.e. it is not convex. Thus, applying a linear scalar convolution in this problem requires additional reasoning. Applying a nonlinear relatively quality criteria scalar convolution $Y(A_i) = \sum_{j=1}^{n} \alpha_j F_j(K(A_i))$.

A nonlinear schema of compromises with weight indices defined on the simplex (8) leads to Pareto optimal decisions according to the lemma of Karlin [22]. The selection of the function $F_j(K(A_i))$ has to be done according to the specifications of the problem, principles used by a making decision person (MDP) as well as accepted schema of compromises. One such a compromise that may be applied in the current problem is a control of compliance to restrictions for component values of following vector quality criterion $K_j \geq B_j$, $j = \overline{1,n}$.

These restrictions are estimated during requirements formulation and analysis to the SWS, as well communication (mapping) of these requirements on the requirements to software architecture [23]. The following may be chosen as a criterion function $Y(A_i) = \sum_{i=1}^{n} \alpha_i f(B_j - K_j(A_i))$, $i = \overline{1,n}$. Let's norm quality criteria by values of restrictions $Y(A_i) = \sum_{i=1}^{n} \alpha_i f(1 - \overline{K}_j(A_i))$, $K_j(A_i) \leq 1$. Now we can select for the function $f(1 - \overline{K}_j(A_i))$ nonlinear scalar convolution offered in [21] for the similar problem of multicriteria selection as following:

$$Y(A_i) = \sum_{j=1}^{m} \alpha_j (1 - K_j(A_i))^{-1}, \; i = \overline{1,n}. \tag{11}$$

This function is nonlinear regarding to the quality criteria. In case when some criteria values will approach to limit of restrictions the minimax model will be accepted for the decision making: $A_{opt} = \arg \min_{A_i \in A} \max_{j=\overline{1,m}} K_j(A_i)$.

For those cases when criteria values are far from restrictions the model of integration optimality can be written as it's shown in the following expression

$$A_{opt} = \arg \min_{A_i \in A} \sum_{j=1}^{m} \alpha_j \left(1 - K_j(A_i)\right)^{-1}.$$

If there are both criteria for minimization and maximization then the integral quality criterion of the selected architecture will be represented as follows:

$$Q(A_i) = \sum_{j \in L_1} \alpha_j \left(1 - \overline{K}_j(A_i)\right)^{-1} + \sum_{j \in L_2} \alpha_j \left(\overline{K}_j(A_i) - 1\right)^{-1}, \quad i = \overline{1,n} \qquad (12)$$

where L_1 is a set of criteria indices for minimization; L_2 is a set of criteria indices for maximization.

It is necessary to define α_j for making the Pareto optimal decision of the problem on the set of quality criteria of the architecture. The method of simplex planning [20] can be used for selection and correction of weights of α_j. We can now solve the problem of mathematical programming with nonlinear criterial function and linear restrictions (9) by substitution the criterion (11) into the ex-pression (8). Thus, we can determine the correction values of the importance criteria (5) for which selected alternative will be the best. The dual procedure of simplex planning [21] can be used for defining the weights αj of criteria in the criterial function (11) or they can be obtained by experimental methods.

3   Case Study for the Method

The correction method in the problem discussed by participants of the international project GlasBox [24] was then applied. The main two requirements to a designed system were scalability and secure access for all project team members to the database. Assessments of some alternatives represented by the project participants are obtained by modified AHP [25, 26] and shown in the Table 1.

There following alternatives are investigated: THTJ – three-tiered architecture on J2EE platform, THTD – three-tiered architecture on .NET platform, TWOT – two-tiered architecture with "fat" client, COAB – the architecture with distributed agent support. We selected the THTJ architecture as the most acceptable for correction and selection

Table 1. Assessments of alternatives regarding to the quality criteria

Quality attributes (criteria)	Alternatives			
	THTJ (A_1)	THTD (A_2)	TWOT (A_3)	COAB (A_4)
1 Modifiability (K_1)	0.521	0.172	0.106	0.210
2 Scalability (K_2)	0.453	0.350	0.064	0.133
3 Performance (K_3)	0.201	0.204	0.347	0.246
4 Cost (K_4)	0.166	0.180	0.427	0.227
5 Portability (K_5)	0.500	0.050	0.050	0.400
6 Ease to install (K_6)	0.268	0.268	0.256	0.218

alternative. As shown in Table 1, the THTJ surpasses all others based on the three criteria K_1, K_2, K_3 while it underperformed on criteria K_3 and K_4.

The assessment of the alternative A_1 needs to corrected such that for each criterion it will not be worse than the other three architectures and it would be better on some criteria.

Here the set $R_i^1 \rightarrow \left(\forall l \in L_i^l, \overline{K}_l^i > \overline{K}_l^j, i \neq j \right) \in \{1; 2; 5\}$ and correspondingly $R_i^2 = \{3; 4\}$. The criterion K_6 does not take part in the substitution. The problem is to use decreasing assessments on 1st, 2nd and 5th criteria to increase assessments on 3rd and 4th but they must stay not worse that for three other alternatives. Since maximum assessment on 1st criterion of second, third and fourth alternatives is 0.210, on 2nd criterion is 0.350, and on 5th criterion is 0.400 so these restrictions look as follows:

$$
\begin{aligned}
&0.521 - \left(\delta\overline{K}_{13} + \delta\overline{K}_{14} \right) \geq 0.21 + y; \\
&0.45 - \left(\delta\overline{K}_{23} + \delta\overline{K}_{24} \right) \geq 0.35 + 0.8y; \ \ 0.5 - \left(\delta\overline{K}_{53} + \delta\overline{K}_{54} \right) \geq 0.4 + 0.8y.
\end{aligned}
\tag{13}
$$

The restrictions where the assessments of third and fourth criteria that are under correction would not be worse than for three other alternatives will look like following:

$$
\begin{aligned}
&0.201 + (1.6 \cdot \delta\overline{K}_{13} + 1.25 \cdot \delta\overline{K}_{23} + 1.3 \cdot \delta\overline{K}_{53}) \geq 0.347 + 0.5y; \\
&0.166 + (2.5 \cdot \delta\overline{K}_{14} + 1.1 \cdot \delta\overline{K}_{24} + 2.0 \cdot \delta\overline{K}_{54}) \geq 0.427 + 0.6y.
\end{aligned}
\tag{14}
$$

The coefficients of substitution C_l^{ilm} in the expression (5) are introduced by experts on the base of criteria importance. The restrictions on maximum change of first, second and fifth criteria look like following:

$$
\delta\overline{K}_{13} + \delta\overline{K}_{14} \leq 0.3; \ \ \delta\overline{K}_{23} + \delta\overline{K}_{24} \leq 0.1; \ \ \delta\overline{K}_{53} + \delta\overline{K}_{54} \leq 0.1.
\tag{15}
$$

Solving the problem of optimization with stated restrictions will lead to:

$$
\begin{aligned}
&\delta\overline{K}_{13} = 0.12; \ \ \delta\overline{K}_{14} = 0.18; \ \ \delta\overline{K}_{23} = 0.04; \ \ \delta\overline{K}_{24} = 0.06; \\
&\delta\overline{K}_{53} = 0.07; \ \ \delta\overline{K}_{54} = 0.03; \ \ y = 0.13.
\end{aligned}
\tag{16}
$$

Corrected assessments of the architectures are represented in the Table 2.

Table 2. Corrected assessments of alternatives.

Quality attributes (criteria)	Alternatives			
	THTJ (A_1)	THTD (A_2)	TWOT (A_3)	COAB (A_4)
1 Modifiability (K_1)	0.221	0.172	0.106	0.210
2 Scalability (K_2)	0.353	0.350	0.064	0.133
3 Performance (K_3)	0.431	0.204	0.347	0.246
4 Cost (K_4)	0.436	0.180	0.427	0.227
5 Portability (K_5)	0.400	0.048	0.052	0.400
6 Ease to install (K_6)	0.268	0.268	0.256	0.218

The alternative A_1 is now the best for all criteria except K_6 for which it is not worse than assessments of other alternatives.

4   Conclusions

Iterative technologies of software design that used widely for development of software products aiming to decrease of development time assume parallel processes running on some stages of SDLC. Thus, change of requirements for the SWS during design process requires amends in the architecture and respectively into all components of the developed software product. It is offered to correct values of the software architecture quality criteria with taking into account the requirements change to minimize the impact of these changes on the final product quality and maximum avoid possible corrections of the project.

The mathematical model is developed on the base of the "substitution – compensation" method. The problems of corrections optimization have been formulated for the SA global quality criterion as linear and nonlinear scalar convolutions and further they were reduced to the problems of linear and nonlinear mathematical programming correspondently. The use case for the developed method is represented to demonstrate its capabilities.

It is planned as further researches to develop a CASE tool on the base of represented models for the processes automation of architecture corrections as a process among other ones of projects management.

References

1. Beck, K., Andres, C.: Extreme Programming Explained: Embrace Change, 2nd edn. Addison Wesley Professional, Boston (2004)
2. Schwaber, K.: Agile Project Management with SCRUM. O'Reilly, Sebastopol (2009)
3. Alotaibi, A.F., Qureshi, M.R.J.: Scrum and temporal distance-based global software development. Int. J. Comput. Netw. Inf. Secur. (IJCNIS) 6(6), 48–53 (2014). https://doi.org/10.5815/ijcnis.2014.06.07
4. Naz, R., Khan, M.N.A., Aamir, M.: Scrum-based methodology for product maintenance and support. Int. J. Eng. Manuf. (IJEM) 6(1), 10–27 (2016). https://doi.org/10.5815/ijem.2016.01.02
5. Rahim, S., Chowdhury, A.Z.M.E., Nandi, D., Rahman, M., Hakim, S.: ScrumFall: a hybrid software process model. Int. J. Inf. Technol. Comput. Sci. (IJITCS) 10(12), 41–48 (2018). https://doi.org/10.5815/ijitcs.2018.12.06
6. Kazman, R., Klein, M., Clements, P.: ATAM: Method for Architecture Evaluation. Software Engineering Institute, Carnegie Mellon University, Pittsburgh, August 2000
7. Ashraf, M.U., Aljedaibi, W.: ATAM-based architecture evaluation using LOTOS formal method. International J. Inf. Technol. Comput. Sci. (IJITCS) 9(3), 10–18 (2017). https://doi.org/10.5815/ijitcs.2017.03.02
8. Karimi, Z., Rashidi, H., Broumandnia, A.: A method to challenge XP agile method through software architecture. Int. J. Latest Trend Comput. 3(3), 73–78 (2012)

9. Stamelos, I.G., Sfetsos, P.: Agile Software Development Quality Assurance. IGI Global, Hershey (2007)
10. Babar, M.A., Brown, A.W., Mistrik, I.: Agile Software Architecture: Aligning Agile Processes and Software Architectures. Morgan Kaufmapp, Burlington (2013)
11. Dobrica, L., Niemela, E.: A survey on software architecture analysis methods. IEEE Trans. Softw. Eng. **28**(7), 638–653 (2002)
12. Zhu, L., Aurum, A., Gorton, L.: Tradeoff and sensitivity analysis in software architecture evaluation using analytic hierarchy process. Softw. Qual. J. **13**, 357–375 (2005)
13. Svahnberg, M., Wholin, C., Lundberg, L.: A quality-driven decision-support method for identifying software architecture candidates. Int. J. Softw. Eng. Knowl. Eng. **13**(5), 547–573 (2003)
14. Al-Naeem, T., Gorton, I., Babar, M.A., Rabhi, F., Benatallah, B.: A quality driven systematic approach for architecting distributed software applications. In: Proceedings of the 27th International Conference on Software Engineering (ICSE), St. Louis, USA, pp. 244–253 (2005)
15. Saaty, T., Vargas, L.: Decision Making with the Analytic Network Process. Springer, New York (2006)
16. Pavlov, A.A., Lishchuk, E.I., Kut, V.I.: Mathematical models of objects weights estimation optimization in the method of pairwise comparisons. System Research and Information Tech-nologies, IPSA, Issue 2, pp.13–21 (2007). (in Russian)
17. Kharchenko, O.G., Galai, I.O., Bodnarchuk, I.O.: Stability of solutions for the optimization problem of software systems architecture. Comput. Sci. Inf. Technol. **771**, 17–24 (2013). Visnyk (Official Gazette) of Lviv Polytechnic National University. (in Ukrainian)
18. Podinovskii, V.V.: Introduction into the Theory of Criteria Importance in Multicriteria Problems of Decision Making. Fismatlit, Moscow (2007). (in Russian)
19. Pavlov, A.A., Lishchuk, K.I.: Operative algorithms of decision making based on the criteria substitution in Saaty's hierarchical system. Bull. NTUU "KPI". Inform. Manag. Comput. Sci. **48**, 78–81 (2008). (in Russian)
20. Nogin, V.D.: Limits of applicability for spreaded methods of scalarization in solution of multicriteria selection problem. In: Methods of Perturbations in Gomological Algebra and Dynamics of Systems. Inter Institute Bulleting of Science Works, Saransk, Publishing of Mordovian University, pp. 59–68 (2004). (in Russian)
21. Voronin, A.N., Ziatdinov, Yu.K.: Theory and Practice of Multicriteria Solutions: Models, Methods. Implementation. Lambert Academic Publishing, Riga (2013). (in Russian)
22. Geimer, Yu.B.: Introduction into the Theory of Operations Research. Nauka, Moscow (1971). (in Russian)
23. Kharchenko, O.G., Yatcyshyn, V.V., Raichev, I.E.: Tool for developing and communicating of quality requirements to software systems. Softw. Eng. **2**, 29–34 (2010). (in Ukrainian)
24. Gorton, I., Haack, J.: Architecting in the face of uncertainty: an experience report. In: Proceedings of the 26th International Conference on Software Engineering (ICSE '04), Edinburgh, Scotland, pp. 543–551 (2004)
25. Kharchenko, A., Bodnarchuk, I., Yatcyshyn, V.: The method for comparative evaluation of software architecture with accounting of trade-offs. Am. J. Inf. Syst. **2**(1), 20–25 (2014)
26. Kharchenko, O.G., Bodnarchuk, I.O., Yatcyshyn, V.V.: Expert system for design of software architecture. Comput. Technol. Publ. **29**, 10–26 (2013). (in Ukrainian)

Effect of the Noise on Generalized Peres Gate Operation

I. M. Yuriychuk[1], Zhengbing Hu[2], and V. G. Deibuk[1(✉)]

[1] Yuriy Fedkovych Chernivtsi National University, Chernivtsi, Ukraine
v.deibuk@chnu.edu.ua
[2] School of Educational Information Technology,
Central China Normal University, Wuhan, China

Abstract. The advantages of quantum computing may be lost due to the presence of various noises in real quantum devices. In this paper, generalized Peres gates are studied using a solid-state model of a linear chain of atoms with nuclear spin one half, which is implanted in a spin-free silicon matrix. The effect of frequency noise on the correctness of gates operation is investigated on the example of the one- and two-step algorithmic transitions. It is shown for the first time that an imbalance by the magnitude of the resonance control frequency leads to a significant decrease in the fidelity of correct gate operation on the digital states. While for the superposition input signals stabilization of the fidelity is observed to a level of 0.4–0.8 with an increase in the imbalance of the resonance frequency. A similar effect takes place at time deviation of the frequency around the resonant value. Minimum values of the frequency imbalance parameters, which ensure correctness of the generalized Peres gate operation, are proposed.

Keywords: Quantum computing · Fidelity · Generalized Peres gate · Ising model

1 Introduction

Real quantum circuits, unlike ideal ones, have a variety of defects [1, 2]. They are associated with both physical and technological factors. One of the peculiarities of quantum systems is the high sensitivity of quantum bits to the interaction with the environment. Such effect inevitably leads to a change in the state of the system, i.e. causes decoherence, and hence errors in the implementation of quantum algorithms. At present, several methods are developed to obtain fault-tolerant fidelity for various physical implementations of qubits. In particular, we note the long-distance entanglement links, the nearest-neighboring topological codes etc. [3, 4]. In the well-known model of a quantum computer proposed by Kane, nuclear spin states of Phosphorus in spin-free silicon matrix enriched by ^{28}Si isotope are considered as qubits [5]. For such system, nuclear spin decoherence time is tens of minutes, and the control error rate does not exceed 10^{-4} [6]. In addition, this system is highly technological in terms of its integration into MOS-structures. Therefore, we consider such a system as promising enough for the construction of a quantum processor [4, 20–24]. Fidelity of the

© Springer Nature Switzerland AG 2020
Z. Hu et al. (Eds.): ICCSEEA 2019, AISC 938, pp. 428–437, 2020.
https://doi.org/10.1007/978-3-030-16621-2_40

one-qubit quantum gates in silicon was discussed in [7, 8]. However, the problem of correct operation of multiqubit gates in the model of nuclear spin qubits in a semi-conductor matrix remains unsolved.

A universal set of reversible one-, two-, and three-qubits gates (NOT, CNOT, Toffoli gates) [9] shown in Fig. 1 is widely used in a reversible quantum design. This set gives a possibility to synthesize an arbitrary reversible Boolean function both at the logical and physical level using different technologies. The generalization of the three-qubit gate is N-qubit Toffoli gate, which can be defined as

$$
\begin{aligned}
Y_i &= X_i, \qquad 1 \leq i \leq N-1, \\
Y_N &= X_1 X_2 \ldots X_{N-1} \oplus X_N
\end{aligned}
\tag{1}
$$

where X_i and $Y_i = X_i$ ($i = 1, \ldots, N-1$) are the input and output control qubits respectively, and X_N and Y_N are the input and output controlled qubits.

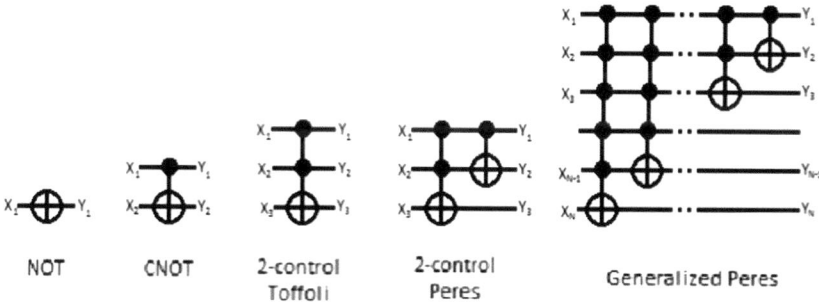

Fig. 1. Symbolic representation of NOT, CNOT, 2-control Toffoli, 2-control Peres, and generalized Peres gates.

A generalized N-qubit Toffoli gate in the case $N = 1$ works as an inverter (NOT), and at $N = 2$ – as a controlled inverter (CNOT). This generalization makes it possible to develop an algorithm for the synthesis of reversible circuits based on the generalized Toffoli gates [10]. The 3-qubit Peres gate combines CNOT and Toffoli gates functions, that is, its output signals (Y_i) can be defined by the input signals (X_i) as follows [11]:

$$
\begin{aligned}
Y_1 &= X_1, \\
Y_2 &= X_1 \oplus X_2, \\
Y_3 &= X_3 \oplus X_1 X_2.
\end{aligned}
\tag{2}
$$

Such 3-qubit gate can be implemented using only 4 elemental quantum gates (the quantum cost is 4), which is the minimal value for all known three-qubit gates and allows reducing the quantum cost of complicated reversible quantum circuits. In the paper [12] the generalized Peres gate was proposed (Fig. 1), for which:

$$Y_i = \bigoplus_{j=1}^{i} X_j \quad i = 1, \ldots, N-1,$$
$$Y_N = X_N \oplus X_1 X_2 \ldots X_{N-1}. \tag{3}$$

The analysis of the generalized Peres gate regarding its circuit design and the quantum cost was made in [12, 13], and the possibility of its physical realization within the framework of the Ising model was presented by us in [14]. In this paper, we have focused on the study of noise effect on the correctness of the generalized Peres gate operation.

The rest of the paper is organized as follows. Section 2 defines the model of the Peres gate based on a chain of coupled nuclear spins in a magnetic field. Section 3 presents the simulation results and discussion of the effect of frequency noise on the correctness of generalized Peres gate operation.

2   The Model of the Peres Gate

The Peres gate can be described by means of a model of interacting nuclear spins (one half) of Phosphorus atoms in the spin-free ^{13}Si silicon matrix, which was discussed in detail in [14]. The Hamiltonian of a chain of coupled nuclear spins in a constant magnetic field can be written in the first approximation taking into account exchange interaction between the nearest (J) and next-nearest (J') neighbors [15]

$$\hat{H}_0 = -\hbar \left(\sum_{k=0}^{N-1} \omega_k I_k^z + 2J \sum_{k=0}^{N-2} I_k^z I_{k+1}^z + 2J' \sum_{k=0}^{N-3} I_k^z I_{k+2}^z \right) \tag{4}$$

The first term describes the interaction of N nuclear spins located at identical distances along the x-axis with a strong constant in time magnetic field $(0, 0, B(x))$, oriented along the z-axis. Here $\omega_k = \gamma B(x_k)$ ($k = 0, 1, 2, \ldots, N-1$) is the Larmore frequency of the k-th spin ($k = 0, 1, 2, \ldots, N-1$); γ is the proton gyromagnetic ratio.

The Hamiltonian (4) is diagonal in the basis of N-spins (qubits) register, which is a tensor product of $|0_k >, |1_k >$ eigenstates of I_k^z operator. The energy spectrum of the system can be found from the solution of the stationary Schrödinger equation [16], and has the form:

$$E_n = E_{i_{N-1} \ldots i_1 i_0} = -\frac{\hbar}{2} \left(\sum_{k=0}^{N-1} (-1)^{i_k} \omega_k + J \sum_{k=0}^{N-2} (-1)^{i_k + i_{k+1}} + J' \sum_{k=0}^{N-3} (-1)^{i_k + i_{k+2}} \right) \tag{5}$$

Transitions between the energy levels of the system are induced by the control pulses of the transverse radiofrequency (RF) magnetic field $(b\cos\omega t, -b\sin\omega t, 0)$ in accordance with Pauli's selection rules. The implementation of the Peres gate in the presented model means providing algorithmic transitions between the energy levels of the system in accordance with the truth-table. We have considered control action of the radiofrequency field in the Hamiltonian of the system as a perturbation:

$$\hat{W}(t) = -\frac{\hbar\Omega}{2}\sum_{k=0}^{N-1}\left[I_k^+ \exp(i\omega t) + I_k^- \exp(-i\omega t)\right] \tag{6}$$

where $I_k^+ = |0_k\rangle\langle 1_k|$ and $I_k^- = |1_k\rangle\langle 0_k|$ are the ascent and descent operators respectively, $\Omega = \gamma b$ is the Rabi frequency. The solution of the nonstationary Schrödinger equation for $H_0 + W(t)$ with the corresponding initial conditions makes it possible to find the probabilities of the allowed transitions. The methodology for solving this problem is described in detail in our previous work [14].

For numerical simulation of the system we consider selective excitation regime, which means realization of such condition:

$$\Omega \ll J \ll \omega$$

Therefore, we set the parameters of the system (in units of 2π MHz) in the following way. Larmore frequencies for N nuclear spins (qubits) are determined by the magnitude of the magnetic field in the z-direction $B(x_k)$ and were chosen as follows

$$\omega_k = 100 \cdot 2^k, \qquad k = 0, 1, \ldots, N-1,$$

which is provided by a certain gradient of the magnetic field along the x-axis. The Rabi frequency of the radiofrequency transverse magnetic field $\Omega = 0.1$. The values of the exchange interaction between the spins of the nearest and next-nearest neighbor qubits are $J = 5$ and $J' = 0.5$ respectively.

One-, two-, and three-qubit gates, namely, NOT, CNOT, and CCNOT gates were tested previously using this model on the basis of ^{12}C-^{13}C diamond crystal [17]. In particular, it was shown that correct operation of these gates is determined by such parameters of the model as magnitude and gradient of the magnetic field, amplitude and frequency of the transverse radiofrequency field, a distance between qubits. An inaccurate setting of the specified parameters can lead to an imbalance of the system and improper operation of the device. To evaluate a measure of good performance of the generalized Peres gate, we will estimate the fidelity parameter [18, 19] as follows:

$$F = \langle \Psi(t)|\Psi_0(t)\rangle \tag{7}$$

where $\Psi_0(t)$ is the ideal wave function and $\Psi(t)$ is the real wave function.

The energy spectrum of the four-qubit system ($N = 4$) can be found from the solution of the stationary problem (5). Generalized Peres gate is realized due to the transitions between energy levels with a rotation of only one spin in accordance with the Pauli principle. Allowed transitions that implement the four-qubit gate are [14]:

$$5|0100> \;\rightarrow\; 7|0110>, \qquad 6|0101> \;\rightarrow\; 8|0111>,$$
$$7|0110> \;\rightarrow\; 5|0100>, \qquad 8|0111> \;\rightarrow\; 6|0101>,$$
$$13|1100> \;\rightarrow\; 9|1000>, \qquad 14|1101> \;\rightarrow\; 10|1001>.$$

Here, the first four digits correspond to the control qubits and the last one to a controlled qubit. For simplicity, we have given both decimal and binary notation. Forbidden transitions

$$9 \rightarrow 15, \ 10 \rightarrow 16, \ 11 \rightarrow 13, \ 12 \rightarrow 14, \ 15 \rightarrow 12, \ \text{and} \ 16 \rightarrow 11$$

can be implemented in two steps, for example, the forbidden transition $9{\rightarrow}15$ can be implemented in two different ways:

$$9|1000> \ \rightarrow 13|1100> \ \rightarrow 15|1110> , 9|1000> \ \rightarrow 11|1010> \ \rightarrow 15|1110> .$$

For allowed transitions, the frequency of the radiofrequency field is determined by the energy of the transition

$$\omega_{mn} = \frac{E_m - E_n}{\hbar}. \tag{8}$$

3 Simulation of the Noise in the Generalized Peres Gate

It is well known that the interaction of the system with the external environment has a destructive effect on the correctness of quantum operations. At the atomic level, in particular, temperature increase leads to the broadening of the energy levels, which is equivalent to the inaccuracy of the setting of the control resonance frequency. In addition, adjustment errors of the magnetic fields lead to a change in the Larmore's and Rabi frequencies, which also causes an imbalance of the system.

Let's consider the effect of frequency noise on the correctness of the generalized Peres gate operation for the digital and superposition quantum states. As quantitative noise characteristic we take relative error η of the transverse radiofrequency magnetic field generator settings at the end of π-pulse ($t_0 = \pi/\Omega$):

$$\omega = \omega_{mn}(1 + \eta). \tag{9}$$

Comparison of the dependence $F(\eta)$ calculated on the digital (blue curve) and the superposition states (red curve) for the algorithmic transition $8 \rightarrow 6$ of the generalized Peres gate ($N = 4$) is presented in Fig. 2. Initial conditions in the superposition case were chosen in the form of uniformly filled states $|k\rangle$ with the probability 1/16:

$$\Psi(0) = \frac{1}{\sqrt{16}} \sum_0^{15} |k\rangle. \tag{10}$$

As follows from Fig. 2, the fidelities for the superposition states are much larger compared to the digital ones, which is particularly true with an increase of the relative error η. In particular, the value of F varies approximately at the level of 0.8. This is the same for two-step transition. Like the last ones, $12 \rightarrow 16 \rightarrow 14$ and $12 \rightarrow 10 \rightarrow 14$

transitions were chosen. As can be seen from Fig. 3, the fidelities for the digital states drop sharply to zero (blue and black curves) as the error of the generator frequency settings increases. At the same time, the value of the fidelity for the superposition states (magenta and orange curves) fluctuates at the level of 0.6 and 0.4 with an increase of relative error η.

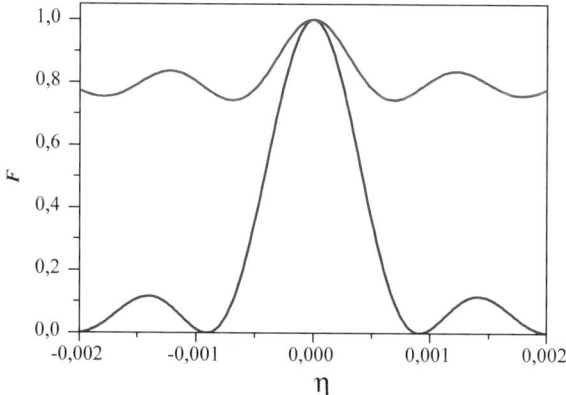

Fig. 2. Fidelity as a function of the relative error η for $8 \to 6$ transition in the four-qubit Peres gate. Superposition states (red), digital states (blue).

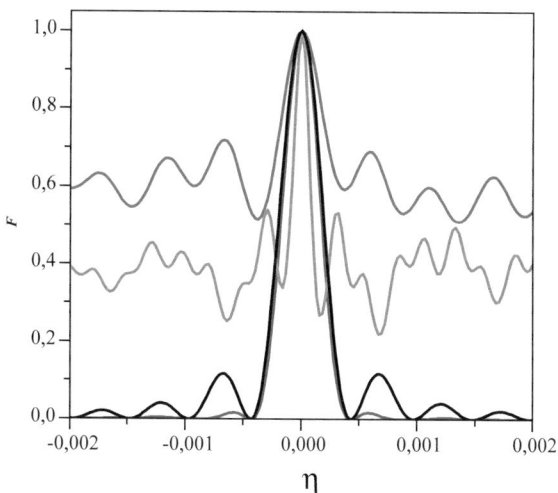

Fig. 3. Fidelity as a function of the relative error η for $12 \to 16 \to 14$ and $12 \to 10 \to 14$ transitions in the four-qubit Peres gate. Digital states (black color – $12 \to 16 \to 14$ transition, blue color – $12 \to 10 \to 14$ transition); superposition states (magenta color – $12 \to 16 \to 14$ transition, orange color – $12 \to 10 \to 14$ transition).

From the analysis, it follows, that more stable operation of the four-qubit Peres gate regarding the error of the generator frequency setting η of the algorithmic transitions is a characteristic feature for the superposition states. Table 1 shows the results of the frequency noise simulation for the Peres gate with different number N of the control signals. Calculated fidelities indicate that the use of the superposition states provides a more stable operation of the generalized Peres gate in respect to an imbalance of the resonant frequency of the transitions. The superposition case is more stable than the digital case due to the nonzero contribution to the fidelity parameter of certain states involved in the transitions. Some terms of uniformly filled initial state always contribute with the constant probability to the final state and the fidelity decays with an increase of the parameter η more slowly than in the digital case.

Table 1. Correctness of the operation of the generalized Peres gate

Number of qubits	Imbalance of the resonance frequency, relative units		Modulation of the resonance frequency, 2π MHz	
	Digital states	Superposition states	Digital states	Superposition states
3	$\|\eta\| \leq 1.4 \cdot 10^{-4}$	$\|\eta\| \leq 6.62 \cdot 10^{-4}$	$\delta \leq 2.22 \cdot 10^{-4}$	$\delta \leq 3.22 \cdot 10^{-4}$
4	$\|\eta\| \leq 7.21 \cdot 10^{-5}$	$\|\eta\| \leq 3.24 \cdot 10^{-5}$	$\delta \leq 1.54 \cdot 10^{-4}$	$\delta \leq 2.16 \cdot 10^{-4}$
5	$\|\eta\| \leq 3.42 \cdot 10^{-5}$	$\|\eta\| \leq 1.47 \cdot 10^{-5}$	$\delta \leq 6.51 \cdot 10^{-5}$	$\delta \leq 8.78 \cdot 10^{-5}$

In addition to the discussed above imbalance of resonance frequency, the known fault of frequency generators is frequency deviations, i.e. small periodic frequency fluctuations over time with respect to the equilibrium value. The effect of time dependence of the generator frequency $\omega = \omega_{mn}\cos(\delta t)$ (δ is the parameter of the time variation of the generator frequency) on the correctness of the Peres gate operation gives similar results (Figs. 4 and 5). The limits of the parameter δ, which ensure correct operation of the gate, were determined from the condition that the fidelity function F reduces to 90%. Besides, as in the case of imbalance of the resonance frequency, the use of the superposition states somewhat extends allowed limits of the time-dependent change of the generator frequency. In addition, the fidelity function in the case of the superposition states takes on values 50–90% in a wide range of the parameter δ. This is especially true for the one-stage transitions (e.g. $8 \rightarrow 6$) where $F \sim 90\%$ virtually regardless of the parameter δ. Calculations for other transitions of the four-qubit gate give similar results, only the value limits of the parameters δ and η are changed (Table 1).

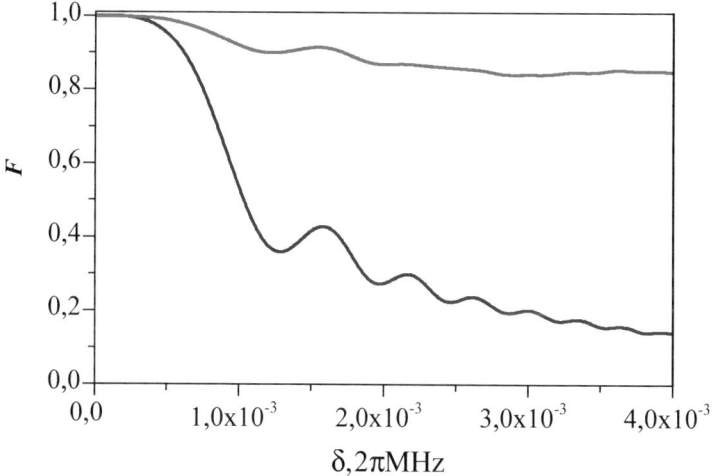

Fig. 4. Fidelity as a function of the parameter δ for 8 → 6 transition in the four-qubit Peres gate for the superposition states (magenta color) and digital states (blue color).

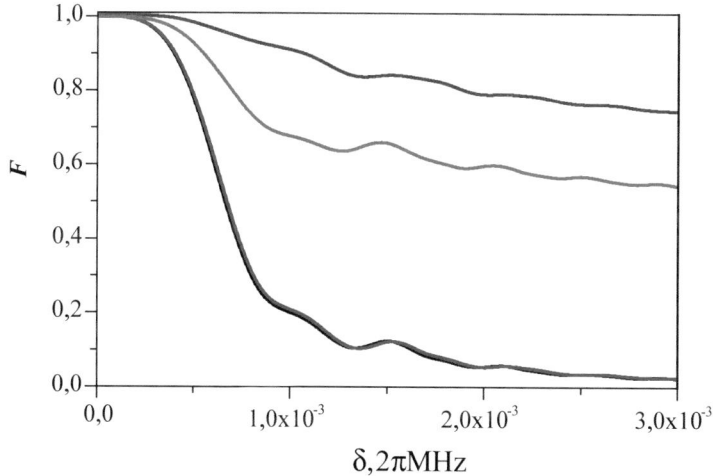

Fig. 5. Fidelity as a function of the parameter δ for 12 → 16 → 14 and 12 → 10 → 14 transitions in the four-qubit Peres gate. Digital states (black color – 12 → 16 → 14 transition, blue color – 12 → 10 → 14 transition); superposition states (red color – 12 → 16 → 14 transition, magenta color – 12 → 10 → 14 transition).

The minimum values of the parameters η and δ, which ensure correct operation of the Peres gates with the number of qubits N = 3, 4, 5 are presented in Table 1. The limits of parameter change necessary for correct gate operation were determined from the analysis of the fidelity function for all possible transitions. For both the digital and superposition states, the limits of correct operation are increased almost linearly with the increase in the number of qubits of the generalized Peres gate. This allows extrapolating the limits to the gate with an arbitrary number of qubits. As follows, although the quantum cost increases linearly with the increase in the number of qubits of the generalized Peres gate [14], the correctness of the gate operation at the same time linearly decreases.

4 Conclusions

A model of generalized (N-qubit) Peres gate based on a chain of nuclear spins one half, which is embedded in a spin-free silicon crystal, is proposed. The effect of the frequency noise on the correctness of the Peres gate operation has been studied for the digital and superposition quantum states. It was shown that an imbalance of the resonance control frequency by magnitude causes a significant decrease in the fidelity of the gate operating on pure digital states. In the case of the superposition states the fidelities are much larger compared to the digital ones. In particular, the fidelity for the one-step transitions fluctuates around the value of 0.8. At the same time, the fidelity for the two-step transitions decreases to 0.6–0.4. A similar effect takes place at a time deviation of the frequency around the resonance. The fidelity for the superposition states takes on values of 0.5–0.9 in a wide range of the parameter δ. This is especially true for the one-stage transitions where the fidelity is of 0.9. The limits of noise parameters that provide correct gate operation of N-qubit Peres gate were determined. It was found a linear reduction of resonance frequency imbalance parameters with an increase in the number of qubits N.

References

1. Gardas, B., Dziarmaga, J., Zurek, W.H., Zwolak, M.: Defects in quantum computers. Sci. Rep. **8**(1), 4539 (2018). https://doi.org/10.1038/s41598-018-22763-2
2. Paler, A., Fowler, A.G., Wille, R.: Reliable quantum circuits have defects. XRDS: Crossroads ACM Mag. Students **23**(1), 34–38 (2016). https://doi.org/10.1145/2983541
3. Knill, E.: Quantum computing with realistically noisy devices. Nature **434**(7029), 39–44 (2005). https://doi.org/10.1038/nature03350
4. Tosi, G., Mohiyaddin, F.A., Schmitt, V., Tenberg, S., Rahman, R., Klimeck, G., Morello, A.: Silicon quantum processor with robust long-distance qubit couplings. Nat. Commun. **8**(1), 11 (2017). https://doi.org/10.1038/s41467-017-00378-x
5. Kane, B.: A silicon-based nuclear spin quantum computer. Nature **393**(6681), 133–137 (1998)
6. Muhonen, J.T., Laucht, A., Simmons, S., et al.: Quantifying the quantum gate fidelity of single-atom spin qubits in silicon by randomized benchmarking. J. Phys.: Condens. Matter **27**(15), 154205 (2015). https://doi.org/10.1088/0953-8984/27/15/154205
7. Pla, J.J., Tan, K.Y., Dehollain, J.P., Lim, W.H., Morton, J.J.L., Zwanenburg, F.A., Jamieson, D.N., Dzurak, A.S., Morello, A.: High-fidelity readout and control of a nuclear spin qubit in silicon. Nature **496**(7445), 334–338 (2013)

8. Linke, N.M., Gutierrez, M., Landsman, K.A., Figgatt, C., Debnath, S., Brown, K.R., Monroe, C.: Fault-tolerant quantum error detection. Sci. Adv. **3**(10), e1701074 (2017). https://doi.org/10.1126/sciadv.1701074

9. Golubitsky, O., Maslov, D.: A study of optimal 4-bit reversible Toffoli circuits and their synthesis. IEEE Trans. Comput. **61**(9), 1341–1353 (2012). https://doi.org/10.1109/TC.2011.144

10. Gupta, P., Agrawal, A., Jha, N.K.: An algorithm for synthesis of reversible logic circuits. IEEE Trans. Comput. Aided Des. Integr. Circ. Syst. **25**(11), 2317–2329 (2006). https://doi.org/10.1109/TCAD.2006.871622

11. Peres, A.: Reversible logic and quantum computers. Phys. Rev. A **32**(6), 3266–3276 (1985). https://doi.org/10.1103/PhysRevA.32.3266

12. Szyprowski, M., Kerntopf, P.: Low quantum cost realization of generalized Peres and Toffoli gates with multiple-control signals. In: Proceedings of the 13th IEEE International Conference on Nanotechnology, pp. 802–807, Beijing (2013)

13. Moraga, C.: Multiple mixed control signals for reversible Peres gates. Electron. Lett. **50**(14), 987–989 (2014)

14. Rozhdov O.I., Yuriychuk I.M., Deibuk V.G.: Building a generalized peres gate with multiple control signals. In: Advances in Intelligent Systems and Computing, vol, 754, pp. 155–164, Springer, Cham (2019)

15. Berman, G.P., Doolen, G.D., Holm, D.D., Tsifrinovich, V.I.: Quantum computer on a class of one-dimensional Ising systems. Phys. Lett. A **193**(5–6), 444–450 (1994)

16. Deibuk, V.G., Yuriychuk, I.M., Yuriychuk, R.I.: Spin model of full adder on Peres gates. Int. J. Comput. **11**(3), 282–292 (2012)

17. Lopez, G.V.: Diamond as a solid-state quantum computer with a linear chain of nuclear spins system. J. Mod. Phys. **5**, 55–60 (2014)

18. Nielsen, M., Chuang, I.: Quantum Computation and Quantum Information. Cambridge University Press, Cambridge (2000)

19. Landauer, R.: Irreversibility and heat generation in the computational process. IBM J. Res. Dev. **5**(1/2), 183–191 (1961)

20. Mendonça, P.E.M., Napolitano, R.D.J., Marchiolli, M.A., Foster, C.J., Liang, Y.-Ch.: Alternative fidelity measure between quantum states. Phys. Rev. A **78**(5), 052330 (2008)

21. Deibuk, V.G., Biloshytskyi, A.V.: Design of a ternary reversible/quantum adder using genetic algorithm. Int. J. Inf. Technol. Comput. Sci. (IJITCS) **7**(9), 38–45 (2015). https://doi.org/10.5815/ijitcs.2015.09.06

22. Bilal, B., Ahmed, S., Kakkar, V.: Optimal realization of universality of peres gate using explicit interaction of cells in quantum dot cellular automata nanotechnology. Int. J. Intell. Syst. Appl. (IJISA) **9**(6), 75–84 (2017). https://doi.org/10.5815/ijisa.2017.06.08

23. Moghimi, S., Reshadinezhad, M.R.: A novel 4 × 4 universal reversible gate as a cost efficient full adder/subtractor in terms of reversible and quantum metrics. IJMECS **7**(11), 28–34 (2015). https://doi.org/10.5815/ijmecs.2015.11.04

24. Sihare, S.R., Nath, V.V.: Analysis of quantum algorithms with classical systems counterpart. Int. J. Inf. Eng. Electron. Bus. (IJIEEB) **9**(2), 20–26 (2017). https://doi.org/10.5815/ijieeb.2017.02.03

Analysis of Propaganda Elements Detecting Algorithms in Text Data

Olena Gavrilenko⑩, Yurii Oliinyk$^{(\boxtimes)}$⑩, and Hanna Khanko⑩

Igor Sikorsky Kyiv Polytechnic Institute, Kyiv 03056, Ukraine
oliyura@gmail.com

Abstract. This article is devoted to the problem of identifying propaganda in text files. Developing methods and techniques that can be used for this analysis is an important task, since the amount of propaganda is enormous. Such amounts of information cannot be analyzed by specialists. A study of how agitation is changing over time is valuable for understanding which areas of our life are particularly covered by propaganda, how its rhetoric is changing, and what impact it will have. All of the above indicates the relevance of research in this area.

The objects of the research are the content of electronic media news, users, and the interrelations between them.

The purpose of this work is to improve the accuracy of the classification of textual information through the appropriate use of existing methods of data mining using the most effective methods of text preprocessing and powerful machine learning algorithms for classification problems.

The methods for solving the problem are considered to solve the problem of classifying text information for spam filtration tasks, contextual advertising, news categorization, creation of subject catalogs.

That is why it is necessary to automate the process of searching, filtering and structuring text data. To solve this problem, the automated classification of texts is used - the task of machine learning from the field of natural language processing. The task of text classification has practical application in many areas, for example, spam filtering, contextual advertising, news categorization, creation of thematic catalogs. Most methods of automatic classification of texts are based on the assumption that the texts of each thematic heading contain certain features, the presence or absence of which indicates the belonging of the text of a rubric. The task of classification methods is to select the best following characteristics and formulate rules, will decide whether to refer the text to a certain category and conduct interactive drilling.

Keywords: Propaganda · Agitation · Convolutional Neural Network · Text data processing · OneClassSVM · Isolation forest · ElipticEnvelope · Association rules

© Springer Nature Switzerland AG 2020
Z. Hu et al. (Eds.): ICCSEEA 2019, AISC 938, pp. 438–447, 2020.
https://doi.org/10.1007/978-3-030-16621-2_41

1 Introduction

Since ancient times propaganda tools have been used to influence the opinion of society. Indeed, propaganda is so powerful that everyone is prone to it. The main channel of accepting propaganda by society is the mass media. The statistics of the media credibility is impressive. Information dissemination is an industry with a turn-over of more than $ 400,000,000,000 per year, of which 206,000,000,000 are spent on mass information, that is, information produced and distributed in identical form to consumers around the world. Everyone thinks verbally and is therefore more or less susceptible to suggestions skillfully used by sales specialists, politicians, journalists, fraudsters, sect organizers, special services and terrorists.

Thanks to the methods of machine and deep learning, it is possible to detect certain patterns in the text data inherent of the propaganda resources. Such technologies will allow users to check content for the presence of special linguistic constructions, words and phrases that contribute to uncritical perception of information [1].

Existing means of text data analysis do not allow obtaining the desired results when solving problems of classification of textual information, so there is a need to develop new algorithms, based on the accumulated information can effectively classify textual information.

This paper discusses the classification algorithms that used to identify propaganda in the news content. The algorithms allowed to divide the data arrays into two classes: "text with propaganda" and "text without propaganda". If you carry out further classification (interactive drilling), you can determine which industry (politics, economy, religion, etc.) the article belongs to. By calculating the ratio of articles that were classified as "text with propaganda" to the total number of articles in this field, it is possible to determine which of them are most susceptible to propaganda.

It should be noted that the proposed method can also be used to analyze posts in social networks.

The article consists chapter "Introduction", "Statement of problem", "Identify the signs", "Discussion and Conclusion". Chapter "Introduction" describes problem and related work. Chapter "Statement of problem" describes mathematical model. Chapter "Identify the signs" describes methods of building signs and test results.

1.1 Related Works

The analysis of propaganda is very important task. Computational propaganda [2] is a term that neatly encapsulates this recent phenomenon—and emerging field of study—of digital misinformation and manipulation. As a communicative practice, computational propaganda describes the use of algorithms, automation, and human curation to purpose-fully manage and distribute misleading information over social media networks [3]. The studies [4] demonstrates both the ease with which malicious actors can harness social media and search engines for propaganda campaigns, and the ability to track and understand such activities by fusing content and activity resources from multiple internet services.

For analysis using many text mining methods and technologies. The work [5] is devoted to the development of a general approach to methods of data mining for the classification of textual information. As the proposed approaches to solving the problem, algorithms TF-IDF, criterion $\chi 2$, gradient boosting algorithm, logistic regression, support vector method and naive Bayesian classifier are considered.

In the paper [6], an innovative similarity estimation method devoted to Kurdish text documents is presented and studied in detail. The method is based on sentences and words, under consideration, and their n-gram phrases. In research [7] provides the accuracy and confidence level of multiple Neural Network architectures, Support Vector Machine and Hidden Markov Model, with the Hidden Markov Model yielding the best results reaching almost 70% accuracy for all languages.

In research [8] using TFT, IDFT transform and proposed approach with SGNB classifier in combination of iterated Lovins stemmer, rainbow stop word removal and word tokenizer. In research [9] proposed model combines several measures in order to model the semantic similarity of words. These measures are: TFIDF, RBF functions, oriented graphs and semantic altruism.

As we see analysis of propaganda is very actual tasks and many methods and technologies can be used for this research.

2   Statement of Problems

The task of classifying text documents can be formulated as an approximation problem for the unknown function Φ: $D \times C \rightarrow \{0,1\}$ (how documents should be classified) via the function K: $D \times C \rightarrow \{0,1\}\}$, which is a classifier, where $C = \{c_1, c_2, \ldots, c_{|C|}\}$ - the set of possible categories, and $D = \{d_1, d_2, \ldots, d_{|D|}\}$- the set of documents

$$\Phi(d_j, c_j) = \begin{cases} 1, d_j \in c_j \\ 0, d_j \notin c_j \end{cases} \tag{1}$$

A document d_j is called a positive example of the category c_i, if $\Phi(d_j, c_j) = 1$, and negative otherwise. If only one category can correspond to each document, then this is a task of unambiguous classification, and if several - of multivalued classification.

Finding of a classifier for a set of categories is considered as a search for binary classifiers.

Description of the data. The Internet Research Agency (IRA) is a well-known Russian "troll" farm whose strategic goal is to cause discord among the American population in the US political system. On February 16, 2018, special adviser Robert S. Muller III accused thirteen Russian citizens and three Russian organizations of interfering in US political and electoral processes. During the hearing, members of the Committee noted the widespread activity of the IRA on Facebook: 3,393 purchased ads, more than 11.4 million US users are exposed to these advertisements; 470 Facebook pages created by IRA employees with an audience of more than 126,000,000 Americans. The Facebook advertising posts used in the study were carefully reviewed by the Minority Committee.

According to the data provided to the Committee by Twitter, an analysis of the relevant activities of the IRA from September 1 to November 15, 2016 shows that more than 36,000 Russian IRA-related tweets about the US election. The bots influence audience is 288 million Americans, the number of tweets linked to IRA accounts is 130, 000.

The data is as follows:

- _unit_id: unique identifier of the entry
- bias: neutral or propaganda
- message:
  - post subject
  - attack: criticism of political persons
  - constituency: criticism of the political constituency
  - information: US and US government news release
  - media: media interaction message
  - mobilization: a message for mobilizing supporters
  - other: other types of messages
  - personal: personal communication, most often with sympathy, support or empathy, or with other personal opinions
  - policy: a message about political policy
  - support: message calling for political support
- embed: HTML message code
- source: the social network where the post was posted: twitter or facebook.
- text: the text of the message.

3 Identify the Signs

Identification of signs is one of the most important steps in learning the machine learning algorithm.

Word2Vec is a model provided by Google that is trained on large text data boxes and represents words as a vector in a 300-dimensional space. Its peculiarity is that words similar in semantics are next to each other. Word2Vec is the "gold standard" for textual data using deep learning. The model studies using two algorithms - the "Continuous Bag-of-Words" (CBOW) model and the Skip-Gram model [10]. Algorithmically, these models are very similar, except that CBOW predicts the target words using words from the context, whereas skip-gram uses the inverse algorithm and predicts the context using the target word. Google provides the ability to use trained Word2Vec in 100-dimensional and 300-dimensional space, you can customize the model for a specific task, touching it on its own body of text.

In solving this problem, the word2vec model in the 300-dimensional space was supported by using political and political posts by Facebook and Twitter users. Case size - 3 GB of text data. Figure 1 shows the generated 300-dimensional vector space of words from the IRA corpus.

The Stanford University research team proposed another method for constructing semantic vector spaces, the model is trained on aggregated global statistics of word use in the body of documents, and the resulting representations demonstrate interesting linear substructures of the resulting vector space. Such a model is called Global Vectors for Word Representation (GloVe) [11]. GloVe can be done on your own corpus of texts, and it was done during the study. A comparative analysis of assessing the efficiency of the classifier using the features of Word2Vec and GloVe is given in the Analysis of Results section.

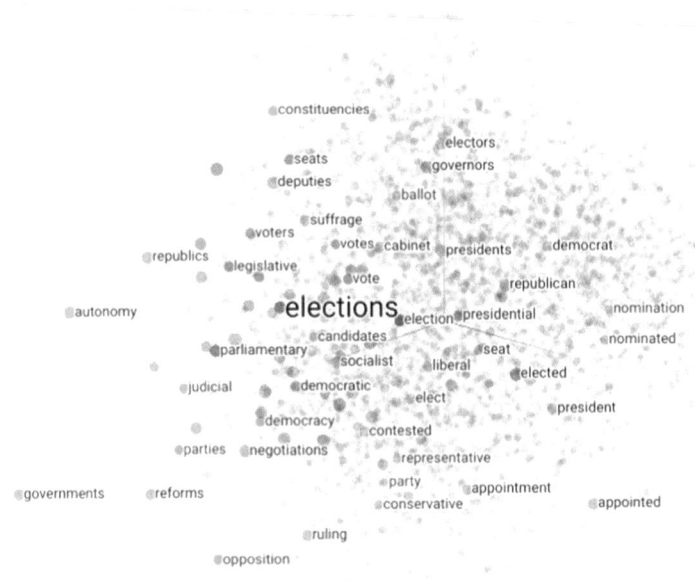

Fig. 1. IRA corpus vector word space

Algorithms Deep Learning. For the task of textual data classification, we used two architectures of neural networks: LSTM (Long Short-Term Memory) recurrent and the multichannel convolutional Neural Network. - CNN (Convolutional Neural Network).

LSTM network is a special case of recurrent neural networks. Information in standard recurrent networks quickly transformed and disappears, does not allow to implement long-term memory. Its architecture uses additional "gates": a residual memory valve, an inlet valve, and an output valve. The residual memory valve implements the forgetting mechanism, which is to determine which information is no longer needed and which one will be used in the future. The input valve determines which input information can be useful and adds it to the long-term memory. The output valve determines the information elements that are most likely to be needed in the near term [12].

To solve the problem, a hierarchical bidirectional LSTM (H-LSTM) is used. Words in vector form are fed to the bidirectional LSTM at the sentence level. The final states of the forward and reverse LSTM are combined together and sent to the two-level LSTM layer. At each stage, the raw data of the direct and inverse LSTM are combined and fed into the final layer, which calculates the probability distribution from the propaganda coloring of the text. The deep network architecture is shown in Fig. 2.

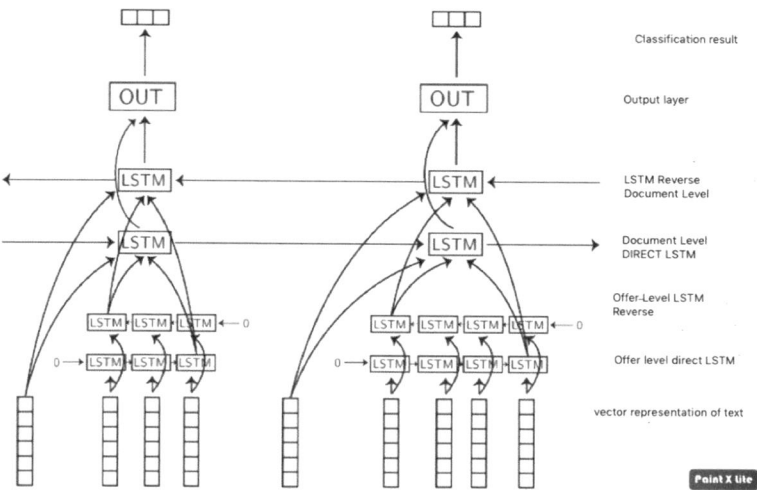

Fig. 2. Hierarchical recurrent network architecture

Consider also used multichannel convolutional Neural Network. Convolutional neural networks are most often used in computer vision problems, but studies show that convolutional neural networks are also successfully used for processing natural language [13].

Each convolution exhibits a specific special pattern. By changing the size of the nuclei and combining their outputs, patterns of any size can be identified [14]. The convolution layer multiplies the filter value by the original values of the vector representations of the words, after which all the results are summarized. Patterns can be expressions, for example, "I hate", "very well", and therefore convolutional networks can identify them in a sentence, regardless of position [15].

It is generally accepted to periodically insert an assembly layer between each other in successive convolution layers in the architecture of ConvNet. Its function is to reduce gradually the number of parameters. The final layer is the fully-connected layer with the decimation method. The main idea of which is instead of teaching a single neural network to teach an ensemble of several neural networks, and then to average the results obtained [16]. Networks for learning are obtained by dropping out neurons with a certain probability q, so the probability that the neuron will remain in the network is p = 1 − q. The "exception" of a neuron means that for any input data or parameters it returns 0 [17].

A fully connected layer outputs an N-dimensional vector to define a class. The work is organized by referring to the feature map obtained in the previous step and determining the properties that are most characteristic of a particular class. The architecture of the convolutional Network is shown in Fig. 3.

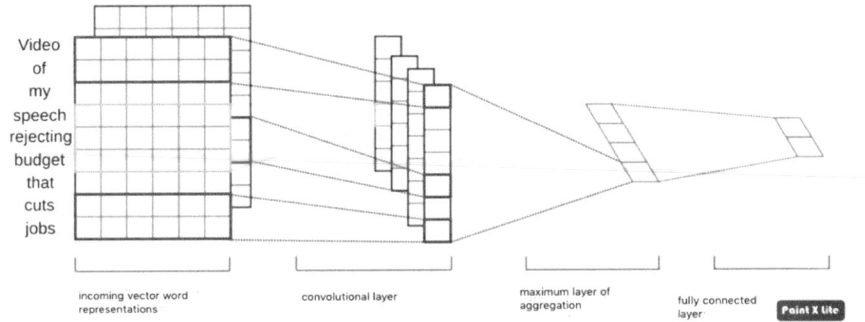

Fig. 3. Architecture of Convolutional Neural Network

Results. The results suggest that the best accuracy and F-score has a model that uses features built by means of the word2vec model and with the Convolutional Neural Network contingent neural network architecture (Table 1).

Table 1. Test results

Method of building signs	Neural network architecture	Accuracy	F-measure
TF-IDF	LSTM	0.752	0.7
TF-IDF	H-LSTM	0.788	0.74
TF-IDF	CNN	0.761	0.72
Word2Vec	LSTM	0.837	0.80
Word2Vec	H-LSTM	0.882	0.82
Word2Vec	CNN	0.877	0.83
Glove	LSTM	0.843	0.77
Glove	H-LSTM	0.858	0.79
Glove	CNN	0.846	0.79

Figures 4, 5 and 6 show growth graphs of the accuracy of models on training data and validation data. Based on the graphs, it can be argued that the model built using CNN has achieved high validation accuracy. Figure 7 shows how long one epoch of the algorithm takes for each architecture.

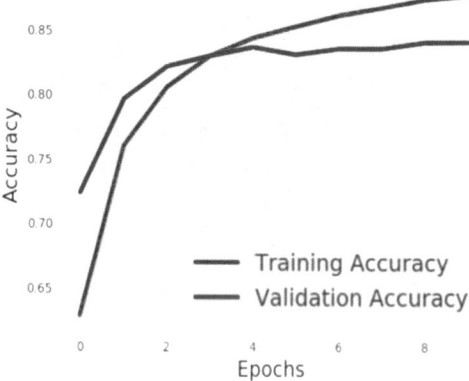

Fig. 4. Accuracy graph for H-LSTM network

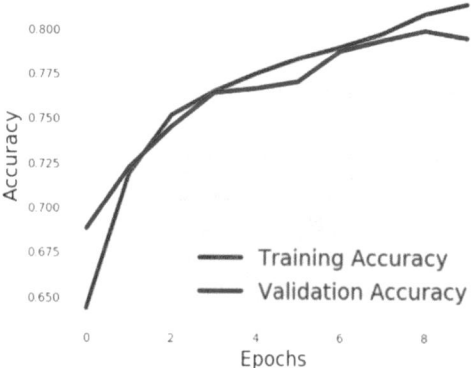

Fig. 5. Accuracy graph for LSTM network

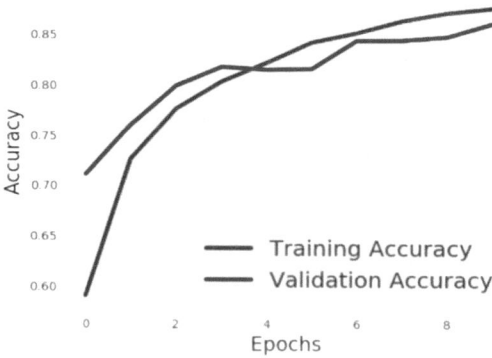

Fig. 6. Accuracy graph for CNN network

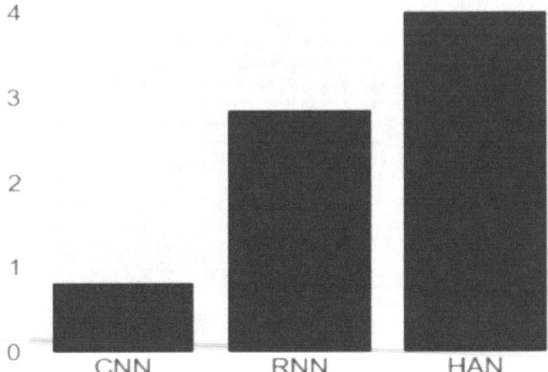

Fig. 7. Dependence of the time of training of one epoch

So, the CNN convolutional neural network algorithm has the best accuracy and the best F-measure result.

4   Discussion and Conclusions

This article is devoted to the problem of identifying propaganda in text files. Developing methods and techniques that can be used for this analysis is an important task, since the amount of propaganda is enormous. Such amounts of information cannot be analyzed by specialists. A study of how agitation is changing over time is valuable for understanding which areas of our life are particularly covered by propaganda (politics, religion, etc.), how its rhetoric is changing, what impact it will have. All of the above indicates the relevance of research in this area.

Modern neural network architectures, such as CNN, LSTM, H-LSTM, are effective for solving text data processing problems. Convolutional network CNN showed the best result −88.2% accuracy.

One of the advantages gained by computerized methods is that large amounts of data can be analyzed, and it is easy to study changes over time. The algorithm allows to realize the right of any person to be informed about the use of forced propaganda. To accept or not this influence is the personal choice of everyone.

A further development of our research is the prediction of the influence of articles or posts classified as propaganda on the opinions and decision-making of readers.

References

1. Agirre, E., Edmonds, P.: Word Sense Disambiguation: Algorithms and Applications, pp. 5–7. Springer Publishing Company Incorporated (2007)
2. Woolley, S.C., Howard, P.N. (eds.): Computational Propaganda: Political Parties, Politicians, and Political Manipulation on Social Media. Oxford University Press (2018)

3. Wooley, S., Howard, P.: Political communication, computational propaganda, and autonomous agents: Introduction. Int. J. Commun. **10**, 4882–4890 (2016)
4. Spangher, A., Ranade, G., Nushi, B., Fourney, A., Horvitz, E.: Analysis of Strategy and Spread of Russia-sponsored Content in the US in 2017. arXiv preprint arXiv:1810.10033 (2018)
5. Gavrilenko, O., Oliynik, Y., Khanko, H.: Review and analysis of algorithms text mining. Project management, systems analysis and logistics, vol. 19, pp. 32–40 (2017)
6. Wakil, K., Ghafoor, M., Abdulrahman, M., Tariq, S.: Plagiarism detection system for the Kurdish language. Int. J. Inf. Technol. Comput. Sci. (IJITCS) **9**(12), 64–71 (2017). https://doi.org/10.5815/ijitcs.2017.12.08
7. Gazeau, V., Varol, C.: Automatic spoken language recognition with neural networks. Int. J. Inf. Technol. Comput. Sci. (IJITCS) **10**(8), 11–17 (2018). https://doi.org/10.5815/ijitcs.2018.08.02
8. Panda, M.: Developing an efficient text pre-processing method with sparse generative Naive Bayes for text mining. Int. J. Mod. Educ. Comput. Sci. (IJMECS) **10**(9), 11–19 (2018). https://doi.org/10.5815/ijmecs.2018.09.02
9. Zaki, T., Bazzi, M.S.E.L., Mammass, D.: An evolutionary model for selecting relevant textual features. Int. J. Mod. Educ. Comput. Sci. (IJMECS) **10**(11), 43–50 (2018). https://doi.org/10.5815/ijmecs.2018.11.06
10. Jurafsky, D., Martin, J.H.: Speech and Language Processing: An Introduction to Natural Language Processing, Computational Linguistics, and Speech Recognition. Pearson Education, Inc., pp. 130–156 (2009)
11. Wang, Y.: Various approaches in text pre-processing. TM Work Paper No, vol. 2, no. 5, pp. 1–3 (2004)
12. Berger, J.: The metronome of apocalyptic time: social media as carrier wave for millenarian contagion. Perspect. Terror. **9**, 2–12 (2015)
13. Vinciarelli, A.: Noisy text categorization, pattern recognition. In: 17th International Conference on ICPR 2004, pp. 554–557 (2004)
14. Reid, S.: 10 misconceptions about neural networks, pp. 2–4, May 2014
15. Olah, C.: Understanding LSTM networks. colah's blog, 27 August 2015. http://colah.github.io/posts/2015-08-Understanding-LSTMs/
16. Han, J., Kamber, M.: Data Mining Concepts and Techniques, pp. 12–16. Morgan Kaufmann (2006)

Digital Filters Optimization Modelling with Non-canonical Hypercomplex Number Systems

Yakiv Kalinovsky[1], Yuliya Boyarinova[2], Iana Khitsko[2(✉)], and Liubov Oleshchenko[2]

[1] Institute for Information Recording, National Academy of Science of Ukraine, Shpaka Street 2, Kyiv 03113, Ukraine
kalinovsky@i.ua
[2] National Technical University of Ukraine "Igor Sikorsky Kyiv Polytechnic Institute", Peremogy Avenue 37, Kyiv 03056, Ukraine
ub@ua.fm, iana.khitsko@gmail.com,
oleshchenkoliubov@gmail.com

Abstract. Recursive digital filter modelling is one of the tasks, which modelling can be improved by using hypercomplex numbers. Existing models are about data representation in canonical hypercomplex number system only. However, canonical number systems have some restrictions. Applying the non-canonical number systems gives more possibilities for filter simulation and its further optimization by its parametric sensitivity since they have more structure constants in Keli table.

The paper proposes a digital filter synthesis method, which is using non-canonical hypercomplex number systems. Use of non-canonical hypercomplex number system with greater number of non-zero structure constants in Keli table can significantly improve the sensitivity of the digital filter.

Keywords: Hypercomplex numbers · Non-canonical number system · Digital filter sensitivity · Filter sensitivity

1 Introduction

Using hypercomplex numerical systems (HNS) for data representation and processing allow to improve the efficiency of various tasks simulation [1–6]. Among such tasks is recursive digital filter modelling with the data representation in hypercomplex numbers. There are recursive filters models with hypercomplex coefficients in canonical HNS which allow to reduce the total filter parametric sensitivity up to 20% comparing to filters with real coefficients [7, 8]. However, the number of systems that can be used for such models is limited by system dimension and multiplication table (Keli table) form [9–11].

Each cell of Keli table for canonical HNS can have only one non-zero constant per structure unit. Non-canonical HNS has more complex multiplication table form. Each table cell can have up to n non-zero constants per structure units where n is HNS dimension. Therefore, non-canonical HNS number of specific dimension is endless

© Springer Nature Switzerland AG 2020
Z. Hu et al. (Eds.): ICCSEEA 2019, AISC 938, pp. 448–458, 2020.
https://doi.org/10.1007/978-3-030-16621-2_42

which gives huge possibilities for research. In particular, it is building the digital filter with the coefficients in non-canonical HNS with better parametric sensitivity performance.

The general approach of using canonical HNS in the amplitude-frequency characteristics construction is already described in [7, 8]. Calculation of filter sensitivity with hypercomplex numbers is investigated in [12–14]. Use of non-canonical HNS makes possible to synthesize the reversible digital filters with better parametric sensitivity performance.

2   Digital Filters with Real and Hypercomplex Coefficients Equivalenting Method

Filter structure synthesis method by converting the digital reversible filter of n order with real coefficients to the digital reversible filter of the first order with hypercomplex coefficients will be used, described in detail in [7].

Consider a digital filter of order 3 with real coefficients, with transfer function:

$$H_R = \frac{\varphi_3 z^{-3} + \varphi_2 z^{-2} + \varphi_1 z^{-1} + \varphi_0}{\phi_3 z^{-3} + \phi_2 z^{-2} + \phi_1 z^{-1} + 1}. \tag{1}$$

Then, following the procedure described in [12], it can be converted to the first-order digital filter which frequency response is:

$$H_\Gamma = \frac{A + Bz^{-1}}{1 + Cz^{-1}} = \frac{(A + Bz^{-1}) \cdot (\overline{1 + Cz^{-1}})}{N(1 + Cz^{-1})}. \tag{2}$$

Hypercomplex coefficients A, B, C belong to some HNS of dimension 3. In (2) conjugate and norm N are defined in accordance with the formulas defined for HNS, which is used.

Let us consider non-canonical HNS of dimension 3 with multiplication table, which has 4 non-canonical non-zero constants. HNS $\Gamma(e, 3)$, which multiplication table is represented in Table 1, is isomorphic to hypercomplex number system $R \oplus C$, which is sum of real and complex numbers with the multiplication table, which is represented in Table 2.

Table 1. Keli table for HNS $\Gamma(e, 3)$

$\Gamma(e, 3)$	e_1	e_2	e_3
e_1	e_1	e_2	e_3
e_2	e_2	$-e_1 + e_3$	$-2e_2$
e_3	e_3	$-2e_2$	$2e_1 - e_3$

Table 2. Keli table for HNS $R \oplus C$

$R \oplus C$	E_1	E_2	E_3
E_1	E_1	0	0
E_2	0	E_2	E_3
E_3	0	E_3	$-E_2$

From the given multiplication tables we can see that calculations in system $R \oplus C$ are much easier. This fact will be used to enhance the digital filter performance.

Then the coefficients of the transfer function H_Γ are of the form: $A = a_1 e_1 + a_2 e_2 + a_3 e_3$, $B = b_1 e_1 + b_2 e_2 + b_3 e_3$, $C = c_1 e_1 + c_2 e_2 + c_3 e_3$, $A, B, C \in \Gamma(e, 3)$.

When substituted to (2) and transformed we obtain the first-order digital filter transfer function with hypercomplex coefficients in $\Gamma(e, 3)$:

$$H_\Gamma = \frac{a_1 + \frac{K}{z} + \frac{M}{z^2} + \frac{L}{z^3}}{1 + \frac{T}{z} + \frac{P}{z^2} + \frac{Q}{z^3}} \tag{3}$$

where

$$K = a_2 c_2 - a_3 c_3 - 3a_1 c_3 + 2a_1 c_1 + b_1;$$

$$M = -2b_3 c_3 + c_2 a_2 c_3 + c_2 a_2 c_1 - 3a_1 c_1 c_3 + c_2 b_2 - 2a_3 c_1 c_3 + 4a_3 c_3^2 + 2a_1 c_2^2 + a_1 c_1^2 + 2a_3 c_2^2 \\ -3b_1 c_3 + 2a_1 c_3^2 + 2b_1 c_1;$$

$$L = c_2 b_2 c_3 + b_1 c_1^2 - 2b_3 c_1 c_3 + c_2 b_2 c_1 + 2b_1 c_2^2 - 3b_1 c_1 c_3 + 2b_1 c_3^2 + 2b_3 c_2^2 + 4b_3 c_3^2$$

$$T = 3c_1 - 3c_2;$$

$$P = -6c_1 c_3 + 3c_2^2 + 3c_1^2;$$

$$Q = 3c_1 c_2^2 + 3c_2^2 c_3 + c_1^3 - 3c_1^2 c_3 + 4c_3^3.$$

Consider a specific example of a third-order filter with real coefficients and the transfer function [7]:

$$H = \frac{0.287589 + 0.6888683 \cdot z^{-1} + 0.6888683 \cdot z^{-2} + 0.287589 \cdot z^{-3}}{1 + 0.418204 \cdot z^{-1} + 0.473048 \cdot z^{-2} + 0.061292 \cdot z^{-3}}$$

Now the coefficient values $a_1, a_2, a_3, b_1, b_2, b_3, c_1, c_2, c_3$ shall be obtained.

Using the method described above, the coefficients of the denominator are equated with the same z^{-i} and the values of hypercomplex coefficients $a_1, a_2, a_3, b_1, b_2, b_3, c_1, c_2, c_3$ are found. Thus, we obtain the following system:

$$\begin{cases} T = 3c_1 - 3c_2 = 0.418204; \\ P = -6c_1c_3 + 3c_2^2 + 3c_1 = 0.473048; \\ Q = 3c_1c_2^2 + 3c_2^2c_3 + c_1^3 - 3c_1^2c_3 + 4c_3^3 = 0.061292, \end{cases} \quad (4)$$

where $c_1 = 0.14032523$, $c_2 = -0.371821$, $c_3 = 0.0009239$.

Substitute these values to the transfer function (3) numerator and equate to the corresponding coefficients of the transfer function of the real filter:

$$\begin{cases} a_1 = 0.287589 \\ 0.2778788a_1 - 0.371821a_2 - 0.0018478a_3 \\ + b_1 = 0.6888683 \\ -0.0018478b_3 - 0.0525194a_2 + 0.295806a_1 \\ -0.3718209b_2 + 0.276246a_3 + 0.2778788b_1 = 0.6888683 \\ -0.0525194b_2 + 0.2958055b_1 + 0.276246b_3 = 0.287589 \end{cases} \quad (5)$$

Express a_1, a_2, b_1, b_3 by a_3, b_2 and obtain the filter parameters as a function of variables a_3, b_2.

$$\begin{aligned} a_1 &= 0.287589 \\ a_2 &= 8.4463122 - 5.3701297a_3 + 7.2213929b_2 \\ b_1 &= 3.7494689 - 1.9948788a_3 + 2.6850649b_2 \\ b_3 &= -2.9738909 + 2.1361279a_3 - 2.685065b_2. \end{aligned} \quad (6)$$

3 Parametric Sensitivity Optimization Method

Digital filter parametric sensitivity is the sensitivity of the digital filter transfer function $|H(w)|$ to the coefficients variations of the filter transfer function [10–12]. Parametric sensitivity function allows us to analyze the impact of coefficients error on the output signal. For the filters with hypercomplex coefficients research we need to consider the possible cumulative error for each of the transfer function coefficients.

Total parametric sensitivity of a first-order filter with hypercomplex coefficients in HNS of dimension 3 is defined by the formula [8]:

$$RCS = \left| \sum_{i=1}^{9} \frac{\alpha_i}{|H|} \cdot \frac{\partial |H|}{\alpha_i} \right| \quad (7)$$

where $\alpha_1 = a_1$, $\alpha_2 = a_2$, $\alpha_3 = a_3$, $\alpha_4 = b_1$, $\alpha_5 = b_2$, $\alpha_6 = b_3$, $\alpha_7 = c_1$, $\alpha_8 = c_2$, $\alpha_9 = c_3$.

Parameters a_3, b_2 can be of any value in system (6). Suppose, that $a_3 = b_2 = 0$. Then, total parametric sensitivity of the filter with hypercomplex coefficients will be built by the formula (7). But overall sensitivity will be greater than for the filter with real coefficients (see Fig. 1) [14].

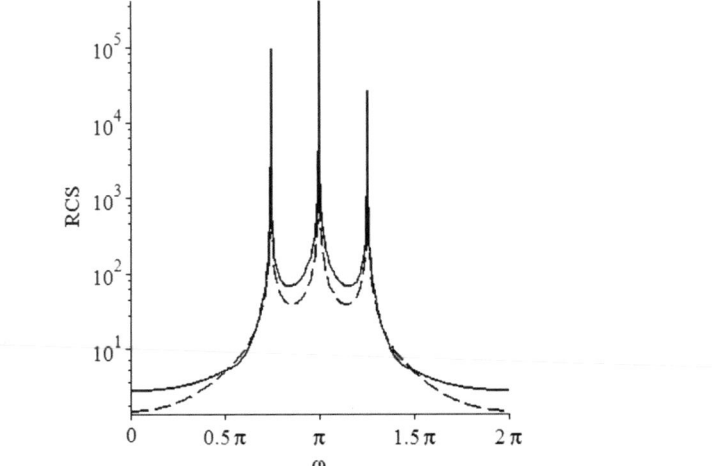

Fig. 1. Total parametric sensitivity for the filters with real (- - -) and hypercomplex coefficients in system $\Gamma(e, 3)$ and $a_3 = b_2 = 0(\text{———})$.

Ratio plot of the total parametric sensitivity of the filter with hypercomplex coefficients to the total parametric sensitivity of the filter with real coefficients is presented in Fig. 2. In this case, the hypercomplex filter sensitivity is much higher than one of a real filter.

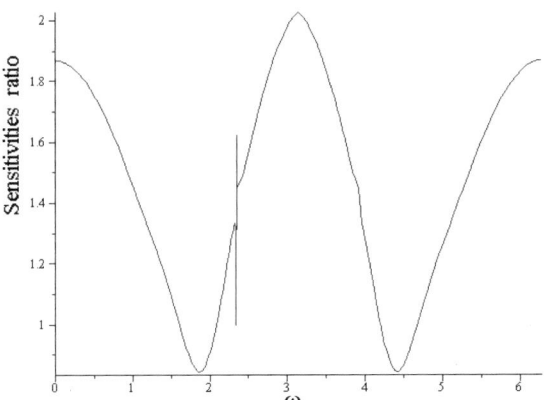

Fig. 2. Total parametric sensitivity RCS of the filter with hypercomplex coefficients to the total parametric sensitivity of the filter with real coefficients ratio.

Filter parameters a_3, b_2 may take different values without altering the transfer function, as we can see from (6). This can be used to optimize the filter parametric sensitivity.

So, values for a_3, b_2 shall be chosen to satisfy conditions (5) and at the same time to optimize a certain criterion. Calculate total sensitivity function to build this criterion, expressing all its components via a_3, b_2.

$$RCS = 2.87589 \cdot 10^5 \left| (z(z^2 + 0.277879z + 0.295806)) / (6.888683 \cdot 10^5 z + 6.888683 \cdot 10^5 z^2 + \right.$$

$$+2.87589 \cdot 10^5 z^2 + 2.87589 \cdot 10^5 - 0.000043zb_2 - 0.0001za_3 - 0.001z^2a_3 - 0.0004b_2) \Big| +$$

$$+3.718209 \cdot 10^5 (8.446331 - 5.37013a_3 + 7.221393b_2) | ((z^3 + 0.418204z^2 + 0.473048z +$$

$$+0.061292)z) / ((z^2 + 0.276955z + 0.433928) \cdot (6.888683 \cdot 10^5 z + 6.888683 \cdot 10^5 z^2 + 2.87589 \cdot 10^5 z^3 +$$

$$+2.87589 \cdot 10^5 - 0.000043zb_2 - 0.0001za_3 - 0.001z^2a_3 - 0.0004b_2)) \Big| + 2 \cdot 10^6 a_3 \cdot |z(0.000924z -$$

$$-0.138123)) / (6.888683 \cdot 10^5 z + 6.888683 \cdot 10^5 z^2 + 2.87589 \cdot 10^5 z^3 + 2.875889 \cdot 10^5 - 0.000043zb_2 -$$

$$-0.0001za_3 - 0.001z^2a_3 - 0.0004b_2) \Big| + 10^6 (3.749469 - 1.994879a_3 + 2.685065b_2) |z^2 + 0.277879z +$$

$$+0.295806) / (6.888683 \cdot 10^5 z + 0.688868 \cdot 10^5 z^2 + 2.87589 \cdot 10^5 z^3 + 2.87589 \cdot 10^5 - 0.000043zb_2 -$$

$$-0.0001za_3 - 0.001z^2a_3 - 0.0004b_2) \Big| + 3.718209 \cdot 10^5 b_2 \left| (z^3 + 0.418204z^2 + 0.473048z + 0.061292) / \right.$$

$$/((z^2 + 0.276955z + 0.433928)(6.888683 \cdot 10^5 z^2 + 6.888683 \cdot 10^5 z^2 + 2.87589 \cdot 10^5 z^3 +$$

$$+2.87589 \cdot 10^5 - 0.000043zb_2 - 0.0001za_3 - 0.001z^2a_3 - 0.0004b_2)) \Big| + 2 \cdot 10^6 (2.97389 +$$

$$+2.1361286a_3 - 2.685065b_2) \cdot |(0.000924z + 0.138123) / (6.888683 \cdot 10^5 z + 6.888683 \cdot 10^5 z^2 +$$

$$+2.87589 \cdot 10^5 z^3 + 2.87589 \cdot 10^5 - 0.000043zb_2 - 0.0001za_3 - 0.001z^2a_3 - 0.0004b_2) \Big| +$$

$$+1.403252 \cdot 10^5 \left| (0.201093z + 0.808015z^2 + 0.396787z^3 - 0.071185 + 2.685065z^4b_2 - \right.$$

$$-1.994878z^4a_3 + 1.502168z^3b_2 - 1.392548z^3a_3 + 0.343983zb_2 + 1.428774z^2b_2 - 1.177149z^2a_3 -$$

$$-0.386364za_3 - 0.287589z^5 + 2.371732 - 0.034218a_3 + 0.023246b_2) / ((6.888683 \cdot 10^5 z +$$

$$+6.888683 \cdot 10^5 z^2 + 2.87589 \cdot 10^5 z^3 + 2.87589 \cdot 10^5 - 0.0001za_3 - 0.001z^2a_3 - 0.000043zb_2 -$$

$$-0.0004b_2)(z^3 + 0.488204z^2 + 0.473048z + 0.06129)) \Big| - 3.71821 \cdot 10^5 \left| ((3.899735z + \right.$$

$$+2.553111z^2 + 0.9986 + 8.446312z^3 - 5.37013z^3a_3 + 3z^2b_2 - 0.6453746a_3 + 0.439284b_2)(z^3 +$$

$$+8.446312z^3 - 5.37013z^3a_3 + 3z^2b_2 - 0.645375a_3 - 0.439284b_2) \cdot (z^3 + 0.418204z^2 +$$

$$+0.473048z + 0.061292)) / ((z^2 + 0.276955z + 0.433928)^2 (6.888683 \cdot 10^5 z + 6.888683 \cdot 10^5 z^2 +$$

$$+2.87589 \cdot 10^5 z^3 + 2.87589 \cdot 10^5 - 0.000043zb_2 - 0.0001za_3 - 0.001z^2a_3 - 0.0004b_2)) \Big| +$$

$$+1847.786738 \left| (-0.4416591z - 1.609719z^2 - 1.098032z^3 - 0.07419 - 2.685065z^4b_2 + \right.$$

$$+1.29972z^4a_3 - 1.502168z^3b_2 + 0.369748z^3a_3 - z^5a_3 - 0.343983zb_2 - 1.428774z^2b_2 +$$

$$+0.803374z^2a_3 + 0.16412za_3 - 3.306609z^4 + 0.007622a_3 - 0.023246b_2) / ((6.888683 \cdot 10^5 z +$$

$$+6.888683 \cdot 10^5 z^2 + 2.875889 \cdot 10^5 z^3 + 2.875889 \cdot 10^5 - 0.000043zb_2 - 0.0001za_3 - 0.001z^2a_3 -$$

$$-0.0004b_2)(z^3 + 0.418204z^2 + 0.473048z + 0.06129)) \Big|$$

Sensitivity function is positive on the whole interval $\omega = 0..2\pi$, so the optimality criterion can take the parametric sensitivity sum for a certain set of values ω, with parameters values a_3, b_2. We select the 33 evenly distributed points on the interval $\{0..2\pi\}$ and calculate the function values at each point, given the fact that $z = \sin(\omega) + i \cdot \cos(\omega)$. Then construct the optimality criterion $S_{RCS}(\omega, a_3, b_2)$ to minimize.

The function $S_{RCS}(\omega, a_3, b_2)$ is too cumbersome and can't be represented in scope of this article. It was also unsuccessful to apply a gradient optimization method, since function differentiation by the components a_3, b_2 is very complex.

It is sufficient to find the approximate optimum to prove the described method efficiency of digital filter synthesis. This became possible with help of function three-dimensional graph construction, which used procedures of analytical calculations MAPLE. At the same time it can be a multistage procedure: first select the wide scope of the search, then it narrows. Accordingly, in Fig. 3 a wide-range search area is presented, in Fig. 4 – narrowed one.

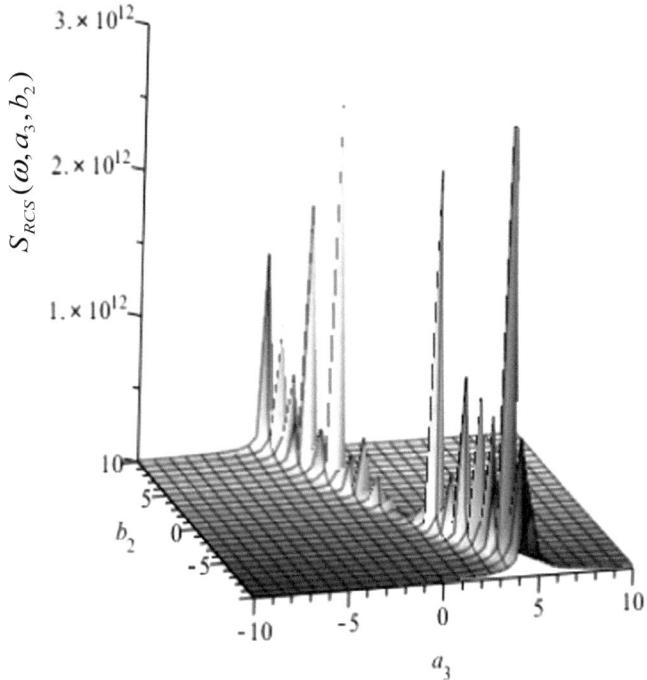

Fig. 3. $S_{RCS}(\omega, a_3, b_2)$ plot wide range search area; $a_3 \in \{-10..10\}, b_2 \in \{-10..10\}$.

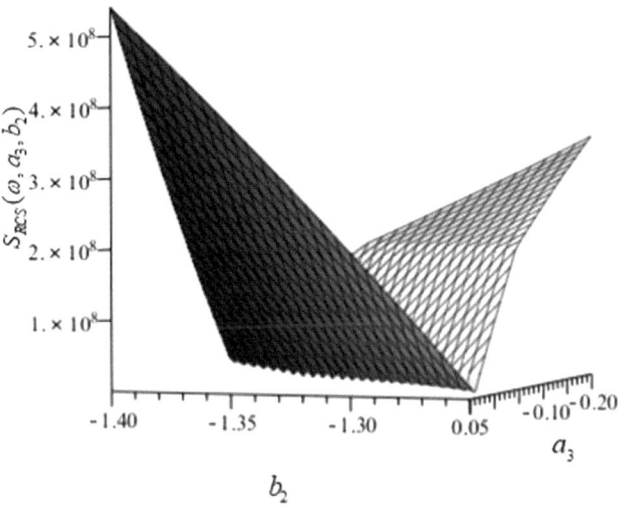

Fig. 4. $S_{RCS}(\omega, a_3, b_2)$ plot for narrow range search area; $a_3 \in \{-0.25..0.05\}$, $b_2 \in \{-1.4.. - 1.25\}$.

Figure 5 is showing the parametric sensitivity changes graph near one of the obtained local minimum with $a_3 = -0.2316615$, $b_2 = -1.2783899677$.

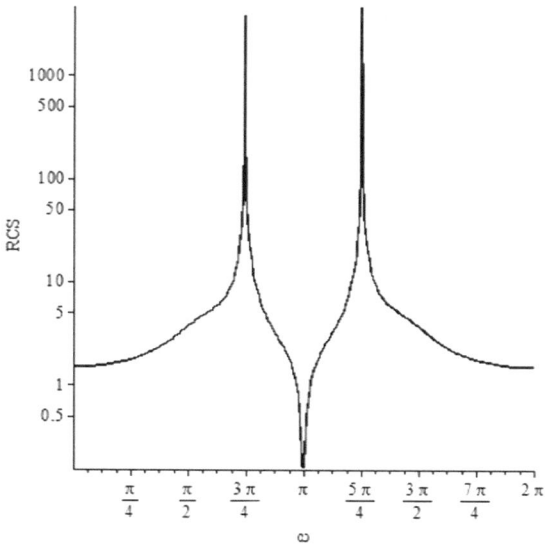

Fig. 5. Total parametric sensitivity *RCS* of the filter with hypercomplex coefficients after optimization procedure depending on the frequency ω.

The sensitivity of the filter with hypercomplex coefficients in $\Gamma(e, 3)$ system to the sensitivity of the filter with real coefficients ratio is shown in the Fig. 6, which shows that the hypercomplex filter sensitivity is lower than the one of the real filter.

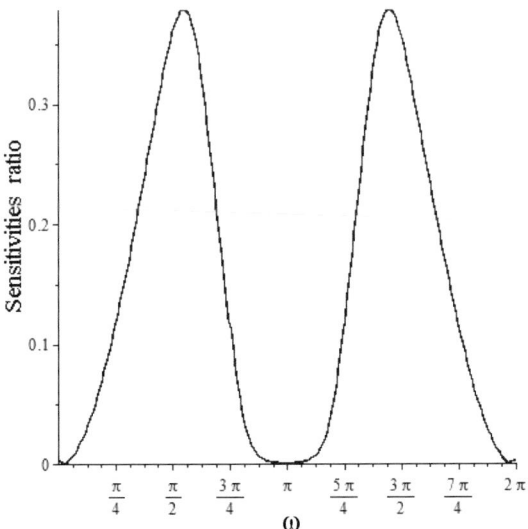

Fig. 6. Ratio of the sensitivities of the filter with hypercomplex coefficients in $\Gamma(e, 3)$ to the one with real coefficients.

In Fig. 7 it is shown the total sensitivity of filters before and after the optimization procedure.

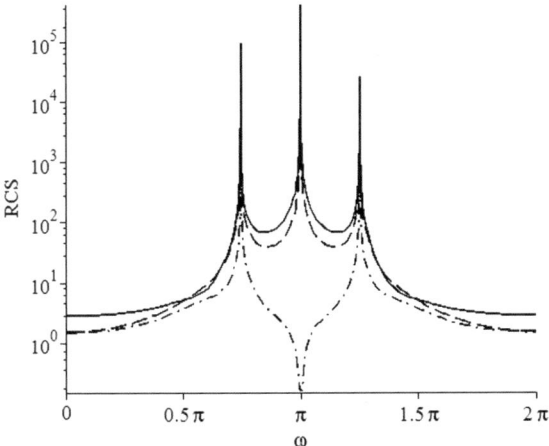

Fig. 7. Total parametric sensitivity for the filters with real (- - -) and hypercomplex coefficients in system $\Gamma(e, 3)$ before (———) and after (-----) optimization.

To implement the digital filter model the analytical and software tools are expanded in the MAPLE system. In particular, these tools include the definition of basic properties and operations for non-canonical HNS; non-canonical HNS structure construction according to given criteria; set of non-canonical HNS determination for representing the digital filter coefficients [15, 16].

The main results of the digital filter sensitivity optimization are represented in Table 3. It shows the averaged parametric sensitivity change of the filter with coefficients in the non-canonical HNS calculated using the optimality criterion compared to the filter with real coefficients. In comparison with the known works on parametric sensitivity optimization of the filters with hypercomplex coefficients [10, 12], we get the maximum decrease in sensitivity up to 43%. It is possible to further optimize parametric sensitivity by searching and analyzing other non-canonical HNS.

Table 3. Total filter parametric sensitivity change with different HNS

Total parametric sensitivities for different filters in comparison	Average change in total parametric sensitivity, %
Sensitivity for filter with real coefficients and filter with coefficients in $\Gamma(e, 3)$ before the optimization	+63%
Sensitivity for filter with real coefficients and filter with coefficients in $\Gamma(e, 3)$ after the optimization	−43%

4   Conclusions

In this article a digital filter synthesis method is proposed with coefficients in non-canonical HNS. At the same time, total parametric sensitivity of such filter is increased for the case when some random coefficients are used.

Total parametric sensitivity decrease method is implemented. It is shown that this method application can decrease filter parametric sensitivity up to 43%. At the same time, precise optimization of the target function $S_{RCS}(\omega, a_3, b_2)$ requires additional research.

References

1. Rajankar, O.S., Kolekar, U.D.: Scale space reduction with interpolation to speed up visual saliency detection. Int. J. Image Graph. Signal Process. (IJIGSP) **7**(8), 58–65 (2015). https://doi.org/10.5815/ijigsp.2015.08.07
2. Khalil, M.I.: Applying quaternion Fourier transforms for enhancing color images. Int. J. Image Graph. Signal Process. (IJIGSP) **4**(2), 9–15 (2012). https://doi.org/10.5815/ijigsp.2012.02.02
3. Hata, R., Akhand, M.A.H., Islam, M.M., Murase, K.: Simplified real-, complex-, and quaternion-valued neuro-fuzzy learning algorithms. Int. J. Intell. Syst. Appl. (IJISA) **10**(5), 1–13 (2018). https://doi.org/10.5815/ijisa.2018.05.01

4. Kumar, S., Tripathi, B.K.: On the root-power mean aggregation based neuron in quaternionic domain. Int. J. Intell. Syst. Appl. (IJISA) **10**(7), 11–26 (2018). https://doi.org/10.5815/ijisa.2018.07.02

5. Kalinovsky, Y.A., Boyarinova, Y.E.: High dimensional isomorphic hypercomplex number systems and their using for calculation efficiency. Infodruk, Kyiv (2012)

6. Kalinovsky, J., Sinkov, M., Boyarinova, Y., Fedorenko, O., Sinkova, T.: Development of theoretical bases and toolkit for information processing in hypercomplex numerical systems. Pomiary. Automatyka. Komputery w gospodarce i ochronie srodowiska **1**, 18–21 (2009)

7. Toyoshima, H.: Computationally efficient implementation of hypercomplex digital filters. IEICE Trans. Fundam. **E85-A**(8), 1870–1876 (2002)

8. Took, C.C., Mandic, D.P.: The quaternion LMS algorithm for adaptive filtering of hypercomplex processes. IEEE Trans. Signal Process. **57**(4), 1316–1327 (2009)

9. Kalinovsky, Y.O., Lande, D.V., Boyarinova, Y.E., Khitsko, I.V.: Infinite hypercomplex number system factorization methods. http://arxiv.org/abs/1401.2844 (2014)

10. Sinkov, M.V., Kalinovsky, Y.A., Boyarinova, Y.E.: Finite-Dimensional Hypercomplex Numerical Systems. Theory Basis. Supplement, Infodruk, Kyiv (2010)

11. Sinkov, M.V., Kalinovskiy, J.A., Boyarinova, Y.E., Sinkova, T.V., Fedorenko, O.V., Gorodko, N.O.: Fundamental principles of effective data presentation and processing on the basis of hypercomplex numerical systems. Data Rec. Storage Process. **12**(2), 62–68 (2010)

12. Kalinosky, Y.A., Fedorenko, O.V.: Principles of constructing digital filters with hypercomplex coefficients. Data Rec. Storage Process. **11**(1), 52–59 (2009)

13. Fedorenko, O.V.: Digital filters with low parametric sensitivity. Data Rec. Storage Process. **10**(2), 87–94 (2008)

14. Kalinovsky, Y.O., Boyarinova, Y.E., Khitsko, I.V.: Reversible digital filters total parametric sensitivity optimization using non-canonical hypercomplex number systems (2015). http://arxiv.org/abs/1506.01701

15. Kalinovsky, Y.O., Boyarinova, Y.E., Khitsko, I.V.: Hypercomplex operations system in Maple. Data Rec. Storage Process. **19**(2), 11–23 (2017)

16. Kalinovsky, Y.O., Boyarinova, Y.E., Khitsko, I.V.: Software complex for hypercomplex computing. Electron. Model. **39**(5), 81–95 (2017)

Approximation Algorithm for Parallel Machines Total Tardiness Minimization Problem for Planning Processes Automation

Alexander Anatolievich Pavlov$^{(\boxtimes)}$ ⓘ, Elena Borisovna Misura ⓘ,
Oleg Valentinovich Melnikov ⓘ, Iryna Pavlovna Mukha ⓘ,
and Kateryna Igoryvna Lishchuk ⓘ

National Technical University of Ukraine
"Igor Sikorsky Kyiv Polytechnic Institute",
37, Prospekt Peremohy, Kyiv 03056, Ukraine
pavlov.fiot@gmail.com, oleg.v.melnikov72@gmail.com,
mip.kpi@gmail.com, elena_misura@ukr.net,
lishchuk_kpi@ukr.net

Abstract. We present an approximation algorithm to solve the NP-hard scheduling problem of minimizing the total tardiness on identical parallel machines with a common due date and release dates of jobs. The algorithm has an estimate of the maximum possible deviation of its approximate solution from the optimum for each individual problem instance. It is based on the PSC-algorithm for the problem with equal release dates of jobs. Sufficient signs of optimality of a feasible solution and the estimate of the deviation of the obtained functional value from the optimum are known for the PSC-algorithm. The functional value obtained by the PSC-algorithm is the lower bound of the deviation of the functional value obtained by the approximation algorithm from the optimum for each individual problem instance. We give the computational data for test instances with dimensions of up to 40,000 jobs and 30 machines. The research shows that the developed algorithm is a very efficient method for the problem solving which allows to solve problems of any practical dimension. The average frequency of an optimal solution obtaining was 29.7%, and the average deviation from the optimum was 6.12%.

Keywords: Planning · Process automation · Scheduling ·
Combinatorial optimization · Approximation algorithm · Total tardiness ·
Identical parallel machines · Common due date · Release dates of jobs

1 Introduction

Scheduling problems play important role in calendar and operational planning processes automation for discrete-type productions. Such productions include, in particular, aircraft manufacturing, shipbuilding enterprises, and small-series productions. Almost all scheduling problems are NP-hard problems of combinatorial optimization. Therefore, creation of efficient approximation algorithms is of great importance for solving a number of practical large dimensional problems.

© Springer Nature Switzerland AG 2020
Z. Hu et al. (Eds.): ICCSEEA 2019, AISC 938, pp. 459–467, 2020.
https://doi.org/10.1007/978-3-030-16621-2_43

Many state-of-the-art scheduling methods are described in the book [1] and also in papers [2–4]. A review of efficient methods for exact exponential algorithms construction can be found in [5]. Some new scheduling methods, with an emphasis on cloud computing, are considered in a recent review [6]. Papers [7–9] research the methods for combinatorial problems solving using parallel computing systems. Parallel and cloud computing [10] and scheduling with restrictions on resources [11] are intensively developing research fields.

We consider in this paper the problem of minimizing the total tardiness on identical parallel machines with a common due date and release dates of jobs (TTP). It is used in the algorithmic ware of the four-level model of planning (including operative planning) and decision making presented in [12]. The interest for tardiness criterion is due to its practical effect in the real life [13]. This criterion is among the most interesting criteria for production systems, especially in the current situation where competitiveness is becoming more and more intensive. Suppliers do ensure their markets and customers. For that, they must have a high service quality while focusing on delivery dates.

The TTP Problem Statement. Given a set J of n jobs. For each job $j \in J$, we know its processing time l_j on one of m identical parallel machines. All jobs have the same due date d. Machines' idle times and interruptions in a job's processing are not allowed. Job j can start no earlier than the specified release date r_j, $0 \leq r_j < d$. We need to build a schedule of jobs $j \in J$ on m machines that minimizes the function:

$$f = \sum_{j \in J} \max\left(0, \ C_j - d\right) \tag{1}$$

where C_j is the completion time of a job j.

This problem belongs to the class of NP-hard problems. It is shown in [14] that this problem is NP-hard already for the case when all jobs arrive simultaneously, i.e. $r_j = 0 \ \forall j \in J$. For this case, [15] gives an estimate of the deviation of a solution from the optimum, and these results were improved in [16] where a PSC-algorithm for the problem solving was given. The theory of PSC-algorithms [17] is a new direction allowing to solve large dimension NP-hard problems in a general formulation exactly and efficiently. When sufficient signs of optimality of current solutions are fulfilled, PSC-algorithms allow to obtain an exact solution by a polynomial subalgorithm. Otherwise, we obtain an efficient exact solution by an exponential subalgorithm or an approximate solution with an estimate of the deviation from the optimum for practical problem instances of large size (up to tens of thousands of jobs).

Unified heuristics and annotated bibliography for a large class of scheduling problems with tardiness criteria are presented in [18]. Kovalyov et al. [19] show that there is no polynomial approximation algorithm with a guaranteed relative error of a solution for the problem in its original formulation unless $P = NP$. Jouglet and Savourey [20] developed a branch and bound algorithm describing new domination rules for the total weighted tardiness problem on parallel machines with distinct release dates and due dates. Their method is based on solving methods for single machine problems. They indicated that Azizoglu and Kirca [21], Yalaoui and Chu [22], as well as Shim and Kim [23], solved the total tardiness problem on identical parallel machines with equal release dates and distinct due dates by branch and bound algorithms and domination rules. Problems with distinct release dates and due dates with total and total weighted tardiness criteria are NP-hard in the strong sense [24]. Several lower bounds were proposed for them in [25]. Tanaka and Fujikuma [26] developed a dynamic programming algorithm with successive sublimation for these problems. It is stated in [24] that the weighted tardiness problem with equal processing times is solvable in pseudo-polynomial time. Lawler et al. [27] indicated that the total tardiness problem on a single machine with a common due date and without release dates is solved in $O(n^2)$ time. Its version with job weights is NP-hard in the ordinary sense [28].

The above analysis of related works shows that the considered problem in original formulation is practically not presented in the scheduling literature. This explains the need for its consideration. The purpose of this paper is to develop, for the first time, an approximation algorithm for the TTP problem solving with an estimate of the maximum possible deviation of the approximate solution from the optimum for each individual problem instance.

The rest of the paper is organized as follows. In Sect. 2, we give theoretical provisions for TTP problem solving in the case of equal release dates of jobs [16]. Also we present the developed approximation algorithm for the problem solving with unequal release dates. In Sect. 3, we present information about experimental studies of the approximation algorithm. Section 4 describes our findings.

2 Approximation Algorithm for the Problem Solving

The Approximation Algorithm AA for TTP problem solving includes two units. In the *first unit*, we solve the problem at equal release dates by the PSC-algorithm (polynomial scheme A1 [16]). The PSC-algorithm includes an approximation algorithm that yields an exact solution if a sufficient sign of optimality is satisfied on the obtained solution. Otherwise, it yields an approximate solution with an estimate of the deviation of the functional value from the optimum. The solution obtained by the PSC-algorithm is the basis for the *second unit*, scheduling for the case when the release dates of jobs are specified.

Let us present some theoretical foundations of the PSC-algorithm for the problem solving with equal release dates [16].

To build an initial schedule, we create a list of jobs in non-decreasing order of their processing times and assign each next job from the list to machine with a minimum *release time* (completion time of all assigned jobs). We assign the first jobs at time zero. As a result of such assignment, jobs of the set J are split into subsets $P_i(\sigma)$, $S_i(\sigma)$, $Q_i(\sigma)$, according to the following notation.

$P_i(\sigma)$ is the set of non-tardy jobs in the schedule of machine i;

$S_i(\sigma)$ is the set of "straddling" tardy jobs in the schedule of machine i for which

$$C_j - l_j < d, C_j > d, \forall j \in S_i(\sigma)$$

where $C_j - l_j$ is the start time of a job j;

$Q_i(\sigma)$ is the set of "fully" tardy jobs in the schedule of machine i for which

$$C_j - l_j \geq d, \forall j \in Q_i(\sigma).$$

Let us denote:

$$P = \bigcup_{i=\overline{1,m}} P_i; \quad S = \bigcup_{i=\overline{1,m}} S_i; \quad Q = \bigcup_{i=\overline{1,m}} Q_i;$$

Here and after, $\overline{p,q}$ means the interval of integers from p to q: $\overline{p,q} = p, p+1, \ldots, q$.

$R_i(\sigma)$ the time reserve of machine i in a schedule σ: $R_i(\sigma) = d - \sum_{j \in P_i(\sigma)} l_j$;

$\Delta_i(\sigma)$ the tardiness of straddling job $j \in S_i(\sigma)$ in regard to the due date:

$$\Delta_i(\sigma) = \sum_{j \in P_i(\sigma) \bigcup S_i(\sigma)} l_j - d.$$

Theorem 1 [15].   There is an optimal schedule that satisfies the conditions:

1. $P \cup S = \{1, 2, \ldots, |P \cup S|\}$;
2. If $|P \cup S| < n$, then $\sum_{j \in P_i \bigcup S_i} l_j \geq d$ and Q_i contains those and only those elements

    which differ from $|P \cup S| + i$ by an amount which is a multiple of m, $i = \overline{1,m}$.

Let Ψ_{PS} denote a class of schedules that correspond to the conditions of Theorem 1. And let $\Psi_{PS}(\mu)$ be a class of schedules corresponding to the condition: $|P(\sigma) \cup S(\sigma)| = \mu$ where μ is a natural number. We can write that $\Psi_{PS}(\mu) \subset \Psi_{PS}$. An optimal schedule σ^* belongs to at least one of the classes $\Psi_{PS}(\mu)$ at some $\mu = \mu^*$. The number of different non-empty classes $\Psi_{PS}(\mu)$ does not exceed m [15].

We distinguish from the class Ψ_{PS} a class of schedules $\Psi_P \subset \Psi_{PS}$ satisfying the following additional conditions:

1. $P = \{1, 2, \ldots, |P|\}$;
2. $\min_{j \in S(\sigma)} l_j > \max_{i=\overline{1,m}} R_i(\sigma)$;
3. If $l_{j_k} \leq l_{j_l}$, then $C_{j_k} - l_{j_k} \leq C_{j_l} - l_{j_l} \; \forall j_k, j_l \in S(\sigma)$.

Definition 1. A schedule with the same number of tardy jobs on all machines is called an even schedule.

The number of tardy jobs on machines differs by a maximum of one in the class Ψ_P [16]. We determine $R_i(\sigma)$ on machines with a smaller number of tardy jobs and $\Delta_i(\sigma)$ on machines with a larger number of tardy jobs.

Two sufficient signs of optimality of a feasible solution were proved in [16]:

1. An even schedule $\sigma \in \Psi_P$ is optimal.
2. If $\Omega_\Sigma(\sigma) = \min(R_\Sigma(\sigma), \Delta_\Sigma(\sigma)) = 0$ in a schedule $\sigma \in \Psi_P$, then the schedule σ is optimal.

Let us introduce the new class $\Psi(\sigma_P) \subset \Psi_{PS}$ that consists from arbitrary schedules σ obtained as a result of sequential directed permutations that we perform in an arbitrary order decreasing $\Delta_\Sigma(\sigma)$ and, consequently, $R_\Sigma(\sigma)$.

Theorem 2 [16]. The following estimate of deviation of the functional value from the optimum is valid for any schedule $\sigma \in \Psi(\sigma_P)$:

$$f(\sigma) - f(\sigma^*) \leq \Omega_\Sigma(\sigma).$$

We need to build a schedule of jobs on the machines in which either $R_\Sigma(\sigma)$ or $\Delta_\Sigma(\sigma)$ takes the minimum value. The PSC-algorithm from [16] is based on permutations of jobs from machines with larger number of tardy jobs to machines with a smaller number of tardy jobs if $\Delta_\Sigma(\sigma) > R_\Sigma(\sigma)$ and vice versa in the opposite case. If at least one of the signs of optimality was fulfilled during the solving process, then we have obtained an optimal schedule. Otherwise, the value of $\Omega_\Sigma(\sigma)$ is the maximum possible deviation from the optimum of the solution obtained by the PSC-algo-rithm.

The Approximation Algorithm AA: Second Unit. Consider the schedule obtained after the first unit execution. Sort all jobs of the sets P and S (together) in non-decreasing order of release dates. Select jobs with a minimum release date and schedule each one to a machine with a minimum release time, but not earlier than their release dates. If there are several options, select a job with a minimum processing time. If the release time on any machine becomes greater than or equal to the due date, then schedule the job to the next machine and stop scheduling jobs from the sets P and S to the current machine. If, as a result of the execution of the algorithm, all the jobs from the sets P and S were scheduled, then arrange the jobs of the set of tardy jobs Q in non-decreasing order of their processing times and schedule each next job of the set Q to the machine with a minimum release time. Denote the obtained schedule by σ_{PS}. If for a current job from the sets P and S we could not find a machine for scheduling without the machine's idle time, then we select a job from the set Q for scheduling, likewise choosing jobs with a minimum release date and scheduling each one to a machine with a minimum release time, but not earlier than their release dates. Then, we return to the scheduling of jobs from the sets P and S repeating the same procedure. If, after performing this procedure and assigning some job, an idle time exists on the machine, then we shift the beginning of all the jobs on this machine to a later time to eliminate the idle time. We perform similar steps until we schedule all jobs.

Theorem 3. The signs of optimality and the estimate of the deviation from the optimum $\Omega_\Sigma(\sigma)$ presented above for the PSC-algorithm, are valid for the schedule σ_{PS}.

Proof is obvious, since the schedule σ_{PS} meets the requirements to the class Ψ_P.

Theorem 4. Any schedule obtained by the Approximation Algorithm has not lesser functional value than the schedule obtained by the PSC-algorithm in the first unit.

Proof. The PSC-algorithm builds a schedule for the job release dates equal to zero. A shift of jobs to a later time, due to the fact that they obtain the release dates, leads to an increase in the total tardiness.

Justification of the Algorithm. The algorithm is built on directed permutations performed according to the following principle. The sets P and S should include the shortest jobs from the entire set of jobs that meet the requirements of the class Ψ_P described in the first unit. The remaining jobs are scheduled according to the rule: the shortest job to the machine with the minimum release time. Theorems 3 and 4 are valid for the resulting schedule. For this cause, we execute the first unit of the algorithm and thereby determine the jobs of the sets P and S. And the functional value obtained in the first unit is an estimate of the maximum possible decrease of the functional value for the approximation algorithm.

The above algorithm also minimizes the start times of jobs, satisfying their release dates as it should be.

3   Computational Studies of the Algorithm

To carry out the computational studies of efficiency of the algorithm, we used instances generator/solver written in C# in Microsoft Visual Studio 2010 environment. We chose processing times of jobs from uniform distribution within interval $[1, 200]$.

The due date was calculated as $0.7 L/m$ where L is the sum of processing times of all jobs. The release dates of jobs were chosen from uniform distribution within $[0, d)$. We researched problems with up to 40,000 jobs and 30 machines. We carried out 100 tests for each pair (n, m) on a personal computer with 2 GB of RAM and a Pentium IV processor that has 3.0 GHz frequency.

We show in Table 1 and in Fig. 1 the average (for 100 generations and runs) solving time by Approximation Algorithm AA depending on the numbers of jobs n and machines m.

Table 1. Average solving time (in milliseconds) by Approximation Algorithm AA depending on the problem size.

n	m				
	5	10	15	20	30
3000	3.73	6.29	6.57	6.23	7.19
5000	51.50	43.61	22.79	22.79	42.44
10,000	177.81	246.39	238.02	179.43	134.88
20,000	980.67	1142.74	998.88	959.86	626.16
40,000	3706.33	4050.73	4716.33	4451.87	2695.27

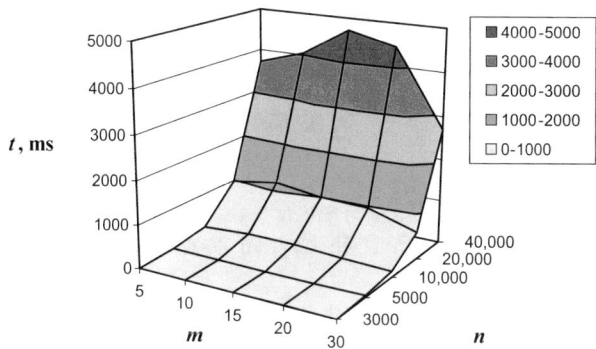

Fig. 1. Solving time for the problem instances of various dimensions (ms)

Table 2. Average percentage of deviation of the functional from the optimum for 100 runs

n	m					
	5	7	10	15	20	30
100	9.09	5.75	5.84	7.39	6.91	8.15
200	5.81	9.41	6.19	6.86	6.14	8.16
400	5.22	5.79	8.40	5.74	6.14	7.12
600	4.95	5.09	8.50	5.74	8.25	7.10
800	4.91	4.84	6.60	5.76	6.44	7.08
1000	4.92	4.76	4.99	6.09	6.45	7.08
1500	4.91	4.70	4.83	5.73	6.20	7.11
2000	4.91	4.70	4.71	5.72	6.15	7.10
2500	4.91	4.69	4.76	5.73	6.14	7.08
3000	4.91	4.69	4.71	5.72	6.14	7.08

Table 2 shows the average percentage of deviation of the functional value from the optimum for 100 generations and runs of the algorithm AA.

The average frequency of an optimal solution obtaining was 29.7%. The average deviation from an optimal solution was 6.12%, the maximum deviation was less than 15%.

The above algorithm also allows to solve much larger dimension instances in a reasonable time. For example, we have obtained a solution to the problem with 1 million jobs and 200 machines in only 6.5 min.

4   Conclusions

For the first time ever, we have developed an approximation algorithm for TTP problem solving with an estimate of the maximum possible deviation of the approximate solution from the optimum for each individual problem instance. The algorithm is based on the PSC-algorithm [16] which yields an exact solution if a sufficient sign of optimality is fulfilled for the obtained solution. Otherwise it yields an approximate solution with an estimate of the deviation of the functional value from the optimum. If the obtained approximate solution meets the requirements of class Ψ_P, then the sufficient signs of optimality and the estimate $\Omega_\Sigma(\sigma)$ of the deviation from the optimum are valid for it. The functional value obtained by the PSC-algorithm is the lower bound of the functional value obtained by the approximation algorithm for each individual problem instance.

We have presented the statistics for solving instances with dimensions up to 40,000 jobs and 30 machines. The average frequency of an optimal solution obtaining was 29.7%. The average deviation from an optimal solution was 6.12%, the maximum deviation was less than 15%.

The above algorithm also allows to solve much larger dimension instances in a reasonable time. For example, we have obtained a solution to the problem with 1 million jobs and 200 machines in only 6.5 min. Thus, the algorithm is very fast, it also does not require a lot of memory resources. Therefore, it is a very efficient method of the problem solving and can be used in production operational control systems.

References

1. Pinedo, M.L.: Scheduling: Theory, Algorithms, and Systems. Springer, Cham (2016). https://doi.org/10.1007/978-3-319-26580-3
2. Heydari, M., Hosseini, S.M., Gholamian, S.A.: Optimal placement and sizing of capacitor and distributed generation with harmonic and resonance considerations using discrete particle swarm optimization. Int. J. Intell. Syst. Appl. (IJISA) 5(7), 42–49 (2013). https://doi.org/10.5815/ijisa.2013.07.06
3. Wang, F., Rao, Y., Wang, F., Hou, Y.: Design and application of a new hybrid heuristic algorithm for flow shop scheduling. Int. J. Comput. Netw. Inf. Secur. (IJCNIS) 3(2), 41–49 (2011). https://doi.org/10.5815/ijcnis.2011.02.06
4. Cai, Y.: Artificial Fish School Algorithm applied in a combinatorial optimization problem. Int. J. Intell. Syst. Appl. (IJISA) 2(1), 37–43 (2010). https://doi.org/10.5815/ijisa.2010.01.06
5. Fomin, F.V., Kratsch, D.: Exact Exponential Algorithms. Springer, Heidelberg (2010). https://doi.org/10.1007/978-3-642-16533-7
6. Soltani, N., Soleimani, B., Barekatain, B.: Heuristic algorithms for task scheduling in cloud computing: a survey. Int. J. Comput. Netw. Inf. Secur. (IJCNIS) 9(8), 16–22 (2017). https://doi.org/10.5815/ijcnis.2017.08.03
7. Garg, R., Singh, A.K.: Enhancing the discrete particle swarm optimization based workflow grid scheduling using hierarchical structure. Int. J. Comput. Netw. Inf. Secur. (IJCNIS) 5(6), 18–26 (2013). https://doi.org/10.5815/ijcnis.2013.06.03
8. Mishra, M.K., Patel, Y.S., Rout, Y., Mund, G.B.: A survey on scheduling heuristics in grid computing environment. Int. J. Mod. Educ. Comput. Sci. (IJMECS) 6(10), 57–83 (2014). https://doi.org/10.5815/ijmecs.2014.10.08
9. Sajedi, H., Rabiee, M.: A metaheuristic algorithm for job scheduling in grid computing. Int. J. Mod. Educ. Comput. Sci. (IJMECS) 6(5), 52–59 (2014). https://doi.org/10.5815/ijmecs.2014.05.07
10. Hwang, K., Dongarra, J., Fox, G.: Distributed and Cloud Computing: From Parallel Processing to the Internet of Things. Morgan Kaufmann, Burlington (2012)
11. Brucker, P., Knust, S.: Complex Scheduling, 2nd edn. GOR-Publications Series. Springer, Heidelberg (2012). https://doi.org/10.1007/978-3-642-23929-8

12. Zgurovsky, M.Z., Pavlov, A.A.: Algorithms and software of the four-level model of planning and decision making. In: Combinatorial Optimization Problems in Planning and Decision Making: Theory and Applications. 1st edn. Studies in Systems, Decision and Control, vol. 173, pp. 407–518. Springer, Cham (2019). https://doi.org/10.1007/978-3-319-98977-8_9

13. Yalaoui, F.: Minimizing total tardiness in parallel-machine scheduling with release dates. Appl. Evol. Comput. **3**(1), 21–46 (2012). https://doi.org/10.4018/jaec.2012010102

14. Tanaev, V.S., Gordon, V.S., Shafransky, Y.M.: Scheduling Theory. Single-Stage Systems. Springer, Dordrecht (1994). https://doi.org/10.1007/978-94-011-1190-4

15. Tanaev, V.S., Shkurba, V.V.: Vvedenie v Teoriju Raspisaniy (Introduction to Scheduling Theory). Nauka, Moscow (1975). (in Russian)

16. Zgurovsky, M.Z., Pavlov, A.A.: The total tardiness of tasks minimization on identical parallel machines with arbitrary fixed times of their start and a common due date. In: Combinatorial Optimization Problems in Planning and Decision Making: Theory and Applications, 1st edn. Studies in Systems, Decision and Control, vol. 173, pp. 265–290. Springer, Cham (2019). https://doi.org/10.1007/978-3-319-98977-8_6

17. Zgurovsky, M.Z., Pavlov, A.A.: Introduction. In: Combinatorial Optimization Problems in Planning and Decision Making: Theory and Applications, 1st edn. Studies in Systems, Decision and Control, vol. 173, pp. 1–14. Springer, Cham (2019). https://doi.org/10.1007/978-3-319-98977-8_1

18. Kramer, A., Subramanian, A.: A unified heuristic and an annotated bibliography for a large class of earliness–tardiness scheduling problems. J. Sched. **20**, 1–37 (2017). https://doi.org/10.1007/s10951-017-0549-6

19. Kovalyov, M.Y., Werner, F.: Approximation schemes for scheduling jobs with common due date on parallel machines to minimize total tardiness. J. Heuristics **8**(4), 415–428 (2002). https://doi.org/10.1023/A:1015487829051

20. Jouglet, A., Savourey, D.: Dominance rules for the parallel machine total weighted tardiness scheduling problem with release dates. Comput. Oper. Res. **38**(9), 1259–1266 (2011). https://doi.org/10.1016/j.cor.2010.12.006

21. Azizoglu, M., Kirca, O.: Tardiness minimization on parallel machines. Int. J. Prod. Econ. **55**(2), 163–168 (1998). https://doi.org/10.1016/s0925-5273(98)00034-6

22. Yalaoui, F., Chu, C.: Parallel machine scheduling to minimize total tardiness. Int. J. Prod. Econ. **76**(3), 265–279 (2002). https://doi.org/10.1016/s0925-5273(01)00175-x

23. Shim, S.-O., Kim, Y.-D.: Scheduling on parallel identical machines to minimize total tardiness. Eur. J. Oper. Res. **177**(1), 135–146 (2007). https://doi.org/10.1016/j.ejor.2005.09.038

24. Lenstra, J., Kan, A.R., Brucker, P.: Complexity of machine scheduling problems. Annals Discr. Math. **1**, 343–362 (1977). https://doi.org/10.1016/s0167-5060(08)70743-x

25. Baptiste, P., Jouglet, A., Savourey, D.: Lower bounds for parallel machine scheduling problems. Int. J. Oper. Res. **3**(6), 643–664 (2008). https://doi.org/10.1504/ijor.2008.019731

26. Tanaka, S., Fujikuma, S.: A dynamic-programming-based exact algorithm for general single-machine scheduling with machine idle time. J. Sched. **15**(3), 347–361 (2011). https://doi.org/10.1007/s10951-011-0242-0

27. Lawler, E.L., Moore, J.M.: A functional equation and its application to resource allocation and sequencing problems. Manag. Sci. **16**(1), 77–84 (1969). https://doi.org/10.1287/mnsc.16.1.77

28. Yuan, J.: The NP-hardness of the single machine common due date weighted tardiness problem. J. Syst. Sci. Complex. **5**(4), 328–333 (1992)

Computer Science for Medicine and Biology

The Ontology as the Core of Integrated Information Environment of Chinese Image Medicine

S. Lupenko[1(✉)] ⓘ, O. Orobchuk[1] ⓘ, and Mingtang Xu[2]

[1] Ternopil Ivan Puluj National Technical University, Ternopil 46000, Ukraine
lupenko.san@gmail.com
[2] Beijing Medical Research Institute "Kundawell", Beijing 100010, China

Abstract. The article is devoted to the improvement of modern onto-oriented information tools for Integrative Medicine (IM), in particular, for its component - Chinese Image Medicine. The architecture of the components integrated onto-oriented information and analytical environment for scientific research, professional healing activities and e-learning of the Chinese Image Medicine is presented. The onto-orientation of the developed environment provides the ability to maintain the necessary level of integration, integrity of knowledge and data in the CIM for various information technologies and systems. The structure of the ontology of Chinese Image Medicine are detailed. Namely, the separate structure of the ontology of CIM is specified. The axiomatic-deductive strategy of organizing the knowledge space of the Chinese Image Medicine is proposed. Developed diagnostic ontology of Chinese Image Medicine, which includes the nosological ontology, the topological diagnostic ontology, the ontology of the diagnostic methods and the ontology of the diagnostic metrics of CIM.

Keywords: Ontology · e-learning systems · Expert systems · Axiomatic-deductive strategy · Integrative Medicine · Chinese Image Medicine

1 Introduction

According to the strategy of the World Health Organization (WHO) in the field of folk medicine [1], an important strategic problem is the development of a scientifically sound approach to the implementation of alternative and complementary medicine in the field of official medicine, both internationally and nationally. Today, in most countries of the world, in particular, in the USA, China, Japan, Korea, Russia, many countries of Europe, Brazil there is a significant revival in the scientific study of non-conventional (alternative, complementary) methods of human health improvement and treatment, which contributes to the formation of such a perspective the direction of medicine as an integrative (integral, holistic) medicine. Integrative (holistic) medicine develops all over the world, dating back to the 90s of the twentieth century. The Academic Consortium for Integrative Medicine and Health, the National Center for Complementary and Integrated Healthcare (NCCIH) were founded in the United States, and in 2001, the Institute for Integrative Medicine was opened in Harvard. In some

© Springer Nature Switzerland AG 2020
Z. Hu et al. (Eds.): ICCSEEA 2019, AISC 938, pp. 471–481, 2020.
https://doi.org/10.1007/978-3-030-16621-2_44

countries of the world there are higher education institutions that train specialists in the field of Integrative Medicine, and many national and international public organizations (associations) have been established, whose activities are aimed at the development of Integrative Medicine around the world. There is a large number of prestigious international journals dedicated to Integrative Medicine. In 2017, Berlin hosted the first World Congress of Integrative Medicine. The Integrative Medicine, synthesizing the best experience of Western, Oriental and traditional medicines, is the key to health in the XXI century. Integrative medicine holistically involves different directions of non-traditional types of medicine. Figure 1 presents a conditional scheme of the strategy for the formation of IM from a set of well-known medical areas, through the scientific selection of qualitative knowledge and methods of each of them and the synthesis of these selected components of Integrative Scientific Medicine.

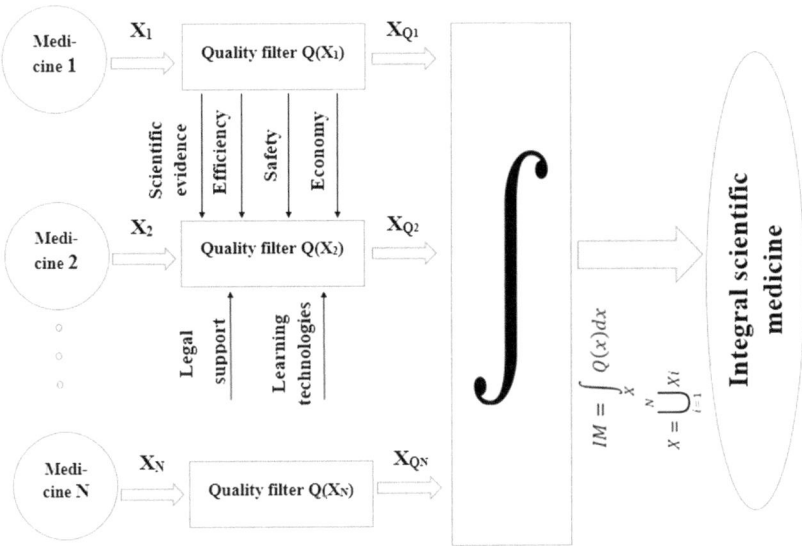

Fig. 1. Conditional scheme of the strategy of formation of integrative medicine from a set of well-known medical directions

In China, IM has become an integral part of public health successfully combining the achievements of Western medicine and traditional Chinese medicine. TCM is rooted in the ancient historical reality and includes naturopathy (treatment with products of natural origin, phytotherapy), qigong, meditation, massage, special diets, acupuncture. The methods and means of Chinese Image Medicine (CIM), which is a part of TCM and its historical roots reach the antiquity of Chinese civilization, are of great interest of scientific research. Nowadays CIM gained a new powerful impetus for its distribution and development worldwide, including the USA, Canada, Germany, Switzerland, China, Russia, Ukraine, Belarus, Brazil, Latvia, Estonia, Czech Republic, Slovakia and Hungary. A world famous centre for studying and research in Chinese Image Medicine is Beijing Medical Research Institute "Kundawell" (China).

Unlike TCM, for which a number of large-scale clinical trials, theoretical scientific substantiation and a range of relevant information and analytical tools (ontologies, expert systems, grid systems for TCM), for Chinese Image medicine (CIM) has almost no similar research and relevant information and analytical tools. The absence of the comprehensive theoretical and experimental researches of the CIM, as well as the lack of technical information and analytical decisions in the field of CIM, is a significant barrier to the creation of a complete scientific CIM paradigm in medical science, as many of the theoretical and experimental aspects and regularities of this area of folk medicine remain unclear.

Given this state of affairs, a Program for the researches of Chinese Image Medicine for 2017–2023 (Program) was developed [5]. The Program is aimed at conducting comprehensive scientific researches of Chinese Image Medicine in order to create theoretical and experimental scientific basis for CIM, which will promote the disclosure of the deep causes and mechanisms of human diseases and will help to create effective methods for their prevention and treatment. According to the CIM research Program, the actual scientific and applied problem is the creation of an integrated onto-oriented information and analytical environment for scientific research, professional healing activities and e-learning of the CIM.

2   Related Work

In [2–4], an overview of the current state, research and further promotion of Integrative Medicine is given. Since traditional Chinese medicine (TCM) is one of the important components of IM, the authentic methods and means of TCM actively are studied, distributed and developed, as evidenced by a series of papers [6–12]. The first results in the direction of creating information intelligent systems in the field of CIM obtained in the works [13–17]. So in [13] architecture of integrated onto-oriented information and analytical environment for scientific research, professional healing activities and e-learning of the CIM was developed. In works [14, 15] the first prototypes of the CIM ontology were developed. In the papers [16, 17], an axiomatic-deductive strategy for organizing knowledge and content in CIM were developed.

The proposed work is devoted to the development and further structuring of the CIM ontology and integrated onto-oriented information and analytical environment for scientific research, professional healing activities and e-learning of the CIM.

3   Results and Discussion

According to Program for the researches of Chinese Image Medicine, the creation of the scientific direction of the CIM is appropriate to realize in four interrelated areas: the theoretical direction, the experimental direction, the direction of clinical researches and the information-analytical direction (see Table 1).

Table 1. Directions of scientific researches of CIM

Direction	Description of the direction
The theoretical direction	Relates to the development of scientific concepts, models, methods, theories of CIM using the theoretical and methodological approach of modern science
The experimental direction	Relates to the organization and conduct out of comprehensive objective instrumental, laboratory and statistical studies of the physical and physiological (biophysical, biochemical, bioinformational) processes in the human body under the influence of the CIM-therapist and the individual practice of ZYQ
Direction of clinical researches	Relates to the development and implementation of clinical research programs for CIM and ZYQ methods in accordance with the requirements and standards of modern evidence-based medicine
Informational-analytical direction	Relates to the development and support of an integrated information and analytical system of scientific research, professional healing activities and training of the CIM, which serves for the organization and coordination of the activities of researchers, CIM-therapists and instructors, the collection, automated statistical and intellectual analysis of the results of treatment by the methods of CIM and the results of ZYQ practice, the creation of a unified database of all theoretical, experimental and clinical research in the field of CIM and ZYQ, the development of the web-oriented system of e-learning of the CIM-specialists, the expert system of support for the adoption of diagnostic and therapeutic decisions in the field of CIM

The actual scientific and applied problem is the creation of an integrated onto-oriented information and analytical environment for scientific research, professional healing activities and e-learning of the CIM. The purpose of developing this information-analytical environment is:

1. Improving the quality (evidence, effectiveness, safety, controllability, reliability, cost-effectiveness, intensity) of professional activity and the exchange of experience of CIM-therapists.
2. Providing effective organization and coordination of the functioning of the CIM-therapists, the researchers of the CIM, the persons studying the CIM.
3. Providing at high scientific, technological and infrastructural levels of collection, automated statistical and intellectual analysis of the results of diagnosis and treatment of CIM methods.
4. Creation of a unificated database and knowledge base for theoretical, experimental and clinical research in the field of CIM.
5. Formation of modern intellectualized information resources in the field of folk, complementary and integrative medicine on both at the national and international levels.

Architecture of the components of the environment integrated onto-oriented information and analytical environment for scientific research, professional healing activities and e-learning of the CIM. General architecture of integrated onto-based information analytical environment of scientific researches, professional healing activities and e-learning of Chinese image medicine is presented in Fig. 2 [13].

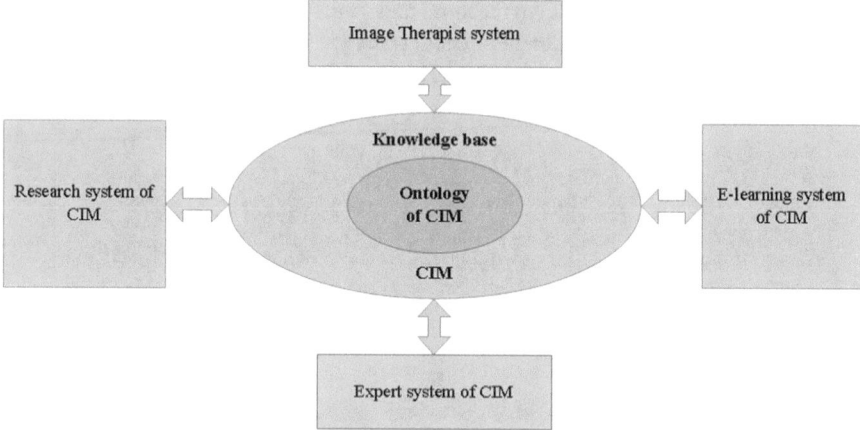

Fig. 2. General architecture of integrated onto-based information analytical environment of scientific researches, professional healing activities and e-learning of Chinese image medicine

Information system of professional healing "Image Therapist" is designed for centralized organization, upgrading (efficacy, safety, controllability, reliability, efficiency, intensity) of professional activities and experience exchange of the existing CIM therapists. General architecture of information system of professional healing "Image Therapist" is presented in Fig. 3 [13].

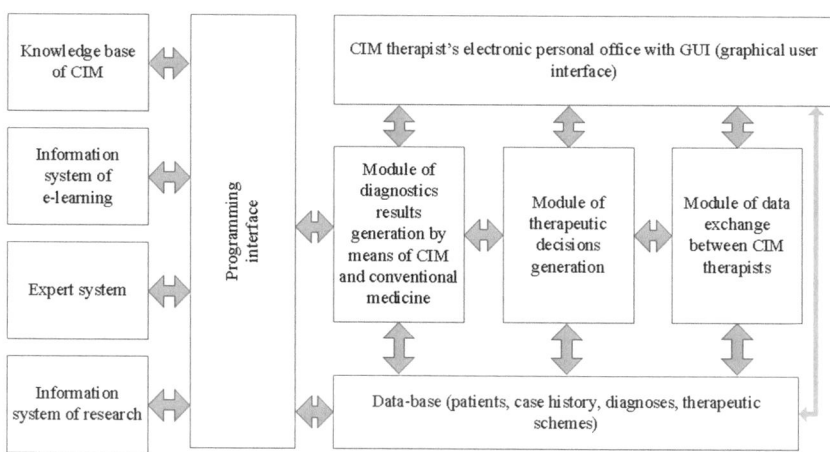

Fig. 3. General architecture of information system of professional healing "Image Therapist"

An important part of the integrated information analytical environment is CIM information system of e-learning. Development of such e-learning system will considerably simplify, intensify and improve quality and availability of educational process in CIM. For implementation of e-learning information system evidence-based standards of CIM learning should be developed firstly; they include educational and professional program for a CIM therapist, educational qualification of a CIM therapist, curricula and steering documents in disciplines, lecture and practice-oriented learning materials, methods of testing and self-assessment testing of CIM specialists. General architecture of the e-learning information system for CIM therapists is presented in Fig. 4 [13].

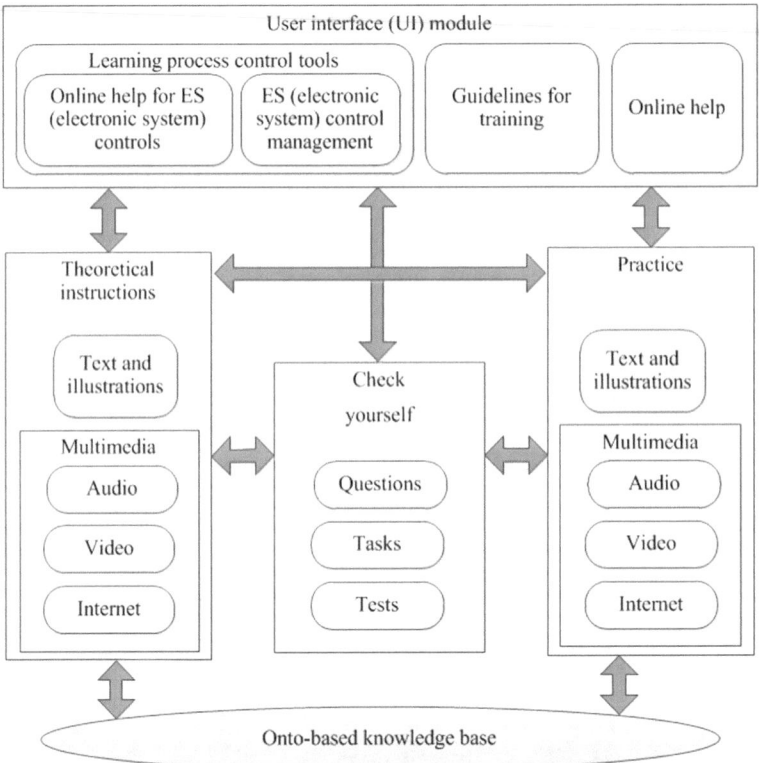

Fig. 4. General architecture of the e-learning information system for CIM therapists

An important part of the developed information analytical environment is the expert system for diagnostic and therapeutic decision-making support in CIM, which will help to improve the skills of CIM therapists. Expert system will issue diagnostic recommendations and personalized patient care scheme by means of CIM, based on data (personal and clinical) about a patient and CIM knowledge-base content. General architecture of the expert system for diagnostic and therapeutic decision-making support in CIM is presented in Fig. 5 [13].

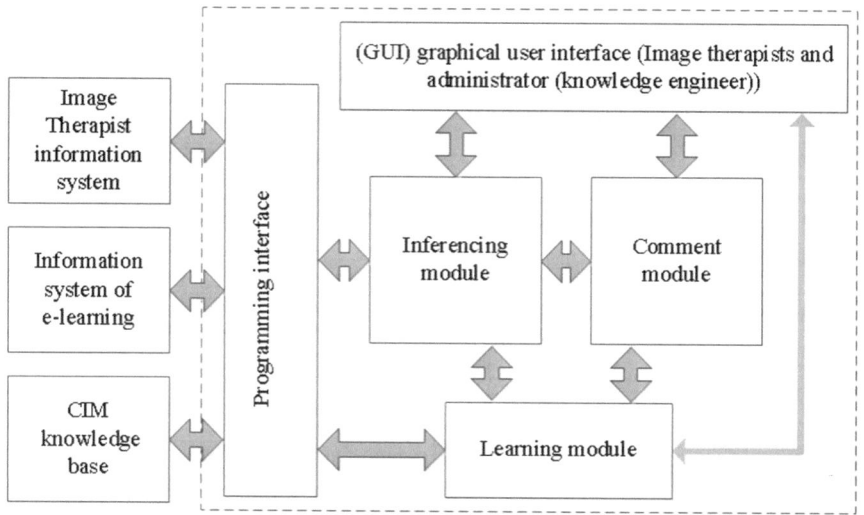

Fig. 5. General architecture of the expert system for diagnostic and therapeutic decision-making support in CIM

Chinese Image Medicine Ontology. The main stage in the development of CIM informational environment, creation of CIM scientific paradigm includes the development of its computer ontology. Besides, CIM ontology development is aimed at the necessity of a comprehensive solving of a range of important tasks of theoretical, clinical, experimental as well as informational and analytical areas of International Research Program, such as: (1) the unification, standardization of information presenting technique (data and knowledge) in CIM, which allows overcoming the problem of semantic heterogeneity of poorly structured and difficult to formalize knowledge of CIM (2) the working out a high-quality dictionary (glossary) and knowledge data-base (thesaurus) in CIM that would have such properties: completeness, consistency, interpretability, unification, integration with other subject areas, including integrative medicine; (3) the multiple reuse of CIM knowledge in various information systems and applications; (4) the necessity of intelligent CIM information search on the Internet in terms of WEB 2.0 semantic search technologies, which ensures a high relevancy and pertinence of the information wanted [14, 15].

Consider the structure of the CIM theory, which defines the general structure of the CIM onology. It is proposed to divide the scientific theory of CIM into two large parts: (1) General Scientific Theory of Integrative Medicine; (2) Special Scientific Theory of Chinese Image Medicine. As the general CIM theory, and the special CIM theory is divided into five of its main sections: (1) the theory of reality and human; (2) the theory of health and diseases; (3) the theory and technology of diagnostics; (4) the theory and technology of therapy; (5) the theory and technology of learning, the professional development of therapists. The sections of the general and special parts of the CIM theory was developed (see Table 2).

Table 2. The sections of the general and special parts of the CIM theory

Sections of theory	Description of section content
The theory of reality and human in IM and CIM	Describes the basic concepts and ideas of the IM (CIM) and serves as a practical-philosophical foundation for the rest of the IM (CIM) sections
The theory of health and diseases in IM (CIM)	Describes the basic concepts of health and diseases in IM (CIM); diagnostic standards of health and diseases for their evaluation by various methods of diagnosis of IM (CIM); the classification and definition of the types of diseases in the IM (CIM)
The theory and diagnostics technology in IM (CIM)	Describes and formalizes theoretical foundations, methods and means of obtaining diagnostic medical information by the methods of the IM (CIM), as well as methods of its interpretation
The theory and technology of therapy in IM (CIM)	Describes and formalizes theoretical foundations, methods and means of conducting therapeutic procedures in the IM (CIM), as well as their interrelations with the corresponding diagnostic information
The theory and technology of training, development of the IM-specialist (CIM-specialist)	Describes educational theoretically and practically oriented content, as well as technologies for its implementation into the educational process for the preparation and improvement of the qualification of the IM-specialist (CIM-specialist)

The development of CIM semantic space should be based on the axiomatic-deductive scheme of CIM theory development that satisfies the requirements for its semantic quality [16, 17]. The first component of the strategy is the deductive-axiomatic strategy of CIM terminology-conceptual apparatus development that involves: (1) the definition of fundamental concepts of CIM theory; this enables the presentation of these data in a format accessible for computerised processing in the form of a hierarchy of classes and relations between them, and further computerised processing of the semantics of the defined informational units by the developed ontology; (2) the definition of derivative concepts of CIM theory, basing on its fundamental concepts. The second component is a deductive-axiomatic strategy for establishing the true statements of CIM theory; it involves: (a) the definition of true statements-axioms, the truth of which is accepted with no proof; (b) the development of rules for logical derivation of all statements of the theory from its axioms. To unify the conceptual apparatus of CIM, the use of a 'top-down' method with a plurality of iterations following the general model of quality data stored in the structured format of computer system according to the ISO/IEC 25012 international standard would be the most effective.

Consider in more detail the diagnostic ontology of CIM. The diagnostic CIM-ontology includes the nosological ontology of CIM, the topological diagnostic ontology of the CIM, the ontology of the diagnostic methods of the CIM and the ontology of the diagnostic metrics of the CIM (see Table 3).

Table 3. Components of the CIM diagnostic ontology

Components of the diagnostic ontology of CIM	Description of components of diagnostic ontology CIM
Nosological ontology of CIM	Nosological CIM ontology reflects knowledge about the types (classes) of diseases that are taken in the diagnostic theory of CIM
Topological diagnostic ontology of CIM	Topological diagnostic ontology CIM reflects the data on the topological localization of diseases related to the physical body, the energy system (field system, Chi system) and information systems (psycho-mental-spiritual system, Shen system), in particular, contains information about body parts, organs, tissues of the physical body, information about bioactive points and energy channels of the human energy system, information about information, psycho-emotional, mental and spiritual topological aspects of a person
Ontology of diagnostic methods of CIM	The ontology of diagnostic methods in the CIM reflects the knowledge about the methods (channels) of obtaining and the specifications of sensory diagnostic information in the CIM
Ontology of the diagnostic metric of CIM	Describes the quantitative characteristics (indicators) of the diagnostic space of the CIM, which determine the degree of manifestation of a particular disease and can be given on a certain number (for example, 1 to 5) or non-numeric (eg, very weak, weak, medium, strong, very strong) scale

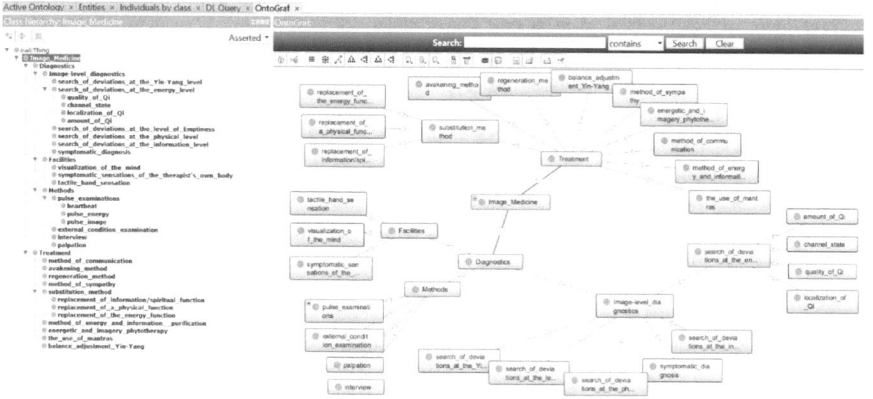

Fig. 6. A fragment of the ontology of CIM

For development and specification analysis of conceptual model description of CIM the OWL language was chosen [14]. A fragment of the ontology of CIM diagnostic methods developed in the Protégé environment using the OWL ontology description language is presented in Fig. 6.

4 Conclusions

The construction of ontology CIM will make possible to unify, standardize the technologies of presentation of information (data and knowledge) in the field of CIM, which will make it possible to overcome the problem of semantic heterogeneity of less structured and difficult formalized knowledge in the field of TCM and CIM, since the use of ontologies eliminates subjective factors, polysemantics, fuzziness of concepts and images that are used explicitly or implicitly by the CIM therapists in the process of diagnostic and therapeutic decision making. Ontology CIM will make it possible to standardize the conceptual-terminology apparatus of the CIM, which will significantly facilitate the CIM-therapists to exchange and accumulate their knowledge and experience in unificated way in an integrated information environment. In addition, the onto-orientation of the developed environment provides the ability to maintain the necessary level of integration, integrity of knowledge and data in the CIM for various information technologies and systems, as well as the possibility of multi repeated use of knowledge in the CIM for various information systems and applications.

References

1. WHO strategy for traditional medicine for 2014–2023 (2013). http://www.who.int/medicines/publications/traditional/trm_strategy14_23/ru/. Accessed 20 Nov 2018
2. Barnes, P., Bloom, B., Nahin, R.: The Use of Complementary and Alternative Medicine in the United States. Findings from the 2007 National Health Interview Survey (NHIS) conducted by the National Center for Complementary and Alternative Medicine (NCCAM) and the National Center for Health Statistics (2008). http://nccam.nih.gov/news/camstats/2007/camsurvey_fs1.htm. Accessed 23 Nov 2016
3. Ananth, S.: Complementary and Alternative Medicine Survey of Hospitals: Summary of Results. Health Forum (American Hospital Association) and the Samueli Institute (2010). http://www.siib.org/news/2468-SIIB/version/default/part/AttachmentData/data/CAM%20Survey%20FINAL.pdf. Accessed 11 Dec 2011
4. Guarneri, E., Horrigan, B., Pechura, C.: The efficacy and cost effectiveness of integrative medicine: a review of the medical and corporate literature. J. Sci. Heal. **5**, 308–312 (2010)
5. International program of scientific research in Chinese image medicine and Zhong Yuan Qigong for 2017–2023. https://kundawell.com/ru/mezhdunarodnaya-programma-nauchnykh-issledovanij-kitajskoj-imidzh-meditsiny-i-chzhun-yuan-tsigun-na-2017-2023-god. Accessed 22 Jan 2018
6. Mukherjee, I., Zain, J.M., Mahanti, P.K.: An automated real-time system for opinion mining using a hybrid approach. Int. J. Intell. Syst. Appl. (IJISA) **8**(7), 55–64 (2016). https://doi.org/10.5815/ijisa.2016.07.06

7. Wang, H.: A computerized diagnostic model based on naive bayesian classifier in traditional chinese medicine. In: Proceedings of the 1st International Conference on BioMedical Engineering and Informatics (BMEI 2008), May 2008, pp. 474–477 (2008)
8. Perova, I., Pliss, I.: Deep hybrid system of computational intelligence with architecture adaptation for medical fuzzy diagnostics. Int. J. Intell. Syst. Appl. (IJISA) **9**(7), 12–21 (2017). https://doi.org/10.5815/ijisa.2017.07.02
9. Huang, M.-J., Chen, M.-Y.: Integrated design of the intelligent web-based Chinese Medical Diagnostic System (CMDS) – systematic development for digestive health. Exp. Syst. Appl. J **32**(2), 658–673 (2007)
10. Mao, Y., Yin, A.: Ontology modeling and development for Traditional Chinese Medicine. In: Proceedings of the 2nd International Conference on Biomedical Engineering and Informatics (BMEI 2009), October 2009, pp. 1–5 (2009)
11. Das, N., Das, L., Rautaray, S.S., Pandey, M.: Big data analytics for medical applications. Int. J. Mod. Educ. Comput. Sci. (IJMECS) **10**(2), 35–42 (2018). https://doi.org/10.5815/ijmecs.2018.02.04
12. Dominic, M., Britto, A.X., Francis, S.: A framework to formulate adaptivity for adaptive e-learning system using user response theory. IJMECS **7**(1), 23–30 (2015). https://doi.org/10.5815/ijmecs.2015.01.04
13. Lupenko, S., Orobchuk, O., Vakulenko, D., Sverstyuk, A., Horkunenko, A.: Integrated onto-based information analytical environment of scientific research, professional healing and e-learning of Chinese Image Medicine. Inf. Syst. Netw. J. **59**, 10–19 (2017)
14. Lupenko, S., Pavlyshyn, A., Orobchuk, O.: Conceptual fundamentals for ontological simulation of Chinese Image Medicine as a promising component of integrative medcine. Sci. Educ. New Dimens. J. **15**, 28–32 (2017)
15. Lupenko, S., Orobchuk, O., Pomazkina, T., Mingtang, X.: Conceptual, formal and software-information fundamentals of ontological modeling of Chinese Image Medicine as an element of integrative medicine. World Sci. **1** (2017). https://doi.org/10.31435/rsglobal_ws
16. Lupenko, S.: Organization of the content of academic discipline in the field of information technologies using ontological approach. In: Proceeding of the International Conference on CSIT. Advances in Intelligent Systems and Computing III, CSIT 2018, 11–14 September, vol. 59, pp. 312–327 (2018)
17. Lupenko, S., et al.: The axiomatic-deductive strategy of knowledge organization in onto-based e-learning systems for Chinese Image Medicine. In: The 1st International Workshop on Information & Data-Driven Medicine (IDDM), November 2018, vol. 59, pp. 126–134 (2018)

Model of the Data Analysis Process to Determine the Person's Professional Inclinations and Abilities

Andrii Bomba[1], Nataliia Kunanets[2], Mariia Nazaruk[1],
Volodymyr Pasichnyk[2], and Nataliia Veretennikova[2(✉)]

[1] Informatics and Applied Mathematics Department,
State Humanitarian University, Rivne, Ukraine
abomba@ukr.net, marinazaruk@gmail.com
[2] Information Systems and Networks Department,
Lviv Polytechnic National University, Lviv, Ukraine
nek.lviv@gmail.com, vpasichnyk@gmail.com,
nataver19@gmail.com

Abstract. In this paper, the authors analyzed the data of vocational guidance tests. Accordingly, a model of data analysis was proposed to determine the person's professional inclinations and abilities, in particular, a methodological approach is described that uses intelligent data analysis to find hidden dependencies in the results of vocational guidance testing (the professional orientation questionnaire by Holland, the questionnaire about professional inclinations by L. Yovashi, the questionnaire about the profession type by E. Klimov, the questionnaire of interests by A. Holomshtok), on the basis of which decisions can be made on the choice of profession. The model of the data analysis process for identifying the professional inclinations and abilities of a person was implemented in the software and algorithmic complex of information and technological support of the profession choice, which allows to optimize the process for choosing the professional direction of a person.

Keywords: Career choice · Vocational guidance test ·
Intellectual data analysis · Methodological approach · Data analysis model

1 Introduction

The formation of a modern high-tech educational social and communication environment in the city is intended to promote maximum satisfaction of the urban population needs, in particular educational ones, which involve the selection and acquisition of a profession, taking into account the personal characteristics of the applicants and the needs of the city community.

The problems of choosing a future profession, professional self-determination and a specialist formation are described in the works by Thomas [1], Nota [2], Eesley [3], Meijers [4], Ceschi [5], Van der Gaag [6], Dimitrakopoulos [7], and others. The researchers argue reasonably that the right profession choice affects the success and productivity of professional activity in the future and the realization of personal

© Springer Nature Switzerland AG 2020
Z. Hu et al. (Eds.): ICCSEEA 2019, AISC 938, pp. 482–492, 2020.
https://doi.org/10.1007/978-3-030-16621-2_45

potentials [8–10]. According to the estimates of foreign scientists, the correct and well-timed profession choice at school age reduces staff turnover by the 2–2.5 times, reduces the cost of their training by 1.5–2 times, increases the labor productivity by 10–15%, therefore the choice of a profession type remains relevant taking into account the desires and opportunities of the applicants.

Successful professional activity is directly related to the physiological and psychological characteristics of a person. Psychophysiological qualities of a person are the features of its psyche, development, body structure, health status. Psychophysiological selection is intended to identify people who according to their abilities and individual psycho and physiological qualities meet the requirements of certain specialties [11, 12].

The theoretical basis for the determination of professional inclinations is the vocational guidance test that is a set of questions answering most truthfully, a person undergoes testing of psychological and emotional characteristics and professional preferences. The test allows you to choose the future profession, helps to identify the person's interests and inclinations to a particular activity field. There are dozens of different vocational guidance tests, but there is some inconsistency between them in terms of scales, evaluation methods and results. In particular, the analysis of several methods for assessing the professional abilities and inclinations of a person (Table 1) was carried out and we distinguish the following methods such as:

- The questionnaire of professional orientation (QPO) by Holland, the essence of which is that success in professional activities depends on the condition of matching a personality type and types of professional environment [13];
- The questionnaire of professional inclinations by Yovashi (QPI), aimed at identifying inclinations for activities in various fields, such as art field (human-artistic image); field of technical interests (human-technician); field of work with people (human-man); field of mental labor (inclinations to mental activity); field of physical labor (inclinations to active physical work); field of production and economy (production and consumption of material goods, planning and economic activity) [14];
- The questionnaire of circle of interests (QCI) by Holomshtok, which allows revealing not only the circle of interests of individuals, but also the degree of their expressiveness, which has a particular importance in creating motivation of choosing a future profession [15];
- The questionnaire of determination of profession type (QPT) by Klimov, which is based on the theory that choosing a profession, a person directs their thoughts firstly with what they will work, especially the work subject, then on what they will do with it, the goal that will be implemented [16].

The main problem is to analyze the data of vocational guidance tests and its results are given in this paper. These data are used to build decision-making rules. Appropriate testing was carried out at secondary schools in order to determine the professional inclinations and abilities of 10–11 grade students to select a future profession. The peculiarity of the analyzed data is that the empirical test data collected by school psychologists is that they are characterized by inaccuracy and subjectivity. In connection with this, there is a problem of objective evaluation and optimization of vocational guidance testing, which is solved by analysis and application of the built decision-making model.

The **purpose** of this work is to develop a model of data analysis process to determine the professional inclinations and abilities of a person based on the results of the system of vocational guidance tests, means of intellectual data analysis to determine the person's professional characteristics.

Table 1. Comparative table of methods

Name	Developer	Number of questions	Taking time (min)	Result
QPO	J. Holland	240	25	Type of professional environment
QPI	L. Yovashi	24	15	Professional inclinations
QCI	A. Holomshtok	96	15	Circle of professional interests
QPT	E. Klimov	30	5	Type of future profession

2 Main Part

In order to identify the general dependencies that help make decisions of a profession choice, it is proposed to use the methodological approach of data mining [17]. This approach provides the possibility of analyzing data and hidden dependencies in them, identifying the mutual influences of object properties, based on information about them, which is stored in databases. In this case, the regularities are defined that are inherent in a certain data set [18, 19]. The technological process of knowledge retrieval may be iterative and consists of four basic steps (Fig. 1).

1. Data retrieval.
2. Data preprocessing.
3. Intellectual data analysis.
4. Evaluation and interpretation of constructed models and found dependencies.

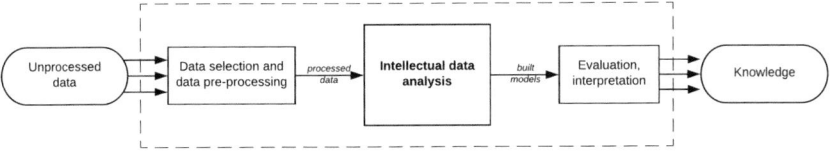

Fig. 1. Scheme of the extracting knowledge process

Within the technology of data mining, a logical method is used to construct decision-making rules. The end result is either a set of logical rules of the form "IF …, THAN", which are mostly obtained by generating matrixes of indeterminacy and reduces, or data templates that can be applied to filter and reduce data. The data that is being studied is presented using examples collected in the decision-making table

$T = (U, V \cup \{d\})$, where U is a non-empty finite set of examples, V is a non-empty finite set of conditional attributes, and d is a decision acceptance attribute with the domain Y_d, $|Y_d| = k$. The value of y and the attribute of d assign each example $x \in U$ to the decision-making class X_i, where $i = 1, 2, \ldots, k$. (the table can have more than one attribute) [19].

Thus, the model of the data analysis process on the determination of professional inclinations and abilities of a person, based on the system results of vocational guidance tests (QPO, QPI, QCI, QPT) will be presented in the form [20]:

$$M = (A, V, R, EscC(v), T, ClasR(v), Evl(v)),$$

where A is a set of persons (agents) involved in the professional guidance testing, V is a set of their properties, R is the test results (by J. Holland, L. Yovashi, E. Klimov and A. Holomshtok), the function $EscC(v)$ eliminates non-essential attributes by construction of redoubts, T is a decision-making table, $ClasR(v)$ is a function that builds a classifier in the form of a set of classification rules, $Evl(v)$ is a function of quality classification assessment.

The set of qualities V according to the above described methods such as QPO, QPI, QCI, QPT is divided into subsets:

$$V = \{V_1, V_2, V_3\},$$

where V_1 is informative properties, V_2 is psychological characteristics, V_3 is personal characteristics.

For each object, we put the set of attribute values in the decision table T.

The V_1 set of informative attributes consists of 5 attributes:

$v_{1,1} =$ «Name, surname».

$v_{1,2} =$ «Year of birth» with a continuous domain Dom_$v_{1,2} \in N$.

$v_{1,3} =$ «Education» is a domain whose Dom_$v_{1,3}$ is defined by the rule:

$$\text{Dom\_}v_{1,3} = \begin{cases} 1, & if & v_{1,3}(a) = "complete \quad secondsry", \\ 2, & if & v_{1,3}(a) = "incomplete \quad secondary", \\ 3, & if & v_{1,3}(a) = "incomplete \ higher", \\ 4, & if & v_{1,3}(a) = "basic \ higher", \\ 5, & if & v_{1,3}(a) = "complete \ higher" \end{cases}$$

$v_{1,4} =$ «Sex» is a domain whose Dom_$v_{1,4}$ is defined by the rule:

$$\text{Dom\_}v_{1,4} = \begin{cases} 0, & if & v_{1,4}(a) = "male", \\ 1, & if & v_{1,4}(a) = "female" \end{cases}.$$

The psychological characteristics of a person are given by the values of the set of attributes $V_2 = \{v_{2,1}, v_{2,2}, v_{2,3}, \ldots, v_{2,24}\}$, where:

$v_{2,1}$ = «concentration of attention»	$v_{2,15}$ = «rationality»
$v_{2,2}$ = «accuracy of information reproduction»	$v_{2,16}$ = «composure»
$v_{2,3}$ = «speed of information reproduction»	$v_{2,17}$ = «endurance»
$v_{2,4}$ = «technical thinking»	$v_{2,18}$ = «patience»
$v_{2,5}$ = «logical thinking»	$v_{2,19}$ = «persistence»
$v_{2,6}$ = «abstract thinking»	$v_{2,20}$ = «determination»
$v_{2,7}$ = «RAM»	$v_{2,21}$ = «courage»
$v_{2,8}$ = «short-term memory»	$v_{2,22}$ = «criticality»
$v_{2,9}$ = «inflexibility»	$v_{2,23}$ = «enthusiasm»
$v_{2,10}$ = «conservatism»	$v_{2,24}$ = «impulsiveness»
$v_{2,11}$ = «dependence»	$v_{2,25}$ = «dominance»
$v_{2,12}$ = «analytical mind»	$v_{2,26}$ = «carefulness»
$v_{2,13}$ = «organization»	$v_{2,27}$ = «originality»
$v_{2,14}$ = «accuracy»	$v_{2,28}$ = «initiativeness»

Personality characteristics and qualities are set by the values of the set of attributes $V_3 = \{v_{3,1}, v_{3,2}, v_{3,3}, \ldots, v_{3,12}\}$, where:

$v_{3,1}$ = «need of achievements»	$v_{3,8}$ = «need of communication»
$v_{3,2}$ = «social status»	$v_{3,9}$ = «desire to teach»
$v_{3,3}$ = «management style»	$v_{3,10}$ = «desire to bring up»
$v_{3,4}$ = «social activity»	$v_{3,11}$ = «desire to treat»
$v_{3,5}$ = «creative activity»	$v_{3,12}$ = «service»
$v_{3,6}$ = «comfort»	$v_{3,13}$ = «independence of decisions»
$v_{3,7}$ = «organizational skills»,	$v_{3,14}$ = «creative abilities»

The set of results of vocational guidance testing R will be in the following way:

$$R_{test} = \{R_{test_1}, R_{test_2}, R_{test_3}, R_{test_4}\},$$

where R_{test_1} is a type of professional environment, R_{test_2} is a range of professional interests, R_{test_3} is a type of profession, R_{test_4} is professional inclinations.

Obviously, the elements of a set R_{test} are classifying attributes (decision-making attributes).

$$R_{test_1} = \{r_{test_1}^1, r_{test_1}^2, r_{test_1}^3, r_{test_1}^4, r_{test_1}^5, r_{test_1}^6\},$$

where $r_{test_1}^1$ is «realistic», $r_{test_1}^2$ is «intellectual», $r_{test_1}^3$ is «social», $r_{test_1}^4$ is «conventional», $r_{test_1}^5$ is «entrepreneurial», $r_{test_1}^6$ is «artistic».

$$R_{test_2} = \{r_{test_2}^1, r_{test_2}^2, r_{test_2}^3, r_{test_2}^4, r_{test_2}^5, r_{test_2}^6, r_{test_2}^7, r_{test_2}^8, r_{test_2}^9, r_{test_2}^{10}\},$$

where $r_{test_2}^1$ is «physics and mathematics», $r_{test_2}^2$ is «chemistry and biology», $r_{test_2}^3$ is «radio engineering», $r_{test_2}^4$ is «mechanics», $r_{test_2}^5$ is «geography and geology», $r_{test_2}^6$ is «literature and art», $r_{test_2}^7$ is «sport and military affairs», $r_{test_2}^8$ is «history and political science», $r_{test_2}^9$ is «pedagogy and medicine», $r_{test_2}^{10}$ is «entrepreneurship».

$$R_{test_3} = \{r_{test_3}^1, r_{test_3}^2, r_{test_3}^3, r_{test_3}^4, r_{test_3}^5\},$$

where $r_{test_3}^1$ is «human – nature», $r_{test_3}^2$ is «human – technology», $r_{test_3}^3$ is «human – human», $r_{test_3}^4$ is «human – sign system», $r_{test_3}^5$ is «human –artistic image».

$$R_{test_4} = \{r_{test_4}^1, r_{test_4}^2, r_{test_4}^3, r_{test_4}^4, r_{test_4}^5, r_{test_4}^6\},$$

where $r_{test_4}^1$ is «work with people», $r_{test_4}^2$ is «mental», $r_{test_4}^3$ is «technical», $r_{test_4}^4$ is «ethics, art», $r_{test_4}^5$ is «physical work», $r_{test_4}^6$ is «planned and economic».

The decision-making table, which is created in the subprocess of the description of the subject area, takes on the form (Table 2):

Table 2. Structure of the decision-making table

Decision-making table						
Conditional attributes			Solution attributes			
$V_1(4)$	$V_2(28)$	$V_3(14)$	$R_{test_1}(6)$	$R_{test_2}(10)$	$R_{test_1}(5)$	$R_{test_1}(6)$
$v_{1,1}, v_{1,2},$	$v_{2,1}, v_{2,2},$	$v_{3,1}, v_{3,2},$	$r_{test_1}^1, r_{test_1}^2,$	$r_{test_2}^1, r_{test_2}^2,$	$r_{test_3}^1,$	$r_{test_4}^1, r_{test_4}^2,$
$v_{1,3}, v_{1,4}$	$v_{2,3}, v_{2,4},$	$v_{3,3}, v_{3,4},$	$r_{test_1}^3, r_{test_1}^4,$	$r_{test_2}^3, r_{test_2}^4,$	$r_{test_3}^2,$	$r_{test_4}^3, r_{test_4}^4,$
	$v_{2,5}, v_{2,6},$	$v_{3,5}, v_{3,6},$	$r_{test_1}^5, r_{test_1}^6$	$r_{test_2}^5, r_{test_2}^6,$	$r_{test_3}^3,$	$r_{test_4}^5, r_{test_4}^6$
	$v_{2,7}, v_{2,8},$	$v_{3,7}, v_{3,8},$		$r_{test_2}^7, r_{test_2}^8,$	$r_{test_3}^4,$	
	$v_{2,9}, v_{2,10},$	$v_{3,9}, v_{3,10},$		$r_{test_2}^9, r_{test_2}^{10}$	$r_{test_3}^5$	
	$v_{2,11}, v_{2,12},$	$v_{3,11}, v_{3,12},$				
	$v_{2,13}, v_{2,14},$	$v_{3,13}, v_{3,14}$				
	$v_{2,15}, v_{2,16},$					
	$v_{2,17}, v_{2,18},$					
	$v_{2,19}, v_{2,20},$					
	$v_{2,21}, v_{2,22},$					
	$v_{2,23}, v_{2,24},$					
	$v_{2,25}, v_{2,26},$					
	$v_{2,27}, v_{2,28}$					

$$T = (A, \{V_1, V_2, V_3\} \cup \{R_{test_1}, R_{test_2}, R_{test_3}, R_{test_4}\}).$$

The described data has several drawbacks. Firstly, there is some redundancy of data. Some attributes contain insignificant information for analysis, so it's necessary to remove attributes that contain excess information at the data extraction stage. Secondly, some attribute values are not known in the data table, due to the inappropriate methods of data analysis, which work only with completed tables. And the general disadvantage of the decisions is their subjectivity. Actually, the subjectivity of the decisions requires solving the problem of finding dependencies in the data and attributes that really influence the decision-making.

The proposed data structure for attribute sets assumes data uncertainty and redundancy. The $EscC(v)$ function is introduced, which eliminates non-essential attributes by constructing reduces (the attributes which influence on decision-making on career aim) to eliminate them, reduce the data size, and shorten the time to execute dependency detection procedures. Reduces were determined using the well-known Johnson algorithm. Above the data presented in the Table 3, a number of experiments were carried out, and a separate selection and removal of unimportant attributes for analysis was performed for each experiment (Table 3).

Table 3. The structure of the results of the decision-making table

№	Attributes used for analysis	Number of conditional attributes	Decision attribute	Number of attributes of a reduce
1.	$V_1, v_{2,15} - v_{2,17}, v_{2,19},$ $v_{3,1}, v_{3,7}, v_{3,12}$	10	$r^1_{test_1}$	4
2.	$V_1, v_{2,1} - v_{2,2}, v_{2,4} -$ $v_{2,5}, v_{2,12}, v_{2,26}, v_{3,12}$	11	$r^2_{test_1}$	5
3.	$V_1, v_{2,5}, v_{2,13} - v_{2,20}, v_{2,28},$ $v_{3,7} - v_{3,14}$	22	$r^3_{test_1}$	12
....	....	....	....	....
27.	$V_1, v_{2,1} - v_{2,5}, v_{2,7}, v_{2,12} -$ $v_{2,19}, v_{2,26}, v_{3,1}, v_{3,3}, v_{3,12}$	29	$r^6_{test_4}$	16

Based on the attributes included in the reduce, the function $ClasR(v)$ builds a dependency classifier between the set of values of the conditional attributes and the decision-making attributes of T table.

The quality of the rule $(\alpha \rightarrow \beta$ (if α, then β) is evaluated according to the following numerical characteristics, namely support is a number of study examples for which both the rule condition α, and the result β are fulfilled; accuracy is a ratio of the number of study examples for which the rule is followed up to the number of study examples for which the rule condition is met; coverage is a ratio of the number of study examples for which the entire rule is executed, to the number of study examples for which the result of the rule is followed.

Thus, based on the result elaboration of vocational guidance testing, a decision is made as to the person's belonging to one of the 6 professional types (Table 4):

$$P\_type = (p\_type_1, p\_type_2, p\_type_3, p\_type_4, p\_type_5, p\_type_6),$$

where p_type_1 = «realistic», p_type_2 = «intellectual», p_type_3 = «social», p_type_4 = «conventional», p_type_5 = «entrepreneurial», p_type_6 = «artistic».

A set of suggestions and recommendations for choosing a future profession is formed depending on which professional type a person.

Table 4. A fragment of the rule base of the decision-making table

IF	Professional inclinations (R_{test_4}) = *work with people* ($r^1_{test_4}$)
AND	Professional interests (R_{test_2}) = *pedagogy and medicine* ($r^9_{test_2}$)
AND	Professional environment (R_{test_1}) = *social* ($r^3_{test_1}$)
AND	Type of profession (R_{test_3}) = *human-human* ($r^3_{test_3}$)
THEN	Professional type (P_type) = $social(p\_type_1)$

3   Results

The results of vocational guidance testing form a comprehensive assessment of the professional orientation of a person. The complex assessment data of a person is stored in the data warehouse. Extracting knowledge from the complex assessment of a person allows to reveal regularities in the results of testing and to solve a problem of professional choice (Fig. 2). The process of extracting knowledge from the complex assessment of a person's data consists of the following steps:

Step 1. To accumulate and consolidate the results of vocational guidance tests.
Step 2. To pre-process data (at this step, the user's responses are analyzed):
    Step 2.1. To format data.
    Step 2.2. To structure and unify data.
    Step 2.3. To anonymize data.
    Step 2.4. To disseminate data.
    Step 2.5. To process incomplete data.
Step 3. To apply the methods of the intellectual analysis to the pre-processed complex estimation data.
Step 4. To evaluate and interpret the results.

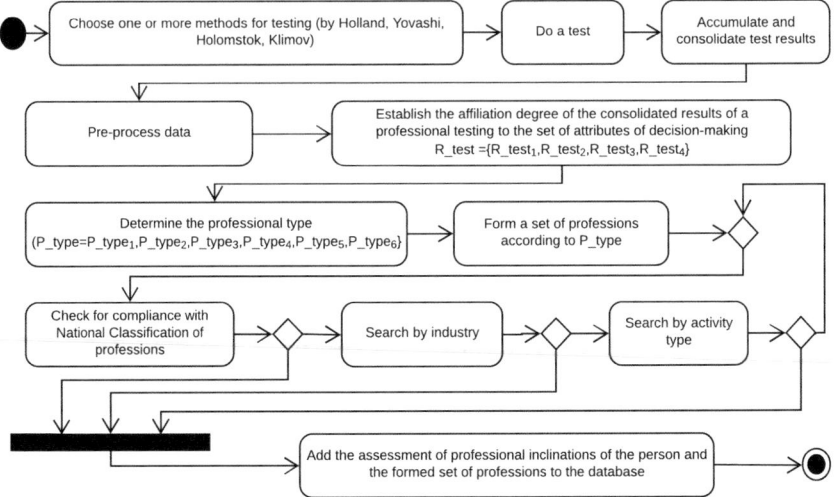

Fig. 2. Recommendation formation for choosing a profession

In order to verify the authenticity of the developed model for identifying the professional inclinations and abilities of a person, experimental studies were conducted, in particular testing pupils (in order to determine the future professional direction), students (to check the compliance of the chosen specialty with the professional abilities, the level of knowledge potential and the requirements of the labor market in the city), and during recruitment at the firm (to establish the correspondence of the professional qualities and inclinations of a candidate to the chosen position) (Fig. 3).

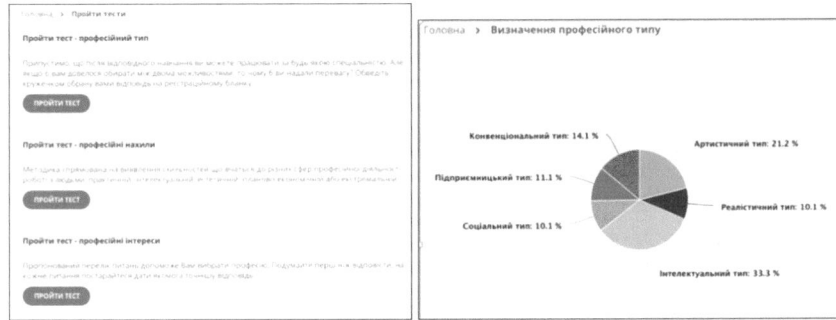

Fig. 3. Vocational guidance testing and its results

In general, the effectiveness of the proposed data analysis model for identifying professional inclinations and abilities of a person, is a significant reduction in the time expenditures for processing the results of vocational guidance testing (Table 5).

Table 5. Comparative table of professional-orientation methods

Name	Result	Test time	Time for processing results
QPO	Professional environment	20 min	7 min
QPI	Professional inclinations	15 min	5 min
QCI	Circle of professional interests	15 min	5 min
QPT	Type of future profession	10 min	3 min
By four techniques together		60 min	20 min
Proposed method	A professional type that gives a comprehensive profile of a person (professional environment and inclinations, type of profession and circle of professional interests)	51 min	1–3 s

The obtained research results are the basis of the software and algorithmic complex of information and technological support of the profession choice [21], which allows to optimize the process of professional orientation of a person. The indicator of the effectiveness of the developed software and algorithmic complex was determined by the ratio of correctly given recommendations on the choice of professional direction of a person to the total number of generated recommendations. It was investigated that in 91% of cases the software and algorithmic complex provides recommendations that coincide with the expert recommendations. This is confirmed by the actual research data and the results of testing pupils and students.

4   Conclusions

The model of the data analysis process for determining the professional inclinations and abilities of a person is based on the results of vocational guidance testing using the methods of J. Holland, L. Yovashi, E. Klimov and A. Holomshtok was developed. The main result of the study was the discovery of a set of attributes by means of intellectual analysis of data, which influence mostly on the decision of a professional type of a person. The obtained research results are the basis of the software complex of informational and technological support of the profession choice, which makes it possible to optimize the process of personalized profession choice and reduce the passage time of the professional orientation tests and the interpretation of their results.

References

1. Thomas, B., Peter, N., Wolter, S.C.: Gender, competitiveness, and study choices in high school: evidence from Switzerland. Am. Econ. Rev. **107**(5), 125–130 (2017)
2. Nota, L., Santilli, S., Soresi, S.: A life-design-based online career intervention for early adolescents: description and initial analysis. Career Dev. Q. **64**(1), 4–19 (2016)

3. Eesley, C., Wang, Y.: Social influence in career choice: evidence from a randomized field experiment on entrepreneurial mentorship. Res. Policy **46**(3), 636–650 (2017)
4. Meijers, F.: A dialogue worth having: vocational competence, career identity and a learning environment for twenty-first century success at work. In: Enhancing Teaching and Learning in the Dutch Vocational Education System, pp. 139–155 (2017)
5. Ceschi, A.: The career decision-making competence: a new construct for the career realm. Eur. J. Train. Dev. **41**(1), 8–27 (2017)
6. Van der Gaag, M.A.E., van den Berg, P.: Modeling the individual process of career choice. In: Advances in Social Simulation, pp. 435–444 (2017)
7. Dimitrakopoulos, I., Karamanis, K.: Decision making using multicriteria analysis: a case study of decision modeling career in education. Case Stud. Bus. Manag. **4**(2), 24 (2017)
8. Bomba, A., Kunanets, N., Nazaruk, M., Pasichnyk, V., Veretennikova, N.: Information technologies of modeling processes for preparation of professionals in smart cities. In: Advances in Intelligent Systems and Computing: Conference Proceedings, pp. 702–712 (2018)
9. Thakar, P., Mehta, A., Manisha: A unified model of clustering and classification to improve students' employability prediction. Int. J. Intell. Syst. Appl. (IJISA) **9**(9), 10–18 (2017). https://doi.org/10.5815/ijisa.2017.09.02
10. Monjurul Alom, B.M., Courtney, M.: Educational data mining: a case study perspectives from primary to University Education in Australia. Int. J. Inf. Technol. Comput. Sci. (IJITCS) **10**(2), 1–9 (2018). https://doi.org/10.5815/ijitcs.2018.02.01
11. Yegorova, O., Durytska, S., Totsenko, V.: The decision support system for choosing a profession. Regist. Storage Process. Data **5**(4), 97–105 (2003). (in Ukrainian)
12. Kaviyarasi, R., Balasubramanian, T.: Exploring the high potential factors that affects students' academic performance. International Journal of Education and Management Engineering (IJEME) **8**(6), 15–23 (2018). https://doi.org/10.5815/ijeme.2018.06.02
13. Holland, J.L., Gottfredson, G.D., Bacer, H.G.: Validity of vocational aspiration and interests of inventories: Extended replicated and reinterpreted. J. Consult. Psychol. **37**, 337–342 (1990)
14. Yovashi, L.: The questionnaire about professional inclinations. http://prevolio.com/tests/. (in Ukrainian)
15. Holomshtok, A.: The questionnaire of interests. http://prevolio.com/tests/. (in Ukrainian)
16. Klimov, E.: How to choose a profession, p. 159 (1990). (in Russian)
17. Almarabeh, H.: Analysis of students' performance by using different data mining classifiers. Int. J. Mod. Educ. Comput. Sci. (IJMECS) **9**(8), 9–15 (2017). https://doi.org/10.5815/ijmecs.2017.08.02
18. Ansari, G.A.: Career guidance through multilevel expert system using data mining technique. Int. J. Inf. Technol. Comput. Sci. (IJITCS) **9**(8), 22–29 (2017). https://doi.org/10.5815/ijitcs.2017.08.03
19. Nikolskyi, Yu.: Model of data analysis process. Comput. Sci. Inf. Technol. **663**, 108–116 (2010). (in Ukrainian)
20. Zavaliy, T., Nikolskyi, Yu.: Data analysis and decision making based on the theory of approximate sets. Inf. Syst. Netw. **610**, 126–136 (2008). (in Ukrainian)
21. Kunanets, N., Nazaruk, M., Nebesnyi, R., Pasichnyk, V.: Information technologies of personalized choice of professionals in smart cities. Inf. Technol. Learn. Tools **65**(3), 277–290 (2018)

The Architecture of Mobile Information System for Providing Safety Recommendations During the Trip

Valeriia Savchuk[1]([⊠]), Yaroslav Vykyuk[2], Volodymyr Pasichnyk[1], Roman Holoshchuk[1], and Nataliia Kunanets[1]

[1] Lviv Polytechnic National University, Lviv, Ukraine
Valeriia.V.Savchuk@lpnu.ua
[2] PHEI "Bukovinian University", Chernivtsi, Ukraine

Abstract. The paper is devoted to the project of mobile information system for providing safety recommendations during the trip. The topicality of the research is caused by the rapid growth of tourism industry all over the world. According to the latest news, Ukrainian aviation companies had fewer plains then needed for 2018 summer vacation period. The deep analysis of dangerous sources with various examples is provided in the paper. In order to develop a project of the system the analysis of modern mobile applications for danger alerts is made. As a result the tasks that should be fulfilled are defined. Some tasks are described in the article, others are still going to be done. The functional features of the system that is being developed by the authors is described and justified. It is being developed in order to define a certain level of danger in any tourist destination and generate recommendations to ensure users safety during the trip that is planned. The paper presents the structure of the system modules and their components basing on UML diagrams. The roles of the users are defined and described. As a result the architecture of the system is developed in order to fulfil all the requirements.

Keywords: Information systems · Mobile applications · Danger sources · Danger predictions · System architecture · System modelling

1 Introduction

The tourism industry is growing with every year all over the world [1]. Ukraine is also concerned. According to the latest news, Ukrainian aviation companies had fewer plains then needed for 2018 summer vacation period.

But when travelling far away tourist is not always aware of danger situations that can occur and their sources.

Every specific danger source needs different way to arrange safety. Theoretically, the government of the country and laws should guarantee safety of every person or society in all spheres of life. Practically, everybody should take care of themselves and know what to do when emergency of any kind happens. It is easy when you are in you native region and know all its features, but it is much harder when you travel far away as because you need to know much more information on the possible sources of danger

© Springer Nature Switzerland AG 2020
Z. Hu et al. (Eds.): ICCSEEA 2019, AISC 938, pp. 493–502, 2020.
https://doi.org/10.1007/978-3-030-16621-2_46

and ways to reduce their influence. So safe state can not arise by its own, as normal, the energy, time and information should be provided for its arrangement.

As it was mentioned above, it is quite difficult to arrange personal safety when travelling, as it is necessary to study a wide range of information on danger level in a region. That is why scientists in Lviv Polytechnic National University and Bukovyna University are working on a project of mobile information system for providing safety recommendations during the trip – "Safe Tourism", main purposes of which is to detect danger sources in a region, count the danger level and provide personalized recommendations to the tourist on where to go and how to behave to avoid emergency.

The topicality of the research is caused by the lack and high need in information technologies to provide safety recommendations to tourists during their trips.

Main parts of the article are devoted to the analysis of danger sources and mobile systems for danger level detection, the developed by authors functionality and structure of the "Safe Tourism".

2   Related Work

In sources [2–13] the analysis of dangerous situations and their results is given. In sources [14–17] the information systems to provide information on dangerous sources are presented and described. In [20, 21] the algorithms to predict natural dangerous situations is described and experimentally. In [18, 19, 22–25] the methods and systems to provide tourist recommendations to the users are described.

2.1   The Analysis of Danger Sources

Danger is a negative property of matter, which manifests itself in its ability to cause damage to certain elements of the world, other words: potential source of harm. If a human being is concerned, the danger sources can be as phenomena, processes, objects, properties that are able under certain conditions to cause damage to the health or life of a person or systems that provide human livelihoods [2].

Danger sources can be: natural, industrial, social.

The general set of danger sources counts more than 150 names, and it is not even quite full [2].

Among natural dangers are bad, or even extreme, weather conditions, natural fires, poisonous plants, dangerous animals, insects, bacteria etc.

Below are given the real examples of worst natural dangerous situations among recent years:

Forest fires in Greece this summer has taken more 90 lives. It was one of the worst fires in Greece among recent years [3]. The fire in Tennessee (USA) in 2016 was the biggest in last 100 years. It took 13 lives, destroyed more than 1000 buildings and lasted nearly one month [4].

The victims of the earthquake in Haiti in 2010 are nearly 3 millions of people, 316 thousands of them died. The natural disaster was getting even worse as it leaded to growth of violence and robbery [5].

Tsunami in the Indian Ocean that was caused by Sumatra-Andaman earthquake was the worst in the human history. It took nearly 230000 lives in 14 countries as the waves were more than 30 m high [5].

As it was mentioned above, animals can also be harmful. For example according to ISAF (International Shark Attack File) before 2012 year 2569 shark attacks were registered: 484 victims died [6]. The most dangerous is Florida cost as there were registered 812 attacks.

Moreover, in Ukraine in Lviv region there were registered 8 victims of snakes this spring (2018) [7].

Among industrial are the dangers caused by use of vehicles, the operation of lifting and transport equipment, the use of combustible, flammable and explosive substances and materials, using processes that occur at elevated temperatures and high pressures, using electric energy, chemicals, different types of radiation etc.

A large number of industrial disasters is associated with nuclear energy usage, both for military and industrial purposes. Thus, by 1993, all nuclear powers carried out at least 2146 explosions: the USA - 1149, the USSR - 715, France - 194, Great Britain - 45, China - 42, India - 1. This led to an increase of the radioactive background on the Earth, which contributed to the growth of the number of cancer patients [8].

One of the most wide known industrial catastrophes is The Chernobyl disaster— caused by 4th explosion reactor unit of the Chernobyl Nuclear Power Plant. Radioactive dust was moved to many European countries by the wind [9].

But the nuclear energy is not the only one that can harm people. For example, pollution in Great Britain is killing 50,000 people a year [10].

To social dangers belong not only terrorist attacks, wars, criminality but cultural diversity and that is why different norms of behaviour, large crowds of people, poverty etc.

It should be mentioned that recently the risk of terrorist attacks is growing according to the registered situations. The biggest and most wide known attack occurred 11th December 2001 year, when passenger planes hit two towers of the World Trade Centre [11].

The other social danger is war, for example Russian aggression in the south region of Ukraine, which is still going on. It leaded to more than 2500 deaths among civilians that are registered [12]. This number is recently growing.

As it was mentioned above, the crowd can also be harmful. More than 100 people died in the crowd in India after religious festival. Another example is situation that has occurred in Cambodia during the festival of water. The crowd has leaded to death of 465 people and hundreds of injured [13].

2.2 Information Systems for Danger Level Detection and Following User Informing

Of course, there is a great amount of user applications that inform users on the danger situations that are likely to occur, but they are mainly taking into account weather conditions and terrorist attacks.

The following mobile information systems provide the user with danger alerts (Table 1):

- Weather dangers alerts: NOAA Weather Radar and Alerts, AccuWeather, Windy
- Forest fire alert: BC Wildfire, Wildfire Analyst Pocket
- Terrorism dangers alert: Terror Alert, TerrorMate, News about Terrorism WTA

These systems are chosen because of their popularity and high rating on Google Play platform (virtual shop of mobile applications for Android operating system) [14].

Below the analysis of mobile information systems that provide the user with weather danger alerts is presented.

NOAA Weather Radar and Alerts is a powerful mobile system for weather predicting and weather danger alerts. The system is developed by one of the leading mobile development companies Apalon [15]. Among the general functions (weather forecast, location search, bookmarks, different maps views etc.) it provides the user severe weather alerts and hurricane tracker.

The AccuWeather is an weather forecasting platform that predicts weather conditions very accurately [16]. The motto of the system is:

"To save lives, protect property, and help people to prosper, while expanding AccuWeather as a healthy and profitable business" [16].

The system provides alerts in the following cases:

- Rain – more than 12.7 mm
- Snow – more than 2.54 mm
- Ice – more than 0.254 mm
- Sustained Wind – more than 48 kph
- Wind Gust – more than 64 kph
- Thunderstorm Probability – 75%

Table 1. Functional comparing of Information systems for weather alerts

Name	Company	Weather alerts	Hurricane predictions	Map	E-mail alert
NOAA weather radar and alerts	Apalon	+	+	+	–
AccuWeather	AccuWeather	+	–	+	–
Windy	Windyty	+	+	+	+

The Windy (or Windyty) is weather forecast visualization tool [17]. Among general weather forecast it provides various weather alerts. The alert consists of the following information: the type of dangerous weather condition, wind, amount of precipitation, temperature, clouds, time/duration etc. Among these functions, the system provides the waves forecast to inform users on if it is safe to do any activities in water (sea, ocean, lakes, rivers, etc.) The other feature that should be mentioned is that AccuWeather provides the users information on various height under the ground, so it is quite useful for pilots, paragliders, skydivers [17].

3 The Functionality and Architecture of the Safety Recommender System

Safe Tourism" is the mobile information system for providing safety recommendations during the trip. Its main functions are to define a certain level of danger in any tourist destination and generate recommendations to ensure user's safety during the trip that is planned (Fig. 1).

The system considers the presence of all the factors that can affect the level of risk in a specific destination, namely: natural, technological and socio-political situations. The main classes of the system users are tourists and administrators (Fig. 2).

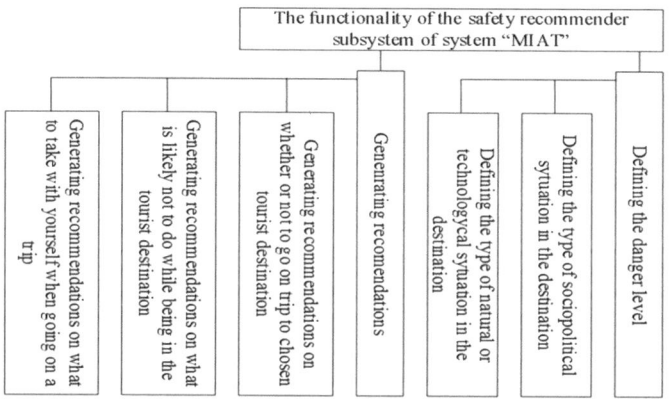

Fig. 1. The functionality of the system "Safe Tourism" [19]

Tourists provide information on the planned period of trip realisation and desirable tourist destinations [18, 19]. The system determines the level of danger according to this input data. If the user does not specify the desired destination or the calculated level of risk is significant the system generates a number of tourist destinations with minimal level of danger in the chosen period of time [20]. System administrators are responsible for its correct functioning and adding new information of the system database. The architecture of the system is complex and extensive (Fig. 3) [20].

The main structural components of the system are:

- "Danger Determination" (DD) is a systems component that is responsible for the analysis of the current natural, technological (industrial), and socio-political situation in the tourist destination today and archives of natural situations for the selected season. Besides these component generates a list of tourist destinations with low level of danger in the chosen period.
- "News Monitoring" (NM) is responsible for finding information in the world news website (namely BBC Word News). It looks for date about present unfriendly political, social, natural and technological situation in a particular region. Search is based on a number of key words from the system database. For example: "fire", "flooding", "rainy season", "revolution", "terrorist", and so on.

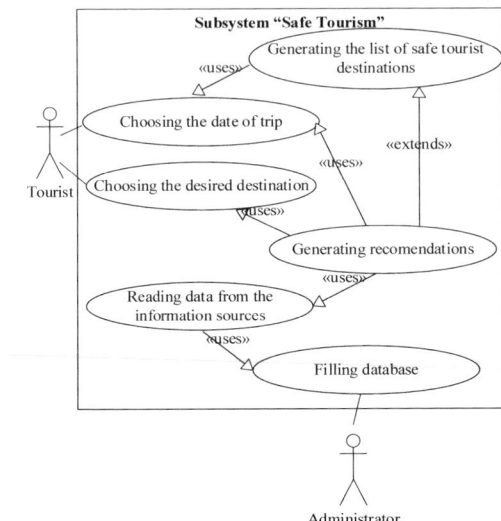

Fig. 2. Users of the system

"Weather Information Extraction" (WIE) is responsible for finding information in world-class weather resource and it forecast for a specific period of time, if it exists.

"Recommendations Generator" (RG) is an important component that generates recommendations based on the result of the DD functioning. Recommendations are divided into two types: the advice on social behaviour and communication and the advice on things to take with yourself on the specific trip.

"Database data input" is a component that is used for "remembering" the results of system work in the database "Danger sources". The database consists of structured and detailed information on danger sources, their peculiarities and ways to minimize the risks.

As it was already mentioned above the system has two types of users, tourists and administrator, which is why there are two types of user interfaces that give the user different opportunities (Fig. 3).

The most complex components of the system are News Monitoring and Danger Determination. Their architectures are described in Figs. 4 and 5.

The News Monitoring component consists of the following subcomponents:

- "Key-words database reading" component – reads information on key words search list that consists in database "Danger sources".
- "Web-site grabbing" is devoted to searching articles according to key-words list through news web-sites and putting it in temporary news.txt file in the cloud.
- Danger sources search" detects the most important articles according to the news popularity in the Internet. The result of Danger source search is the result of the component functioning.

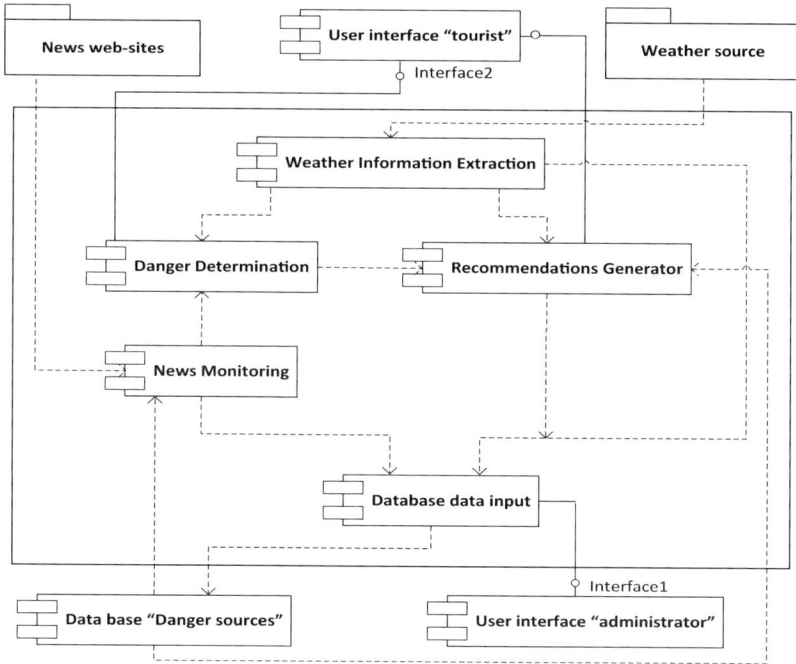

Fig. 3. Architecture of the system "Safe Tourism"

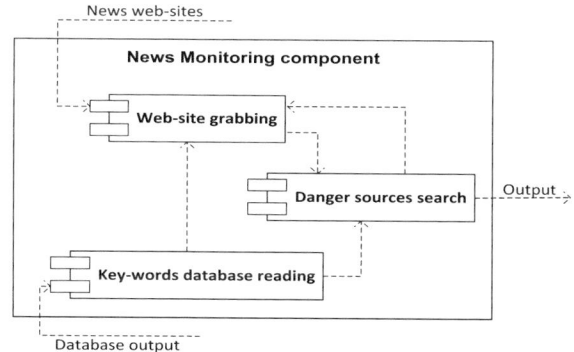

Fig. 4. Architecture of the News Monitoring component

The Danger Determination component consists of the following subcomponents:

- "Attractivety computing" is devoted to defining most popular regions for the tourist trips, taking into account seasonal issues and destination features.
- "Danger predictioning" component functioning is based on developed methods to predict natural dangers [21].
- "Danger sources determination" component defines which emergencies are more likely to occur in a region and what are the danger sources.

- "Danger level computing" counts the influence of every danger source and possible level of threat.
- As a result the beta version of the system is developed (Fig. 6).

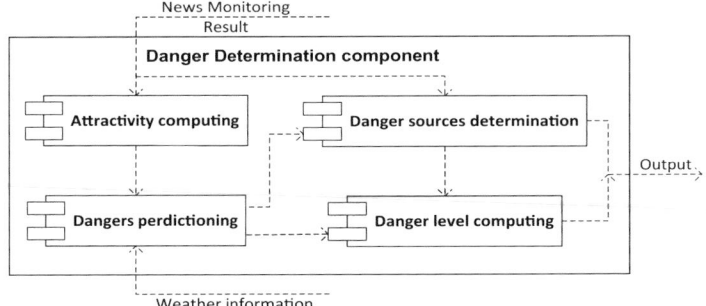

Fig. 5. Danger Determination component structure

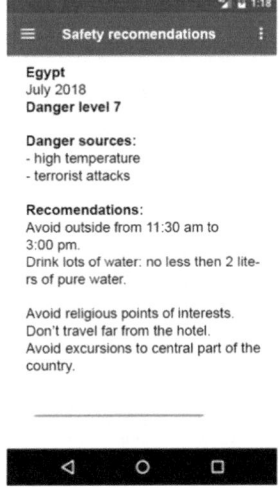

Fig. 6. "Safe tourism" recommendations user interface

4   Conclusions

Scientists in Lviv Polytechnic National University and Bukovyna University are working on a project of mobile information system for providing safety recommendations during the trip – "Safe Tourism", main purposes of which is to detect danger sources in a region, count the danger level and provide personalized recommendations to the tourist on where to go and how to behave to avoid emergency.

As a result the authors developed and presented the architecture of the system "Safe tourism" to give personalized recommendation on how to avoid danger situations or minimize the risk taking into account various dangerous situations using several kinds of information sources: database with archived information on past dangerous situations and ways to minimize risks, up to date internet news web-sites and up-to-date weather conditions sources. The structure of the system is presented with the help of UML diagrams.

References

1. Kuzyk, S.P.: Tourism Geography: Schoolbook. Knowledge, Kiev (2011). (in Ukrainian)
2. The sources of danger and dangerous factors. Osvita.Ua. (in Ukrainian). http://osvita.ua/vnz/reports/bjd/22791/. Accessed 18 June 2018
3. Massive forest fires in Europe: suffered Greece, France and Portuguese. Unian (2017). (in Ukrainian). https://www.unian.ua/world/2080404-masshtabni-lisovi-pojeji-u-evropi-naybilshe-strajdayut-gretsiya-frantsiya-ta-portugaliya.html. Accessed 18 June 2018
4. Record count of victims: in USA because of forest fires 13 people died. (in Ukrainian). https://www.5.ua/svit/rekordni-vtraty-u-ssha-vid-lisovykh-pozhezh-zahynuly-13-osib-132723.html. Accessed 18 June 2018
5. 10 worst natural disasters of 21 century. Rate1. (in Ukrainian). http://www.rate1.com.ua/ua/dovkillja/priroda/2391/. Accessed 18 June 2018
6. 1580-2012 Map of World's Confirmed Unprovoked Shark Attacks (N = 2,569). International Shark Attack File. (in Ukrainian). http://www.webcitation.org/6HdYjpDyI. Accessed 18 June 2018
7. This year in Lviv region 8 people were hospitalised with snake bites. Lviv City News. (in Ukrainian). https://city-adm.lviv.ua/. Accessed 18 June 2018
8. Miahchenko, O.P.: The safety of live of human and society. Schoolbook. The educational literature center, Kiev (2010). (in Ukrainian)
9. Ganotsky, V.: Chornobyl – our pain. The requiem story. Central Ukrainian Publishing House, Kirovograd (2006)
10. Griffin, A.: Pollution is killing millions of people a year and the world is reaching 'crisis point', experts warn. Independent (2017). (in Ukrainian). https://www.independent.co.uk/environment/pollution-air-clean-water-vehicles-diesel-car-tax-lancet-report-deaths-fatal-disease-a8009751.html. Accessed 18 June 2018
11. The 9/11 terrorist attacks. BBC History. (in Ukrainian). http://www.bbc.co.uk/history/events/the_september_11th_terrorist_attacks. Accessed 18 June 2018
12. The war in Donbas: UNO told the number of civil victims. Segodnya. (in Ukrainian). https://ukr.segodnya.ua/regions/donetsk/voyna-na-donbasse-v-oon-nazvali-chislo-zhertv-sredi-mirnyh-zhiteley-1029489.html. Accessed 20 June 2018
13. More than 100 dead in crowd on festival in India. TSN. (in Ukrainian). https://tsn.ua/svit/bilshe-100-lyudey-zadavili-na-smert-na-festivali-v-indiyi.html. Accessed 20 June 2018
14. Google Play. https://play.google.com/store. Accessed 20 June 2018
15. NOAA Weather Radar. Apalon. https://www.apalon.com/noaa_radar_free.html. Accessed 20 June 2018
16. AccuWeather. AccuWeather. https://www.accuweather.com/en/about. Accessed 20 June 2018
17. Windy.com. AppStore Preview. https://itunes.apple.com/us/app/windy-com/id1161387262?mt=8. Accessed 20 June 2018

18. Power, D.J., Sharda, R., Burstein, F.: Decision Support Systems. Wiley Encyclopedia of Management (2015)
19. Huang, C.D., Goo, J., Nam, K., Yoo, C.W.: Smart tourism technologies in travel planning: the role of exploration and exploitation. Inf. Manag. **54**(6), 757–770 (2017). https://doi.org/10.1016/j.im.2016.11.010
20. Savchuk, V.V., Kunanec, N.E., Pasichnyk, V.V., Popiel, P., Weryńska-Bieniasz, R., Kashaganova, G., Kalizhanova, A.: Safety recommendation component of mobile information assistant of the tourist. In: Proceedings on SPIE 10445, Photonics Applications in Astronomy, Communications, Industry, and High Energy Physics Experiments 2017, 104455Z, 7 August 2017. https://doi.org/10.1117/12.2280833
21. Vyklyuk, Y., Radovanović, M.M., Stanojević, G.B., Milovanović, B., Leko, T., Milenković, M., Petrović, M., Yamashkin, A.A., Pešić, A.M., Jakovljević, D., Milićević, S.M.: Hurricane genesis modelling based on the relationship between solar activity and hurricanes II. J. Atmos. Solar-Terr. Phys. (2017). https://doi.org/10.1016/j.jastp.2017.09.008
22. Barabash, O., Shevchenko, G., Dakhno, N., Neshcheret, O., Musienko, A.: Information technology of targeting: optimization of decision making process in a competitive environment. Int. J. Intell. Syst. Appl. (IJISA) **9**(12), 1–9 (2017). https://doi.org/10.5815/ijisa.2017.12.01
23. Dennouni, N., Peter, Y., Lancieri, L., Slama, Z.: Towards an incremental recommendation of POIs for mobile tourists without profiles. Int. J. Intell. Syst. Appl. (IJISA) **10**(10), 42–52 (2018). https://doi.org/10.5815/ijisa.2018.10.05
24. Hadiwijaya, N.A., Hamdani, H., Syafrianto, A., Tanjung, Z.: The decision model for selection of tourism site using analytic network process method. Int. J. Intell. Syst. Appl. (IJISA) **10**(9), 23–31 (2018). https://doi.org/10.5815/ijisa.2018.09.03
25. Alemu, T.A., Tegegne, A.K., Tarekegn, A.N.: Recommender system in tourism using case based reasoning approach. Int. J. Inf. Eng. Electron. Bus. (IJIEEB) **9**(5), 34–43 (2017). https://doi.org/10.5815/ijieeb.2017.05.05

Research and Determination of Personal Information Security Culture Level Using Fuzzy Logic Methods

Mariia Dorosh[✉], Mariia Voitsekhovska, and Iryna Balchenko

Chernihiv National University of Technology, Chernihiv, Ukraine
mariyaya5536@gmail.com

Abstract. The tasks of determining the state of the organizational information security culture are considered. Despite the dozens of technical and technological means to provide an information security, the issue of quantifying the level of information security culture (ISC) in the organization remains inadequately investigated. The personal information security culture of employee becomes the basis of organizational ISC. The model for identifying a personal ISC using the fuzzy logic for the formalized assessment of personnel ISC in the overall quantitative assessment of the organization's IS was proposed. The need for this approach caused with difficulty of obtaining quantitative evaluation indicators of personnel ISC in assessing the overall security of the organization.

As an example of using this model, the assessment of the user's personal cybersecurity culture is considered. It was carried out by collecting the input data by questionnaire and represented as the "inputs-output" surfaces of the fuzzy hierarchical system. The results of the assessment show the problems in ISC wich must be corrected with training, motivation and additional instructing. The presented model can be considered as a part of the ISMS audit, which assessing the awareness of employees as one of aspects of the organization's ISC.

Keywords: Information security culture · Fuzzy logic · Personal culture

1 Introduction

The development of a network society and widening of information technologies have led to fundamental changes, but opened a new opportunities for success achievement in business, high-tech goods and services production and social interaction. Nevertheless, there are some new threats appeared (or showed themselves) with new opportunities related with Internet of Everything "digital universe".

Taking into account the rapid development of technical systems for IS it is still almost impossible to exclude a person from the process of creating, processing, transmitting information, which maintains a high degree of vulnerability to attacks based on social engineering [1]. Also, the effective implementation of integrated information security systems is a big enough problem for small under-funded and developing private companies and start-ups with security implementation standards. The reason is the high cost of software and hardware and the need to attract a highly specialized staff with appropriate pay.

© Springer Nature Switzerland AG 2020
Z. Hu et al. (Eds.): ICCSEEA 2019, AISC 938, pp. 503–512, 2020.
https://doi.org/10.1007/978-3-030-16621-2_47

Also, the importance of strengthening the human factor by raising the level of ISC is caused by the possibility of catastrophic consequences as a result of the occurrence of IS incidents on critical infrastructure objects.

One of the key issues in the design and implementation of the Information Security Management System (ISMS) is to determine the current ISC level of the employees and participants of the organization's information processes. The requirements for the technical IS aspect are devoted to many industry standards, recommendations and techniques that are intended to justify, evaluate and improve the system of policies and tools for IS providing.

During the audit of the organization ISMS, there are weakly formalized methods for evaluating the ISC of employees in the overall IS system of the organization. It should be noted that hardware protection systems are aimed at responding and eliminating the effects of IS incidents on known attacks only, and detecting and responding to attacks of new types is a function of personnel.

Thus, the indivisibility of the personal ISC of the employees (as end-users of the internal information space of the organization) and the organizational ISC is the basis for further study of the personal information security culture.

2   The Aim of Research

As a component of the expert system for assessing the state of the organizational ISC, is considered the block for assessing the state of the personal cybersecurity culture of the end-user, which employees are. Since personal skills and knowledge in the field of information security directly affects the ISC, identifying the "weaknesses" of direct participants in the information space of an organization will help to further determine the set of necessary measures to eliminate spread the basics of ISC among employees, which will enhance the overall ISC of organization. So, the aim of the study is researching of personal information security culture by forming an expert system using the fuzzy logic algorithm.

3   Related Works

Problems of using a culture as a carrier and translator of safe and informed behavior are of interest to many potentially dangerous knowledge areas, such as nuclear energy, ecology, medicine, – in areas where human activities can have destructive effects. Culture is considered as a source of wealth and rarities factors in the economic behavior modeling [2].

Problems of culture evaluating in various fields have troubled many scientists. The culture as a factor of ensuring an environmental safety was investigated by I.V. Lesnikova and N.M. Yastrebova. The issues of the safety culture formation in the nuclear energy field are dedicated to the work of Ukrainian and foreign researchers: V.V. Begun, G.A. Novikova, etc.

The works of Schlienger and Teufel [3] are devoted to the consideration of the ISC from the standpoint of the influential instrument of the existing corporate culture.

In more detail, organization's ISC aspects that directly related to the source of risk - the staff, are presented in the works Krombholz et al. [4], Mouton et al. [5], Okere, van Niekerk and Carroll [6], Alhogail and Mirza [7], Ochang et al. [8]. Such increased attention to the human factor is due to the impossibility of excluding a person as a participant in the process of information processing. Consequently, gaps in information security of direct implementers will be sources of IS threats to the organization.

Effective models for assessing the level of culture in general and the culture of information security, in particular, recognize the models of indicators considered in [9], and as the most appropriate ones are presented in the models of latent variables, determined by causal and effect-indicators [10].

Recently, the attention of sociologists and other scientists is attracted to a socio-technological culture phenomenon. Socio-technological culture of employees is an organic part of the corporate culture aimed for integrating the achievements of technical and human sciences, applying integrated principles to studying the company's social space, its "healthy" functioning in a competitive market and its active development in accordance with the goals of the organization's development [11].

According to assertion [7], an ISC can be defined as collection of perceptions, attitudes, values, assumptions and knowledge that guide the human interaction with information assets in organization with the aim of influencing employees' behavior to preserve information security. Also, one of the concept definitions of ISC can be noted one that reflects the attitude towards information and information space as one of the most dangerous and most influential environments. Thus, the exemplary attitude to security has been formed in the nuclear power industry: "The security culture is a set of characteristics and peculiarities of organizations' activities and individual behavior establishing the security problems of nuclear power plants as having a higher priority are given attention that is determined their significance" [10].

The above definition covers not only the hardware and technology problems of information resources, but also a set of behavioral templates guided by the personnel (operators, administrators, other employees of the organization, management) in the course of inside and outside professional activities. The importance of continuous compliance with the IS principles is due to the scale and continuity of information threats, the multi-vector attacks on information resources, complicated understanding of the hidden value of information, which is perceived as not important.

4   Research Methods and Data Collection

For the quantitative assessment of the ISC participants level for the organization's information processes and it's supporting the using of artificial intelligence methods in particular are proposed. The use of expert systems (ES) (as opposed to decision support systems) allows not only a simulation of an expert's considerations but also provides an explanation for the findings made [12]. The following recommendation is based on the analysis of the existing ES, taking into account the advantages and disadvantages of their different types. For a given task, it is worth developing an ES based on rules. Because this makes it easy to create and adjust a rule base in the event of a change in the characteristics of the real system, the addition of a linguistic model or new solutions options. A set of fuzzy production rules has been adopted as a basis when developing the ES based on a fuzzy inference system (FIS).

This paper aims to determine the level of personal ISC of employees. Since the physical measurement of ISC indicators is impossible, it was proposed to resort to questioning respondents with subsequent processing of the results. To determine the level of personal ISC, a questionnaire has been developed and contains clear and concise answers. To preserve the interest of the participants, the survey takes no more than 5–7 min for the respondents. Increasing the number of questions or a proposal to form a detailed answer will result in a loss of interest among participants and will have a negative impact on the results of the final questions. Questioning was carried out on a voluntary basis and with the preservation of anonymity (without collecting e-mail addresses). To compile the questionnaire Recommendations [13] were used.

The target audience is represented with students of the 121-Software Engineering specialty and participants of projects that are implemented by the Information Technologies and Software Engineering department of CNUT.

The purpose of the survey is to obtain data from primary sources regarding the use of cyber defense tools and methods by end-users in their daily activities (studies, work, project activities and recreation). At the same time, participants' questioning is justified by the fact that they are active users of the internal information space during activity, and the formed foundations of the personal cybersecurity, as a fragment of a personal ISC, affect work and project activities.

The questionnaire contains 9 questions that cover the basic provisions regarding the basics of personal cyber defense. The probability of error when filling out the questionnaire is minimized by providing a variety of and the most clearly compiled response options. Questions are logically grouped into 3 blocks and focus on ensuring the protection of accounts, personal gadgets and OS.

According to the results of the survey, 39.5% of respondents prefer to work with administrator rights; 52.6% use one common account for business and personal use. 18.4% of participants are highly active in social networks and regularly update information and photos. The same part of the respondents use personal pages in social networks to collect and store an interesting for them information. 39.5% of respondents are rare visitors to their own pages, and 21.1% have lost interest in social networks.

47.4% of respondents monitor the relevance of software using automatic OS and software updates. 44.7% prefer a selective upgrade approach. At the same time, 44.7% of respondents did not installed any anti-virus software, and only 21.1% of participants carry out regular anti-virus checks of their own systems.

Half (50%) of respondents prefer to protect their gadgets with a biometric key, 28.9% of respondents use symbolic or graphical blockers.

As for the password policy, 15.8% of respondents use "strong" passwords for each account and then save them with the help of the password manager, 21.1% store unique passwords on physical media (notebooks, individual files). The majority of respondents (44.7%) prefer to use several memorable passwords for most registration cases. Other respondents use the same password for all accounts or use password autosave in the browser.

5 Assessment of the Level of Personal Information Security Culture

Since the model of personal ISC of the end-user (employee, digital partner, intermediary) cannot be formalized with exact mathematical methods, intellectual methods based on a humanlike inference system can be used to solve this problem.

As a mathematical apparatus for determining the level of awareness, we will use the provisions of fuzzy logic [14] and the fuzzy sets theory. This approach avoids the need for quantitative assessment of indicators, replacing them with qualitative characteristics in the form of phrasal statements. This method has shown good formalization results in the works of Singh and Tomar [15], Khare, Rana and Jain [16], Atlam et al. [17].

Thus, the use of fuzzy logic solves the problems associated with the inaccuracy of quantitative characteristics determination, uncertainty in describing the situation, etc.

We will form the input data for constructing a hierarchical fuzzy knowledge base to determine the basic level of awareness in the issues of personal cybersecurity (Table 1). Thus, 9 linguistic variables grouped logically into three blocks were formed.

We see that it is extremely difficult to obtain accurate or at least numeric input variables in such field as an ISC and also to establish formal dependencies between inputs and the resulting variable, so it was decided to take advantage of the fuzzy logic methods built on the Mamdani algorithm [18]. The rulebase is transparent and easy to understand, and makes the rules accessible for simplified correction. This makes it easy to interpret in terms that are understood not only for developers, but also for users.

To form a fuzzy system for assessing the base level of end-user personal cybersecurity culture, we used the MATLAB computing environment Fuzzy Logic Designer. When forming a fuzzy model, it was decided to resort to the hierarchical principle of construction, which makes it possible to reduce the number of rules and easily operate with dependencies of intermediate variables on several inputs [19]. This model is represented by three subsystems that correspond to thematic triples of incoming variables and the resulting subsystem based on the three outputs of the mentioned subsystems of a lower hierarchical level (Fig. 1).

Figure 1 shows a hierarchical system that models the dependence $y = f(x_1, x_2, x_3, x_4, x_5, x_6, x_7, x_8, x_9)$ using four knowledge bases. These knowledge bases describe such dependencies: $y_1 = f_1(x_1, x_2, x_3)$, $y_2 = f_2(x_4, x_5, x_6)$, $y_3 = f_3(x_7, x_8, x_9)$ and $y_4 = f_4(y_1, y_2, y_3)$.

To describe the variables a triangular membership functions selected. The variables $x_1, x_2, x_4, x_5, x_7, x_8$ we decided to characterize with three terms. The variables $x_3, x_6, x_9, y_1, y_2, y_3, y_4$ were proposed to describe with five terms.

Table 1. Input data of the hierarchical fuzzy knowledge base

Name of variable			Linguistic variable	Possible answer	Term
Resultant	Intermediate	Input			
y_4	y_1	x_1	Accounts for work and interests	Single account with device sync	L
				Partially separated accounts and gadgets	M
				Strictly separated personal and work accounts and gadgets	H
		x_2	Account in OS	Administrator account	M
				User account	H
				No account	L
		x_3	Activity in social networks	Accounts in main social networks (personal information, photos, preferences)	L
				Profile for notes and multimedia	ML
				Accounts with a rare update	M
				Outdated account	MH
				No accounts	H
	y_2	x_4	OS and software updates	Auto update	H
				Selective installation of updates	M
				Ignoring updates	L
		x_5	Antivirus software	Installed on each computing device	H
				Installed on a laptop/computer	M
				Antivirus is not installed/activated	L
		x_6	System scan with antivirus	Once a week	H
				Once every two weeks	MH
				Once a month	M
				Even less often/even more rarely	ML
				Never (antivirus is not installed)	L
	y_3	x_7	Online banking	High/regular activity	L
				Low/rare activity	M
				No activity	H
		x_8	Lock phone and laptop	Symbol or graphic blocker	M
				Biometric screen blocker (fingerprint)	H
				Absence/ignore lock function	L
		x_9	About passwords	Separate password for each account + storage on media	MH
				"Strong" passwords + password manager	H
				Several memorable passwords	M
				Autosave of passwords in the browser	ML
				Single password for most accounts	L

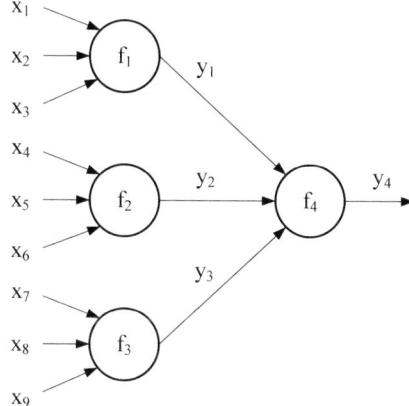

Fig. 1. Hierarchical fuzzy knowledge base

Thus, for example, for the variable x_1:

- Term "*L*" corresponds to the answer to the question in the form – "I use single account, laptop and smartphone, I appreciate the mobility and advantages of synchronization";
- Term "*M*" corresponds to the answer to the question in the form – "The line between work and personal gadgets is very conditional but I try to separate them";
- Term "*H*" corresponds to the answer to the question in the form – "I use a working laptop, telephone and a separate mailing address".

By means of formation of the hierarchical fuzzy rulebased systems, there were compiled 45 rules for the FIS_y1.fis subsystem, 31 for the FIS_y2.fis, 45 for the FIS_y3.fis and 105 for the FIS_y4.fis top-level subsystem.

The control of the correctness and completeness of the created rule base of the resulting subsystem is possible due to the modeling of the "inputs – outputs" surface for $y_4 = f_4 (y_1, y_2, y_3)$ (Fig. 2) and others according to modelling system.

The data obtained from the surveys gave us an opportunity to formulate the level of ISC based on awareness of basic cybersecurity issues for each participant (Fig. 3). Based on these results, definitions of the general level of basic awareness of the personal cybersecurity issues within this group was determined.

The results are shown in Fig. 3 indicate that the existing level of the personal ISC for respondents can be described as "below average (*ML*)". Such an estimate obtained with the help of fuzzy logic methods makes it possible to substantially simplify the formalization of an expert assessment. And also move from qualitative evaluation to quantitative and increase its accuracy.

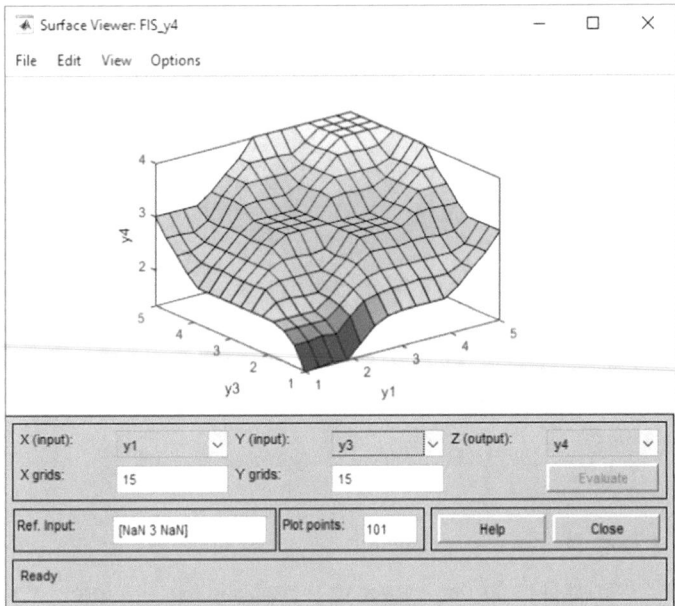

Fig. 2. The "input-output" surface of the resulting fuzzy system FIS_y4.fis

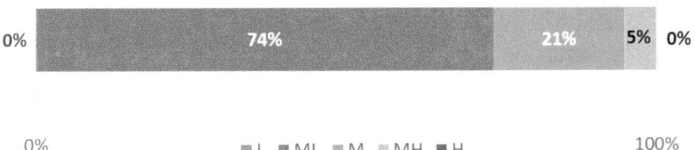

Fig. 3. The results of survey: awareness level of basic cybersecurity issues of participants

It is clear that it is necessary the ISC introduction implement using project-based approach and taking into account the need for continuous improvement of this process' implementation. It can be realized with the project or/and process approaches in the formation, implementation and realization of the organization's ISC.

Thus, if a problem arises with the determination, formation and further improvement of the ISC of employees and participants in the organization's information processes by increasing awareness of IS issues, the introduction of ISC into the socio-technological culture a complex project approach can be used. The main stages can be:

- Substantiation of expediency and necessity of implementation or improvement of ISC in the organization;
- Analysis of the current ISC level in the organization and planning measures for its implementation or improvement;
- Realization of planned activities and ISC implementation for all participants in the organization's information processes.

The question of ISC formation and wide-scale introduction in organizations should be considered as a standardized procedure. It is clear, it is necessary the ISC introduction implement using project-based approach and taking into account the need for continuous improvement of this process' implementation.

6   Conclusions and Perspectives

As a set of basic recommendations of the providing appropriate level of the organization's ISC, it is proposed to develop an expert system that will determine the level of the existing organization's ISC including the assessment of the level of the personal ISC. A model for estimating the level of personal culture is constructed with using the fuzzy logic.

The advantages of this approach to assessing the level of the basic culture of end-user cybersecurity can be seen in terms of clarity and simplicity of implementation, the convenience of adjusting the rule-base in the event of the need case to make changes. The ability to use AI in AI stands out with its ability to be used for assessing the ISC that is not susceptible to machine logic. Among the disadvantages we can notice the expert subjectivity when compiling the rulebase. This lack can be reduced involving a group of experts.

The main perspective of our research is including the inputs weights of the expert model and taking into account a risk factors.

Also discussion questions can be:

- Formation of a sufficient amount of input data, and the establishment of interconnections;
- Determination of the importance of various factors;
- Construction of an expert information system for assessing the level of culture of information security of the organization;
- Use of neural networks for the formulation of recommendations based on evaluation results.

References

1. Fan, W., Kevin, L., Rong, R.: Social engineering: I-E based model of human weakness for attack and defense investigations. Int. J. Comput. Netw. Inf. Secur. (IJCNIS) 9(1), 1–11 (2017). https://doi.org/10.5815/ijcnis.2017.01.01
2. Beugelsdijk, S., Maseland, R.: Culture in Economics: History, Methodological Reflections and Contemporary Applications. Cambridge University Press, Cambridge (2010)

3. Schlienger, T., Teufel, S.: Information Security Culture: The Socio-Cultural Dimension in Information Security Management, pp. 191–202 (2002)
4. Krombholz, K., Hobel, H., Huber, M., Weippl, E.: Social engineering attacks on the knowledge worker. In: Proceedings of the 6th International Conference on Security of Information and Networks, Aksaray, Turkey, 26–28 November 2013, pp. 28–35 (2013). https://doi.org/10.1145/2523514.2523596
5. Mouton, F., Leenen, L., Venter, H.S.: Social engineering attack examples, templates and scenarios. Comput. Secur. **59**, 186–209 (2016)
6. Okere, I., van Niekerk, J., Carroll, M.: Assessing information security culture: a critical analysis of current approaches. In: Proceedings of the 2012 Information Security for South Africa, pp. 1–8 (2012). https://doi.org/10.1109/issa.2012.6320442
7. Alhogail, A., Mirza, A.: Information security culture: a definition and a literature review. In: Proceedings of IEEE World Congress on Computer Applications and Information Systems, pp. 1–7 (2014)
8. Ochang, P.A., Irving, P.J., Ofem, P.O.: Research on wireless network security awareness of average users. Int. J. Wirel. Microwave Technol. (IJWMT) **6**(2), 21–29 (2016). https://doi.org/10.5815/ijwmt.2016.02.03
9. Tolstova, Y.N.: Izmerenie v sotsiologii: Kurs lektsiy [Measurement in sociology: Course of lectures]. INFRA-M, Moscow (1998). (in Russian)
10. Begun, V.V., Shirokov, S.V., Begun, S.V., Pismenniy, E.M., Litvinov, V.V., Kazachkov, I.V.: Kultura bezpeki v yadernIy energetitsi [Culture of safety in nuclear power] Kyiv (2012). (in Ukrainian)
11. Hromtsov, A.V.: Sotsialno-tehnologicheskaya kultura personala, kak faktor formirovaniya konkurentosposobnosti firmyi [Socio-technological culture of the personnel as a factor of formation of the firm's competitiveness]. Paper presented at the «Lomonosov» Conference (2007). https://lomonosov-msu.ru/archive/Lomonosov_2007/17/hromcov_av.doc.pdf
12. Angeli, C.: Diagnostic expert systems - from expert's knowledge to real-time systems. In: Sajja, P., Akerkar, R. (eds.) Advanced Knowledge Based Systems: Model, Applications & Research, vol. 1, pp. 50–73 (2010)
13. Tokarev, B.: Printsipy sostavleniya oprosnikov dlya marketingovykh issledovaniy [Principles of compiling questionnaires for marketing research]. https://www.marketing.spb.ru/lib-research/methods/poll_questionnaire.htm. (this page was last modified on 2018)
14. Zadeh, L.A.: Fuzzy sets as basis for a theory of possibility. Fuzzy Sets Syst. **1**, 3–28 (1978)
15. Singh, A.P., Tomar, P.: Web service component reusability evaluation: a fuzzy multi-criteria approach. Int. J. Inf. Technol. Comput. Sci. (IJITCS) **8**(1), 40–47 (2016). https://doi.org/10.5815/ijitcs.2016.01.05
16. Khare, A.K., Rana, J.L., Jain, R.C.: Detection of wormhole, blackhole and DDOS attack in MANET using trust estimation under fuzzy logic methodology. Int. J. Comput. Netw. Inf. Secur. (IJCNIS) **9**(7), 29–35 (2017). https://doi.org/10.5815/ijcnis.2017.07.04
17. Atlam, H.F., Alenezi, A., Hussein, R.K., Wills, G.B.: Validation of an adaptive risk-based access control model for the internet of things. Int. J. Comput. Netw. Inf. Secur. (IJCNIS) **10**(1), 26–35 (2018). https://doi.org/10.5815/ijcnis.2018.01.04
18. Pegat, A.: Nechetkoe modelirovanie i upravlenie [Fuzzy modeling and control]. Binom. Laboratoriya znaniy, Moscow (2013). http://padaread.com/?book=89681&pg=1
19. Shtovba, S.D.: Vvedenie v teoriyu nechetkikh mnozhestv i nechetkuyu logiku [Introduction to the theory of fuzzy sets and fuzzy logic]. MATLAB. Exponenta (2001). http://matlab.exponenta.ru/fuzzylogic/book1/4_6.php

Anomalies Detection Approach in Electrocardiogram Analysis Using Linguistic Modeling

Igor Baklan$^{(\boxtimes)}$ ⬤, Iryna Mukha⬤, Yurii Oliinyk⬤,
Kateryna Lishchuk⬤, Evgenii Nedashkivsky⬤,
and Olena Gavrilenko⬤

Igor Sikorsky Kyiv Polytechnic Institute, Kyiv 03056, Ukraine
iaa@ukr.net

Abstract. This work proposes a new approach for identifying heart anomalies on electrocardiograms data using linguistic modeling. The process of identifying anomalies in the proposed approach consists of the following subtasks: the subtask of interval splitting, the subtask of linguistics, the subtask of anomalies searching. The approach includes: the creation of a linguistic pattern database, represent ECG as well as linguistic chain, use linguistic pattern database to search for linguistic chain parts based on abnormal patterns. Linguistic model is suggested for the creation of an anomalies database for further detection in a cardiogram, reproduced in the form of a linguistic chain. Storing ECG as a linguistic model facilitates is easy for data storing and data search in patient history. Linguistic pattern database has filling stages: signal conversion in digital time series, interval splitting, matching interval with an alphabet symbol, creation of alphabetic symbol time series. Anomalies search based on seeking abnormal linguistic patterns in ECG linguistic chains.

Keywords: ECG · Linguistic modeling · Linguistic model · Linguistic chain

1 Introduction

Electrocardiography is one of the basic methods of heart research and diagnostics of diseases of the cardiovascular system. An ECG is a curve that displays the electrical activity of the heart. Essentially, an ECG is a record of variations in the potential difference occurring in the heart during its excitation. All ECGs are classified by physicians as normal and abnormal, which contain anomalies in the work of the heart. Anomalies detection in ECG is the definition of areas on the curve that do not correspond to the normal work of the heart and contains one of the anomalies in the work of the heart (for example, arrhythmia, ischemic disease, etc.).

The modern approach to cardiology is to analyze the ECG automatically. It allows to provide information support for diagnostic decisions of a doctor, and also to provide an increase in accuracy and reliability of diagnostic findings, which ultimately contributes to increasing the effectiveness of diagnosis and treatment of pathologies of the cardiovascular system of a person.

© Springer Nature Switzerland AG 2020
Z. Hu et al. (Eds.): ICCSEEA 2019, AISC 938, pp. 513–522, 2020.
https://doi.org/10.1007/978-3-030-16621-2_48

Methods of intellectual decision-making support for physicians based on computer processing and data analysis are not yet widely used, although automated information analysis already allows solving many medical tasks, in particular, related to the diagnosis of the functional state of the cardiovascular system.

In the article proposes new approach to recognizing anomalies using linguistic modeling and linguistic chains.

The article consists section "Introduction", "Mathematical approach to recognize anomalies", "Abnormal patterns from LPDB", "Discussion" and "Conclusion". Section "Introduction" describes related work, researches tasks, linguistic model and ECG methods. Section "Mathematical approach to recognize anomalies" describes three subtasks of construction a linguistic ECG model. Section "Abnormal patterns from LPDB" describes examples of graphic images of an anomaly and the corresponding linguistic chains.

1.1   Related Work

The analysis of existing software tools for the analysis of anomalies based on ECG data showed that many of them are based on the use of neural networks. The use of an artificial neural network is conditioned by its ability to process fuzzy and complex output data for their classification [1].

Currently, several types of neural networks are used: full-fledged [2], convolutional [3], recurrent [4]. Typically, for working with sequences, convolutional or recurrent neural networks are used, since the usual ones do not take into account the temporal bonds in sequence, as they do, for example, the convolution (and this can be controlled through the length of the convolution nucleus) or recurrent where the mechanism of the bundle between adjacent states is embedded in the structure of the neural network.

For a simulation through recurrent networks, more memory is needed than by modeling their convolution. In addition, recurrent neural networks are difficult to learn in sequences of a great length (more than a few tens of steps) due to "explosions" or "fading" of gradients.

The advantage of neural network technology is an ability to classify events that are not in the training set, summarizing the previous experience and applying it in new cases. However, to train an artificial neural network, one needs to use a large amount of accumulation information (cardiograms) as training and, as a result, significant hardware and software resources.

In addition to the approach mentioned above to detecting and analyzing anomalies (emissions or novelty), there are other ones.

The statistical approach is based on identifying extremes for certain types of data. This approach is easily visualized using a span chart. In this case, the anomaly is not always characterized by extreme values, which is a disadvantage of this approach.

The idea of the model approach is the following: building model that describes the data (for example, regression model, F model, SVD, linguistic model, etc.). Points that are very different from the model marked as abnormal. The advantage of this method is the ability to consider nature and specificity of the data. The disadvantages include the fact that the method is more suitable for detecting novelty than for detecting emissions.

The iterative approach consists the detection and deleting abnormal objects on each iteration step. Complexity and accuracy are disadvantages of this approach.

The describing methods above allow determining the anomalies of one parameter (one indicator on a cardiogram).

In case of a metric approach will be few neighbors elements in emissions. That is a clustering task. As a measure of an anomaly the "distance to the k-neighbor", the distance of Mahalanobis is being used. It is worth noting that this approach is the most common. For example, the following machine learning methods are being used for anomalies detection:

- OneClassSVM;
- Isolation Forest;
- ElipticEnvelope;
- Association rules.

These methods can be widely used both for training neural networks and in the linguistic model, that is presented in this research.

In paper [5] proposed to modify the methods of selecting Initial centroids. Instead of a random selecting of cluster point in K-means, the enhanced approach is using the Weighted Average Mean as the basis for selecting initial configurations or initial conditions. In paper [6] proposed f novel singleton Apriori has been proposed in this paper that scans all items in a transaction in a database only once which results in minimizing the scanning time. In paper [7] proposed novel fuzzy-based multi-fever symptom classifier, that has two stages. The first stage is fever type confirmation using common fever symptoms, leading to five major fuzzy rules and the second phase is determining the level of infection (severe or mild) of the confirmed type of fever using unique fever symptoms. In [8] proposed Priority Based New Approach for Correlation Clustering method, that to solve the problem of chromatic correlation clustering where data objects as nodes of a graph are connected through color-labeled edges representing relations among objects.

It is recommended to use a joint method that includes several types of different methods to improve the accuracy of determining anomalies.

1.2 Researches Tasks

Primary goal: increasing the accuracy and reliability of diagnostic conclusions about the state of the cardiovascular system due to the development of new computer methods of ECG data procession.

The research purpose:

- Creation of a linguistic pattern database (LPDB);
- Represent ECG as well as a linguistic chain;
- Use LPDB to search for linguistic chain parts based on abnormal patterns.

1.3 Linguistic Model

Linguistic model – set of symbolic (linguistic) sequences based on the same linguistic parameters and formal grammar restored from this set [9].

LPDB filling stages:

- Signal conversion in digital time series;
- Interval splitting. Split the time series value domain into intervals of one of four types:
  - Equivalent intervals;
  - Logarithmic intervals;
  - Equiprobable interval;
  - Intervals for a specified distribution (Poisson, Dirichlet, normal, beta distribution, etc.). The number of intervals should correspond to the alphabet chosen;
- Lnguistization. Matching interval to alphabet symbol;
- Creation alphabet symbol time series.

1.4   ECG Methods

ECG methods are described in [10].

The curve, which reflects the electrical activity of the heart, is called an (ECG). Thus, an ECG is a record of the variations in the potentialities that arise in a heart during its excitation. When registering an ECG in each heart cycle, a series of peak connected to each other is struck out. Vertically, the voltages of the teeth are displayed, horizontally - their duration. Einthoven called them the Latin letters P, Q, R, S, T - in the order they appear from left to right.

In Ukraine, the standard of a tape's speed is 50 mm/s (abroad is 25 mm/s). At a 50 mm/s speed of the tape, each small cell, set between the adjacent vertical lines (distance 1 mm), corresponds to an interval of 0.02 s. Each fifth vertical line on an electrocardiographic tape is thicker. The constant speed of the ribbon and the millimeter net on the paper allow measuring a duration of the teeth and the intervals of the ECG, as well as an amplitude of these teeth.

The electrocardiogram shows schematically three primary indicators of Fig. 1:

- Teeth - bulges with a sharp angle up or down, which are denoted by P, Q, R, S, T;
- Segments - the distance between neighboring teeth;
- Interval - a gap that includes both a tooth and a segment.

Based on the indicators mentioned above, the cardiologist determines the level of reduction and restoration of a heart muscle. In addition to these indicators, the electric axis of the heart can also be determined during the electrocardiogram take off, indicating the approximate location of the organ in the chest cavity. The latter depends on the design of the human body and chronic pathology. The electric heart axis may be normal, vertical and horizontal.

When decoding the ECG, the norm will be as follows:

- The distance between the teeth R and R should be equal throughout the cardiogram;
- The intervals between PQRST should be 120 to 200 m/s (figuratively it is determined by 2–3 squares). This is an indicator of the passage of the pulse in all cardiac departments from the atrium to the ventricles;
- The interval between Q and S indicates the passage of the pulse through the ventricles (60–100 m/s);

- The duration of the contractile capacity of the ventricles is determined by Q and T (in the norm 400–450 m/s).

The least deviation from these parameters may indicate the onset or development of a pathological process in the cardiac muscle.

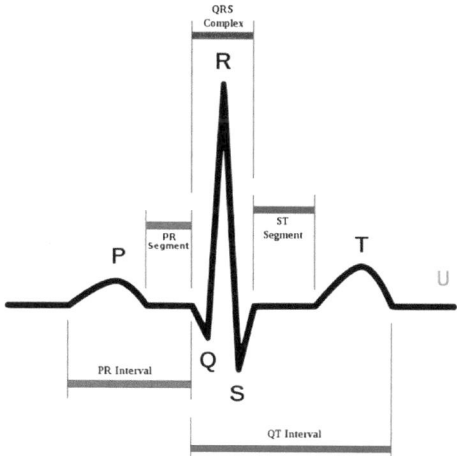

Fig. 1. Schematic elements of ECG

2   Mathematical Approach to Recognizing Anomalies

According to the phases of construction of a linguistic ECG model, the initial task will be divided into the following subtasks:

- The subtask of interval splitting;
- The subtask of linguistics;
- The subtask of anomalies searching.

2.1   The Subtask of Interval Splitting

Assignment of the given subtask is construction a user alphabet by splitting a sorted numerical series of ECGs into a plurality of intervals, each element of which characterizes a particular letter of the alphabet.

It is given:

- Hypothetic complexity of the alphabet a;
- The vector of integers $\bar{D}$ with capacity $k = |\bar{R}|$.

The results:

- The vector of pairs of integer values $\bar{I}$ with capacity $n = |\bar{I}|$.

Restrictions: defined by the maximal amplitude of ECG device.

An interval approach can be used to break up the multitude of numerical data of a series. We will arrange the intervals of the received sequences. We will consider intervals which are not degenerate, that is, the upper and lower boundaries of the interval do not coincide.

The following types of interval splitting are possible:

– Equivalent intervals;
– Logarithmic intervals;
– Equiprobable interval;
– Intervals for a specified distribution (Poisson, Dirichlet, normal, beta distribution, etc.).

As a result of an interval splitting, we receive two sets of intervals:

$$I_{0,1} = [a_0, a_1], I_{1,2} = [a_1, a_2], \ldots, I_{N-1,N} = [a_{N-1}, a_N],$$

where $a_0 = min(a(k)), a_n = 0$;

$$J_{0,1} = [b_0, b_1], J_{1,2} = [b_1, b_2], \ldots, J_{N-1,N} = [b_{N-1}, b_N],$$

where $b_0 = 0, b_n = max(b(k))$;

There are two ways to represent time series: in absolute values or by using normalization.

After choosing the alphabet (for example, Latin), we assign to each member of the sequences $(a(k))$ and $(b(k))$ characters of the alphabet a^i and $b^i i = 1, K, j = 1, L$. We will note that the capacity of the alphabet should be not much smaller than the sequence a (k) $\oplus$ b (k). As a result of these transformations, a linguistic sequence c_i has been obtained.

2.2   The Subtask of Linguistics

Assignment of the given subtask is to obtain a linguistic chain by finding the appropriate letter of the alphabet for each value of the numerical number of the ECG. The letter of the alphabet uniquely corresponds to a certain interval from the set of intervals obtained as a result of the solution of the previous problem.

It is given:

– The vector of integers $\bar{R}$ with capacity $k = |\bar{R}|$;
– The vector of pairs of integer values $\bar{I}$ with capacity $n = |\bar{I}|$.

The results:

– The vector of real $\bar{A}$ with capacity k.

Restrictions:

$$\forall x_i \in \bar{A} : \exists d_i \in \bar{R}, \exists y_j \in \bar{I}, d_i \in \left[y_j^1; y_j^2 \right], x_i = j, \text{where } i \in [0; k), j \in [0; k) \quad (1)$$

2.3   The Subtask of Anomalies Searching

The task purpose – to find a pattern (word) in the linguistic chain (text). In the future, short chains of patterns will be names as words, and an ECG in the form of linguistic chains will be named as a text. To search for words in the text, there are classical methods that are not the purpose of this study. For example, Isolation Forest [11] and text mining algorithms [12] can be used.

3   Abnormal Patterns from LPDB

Below are examples of graphic images of an anomaly and the corresponding linguistic chains.

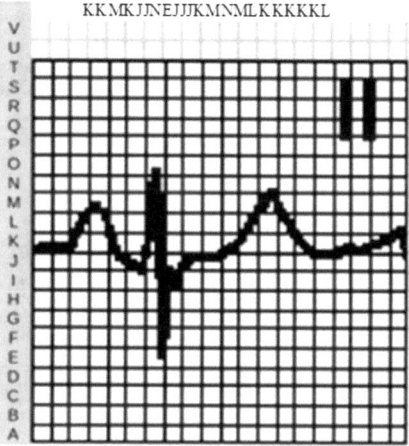

Fig. 2. High peak P in lead II

In Fig. 2 is being shown an enlarged right atrium. Peak amplitude P in lead II > 2,5 mm (P pulmonale). Specific has only 50%, in 1/3 cases P pulmonale called by increase left atrium. Marked by COPD, congenital defects of the heart, congestive heart failure, CHD. Template value = 'KKMKJJNEJJJKMNMLKKKKKL'.

In Fig. 3 shows the following:

- **Dextrocardia.** Negative peaks P and T, inverted complex QRS in lead I without increase peak of amplitude R in chest leads. Dextrocardia can be one of the situs inversus display or isolation. Isolated dextrocardia often combines with other congenital abnormalities, including corrected transposition of main head arteries, pulmonary artery stenosis, ventricular septal and interatrial defects.
- **Wrong electrode attaching.** If electrode, destined for a left hand, was attached on right hand, then registered negative peaks P and T, inverted complex QRS with a normal location transition zone in chest leads. Template value = 'LMLLLMNCMLMLLKKKLMMMM'.

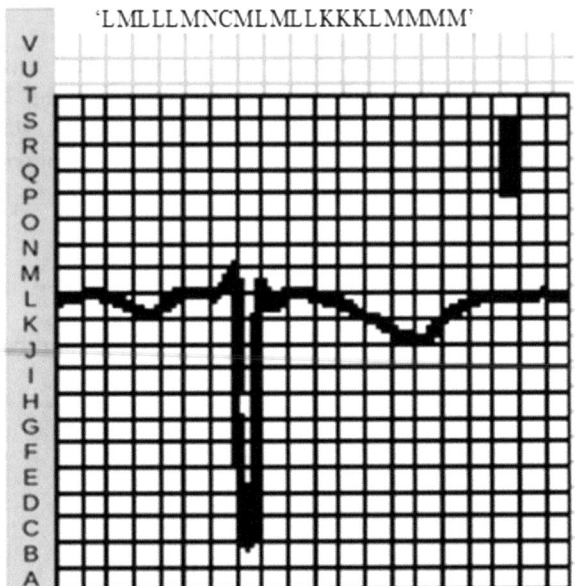

Fig. 3. Negative P in I lead

In Fig. 4 the increased left atrium is being shown. P mitrale: in lead V1 last part (rising knee) peak P expanded (>0,04 c), amplitude >1 mm, peak P expanded in II lead (>0,12 c). Observer in mitral and aortic defects, cardiac insufficiency, myocardial

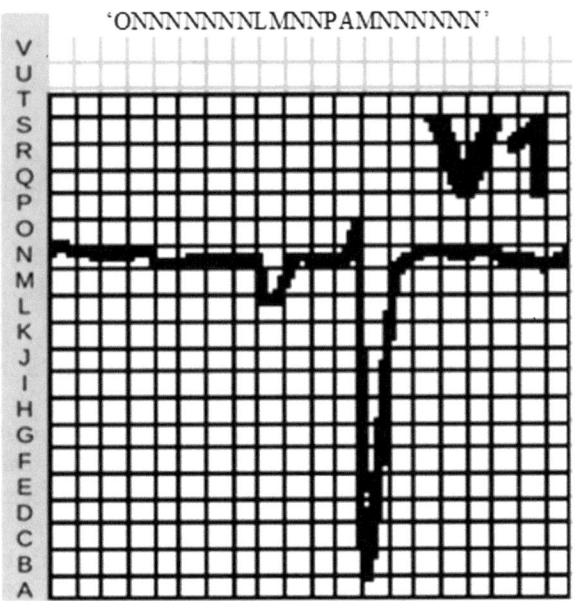

Fig. 4. Deep negative P in lead V1

infarction. Appear specificity of these indicators—more than 90%. Template value = 'ONNNNNNNLMNNPAMNNNNNN'.

Storing ECG as a linguistic model facilitates secure data storing, and data search in patient history.

4 Discussion

Suggested approach has some ECG specialties: proposed approach improves existing methods ECG procession, clearly structured data storing allows to avoid redundancy of data and requires less storage volume, improved ECG analysis algorithm allows to define violation in heart work.

The described approach can be a part of the modern medical reform realization, that include electronic storing and data processing patient history.

5 Conclusions

Describing the creation of linguistic pattern database allows data storing in a linguistic chain view. Anomalies pattern building are suggested. Anomalies search based on seeking abnormal linguistic patterns in ECG linguistic chains. Data representation is easy for procession and viewing.

References

1. Blatter, C.: Wavelet analysis. Theory fundamentals, Tekhnosfera (2004)
2. Hsu, K.Y., Li, H.Y., Psaltis, D.: Holographic implementation of a fully connected neural network. Proc. IEEE **78**(10), 1637–1645 (1990)
3. Faust, O., Ng, E.Y.: Computer-aided diagnosis of cardiovascular diseases based on ECG signals: a survey. J. Mech. Med. Biol. **16**(01), 1640001 (2016)
4. Luz, E.J.D.S., Schwartz, W.R., Cámara-Chávez, G., Menotti, D.: ECG-based heartbeat classification for arrhythmia detection: a survey. Comput. Methods Programs Biomed. **127**, 144–164 (2016)
5. Fabregas, A.C., Gerardo, B.D., Tanguilig III, B.T.: Enhanced initial centroids for k-means algorithm. Int. J. Inf. Technol. Comput. Sci. (IJITCS) **9**(1), 26–33 (2017). https://doi.org/10.5815/ijitcs.2017.01.04
6. Mani, K., Akila, R.: Enhancing the performance in generating association rules using singleton apriori. Int. J. Inf. Technol. Comput. Sci. (IJITCS) **9**(1), 58–64 (2017). https://doi.org/10.5815/ijitcs.2017.01.07
7. Ajenaghughrure, I.B., Sujatha, P., Akazue, M.I.: Fuzzy-based multi-fever symptom classifier diagnosis model. Int. J. Inf. Technol. Comput. Sci. (IJITCS) **9**(10), 13–28 (2017). https://doi.org/10.5815/ijitcs.2017.10.02
8. Jain, A., Tyagi, S.: Priority-based new approach for correlation clustering. Int. J. Inf. Technol. Comput. Sci. (IJITCS) **9**(3), 71–79 (2017). https://doi.org/10.5815/ijitcs.2017.03.08
9. Baklan, I.V.: Interval approach to linguistic model creation. System technologies. Regional intercollegiate collection of scientific papers, Dnepropetrovsk, vol. 86, no. 3, pp. 3–8 (2013)

10. Clifford, G.D., Azuaje, F., McSharry, P.E.: Advanced Methods and Tools for ECG Data Analysis. Engineering in Medicine & Biology Series. Artech House, Inc. (2006)
11. Tomashevskii, V.M., Oliynik, Y.O., Yaskov, V.V., Romanchuk, V.M.: Realtime text stream anomalies analysis system. Visnyk of Kherson National Technical University, vol. 66, no. 3, pp. 361–366 (2018)
12. Olena, G., Yuri, O., Hanna, K.: Review and analysis of algorithms TEXT MINING [Project management, systems analysis, and logistics], no. 19, pp. 32–40 (2017). (in Ukrainian)

Information Technologies for Teaching Children with ASD

Vasyl Andrunyk⬤, Tetiana Shestakevych$^{(\boxtimes)}$⬤,
Volodymyr Pasichnyk⬤, and Nataliia Kunanets⬤

Lviv Polytechnic National University, S. Bandery Street, 12, Lviv, Ukraine
{Vasyl.A.Andrunyk,Tetiana.V.Shestakevych}@lpnu.ua

Abstract. At the education of children with autistic spectrum disorder, special attention should be paid to the correction and maintenance of certain skills. In a variety of assistive technologies and application for students with autism, there is a problem of choice of the relevant product that will fully satisfy users' needs. Such IT product should have appropriate educational characteristics, as well as take into account users' needs and strengths. A recommender system to support such decision making is one of the approaches to deal with the problem. The modeling of such a system is the first stage of its design.

Keywords: Student with ASD · Universal design · WCAG · Social skills · Communication assistance for student with ASD

1 Introduction

The deviation of the autistic spectrum is can be detected at the age of 2–4 years, when specialists can specify abnormalities in communication with other people and the outside world i.e. difficulties with social communication, interaction, and imagination. Since 2013, all autism-related diagnoses are described by the term *Autism Spectrum Disorder* (ASD). Autism spectrum disorder is a common term used to group some brain disorders, including autistic disorders, Asperger's syndrome, children's disintegration disorder, and other widespread developmental disorders, including intellectual. A child with ASD has special needs in organizing of the educational process.

In addition to medical and pedagogical support of the education of children with ASD, it is undoubtedly that various information technologies became an important modern tool for the improvement of educational process. The goal of this paper is to answer the questions: *How to choose the most appropriate application for each student with ASD in a variety of existing ITs? How to support such decision? What are the most relevant features of applications for students with ASD? What should be the structure of a recommender system, that will help in choosing such application?*

The structure of paper is organized as follows. Section 2 briefly describes the related works on available information technologies, designed for students with ASD to find the most relevant features of such technologies. All the technologies were divided into groups. The technologies in each group are aimed to improve some specific psychophysiological feature of a student with ASD. Section 3 investigates the features

© Springer Nature Switzerland AG 2020
Z. Hu et al. (Eds.): ICCSEEA 2019, AISC 938, pp. 523–533, 2020.
https://doi.org/10.1007/978-3-030-16621-2_49

of the ICT evaluation, designed for the education of children with ASD. Section 4 presents a model of a recommender system, that will allow to support the decision of which application should be chosen to fulfill the educational demands of a precise student with ASD. Section 5 addresses the conclusions and future work directions.

2   Related Works

2.1   Features of Teaching Children with ASD

Education is said to be one of the most effective ways of socialization of people with special needs. For people with ASD, learning has some specifics and should be focused on correcting the relevant skills. Researchers identify similar skills, the correction and maintenance of which is important for the education of children with ASD.

In [1], it was emphasized that it is necessary to improve communicating, social skills and traditional education activity. Aresti-Bartolome [2] underlines the importance of improving communication and interaction, social learning and imitation skills, and other related conditions for the education of such children. On the basis of the above mentioned researches, and taking into account the professional advices of the specialists of governmental educational institution for children with ASD (*Dovira* center at Lviv, Ukraine, founded in 1974, now has about 250 students), we shall analyze the existing informational technologies for student with ASD concerning whether it assists some academic discipline, or some special skills, relevant for such student. I.e., we shall divide the information technologies into three groups, the Communication skills assisting group, the Social skills assisting group, and the Academic education assisting group.

2.2   The Variety of Information Technologies for Students with ASD

Although most children are now considered as "digital hostages", children with ASD, on the contrary, feel very comfortable, more willingly "communicating" with modern gadgets and other technological devices. In the article for the Indiana Resource Center for Autism, authored by Kristie Brown Lofland, children with ASD are "visual learners", which means that modern information technology, including assistive technologies, can be a valuable tool in the educational process [3, 27, 33, 34].

Researchers claim that people with ASD consider interaction with a computer or tablet less stressful and more engaging than interaction with other people. Information technology helps children with ASD to become more confident in social situations [31]. In "Huffington Post 2015" [4], it is explained how modern technology helps children with ASD, and special teacher Kathryn deBros claims that modern technology is a powerful auxiliary tool for students who need to be socialized.

The University of Bath (https://www.bath.ac.uk/) is one of the active researchers in the field of life support and education for people with ASD. This institution offers the "SMART-ASD: Matching Autistic People with Technology Resources" course for those who want to improve their theoretical and practical knowledge related to autism [5]. In this course, the University staff recalls the technologies used to support an education of students with ASD. Such technologies are augmented reality, virtual reality, robots, mobile + tablet (sometimes called the iPad and mobile technologies), Tangible technologies [5]. The course highlights the advantages and disadvantages of applications developed with the use of such technologies, their features, etc.

The researcher Aresti-Bartolome [2], in the list of the most relevant technologies for application development, suggests by mixed reality, virtual reality, augmented reality, robots, and telehealth.

2.3 ICT in Support of Communication for Students with ASD

The *Autismspeaks* organization [6], founded in the USA to find a cure for autism, claims that around 25% of ASD children are non-verbal, the ability of others to communicate is low. Difficulties in communication significantly affect the quality of life, education, development of social relations.

With the exception of verbal language, all communication is considered to be assistive and augmentative/alternative communication (AAC). Sue Fletcher-Watson [7] advices to communicate with ASD children using applications for AAC communications to enhance its quality. The *Proloquo2Go* application, suggested in [8, 9], allows children with communication problems to work with images and text. The *Claro* reader with text messaging is able to speak any available text with a wide range of voices, and works at iOS, Android and Windows 10 [10]. There is also an alternative AAC applications for adult users with ASD [11].

2.4 Assistive Technologies for the Improvement of Social Skills

One of the modern technology of assisting children with ASD is augmented reality (AR). This technology combines virtual objects, generated by the computer, with the real environment [12]. Researchers at Cambridge University have developed an AR system, designed to help children and students with ASD by offering additional visualization for learning or playing [13]. According to one of the developers of such a system, the technology of AR can help autistic children to incorporate what they learn from the computer system into their reality [14].

At *Aspect Hunter* school, Australia [15], children with ASD, while educated, are using *Sphero* robots [16]. Such technologies are used to alleviate the unpleasant social aspects.

The VR technology relates to the disability field, usually because of its ability to overcome the physical and cognitive barriers to social integration. Researchers Nyaz Didehbani et al. described virtual reality as a very promising and motivational platform for safe practice and rehearsal of social skills for children with ASD [17].

2.5 Assistive Technology for Academic Learning

In addition to the listed above, one of the promising advanced technologies that can help children with ASD, is a *3D holograms*. The technology, based on 3D holograms, is called *mixed reality* (MX). With the help of holography, educational institutions can help each other to overcome a shortage or lack of a particular specialist or teacher [18].

Microsoft *HoloLens* is the first self-contained holographic computer, that allows communication with visual content and interaction with holograms around. *HoloStudy* is a series of lessons in geology, physics, chemistry, and biology; the knowledge is visualized in an unusual way, using all the features of mixed and augmented reality [19].

Thus, we came into a conclusion that a huge line of existing information technologies and applications, designed to correct and support the educational abilities of students with ASD. In general, adjustments take place in three main areas: to support communication for such children, to develop their social skills, and to support the process of academic education.

3   Features of the ICT Evaluation, Designed for the Education of Children with ASD

In addition to the abovementioned, hundreds and thousands of applications, both online and mobile, have been developed for the education of children with ASD. The authors of such ICTs are not only companies and corporations that create commercial products for the education of children with ASD, but also inclusive educators who develop applications for their own professional needs, developers of free software, enthusiasts. There is an indisputable need to support the process of choosing the best applications, which would take into account the needs of all participants in the child with ASD education.

3.1   The Digests of Applications for Children with ASD

There is a number of researches and digests [2, 20] that aim to discuss applications for children with ASD training. The ASD experts offer ready-made digests to help to choose such application [7, 8], but without specifying the parameters and characteristics behind such applications election. In such digests, applications are grouped according to different characteristics, mainly by the types of skills such applications are to correct.

The DART project of the University of Edinburgh has conducted a research on applications for people with ASD (http://dart.ed.ac.uk/app-reviews-fir/). It was selected a 100 of the most relevant applications, that were evaluated for the following parameters: look and feel; accessibility; entertainment; meeting the needs of people with ASD; and educational potential, etc.

In the above mentioned course offered by the University of Bath, it is suggested to consider the following options for comparing applications [5]: size of the dictionary, the size of the button, the availability of panels or dynamics, the level of distractors, the level of availability of customization, the platform on which the application is implemented, the cost.

The *Autism Association of Western Australia* (founded in 1967) offers a subjective approach to choosing applications for children with ASD [20]. While choosing an application, one should take into account the characteristics of the child that will use it, its preferences, strengths, and weaknesses, the application environment; identify the desired features of such an application: the purpose of its use, the understanding of the skills that can be improved with the application, and whether it may be used as a reward or for motivation; some features of such applications are listed in Fig. 1, and even more, it is necessary to check whether it is possible to print out a document in such an application, the amount of resources of the computer or tablet it uses, whether it is possible to use such an application individually, or it needs someone else to be near.

The variety of appraisal characteristics of applications for teaching children with ASD reveals the subjectivity of such assessments, while the best result should be achieved with the objective evaluation of the applications by experts in ASD, inclusive educators, teachers, IT professionals, and, undoubtedly, parents of children with ASD.

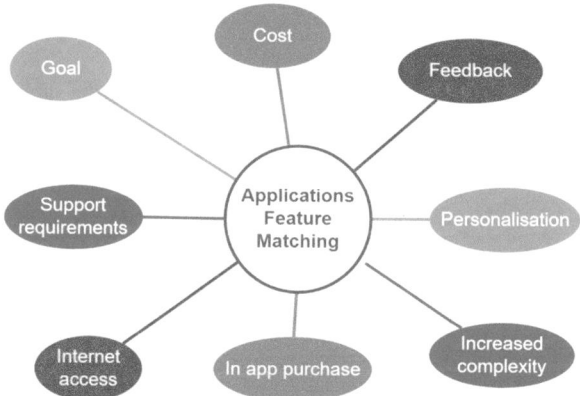

Fig. 1. Some characteristics of applications for children with ASD (after https://www. autismapps.org.au)

3.2 Accessibility and Usability Demands

The requirements for the development of the information technologies are set out in standards and manuals (ISO/IEC 9126-1, RAMS and FURPS). The importance of developing high-quality software is of particular meaning when it comes to supporting the education and socialization of people with special needs. The accessibility and usability of information technologies [21] are important features for users with ASD.

The demands of web accessibility and usability, for the most, pursue similar goals. Web accessibility is considered as a component of universal design, the purpose of which is to lower the discrimination level in access to the living environment for people with special needs. Usability of web technologies is the availability of effective design of websites and applications [21]. The main characteristics of accessibility and usability of online resources for children with autism were discussed in [21]. In the paper, the authors summarize different guidelines and proposed four main areas, the specifics of which should be taken into account when creating an application for children with autism (Fig. 2).

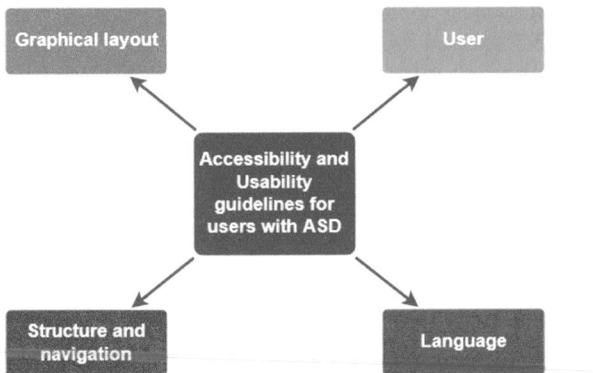

Fig. 2. Main areas of Accessibility and Usability guidelines for users with ASD

In [21], as well as in [22], a close attention is payed to the area of the graphical layout of applications for users with ASD. As a result, a common and different in an understanding of the graphical layout is presented at Fig. 3.

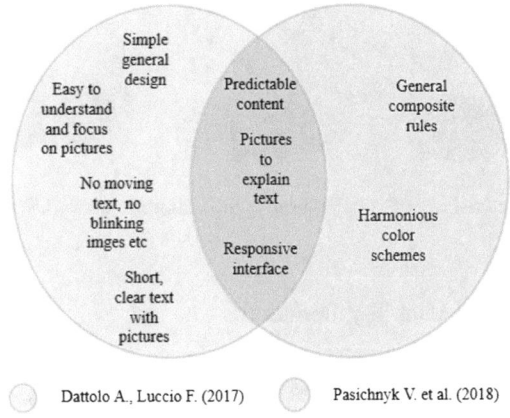

Fig. 3. The graphical layout approach of applications for users with ASD [21, 25]

The widely used web content accessibility recommender system WCAG 2.0 is offered by the World Wide Web Consortium (W3C) and was implemented as a standard ISO/IEC40500:2012 in 2012. These recommendations are aimed at ensuring the availability of web content to a wider range of people with special needs, including blindness and poor eyesight, deafness and hearing impairment, cognitive features, movement restrictions, linguistic features and their combinations [23]. Following such guideline will also make web content more usable for users in general. The model of the system of such recommendations is given in [24]. The studies [21, 32] examines a number of approaches and guidelines offered by various researchers to match web accessibility.

A series of information technologies, such as AChecker (https://achecker.ca/checker/index.php), has been developed to automatically evaluate websites to comply with WCAG 2.0 requirements at various levels. Investigating websites that are tangent to inclusive learning by such a means is proposed in [24]. Unfortunately, this resource does not allow you to automatically evaluate applications.

Thus, the authors concluded that it was necessary to develop a recommender system for identifying the best application that would enable taking into account the demands of all those involved in the inclusive education process. One of the basic requirements is that an application should be in compliance with WCAG 2.0 (ISO/IEC 40500:2012).

4 The Model of the Recommender System

Recommender systems nowadays are implemented in various fields, including education [25, 26, 28, 35]. The model of the recommender system that helps choosing the best application for a student with ASD, is presented by means of the UML. The use case diagram gives an idea, in particular, of the participants of the education process of a student with ASD (Fig. 4).

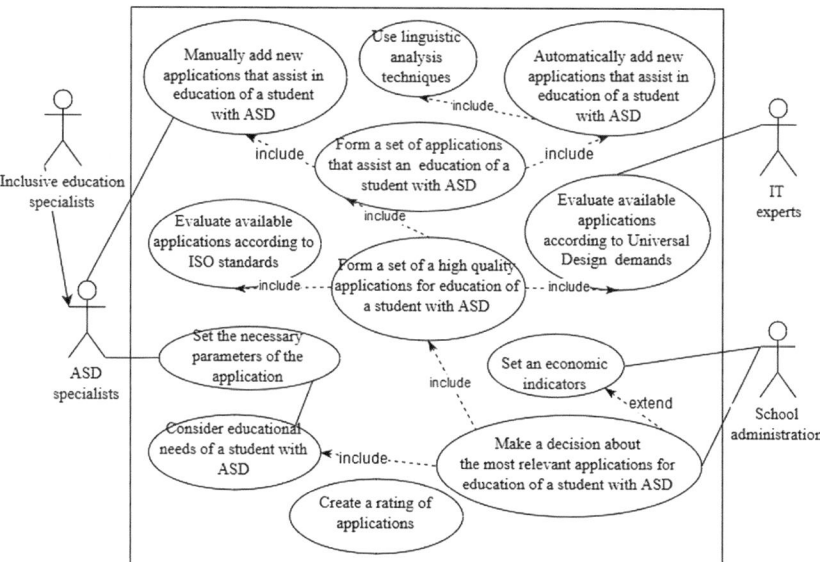

Fig. 4. The use case diagram of the recommender system of choosing an application to support an education of a student with ASD

To identify a better application, there are specialists of the child's with ASD education, who are involved in determining the correctional and educational needs of such a child. The school administration sets financial and territorial requirements and restrictions to the requested application. Specialists in teaching children with ASD make the software requirements according to the educational needs and personal abilities of such a student. Experts, i.e. specialists in ASD and IT, evaluate the proposed applications. The recommender system takes into account all the requirements and limitations and offers the best option of the necessary application. The automatic search [29, 30] for new applications will allow the dynamical search on the Internet for current software that is useful for teaching children with ASD. Creating automatic application ratings for students with ASD and other special needs will be useful when selecting it in educational needs.

To present the overall workflows, data flows, and other related activities in the designed recommender system, the activity diagram was used (Fig. 5).

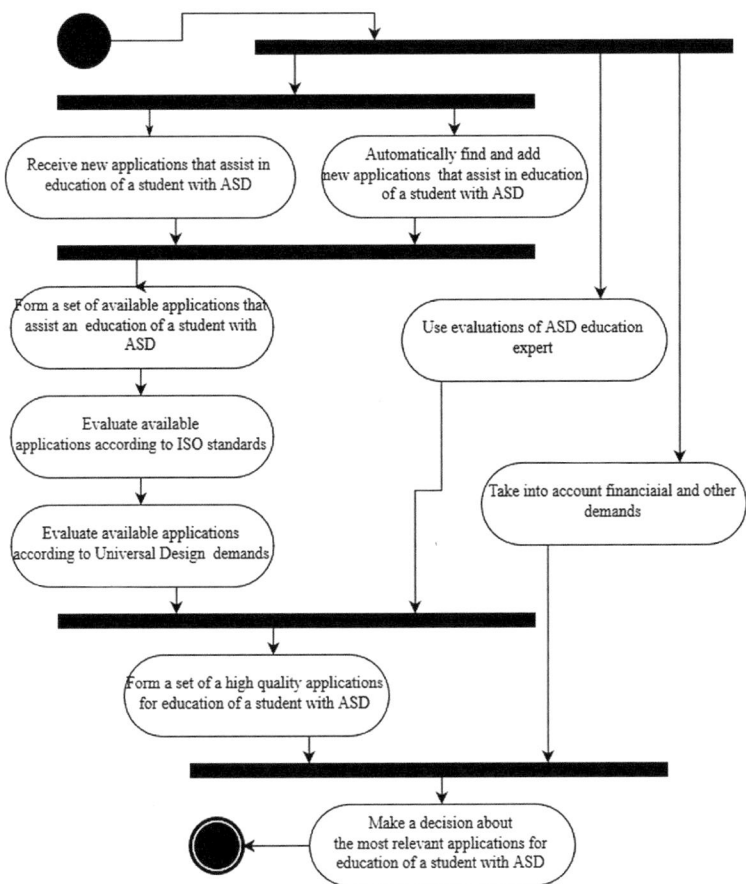

Fig. 5. The activity diagram of the recommender system of choosing an application to support an education of a student with ASD

The development of the recommender system with the specified functions will enable to improve and support the application search process, taking into account the requirements and restrictions of the participants in the process of educating of the student with ASD.

5 Conclusions

The most relevant features of the information technology, designed for an education of a student with ASD, are the types of the skills, such technology should assist with. Such information technologies can be divided into three groups:

- The Communication skills assisting group,
- The Social skills assisting group, and
- The Academic education assisting group.

Choosing the best application for teaching a child with ASD involves taking into account both the needs of the autistic student and the needs of the relevant training and correctional facilities. Moreover, when choosing the best applications, it is necessary to take into account the professional, financial and technological capabilities of such training and correctional facilities. The structure of the appropriate recommender system, should be designed for comfortable use of the specialists of students with ASD education, school administration, IT specialists. Such system will help the school administration to support the decision making and consider the available economic indicators, expert evaluation, and the needs of each student with ASD.

There is no solid approach in evaluation of an information technology, that is designed for education of a student with ASD. The development of a universal technique, that will help to evaluate any IT for education of students with ASD, is a direction of the next investigation. Having such technique will improve the decision making process in education of students with ASD.

References

1. Sailers, E., Coopin, M., Marden, J.: iPhone, iPad and iPod touch Apps for (Special) Education. https://emedea.it/centro-ausili/images/pdf/24470331-iPhone-iPad-and-iPod-touch-Apps-for-Special-Education.pdf
2. Aresti-Bartolome, N., Garcia-Zapirain, B.: Technologies as support tools for persons with autistic spectrum disorder: a systematic review. Int. J. Environ. Res. Public Health **11**(8), 7767–7802 (2014). https://www.ncbi.nlm.nih.gov/pmc/articles/PMC4143832/
3. EdTech. https://edtechmagazine.com/k12/article/2016/08/3-ways-technology-can-help-students-autism
4. HUFFPOST. https://www.huffingtonpost.com/2015/04/20/teaching-technologyautism_n_6865030.html
5. FutureLearn. https://www.futurelearn.com/courses/supporting-autism
6. AutismSpeaks. from https://www.autismspeaks.org/
7. Kindy Segovia's AT Tidbits. https://dart.ed.ac.uk/app-wheel-update/
8. The University of Edinburgh. www.kindysegovia.com/318-2/

9. AssistiveWare. from https://www.assistiveware.com/products/proloquo2go
10. ClaroSoftware. from https://www.claro-apps.com/
11. Aacorn. https://www.aacornapp.com/
12. He, Z., Peng, L., Han, H., Xu, M., Wang, G., Bao, X., Yu, H., Hou, Z., Wang, H., Zhu, L., Zhang, Z.: Design and implementation of augmented reality cloud platform system for 3DEntity objects. In: 8th International Congress of Information and Communication Technology (ICICT-2018). Procedia Comput. Sci. **131**, 108–115 (2018)
13. Bhatt, S.K., De Leon, N.I., Al-Jumaily, A.: Augmented reality game therapy for children with autism spectrum disorder. Int. J. Smart Sensing Intell. Syst. **7**(2) (2014). https://augmentedrealitynews.org/games/augmented-reality-helps-children-with-autism/
14. University of Cambridge. https://www.enterprise.cam.ac.uk/news/the-land-of-make-believe/
15. Autism Spectrum Australia. https://www.autismspectrum.org.au/school/aspect-hunter-school
16. Sphero. https://www.sphero.com/
17. Didehbani, N., Allen, T., Kandalaft, M., Krawczyk, D., Chapman, S.: Virtual reality social cognition training for children with high functioning autism. Comput. Hum. Behav. **62**, 703–711 (2016)
18. Aina, O.: Application of holographic technology in education. Bachelor's thesis of Degree Programme in Business Information Technology. Tornio University of Applied Sciences, 67 p. (2010)
19. HoloStudy. http://www.holo.study/
20. AutismApps. https://www.autismapps.org.au/making-it-work/
21. Dattolo, A., Luccio, F.L.: A review of websites and mobile applications for people with autism spectrum disorders: towards shared guidelines. In: Gaggi, O., Manzoni, P., Palazzi, C., Bujari, A., Marquez-Barja, J. (eds.) Smart Objects and Technologies for Social Good, GOODTECHS 2016. Lecture Notes of the Institute for Computer Sciences, Social Informatics and Telecommunications Engineering, vol. 195. Springer, Cham (2017)
22. Pasichnyk, V., Shestakevych, T., Kunanets, N., Andrunyk, V.: Analysis of completeness, diversity and ergonomics of information online resources of diagnostic and correction facilities in Ukraine. In: Proceedings of the 14th International Conference on ICT in Education, Research and Industrial Applications. Integration, Harmonization and Knowledge Transfer. Volume I: Main Conference, pp. 193–208 (2018)
23. ISO. https://www.iso.org/standard/58625.html
24. Shestakevych, T., Pasichnyk, V., Kunanets, N., Medykovskyy, M., Antonyuk, N.: The content web-accessibility of information and technology support in a complex system of educational and social inclusion. In: 2018 13th International Scientific and Technical Conference on Computer Sciences and Information Technologies (CSIT), Lviv, pp. XXVI–XXXI (2018)
25. Adam, N.L., Zulkafli, M.A., Soh, S.C., Kamal, N.A.M.: Preliminary study on educational recommender system. In: 2017 IEEE Conference on e-Learning, e-Management and e-Services (IC3e), pp. 97–101 (2017)
26. Rawat, B., Dwivedi, S.K.: An architecture for recommendation of courses in e-learning system. Int. J. Inf. Technol. Comput. Sci. (IJITCS) **9**(4), 39–47 (2017). https://doi.org/10.5815/ijitcs.2017.04.06
27. Fetaji, B., Fetaji, M., Ebibi, M., Kera, S.: Analyses of impacting factors of ICT in education management: case study. Int. J. Modern Educ. Comput. Sci. (IJMECS) **10**(2), 26–34 (2018). https://doi.org/10.5815/ijmecs.2018.02.03
28. Ehimwenma, K.E., Crowther, P., Beer, M.: Formalizing logic based rules for skills classification and recommendation of learning materials. Int. J. Inf. Technol. Comput. Sci. (IJITCS) **10**(9), 1–12 (2018). https://doi.org/10.5815/ijitcs.2018.09.01

29. Chyrun, L., Kis, I., Vysotska, V., Chyrun, L.: Content monitoring method for cut formation of person psychological state in social scoring. In: Proceedings of 2018 IEEE 13th International Scientific and Technical Conference on Computer Sciences and Information Technologies, CSIT 2018, vol. 2, pp. 106–112 (2018)

30. Vysotska, V., Lytvyn, V., Hrendus, M., Kubinska, S., Brodyak, O.: Method of textual information authorship analysis based on stylometry. In: Proceedings of 2018 IEEE 13th International Scientific and Technical Conference on Computer Sciences and Information Technologies, CSIT 2018, vol. 2, pp. 9–16 (2018)

31. Khan, A., Madden, J.: Active learning: a new assessment model that boost confidence and learning while reducing test anxiety. Int. J. Modern Educ. Comput. Sci. (IJMECS) **10**(12), 1–9 (2018). https://doi.org/10.5815/ijmecs.2018.12.01

32. Abduganiev, S.G.: Towards automated web accessibility evaluation: a comparative study. Int. J. Inf. Technol. Comput. Sci. (IJITCS) **9**(9), 18–44 (2017). https://doi.org/10.5815/ijitcs.2017.09.03

33. Veretennikova, N., Lozytskyi, O., Kunanets, N., Pasichnyk, V.: Information and technological service for the accompaniment of the educational process of people with visual impairments. In: ICT in Education, Research and Industrial Applications. Integration, Harmonization and Knowledge Transfer. Proceedings of the 14th International Conference on ICT in Education, Research and Industrial Applications. Integration, Harmonization and Knowledge Transfer (ICTERI 2018). Volume I: Main Conference, pp. 290–301 (2018)

34. Nosenko, Y., Matyukh, Z.: The implementation of multimedia technology in ukrainian inclusive pre-school education. In: ICT in Education, Research and Industrial Applications. Integration, Harmonization and Knowledge Transfer. Proceedings of the 13th International Conference on ICT in Education, Research and Industrial Applications. Integration, Harmonization and Knowledge Transfer (ICTERI 2017), pp. 459–466 (2017)

35. Lytvyn, V., Vysotska, V., Dosyn, D., Lozynska, O., Oborska, O.: Methods of building intelligent decision support systems based on adaptive ontology. In: Proceedings of the 2018 IEEE 2nd International Conference on Data Stream Mining and Processing, DSMP 2018. pp. 145–150 (2018)

Connections Between Long
Genetic and Literary Texts.
The Quantum-Algorithmic Modelling

Sergey V. Petoukhov$^{(\boxtimes)}$

Mechanical Engineering Research Institute, Russian Academy of Sciences,
M. Kharitonievsky pereulok, 4, Moscow, Russia
spetoukhov@gmail.com

Abstract. The article is devoted to a development of quantum biology as a known scientific direction, which can give new decisions for various fields of science and technology. More precisely, the article describes some deep analogies between the structural organisation of long DNA sequences and of long literary texts; in addition, applications of formalisms of quantum informatics for modeling phenomenological results are shown. The described results support thoughts of some authors that linguistic languages are a continuation of the genetic language and that formalisms of quantum informatics can be effectively used for modelling and understanding biological structures. The author believes that the further usage of the concepts and formalisms of quantum informatics in genetics and bioinformatics will lead to the development of quantum-algorithmic bioinformatics, achievements of which will be useful, in particular, for a creation of systems of artificial intelligence.

Keywords: Algorithms · Quantum computing · Alphabet · DNA · Literary text

1 Introduction

Quantum biology having deep roots in works of pioneers of quantum mechanics is intensively developed in modern science [1]. This scientific direction is interesting for deeper understanding living nature and for different technological applications including systems of artificial intelligence, where applications of algorithms of quantum informatics are considered as very perspective [2]. In the historically first scientific paper on quantum biology, one of founders of quantum mechanics P. Jordan claimed that life's missing laws were the rules of chance and probability (the indeterminism) of the quantum world that were somehow scaled up inside living organisms [1].

One of possible ways to reveal these rules is connected with studying structural analogies between long genetic and literary texts concerning probabilities of elements in them. Some results of this study are presented in this article. The author tries to reveal such hidden rules of probabilities taking into account the following thoughts of many scientists: languages of human dialogue are formed not from an empty place, but they are continuation of the genetic language or, anyhow, are closely connected with it,

© Springer Nature Switzerland AG 2020
Z. Hu et al. (Eds.): ICCSEEA 2019, AISC 938, pp. 534–543, 2020.
https://doi.org/10.1007/978-3-030-16621-2_50

confirming the idea of information commonality of organisms. For example, Jakobson [3, 4], who is one of the most famous experts and who is the author of a deep theory of binary oppositions in linguistics, stated that the genetic code system is the basic simulator, which underlies all verbal codes of human languages. According to Jakobson, all relations among linguistic phonemes are decomposed into a series of binary oppositions of elementary differential attributes by analogy with binary-oppositional feature of the genetic code systems having double helixes of DNA molecules with their complementary pairs of nitrogenous bases, etc.

As Jakobson writes, *"The heredity in itself is the fundamental form of communications ... Perhaps, the bases of language structures, which are imposed on molecular communications, have been constructed by its structural principles directly"* [5, p. 396]. *"Jakobson reveals distinctly a binary opposition of sound attributes as underlying each system of phonemes.... Jakobson was interested especially in the general analogies of language structures with the genetic code, and he considered these analogies as indubitable"* [6]. These questions have arisen at Jakobson as consequence of its long-term researches of connections of linguistics, biology and physics. Such connections were considered at a united seminar of physicists and linguists, which was organized by Niels Bohr and Roman Jakobson jointly in Massachusetts Institute of Technology. The book title «On the Yin and Yang Nature of Language» [7] emphasises the important role of binary oppositions.

In addition many researchers perceive a linguistic language as a living organism. The book "Linguistic genetics" [8] says: *"The opinion about language as about a living organism, which is submitted to the laws of a nature, ascends to a deep antiquity ... Research of a nature, of disposition and of reasons of isomorphism between genetic and linguistic regularities is one of the most important fundamental problems for linguistics of our time".*

The Nobel Prize winner in molecular genetics Jacob was also thinking about deep relations between genetics and linguistic languages in a connection with the principle of binary oppositions systematically described in the Ancient Chinese book "I-Ching". He wrote: *«C'est peut-être I Ching qu'il faudrait étudier pour saisir les relations entre hérédité et langage»* (it means in English: *perhaps, for revealing of relations between genetics and language it would be necessary to study them through the Ancient Chinese "I Ching"*) [9]. These questions are connected with the theme of archetypes of the unconscious introduced by the creator of analytical psychology Carl Jung; a prominent contribution to the development of the concept of archetypes was made by Wolfgang Pauli as a result of his many years of cooperation on this topic with Jung [10].

The described results were received as an interesting scientific addition in the course of author's study of structural connections between molecular genetics and mathematical formalisms of quantum informatics [11–13]. Below we specially note those results about peculiarities of sequences of hydrogen bonds in long DNA sequences, which, firstly, have allowed discovering a new class of symmetries in long DNA sequences and, secondly, have led to revealing the connections between long genetic and literary texts.

One should mention suppositions of many authors that formalisms of quantum informatics can be effectively used for deep understanding and modelling biological bodies [11–20]. Our work belongs to this direction of thoughts and we believe that the described structural connections between long genetic and literary texts reflect deep quantum algorithmical aspects of organization of biological bodies.

2   On the Principle of Binary Oppositions in the DNA Alphabet and in the Alphabet of Russian Language

As known the genetic coding systems are constructed by the nature on the principle of binary oppositions including double helixes of DNA molecules with their complementary pairs of nitrogenous bases connected by two types of hydrogen bonds, etc. The binary-oppositional structure of the DNA alphabet of nitrogenous bases is used in mathematical models of genetic informatics in some works, which show connections of genetic texts with binary language of computers and also with binary schemes and tables of the ancient Chinese book "I Ching" [21–23].

But the Russian alphabet (and alphabets of some other languages) also has a binary-oppositional structure since it has two binary-oppositional sub-alphabets: the sub-alphabet of vowels and the sub-alphabet of consonants. Each of these sub-alphabets also has its own binary-oppositional structure: the sub-alphabet of vowels consists of the sub-sub-alphabet of long vowels and the sub-sub-alphabet of short (or iotated) vowels; the sub-alphabet of consonants consists of the sub-sub-alphabet of voiced consonants and the sub-sub-alphabet of deaf consonants (Fig. 1). The soft sign "ь" and the hard sign "ъ" in the Russian alphabet do not convey any sound and therefore they are not taken into account in its phonologic structure.

The Russian alphabet can be considered as consisting of the following two class of equivalency (in Fig. 1, the first class is marked by yellow and the second class is marked by green):

(1) The first class of equivalency combines all short (iotated) vowels and all deaf consonants: е, ё, ю, я, п, ф, к, т, ш, с, х, ц, ч, щ. Any of 14 members of this class we denote by the general symbol 0;

(2) The second class of equivalency combines all long vowels and all voiced consonants: а, и, о, у, ы, э, б, в, г, д, ж, з, й, л, м, н, р. Any of 17 members of this part we denote by the general symbol 1.

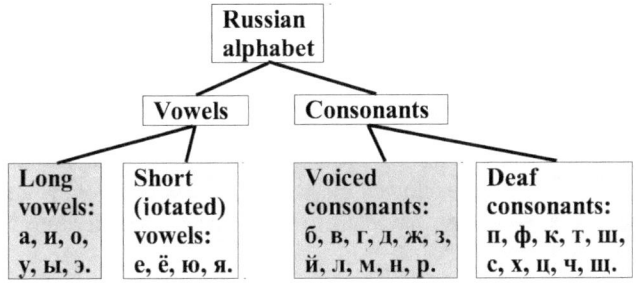

Fig. 1. The binary-oppositional structure of the Russian alphabet.

Any Russian text can be converted into the corresponding binary sequence of these two symbols 0 and 1 (like as 100110110...) by the following procedure:

(1) all punctuation marks and the spacings between words, as well as all soft signs « ь » and hard signs «ъ», are deleted from the text;
(2) each of the remaining letters is replaced by the corresponding symbol 0 or 1 depending its belonging to one of these two classes of equivalency.

For example, as a result of this procedure, the Russian text "Лев Толстой – великий русский писатель" (in English: «Leo Tolstoy - a great Russian writer») turns into a binary sequence 10101100111011011111000110101001. The computer program for our analysis of literary texts was created by our graduate student V.I.Svirin in line with a technical task formulated by the author.

For further description one should remind about the mathematical operation of the tensor product. This operation is known in mathematics, physics and informatics, where it gives a way of putting vector spaces together to form larger vector spaces. The tensor product is the crucial operation to understanding the quantum mechanics of multiparticle systems [24] and is one of basic instruments in quantum informatics. The following quotation speaks about the tensor product: «*This construction is crucial to understanding the quantum mechanics of multiparticle systems*» [24, p. 71] since in line with the postulate of quantum mechanics: the state space of a composite system is the tensor product of the state spaces of its components. By definition, under the tensor product of two vectors, each of components of the first vector is multiplied with all components of the second vector. The expression (1) shows an example of the tensor product (denoted by the symbol $\otimes$) of two 2-dimensional vectors $[x, y]$ and $[v, w]$, which gives in the result one 4-dimensional vector $[xv, xw, yv, yw]$:

$$[x, y] \otimes [v, w] = [x[v, w], y[v, w]] = [xv, xw, yv, yw] \tag{1}$$

We analyze the named binary representations of Russian long literary texts by means of the method, which was used for analyzing long DNA sequences of numbers of hydrogen bonds 3 and 2 [11]. More precisely, each of long binary sequences of numbers 0 and 1 (for example the sequence 1-0-1-0-1-1-0-0-1-1-1-0-...) can be represented also a sequence of binary doublets (10-10-11-00-11-10-...), or a sequence of binary triplets (101-011-001-110-...) or, in a general case, as a sequence of binary n-plets ($n = 1, 2, 3, 4, 5, ...$). We call such different representations of a long literary text in the form of sequences of binary n-plets as its "binary n-plet representations".

A complete alphabet of binary n-plets under fixed value n is called "the alphabet of binary n-plets" and it contains 2^n members. For example, the alphabet of binary monoplets contains 2 members (0 and 1); the alphabet of binary doublets contains 4 members (00, 01, 10, 11), which only exist in binary doublet representation of any long literary text; the alphabet of binary triplets contains $2^3 = 8$ members (000, 001, 010, 011, 100, 101, 110, 111), which only exist in the binary triplet representation of any long literary text, etc. One can mention that the tensor family of vectors $[0, 1]^{(n)}$ (where (n) means the tensor power; $n = 1, 2, 3, 4, ...$) contains vectors, whose sets of components coincide with corresponding alphabets of binary n-plets. For example, $[0, 1]^{(2)} = [00, 01, 10, 11]$; $[0, 1]^{(3)} = [000, 001, 010, 011, 100, 101, 110, 111]$, etc.

In binary n-plet representations of long literary texts in Russian, we analyze probabilities (or frequencies or percentage) of each of members of concrete alphabets of binary n-plets in concrete long texts in Russian. Under a fixed value n, each of these probabilities is equal to the ratio: (the total quantity of a corresponding member of an alphabet of binary n-plets) divided by (the total quantity of these binary n-plets).

For example, in the text of the work "Anna Karenina" by Leo Tolstoy in its binary doublet representation there exist the total number 654523 of doublets 00, 01, 10 and 11, including 75895 doublets 00, 142504 doublets 01, 142547 doublets 10 and 293577 doublets 11. Correspondingly the probability of doublets 00 is equal to 75895/654523 = 0,115954672; the probability of doublets 01 is equal to 142504/654523 = 0,217721914; the probabilty of doublets 10 is equal to 142547/654523 = 0,21778761; the probability of doublets 11 is equal to 293577/654523 = 0,448535804.

In addition to calculation of probabilities of all members of alphabets of n-plets (n = 1, 2, 3, 4) in long literary Russian texts, the author models all these probabilities with a good level of accuracy by means of the same quantum-algorithmic method, which was used for modeling probabilities in long DNA sequences of hydrogen bonds 3 and 2 in his article [11]. This method explores some classical formalism from quantum informatics [24] and is based on the notion of the genetic qubit [11, 12].

In our model approach, phenomenologic probabilities of all members of binary n-plet alphabets $[0, 1]^{(n)}$ (under fixed n, where n = 1, 2, 3, 4, … is not too large) are modeled by numeric values of appropriate coordinates of 2^n-dimensional vector from the tensor family of vectors of probabilities $[0_P, 1_P]^{(n)}$ where 0_P and 1_P denote probabilities of the symbol 0 and of the symbol 1 in the binary monoplet representation of the analyzed long Russian text; these probabilities are shown in the first column of tables in Fig. 2.

Correspondingly, in this model approach, to get model values of probabilities of all 2^n members of the binary n-plet alphabet in a considered binary n-plet representation of the analyzed literary text, one should calculate the following 2^n coordinates of vectors of probabilities from their tensor family:

- in the case of the alphabet of binary doublets, there exist 4 coordinates of the vector $[0_P, 1_P]^{(2)} = [0_P0_P, 0_P1_P, 1_P0_P, 1_P1_P]$; these 4 numeric coordinates correspond to 4 appropriate coordinates of the vector of all members of the alphabet of binary doublets [00, 01, 10, 11];
- in the case of the alphabet of binary triplets, there exist 8 coordinates of the vector $[0_P, 1_P]^{(3)} = [0_P0_P0_P, 0_P0_P1_P, 0_P1_P0_P, 0_P1_P1_P, 1_P0_P0_P, 1_P0_P1_P, 1_P1_P0_P, 1_P1_P1_P]$; these 8 numeric coordinates correspond to 8 appropriate coordinates of the vector of all members of the alphabet of binary triplets [000, 001, 010, 011, 100, 101, 110, 111], etc.

One should emphasise here that all coordinates of these vectors $[0_P, 1_P]^{(n)}$, where n = 2, 3, 4, … are expressed by means of only two values 0_P and 1_P. Briefly speaking, in line with the proposed model approach, for an approximate prediction of values of probabilities of all members of binary n-plet alphabets in a binary n-plet representation of a long Russian literary text - on the basis of knowledge about only two probabilities 0_P and 1_P - it is enough to do the following:

- Calculate probabilities 0_P and 1_P of binary monoplets 0 and 1 in the sequence;
- Calculate the product of these probabilities 0_P and 1_P in coordinates of vectors $[0_P, 1_P]^{(n)}$, where $n = 2, 3, 4, \dots$.

The expression (2) shows an example of such calculation of the probability $0_P0_P1_P$ for the member 001 of the alphabet of binary triplets under $0_P = 0,36$ and $1_P = 0,64$ $(0_P + 1_P = 1)$:

$$0_P0_P1_P = 0,36 * 0,36 * 0,64 = 0,0829 \qquad (2)$$

Figure 2 represents graphs of results of our study of the probabilities in the novel by L.N. Tolstoy «Anna Karenina» containing 1309047 letters (in Russian edition). One can see from Fig. 2 that model values (red points) of the probabilities turned out to be almost exactly superimposed on the phenomenologic points of blue color.

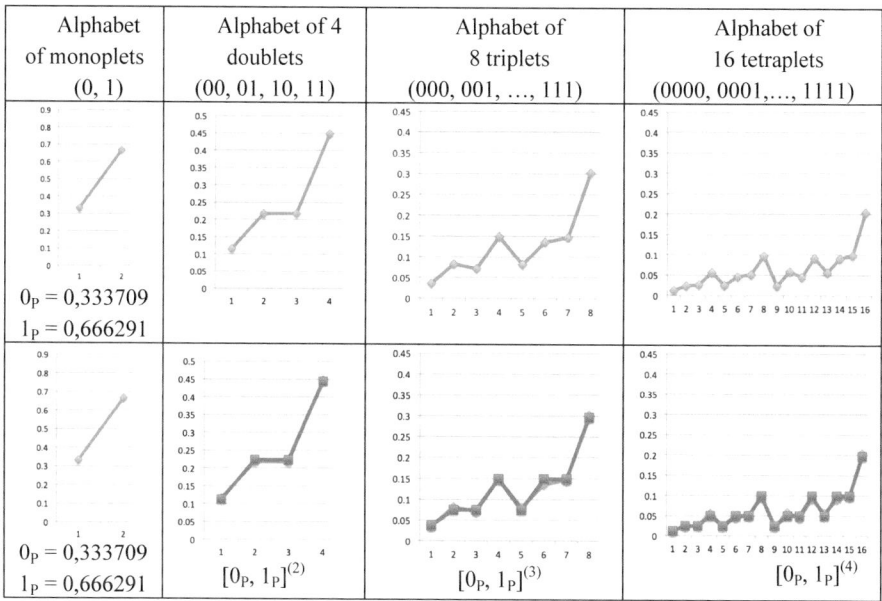

Alphabet of monoplets (0, 1)	Alphabet of 4 doublets (00, 01, 10, 11)	Alphabet of 8 triplets (000, 001, ..., 111)	Alphabet of 16 tetraplets (0000, 0001,..., 1111)
$0_P = 0,333709$ $1_P = 0,666291$			
$0_P = 0,333709$ $1_P = 0,666291$	$[0_P, 1_P]^{(2)}$	$[0_P, 1_P]^{(3)}$	$[0_P, 1_P]^{(4)}$

Fig. 2. Graphical representations of probabilities of members of the alphabets of binary n-plets (n = 1, 2, 3, 4) in the Russian novel by L.N. Tolstoy "Anna Karenina" (the original literary text is taken from http://samolit.com/books/62/). Blue points correspond to phenomenologic values of the probabilities for cases of alphabets named at the top of the columns, and red points correspond to model values of the probabilities calculated as components of the vectors $[0_P, 1_P]^{(n)}$, where 0_P and 1_P are probabilities of binary monoplets 0 and 1; (n) means tensor powers.

So, knowing only two probabilities 0_P and 1_P of binary monoplets 0 and 1 in the binary *n*-plet representation of this novel, one can predict - with good level of accuracy - dozens of probabilities of all members of the alphabets of binary n-plets for this Russian novel with a qood level of accuracy. Very similar results have been received in our analogic study of many long literary works in Russian language: L.N. Tolstoy «War and Peace»; F.M. Dostoevsky «Crime and Punishment» and «Idiot»; A.S. Pushkin «Evgenij Onegin» and «Dubrovsky»; the Biblie in Russian [11]. All these results confirm that studied probabilities of members of different alphabets of binary n-plets ($n = 1, 2, 3, 4$) are interrelated each other in a certain degree and that this interrelation can be modeled on the basis of the tensor family of vectors $[0_P, 1_P]^{(n)}$ ($n = 1, 2, 3, 4, \ldots$ is not too large). We continue such study of long literary texts also in different languages (English, German, etc.) taking into account peculiarities of their alphabets and phonetics.

3 On Similar Properties in Long DNA Sequences

In DNA double helixes, complementary letters A and T are connected by 2 hydrogen bonds and complementary letters C and G are connected by 3 hydrogen bonds. Correspondingly, any DNA can be considered as a chain of 2 and 3 hydrogen bonds, for example, 33223223233.… We analysed properties of such hydrogen bonds sequences for many long DNA of different organisms (here the term "long" means that DNA contains ≥ 100000 letters). Below results of our study are briefly presented, which confirm the existence of structural connections between long genetic and literary texts.

By analogy with the described above approach to binary n-plet representations of literary texts, each of long sequences of numbers 2 and 3 of hydrogen bonds (for example, the H-sequence of monoplets 3-2-3-3-3-2-2-2-3-2-3-3-…) can be represented also as a sequence of H-doublets (32-33-32-22-32-33-…), or a sequence of H-triplets (323-332-223-233-…), etc. A complete set of H-n-plets under fixed value n is called "the alphabet of H-n-plets" and it contains 2^n members: for example, the alphabet of H-monoplets contains 2 members (3 and 2); the alphabet of H-doublets contains 4 members (33, 32, 23, 22); the alphabet of H-triplets contains $2^3 = 8$ members (333, 332, 323, 322, 233, 232, 223, 222); etc. It is obvious that, if the number 3 is denoted by the symbol "0" and the number 2 is denoted by the symbol "1", then these H-n-plets are represented in forms of ordinary binary sequences like 011, etc.; but here we prefer to use the representations for H-sequences and H-n-plets on the basis of numbers 3 and 2 directly without the additional denotations.

The preprint [11] describes in details those hidden regularities in many tested H-sequences of long DNA of different organisms, which are connected with members of the alphabets of hydrogen n-plets. These regularities concern probabilities (or percentage) of members of the alphabets of hydrogen n-plets (briefly, H-n-plets). The author has revealed that for a concrete long DNA sequence, probabilities of members of different alphabets of H-n-plets ($n = 1, 2, 3, 4, \ldots$ is not too large) are interrelated each other. This interrelation can be modeled by the same method, which was described above for the analysis of literary texts: if 3_p and 2_p are probabilities of numbers 3 and 2 of hydrogen bonds in the considered DNA, then tensor powers (n) of the 2-dimensional vector $[3_p, 2_p]$ gives 2^n-dimensional vectors $[3_p, 2_p]^{(n)}$, components of which are model values of probabilities of members of the corresponding alphabet of H-n-plets in this DNA. Figure 3 shows relevant data for the case of the genome of microorganism *Chlamydia trachomatis* that causes chlamydia - illness of a billion people. One can see strong analogies between data in Fig. 2 and in Fig. 3 and also a possibility of modelling the described probabilities in both cases with good levels of accuracy. We have many other similar results that testify in favor of structural connections between long Russian literary texts and long DNA texts.

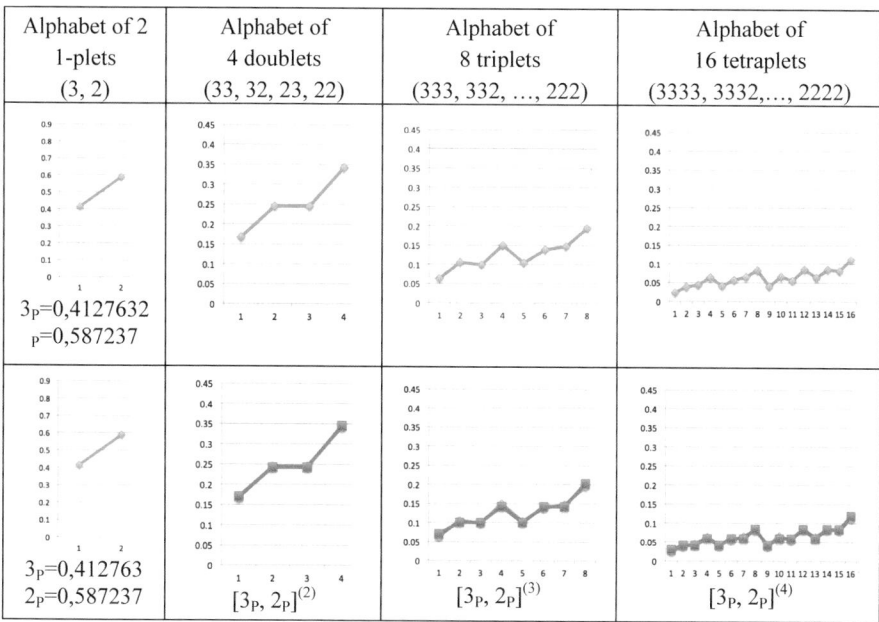

Fig. 3. Graphical representations of probabilities of members of the alphabets of hydrogen n-plets (n = 1, 2, 3, 4) in the complete genome *Chlamydia trachomatis* strain QH111L, 1025839 bp, GenBank: CP018052.1. Blue points correspond to phenomenologic values of the probabilities and red points correspond to model values of the probabilities calculated as components of the vectors $[3_p, 2_p]^{(n)}$, where 3_P and 2_P are probabilities of hydrogen bonds 3 and 2; (n) means tensor powers.

4   Conclusions

The presented results support the existing hypothesis of some authors about a structural origin of linguistic languages on the basis of the genetic language. The principle of binary oppositions plays important role both in organization of genetic and linquistic languages. Our work develops knowledge about possibilities of applications of mathematics of quantum informatics for modeling biological systems.

The author hopes that the further usage of the concepts and formalisms of quantum informatics in genetics will lead to the development of substantial quantum-algorithmic genetics. Consideration of biological phenomena (including the phenomena of inheritance of the intellectual abilities of biological bodies) from the standpoint of the theory of quantum computers gives many valuable opportunities for their comprehension and also for development of artificial intelligence systems and of genetic analysis methods [25–28] (the work [2] contains a valuable review about quantum computing and problems of artificial intelligence). For example, an adult human organism has around 100 trillion (10^{14}) human cells and each of cells containts an identical complect of DNA, whose genetic information is used for physiological functioning organism as the holistic system of cells. How such huge numbers of cells can reliably functioning as a cooperative whole? Associations with quantum computing can help to model and to understand such holistic biological systems with their ability of computing complex tasks and transferring genetic information from one generation to another. The described results are connected additionally with the themes of Jungian's archetypes and of the important role of hydrogen atoms and hydrogen bonds in living matter.

Acknowledgment. Some results of this paper have been possible due to a long-term cooperation between Russian and Hungarian Academies of Sciences on the topic "Non-linear models and symmetrologic analysis in biomechanics, bioinformatics, and the theory of self-organizing systems", where S.V. Petoukhov was a scientific chief from the Russian Academy of Sciences. The authors are grateful to G. Darvas, E. Fimmel, M. He, Z.B. Hu, I. Stepanyan and V. Svirin for their collaboration. Special thanks to the German Academic Exchange Service (DAAD) for providing the very useful internship for S.V. Petoukhov in autumn 2017 at the Institute of Mathematical Biology of the Mannheim University of Applied Sciences (Germany) where his host was Prof. E. Fimmel.

References

1. McFadden, J., Al-Khalili, J.: The origins of quantum biology. In: Proceedings of the Royal Society A: Mathematical, Physical and Engineering Sciences, 12 December 2018. https://royalsocietypublishing.org/doi/full/10.1098/rspa.2018.0674
2. Biamonte, J., Wittek, P., Pancotti, N., Rebentrost, P., Wiebe, N., Lloud, S.: Quantum machine learning. Nature **549**, 195–202 (2017). https://doi.org/10.1038/nature23474
3. Jakobson, R.: Language in Lirature. MIT Press, Cambridge (1987)
4. Jakobson, R.: Texts, Documents, Studies. RGGU, Moscow (1999, in Russian)
5. Jakobson, R.O.: Selected Works. Progress (1985 in Russian)
6. Ivanov Viach Vs. Linguistic way of Roman Jakobson. In: Jakobson, R.O. (ed.) Selected Works. Progress (1985, in Russian)
7. Bailey, Ch.-J.N.: On the Ying and Yang Nature of Language. Karoma, Ann Arbor (1982)
8. Makovsky, M.M.: Linguistic Genetics. Nauka (1992, in Russian)
9. Jacob, F.: Le modele linguistique en biologie. Critique Mars **30**(322), 197–205 (1974)
10. Meier, A. (ed.): Atom and Archetype. The Pauli/Jung Letters, 1932–1958. Princeton University Press, Princeton (2001). ISBN 978-0-691012-07-0
11. Petoukhov, S.V.: The Genetic Coding System and Unitary Matrices. Preprints 2018, 2018040131. The second version. (https://doi.org/10.20944/preprints201804.0131.v2)
12. Petoukhov S.V. The rules of long DNA-sequences and tetra-groups of oligonucleotides. arXiv:1709.04943v5, 5th version, 8 October 2018, 159 p
13. Petoukhov, S.V., Svirin, V.I.: The New wide class of symmetries in long DNA-texts. Elements of quantum-information genetics. Biologia Serbica **40**(1), 51 (2018). Special Edition, ISSN 2334-6590, UDK 57(051). Book of Abstracts, Belgrade Bioinformatics Conference 2018, 18–22 June 2018

14. Penrose, R.: Shadows of mind. Oxford University Press, New York (1994)
15. Igamberdiev, A.U.: Quantum mechanical properties of biosystems: a framework for complexity, structural stability, and transformations. Biosystems **31**(1), 65–73 (1993)
16. Matsuno, K.: Cell motility as an entangled quantum coherence. BioSystems **51**, 15–19 (1999)
17. Matsuno, K., Paton, R.C.: Is there a biology of quantum information? BioSystems **55**, 39–46 (2000)
18. Patel, A.: Why genetic information processing could have a quantum basis. J. Biosci. **26**(2), 145–151 (2001)
19. Abbott, D., Davies, P.C.W., Pati, A.K. (eds.): Foreword by Sir Roger Penrose: Quantum Aspects of Life (2008). ISBN-13: 978-1-84816-253-2
20. Petoukhov, S.V.: Matrix genetics, algebras of the genetic code, noise immunity. Moscow, RCD, Russia, 316 p. (2008, in Russian)
21. Petoukhov, S.V., He, M.: Symmetrical Analysis Techniques for Genetic Systems and Bioinformatics: Advanced Patterns and Applications. IGI Global, Hershey (2010)
22. Hu, Z.B., Petoukhov, S.V.: I-Ching, dyadic groups of binary numbers and the geno-logic coding in living bodies. Prog. Biophys. Mol. Bio. **131**, 354–368 (2017)
23. Nielsen, M.A., Chuang, I.L.: Quantum Computation and Quantum Information. Cambridge University Press, New York (2010)
24. Singh, D.A.A.G., Leavline, E.J., Priyanka, R., Priya, P.P.: Dimensionality reduction using genetic algorithm for improving accuracy in medical diagnosis. Int. J. Intell. Syst. Appl. (IJISA) **8**(1), 67–73 (2016). https://doi.org/10.5815/ijisa.2016.01.08
25. Chawda, B.V., Patel, J.M.: Investigating performance of various natural computing algorithms. Int. J. Intell. Syst. Appl. (IJISA) **9**(1), 46–59 (2017). https://doi.org/10.5815/ijisa.2017.01.05
26. Mousa, H.M.: DNA-genetic encryption technique. Int. J. Comput. Netw. Inf. Secur. (IJCNIS) **8**(7), 1–9 (2016). https://doi.org/10.5815/ijcnis.2016.07.01
27. Abo-Zahhad, M., Ahmed, S.M., Abd-Elrahman, S.A.: Genomic analysis and classification of exon and intron sequences using DNA numerical mapping techniques. Int. J. Inf. Technol. Comput. Sci. (IJITCS) **4**(8), 22–36 (2012). https://doi.org/10.5815/ijitcs.2012.08.03
28. Srivastava, P.C., Agrawal, A., Mishra, K.N., Ojha, P.K., Garg, R.: Fingerprints, Iris and DNA features based multimodal systems: a review. Int. J. Inf. Technol. Comput. Sci. (IJITCS) **5**(2), 88–111 (2013). https://doi.org/10.5815/ijitcs.2013.02.10

On Symmetries Inside Complete Sets of Chromosomes

Sergey Petoukhov$^{(\boxtimes)}$, Elena Petukhova, and Vitaliy Svirin

Mechanical Engineering Research Institute,
Russian Academy of Sciences, M. Kharitonievsky pereulok, 4, Moscow, Russia
spetoukhov@gmail.com

Abstract. Many genetic problems in fields of medicine, biotechnology and bio-medical engineering are connected with secrets of complete sets of chromosomes (karyotypes) in human and other eukaryotic organisms. The article is devoted to new symmetries inside complete sets of chromosomes, which add important knowledge into the field of cytogenetics. Simultaneously the article describes new methods of analysis and modeling long sequences of hydrogen bonds in DNA. The proposed model approach is connected with the tensor product of matrices and with formalisms of quantum informatics. The described results can be used in development of new approaches for modeling genetic informatics from standpoint of quantum informatics.

Keywords: DNA · Symmetry · Hydrogen bond · Probability · Quantum informatics

1 Introduction

The achievements of molecular genetics and biotechnology give many useful tools and methods for our life including new methods in biotechnology, engineering devices for medicine, new knowledge for education, etc. One of the important directions of scientific and technological progress in this connection is the development of so called "personal genetics". The American journal «Time» in 2008 year has published a list of «the best inventions of the year». The first place in this list was given to «personalized genetics» from the company «23andMe» [1]. This innovation was recognized much more important then many others, including the Large Hadron Collider from the field of nuclear physics. The company «23andMe» proposes information about genetic peculiarities of persons with a low prise $399 only.

The personalized genetics leads to personalized pharmacology, in particular. The current pharmacology, which makes no distinction between people, has defects. Firstly, because the same disease can be caused by different causes, and secondly, which is much more important, because different people tolerate medications in different ways. For one person a new medicine promises deliverance from all ills, for another person it leads to an anaphylactic shock. We need to know about each patient what is useful for

© Springer Nature Switzerland AG 2020
Z. Hu et al. (Eds.): ICCSEEA 2019, AISC 938, pp. 544–554, 2020.
https://doi.org/10.1007/978-3-030-16621-2_51

his individual organism. In addition, knowing his genetic characteristics, a person could, say, be screened for cancer or heart disease in advance - and detect the disease in its early stages. All these and many other questions need a deep knowledge about molecular-genetic regularities.

Each eukaryotic organism has its own set of chromosomes, which differ greatly from each other by their molecular dimensions, their sequences of letters, kinds and quantities of genes in them, cytogenetic bands (which shows biochemical specifity of different parts of chromosomes), etc. For example, human organisms contain 22 autosomes and 2 sex chromosomes X and Y (Fig. 1). These chromosomes contain long DNA molecules, the lengths of texts in which lie in the range from 50 to 250 million letters approximately.

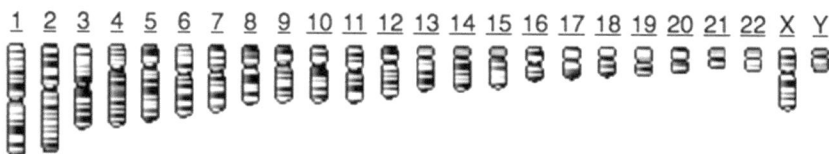

Fig. 1. Human chromosomes (from https://www.ncbi.nlm.nih.gov/genome/51)

Each chromosome has DNA with its own history of mutations, the influence of natural selection factors, gene drifts, etc. Taking all these into account, it seems unlikely that there are common quantitative regularities in the DNA sequences of such different chromosomes of one organism. However, the results of our research reveal unexpectedly the existence of such general regularities of symmetry. It means that all chromosomes inside its set for a concrete organism are not completely individual objects but they are closely interrelated each other in relation to the described symmetry of their DNA sequences. This article is devoted to the description of these results and their modeling capabilities based on quantum-algorithmic formalisms.

2   Chains of Hydrogen Bonds of DNA in Different Chromosomes

In this Section we will study hidden regularities in long sequences of double helixes of DNA. The term "long DNA sequences" means that each of sequences contains no less than 100000 nucleotides. We will consider any of long DNA sequences as a chain of numbers 2 and 3 of hydrogen bonds, which connect complementary nucleotides A-T (adenine and thymin) and C-G (cytosine and guanine) correspondingly (one can denote conditionally A = T = 2 and C = G = 3). We call such chains of numbers of hydrogen bonds as "hydrogen bonds sequences" or briefly "H-sequences".

Each of long chains (or sequences) of separate numbers 2 and 3 of hydrogen bonds (for example, the H-sequence 3-2-3-3-3-2-2-2-3-2-3-3-...) can be represented also as a sequence of H-doublets (32-33-32-22-32-33-...), or a sequence of H-triplets (323-332-223-233-...) or, in a general case, as a sequence of H-n-plets ($n = 1, 2, 3, 4, 5,...$). We call such different representations of a long DNA sequence of nucleotides A, T, C, G in forms of sequences of H-n-plets as its "H-n-plet representations".

A complete set of H-n-plets under fixed value n is called "the alphabet of H-n-plets" and it contains 2^n members. For example, the alphabet of H-monoplets contains 2 members (3 and 2); the alphabet of H-doublets contains 4 members (33, 32, 23, 22), only which exist in the H-doublet representation of DNA sequences; the alphabet of H-triplets contains $2^3 = 8$ members (333, 332, 323, 322, 233, 232, 223, 222), only which exist in the H-triplet representation of DNA sequences; etc. In any H-n-plet representation of a long DNA sequence, each of the members of the alphabet of n-plets occurs with a certain probability (or percentage, or frequency). For species of organisms, which are classical model organisms in genetics, we study these probabilities of all members of the alphabets of H-n-plets in the DNA of each of the chromosomes belonging to the same species of organism (we study the cases with $n = 1, 2, 3, 4, 5$). In this study, we find that all these chromosomes are symmetrical each other in relation to values of these probabilities with a good level of accuracy. In other words, values of these probabilities are approximate invariants inside the concrete set of chromosomes.

One can mention that the tensor family of vectors $[3, 2]^{(n)}$ (where (n) means the tensor power, $n = 1, 2, 3, 4, ...$) contains vectors, whose sets of components coincide with corresponding alphabets of H-n-plets. For example, $[3, 2]^{(2)} = [33, 32, 23, 22]$; $[3, 2]^{(3)} = [333, 332, 323, 322, 233, 232, 223, 222]$, etc. By this reason, one can note conditionally that all alphabets of H-n-plets belong to the same tensor family of vectors. As one of our results, Fig. 2 shows probabilities of members of H-n-alphabets ($n = 1, 2, 3, 4$) in DNA sequences of all 5 chromosomes of a plant *Arabidopsis thaliana*; initial data about DNA sequences of these chromosomes were taken from the GenBank: https://www.ncbi.nlm.nih.gov/genome/4. The lengths of these DNA sequences are the following: in the chromosome #1, the length is equal to 30427671 bp; in the chromosome #2 – 19698289 bp; in the chromosome #3 – 23459830 bp; in the chromosome #4 – 18585056; in the chromosome #5 – 26975502 bp.

Members of all these H-n-alphabets in Fig. 2 are given in a certain arrangement corresponding to arrangements of coordinates of vectors $[3, 2]^{(n)}$, where (n) means the tensor power. For example, the vector $[3, 2]^{(2)} = [33, 32, 23, 33]$ have the arrangement of its coordinates 33, 32, 23, 22, which is used in Fig. 2 for the appropriate graph in the case of the alphabet of H-doublets. Similar correspondences hold true for vectors $[3, 2]^{(3)}$, $[3, 2]^{(4)}$ and appropriate graphics of probabilities of members of other H-n-alphabets in Fig. 2.

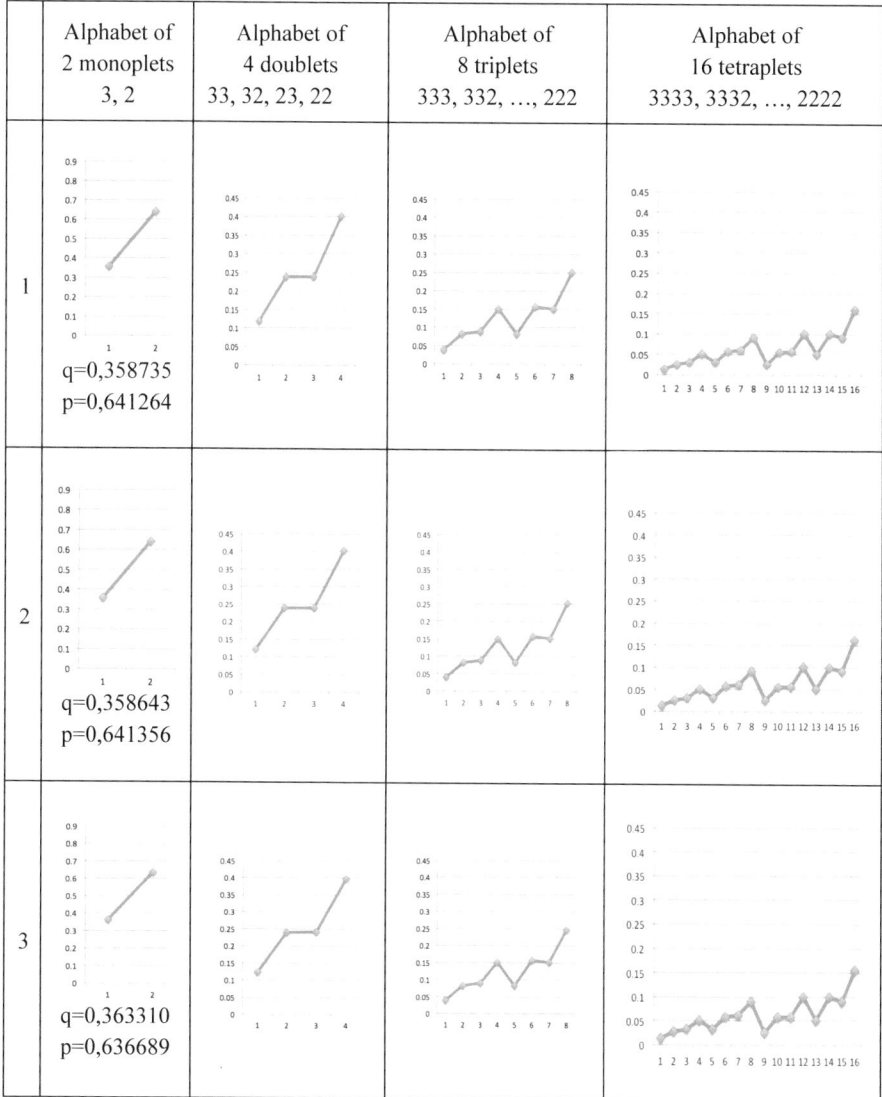

Fig. 2. Probabilities of members of H-*n*-alphabets (*n* = 1, 2, 3, 4, 5) in DNA sequences of all 5 chromosomes of a plant *Arabidopsis thaliana*. The left column contains the sequence numbers of the chromosomes. Each of points in graphs shows a probability of an appropriate member of alphabets of H-*n*-plets. For each of chromosomes, its probabilities q and p of H-monoplets 3 and 2 are shown in the second column. See explanations in the text.

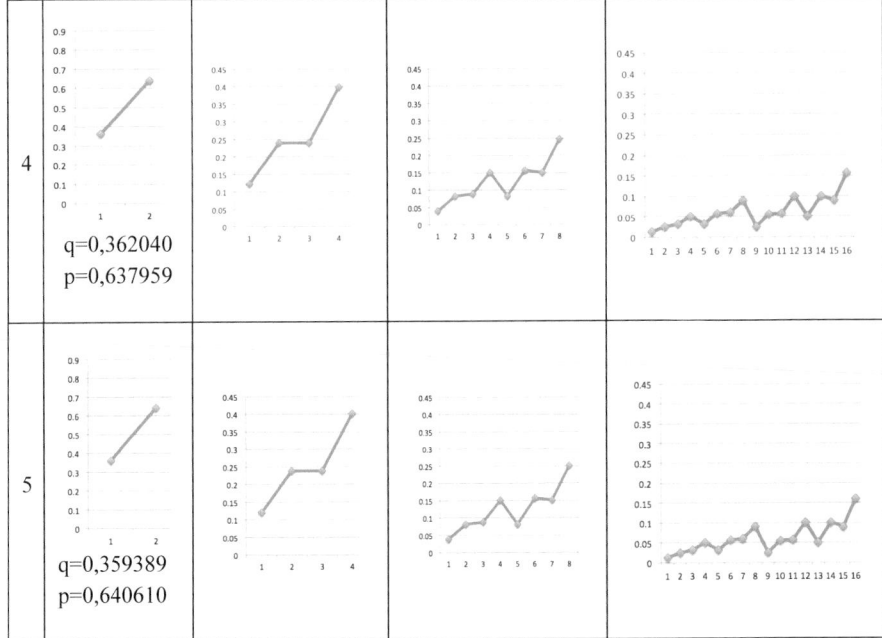

Fig. 2. (*continued*)

In addition to graphs in Fig. 2, Table 1 shows numeric values of probabilities of all members of these H-*n*-alphabets for all 5 chromosomes and for *n* = 1, 2, 3 (more details for cases *n* = 4, 5 are given in the preprint [2]. For statistical analysis of these phenomenolgical data from all chromosomes we use Excel with denotations and formulas from the website on "Real statistics using Excel" (http://www.real-statistics.com/descriptive-statistics/measures-variability/).

One can see from Table 1, that the probability of each of members of these H-*n*-alphabets has approximately the same value in all 5 chromosomes or, in other words, it serves as a special invariant inside the complete set of chromosomes *Arabidopsis thaliana*. It is obvious that these regularities of probabilities of the hydrogen bonds 3 and 2 and of their ensembles define in long DNA sequences a special limitation for arrangements of nucleotides A, T, C, G, which are connected with hydrogen bonds in DNA sequences (A = T = 2, C = G = 3): only special classes of locations of the nucleotides in a long DNA sequence can provide such regularities.

Some data from Table 1 about probabilities of the H-monoplet 3 and of the H-monoplet 2 remind the following known data: the article [3] shows an approximate equality of probabilities of each of nucleotides A, T, C and G in all human nuclear chromosomes with a precision in a few percentage. Correspondingly a summary probability of nucleotides A and T (and also C and G), which is directly connected with the frequency of the H-monoplet 2 (the H-monoplet 3), has approximately the same values in all human nuclear chromosomes.

Similar results (with some other levels of accuracy) have been obtained in our analysis of the following complete sets of nuclear chromosomes on the basis of initial

Table 1. Probabilities of all members of the H-*n*-alphabets (n = 1, 2, 3) in DNA sequences of all 5 chromosomes of Arabidopsis thaliana are shown. Symbols q and p denote probabilities of H-monoplets 3 and 2 correspondingly and are shown in the left column separately for each chromosome. The probability values are rounded to the fourths decimal place. Traditional statistical characteristics are also shown: $\bar{X}$ denotes the mean of the sample $S =\{x_1, x_2, ..., x_n\}$; s – sample standard deviation; V – coefficient of variation (see their formulas in http://www.real-statistics.com/descriptive-statistics/measures-variability/).

	Probabilities q and p of H-monoplets: 3 and 2		Probabilities of 4 H-doublets: 33, 32, 23, 22			
	3 (q)	2 (p)	33	32	23	22
1	0,3587	0,6413	0,1198	0,2390	0,2389	0,4023
2	0,3586	0,6414	0,1206	0,2381	0,2380	0,4033
3	0,3633	0,6367	0,1230	0,2402	0,2405	0,3964
4	0,3620	0,6380	0,1219	0,2401	0,2401	0,3978
5	0,3594	0,6406	0,1204	0,2391	0,2389	0,4016
$\bar{X}$	0,3604	0,6396	0,1211	0,2393	0,2393	0,4003
s	0,0019	0,0019	0,0012	0,0008	0,0009	0,0027
V	0,0053	0,0030	0,0095	0,0033	0,0037	0,0067

	Probabilities of 8 H-triplets: 333, 332, 323, 322, 233, 232, 223, 222							
	333	332	323	322	233	232	223	222
1	0,0385	0,0811	0,0880	0,1507	0,0812	0,1577	0,1514	0,2512
2	0,0392	0,0813	0,0876	0,1504	0,0813	0,1568	0,1506	0,2527
3	0,0402	0,0826	0,0897	0,1508	0,0828	0,1572	0,1509	0,2457
4	0,0396	0,0827	0,0890	0,1508	0,0822	0,1577	0,1509	0,2470
5	0,0390	0,0814	0,0881	0,1509	0,0814	0,1578	0,1508	0,2506
$\bar{X}$	0,0393	0,0818	0,0885	0,1507	0,0818	0,1575	0,1509	0,2495
s	0,0006	0,0007	0,0008	0,0002	0,0006	0,0004	0,0002	0,0027
V	0,0145	0,0082	0,0089	0,0013	0,0077	0,0024	0,0016	0,0107

data about their DNA sequences from the GenBank: a nematode *Caenorhabditis elegans*, fruit fly *Drosophila melanogaster*, house mouse *Mus musculus*, Homo sapiens.

These results allow assuming existence of the following general rule for eukaryots about probabilities of hydrogen bonds in complete sets of nuclear chromosomes:

- In complete sets of nuclear chromosomes of different organisms, a probability of any member of the alphabets of n-plets has approximately the same value in all chromosomes ($n = 1, 2, 3, 4, 5 \ldots$ is not too large).

Of course, further researches are needed to define a degree of universality and precision of this rule. This rule has mathematical analogies with the main law of population genetics - the Hardy–Weinberg law as it is described in the preprint [2].

3 On the Modeling Approach with the Tensor Product of Vectors

Our conception of the quantum-algorithmic genetics (or the quantum-computing genetics) [4, 5] has led the authors to the useful mathematical model for described phenomenological probabilities. This model uses the notion of genetic qubits [2, 4] and also some classical formalism from quantum informatics [6].

In our proposed approach, phenomenological probabilities of all members of alphabets of H-n-plets $[3, 2]^{(n)}$ in a long DNA sequence under fixed n are modeled by numeric values of appropriate coordinates of 2^n-dimensional vector from the tensor family of vectors of probabilities $[q, p]^{(n)}$ where q and p mean probabilities of the H-monoplet 3 and the H-monoplet 2 correspondingly (these probabilities q and p are shown for each of chromosomes in the second columns in Fig. 3). Correspondingly, in this model approach, to get model values of frequencies of all 2^n members of the H-n-alphabet in a considered long DNA, one should calculate the following 2^n coordinates of frequency vectors from their tensor family:

– In the case of H-doublets-alphabet, 4 coordinates of the vector $[q, p]^{(2)} = [qq, qp, pq, pp]$;
– In the case of H-triplets-alphabet, 8 coordinates of the vector $[q, p]^{(3)} = [qqq, qqp, qpq, qpp, pqq, pqp, ppq, ppp]$; etc.

Briefly speaking, in line with the proposed model approach, for a prediction of approximate values of probabilities "f" of many members of H-n-alphabets in a long DNA sequence on the basis of knowledge about only two probabilities q and p, it is enough to do the following:

- Calculate probabilities q and p of H-monoplets 3 and 2 in the sequence;
- Replace digits 3 and 2 in H-symbols of these H-alphabetic members by their probabilities q and p as factors;
- Calculate the product of these factors.

The expression (1) shows an example of such calculation of the probability f(332) for the member 332 of the alphabet of H-triplets under q = 0,36 and p = 0,64 (q + p = 1):

$$f(332) = qqp = 0,36 * 0,36 * 0,64 = 0,0829. \tag{1}$$

Figure 3 shows nice correspondences between these model values (red points) of components of probability vectors $[q, p]^{(n)}$ and phenomenological values (blue points)

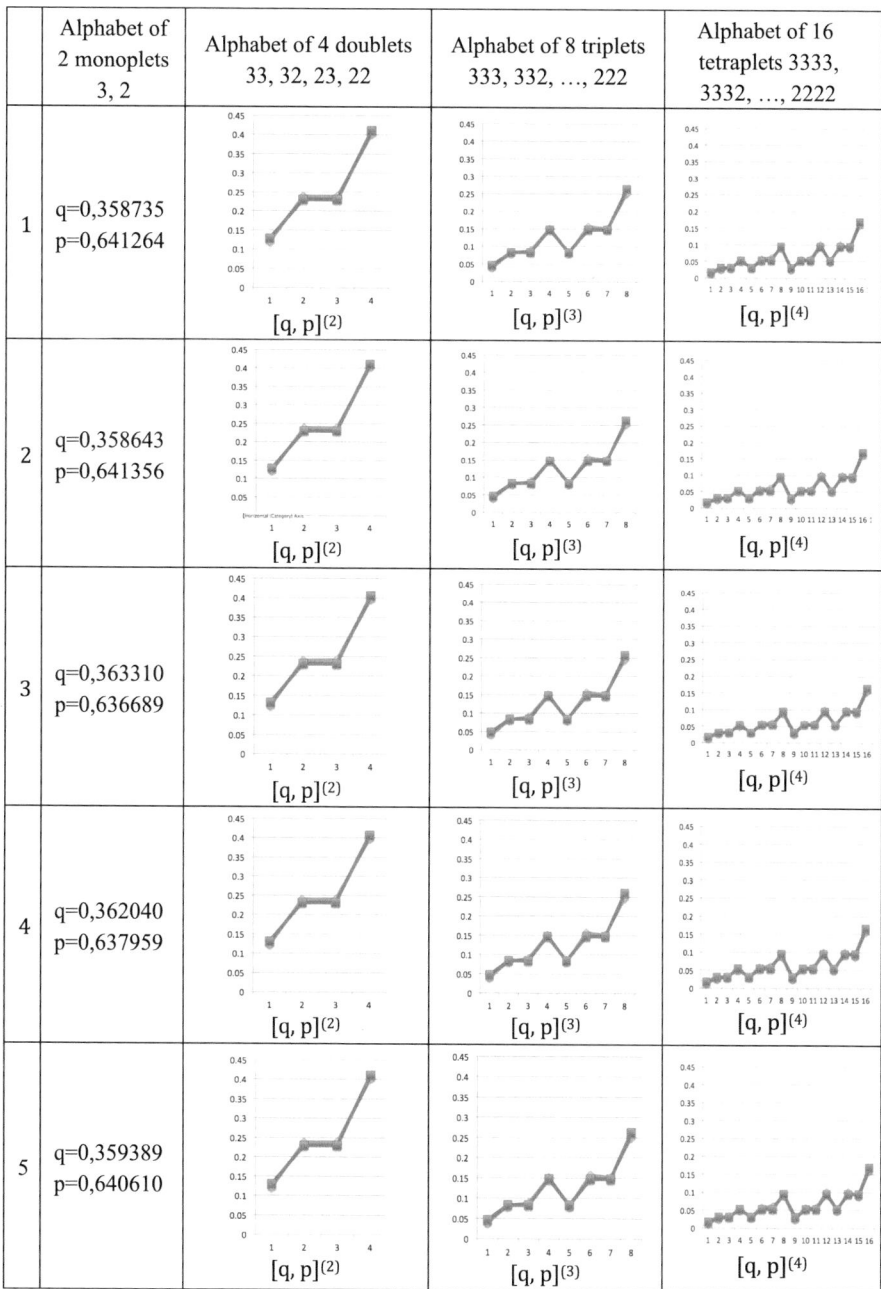

	Alphabet of 2 monoplets 3, 2	Alphabet of 4 doublets 33, 32, 23, 22	Alphabet of 8 triplets 333, 332, …, 222	Alphabet of 16 tetraplets 3333, 3332, …, 2222
1	q=0,358735 p=0,641264	$[q, p]^{(2)}$	$[q, p]^{(3)}$	$[q, p]^{(4)}$
2	q=0,358643 p=0,641356	$[q, p]^{(2)}$	$[q, p]^{(3)}$	$[q, p]^{(4)}$
3	q=0,363310 p=0,636689	$[q, p]^{(2)}$	$[q, p]^{(3)}$	$[q, p]^{(4)}$
4	q=0,362040 p=0,637959	$[q, p]^{(2)}$	$[q, p]^{(3)}$	$[q, p]^{(4)}$
5	q=0,359389 p=0,640610	$[q, p]^{(2)}$	$[q, p]^{(3)}$	$[q, p]^{(4)}$

Fig. 3. The nice correspondence between coordinate values (red points) of model vectors $[q, p]^{(n)}$ and phenomenological values (blue points are taken from Fig. 2) of probabilities of all members of alphabets of H-n-plets ($n = 2, 3, 4$) for DNA sequences of all 5 chromosomes of the plant *Arabidopsis thaliana*. The left column contains chromosome numbers.

Table 2. Values of coordinates of vectors [q, p]$^{(n)}$ are shown, which are used as models of values of frequencies of all members of the H-n-alphabets (n = 1, 2, 3, 4) in DNA sequences of all 5 chromosomes of *Arabidopsis thaliana*. Symbols q and p denote frequencies of H-monoplets 3 and 2 correspondingly and are shown in the left column separately for each chromosome. The model values are rounded to the sixth decimal place. Traditional statistical symbols are also shown: $\bar{X}$ denotes the mean of the sample S = $\{x_1, x_2, ..., x_n\}$; s – sample standard deviation; V – coefficient of variation (see their formulas in http://www.real-statistics.com/descriptive-statistics/measures-variability/).

	Probabilities q and p of H-monoplets 3 and 2		Probabilities of 4 H-doublets 33, 32, 23, 22			
	MODEL VALUES:					
	3 (q)	2 (p)	33 (qq)	32 (qp)	23 (pq)	22 (pp)
1	0,3587	0,6413	0,1287	0,2300	0,2300	0,4112
2	0,3586	0,6414	0,1286	0,2300	0,2300	0,4113
3	0,3633	0,6367	0,1320	0,2313	0,2313	0,4054
4	0,3620	0,6380	0,1311	0,2310	0,2310	0,4070
5	0,3594	0,6406	0,1292	0,2302	0,2302	0,4104
$\bar{X}$	0,3604	0,6396	0,1299	0,2305	0,2305	0,4091
s	0,0019	0,0019	0,0014	0,0005	0,0005	0,0024
V	0,0053	0,0030	0,0106	0,0023	0,0023	0,0059

Probabilities of 8 H-triplets: 333, 332, 323, 322, 233, 232, 223, 222							
333	332	323	322	233	232	223	222
MODEL VALUES:							
qqq	qqp	qpq	qpp	pqq	pqp	ppq	ppp
0,0462	0,0825	0,0825	0,1475	0,0825	0,1475	0,1475	0,2637
0,0461	0,0825	0,0825	0,1475	0,0825	0,1475	0,1475	0,2638
0,0480	0,0840	0,0840	0,1473	0,0840	0,1473	0,1473	0,2581
0,0475	0,0836	0,0836	0,1473	0,0836	0,1473	0,1473	0,2596
0,0464	0,0827	0,0827	0,1475	0,0827	0,1475	0,1475	0,2629
0,0468	0,0831	0,0831	0,1474	0,0831	0,1474	0,1474	0,2616
0,0007	0,0006	0,0006	0,0001	0,0006	0,0001	0,0001	0,0023
0,0158	0,0076	0,0076	0,0007	0,0076	0,0007	0,0007	0,0089

Row labels for the second table: 1, 2, 3, 4, 5, $\bar{X}$, s, V.

of probabilities of all members of H-n-alphabets (n = 2, 3, 4). One can see in these graphs that the model points of red color turned out to be almost exactly superimposed on the phenomenological points of blue color. It gives evidences in favor that knowing only two probabilities q and p of H-monoplets 3 and 2 in a long DNA sequence, one can predict dozens of probabilities of all members of alphabets of H-n-plets for this DNA sequence with a good level of accuracy. We presume that a similar model correspondence holds true also for n = 6, 7, 8, … (if n is not too large) but this should be studied in future researches.

In addition to graphs in Fig. 3, Table 2 shows numeric values of coordinates of vectors [q, p]$^{(n)}$ for all 5 chromosomes under n = 1, 2, 3. By a comparison of Tables 1 and 2, one can see that the proposed model approach gives nice results in modeling the phenomenological probabilities. More details about this model approach and its results for n = 1, 2, 3, 4, 5 one can see in the preprint [2].

4 Conclusions

Many genetic problems in fields of medicine, biotechnology, agriculture and bio-medical engineering are connected with secrets of sets of chromosomes in human and other eukaryotic organisms. Complete set of chromosomes (karyotypes) in species or in individual organisms are intensively studied in cytogenetics by means more and more effective technologies taking into account their importance. Karyotypes can be used for many purposes; such as to study chromosomal aberrations, cellular function, taxonomic relationships, and to gather information about past evolutionary events (https://en.wikipe dia.org/wiki/Karyotype). Chromosomal abnormalities lead to many diseases in humans including Down syndrome (https://en.wikipedia.org/wiki/Karyotype#Chromosome_ abnormalities).

The described discovery of new symmetries inside complete sets of chromosomes gives deeper scientific knowledge in cytogenetics for many medicine and biotechno-logical problems. We believe that this will lead to new tools and methods in cytoge-netics for medical and biotechnological purposes. It will enrich mathematical bioinformatics presented by very different publications [7–11].

In this article we described the new method of studying cooperative properties in long DNA sequences of hydrogen bonds. The effectiveness of the proposed model approach using the tensor product of vectors gives additional evidence in favor that formalisms of quantum informatics can be used in mathematical bioinformatics. The tensor product is the crucial operation to understanding the quantum mechanics of multiparticle systems and is one of basic instruments in quantum informatics. The following quotation speaks about the tensor product: «*This construction is crucial to understanding the quantum mechanics of multiparticle systems*» [6, p. 71] since in line with the postulate of quantum mechanics: the state space of a composite system is the tensor product of the state spaces of its components. The materials of our article support thoughts of some authors about functioning biological bodies on the principles of quantum computers [12–19].

References

1. Hamilton, A.: The retail DNA test. Time, 29 October 2008
2. Petoukhov, S.V.: The genetic coding system and unitary matrices. Preprints 2018, 2018040131. https://doi.org/10.20944/preprints201804.0131.v2
3. Okamura, K., Wei, J., Scherer, S.: Evolutionary implications of inversions that have caused intra-strand parity in DNA. BMC Genom. **8**, 160–166 (2007)
4. Petoukhov, S.V.: The rules of long DNA-sequences and tetra-groups of oligonucleotides. arXiv:1709.04943v5, 5th version, 8 October 2018, 159 p.
5. Petoukhov, S.V., Svirin, V.I.: The new wide class of symmetries in long DNA-texts. Elements of quantum-information genetics. Biologia Serbica **40**(1), 51 (2018)
6. Nielsen, M.A., Chuang, I.L.: Quantum Computation and Quantum Information. Cambridge University Press, New York (2010)
7. Singh, D.A.A.G., Leavline, E.J., Priyanka, R., Priya, P.P.: Dimensionality reduction using genetic algorithm for improving accuracy in medical diagnosis. Int. J. Intell. Syst. Appl. (IJISA) **8**(1), 67–73 (2016). https://doi.org/10.5815/ijisa.2016.01.08
8. Chawda, B.V., Patel, J.M.: Investigating performance of various natural computing algorithms. Int. J. Intell. Syst. Appl. (IJISA) **9**(1), 46–59 (2017). https://doi.org/10.5815/ijisa.2017.01.05
9. Mousa, H.M.: DNA-genetic encryption technique. Int. J. Comput. Netw. Information Secur. (IJCNIS) **8**(7), 1–9 (2016). https://doi.org/10.5815/ijcnis.2016.07.01
10. Abo-Zahhad, M., Ahmed, S.M., Abd-Elrahman, S.A.: Genomic analysis and classification of exon and intron sequences using DNA numerical mapping techniques. Int. J. Inf. Technol. Comput. Sci. (IJITCS) **4**(8), 22–36 (2012). https://doi.org/10.5815/ijitcs.2012.08.03
11. Srivastava, P.C., Agrawal, A., Mishra, K.N., Ojha, P.K., Garg, R.: Fingerprints, Iris and DNA features based multimodal systems: a review. Int. J. Inf. Technol. Computer Sci. (IJITCS) **5**(2), 88–111 (2013). https://doi.org/10.5815/ijitcs.2013.02.10
12. Penrose, R.: Shadows of Mind. Oxford University Press, New York (1994)
13. Igamberdiev, A.U.: Quantum mechanical properties of biosystems: a framework for complexity, structural stability, and transformations. Biosystems **31**(1), 65–73 (1993)
14. Matsuno, K.: Cell motility as entangled quantum coherence. BioSystems **51**, 15–19 (1999)
15. Matsuno, K., Paton, R.C.: Is there a biology of quantum information? BioSystems **55**, 39–46 (2000)
16. Patel, A.: Quantum algorithms and the genetic code. Pramana – J. Phys. **56**(2–3), 367–381 arXiv:quant-ph/0002037 (2001a)
17. Patel, A.: Testing quantum dynamics in genetic information processing. J. Genet. **80**(1), 39–43 (2001b)
18. Patel, A.: Why genetic information processing could have a quantum basis. J. Biosci. **26**(2), 145–151 (2001c)
19. Abbott, D., Davies, P.C.W., Pati, A.K., foreword by Sir Roger Penrose: Quantum Aspects of Life (2008). ISBN-13: 978-1-84816-253-2

Elastic and Non-elastic Properties of Cadherin Ectodomain: Comparison with Mechanical System

I. V. Likhachev[1,2], N. K. Balabaev[1], and O. V. Galzitskaya[2(✉)]

[1] Institute of Mathematical Problems of Biology, Russian Academy of Sciences,
Keldysh Institute of Applied Mathematics of Russian Academy of Sciences,
Pushchino, Moscow Region 142290, Russia
[2] Institute of Protein Research, Russian Academy of Sciences,
4, Institutskaya Street, Pushchino, Moscow Region 142290, Russia
ogalzit@evga.protres.ru

Abstract. Cadherins are predominantly homotypically acting cell-cell adhesion molecules that play an essential role in animal differentiation and maintenance of tissue integrity. Their cell-cell adhesive function relies on the presence of Ca^{2+} but is also influenced by other ions, in particular Mg^{2+}. Initial studies were devoted to analysis of the influence of various cations on cadherin-mediated cell-cell adhesion, and the subsequent work is focused predominantly on the role of Ca^{2+}, disregarding other physiologically relevant ions that are also able to form water complexes and stabilize protein conformations, i.e. Mg^{2+}. The amino acid sequence analysis revealed earlier the presence of several putative acidic Ca^{2+} binding motifs that were later confirmed. Using molecular dynamics simulations we demonstrated that the cadherin ectodomain acts as a spring but also dampens pulling forces. Both the simple model system (Glu-ion-Glu) and the complex cadherin macromolecule equally demonstrate two types of behavior: the systems with potassium and sodium ions possess less mechanical stability for external force action than the systems with calcium and magnesium.

Keywords: C-cadherin · Molecular dynamics · Divalent and monovalent ions

1 Introduction

Cadherins are proteins of cell adhesion, providing a calcium-dependent compound of cells in dense tissues of the body. The specificity of the formation of contacts between cells is very important for the development of the organism, in particular for the formation of tissues from cells. Cadherin consists of extracellular, small cytoplasmic component and membrane parts. The intercellular part consists of five domains from EC1 to EC5, each of 110 amino acid residues (Fig. 1). EC2–EC5 domains are similar in their amino acid sequence, while EC1 is less conserved; it is maximally removed from the cell and is responsible for the specificity of contact formation. Thus, cells can only come into contact with cells that have identical cadherin [1–3], but in some cases heterophilic contacts between classical cadherins are also possible [4–6]. Other EC domains can interact with different partners, thereby providing unique functionality for

© Springer Nature Switzerland AG 2020
Z. Hu et al. (Eds.): ICCSEEA 2019, AISC 938, pp. 555–566, 2020.
https://doi.org/10.1007/978-3-030-16621-2_52

cadherins. For example, the EC4 domain can interact with the fibroblast growth factor receptor (FGFR) [7]. Cadherins play an important role in maintaining the integrity of tissues. It is shown that the function of gluing cells in tissue depends on the presence of Ca^{2+} ions [8, 9], and is also influenced by other ions, in particular, Mg^{2+} [10].

C-cadherin (also known as EP-cadherin) is a typical representative of cadherins. Electron microscopy studies have shown that the intercellular part of E-cadherin (very similar to C-cadherin) adopts an elongated, rod-like conformation in the presence of Ca^{2+} [11]. Without divalent ions, molecules adopt arbitrary conformations, losing their physiological functions, i.e., the functions of maintaining the integrity of tissues. The studies, carried out using the results of analyzes of the crystal structure and NMR data, show that an increase in the Ca^{2+} concentration leads to stabilization of the elongated conformation of cadherin. In the rod-shaped conformation, cadherin is capable of trans-interaction with cadherin of another cell (Fig. 1). Trans-interaction cadherins of different cells is due to the contact of the first domains of EC1 in each of the proteins (most distant from the cell membrane) [12].

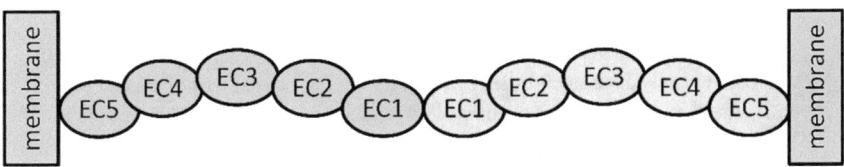

Fig. 1. Interaction of two cadherins.

The knowledge of the mechanical characteristics of cell adhesion proteins is important for assessing the ability of intercellular contacts. In the papers [10, 13] the authors showed that the mechanical properties of cadherin depend on the presence or absence of ions in the structure of the molecule, as well as their type (Ca^{2+}, K^+ or Na^+). It was demonstrated that cadherin is more stable and retains a rod-shaped conformation in the presence of Ca^{2+} ions rather than with K^+ and Na^+ ions. In the absence of ions, the structure of the protein is unstable.

The experiments on the full unfolding of cadherin with ions Ca^{2+}, Mg^{2+}, K^+, Na^+ by using atomic force microscopy (AFM) [10] are described. In our paper, conclusions are drawn on greater stability of the macromolecule in the presence of Ca^{2+} ions compared to Mg^{2+} ions solely from the large peaks of force corresponding to the stretching of 350 Å. We do not question these data, but rather confirm the presence of large peaks of force at this stage of unfolding. However, in addition to these peaks, in the MD simulations, in view of a higher available resolution, smaller peaks of force are observed, in which cadherin with ions such as Ca^{2+} and Mg^{2+} has approximately the same strength reaction, which is significantly higher than cadherin with K^+, Na^+ ions. We focus on the fact that the studied molecule under the conditions of the intercellular space should not violate the secondary structure. And this structure is already broken even before the peak is observed in the experiments of atomic force microscopy. While works on molecular dynamics consider small deformations with ions Ca^{2+}, K^+, Na^+.

The question remains open about the behavior of C-cadherin with Mg^{2+} ions under small deformations that a protein can undergo in biological conditions.

In this paper, it is shown that for small deformations cadherin with Mg^{2+} ions has plasticity and elasticity similar to cadherin with Ca^{2+} ions. To attain a better understanding of the interaction of proteins and ions in water, we performed the series of experiments on the stretching of the Glu-Ion-Glu system (two molecules of glutamic acid and an ion between them). This simple model clearly shows that in water the ions Ca^{2+} and Mg^{2+} interact with the protein more strongly than the ions K^+ and Na^+. It is also concluded that only explicit water is suitable for experiments with cadherin.

2 Methods

2.1 Cadherin Protein Model with Different Ions

We used the cadherin protein, from the embryo of the frog (*Xenopus laevis*) [15, 16], whose structure was obtained with a resolution of 3.08 Å. The initial coordinates of the atoms were taken from the Protein Data Bank (protein code: 1L3W). The structure of cadherin with Mg^{2+} ions was obtained by arranging 12 magnesium ions in the places where calcium ions were located. We also replaced these 12 calcium ions with potassium and sodium ions with a single positive charge, reducing the charge of the system by 12 elementary electric charges. To compensate for the charge, we replaced 3 water molecules at the four contact points of EC1–EC5 domains with potassium and sodium ions, respectively.

The protein was enclosed in a water shell. To this end, the protein was placed into a parallelepiped filled with water molecules. The water molecules, which overlap with the protein atoms, were removed, the distance between atoms water and protein atoms being less than 3 Å. Then water molecules located more than 7 Å from the protein were removed. This step is important for the analysis. Thus, we received 3478 water molecules for C-Cadherin. The whole system of protein and water molecules was enclosed into a sufficiently large sphere-cylinder; the diameter of the sphere-cylinder is 100 Å and the length is 500 Å with impenetrable repulsive walls. This sphere-cylinder did not affect the dynamics of protein and water and, at the same time, did not allow water molecules go to the infinity, returning them into the modeling region. During preparation of the initial data for the first time, random velocities were assigned to all atoms and relaxation of the system was performed.

We think this water model is a good approximation for observation of the mechanical properties of this protein. In the presence of the sorbate layer of water, all atoms of the protein are surrounded by water molecules such as in bulk water. If the protein structure changes the water follows up these changes, and it is adapted to protein such as in bulk water. Certainly, the presence of surface tension may distort the protein characteristics. But we think its influence is in agreement with important precision. In addition, it is possible to evaluate such influence by obtaining a thicker layer of water in one of the experiments and comparing the results.

2.2   Molecular Dynamics Simulations

The study was carried out with the help of the method of molecular dynamics using the program PUMA, developed at the IMPB, RAS [19–21]. The system of classical motion equations of atoms was resolved in the all-atom force field AMBER-99 [22]. Ion parameters were added to the force field for Ca^{2+}, Mg^{2+}, Na^+ and K^+ from [23]. Unlike [24], full atoms force field was used.

Parameters of Van-der-Waals interactions are given in Table 1.

Table 1. Force field parameters for ions

Name	Epsilon	R	Full name
Ca^{2+}	0.3353	1.7433	Calcium ion
Mg^{2+}	0.4677	1.4394	Magnesium ion
Na^+	0.0263	2.0484	Sodium ion
K^+	0.0091	2.4934	Potassium ion

The TIP3P model was used for water molecules, bonds and angles being not fixed. To maintain constant temperature, a collisional thermostat was used [25, 26]. The mean collision frequency of atoms with virtual particles was 10 ps^{-1}, the mass of each virtual particle was 1 atomic mass unit. Equations of motion were integrated numerically using the velocity version of the Verlet algorithm [27] with a time step 1 fs (10^{-15} s). This time step is correct because we used the collisional thermostat to maintain constant temperature during the simulations.

Also we have introduced the 10.5 Å shielding the radius of Coulomb interaction. So we guaranteed the electro neutrality of the whole molecular system.

On the one hand, the shielding (with the finite interaction radius) was introduced for decreasing the volume of the calculations. On the other hand, it is an important step in our simulation model. The goal is to define implicitly the salt solution. The salt solution efficiently shields Coulomb interactions of charged protein groups and explicit ions at long distances. The shielding radius is similar to the Debye radius in the electrolyte solution.

3   Results

3.1   Steered Molecular Dynamics Simulations

For the understanding of elastic and non-elastic properties, we used the method of molecular dynamics. It is a common method for observation of nano-systems. We have performed a free relaxation experiments as steered molecular dynamics simulations. Applying the force in our simulations is analogous to AFM experiments. But on the one hand, we are not limited by biological conditions (seconds of time during slow expanding). On the other hand, we wish to get some characteristics like under biological conditions. Thus we wish to use some parameters to decrease the computational complexity of our model and define realistic properties. Sometimes we may only obtain an order of the desired value.

We performed the series of free relaxation experiments with Ca^{2+} from different initial states. Different moments of time of cyclic stretching experiments were studied. Initial states were the following: the distances of 175, 197, 207 and 215 Å between the ends of protein. The goal of these experiments is to disclose the fluctuation motions as well as to determine what the unstrained state is for cadherin.

We have performed free relaxation experiments with all types of ions. Also we compared the results of all free relaxation experiments.

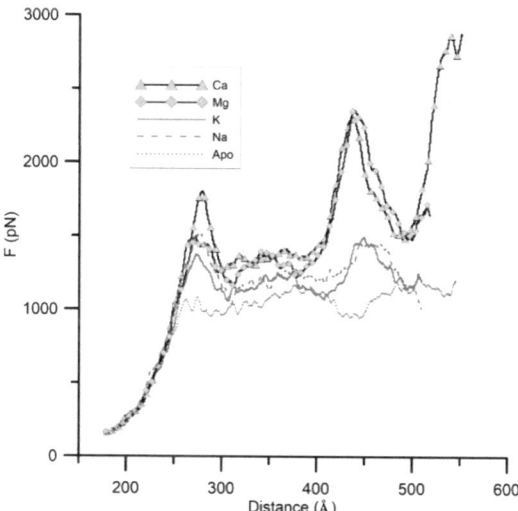

Fig. 2. Steered MD simulations with constant velocity. Force dependence [pN] on distance [Å].

MD simulations of stretching C-Cadherin with some force were performed, which guarantied constant stretching speed. The velocity of expansion is 0.1 Å/ps. The start point of each experiment is the relaxed structure of each variant of C-Cadherin – C-Cadherin with Ca^{2+}, Mg^{2+}, Na^+, K^+. In this case, we are interested only in the biological conditions. That is why long distance experiments were not conducted. The first peak of force is of interest and the time when the protein is not unfolded (Fig. 2). To check the latter, we used the TAMD program [14]. The 3D structure was analyzed as well as the dynamic contact map.

Our experiments display that unfolding did not occur during the stretching. **There were two different classes of behavior: (1) with ions of Ca^{2+} and Mg^{+2}, and (2) with ions of K^+ and Na^+.** There were some fluctuations in all experiments. We are sure that one of the experiments is only a private case of all experiments. Concrete values of force can be different from experiment to experiment. But we can observe the common tendency of the C-Cadherin behavior.

3.2   Properties of Elasticity and Plasticity for Cadherin with Ions

Cyclic Experiments

Force-distance characteristics show that Cadherin can be described like a spring and plasticine (or piston). Ions made the structure harder. But even if ions are present cadherin has in part plasticity properties. We assume that small force can stretch the C-Cadherin molecule with Ca^{2+}, but only at some distance as "spring" properties are present.

First resume. Force-distance characteristics show that Cadherin-Ca^{2+} and Cadherin-Mg^{2+} have the same elastic properties (yielding flow). **Second resume.** Both experiments show that Cadherin (either with Ca^{2+} or Mg^{2+}) has partially elastic spring properties and partially plasticity properties.

We have performed the series of experiments with cyclic stretching/pulling at different values of amplitude and periods. Experiments are more interested with stretching by 50 Å during 1 ns (Fig. 3).

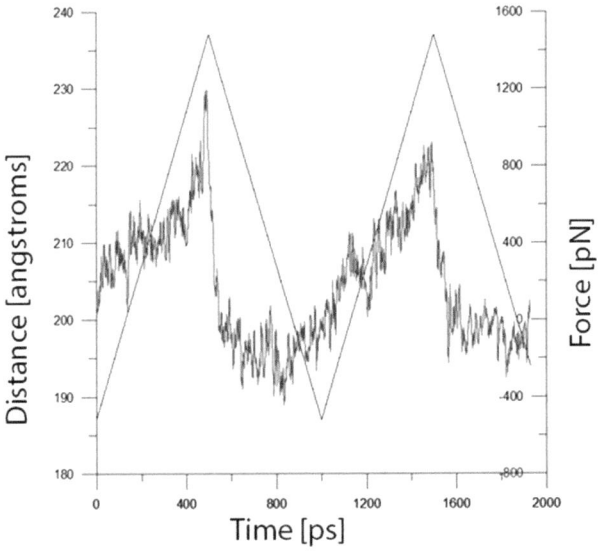

Fig. 3. Cyclic constant velocity extension MD experiments with Mg^{2+}. Distance dependence [Å] on time [ps] (black, right axis) and force dependence on time (red, left axis).

The beginning of the periodic function corresponds to an initial relaxed state of C-Cadherin.

C-Cadherin with Ca^{2+} and Mg^{2+} ions has the same elastic properties as shown by the cyclic experiments.

3.3   Comparison of Cadherin Simulations with Mechanical Piston-Spring System

We have plotted the force dependence on distance. The hysteresis behavior was observed in cyclic experiments.

For better understanding of the force characteristics, we suggest to consider the protein like a spring and like plasticine in the regions of force characteristics when it is possible (Fig. 4).

What is the method of connection between the spring and the piston? Is it a series or parallel connection? If we can see the figure of hysteresis, the connection is parallel.

If we stretch the system (Fig. 5, top red line)

$$F_{exp} = k * \Delta x + \eta * V, \tag{1}$$

F_{exp} is the stretching force, k is the spring constant, η is the piston viscosity, Δx is the change in the length of the spring, V is the speed.

If we pull the system (Fig. 5, bottom red line)

$$F_{pul} = k * \Delta x - \eta * V, \tag{2}$$

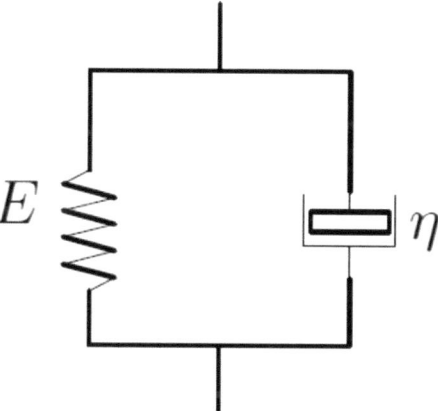

Fig. 4. Mechanical system "piston-spring" (E is the energy of spring, η is the piston viscosity).

F_{pul} is the force of pulling. The velocity is constant, 0.1 Å/ps.
The distance between two red lines is the double force of the piston:

$$F_{exp} - F_{pul} = 2\eta * V$$
$$\eta = (300 - (-200) \, pN/(2 * 0.1 \, \text{Å}/ps) = 2500 \, N * s/nm$$
$$= 2.5 * 10^{-11} N * s/m = 25 \, pN * s/m$$
$$= 2.5 * 10^{-11} kg/s$$

Spring force: $F_{sping} = k*\Delta x$. $k = (400 - 0)/(200 - 170)$ pN/ Å $= 133$ N/m.

On the green part of the graph we can suppose that the friction force is the same (the velocity is the same), but the elastic force is not decreasing to zero. We performed the series of constant force experiments. They showed that the elasticity force dependence is not linear, the elasticity coefficient is different upon application of different force.

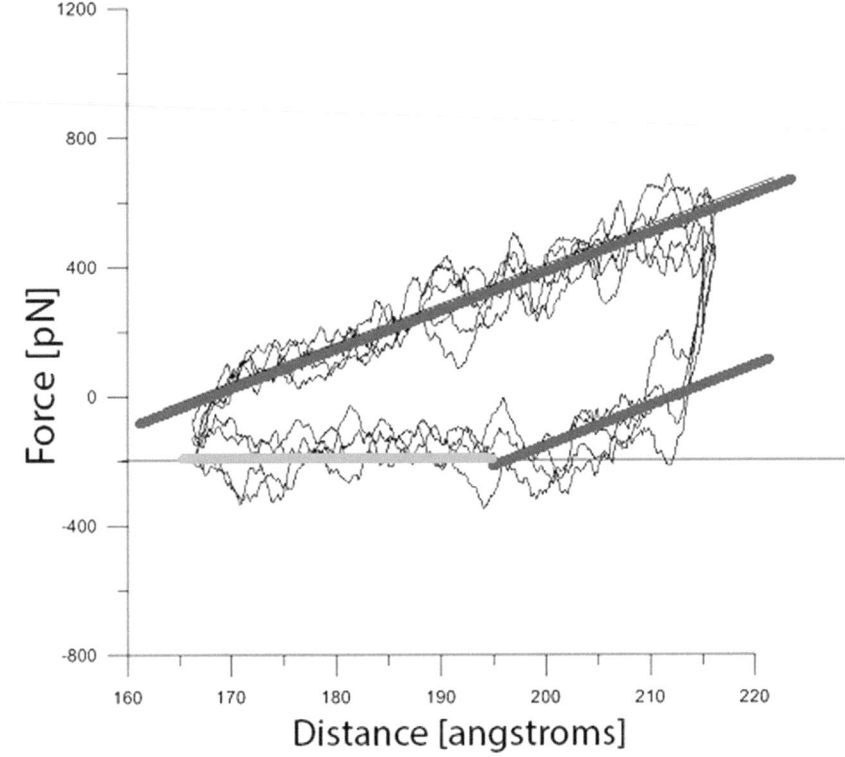

Fig. 5. MD-experiments with calcium ions: hysteresis. Dependence of force [pN] on distance [Å].

Hysteresis picture of C-Cadherin with 12 Mg^{2+} cations looks the same.

3.4   Model Glu-cation-Glu

Since Ca^{2+} ions in the cadherin molecule interact with the aspartic and glutamic acid residues, a better perception of their role in the mechanical properties and stability of the protein was achieved by modeling the stretching of two glutamic acid molecules with various metal ions placed between them. The stretching rate was 0.1 Å/ps. The force was applied to the carbon atoms of the COOH groups of glutamic acids. For each type of ion MD simulations were carried out both in water and in vacuum at the temperature of 300 K.

For all numerical experiments, the dependence of the reaction force on the distance between the points of force application was constructed (Fig. 6).

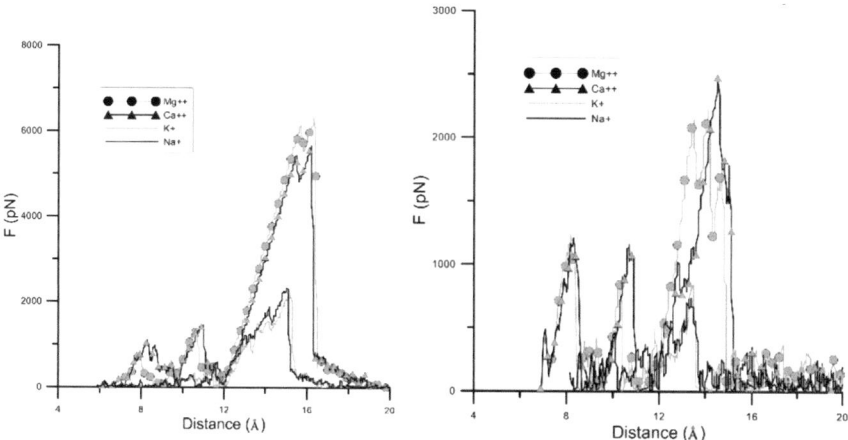

Fig. 6. MD Experiments on stretching the model system of two amino acids at the constant rate of 0.1 Å/ps with different ions. On the left, experiments in vacuum, on the right, in water. Force dependence [pN] on distance [Å].

In the initial configuration, the metal atom is bound to four oxygen atoms of glutamic acids. The analysis of the behavior of molecular dynamics trajectories allow us to draw the following conclusions. The first peaks of the smaller amplitude correspond to the sequential separation of two oxygen atoms of different amino acids from the metal ion. The third peak corresponds to the loss of the bond between the amino acids. In the experiments with water, the magnitude of the force corresponding to the third peak is much less than its value in vacuum. This is explained by the fact that the change of the contact of the ion and the glutamic acid molecule for the contacts of the ion and glutamic acid with the surrounding water molecules is energetically more advantageous than breaking the contact in vacuum. The dependences in Fig. 6 also indicate that the forces obtained in the experiments with calcium and magnesium, both in vacuum and in the water environment, far exceed the forces in experiments with potassium and sodium ions.

Thus, both the simple model system and the complex cadherin macromolecule equally demonstrate two types of behavior: the systems with potassium and sodium ions possess less mechanical stability for external force action than the systems with calcium and magnesium. Experiments with simple model systems reveal the key mechanism underlying the cadherian mechanical stability.

4 Conclusion

According to [13] the stability of the rod-like conformation inherent in cadherin in the intercellular space depends on the presence of calcium ions in the macromolecule. In their absence, cadherin assumes an accidental collapsed state. In the present work, the stretching of the cadherin molecule is simulated both in the presence of Ca^{2+}, Mg^{2+}, Na^+, K^+ ions, and in the apo-form at a constant rate. Experiments on stretching the cadherin beyond the ends at a constant rate with different ions revealed two types of system behavior: Ca^{2+} and Mg^{2+} ions impart greater mechanical stability to the protein than Na^+ and K^+ ions. For 3 ns of simulations, in all cases, two peaks of force were observed associated with the unfolding of the polypeptide chain of the first domain and with the increase in distances between domains. For sets of independent trajectories obtained for systems with each type of ions, the deployment scenarios are repeated.

When comparing molecular dynamics simulation with the atomic force microscopy experiments, only a qualitative assessment of the results is possible. The rates of stretching of the macromolecule in computer experiments exceed by orders of magnitude those obtained in the experiment. This leads to the fact that the force influence on cadherin during simulation is large (1500–2500 pN versus 100–200 pN in the experiment). Nevertheless, it can be assumed that the regularities inherent to the system obtained by simulations qualitatively correspond to the experimental data. The performed computer simulation and the results of the AFM experiments indicate the important role of metal ions in the mechanical properties of cadherin.

Also we have compared C-Cadherin in the cycle experiments with a popular physical system – piston and spring. Most potentials of Amber Force Field near the minimum potential energy look like a parabola. The potential of Hook's spring is a parabola too. But we have shown that various biology systems behave differently under tension and compression. The name of this phenomenon is hysteresis. This protein works like a piston and a spring. And we have shown how to calculate the parameters of similar systems.

May be in future the problem with storage of MD trajectories will occur. But with new novel cloud architecture such problem may be resolved [30–32].

Funding. This research was funded by the RUSSIAN SCIENCE FOUNDATION.

References

1. Takeichi, M., Hatta, K., Nose, A., Nagafuchi, A.: Identification of a gene family of cadherin cell adhesion molecules. Cell Differ. Dev. **25**, 91–94 (1988)
2. Chen, J., Ruan, H., Ng, S.M., Gao, C., Soo, H.M., Wu, W., Zhang, Z., Wen, Z., Lane, D.P., Peng, J.: Loss of function of def selectively up-regulates Delta 113p53 expression to arrest expansion growth of digestive organs in zebrafish. Genes Dev. **19**, 2900–2911 (2005)
3. Patel, N.A., Curiel, S., Zhang, Q., Sridharan, T.K., et al.: Torrelles submillimeter array observations of 321 GHz water maser emission in cepheus A. Astrophys. J. **658**, L55–L58 (2007)

4. Niessen, C.M., Gumbiner, B.M.: Cadherin-mediated cell sorting not determined by binding or adhesion specificity. J. Cell Biol. **156**, 389–399 (2002)
5. Foty, R.A., Steinberg, M.S.: The differential adhesion hypothesis: a direct evaluation. Dev. Biol. **278**, 255–263 (2005)
6. Shimoyama, Y., Tsujimoto, G., Kitajima, M., Natori, M.: Identification of three human type-II classic cadherins and frequent heterophilic interactions between different subclasses of type-II classic cadherins. Biochem. J. **349**, 159–167 (2000)
7. Williams, E.J., Williams, G., Howell, F.V., Skaper, S.D., Walsh, F.S., Doherty, P.: Identification of an N-cadherin motif that can interact with the fibroblast growth factor receptor and is required for axonal growth. J. Biol. Chem. **276**, 43879–43886 (2001)
8. Nagar, B., Overduin, M., Ikura, M., Rini, J.M.: Structural basis of calcium-induced E-cadherin rigidification and dimerization. Nature **380**, 360–364 (1996)
9. Haussinger, D., Ahrens, T., Aberle, T., Engel, J., Stetefeld, J., Grzesiek, S.: Proteolytic E-cadherin activation followed by solution NMR and X-ray crystallography. EMBO J. **23**, 1699–1708 (2004)
10. Oroz, J., Valbuena, A., Vera, A.M., Mendieta, J., Gomez-Puertas, P., Carrion-Vazquez, M.: Nanomechanics of the Cadherin ectodomain. J. Biol. Chem. **286**, 9405–9418 (2011)
11. Pokutta, S., Herrenknecht, K., Kemler, R., Engel, J.: Conformational changes of the recombinant extracellular domain of E-cadherin upon calcium binding. Eur. J. Biochem. **223** (3), 1019–1026 (1994)
12. Zhang, Y., Sivasankar, S., Nelson, J., Chu, S.: Resolving cadherin interactions and binding. PNAS **106**, 109–114 (2009)
13. Sotomayor, M., Schulten, K.: The allosteric role of the Ca^{2+} switch in adhesion and elasticity of C-cadherin. Biophys. J. **94**, 4621–4633 (2008)
14. Likhachev, I.V., Balabaev, N.K., Galzitskaya, O.V.: Available instruments for analyzing molecular dynamics trajectories. Open Biochem. J. **10**, 1–11 (2016)
15. Choi, Y.S., Sehgal, R., McCrea, P., Gumbiner, B.M.: A cadherin-like protein in eggs and cleaving embryos of Xenopus laevis is expressed in oocytes in response to progesterone. J. Cell Biol. **110**, 1575–1582 (1990)
16. Ginsberg, D., DeSimone, D., Geiger, B.: Expression of a novel cadherin (EP-cadherin) in unfertilized eggs and early Xenopus embryos. Development **111**, 315–325 (1991)
17. Ueda, M.V., Takeichi, M.: Two mechanisms in cell adhesion revealed by effects of divalent cations. Cell Struct. Funct. **1**, 377–388 (1976)
18. Boggon, T.J., Murray, J., Chappuis-Flament, S., Wong, E., Gumbiner, B.M., Shapiro, L.: C-cadherin ectodomain structure and implications for cell adhesion mechanisms. Science **296**, 1308–1313 (2002)
19. Glyakina, A.V., Likhachev, I.V., Balabaev, N.K., Galzitskaya, O.V.: Comparative mechanical unfolding studies of spectrin domains R15, R16 and R17. J. Struct. Biol. **201**, 162–170 (2018)
20. Glyakina, A.V., Likhachev, I.V., Balabaev, N.K., Galzitskaya, O.V.: Mechanical stability analysis of the protein L immunoglobulin-binding domain by full alanine screening using molecular dynamics simulations. Biotech. J. **10**, 386–394 (2015)
21. Glyakina, A.V., Likhachev, I.V., Balabaev, N.K., Galzitskaya, O.V.: Right-and left-handed three-helix proteins. II. Similarity and differences in mechanical unfolding of proteins. Proteins **82**, 90–102 (2014)
22. Wang, J., Cieplak, P., Kollmann, P.A.: How well does a restrained electrostatic potential (RESP) model perform in calculating conformational energies of organic and biological molecules? J. Comp. Chem. **21**, 1049–1074 (2000)
23. Aqvist, J.: Ion-water interaction potentials derived from free energy perturbation slmulations. J. Phys. Chem. **94**, 8021–8024 (1990)

24. Chawda, B.V., Patel, J.M.: Investigating performance of various natural computing algorithms. Int. J. Intell. Syst. Appl. (IJISA) **9**(1), 46–59 (2017). https://doi.org/10.5815/ijisa.2017.01.05

25. Lemak, A.S., Balabaev, N.K.: A comparison between collisional dynamics and Brownian dynamics. Mol. Simul. **15**, 223–231 (1995)

26. Lemak, A.S., Balabaev, N.K.: Molecular dynamics simulation of a polymer chain in solution by collisional dynamics method. J. Comp. Chem. **17**, 1685–1695 (1996)

27. Allen, M.P., Tildesley, D.J.: Computer Simulation of Liquids. Clarendon Press, Oxford (1989)

28. Piplani, S., Saini, V., Niraj, R.R., Pushp, A., Kumar, A.: Homology modelling and molecular docking studies of human placental cadherin protein for its role in teratogenic effects of anti-epileptic drugs. Comput. Biol. Chem. **60**, 1–8 (2016)

29. Alaofi, A., Farokhi, E., Prasasty, V.D., Anbanandam, A., Kuczera, K., Siahaan, T.J.: Probing the interaction between cHAVc3 peptide and the EC1 domain of E-cadherin using NMR and molecular dynamics simulations. J. Biomol. Struct. Dyn. **35**, 92–104 (2017)

30. Althagafy, E., Qureshi, M.R.J.: Novel cloud architecture to decrease problems related to big data. Int. J. Comput. Netw. Inf. Secur. (IJCNIS) **9**(2), 53–60 (2017). https://doi.org/10.5815/ijcnis.2017.02.07

31. Kaur, J., Kaur, K.: Internet of Things: a review on technologies, architecture, challenges, applications, future trends. Int. J. Comput. Netw. Inf. Secur. (IJCNIS) **9**(4), 57–70 (2017). https://doi.org/10.5815/ijcnis.2017.04.07

32. Akomolafe, O.P., Abodunrin, M.O.: A hybrid cryptographic model for data storage in mobile cloud computing. Int. J. Comput. Netw. Inf. Secur. (IJCNIS) **9**(6), 53–60 (2017). https://doi.org/10.5815/ijcnis.2017.06.06

Multidimensional Approach for Analysis of Chromosomes Nucleotide Composition

Ivan V. Stepanyan[(✉)]

Mechanical Engineering Research Institute, Russian Academy of Sciences,
M. Kharitonievsky pereulok, 4, Moscow, Russia
neurocomp.pro@gmail.com

Abstract. A chromosome includes a DNA molecule with a part or all of the genome of an organism. Statistically known that chromosomes nucleotide compositions are different for different biological species. Special comparative method of visualization the chromosomes nucleotide composition of various organisms is described. This analysis is conducted by means of a metric space of binary orthogonal functions taking into account physical-chemical parameters of nitrogenous bases of the genetic code. In consideration that genetic algebra and geometry are connected with a relations purposed in this article algorithms allows to display a statistical chromosome nucleotide composition data in a metric spaces using multidimensional analysis.

Keywords: Nucleotide composition · DNA symmetries ·
Multidimensional analysis

1 Introduction

Genetic nucleotides sequences are one-dimensional linear data consist of 4-letters alphabet of complimentary pairs of nucleotides. For the analysis of long nucleotide sequences, various mathematical methods are applied, including statistical. At the same time, it is actual to develop a system of new methods that would allow to visualize differences between nucleotide sequences and their characteristics clearly and without the using of complex programs and special skills. The paper structure consists of observation, mathematical description of physical chemistry of chromosomes and genetic sub-alphabets. The following are examples of multidimensional visualizations of chromosomes. The goal of research is the development and discussion of algorithms that can help facilitate the perception and visualization of genetic information and genetic fenomena.

2 Mathematical Description of Physical Chemistry of Chromosomes and Symmetric Hadamard Matrices of Genetic Sub-alphabets

2.1 Related Works

Let us recall some approaches to geometrical representations of molecular-genetic alphabets. In [1] described that every genetic nucleotide letter has a binary sub-alphabets

© Springer Nature Switzerland AG 2020
Z. Hu et al. (Eds.): ICCSEA 2019, AISC 938, pp. 567–575, 2020.
https://doi.org/10.1007/978-3-030-16621-2_53

of the genetic code (alphabets inside alphabets). Binary sub-alphabets of genetic coding are represented according to kinds of binary attributes in the set of nitrogenous bases. Three of them are represented according to kinds of binary opposite attributes and one is constant: pyrimidine or purine; amino or keto; three hydrogen bonds or two hydrogen bonds; phosphate residue (non-opposite parameter). In nucleotide composition of DNA molecules of different species of living organisms Chargaff noticed regularities that have verbal and mathematical equations representation [2]. These features features are the characterizing properties of the DNA of many living organisms and can be visualized algorithmically. For this visualization, the methods of parametrization the using of binary sub-alphabets and binary geometry will be applied.

Work [3] presents a gene retrieval system, which is based on feature dimensionality minimization and classification of the microarray gene data. The epipolar geometry [4] is the intrinsic projective geometry with the algebraic representation. Recovery of epipolar geometry is a fundamental problem in computer vision. The work [5] elucidates approach a new distance measure using DNA hybridization melting temperature that gives approximate solutions for the multiple sequence alignment problem. The main application of this multiple sequence alignment is to identify the sub-sequences for the functional study of the whole genome sequences. The paper [6] presents the method of medical images similarity estimation based on feature extraction and analysis. The described method of images similarity estimation based on feature analysis consists of several stages and feature analysis. Automatic method of karyotype analysis in cytogenetics description is given at paper [7] with locally adaptive thresholding method to segment chromosome clusters. The CGR-method is described at [8]. Structured alphabets of DNA and RNA in their matrix form of representations are connected with Walsh functions and a new type of systems of multidimensional numbers [9, 10]. Some related investigations were discussed at [11–13].

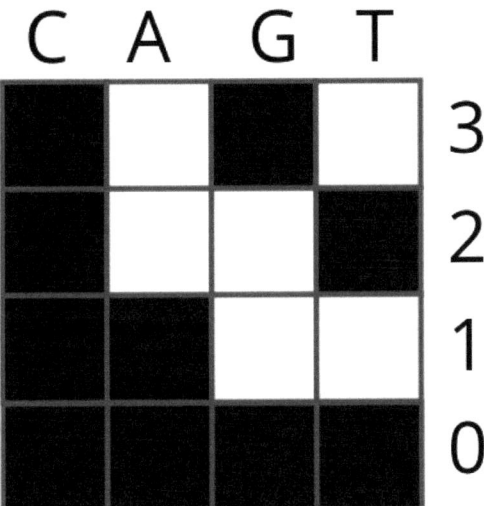

Fig. 1. Variant of Hadamard matrix based on nucleotide sub-alphabets coding. Symbols of a genetic sub-alphabets from a viewpoint of the binary attributes are 1, 2, 3 and 0. Blacks cells are 1, white cells are −1 or wise versa depending on coding.

2.2   Genetic Coding as a System of Orthogonal Walsh Functions

System of genetic sub-alphabets can be represented as Hadamard matrix shown on Fig. 1. Each row of Hadamard matrix is a Walsh function. We note, that present Hadamard matrix is symmetrical [14]. Every nucleotide can be replaced with relevant sub-alphabet without changing the structure of matrix.

Walsh functions in mathematical harmonic analysis is a complete orthogonal set of functions that can be used to represent any discrete function [15] as well as using Fourier analysis with trigonometric functions can represent any continuous function. More information about relations of genetic coding to Hadamard matrices is described in Petoukhov's articles, for example in [1].

2.3   Scale-Free Parametrization of Chromosomes Nucleotide Composition

Multidimensional analysis of chromosomes nucleotide composition is based on a splitting of genetic nucleotide sequence by equal parts with n nucleotides in each where n is a scaling free parameter. Thus, genetic nucleotide sequence is a 3-channel binary-opposite code divided by parts by n nucleotides (n-plets). Converting binary n-symbol words into decimal numbers allows to use them as coordinates at Cartesian spaces of different dimensions.

2.4   Geometrization

Set of three binary-opposite sub-alphabets can be matched to axes {X, Y, Z} of Cartesian coordinate system. This serves to represent genetic nucleotide DNA sequence at Cartesian space using described scale-free parametrization of chromosomes nucleotide composition. Algorithm's scale-free parameter of coefficient n works as a resolution of geometrical visualization: too big n gives small quantity of points, too small n gives small coordinate mesh. We found good results for n lays near range between 6 and 24 nucleotides.

We found that chromosomes of different species of organisms can have individual patterns, that are often fractals or fractal-like mosaics.

Described algorithm of a geometrization of bioinformatic data could be applied not only to chromosomes, but to any genetic nucleotide sequences including RNA, viruses, mitochondries DNA. Examples are shown at figures below.

3   Examples of Multidimensional Analysis of Chromosomes Nucleotide Composition

3.1   3-D Visualization of Chromosomes Nucleotide Composition

Using the set {X, Y, Z} as orthogonal basis of 3-dimensional Cartesian coordinate system gives graphical results of 3D mapping of chromosome obtained on the base of described method. Resulting Sierpinski tetrix that is the 3-D analogue of the Sierpinski triangle is shown at Fig. 2. Demonstrated 3-D visualization is not very suitable for analysis itself but 2-D projections of this tetrix can display geometrically wide differences between species of living organisms.

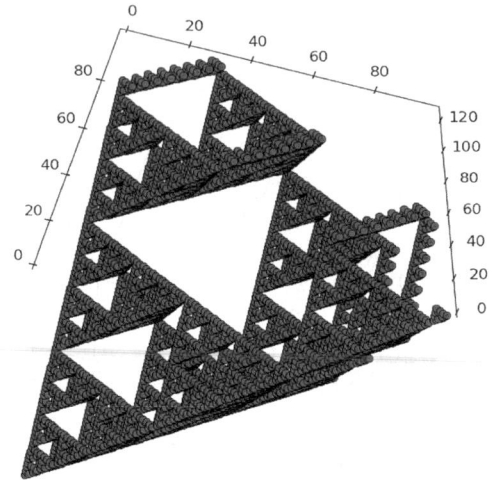

Fig. 2. Illustration of typical 3-dimensional representation of nucleotide composition of example chromosome. Axises X, Y and Z are corresponding to decimal representations of binary coding of each n-plet using a three binary-opposite sub-alphabets.

3.2   2-D Analysis of Chromosomes Nucleotide Composition

To reach 2-dimensional analysis of chromosomes nucleotide composition any pair of genetic binary sub-alphabets as orthogonal basis can be used. We note, that only two of three sub-alphabets are enough for 2-D visualization because addition modulo 2 of any pair of sub-alphabets gives 3-rd sub-alphabet as a property of Hadamard matrix.

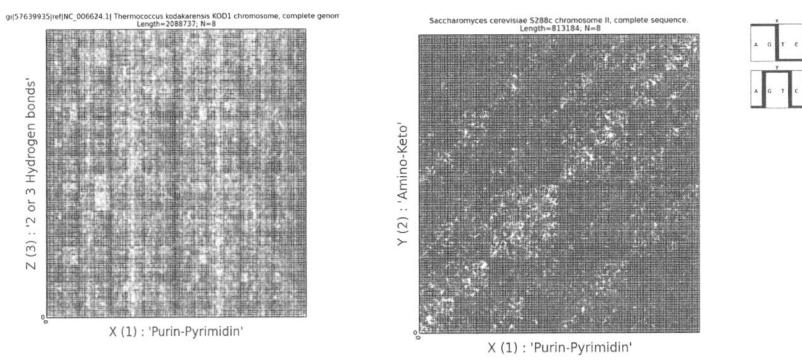

Fig. 3. Illustration of 2-dimensional representation of chromosome nucleotide composition. A pair of Walsh-functions used for parametrization is shown on top right corner. Axises of abscissae and ordinates are corresponding to decimal representations of binary coding of each n-plet using a pair of sub-alphabets. More info in text.

Thus, the using 2-dimensional projections {X, Y}, (X, Z} and {Y, Z} of 3-D Sierpinski tetrix into Cartesian 2-D coordinate system is giving three different mappings using corresponding sub-alphabets of physical chemistry properties for analyzing chromosome (Figs. 3 and 4).

Such mosaic pattern shows phenomenology of «presence-and-absence» of different n-plets. Possible way of an unambiguous representation of a long nucleotide sequence in a case of its division with a certain value n (for example, with n = 8) is connected with a construction of additional visual patterns, which reflect an order of n-plets in the sequence.

2-D patterns, which are obtained by means of the described method, sometimes resemble fractal patterns of long nucleotide sequences and amino acid sequences, which were previously obtained by means of the known method "Chaos Game Representation" (CGR-method) in works [16] though both methods are quite different in their algorithmic essence.

The preliminary results includes the stability of resulting fractal-like and other patterns in a case of shifts of reading frame of such sequences, or in a case of reversing of sequences, or in a case of the permutation of fragments of a sequence, or in case of a removal of certain parts of sequences; these results are mainly similar to the results of studies of fractal genetic networks for long nucleotide sequences [17]. In particularly, we saw a stability of mosaic patterns in cases of transformations of examined nucleotide sequences by means of removal of the every second nucleotide in sequences, or removal of the every third nucleotide in sequences, etc. Adjacent variants can be added to the described method for deeper research of long genetic sequences by means of their binary presentation.

Different random nucleotide sequences with 100000 nucleotides in cases of its division into n-plets with n = 8, 16, 28 were generated by a computer program. Visual patterns were produced by the described method for a long random sequences of nucleotides. Appropriate visual patterns have non-regular chaotic characteristics in contrast to cases of real genomes.

Additionally we have tested different kinds of penicillins to receive binary patterns. The results testify that in this group of antibiotics their long nucleotide sequences usually generate non–regular mosaics, which resemble mosaics of random nucleotide sequences. Perhaps medical meaning of penicillins is connected with this their peculiarity.

A random sequence with the second Chargaff's rule was algorithmically generated using the second Chargaff's rule [18]. Special kind of regularities in a form of squares for this random sequence were visualized with n = 6 at Fig. 4.

From the other side, algorithmically generated random sequences without Chargaff rule have irregular chaotic characteristics without any regularities. This allows to say that geometrization of nucleotide states of chromosomes has relations with Chargaff rules and can help to visualize the statistical characteristics of genomes of living organisms.

Quantities of elements 0 and 1, which are met in n-plets of two kinds of the n-bit binary presentation of a long nucleotide sequence was used to construct a new type of a visual pattern of the sequences (Fig. 5).

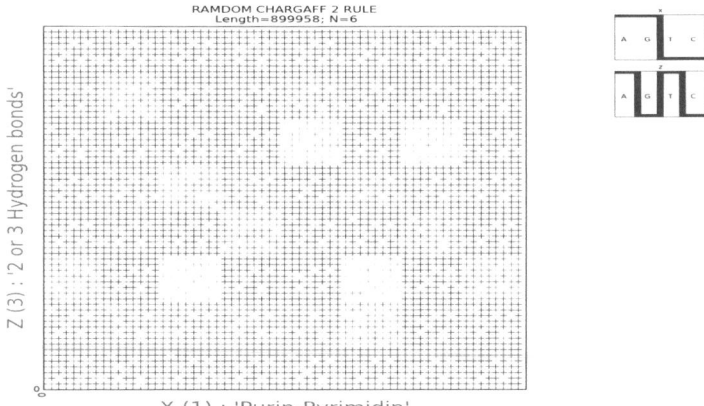

Fig. 4. Illustration of 2-dimensional representation of randomly generated chromosome with 2-nd Chargaff rule. A pair of Walsh-functions used for parametrization is shown on top right corner. Axises of abscissae and ordinates are corresponding to decimal representations of binary coding of each n-plet using a pair of sub-alphabets. More info in text.

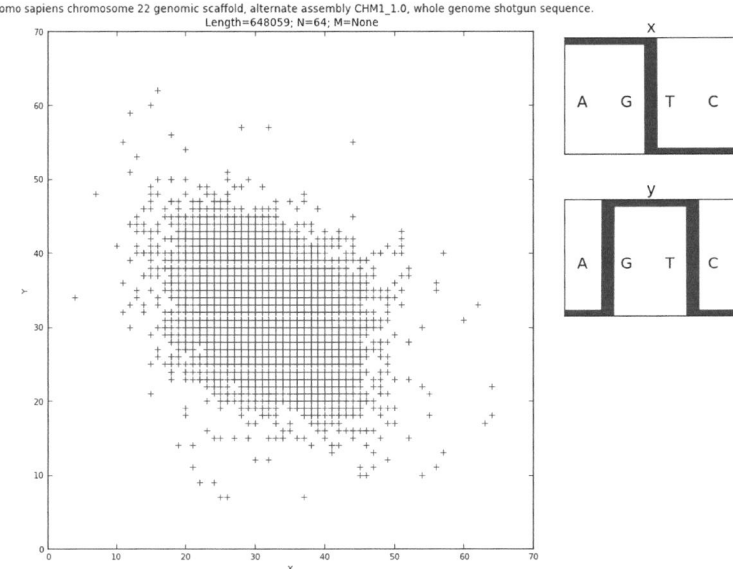

Fig. 5. Illustration of 2-dimensional representation of chromosome nucleotide composition. A pair of Walsh-functions used for parametrization is shown on top right corner. Axises of abscissae and ordinates are corresponding to quantity of ones of each n-plet using a pair of sub-alphabets. More info in text.

The described method of two-dimensional analysis seems to be useful for the study of hidden regularities in chromosomes and also for classification and comparative analysis of different genetic sequences with possible applications in biotechnology and medicine.

Multidimensional structure of chromosomes is represented in a computer binary format suitable for analysis using binary artificial neural networks. Multidimensional analysis of chromosomes nucleotide composition could help in understanding deeper genetic phenomena. For effective resistance to the new and dangerous diseases and viruses science needs an innovations in the field of analysis of genetic coding and corresponding artificial intelligence systems.

The discovery of new binary fractal-like patterns, which are revealed from long genetic sequences of chromosomes provokes many questions about relations between the genetic system and those fields of science and technology, where digital binary fractals are used, for example, fields of radiophysics, technology of fractal antenna, digital fractal codes, etc.

3.3   1-D Analysis of Chromosomes Nucleotide Composition

To reach 1-dimensional analysis of chromosomes nucleotide composition all of genetic binary-opposition sub-alphabets can be used. Thus, the using of separated 1-dimensional X, Y and Z Cartesian axises giving three different mappings using corresponding sub-alphabets of physical chemistry properties of chromosome. On a resulting figure three rows for three sub-alphabets are shown for a human chromosome.

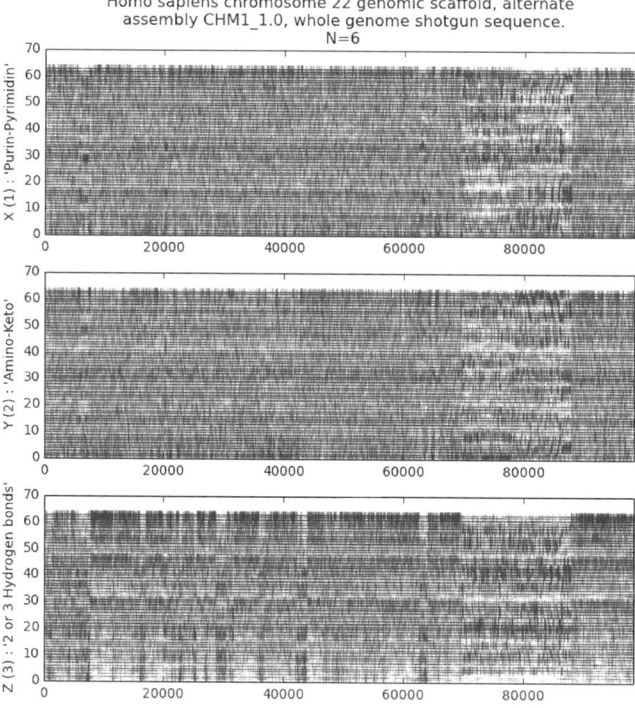

Fig. 6. Illustration of 1-dimensional 3-channel representation of Homo sapiens 22 chromosome nucleotide composition. Each of three rows corresponds to binary-opposite sub-alphabet. Axis of abscissae is an index number of n-plet, axis of ordinates is a decimal representation of binary number of n-plet.

Special regions with different nucleotide composition has a clear character and well seen on Fig. 6 with n = 6. This special regions are having distinguishable nucleotide state that can be visualized in 2D parametric spaces algorithmically.

4   Discussion and Results

General scientific methods for studying nucleotide sequences usually concentrate their attention on those fragments (or N-mers, or N-plets), which exist inside the sequences. The described method investigates a deficit of different types of N-plets in nucleotide sequences with geometrical representation. We found the high noise stability of geometrical patterns on the base of Walsh functions: changing parameter n, reversing, different types of remixing genetic sequence in chromosome are saving stable the general visualization of structure. In our research we found that even small parts of chromosome are repeating general pattern of all chromosome that is correlated with theory of noise-immunity coding.

In's interesting to note, that some fractals are similar for different organisms and some are different. This obsrevation is not always corresponds to the logic of known phylogenetic tree [19]. Comparison analysis of chromosomes of different species including humans, different monkeys, crocodile, bacterias was carried out. We also note that it's possible to display DNA as 2-D Sierpinski triangles with condition of using of Walsh functions with special kinds of errors. Results are close to non linear dynamics, chaos & fractals theory, binary code principle and regulation, modulo two addition, p-adic number systems, holography etc. As mentioned, only two of three sub-alphabets are enough for 2-D visualization that is regarded to property of Hadamard matrix. In this connection the most interesting mosaics were constructed on the coordinate system of pyrimidine/purine and amino/keto properties – the binary-opposite sub-alphabets not related to hydrogen bonds information.

5   Conclusions

Scientific novelty among other things lies in the multidimensional approach to visualization of the genome. We assume that multidimensional representations are associated with the immune system, which operates on the principle of recognition of patterns of nucleotide composition, which can be different for different types of living organisms, including parasites and their carriers, etc.

Acknowledgments. Part of the calculations was performed on the supercomputer "MVS-10P" (JSCC RAS). Author thanks Sergey Petoukhov and Vitaly Svirin for scientific discussions.

References

1. Petoukhov, S.V., He, M.: Symmetrical Analysis Techniques for Genetic Systems and Bioinformatics: Advanced Patterns and Applications, 271 p. IGI Global, Hershey (2010)
2. Chargaff, E., Lipshitz, R., Green, C.: Composition of the deoxypentose nucleic acids of four genera of sea-urchin. J. Biol. Chem. **195**(1), 155–160 (1952). PMID 14938364

3. Scaria, T., Christopher, T.: Microarray gene retrieval system based on LFDA and SVM. Int. J. Intell. Syst. Appl. (IJISA) **10**(1), 9–15 (2018). https://doi.org/10.5815/ijisa.2018.01.02

4. Zhang, D., Wang, Y., Tao, W.: Epipolar geometry estimation for wide baseline stereo. Int. J. Eng. Manuf. (IJEM) **2**(3), 38–45 (2012). https://doi.org/10.5815/ijem.2012.03.06

5. Jayapriya, J., Arock, M.: A novel distance metric for aligning multiple sequences using DNA hybridization process. Int. J. Intell. Syst. Appl. (IJISA) **8**(6), 40–47 (2016). https://doi.org/10.5815/ijisa.2016.06.05

6. Hu, Z., Dychka, I., Sulema, Y., Valchuk, Y., Shkurat, O.: Method of medical images similarity estimation based on feature analysis. Int. J. Intell. Syst. Appl. (IJISA) **10**(5), 14–22 (2018). https://doi.org/10.5815/ijisa.2018.05.02

7. Moallem, P., Karimizadeh, A., Yazdchi, M.: Using shape information and dark paths for automatic recognition of touching and overlapping chromosomes in G-band images. Int. J. Image Graphics Sign. Process. (IJIGSP) **5**(5), 22–28 (2013). https://doi.org/10.5815/ijigsp.2013.05.03

8. Feldman, D.P.: "17.4 The chaos game", Chaos and Fractals: An Elementary Introduction, pp. 178–180. Oxford University Press (2012). ISBN 9780199566440

9. Petoukhov, S.V.: Genetic coding and united-hypercomplex systems in the models of algebraic biology. Biosystems **158**, 31–46 (2017)

10. Stepanian, I.V., Petoukhov, S.V.: The matrix method of representation, analysis and classification of long genetic sequences. http://arxiv.org/pdf/1310.8469.pdf

11. Mousa, H.M.: DNA-genetic encryption technique. Int. J. Comput. Netw. Inf. Secur. (IJCNIS) **8**(7), 1–9 (2016). https://doi.org/10.5815/ijcnis.2016.07.01

12. Chandel, A., Sood, M.: A genetic approach based solution for seat allocation during counseling for engineering courses. Int. J. Inf. Eng. Electron. Bus. (IJIEEB) **8**(1), 29–36 (2016). https://doi.org/10.5815/ijieeb.2016.01.04

13. Wu, J., Zhang, W., Jiang, R.: Prioritization of candidate nonsynonymous single nucleotide polymorphisms via sequence conservation features. Int. J. Eng. Manuf. (IJEM) **1**(5), 66–72 (2011). https://doi.org/10.5815/ijem.2011.05.09

14. Balonin, N.A., Balonin, Y.N., Djokovic, D.Z., Karbovskiy, D.A., Sergeev, M.B.: Construction of symmetric Hadamard matrices. https://arxiv.org/abs/1708.05098

15. Georgiou, S., Koukouvinos, C., Seberry, J.: Hadamard matrices, orthogonal designs and construction algorithms. In: Designs 2002: Further Computational and Constructive Design Theory, pp. 133–205. Kluwer, Boston (2003). ISBN 1-4020-7599-5

16. Jeffrey, H.J.: Chaos game representation of gene structure. Nucleic Acids Res. **18**(8), 2163–2170 (1990)

17. Petoukhov, S.V., Svirin, V.I.: Fractal genetic nets and symmetry principles in long nucleotide sequences. Symmetry Cult. Sci. **23**(3–4), 303–322 (2012)

18. Rudner, R., Karkas, J.D., Chargaff, E.: Separation of B. SubtilisDNA into complementary strands. 3. Direct analysis. In: Proceedings of the National Academy of Sciences of the United States of America, vol. 60, no. 3, pp. 921–922 (1968). https://doi.org/10.1073/pnas.60.3.921. PMC 225140

19. Townsend, J.P., Su, Z., Tekle, Y.: Phylogenetic signal and noise: predicting the power of a data set to resolve phylogeny. Genetics **61**(5), 835–849 (2012). https://doi.org/10.1093/sysbio/sys036. PMID 22389443

Computer Science and Education

The Student Training System
Based on the Approaches of Gamification

Nataliya Shakhovska$^{(\boxtimes)}$ ⓘ, Olena Vovk, Roman Holoshchuk ⓘ,
and Roman Hasko

Lviv Polytechnic National University, Lviv, Ukraine
{nataliya.b.shakhovska,olena.b.vovk,
roman.o.holoshchuk}@lpnu.ua, r.hasko@gmail.com

Abstract. The paper discusses a virtual computer game laboratory named SAUDAI as an innovative method of teaching IT students that will encourage them to learn and generate students' interest in learning and research. The system is designed to provide students with convenient environment for development and testing of artificial intelligence and motivate students to research, develop, and deepen their knowledge of artificial intelligence, theory of algorithms and decision support systems. The essence of this work is gamification of studying information technology, specifically, the development of a system that allows the user to develop a software solution that can act as a player in a number of computer games. While developing artificial intelligence for each game, the student acquires knowledge of relevant information technology, each game system set up so that the student can acquire and use the skills on the specific topic.

Keywords: Training system · Artificial intelligence · IT · Game

1 Introduction

Recently, in Ukraine and European countries, there is a tendency for students to lose interest in research and learning in general [1, 2]. The main reason for this development is lack of motivation. Students do not have the opportunity to concentrate on their studies. Moreover, development of information technology is changing rapidly. First, this applies to students from areas related to information technology. The scope of IT is innovative, and it is natural that traditional teaching methods such as lectures and books are not always effective. It is to encourage such students to study and research that it is necessary to update the traditional system, making it more interesting and relevant. In this complex issue, gamification comes to the help.

Methods of gamification are a good way of student motivation. In addition, it can build up such human traits as confidence and self-expression. The article considers a system in which students can learn the curriculum by creating so-called bots for games. Thus, we developed the software product, which realizes a competition of bots under the given circumstances. The education process is given in a form of game and bots' competition organization. It allows us to improve soft skills and organize teamwork.

Z. Hu et al. (Eds.): ICCSEEA 2019, AISC 938, pp. 579–589, 2020.
https://doi.org/10.1007/978-3-030-16621-2_54

2 Theoretical Background

To begin with, our goal is the encouragement of modern youth to conduct independent scientific research, educate among them highly qualified specialists [2–5], as well as the use of innovative information technologies for distance learning, for example, artificial intelligence systems [6–8].

One method for students' evaluation is fuzzy approach [7], however the possibility to use such approach at high school is not described.

Works [12, 13] describe agent technologies for education process, but it is more efficient to allow students to create their own agent or bot.

Education process can be presented as ontology [14]. The main issue in this case is evaluation process describing.

In this paper, a virtual computer lab of interactive games system is considered that allows users to test their programming and algorithmization skills. The main application of the system is to replace the familiar scheme of checking students' computer knowledge, making the learning interesting and effective. The soft-skills and teamwork should be improved. For this purpose, we use gamification.

Solution of this problem is a well-known method to use games in education. According to the Robeson "Is it all a game? Understanding the principles of gamification" [9, 10], this method improves such human traits as confidence and self-expression.

Competition is one of the main elements of gaming. In addition, it is the best way of encouraging students to study.

In particular, popular and understandable examples of gaming are honor boards, competitions between classes for a symbolic "currency", use of game elements directly during the lessons and the like.

The core of the gamification strategy is a reward for the completed tasks. In detail, [11], discuss the issue of gaming in a study. In this study, various methods of gaming and their effectiveness are analyzed in detail. The results showed that gaming is positively affecting the work of students, but the conditions are very important, as well as the potential of users (students' specialties).

The idea of writing bots is not new [15, 16]. For example, there are qualitative resources like theaigames.com, where users can write bots to various games, after which they will compete among themselves. Also there is a resource www.botchief. com, where the user is provided with tools for creating Internet bots that reproduce the actions of a person in social networks or on other sites. Nevertheless, these resources have nothing to do with education; the idea of combining the writing of bots with training is relevant.

Now, the main analogue of the traditional training system is SAUDAI - virtual computer lab of interactive games and student evaluation.

In general, students have access to all necessary information like lectures, theory to laboratory work. However, practice shows that very rarely students attend all lectures, write down theory and get prepared for labs. The defense of the labs is also extremely undesirable for students. Some students cannot show their real knowledge or abilities in oral defenses, because of their shyness, or the inability to formulate a thought clearly. Even if a student has good programming skills and a good knowledge of the topic, the teacher does not always have the opportunity to properly estimate it. In addition, the defense of the labs takes a very long time, whereas students can spend it on something more productive.

The **objective of this paper** is discussing key features of the architecture of student training system demonstrated in the form of virtual computer lab games.

3 The Results and Discussion

3.1 The Education System as a Set of Agents

The education system is presented as $\langle L_t, L_s, C, E \rangle$, where L_t is model of knowledge, presented as semantic network; L_s is model of knowledge, interpreted by the teacher; C is algorithm of knowledge management; E is the goal of education.

The education process is built as competitions of students' agents or bots. We have such characteristics of agents (bots) as

- Performance Measure of Agent is the criterion which determines how successful an agent is (degree of similarity to L_t);
- Behavior of Agent is the action that agent performs after any given sequence of percepts;
- Percept is agent's perceptual inputs at a given instance;
- Percept Sequence is the history of all that an agent has perceived till date;
- Agent Function is a map from the precept sequence to an action.

It is necessary to develop a common architecture of the system that will complement the existing system in the university, and which will provide users with a constant and quick access to necessary materials, for example lecture notes, articles, video lectures, and simplify, speed up and make more objective assessment of knowledge, increase students' interest in education. The basis of such a system should be the analysis of the competences of users on the basis of methods of processing information resources.

In contrast to the existing educational portals such as UDACITY, Coursera, EDX, combining video explanation of the teacher, homework and tests to check students' knowledge, we propose an opportunity to get research experience in finding new ways of solution of tasks and interactive work between team members and their mentors or other participants. There are such teaching programs, which use gamification:

- Codecademy – teaching programming in JavaScript, HTML, Python, Ruby.
- Motion Math Games – Mobile Games make learning math fun and exciting.
- Mathletics – program for schools aimed at attracting children to mathematics through games and the Challenge.
- Khanacademy – free video courses on various subjects.
- Spongelab – personalized platform for science education.
- Foldit – the solution of scientific problems as puzzles.

We propose to combine approaches from Codeacademy, Mathletics and Spongelab. Education process will be conducted in the form of a game. Therefore, we propose to use gamification for practical work (see Fig. 1). The first stage of game is to study known methods (game with mentor). For example, we propose to use labyrinth for deep search learning. Second level (exploratory level or level of fire between students' teams) is used to create game elements and use them to race with other students from different universities. We developed BrainRacing game, which consists of such steps:

1. There is a car on the road. Two players should solve different tasks to move the car. When the car gets to one player's base – the player wins.
2. Players program a car to move in their direction and rewrite it, if the car is already programmed.

The games platform architecture is shown in Fig. 1. The "bot" means algorithm created by student.

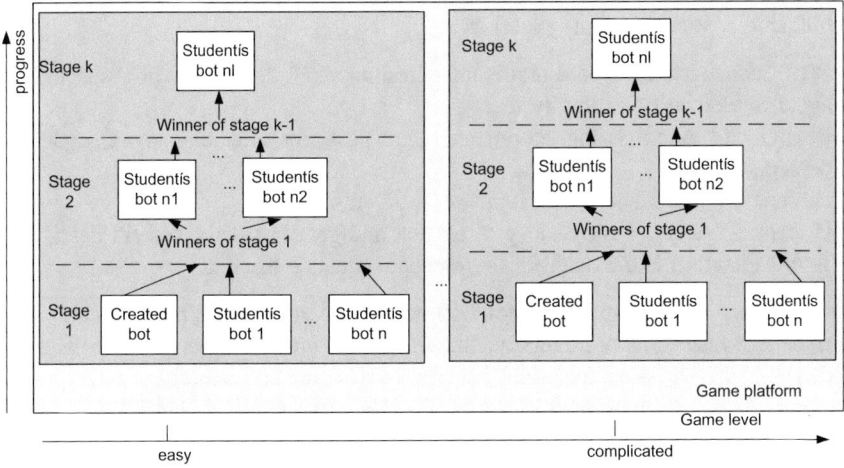

Fig. 1. The explanation of gamification usage

The technical approach and comparison with other applications is shown in Fig. 2.
The application of AND-OR graph, expert system and methods of neural networks will allow to evaluate the students' knowledge and identify the most appropriate training trajectory in the form of a game.

3.2   The System Description

The key concepts of created system are competence and learning outcomes in the process of creating a new educational program.

Stakeholders of the precedent and their requirements are follows:

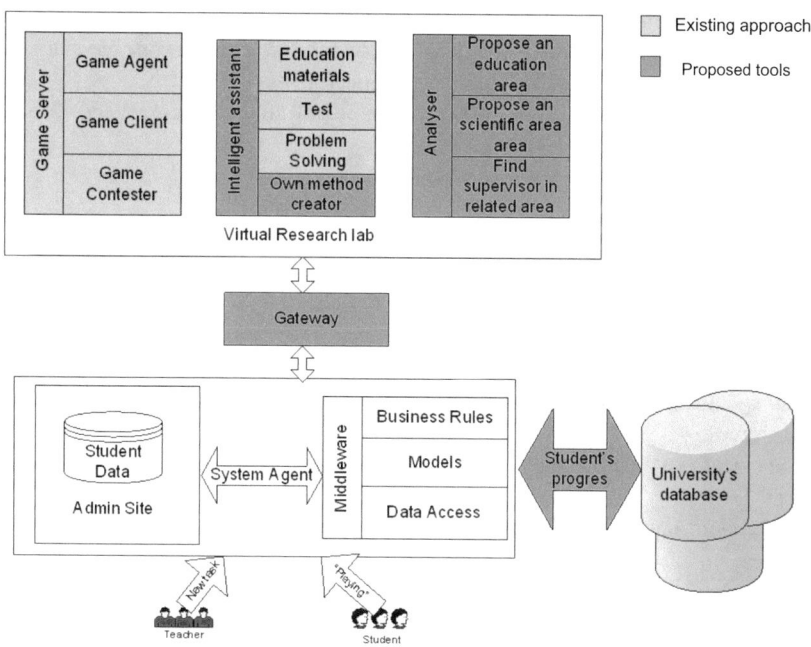

Fig. 2. Differences between project design and existing solutions

- Administrator: should post games on the site, and create competitions.
- User: sends a software solution (bot) to the game and to the competition.
- The system must send the bot to the agent, then get the results of the games, display these results, and show the playback.
- Agent: after receiving the game and bots, it should hold it with the received bots, get the results and return the system to them, and save each game progress so that the system can play it.

Program agent (PA) user is the main actor of this precedent. This separated system interacts with all other actors:

- Receives the game from the administrator, the bot from the user,
- Sends them to the agent,
- Receives the results of the game from him,
- Displays them, and reproduces the game.

Preconditions: the system must be active; the game and the agent must be well defended and run without errors, regardless of which bots the user directs; the user is registered, and the bot must pass the plagiarism check.

The main successful scenario:

1. The administrator adds the game to the system;
2. The administrator sets the rules of the game;
3. The system adds the game and rules to it in the list of games on the site;
4. The administrator creates a new competition for the game;
5. The system adds a new competition to the list of competitions on the site;
6. The user adds bot to the system;
7. The user sends the bot to the competition;
8. The system receives the bot from the user;
9. The system looks for the active agent;
10. The system sends the game and received bots to a free agent;
11. The agent receives the game and bots;
12. The agent launches the game, hands over the bots to it;
13. The agent receives the result of the game and a file with the record of all moves;
14. The agent transfers to the system the result of the game and a file with the record of all moves;
15. The system receives data from the agent and displays the results of the game on the site;
16. The system reproduces the course of the game on the site on a file with recorded moves;
17. The system records the statistics of the results of the games of various users and analyzes the level of competences of the participants in the project.

Expansion of the main scenario or alternative flows:

- A negative check on plagiarism - a bot that is not unique is not downloaded to the site, an attempt to upload it is recorded in the user profile, the user receives a corresponding message;
- When adding a bot to the competition:

a. The user tries to send the bot to the competition, which is closed: the PA notifies the user about the impossibility to add the bot to the selected competitions, since it is not open and returns it to the contest selection page.
b. The mismatch between the bot and the game for which the competition is created: the PA notifies the user about the impossibility of adding the bot to the selected competition, since it concerns the wrong game and returns it to the contest selection page.

- Cannot find an available agent:

a. No active agent: the system adds an agent request to the waiting queue, where it will stay until the active agent appears.

b. No free agent: the system waits until one of the agents finishes the game and sends to it a new request.

- Unable to get results: the administrator reports an agent error; the system displays information about the temporary unavailability of the results;
- Cannot build the playback:

a. There is no script to build this game: instead of playing, a message is displayed about the temporary impossibility of displaying the game.
b. Playback is not built: instead of playing, a message is displayed about the temporary impossibility to show the game, the administrator reports an error.

Post conditions:

- The result of the competition is found on the game's play site - it can be viewed using the script at any time;
- The evaluation of the efficiency of each user's bot is stored in the database, displayed in the user's profile.

Special conditions:

- The agent must complete the game and send the results regardless of whether the bot is working correctly or not;
- It is necessary to establish parallel connection of the system with several agents.

List of necessary technologies and additional devices:

- The PA must be designed as a Web-based system;
- Agents are on separate servers and connect to the system via the Internet;
- The database is located on a separate cloud service, the site always has access to it;
- The site is located on the cloud service;
- The user uses a computer, a browser.

The use case diagram gives the most general idea of the functional purpose of the system (Fig. 3).

The implementation of a separate use case requires the participation and interaction of certain instances of actors and classes. The most suitable tool for describing such an interaction is the sequence diagrams (Fig. 4) [6].

The user is authorized, adds the solution and sends the bot, and then waits for the results. While waiting, the system sends the bot and game to the agent, the agent starts the game and passes it to the bot. The game works based on the agent communicates with the bot and returns to the agent the results of the bot's activity. After that, the agent returns these results to the system, and the system displays them to the user.

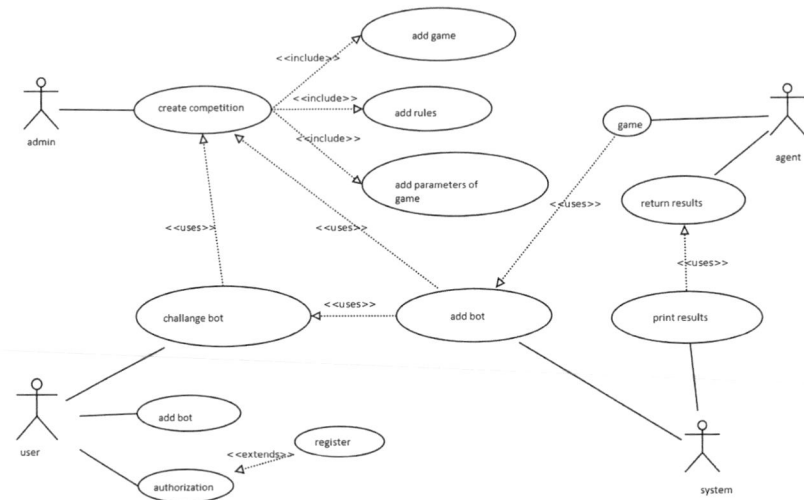

Fig. 3. Diagram of system usage options

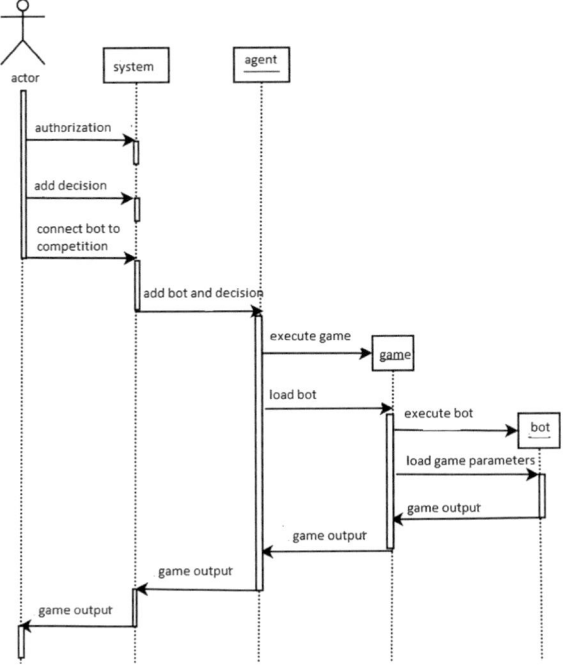

Fig. 4. Diagram of the sequence of participation in the competition

4   Results and Discussion

Competitions were held between the students of Lviv Polytechnic National University and based on the results the students received grades (Figs. 5 and 6).

Two different competitions were organized. The fight between students was organized in the first round. The second round was competition between groups.

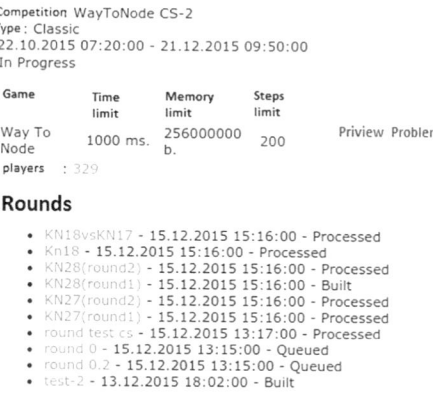

Competition WayToNode CS-2
Type : Classic
22.10.2015 07:20:00 - 21.12.2015 09:50:00
In Progress

Game	Time limit	Memory limit	Steps limit	
Way To Node	1000 ms.	256000000 b.	200	Priview Problem

players   : 329

Rounds

- KN18vsKN17 - 15.12.2015 15:16:00 - Processed
- Kn18 - 15.12.2015 15:16:00 - Processed
- KN28(round2) - 15.12.2015 15:16:00 - Processed
- KN28(round1) - 15.12.2015 15:16:00 - Built
- KN27(round2) - 15.12.2015 15:16:00 - Processed
- KN27(round1) - 15.12.2015 15:16:00 - Processed
- round test cs - 15.12.2015 13:17:00 - Processed
- round 0 - 15.12.2015 13:15:00 - Queued
- round 0.2 - 15.12.2015 13:15:00 - Queued
- test-2 - 13.12.2015 18:02:00 - Built

Fig. 5. Competitions

KN28(round2)

Processed
15.12.2015 15:16:00
Games processed: 110 / 110

Vova(Goein lab 1.2)	475
Stepan(not rand 00)	453
Юлія(Solution 13:01 13.11.2015)	267
skittles(No Solution)	67
Lyudmula(No Solution)	53
Влад(No Solution)	0
Sergiy(No Solution)	0
Kimmi76 (No Solution)	0
ZaraRk(No Solution)	0
zagrava123(No Solution)	0
Genek0133(Genek0133)	0

Fig. 6. Results of the competitions of one of the groups

5   Conclusions and Prospects for Further Research

The architecture of training system is developed. The education process is given in a form of game and bots' competition organization. It allows us to improve soft skills and organize teamwork. The main users and activities are described. By creating artificial intelligence for each game, the student learns from the relevant field of information technology. The Advantage of the online system is constant access to all materials in different forms. The prospects of further research are organization of bots' competition for all technical universities in Ukraine and for triple-learning [17–22].

References

1. Shakhovska, N., Vovk, O., Hasko, R., Kryvenchuk, Y.: The method of big data processing for distance educational system. In: Conference on Computer Science and Information Technologies, pp. 461–473. Springer, Cham, September 2017
2. Rashkevych, Y.M.: The Bologna Process and the New Paradigm of Higher Education. Vydavnytstvo Lvivska Politechnika, Lviv (2014). (in Ukr.)
3. System administrator leveling matrix. https://docs.google.com/spreadsheets/d/1FBr20VIOePQH2aAH2a_6irvdB1NOTHZaD8U5e2MOMiw/pub?output=html
4. Table of levels for the system administrator. http://habrahabr.ru/post/145148/
5. Matrix of programmer's competence. http://habrahabr.ru/post/37707/
6. Beltadze, G.N.: Game theory - basis of higher education and teaching organization. Int. J. Mod. Educ. Comput. Sci. (IJMECS) 8(6), 41–49 (2016). https://doi.org/10.5815/ijmecs.2016.06.06
7. Brunnet, N., Portugal, C.: Digital games and interactive activities: design of experiences to enhance children teaching-learning process. Int. J. Mod. Educ. Comput. Sci. (IJMECS) 8(12), 1–9 (2016). https://doi.org/10.5815/ijmecs.2016.12.01
8. Doman, M., Sleigh, M., Garrison, C.: Effect of gamemaker on student attitudes and perceptions of instructors. Int. J. Mod. Educ. Comput. Sci. (IJMECS) 7(9), 1–13 (2015). https://doi.org/10.5815/ijmecs.2015.09.01
9. Robson, K., Plangger, K., Kietzmann, J.H., McCarthy, I., Pitt, L.: Is it all a game? Understanding the principles of gamification. Bus. Horiz. 58(4), 411–420 (2015)
10. Bobalo, Y., Stakhiv, P., Shakhovska, N.: Features of an eLearning software for teaching and self-studying of electrical engineering. In: 2015 16th International Conference on Computational Problems of Electrical Engineering (CPEE), pp. 7–9. IEEE, September 2015
11. Hamari, J., Koivisto, J., Sarsa, H.: Does gamification work?–a literature review of empirical studies on gamification. In: 2014 47th Hawaii International Conference on System Sciences (HICSS), pp. 3025–3034. IEEE, January 2014
12. Yi, Z., Zhao, K., Li, Y., Cheng, P.: Remote intelligent tutoring system based on multi-agent. In: 2010 2nd International Conference on Information Engineering and Computer Science (ICIECS), pp. 1–4. IEEE, December 2010
13. Sungkur, R.K., Pudaruth, S., Mahomudally, J., Sookaloo, A.: Measuring students' uncertainty in a tutoring system using agent technology. In: 2014 IEEE International Advance Computing Conference (IACC), pp. 937–940. IEEE, February 2014
14. Salukvadze, M.E., Beltadze, G.N.: The optimal principle of stable solutions in lexicographic cooperative games. Int. J. Mod. Educ. Comput. Sci. (IJMECS) 6(3), 11–18 (2014). https://doi.org/10.5815/ijmecs.2014.03.02
15. Albilali, A.A., Qureshi, R.J.: Proposal to teach software development using gaming technique. Int. J. Mod. Educ. Comput. Sci. (IJMECS) 8(8), 21–27 (2016). https://doi.org/10.5815/ijmecs.2016.08.03
16. Korkmaz, Ö.: The effects of scratch-based game activities on students' attitudes, self-efficacy and academic achievement. Int. J. Mod. Educ. Comput. Sci. (IJMECS) 8(1), 16–23 (2016). https://doi.org/10.5815/ijmecs.2016.01.03

17. Hasko, R., Shakhovska, N.: Tripled learning: conception and first steps applications. In: CEUR Workshop Proceedings 2105, pp. 481–484, May 2018
18. Oleksiv, I., Izonin, I., Kharchuk, V., Tkachenko, R., Doroshenko, A.: Identification of IT sector stakeholder's requirements to masters program in information system in Lviv region. In: CEUR Workshop Proceedings 2105, pp. 429–437, May 2018
19. Kunch, Z., Kharchuk, L., Syerov, Y., Fedushko, S.: Development of concept of terminological online assistant for electric power engineering specialists. In: 2017 12th International Scientific and Technical Conference on Computer Sciences and Information Technologies (CSIT), vol. 1, pp. 83–86. IEEE, September 2017
20. Maslovskyi, S., Sachenko, A.: Adaptive test system of student knowledge based on neural networks. In: 2015 IEEE 8th International Conference on Intelligent Data Acquisition and Advanced Computing Systems: Technology and Applications (IDAACS), vol. 2, pp. 940–944. IEEE, September 2015
21. Lendyuk, T., Rippa, S., Sachenko, S.: Simulation of computer adaptive learning and improved algorithm of pyramidal testing. In: 2013 IEEE 7th International Conference on Intelligent Data Acquisition and Advanced Computing Systems (IDAACS), vol. 2, pp. 764–769. IEEE, September 2013
22. Lendyuk, T., Rippa, S., Bodnar, O., Sachenko, A.: Ontology application in context of mastering the knowledge for students. In: 2018 IEEE 13th International Scientific and Technical Conference on Computer Sciences and Information Technologies (CSIT), vol. 2, pp. 123–126. IEEE, September 2018

Electrical Engineering Disciplines Teaching System for Students with Special Needs

Yuriy Bobalo[1], Petro Stakhiv[1,2], Nataliya Shakhovska[1(✉)] [iD],
and Orest Hamola[1]

[1] Lviv Polytechnic National University, Lviv, Ukraine
rector@lp.edu.ua, petro.stakhiv@p.lodz.pl,
{nataliya.b.shakhovska, orest.y.hamola}@lpnu.ua
[2] Technical University of Lodz, Lodz, Poland

Abstract. This paper presents the features of an educational and methodological resource complex for students with special needs (disabilities). Approaches to the organization of the learning process of students with disabilities are considered. Different methods and techniques are proposed to deliver lectures, conduct practical and laboratory classes. The administrative environment of the virtual lab is given. The structure of a laboratory-working place and the schema of laboratory practice are build. A system of testing, which allows the generation of questions and selection of teaching material on the basis of analysis of answers and previous knowledge of the students has been developed.

Keywords: Students with special needs · Distance learning · Self-studying · Big data technologies · Tutors control

1 Introduction

For distance learning, a teacher-student relationship is very important. An interaction between participants of the educational process takes place online on scheduled days and at determined times. This enables students with disabilities to find answers to unclear questions.

The distance learning systems widely used by engineering students with special needs must meet the additional requirements, as the fulfillment of laboratory bench exercises is an integral part of the training of engineers, especially in electrical engineering disciplines.

Additionally, students with special educational needs have the opportunity to move forward on the individual learning trajectory, set their own goals, choose the optimal forms and pace of learning, apply the learning methods most relevant to each individual's characteristics.

Each disabled student suffers from particular features and limitations that make their learning difficult. To ease the way, it is necessary that the work with the students be supported by special methods, pedagogical techniques and adaptive technical teaching aids introduced into the process of learning. Typically, such students are very inquisitive and diligent, but experience certain problems: knowledge gaps, increased

Z. Hu et al. (Eds.): ICCSEEA 2019, AISC 938, pp. 590–599, 2020.
https://doi.org/10.1007/978-3-030-16621-2_55

fatigability, reserved demeanor, low self-esteem, and vulnerability. At the same time, they cannot slow down the pace of training or reduce the number of classes, because this reduces the quality of professional training. Therefore, the system of supervision of their training, comprehensive assistance and support is being implemented.

It is often difficult for students with hearing or the musculoskeletal system disorders to participate in the educational process in higher school. This especially concerns students of engineering specialties, since the training of engineers involves a large number of practical work in laboratories, assumes giving permission for exercises fulfillment, and making results presentation.

The training support of students with special needs should be carried out in the following areas: technical, methodical, and social. In the article, we shall focus on the technical and methodical one.

The technical support involves adaptive technological tools and special training technologies that enable one to build one's own learning trajectory. To do this, the article proposes the development of special laboratory benches, a virtual laboratory and system for individual training and self-control of knowledge.

The methodical support involves the optimization of delivery of teaching material in the form most convenient for a student with special needs. For students who have a decreased ability to see, technical teaching means should be equipped with voice interface cards (modules). Students with hearing disorders should be definitely provided with a textual presentation of information; for those with limitations of the musculoskeletal system, the fulfillment of laboratory bench works can be ensured by replacing them with virtual ones likely with elements of complemented and virtual reality.

The complex approach that consists of lectures providing, practice work organization and exams supporting is the purpose of this paper. The novelty of this approach is combining of traditional lections with virtual labs. In addition, it is possible to use augmented reality for materials presentation. 128 students including three students with disabilities are attended to experiment.

The structure of paper consists of related works overview, main methods description and results presentation.

2 Related Works

Nowadays, self-study as a form of acquiring knowledge is gaining its popularity both in Ukraine and in the world. It can be defined as an individual-directed activity towards the independent acquisition of knowledge and/or experience [1, 2]. In the Western world, this form has been existing for quite a long time, and is very popular among students and teachers because of its cost and educational efficiency.

The following technologies came to be used for self-study.

Moodle is a free, open system of distance learning. The system implements the philosophy of "pedagogy of social constructivism" and focuses primarily on organizing the interaction between the teacher and the students, although it is also suitable for organizing traditional distance courses, as well as for supporting a classroom-based mode of learning. Moodle is translated into dozens of languages, including partial translation into Ukrainian. The system is used in 175 countries of the world, as well as in

Ukraine. Lviv Polytechnic National University widely uses this system for the placement of teaching materials. What we should also note is that this project is open [3].

ATutor is a Learning Content Management System (LCMS). The program is easy to install, configure, and maintain by system administrators; teachers (tutor) can easily create and transfer teaching materials and run their online courses. And since the system is modular, it is open for the modernization and expansion of functional capabilities. It is used in Spain, Bulgaria, Serbia [4].

Claroline is a collaborative e-learning platform (Learning Management System) released under the Open Source license. The platform is used in more than 80 countries and translated into more than 30 languages [5].

Work [6] describes agent technologies for education process for students with disabilities. However, the disadvantage of the above technologies is that they work with "standard tests", i.e. an automatic generation of input data for tasks [7, 8] is not allowed. This leads to the fact that with a limited set of tasks and multiple self-study, students are likely not to solve problems, but guess the correct answers, which does not contribute to improving their skills. Thus, for today, there is not any software designed to deal effectively with the problem of self-study, i.e. to enable the students provided with a limited set of test tasks to solve new ones, the input data for which do not repeat.

The presented in [10–17] approaches describe local problems connected to e-learning, but they can't be used for electronics engineering disciplines studies without previous processing.

Education process can be presented as ontology using Big data approach [18]. The main issue in this case is evaluation process describing.

Therefore, the development of methodological and instrumental foundations for the modelling of automated systems of learning management and knowledge control, appropriate to modern trends in the development of information technologies and to didactic principles of the organization and conduct of teaching activity, becomes relevant.

The purpose of the paper is to develop the architecture of an educational and methodological complex for students with special needs based on big data.

3 Materials and Methods

The main tasks of the information technology of the distance training and counselling center for the disabled based on the Big data technology are

- The organization and implementation of a complete educational process,
- Combining information about the students who are to perform physical experiments and laboratory exercises in various laboratories, their level of theoretical knowledge and self-control measures using an electronic textbook.

It is also necessary to receive information from medical centers to identify the peculiarities of working with this or that student and to provide them with the appropriate academic facilities (sound, image, text, tactile means, etc.). Additionally, the system can be used for remote control of knowledge and become a didactic tool for self-study using Big data technology [18] or virtual lab [9, 10]. The database of teaching materials combines information from electronic textbooks, instructions to laboratory, practical or independent works.

To study electrical engineering disciplines, it is necessary to provide the availability of all forms of classes:

- Lectures,
- Seminars,
- Practical classes,
- Projects (individual or in team),
- Laboratory practicum,
- Final classes.

Lectures are intended either for a complete presentation of learning material (study) or for a brief statement (independent work), and can be presented in the form of text, videos, recordings, etc.

Practical classes include some theoretical material to explain the problem-solving principles, to identify regularities, etc., and the application of automated means for the generation of tasks and the check as to the correct students' solutions.

Laboratory practicum include the use of computer simulation software, a development environment medium, etc. for the student to complete a practical task, and the automated check as to the correct results obtained.

Seminars and project must be evaluated after oral presentation and can't be simulated.

Final classes are used to evaluate the student's level of theoretical and practical training. Accordingly, test assignments of different levels of complexity are used (including multiple-choice tasks), tasks (entering answers by using the keyboard), graphical tasks (graphics package to create diagrams, charts, etc.).

In order to ensure that the number of allocated credits is consistent with the amount of learning activities, it is necessary to determine:

- Time to be taken for performing each type of task within each topic;
- Time to learn theoretical material;
- Time to carry out laboratory works (practical works).

An important component of the learning process is lectures as they in combination with knowledge self-assessment tools lay the foundation for studying a discipline. These facilities are likely to be universal for both academic disciplines and categories of students with disabilities. Based on the three categories of students experiencing movement disorders, speech and hearing disorders, vision deficiency, we can recommend the following approaches.

For the students with movement disorders (first category) to study theoretical material, the only requirement is to have access to the Internet or other digital storage media (flash memory cards, disks, etc.). At the same time, appropriate software, such as described in [1], is required to assess or self-access knowledge, as well as the possibility to share information online in order to visualize a student-teacher dialogue.

The second category of disabled students (speech and hearing disorders) can succeed in the mastering of theoretical knowledge using standard computer input/output devices (keyboard, display) complemented by Internet procedures and video surveillance equipment. In some cases, when evaluating the level of knowledge, one can use special computer graphics-based tests, and in this way speed up the evaluation procedure.

The third category of students with disabilities, i.e. having vision deficiency, when learning material, requires the use of special equipment, in the first place speech analyzers and synthesizers, and a special set of tests for assessing knowledge. These devices, supplemented by the equipment recommended for the first category of students are sure to enable the student to get full access to theoretical knowledge and the level of their assessment. The complexity of implementing the proposed approach is primarily caused by the high cost of the above equipment. However, this disadvantage can be eliminated with the help of an assistant present in studying the material or assessing the level of knowledge.

People with special needs should be provided with necessary facilities for doing laboratory exercises and practical works when studying electrical engineering disciplines both in laboratories and remotely. When teaching disabled students, to conduct laboratory practicums is perhaps the most difficult task both in terms of technical support and in terms of methodological support. Equipment for deaf and blind, adaptive appliances and specialized training technologies are the key to the successful training of students with various physical disabilities (disability nosology). The appropriate technical support should compensate for the functional limitations of students with special needs for their perceiving and mastering learning material. The implementation of a set of activities for organizing a laboratory practicum in electrical engineering disciplines for students with special needs is carried out in the following directions:

- Provision of appropriate adaptive appliances;
- Introduction of special technologies and teaching methods;
- Individualization of learning with the participation of a teacher (tutor's assistance).

The structure of a laboratory working place contains a laboratory bench with cells for a universal power supply, a signal generator, and physical models of the objects under study, a computer and a oscilloscope's universal USB, web cameras (Fig. 1). Such a structure allows the strategy of a reasonable combination of physical and virtual experiments with the use of modern means of research into electromagnetic processes to be implemented.

The technical support of a laboratory practicum for students with special needs includes a shared classroom training technology, an individual classroom training technology, technical aids for distance work, and software. The organization of a laboratory practicum is carried out taking into account the peculiarities of the student's physical disabilities, namely, a convenient way of perceiving the educational material (visually, aurally, and tactilely). Based on this, students with special needs are divided into three groups: with movement disorders; with hearing disorders; with visual impairment. For each of these groups, an appropriate technical and methodological support is being developed.

Taking into account the physical disabilities of the students, a laboratory practicum is carried out in the computerized laboratory of electrical engineering and electronics of the Department of Theoretical and General Electrical Engineering LPNU or remotely.

For this purpose, there are several working places provided with convenient access, and equipment for hearing and visually impaired (deaf and blind) students.

In the case of a remotely-conducted laboratory practicum, the students install the necessary software for simulating electronic devices (Multisim, Orcad) on a home computer or so does the system administrator on a computer in a specialized room, or provides remote access to a computer in the laboratory of the Department of Theoretical and General Electrical Engineering, Lviv Polytechnic National University (LPNU).

The laboratory practicum is carried out with the participation of a teacher who provides an individual approach and differentiated training, i.e. takes into account the differences between students regarding their level of knowledge and limited physical abilities. The teacher organizes training under a certain scenario (plan), controls the comprehension of the learned material both personally and applying pre-programmed test tasks.

A virtual laboratory practicum consists in the interaction of the student with virtual laboratory equipment based on the simulation models of the studied physical processes. The program interface provides an interactive interaction between the student and the objects being studied.

To implement the proposed structure of the laboratory practicum, a computerized laboratory has been created. In the similar mode, laboratories with installed hardware and software for simulating analogue electronic circuits, a training materials database, a

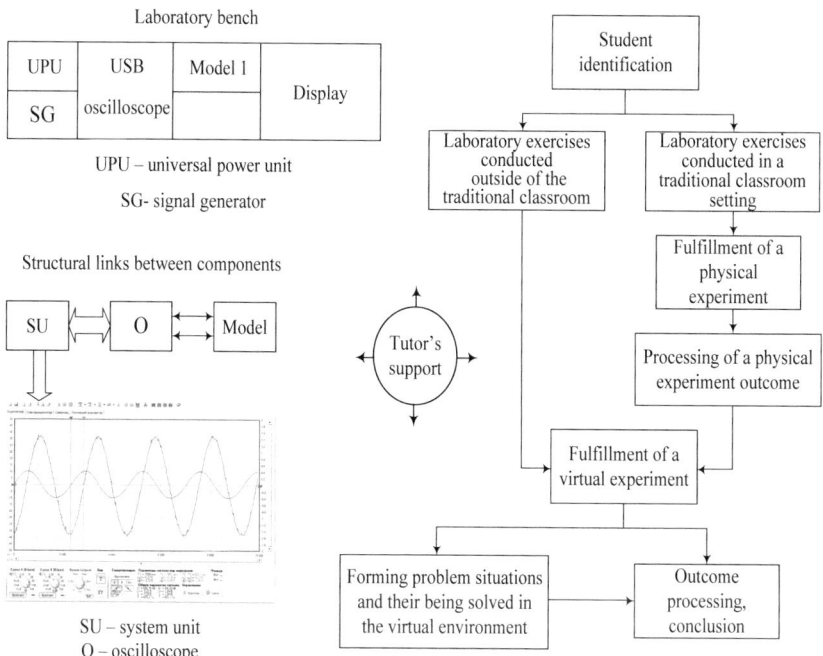

Fig. 1. Structure of a laboratory working place

training and content management server and a Web server are involved. The availability of laboratory Web-cameras (one per 2–3 working places) enables the teacher to observe the student's work.

The administrative environment of the virtual lab enables teaching materials and tests to be independently created and administrated by teachers [11, 12]. The designed system generalizes the specific character of the educational institution and is a means for effective implementation of personalized distance learning and remote knowledge control. The peculiarity of the designed system is the selection of ergonomic indicators of a student's working place according to his or her personal data, namely, medical grounds. Therefore, in the data warehouse of the system, it is necessary to provide data structures for storing the characteristics of the student's disease history. Another important factor of the system is to take into account students' learning outcomes when designing a procedure of delivering the teaching material to them.

The subsystem of testing is based on the principle of the "Question and Answer" dialogue, with the answer being chosen from a list of analogous ambiguously distributed options. Some questions are likely to be provided with the necessary references and assistance.

To implement the proposed structure of the laboratory practice a computerized laboratory was created, where the workplace is equipped with a computer, a universal USB oscilloscope and physical layouts of the objects under study (Figs. 2 and 3).

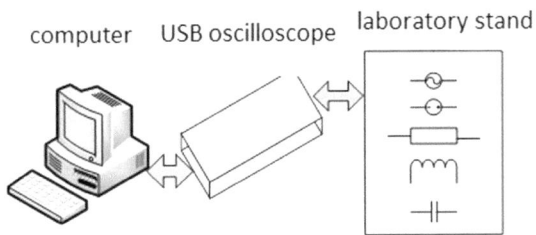

Fig. 2. The schema of laboratory practice

The computer is used to solve the following tasks:

- Realization of virtual laboratory works in the instrumental environment of applied programs for modeling of electrical devices;
- Realization of measurements by means of interaction through the program interface of a universal USB oscilloscope.

The universal USB oscilloscope is designed to measure analog signals with the ability to save measurement results in the form of vector or raster images or in a data file (binary or textual) for further analysis (Fig. 4). It has two channels, which allows simultaneous measurement of the voltages on the two components of the electric circuit relative to the common base node. The signals of these channels, which correspond to the measured voltages, are programmed, resulting in their parameters, including frequency, phase shift, are determined. To measure the USB current with an oscilloscope, it is necessary to

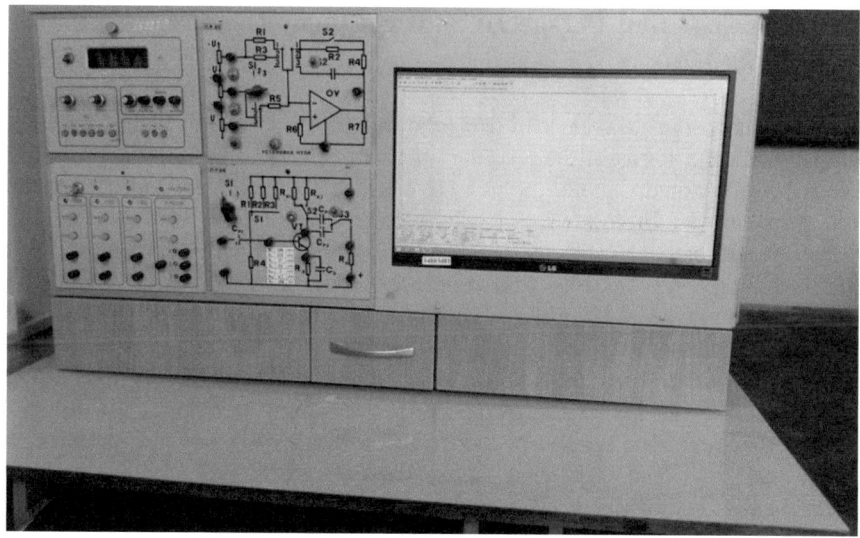

Fig. 3. The work place in laboratory room

activate a reference resistor with known resistance in the circuit. Thus, such oscilloscope is simultaneously used to measure voltage, current, phase shift, frequency, which will save on the purchase of voltmeter, ammeters, wattmeter, phase meters.

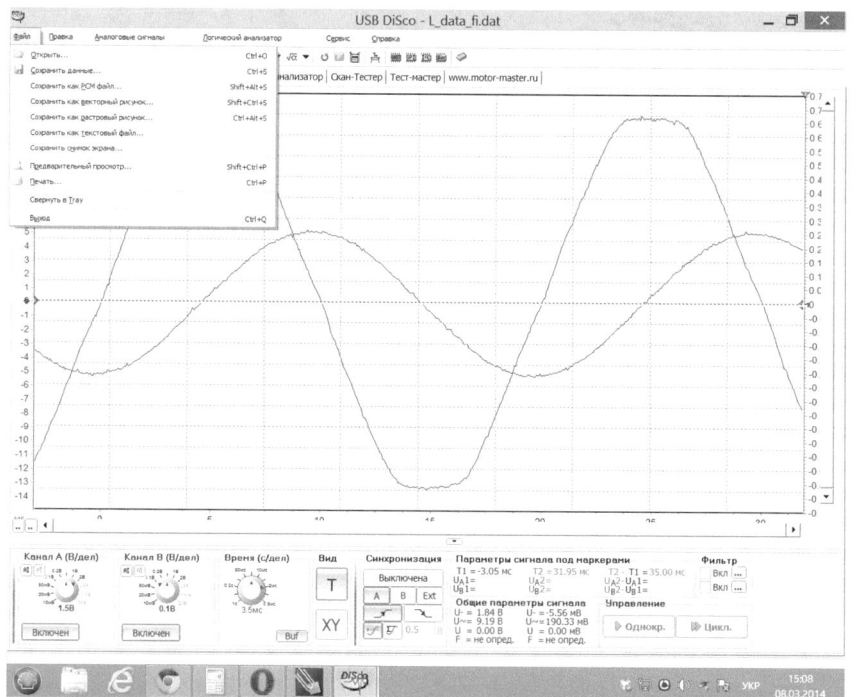

Fig. 4. The interface USB Disco 2 of universal oscilloscope

4 Conclusions and Perspectives of Further Scientific Developments

This paper shares our experience of using the basic approaches and technical means to arranging different forms of the learning process in studying the academic discipline "Theory of Electronic Circuits", which are oriented to certain categories of students with special needs. In contrast to traditional case studies, our approach allows to organize virtual practice work and can be used for students with disabilities. 128 students including three students with problems of the musculoskeletal system were involved to such type of teaching. A system of selection of ergonomic indicators for different categories of nosologies is developed. A system of testing, which allows the generation of questions and selection of teaching material on the basis of analysis of answers and previous knowledge of the students has been developed. The virtual lab based on the simulation models of the studied physical processes allows organizing practice work in augmented mode. The workplace is equipped with a computer, a universal USB oscilloscope and physical layouts of the objects under study.

Further research will be aimed at developing a laboratory workshop for people with reduced visual acuity.

References

1. Luppicini, R.: A systems definition of educational technology in society. Educ. Technol. Soc. **8**(3), 103–109 (2005)
2. Oleksiv, I., Izonin, I., Kharchuk, V., Tkachenko, R., Doroshenko, A.: Identification of IT sector stakeholder's requirements to masters program in information system in Lviv region. In: Proceedings of the Ph.D. Symposium at 14th International Conference on ICT in Education, Research, and Industrial Applications ICTERI 2018, Kyiv, Ukraine, 16 May 2018, pp. 429–437 (2018)
3. Galeev, I., Sosnovsky, S., Chepegin, V.: MONAP-II: the analysis of quality of the learning process model. In: Proceedings of IEEE International Conference on Advanced Learning Technologies (ICALT 2002), pp. 9–12, September 2002
4. Ananthanarayan, A., Balachandran, R., Grossman, R., Gu, Y., Hong, X., Levera, J., Mazzucco, M.: Data webs for earth science data. Paral. Comput. **29**(10), 1363–1379 (2003)
5. Greenagel, F.L.: The Illusion of e-Learning: Why We Are Missing Out on the Promise of Technology. League White Papers (2002)
6. Kasperskyi, A.: System of Knowledge Development in the Field of Electronics in a Secondary and Higher Pedagogical School. Drahomanov National Pedagogical University, Kyiv (2002). (Ukrainian)
7. Vysotska, V.: Linguistic analysis of textual commercial content for information resources processing. In: 2016 13th International Conference on Modern Problems of Radio Engineering. Telecommunications and Computer Science (TCSET), pp. 709–713. IEEE, February 2016
8. Roussel, N., Geiker, M.R., Dufour, F., Thrane, L.N., Szabo, P.: Computational modeling of concrete flow: general overview. Cement Concr. Res. **37**(9), 1298–1307 (2007)

9. Shakhovska, N., Vovk, O., Hasko, R., Kryvenchuk, Y.: The method of big data processing for distance educational system. In: Conference on Computer Science and Information Technologies, pp. 461–473. Springer, Cham, September 2017

10. Bobalo, Y., Mandziy, B., Stakhiv, P., Pysarenko, L., Yakymenko, Y.: Fundamentals of the theory of electronic circuits, 332 p. Publishing House of Lviv Polytechnic National University (2008). (Ukrainian)

11. Hasko, R., Shakhovska, N.: Tripled learning: conception and first steps applications. In: Proceedings of the Ph.D. Symposium at 14th International Conference on ICT in Education, Research, and Industrial Applications, ICTERI 2018, Kyiv, Ukraine, 16 May, pp. 481–484, May 2018

12. Nadeem, F., Mahgoub, S.: Student-centered role-based case study model to improve learning in decision support systems. Int. J. Modern Educ. Comput. Sci. (IJMECS) 6(10), 16–22 (2014). https://doi.org/10.5815/ijmecs.2014.10.03

13. Bobalo, Y., Stakhiv, P., Shakhovska, N.: Features of an eLearning software for teaching and self-studying of electrical engineering. In: 2015 16th International Conference on Computational Problems of Electrical Engineering (CPEE), pp. 7–9. IEEE, September 2015

14. Lashayo, D.M., Johar, M.G.M.: Preliminary study on multi-factors affecting adoption of e-learning systems in universities: a case of open University of Tanzania (OUT). Int. J. Modern Educ. Comput. Sci. (IJMECS) 10(3), 29–37 (2018). https://doi.org/10.5815/ijmecs.2018.03.04

15. Mohamed, S., Chebbi, M., Behera, S.K.: AMMAS: ambient mobile multi-agents system: simulation of the m-learning. Int. J. Modern Educ. Comput. Sci. (IJMECS) 9(1), 36–42 (2017). https://doi.org/10.5815/ijmecs.2017.01.04

16. Pal, S.K., Bhowmick, S.: Evaluation framework for disabled students based on speech recognition technology. Int. J. Modern Educ. Comput. Sci. (IJMECS) 9(10), 10–17 (2017). https://doi.org/10.5815/ijmecs.2017.10.02

17. Dominic, M., Xavier, B.A., Francis, S.: A framework to formulate adaptivity for adaptive e-learning system using user response theory. Int. J. Modern Educ. Comput. Sci. (IJMECS) 7 (1), 23–30 (2015). https://doi.org/10.5815/ijmecs.2015.01.04

18. Shakhovska, N.: Consolidated processing for differential information products. In: 2011 Proceedings of VIIth International Conference on Perspective Technologies and Methods in MEMS Design (MEMSTECH), pp. 176–177. IEEE, May 2011

Method and Model of Analysis of Possible Threats in User Authentication in Electronic Information Educational Environment of the University

V. A. Lakhno$^{(\boxtimes)}$ ⓘ, D. Y. Kasatkin ⓘ, A. I. Blozva ⓘ,
and B. S. Gusev ⓘ

Department of Computer Systems and Networks,
National University of Life and Environmental Sciences of Ukraine,
Kyiv, Ukraine
valss2l@ukr.net

Abstract. The article presents the research results of the development of a method and mathematical model of subject authentication in the electronic information educational environment of universities (EIEEU). The problem was solved on the basis of processing updated data sets with software and hardware detection tools. It is proposed to separate the verification procedures for new threats and precedents that have already been entered into the knowledge base of protection and cybersecurity systems of EIEEU. A method for analyzing threats is described, which involves the use of information sets characterizing each of the threats by features typical of their implementation, and takes into account the information content of each of the feat user. It is shown that the use of the proposed method and mathematical model of the subject authentication in the EIEEU, allows allocating resources of information protection systems and cybersecurity of educational institutions effectively. A reduction in the number of erroneous and false reactions while identifying threats, as well as an increase of the probability of detection of new cyber threats in the EIEEU, are confirmed.

Keywords: Cybersecurity · Information security ·
Electronic information educational environment of universities ·
Method · Model · Subject authentication

1 Introduction

The achieved level of development and scale of the use of information technologies (IT) and systems (ITS) in educational institutions (EI), and first of all in large universities, makes experts think about information protection issues (IP) and cybersecurity (CrS) of such systems at the stage of their design [1, 2]. The papers [2–4] note that the cybersecurity systems (CrSS) of educational institutions, in particular, the electronic information educational environment of universities (EIEEU), cannot guarantee the impossibility of external unauthorized penetration into the EIEEU. The protected data that are stored and circulate in the EIEEU can include [2, 4–6]: personal

© Springer Nature Switzerland AG 2020
Z. Hu et al. (Eds.): ICCSEEA 2019, AISC 938, pp. 600–609, 2020.
https://doi.org/10.1007/978-3-030-16621-2_56

data of students (students), teachers, employees; digitized information representing the intellectual property of the educational institution; information arrays providing the learning process (for example, multimedia content, databases, training programs); etc.

This information can be an object of theft or distortion from the side of external (internal) computer intruders or from students' or employees' hooligan motives.

As many authors show [7–9], the authentication of subscribers (users) in the EIEEU still remains a reliable and relatively affordable information protection tool (IPT) in EI from unauthorized access (UAA) and intervention in work. It should be noted that the existing IPTs are often built on fairly complex algorithms analyzing the characteristic realizations of subject features (as an object of observation), in particular in the EIEEU. In order to increase the security of EIEEU, strict verification rules are often used. According to this approach to the security policy (SP) in the EIEEU, a sufficiently high level of security and reliability of the system can be achieved, but a probability of false alerts and reactions to false threats to the EIEEU will also increase. This, in turn, will require additional material and other resources to enhance protection.

All of the above caused the relevance of the presented research results on the development of a method and model for identifying threats of UAA to the electronic information educational environment of universities. In this case, the focus is on minimizing the erroneous results of testing subscribers, which will allow one to increase the effectiveness in identifying new threats in the EIEEU.

2   Literature Review and Problem Posing

The papers [7–9] present the results of studies on the development of methods and models for increasing the effectiveness of the classification of subscribers' behavior in the networks of organizations and enterprises.

The authors [10, 11] note that the possibility of theft or misuse of legitimate subscriber credentials is a potential weakness of cybersecurity systems of many computer networks and systems of enterprises and organizations. However, the authors admit that solving the problem is possible by tracking patterns of the network traffic. This will give a potential to detect unauthorized use of subscriber credentials. But these studies are largely of an initial nature and need experimental verification.

The work [12] presents the results of research in the area of subscriber authentication as a set of graphics specific for subscriber. In particular, an aspect of attribute analysis with a time limit is considered, which allows using new possibilities of analyzing the subscriber's behavior in the network. The research is still ongoing and requires extensive experimental testing.

Thus, a problem of improvement of existing and development of new methods and models for identifying threats of unauthorized access to the electronic information educational environment of universities, with focusing on the tasks of minimizing the erroneous results of testing subscriber actions remains still relevant. This should increase the effectiveness of identifying new threats in the EIEEU and effectively allocate the resources of IPT and cybersecurity of educational institutions.

3 Purpose and Objectives

The purpose of the research is to develop methods and mathematical models that are applicable to the procedures for detecting threats of unauthorized access when authenticating subscribers in the electronic information educational environment of university.

To achieve the goal, a method for analyzing data on possible threats in the EIEEU, which will allow minimizing the time of their recognition, and a mathematical model of authentication of subscribers of EIOSU, which allows reducing the number of false reactions and reports of false threats, were developed.

4 Methods and Models

While forming the sets involved in the detection of threats (hereinafter SDT – a set for detecting threats) for the EIEEU, the following was taken into account:

1. The detector should not be activated in legitimate actions of subscribers (users in the EIEEU). Also the detector should not be activated in legitimate actions of objects and subjects of IPT and CrS in the EIEEU;
2. The intervals of values of implementations of signs that correspond to objects of detection should be sufficient to minimize identities. We assume that the objects used for training the detection systems were obtained using the methods and models described in [13–15].
3. If the copy of the set participating in detecting a threat of EIEEU, has successfully recognized it, then this copy is stored in the IPT knowledge base and then participates in producing new generations of objects used for training the CrP systems.

After analyzing the existing methods and algorithms of authentication of subscribers of the EIEEU, an improved scheme was developed, which is shown in Fig. 1.

Two streams are used on the scheme (see Fig. 1). Stream №1 is designed to search in the database of threats that have already been encountered in the IPT knowledge base of a specific EIEEU. Stream №2 is designed to identify new threats. New threats are threats that have not been previously encountered when authenticating the subject of EIEEU. The distinctive feature of the proposed scheme is the ability to use a mechanism certifying verification. This approach is based on receiving an activation signal from the "nearest" SDT. At the same time, when analyzing a signal, the entire data array is used, which are provided by the same subject of the EIEEU.

The following equations were used to calculate the speed with which attempts of penetration into the EIEEU can be found out (for the proposed scheme of authentication of the subject based on updated SDT (further USDT), Fig. 2):

$$\frac{da}{dt} = rev \cdot \zeta \cdot |Tr| - \alpha \cdot |Tr(MP \backslash W(MP))| \tag{1}$$

$$W(MP) = \{w | \forall w \exists mp \in MP : (w) \ Con \ MP |\} \tag{2}$$

where $Tr = Tr(S)-$ is a variety of threats for EIEEU; $MP-$ is IPT in EIEEU; $W(MP)-$ is IPT involved in verification of subscribers; $rev-$ is a coefficient of delay in including of IPT in EIEEU; $\zeta-$ is a coefficient characterizing the increase in the number of threats; $Tr(PR)-$ is threats neutralized by IPT; $\alpha-$ is a coefficient characterizing the degree of failure of the workability of EIEEU caused by an attack; $Con-$ is conditions under which threats to an EIESS are neutralized by the existing protection system.

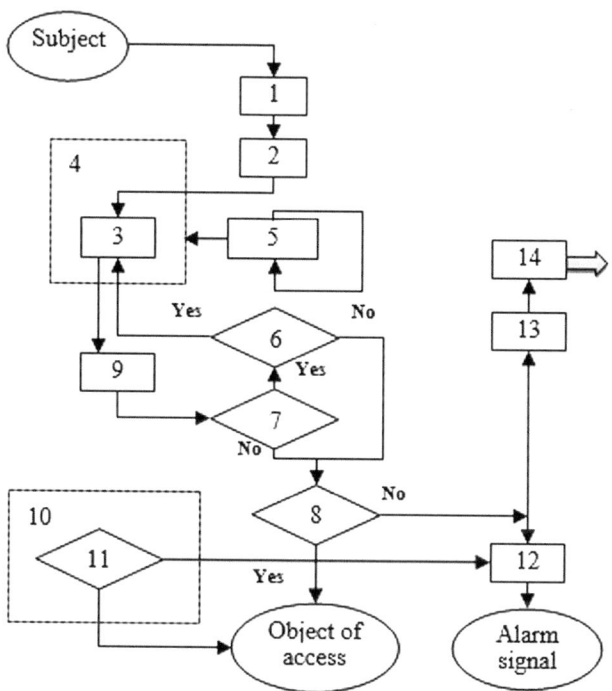

Designations adopted on the scheme: *1 – to count identifying data of the subject; 2 – to form a representative set for searching; 3 – implementation of search procedures based on a variety of sets used for detection (VSDT); 4 – a variety of sets used for detecting threats of EIEEU; 5 – updating of VSDT; 6 – verification of signal; 7 – waiting for a confirmation signal; 8 – verification is passed; 9 – formation of resulting data; 10 – a variety of sets recovered from memory, which are used to detect threats; 11 – search among a variety of sets recovered from memory, which are used to detect threats; 12 – blocking the actions of the subject; 13 – recording in the memory of EIEEU protection system; 14 – updating memory.*

Fig. 1. Scheme of subject authentication in EIEEU based on updated SDT.

The proposed method for analyzing data on possible threats in the EIEEU is schematically shown in Fig. 2.

Taking into account the results of the works [13–15], it was proposed to apply weighting values for USDT with a particularly pronounced "cause-and-effect"

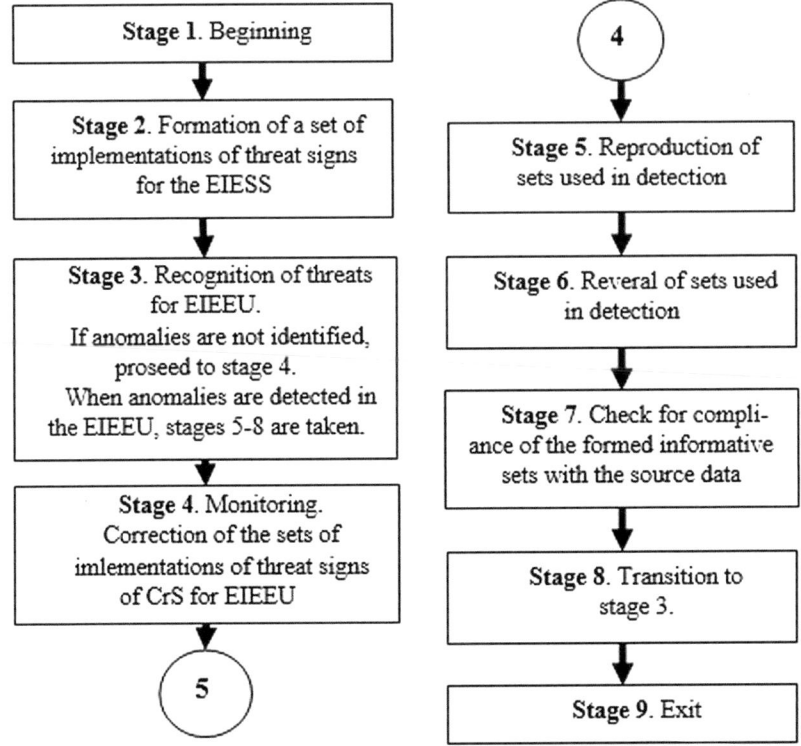

Fig. 2. Algorithm of generating sets involved in detecting threats for EIEEU.

relationship in the process of "training" USDT. It is shown that from the view of priority of increasing the effectiveness of identifying threats of EIEEU, it is obligatory to exclude SDT that are not included in the allowable field. In the end, as a result of the "training" of the threat recognition system and the corresponding algorithm, we obtain the resulting subset capable of further training.

We suppose that the authentication process involves data obtained from authorized subjects. In addition, SDT, service data about subscribers and regulatory rules are involved in the process of analyzing data on possible threats in the EIEEU. The input of the algorithm uses data obtained from IPT of EIEEU (for example, the initial information from the subject). Next, SDT verify this information and generate the appropriate solution.

An example of the regulatory rules for possible threats analysis subsystem in users' authentication in the electronic information educational environment of university is below.

The following initial data are given: $DS-$ is a variety of SDT $(ds \subset DS)$; $ID-$ an input information from the subject of EIEEU $(id \subset ID)$; $SS-$ a variety of SDT's signs implementations or initial data of the subject; $VE-$ results of verification over a period of time t; $SP-$ service parameters of IPT in EIEEU.

The following list of regulatory rules for EIEEU DBMS (authentication model of subscriber) has been received:

$$VE(ds(ID)) = \sum (all\ SS(id) > tv(SS)) \tag{3}$$

$$if \quad VE(ds(ID)) > tvr \quad \& \quad nc = isNull \quad then \quad VE = 1 \quad \& \quad new\ ds(ID) \tag{4}$$

$$if \quad VE(ds(ID)) > tvr \quad \& \quad nc == 0 \quad then \quad VE = 0 \quad \& \quad Stop \tag{5}$$

$$if \quad VE(ds(ID)) > nc \cdot tvr \quad then \quad ds(nit) \subset new\ DS \tag{6}$$

where $tv-$ a threshold level of similarity of comparison of the signs implementation set (SS) with the original data; $tvr-$ a threshold level of the resulting SDT data with fixing a warning about the threat for EIEEU; $nc-$ the minimum required amount of reliable evidence from other SDTs (it is mandatory for classifying the subject as a potential danger); $nit-$ the number of iterations in the loop.

The following types of elements that can be used in the proposed authentication scheme were considered:

Group №1 includes sets involved in the detection process (for example, binary matrices used as objects of learning of sets). Each set corresponds to a specific class of implementations of observation objects signs [13–15]. The first group is responsible for processing information provided by the subject in the EIEEU.

Group № 2 includes the basic objects (BO) of EIEEU. For each element of protected EIEEU (in some cases for the entire system), a set of internal BOs is formed. BO data are intended to implement service functions. The BO of typical EIEEU can include: information arrays, as well as an object that will contain, record and accumulate data by its own characteristics of the protected object. Later this information will be used as the basis for the synthesis of sets involved in the detection of threats. The service configurations of the EIEEU protection systems should be included into the second group. These configurations contain the data required to adjust the sets involved in the detection of threats.

It is necessary to mention the control subsystem of EIEEU. This subsystem directly implements the control over the process of authentication of subjects in the process of keyboard recognition in the EIEEU based on USDT.

The proposed method allowed: (1) increasing the validity and reliability of verification results (this was achieved through additional checks during the use of USDT); (2) increasing the information content of the SDT sets for specific cyber threats of EIEEU (this was achieved by preserving the values of threats signs with a high degree to minimize the feature space allocated under the SDT).

5   Simulation Experiment

Table 1 shows a fragment for a set of output data, obtained during the comparison of computational experiments and experimental verification, of the proposed authentication scheme for the problem of analyzing the behavior of subject in EIEEU.

Table 1. Fragment for a set of output data during experiments

$rev \cdot \zeta \cdot |Tr|$	t, s	Probability of detection of potentially dangerous subjects of EIEEU									
		Standard authentication of subject in EIEEU, 100%		Authentication of subject in EIEEU based on USDT, 100%							
				$\alpha = 0,1$				$\alpha = 0,9$			
		$\alpha = 0,1$	$\alpha = 0,9$	P_m – probability that SDT data will be mistakenly identified with data represented by the subject. (the threshold of similarity of sets was determined in advance)							
				0,7		0,98		0,7		0,98	
				ξ – a coefficient characterizing the ability to apply test results based on a specific SDT to several sub-types of threats							
				0,1	0,3	0,1	0,3	0,1	0,3	0,1	0,3
1	50	59,2	0,2	100	99,9	99,3	99,2	77,2	71	50,3	45,4
	160	59,2	0,1	60,1	59,9	59,1	59,1	0,8	0,4	0,2	0,1
10	50	61,3	0,4	99,9	98,1	98,4	96	77,3	70,9	50,3	49,8
	160	60,1	0,4	60, 2	59,1	60,2	59,7	0,6	0,3	0,3	0,1

Figure 3 shows the results of testing the proposed authentication scheme for the problem of analyzing the behaviour of subject in the EIEEU. The dependencies for the EIEEU password protection system module are shown in the graphs of Fig. 3. At the

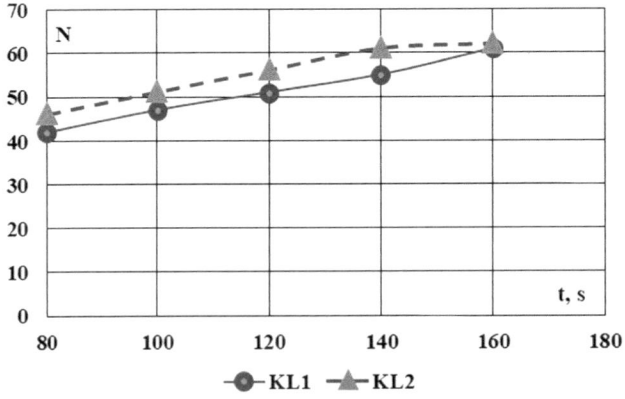

KL1 – authentication of subject in the process of keyboard recognition in EIEEU based on the application of USDT; KL2 – protection by ordinary passwords for the subject; N – the number of transitions when the subject operates in EIEEU on the keyboard.

Fig. 3. Results of testing of proposed authentication scheme for the problem of analyzing the behaviour of subject in EIEEU (on the example of a distance learning system of European University (Ukraine)).

same time, the speed of recognition of the subject in EIEEU from the number of characters entered by the subject is analyzed. In the experiments, it was assumed that the password length was 8 characters, and the number of possible attempts to enter a password was 3. The number of sets involved in the detection was 250.

Thus, it was shown that even if an attacker found out the password, it is difficult for him to fake the machine manner of an authorized subscriber, it is necessary to keep an eye on the subscriber for a long time and then train in EIEEU. Consequently, the proposed authentication scheme of subject in EIEEU based on USDT, is sufficiently effective for the tasks of subscriber identification in the system.

6 Discussion

The advantages of the study include the fact that the proposed solutions, in particular, the developed software modules for authentication, compared with the results of the studies presented in [7–9, 16–19], showed a greater probability of detecting potentially dangerous subjects in information systems and networks of enterprises, and a lower probability that the SDT will be mistakenly identified with the data provided by the network subscriber.

The software products created on the basis of the proposed solutions allowed one to automate the control, maintenance and change of subscribers' accounts of the networks of two large universities in Ukraine. At the same time, a possibility to adjust the subscriber access levels to information resources and automate user authentication in the EIEEU was in the basis of the software product "Threat Analyzer" [14].

At the present stage of research, the definite drawback of the work is insufficient approbation of the proposed solutions. The experiments have so far been carried out only on the platforms of two universities: The National University of Life and Environmental Sciences of Ukraine and European University.

The prospect of further research is determined by the possibilities of applying the obtained results for the subsequent algorithmization of processes associated with the analysis of EIEEU protection. It is also possible to programmatically automate the processing of data on possible cyber threats in the process of applying the selective algorithm for processing updated data sets from software and hardware for detecting IPT and CrSB of an educational institution.

In this context, our work continues previous publications [14, 15].

7 Gratitude

The research and the article were done within the framework of promising scientific and technical programs of the Department of Computer Systems and Networks of the National University of Life and Environmental Sciences of Ukraine, as well as the grant of the Republic of Kazakhstan, "Development of adaptive expert systems in the area of cybersecurity of critical objects of informatization".

8 Conclusions

The following results have been obtained:

A method for analyzing data on possible cyber threats in electronic information educational environment of universities and other large educational institutions, which differs from the existing ones by the possibility of minimizing time costs in the process of detection of new threats when authenticating subscribers to EIEEU was suggested;

A mathematical model of subscriber authentication in the electronic information educational environment of universities was proposed. The model differs from the existing ones by the possibility of using a selective algorithm in processing updated data sets from software and hardware for detecting information protection systems and the cybersecurity of an educational institution, which reduces the number of false positives and false reports.

References

1. Fan, W., Kevin, L., Rong, R.: Social engineering: I-E based model of human weakness for attack and defense investigations. Int. J. Comput. Netw. Inf. Secur. (IJCNIS) 9(1), 1–11 (2017). https://doi.org/10.5815/ijcnis.2017.01.01
2. Hu, Z., Khokhlachova, Y., Sydorenko, V., Opirskyy, I.: Method for optimization of information security systems behavior under conditions of influences. Int. J. Intell. Syst. Appl. (IJISA) 9(12), 46–58 (2017). https://doi.org/10.5815/ijisa.2017.12.05
3. Jin, G., et al.: Evaluation of game-based learning in cybersecurity education for high school students. J. Educ. Learn. 12(1), 150–158 (2018)
4. Diaz, L.J., et al.: The risks and liability of governing board members to address cyber security risks in higher education. JC & UL 43, 49 (2017)
5. Ghernaouti, S., Wanner, B.: Research and education as key success factors for developing a cybersecurity culture. In: Cybersecurity Best Practices, pp. 539–552 (2018). Springer, Wiesbaden. https://doi.org/10.1007/978-3-658-21655-9_38
6. Caelli, W.J., Liu, V.: Cybersecurity education at formal university level: an Australian perspective. J. Colloq. Inf. Syst. Secur. Educ. 5(2), 26–44 (2018)
7. Krishnamoorthy, S., Rueda, L., Saad, S., Elmiligi, H.: Identification of user behavioral biometrics for authentication using keystroke dynamics and machine learning. In: Proceedings of the 2018 2nd International Conference on Biometric Engineering and Applications, pp. 50–57. ACM (2018). https://doi.org/10.1145/3230820.3230829
8. Turkanović, M., Brumen, B., Hölbl, M.: A novel user authentication and key agreement scheme for heterogeneous ad hoc wireless sensor networks, based on the Internet of Things notion. Ad Hoc Netw. 20, 96–112 (2014). https://doi.org/10.1016/j.adhoc.2014.03.009
9. Amin, R., Biswas, G.P.: A secure light weight scheme for user authentication and key agreement in multi-gateway based wireless sensor networks. Ad Hoc Netw. 36, 58–80 (2016). https://doi.org/10.1016/j.adhoc.2015.05.020
10. Kent, A.D., Liebrock, L.M., Neil, J.C.: Authentication graphs: analyzing user behavior within an enterprise network. Comput. Secur. 48, 150–166 (2015). https://doi.org/10.1016/j.cose.2014.09.001
11. Heard, N.A., Palla, K., Skoularidou, M.: Topic modelling of authentication events in an enterprise computer network (2016). https://doi.org/10.1109/isi.2016.7745466

12. Singhal, A., Ou, X.: Security risk analysis of enterprise networks using probabilistic attack graphs. In: Network Security Metrics, pp. 53–73 (2017). Springer, Cham. https://doi.org/10.1007/978-3-319-66505-4_3
13. Lakhno, V., Petrov, A.l., Petrov, A.: Development of a support system for managing the cyber security of information and communication environment of transport. In: Information Systems Architecture and Technology: Conference 2017, ISAT, pp. 113–127 (2017). Springer. https://doi.org/10.1007/978-3-319-67229-8_11
14. Lakhno, V., Akhmetov, B., Korchenko, A., et al.: Development of a decision support system based on expert evaluation for the situation center of transport cybersecurity. J. Theor. Appl. Inf. Technol. **96**(14), 4530–4540 (2018)
15. Akhmetov, B., Lakhno, V., Akhmetov, B., Alimseitova, Z.: Development of sectoral intellectualized expert systems and decision making support systems in cybersecurity. In: Silhavy, R., Silhavy, P., Prokopova, Z. (eds.) Intelligent Systems in Cybernetics and Automation Control Theory. CoMeSySo 2018. Advances in Intelligent Systems and Computing, vol 860, pp. 162–171 (2019). https://doi.org/10.1007/978-3-030-00184-1_15
16. Hasani, S.M., Modiri, N.: Criteria specifications for the comparison and evaluation of access control models. Int. J. Comput. Netw. Inf. Secur. (IJCNIS) **5**(5), 19–29 (2013). https://doi.org/10.5815/ijcnis.2013.05.03
17. Hu, Z., Gnatyuk, S., Koval, O., Gnatyuk, V., Bondarovets, S.: Anomaly detection system in secure cloud computing environment. Int. J. Comput. Netw. Inf. Secur. (IJCNIS) **9**(4), 10–21 (2017). https://doi.org/10.5815/ijcnis.2017.04.02
18. Tereykovska, L., et al.: Encoding of neural network model exit signal, that is devoted for distinction of graphical images in biometric authenticate systems. In: News of the National Academy of Sciences of the Republic of Kazakhstan, Series of Geology and Technical Sciences, vol. 6, no. 426, pp. 217–224 (2017)
19. Dychka, I., Tereikovskyi, I., Tereikovska, L., et al.: Deobfuscation of computer virus malware code with value state dependence graph. In: Advances in Intelligent Systems and Computing, pp. 370–379 (2018). https://doi.org/10.1007/978-3-319-91008-6

Implementation of Active Learning
in the Master's Program on Cybersecurity

Volodymyr Buriachok and Volodymyr Sokolov(✉)

Borys Grinchenko Kyiv University, Kyiv, Ukraine
{v.buriachok, v.sokolov}@kubg.edu.ua

Abstract. The paper examines the possibility and practical approach of combining the higher education standards in Ukraine and the best international practices in the training "Cybersecurity" specialists, since the process of transformation and formation of the world information society will gain momentum. The National Sustainable Development Strategy (see 3.1 "National Security and Defense Reform" [1]) for the period up to 2020 gives opportunity for a competitive global environment and the implementation of a number of innovations at the national level. It is impossible to handle these issues without assurance of having access to the modern infrastructure by all population sectors in order to effectively use it in practice. Essentially, the approach proposed in the CDIO standard, which creates conditions for active learning of technical students, is applied. The approach has been tested at two higher educational institutions of Ukraine (State University of Telecommunications and Borys Grinchenko Kyiv University) and described through of the program of the course "Wireless and Mobile Security." The paper presents the results of master's research.

Keywords: Cybersecurity · Information and communication systems ·
Active learning · Professional competences · Learning outcomes ·
Wireless security

1 Introduction

Since Ukraine was part of the Soviet Union, the technical high school was industrially oriented. All students who graduated from a higher educational institution were distributed, as a rule, to industrial enterprises and design offices. Thus, changing industry requirements effected curricula.

After the collapse of the Soviet Union, many industrial relations have been broken; therefore, there was an imbalance in the training of specialists in many technical specialties. For example, the level of specialists training in aviation technology in Kyiv and Kharkiv exceeded by many times over the needs of the state for such specialists.

During the 90s, the decline in industrial production led to a glut of the labor market with technical personnel, and as a result, the collapse of training, a decline in the popularity of technical specialties as a whole, as well as the outflow of highly qualified teaching staff to adjacent areas and abroad. As a result, over two decades, the practical

Z. Hu et al. (Eds.): ICCSEEA 2019, AISC 938, pp. 610–624, 2020.
https://doi.org/10.1007/978-3-030-16621-2_57

component of education has been lost, the connection between practical industrial productions has been lost, and the continuity of pedagogical and scientific personnel has been broken.

To restore the links of higher technical schools with industrial enterprises, a transition from academic activities to real challenges of modern industrial production is required. As one of the methods of convergence, we propose to use *active learning*.

The content of the paper is organized as follows. In Sect. 2 "Related Works" provides an analysis of the issues of safety training, methods of assessment and the formation of training materials using active training. Section 3 "Active Learning in the Cybersecurity" provides an overview of the rationale for the implementation of practice-oriented learning approaches. Our approaches to the implementation and development of cybersecurity training are given in Sect. 4 "Cybersecurity Master's Program." The current state of one of the master's courses is given in Sect. 5 "Course Implementation Example." Research results of the introduction of active learning and the results of scientific work of students are given in Sects. 6 "Experimental Research of the Program Implementation" and 7 "Experimental Research in Master's Diplomas." The paper end with Sects. 8 "Gratitude" and 9 "Conclusions and Future Work."

2 Related Works

The widespread use of e-learning [2] and even m-learning courses [3] reduces the amount of time when a student has the opportunity to work with real software and hardware systems under the direct supervision of a teacher, share experiences in a group of students, and realize the gaps in his areas of knowledge.

As the number and quality of attacks grows, approaches to ensuring information security are also actively developing. For example, penetration testing [4] and forensic [5] training courses outdated by more than 20% over the year due to their active research. Therefore, it is necessary to involve undergraduates in scientific and practical activities, and publish the results in the form of scientific articles, abstracts of conferences and educational materials, as shown in [6]. In addition, at the same time, students themselves can check and evaluate newly published teaching materials, find errors and improve them [7].

In [8] it is proposed to rank students into three groups, but in this case, for each of them it is required to develop separate tasks. The method of active learning allows minimizing differences between students. Even backward students can get involved in work at any stage of its implementation. Tracking student performance can be carried out according to the methods presented in [9] but the role of examinations with active learning decreases, which leads to a decrease in anxiety among students [10].

3 Active Learning in the Cybersecurity

Active learning implies a close relationship between the knowledge and skills acquired and the requirements of employers. Figure 1 shows a looped diagram in which the employer acts as the customer of the technical and personal requirements. Based on these requirements, lists of hard and soft skills are formed, and they, in turn, form *professional competences* (PCs) and *learning outcomes* (LOs). Moreover, the courses are based on PCs and LOs. Therefore, if the requirements of the employer change over time, then the syllabus changes along the chain: requirements of the employer → skills → PCs and LOs → syllabus.

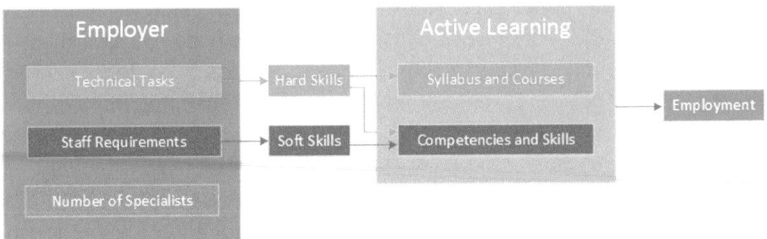

Fig. 1. Master skills and curriculum formation scheme.

For large technical universities, it is possible to create their own curricula for active learning, for example, the "Technology-Enabled Active Learning" program at the Massachusetts Institute of Technology [11]. But the majority of higher education institutions join the initiatives of national agencies, for example, in the United States there are standards K-4 and 9–12 "Science as Inquiry. Developing Student Abilities and Under Standing" [12], or to international initiatives, for example, the CDIO Initiative [13] (anagram from "Conceiving, Designing, Implementing and Operating").

4   Cybersecurity Master's Program

This report presents the results of the implementation of the master's practical program on the security of wireless and mobile systems in two higher educational institutions of State University of Telecommunications (2016–2018) and Borys Grinchenko Kyiv University (2018–2019). This program was developed in the framework of the Tempus project #544455-TEMPUS-1-2013-1-SE-TEMPUS-JPCR "Educating the Next Generation Experts in Cyber Security: the New EU-Recognized Master's Program" (ENGENSEC) [14], which is funded by the European Union, significantly improved the quality of the student's training of the second (master) educational level in 125 "Cybersecurity" specialty. During its development, recommendations of CDIO standards [13] were taken into account, as well as the practice of presenting courses on the physical basics of information security and information security for graduate students.

The following courses were developed as part of the project: "Wireless and Mobile Security," "Network and Cloud Security," "Web Security," "Penetration Testing and Ethical Hacking," "Malware," "Secure Software Development," and "Digital Forensic." But among them there are no courses on cryptography, therefore the courses "Construction and Analysis of Cryptosystems" and "Mathematical Methods of Cryptography" were introduced into the program, and also for fixing the developed material at industrial enterprises, scientific research and industrial (technological) internship are provided. The relationship between the knowledge gained in the bachelor's course and the courses inside the master's program is shown in Fig. 2.

Master's work is based on one of the courses and does not require a comprehensive final exam. Protection is carried out on the written master's diploma. Approbation of the results of work is carried out at scientific conferences. It is recommended to submit publications to international conferences that are part of scientific abstract and citation databases (SciVerse Scopus, Web of Science, etc.). The results of master's studies should be presented in one of the scientific journals approved by the Higher Attestation Commission of Ukraine [15]. Since the process of reviewing and publishing materials may take a lot of time, the choice of topic, source analysis, concept development, research and publication of the results should be carried out as soon as possible, almost from the first weeks of the start of studies in the magistracy. One and a half years of study in the magistracy may not be enough for the full publication of the results of the study.

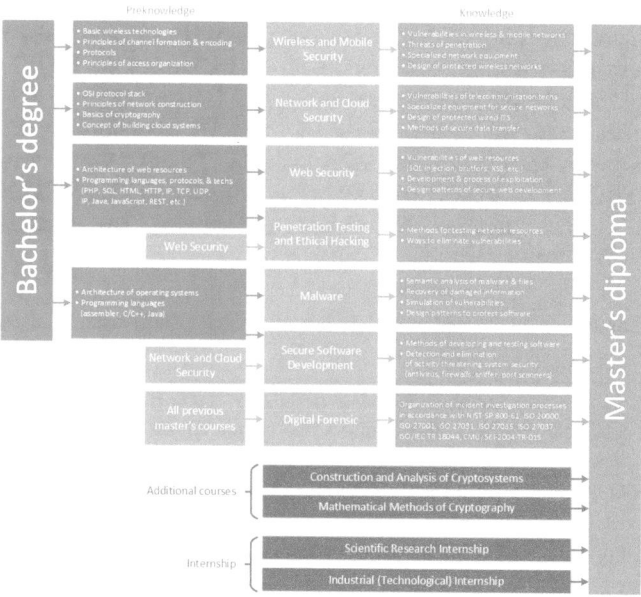

Fig. 2. Block diagram of the master's program in 125 "Cybersecurity" specialty.

5 Course Implementation Example

Consider the relationship between courses and PCs/LOs on the example of the course "Wireless and Mobile Security." For completeness, Fig. 3 shows not only the relationship between the master's courses [16], but also the bachelor's [17, 18]. In the figure, the *b* and *m* indices denote bachelor and master PCs and LOs. Detailed decoding of the indices is given in Tables 1 and 2.

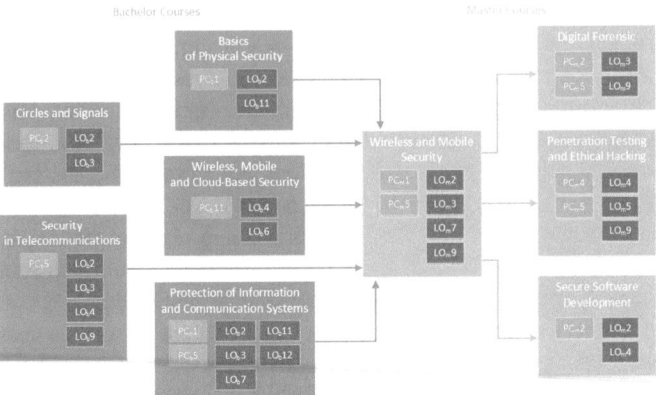

Fig. 3. Continuity of competencies for the course "Wireless and Mobile Security."

The laboratory workshop contains issues of creating a wireless access point, installing penetration testing OS for embedded systems (PwnPi and Kali), building map of wireless networks, monitoring of network traffic, hacking of Wi-Fi (WEP and WPS), researching of radio frequency resources in 2.4–2.5 GHz range, DoS attacking, Wi-Fi fuzzing, wireless networking, and RFID sniffing. All laboratory works use widely available hardware (Raspberry Pi, GPS module, Pololu Wixel, Arduino Nano, NodeMCU, TI CC2500, RDM6300, USB charger meter, OLED SSD1306, microSD cards, adapters, power supplies, breadboards and cables). The price of one set of equipment is €200–300.

Table 1. Professional competencies of bachelor and master courses.

PCs	Description
PC_b1	Ability to apply the legislative and regulatory framework, as well as state and international requirements, practices and standards for the purpose of carrying out professional activities in the field of information and/or cybersecurity
PC_b2	Ability to use information and communication technologies, modern methods and models of information and/or cybersecurity
PC_b5	Ability to provide protection in information and telecommunication (automated) systems in order to implement the established information and/or cybersecurity policy
PC_b11	Ability to monitor the processes in information and telecommunication systems in accordance with the established information and/or cybersecurity policy
PC_m1	Ability to use modern information and security technologies in the field of information security
PC_m2	Ability to detect vulnerabilities and secure wired and wireless networks, investigate incident information and/or cybersecurity and counteract malware
PC_m4	Ability to secure network resources and cryptographic protection of information in information and/or cybersecurity systems
PC_m5	Ability to ensure the protection of information processed in information and communication systems, the administration of such systems and their exploitation

Table 2. Learning outcomes of bachelor and master courses.

LOs	Description
LO_b2	1. Carry out professional activity on the basis of knowledge of modern information and communication technologies 2. Develop and analyze of information and communication system (ICS) projects based on standardized technologies and data transfer protocols 3. Apply in the professional activity knowledge, skills and practice, on the structures of modern computing systems, methods and means of information processing, architectures of operating systems 4. Protect resources and processes in ICS based on security models (end-to-end automation, flow management, Bell-LaPadula, Biba, Clark-Wilson, and others), as well as established modes for the safe operation of ICS 5. Perform software analysis to assess compliance with established information and/or cybersecurity requirements in ICS
LO_b3	1. Provide processes for the protection of information and telecommunication systems by installing and correct operation of software and hardware and software complexes of protection means 2. Provide the operation of special software for data protection against destructive software effects, destructive codes in information, information and telecommunication systems 3. Develop the operational documentation according to standards for integrated security systems
LO_b4	1. Solve the maintenance tasks (including review, testing, and accountability) of the access control system in accordance with the principles, criteria of access and established security policy in information and telecommunication systems 2. Implement measures to counteract the unauthorized access to information resources and processes in information, information and telecommunication systems 3. Solve the problems of managing access to information resources and processes in information and information-telecommunication systems based on models of access control (mandate, discriminatory, role-playing) 4. Solve the tasks of centralized and decentralized administration of access to information resources and processes in information, information and telecommunication systems 5. Ensure the accountability of the access control system for information resources and processes in the ICS
LO_b6	1. Solve tasks of managing processes ensuring business continuity using software backup procedures and directly information resources 2. Solve tasks of correction of goals, strategies, and plans of ensuring the non-integrity of business after cyber-attacks, crashes and failures of different classes 3. Create and implement plans for ensuring business continuity, including critical infrastructure 4. Analyze the settings of the elements of information systems and communication equipment
LO_b7	1. Solve the problems of support and implementation of integrated information security systems, as well as to counteract unauthorized access to resources and processes in information and information and telecommunication systems 2. Assess the security level of information processed in the ICS using tools for assessing potential vulnerabilities 3. Solve the problems of management of the complex information security system in information, information and telecommunication 4. Solve the tasks of examination, testing of complex information security systems

(*continued*)

Table 2. (*continued*)

LOs	Description
LO$_b$9	1. Ensure continuity of business processes of the organization based on the information security management system, in accordance with domestic and international requirements and standards 2. Ensure the functioning of the information and/or cybersecurity management system of the organization on the basis of information risk management, the implementation of procedures for their quantitative and qualitative assessment
LO$_b$11	1. Provide monitoring processes for access to ICS resources and processes 2. Provide configuration and operation of monitoring systems of resources and processes in ICS
LO$_b$12	1. Implement and support intrusion detection systems and use protection complexes to provide the necessary level of information security in information, information, telecommunication and automated systems 2. Analyze the effectiveness of systems for detecting and counteracting unauthorized access to resources and processes in ICS 3. Analyze and implement anti-malware security systems
LO$_m$2	1. Be able to identify and formulate actual scientific problems, generate and integrate new ideas and new knowledge in the field of information and/or cybersecurity protection 2. Be able to use specialized software packages, modern information and/or security technologies in the field of information security 3. Know the vulnerability and methods of their application in various telecommunication technologies 4. Know how to deal with these vulnerabilities, as well as specialized network equipment used to secure corporate networks 5. Be able to design protected (including threats) wired telecommunication systems 6. Know the methods of organizing secure data transmission in an unprotected environment
LO$_m$3	1. Know vulnerabilities and methods of their application in wireless and mobile networks 2. Be able to detect threats of penetration or access of malicious people to such networks 3. Know the specialized network equipment used to ensure the security of wireless and mobile networks 4. Be able to design protected (including threats) wireless networks
LO$_m$4	Know the methods and methods of developing and testing software for detecting and eliminating activities that threaten system security (antiviruses, firewalls, sniffers, port scanners)
LO$_m$5	1. Be able to carry out semantic analysis of files 2. Be able to detect malignant software and files according to their structure and behavior 3. Be able to recover damaged information 4. Be able to model software vulnerabilities and use design patterns to protect software
LO$_m$7	1. Know the methods and methods of testing network resources for security vulnerabilities 2. Be able to find ways to eliminate them
LO$_m$9	1. Have practical skills for conducting safety audits of information and communication systems, their administration and exploitation 2. Be able to design perspective cryptosystems and apply modern technologies of cryptographic protection of information in information and/or cybersecurity systems

In addition, the course has developed a training laboratory workshop "Wireless and Mobile Security" in English [19] and Ukrainian languages [20]. The course was tested at *State University of Telecommunications* and *Borys Grinchenko Kyiv University*. Parts of the course was evaluated in *Blekinge Institute of Technology* (Sweden), *Wroclaw University of Science and Technology* (Poland), *Bonch-Bruevich Saint-Petersburg State University of Telecommunications* (Russian Federation), *International Coordination, Training and Research Center of the Federal Criminal Police Office* (Germany), *Kharkiv National University of Radioelectronics*, and *Lviv Polytechnic National University*.

An active learning approach is implemented in this course. Thus, the topics of the course are disclosed in the laboratory. Each lab contains a *purpose* that details the practical task. Knowledge and skills of the student are formalized. Also included are recommended *hardware* and *software*, *safety instructions*, a list of recommended *literature* and *references*. The student gets the opportunity to consolidate his theoretical knowledge into practice, and a brief theoretical justification helps to quickly get up to speed.

Such an approach in structuring educational information makes it easy to combine theoretical and practical knowledge, forms design thinking for a student, and reduces the time for performing laboratory work approximately one and a half times.

6   Experimental Research of the Program Implementation

To determine the effectiveness of introducing active learning into the educational process, we used data on student assessments for the course "Wireless and Mobile Security" for the period from 2016 to 2018. The number of students who studied before the implementation of active training was 45 people, and after its introduction—57. An alphabetic scale was used for averaging the results (see Table 3) instead of a 100-point scale. Figure 4 shows the distribution of marks obtained by analyzing student success reports.

Table 3. Grading scale.

Mark	Number of points	Mark	Number of points
A	≥ 90	D	≥ 69
B	≥ 82	E	≥ 60
C	≥ 75	F	≥ 35

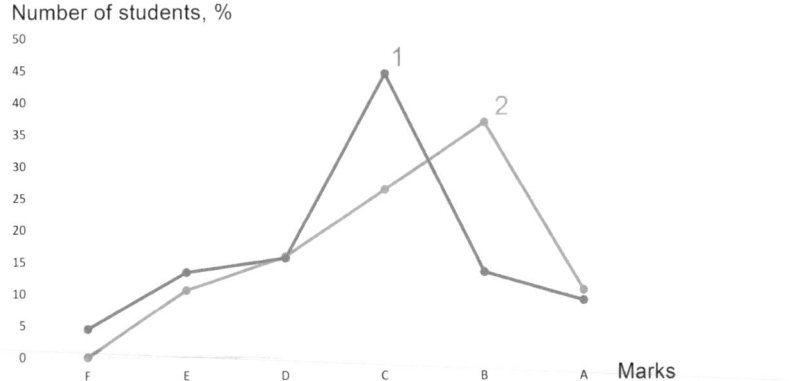

Fig. 4. Student success before (1) and after (2) implementation of active learning.

The distribution of marks has become uniform: the curve of distribution before the introduction of active learning was subordinate to the Laplace distribution, and after implementation, it began to approach the χ^2-distribution with 4 degrees of freedom. The average score increased by 4 points (from 76.3 to 79.3).

Of course, students' assessments do not fully reflect the success of the learning material, but allow us to evaluate the qualitative characteristics of the processes.

The quality of the implementation of active learning has been verified through external testing at Cisco Networking Academy. The most familiar material is the experimental course "IoT Fundamentals: IoT Security" (version 1.0). The graduates were given a week to get acquainted with the materials of the course and pass the exam (passing score was 75, so students who successfully passed the exam, received certificates).

7   Experimental Research in Master's Diplomas

The quintessence of training in the master's program is the writing of a scientific work (in the form of a master's degree, scientific articles and a speech at scientific conferences). Below are six examples of master's diplomas, which can be divided into three categories: hardware, software and at the same time hardware and software.

1. Hardware implementation:

   - *Analysis of Integrity of Data Transmission in 2.4–2.5 GHz Wireless Communication Channels using the Hardware Spectrum Analyzer* (production of printed circuit boards for three spectrum analyzers shown in Fig. 5 and testing of real devices) [21].

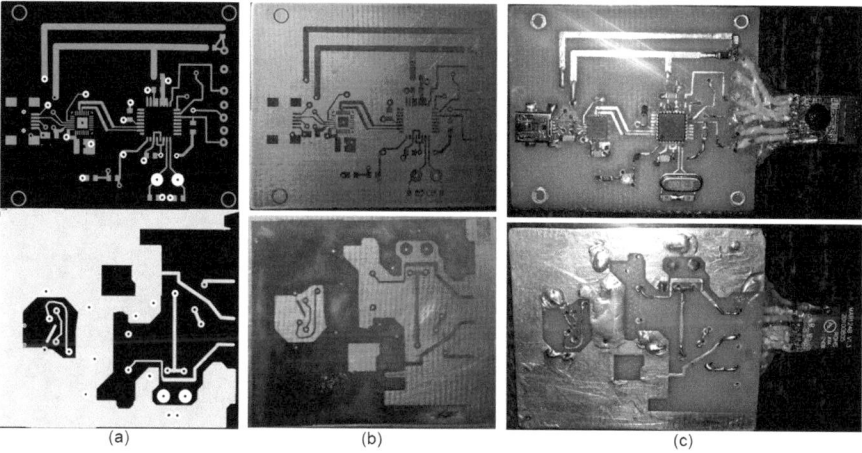

(a) (b) (c)

Fig. 5. An example of a spectrum analyzer circuit board on the CC2500 module in graphic editor (a), etched (b) and in the finished device (c).

- *Investigating Wireless Botnets and Making Recommendations on Their Use for Implementing Denial-of-Service Attacks* (manufacturing of wireless bots shown in Fig. 6 and conduct an experiment) [22].

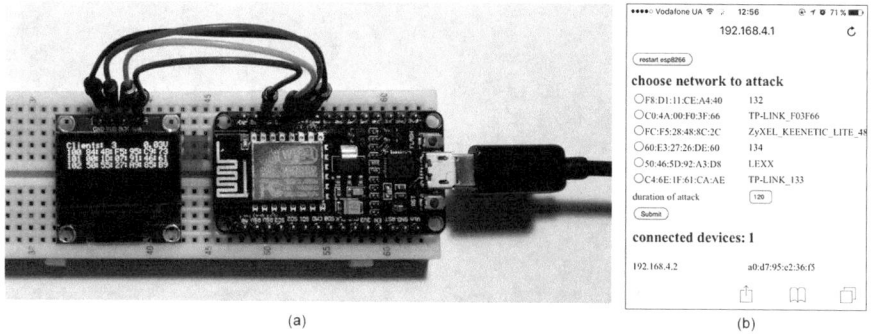

(a) (b)

Fig. 6. Botnet system controller based on the NodeMCU 1.0 board (a) and its web interface (b).

2. Software implementation:

- *Software Complex for Comparative Analysis of Integrity of Data Transmission in 2.4–2.5 GHz Wireless Channels* (development of protocols and unification of work with modules, see Fig. 7) [23].

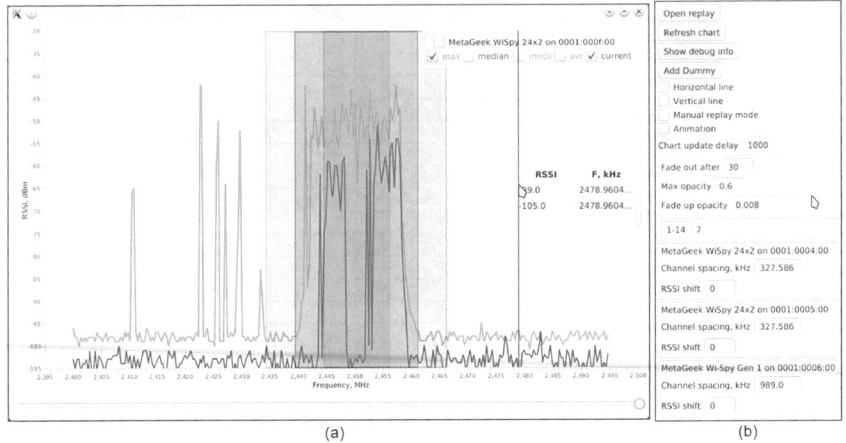

(a)   (b)

Fig. 7. Interfaces of a software analysis system: the main window (a) and settings (b).

- *Methodology of Counteraction to Social Engineering at Objects of Information Activity* (fishing page design and collecting statistics, see Fig. 8) [24].

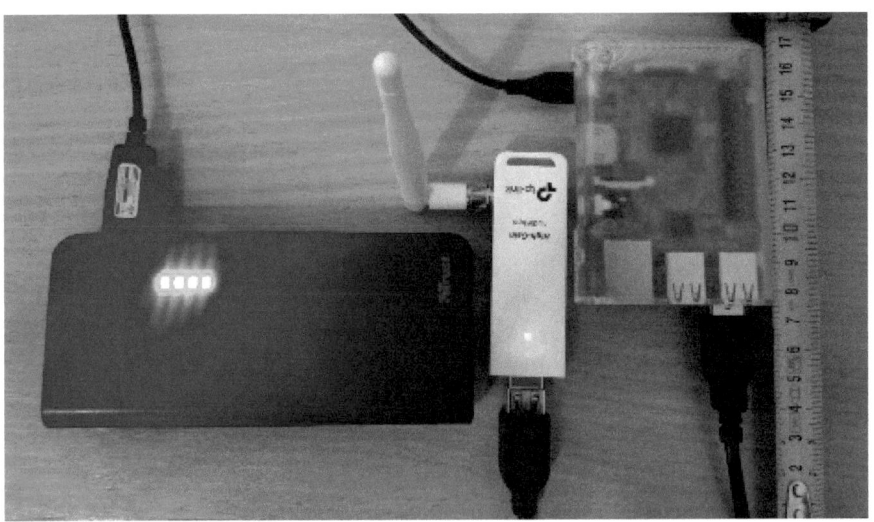

Fig. 8. Experimental setup based on the Raspberry Pi 3 board.

3. Hardware and software implementation:

- *Research on the Security of Low-Power Wireless Technologies* (collection of IEEE 802.15.4/802.16 data and analysis of received packages, see Fig. 9) [25, 26].

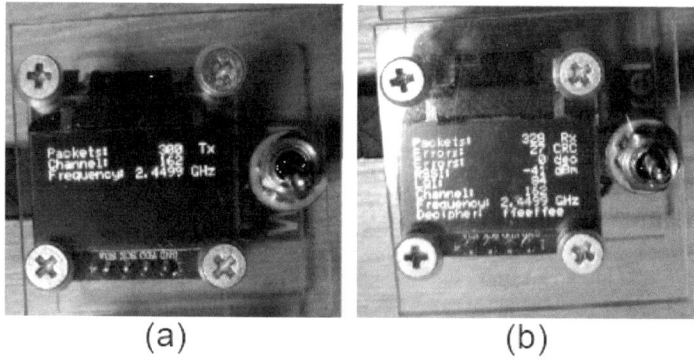

Fig. 9. Hardware emulator transmitter (a) and receiver (b) based on the Pololu Wixel board.

- *Investigation of Ways and Recommendations on Safety of Monitoring Systems of Wireless Ad Hoc Networks in Conditions of Third-Party Influence* (spectrum research and programming services, see Fig. 10) [27].

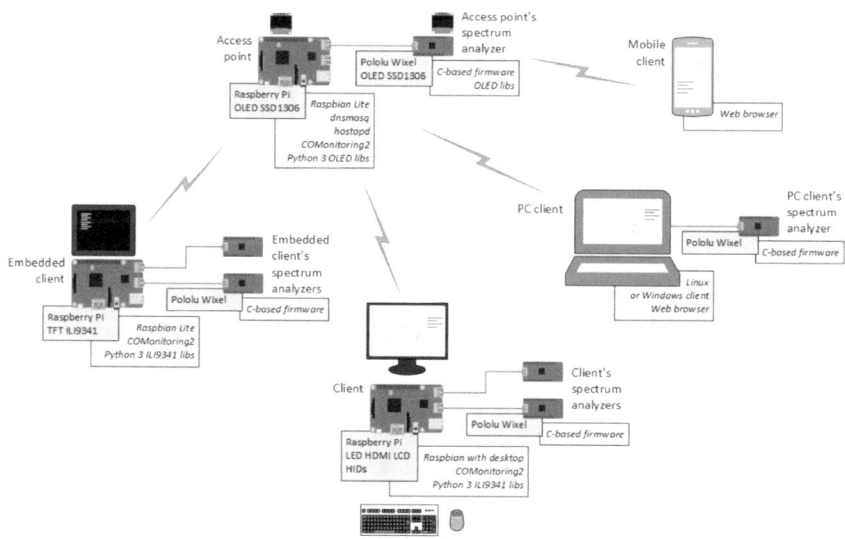

Fig. 10. General spectrum analysis subsystem.

Thus, each master's diploma is supported by at least one scientific work.

8 Gratitude

The authors of the article are grateful to the EU ENGENSEC project chief Anders Carlsson from Department of Computer Science and Engineering (Faculty of Computing, Blekinge Institute of Technology, Karlskrona, Sweden) for helping to form a master's program in cybersecurity [28].

9   Conclusions and Future Work

Formation of the regulatory framework that would meet the current requirements of world standards for ensuring the cybersecurity of the state and its harmonization with international law. The authors of the publication are preparing to release the translation of the current version of the CDIO standard [13], which can quickly improve the level student's preparation in cybersecurity. This, in turn, will assure the security of state information resources and the reliability of the protection of critical infrastructure objects, increase the competence of specialists in cybersecurity and information security and organize international cooperation in the field of masters training.

For the first time, an integrated approach to the training of graduate students in security is offered: the student is determined with the research topic in the first months of the magistracy and all subsequent active training is aimed at obtaining scientific and practical results.

According to the results of this international project and educational programs, it is planned to create a project of a master's standard (see [13] for the current national bachelor's program) for the training of specialists in the field of cybersecurity.

References

1. Lozhkin, B.: The strategy for sustainable development "Ukraine 2020" (2015). http://zakon1.rada.gov.ua/laws/show/5/2015. Accessed 6 Nov 2018. [Publication in Ukrainian]
2. Rawat, B., Dwivedi, S.K.: An architecture for recommendation of courses in e-learning system. Int. J. Inf. Technol. Comput. Sci. (IJITCS) 9(4), 39–47 (2017). https://doi.org/10.5815/ijitcs.2017.04.06
3. Adejo, O.W., Ewuzie, I., Usoro, A., Connolly, T.: E-learning to m-learning: framework for data protection and security in cloud infrastructure. Int. J. Inf. Technol. Comput. Sci. (IJITCS) 10(4), 1–9 (2018). https://doi.org/10.5815/ijitcs.2018.04.01
4. Sharma, S., Mahajan, S.: Design and implementation of a security scheme for detecting system vulnerabilities. Int. J. Comput. Netw. Inf. Secur. (IJCNIS) 9(10), 24–32 (2017). https://doi.org/10.5815/ijcnis.2017.10.03
5. Lazzez, A., Slimani, T.: Forensics investigation of web application security attacks. Int. J. Comput. Netw. Inf. Secur. (IJCNIS) 7(3), 10–17 (2015). https://doi.org/10.5815/ijcnis.2015.03.02
6. Ehimwenma, K.E., Crowther, P., Beer, M.: Formalizing logic based rules for skills classification and recommendation of learning materials. Int. J. Inf. Technol. Comput. Sci. (IJITCS) 10(9), 1–12 (2018). https://doi.org/10.5815/ijitcs.2018.09.01
7. Almeida, F.: Framework for software code reviews and inspections in a classroom environment. Int. J. Modern Educ. Comput. Sci. (IJMECS) 10(10), 31–39 (2018). https://doi.org/10.5815/ijmecs.2018.10.04
8. Kaviyarasi, R., Balasubramanian, T.: Exploring the high potential factors that affects students' academic performance. Int. J. Educ. Manag. Eng. (IJEME) 8(6), 15–23 (2018). https://doi.org/10.5815/ijeme.2018.06.02
9. Kumar, M., Singh, A.J.: Evaluation of data mining techniques for predicting student's performance. Int. J. Modern Educ. Comput. Sci. (IJMECS) 9(8), 25–31 (2017). https://doi.org/10.5815/ijmecs.2017.08.04

10. Khan, A., Madden, J.: Active learning: a new assessment model that boost confidence and learning while reducing test anxiety. Int. J. Modern Educ. Comput. Sci. (IJMECS) **10**(12), 1–9 (2018). https://doi.org/10.5815/ijmecs.2018.12.01
11. Dori, Y.J., Belcher, J., Bessette, M., Danziger, M., McKinney, A., Hult, E.: Technology for active learning. Mater. Today **12**, 44–49 (2003). https://doi.org/10.1016/S1369-7021(03) 01225-2
12. National Research Council: National Science Education Standards, pp 69–91. The National Academies Press, Washington (1996). https://doi.org/10.17226/4962
13. Crawley, E.F., Malmqvist, J., Östlund, S., Brodeur, D.R., Edström, K.: The CDIO approach. In: Rethinking Engineering Education, pp. 11–45. Springer, Cham (2014). https://doi.org/10. 1007/978–3-319-05561-9_2
14. Carlsson, A.: Educating the next generation experts in cyber security: The new EU-recognized master's program (2016). http://engensec.eu/about-the-project/. Accessed 6 Nov 2018
15. Ministry of Education and Science of Ukraine, Scientific specialized publications (2018). https://mon.gov.ua/ua/nauka/nauka/atestaciya-kadriv-vishoyi-kvalifikaciyi/naukovi-fahovi-vidannya. Accessed 6 Nov 2018. [Publication in Ukrainian]
16. Buryachok, V.L., Bessalov, A.V., Tolyupa, S.V., Abramov, V.O.: Security of Information and Communication Systems. Educational and Professional Program of the Second (Master) Level of Higher Education, pp. 2–12. KUBG Publication, Kyiv (2018). [Publication in Ukrainian]
17. Semko, V.V., Bessalov, A.V., Melnyk, I.Y., Yermoshyn, V.V.: Security of Information and Communication Systems. Educational and Professional Program of the First (Bachelor) Level of Higher Education, pp. 2–14. KUBG Publication, Kyiv (2018). [Publication in Ukrainian]
18. Yudin, O.K., Oksiyuk, O.H., Babenko, T.V., Buryachok, V.L., Voronov, V.R., Kachynskyy, A.B., Kurnyetsov, O.O., Maksymovych, V.M., Chechelnatskyy, V.Y.: Standard of Higher Education in Specialty 125 "Cybersecurity" for the First (Bachelor) Level of Higher Education, pp. 2–18. Ministry of Education and Science Publication, Kyiv (2018). [Publication in Ukrainian]
19. Sokolov, V., TajDini, M., Buryachok, V.: Wireless and Mobile Security: Laboratory Workshop, pp. 6–124. SUT Publication, Kyiv (2017). https://doi.org/10.5281/zenodo. 2528820
20. Sokolov, V.Y., TajDini, M., Raiter, O.P.: Wireless and Mobile Security: Laboratory Workshop, pp. 6–122. SUT Publication, Kyiv (2018). https://doi.org/10.5281/zenodo. 2528822. [Publication in Ukrainian]
21. Sokolov, V.Y.: Comparison of possible approaches for the development of low-budget spectrum analyzers for sensory networks in the range of 2.4–2.5 GHz. Cybersecur.: Educ. Sci. Technol. **2**, 31–46 (2018). https://doi.org/10.28925/2663-4023.2018.2.3146. [Publication in Ukrainian]
22. Buryachok, V.L., Sokolov, V.Y.: Using 2.4 GHz wireless botnets to implement denial-of-service attacks. Web Sch. **24**, 14–21 (2018). https://doi.org/10.31435/rsglobal_wos/ 12062018/5734
23. Buryachok, V., Sokolov, V.: Miniaturization of wireless monitoring systems 2.4–2.5 GHz band. In: Shevchenko, V.L., Nakonechnyy, V.S. (eds.) 2nd International Scientific-Technical Conference on Actual Problems of Science and Technology (APST), pp. 41–43. SUT, Kyiv (2015)
24. Kurbanmuradov, D.M., Sokolov, V.Y.: Methodology of counteraction to social engineering at objects of information activity. Cybersecur.: Educ. Sci. Technol. **1**, 6–25 (2018). https:// doi.org/10.28925/2663-4023.2018.1.616. [Publication in Ukrainian]

25. TajDini, M., Sokolov, V.Y.: Internet of things security problems. Mod. Inf. Prot. **1**, 120–127 (2017). https://doi.org/10.5281/zenodo.2528814
26. TajDini, M., Sokolov, V.Y.: Penetration tests for Bluetooth low energy and Zigbee using the software-defined radio. Mod. Inf. Prot. **1**, 82–89 (2018). https://doi.org/10.5281/zenodo. 2528810
27. Bogachuk, I., Sokolov, V., Buriachok, V.: Monitoring subsystem for wireless systems based on miniature spectrum analyzers. In: Ageyev, D.V., Lemeshko, O.V. (eds.) 5th International Scientific and Practical Conference Problems of Infocommunications, Science and Technology (PICST), pp. 581–585. IEEE, Kharkiv (2018). https://doi.org/10.1109/ infocommst.2018.8632151
28. Gustavsson, R., Truksans, L., Carlsson, A., Balodis, M.: Remote security labs in the cloud. In: Rüütmann, T., Auer, M.E. (eds.) Engineering Education towards Excellence and Innovation (EDUCON), pp. 1–8. IEEE, Tallin (2015). https://doi.org/10.1109/educon.2015. 7095971

Predicting Pupil's Successfulness Factors Using Machine Learning Algorithms and Mathematical Modelling Methods

Solomia Fedushko$^{(\boxtimes)}$ and Taras Ustyianovych

Lviv Polytechnic National University, Lviv 79013, Ukraine
{solomiia.s.fedushko,
taras.ustyianovych.dk.2017}@lpnu.ua

Abstract. Taking into account the challenges and problems that are faced by the modern educational process, it is considered to use modern intelligent systems and algorithms to improve the education and teaching levels in educational institutions. The article describes an algorithm of actions on machine learning using, determining the students success level and analyzing the obtained data. This research can be efficiently used to find out and detect the modern educational problems, and individual and collective pupils sample features, implement the classification process and regression analysis of the data set. Results obtained from the algorithms usage, data analysis are described and demonstrated. The main features, knowledge and insights obtaining methods from the dataset are determined. The applied method is quite efficient and is capable of assessing pupil's performance metrics. Predicting student's and pupil's characteristics will help to segment and divide them into different classes so that it will allow pupils to develop communication, leadership, and self-management skills while studying at school or university. The results show that performance metrics assessment is an integral part of modern education process that is slightly crucial for its improvement and pupil's trends in education exploration.

Keywords: Machine learning · Intelligent systems · Data processing · EDA · Education · Modern educational system · School educational process

1 Introduction

The modern educational system needs a new qualitative teaching and studying arrangement model. For this, there is a need for an efficient information technology implementation in the scientific and educational sphere. A large number of data have a significant impact on the various processes formation in many areas of society. However, in order to benefit from their usage, a need to create a specific automation data processing model is crucial. Machine learning is able to solve a plenty of educational process tasks at once, in particular: pupils/students clustering; their classification, based on certain features and on certain groups; success and its factors prediction; improvement of the educational process level; finding an specific approach to each individual of educational institutions, etc. This article discusses the machine learning usage to predict and model the success level on the basis of a sample of pupils,

Z. Hu et al. (Eds.): ICCSEEA 2019, AISC 938, pp. 625–636, 2020.
https://doi.org/10.1007/978-3-030-16621-2_58

school graduates, finding factors that promote a certain pupil successfulness level, and regularity identification between student groups. The purpose of the study is to determine the performance metrics, using methods of mathematical modeling, machine learning, automate student's classification process, optimize educational and academic activities in schools and institutions of higher education, and find those factors that are most clearly shown in entrants. The methods described above were used in the following educational and research processes: the intellectual analysis of student sample data, exploration of hidden student individual characteristics, and optimization of various processes in higher education. This study will improve the approach of teachers to students, accomplish communication between them, enable various skills development in youth by identifying their interests, and contribute to improving the education level. The applied methods will help to accomplish the activity of the teachers, psychologists and others responsible for the educational process. The research methods will help to take new information about each student, taking into account pupil's personal characteristics. The educational sector will benefit from this, as it optimizes a plenty of processes at school or university, will enable more active skills development, provide young generation formation.

2   Related Work Analysis

AI (artificial intelligence) algorithms implementation is becoming widespread today. Their functions, namely machine learning and processes, can be integrated into the educational and teaching processes, its detailed forecasting and improvement. In 1959, Samuel [1] gave a clear definition of this phenomenon. Machine learning (ML) is a field of artificial intelligence that uses statistical techniques to give computer systems the ability to "learn" (e.g., progressively improve performance on a specific task) from data, without being explicitly programmed. For its implementation in the educational field both machine learning with a teacher (Supervised Learning) and without it (unsupervised learning) can be used. It can be well-used to process a large number of various data types - Big Data [2–4]. Accordingly, the result's quality is predetermined by the amount of data. Zhuang, Gan [5] proposed their machine learning usage method in the educational process in order to get a probability of student's enrollment at a high school. They used the logistic regression algorithm, adapting it to their developed model for assessing student admission at new educational institutions. The study of du Boulay [6] describes the further artificial intelligence techniques usage and improvement as means of helping students to gain new knowledge and certain skills. The teacher's activity and artificial intelligence systems meta-analysis, their comparison and probable efficiency during educational tasks execution and working with students are considered. The process will be implemented through data processing using artificial neural networks and algorithms. Narayanan, Kommuri, Subramanian, Bijlani suggested a methodology for testing questions formulation, analyzing the student efficiency metrics. They handled about 2,000 university student's score data, which allowed them to gain new knowledge and information about students. That method allowed an assessment question calibration for its further usage in educational process improvement. Kumar, Handa [7, 8] suggested methods for determining student's success using Data Mining techniques, reviewing the most important data analysis method references.

Research Hypothesis

Hypothesis 1: pupil's intereses influence on their performance metrics, determine successfulness level and make them better at studying. Exploratory data analysis methods were used in order to confirm or reject the hypothesis.

Hypothesis 2: machine learning models can classify pupils into 2 classes (successful, unsuccessful) using the following characteristics: gender, English level, intereses, productivity, average score, and desire to study abroad. Such algorithms were used: Decision tree, logistic regression, support vector machine, naive Bayes.

3 Main Part of Research

The article reveals factors that facilitate a particular student success, finds out hidden patterns between data of successful and unsuccessful students. The equation for determining the Ukrainian pupil success is proposed. Classification and application of machine learning algorithms on this sample allowed us to form the school graduate preferences and inclinations. Machine learning has become widely-used recently and that is why pupil and student classification and clusterization methods should be used. The issue of the university entrant regularities and success factors was quite mutable, as during the last decades, with the education change, teaching methods and a teaching process itselves, the young generation, especially adolescents, have changed markedly. The study/research goals and objectives are to propose an equation for determining the student achievement level; apply machine learning algorithms for classification process automation; find patterns in the dataset. The research will make it easier to find an approach to each pupil, to wit, a data aggregate will help to identify the main Ukrainian school development aspects improve the learning process; assist students in self-determination, using machine learning algorithms.

This problem solution helps to form a conscious young generation, make it easier to them to determine their goals. The data processing methods can accomplish the educational and teaching levels, which is most important in the Ukrainian and international education development context. Mathematical and statistical methods were used during the research process: online sociological survey; checking data for normal distribution (Gauss's law); machine learning algorithms (k-nn, decision tree, logistic regression, Naïve Bayes for classification). This research can be used globally, not only for Ukrainian educational institutions, but also abroad, as the young generation and adolescent trends are in some kind similar due to the globalization process. It will answer the question: which factor support particular student success [9]; allow determining how successful a student can be in advance by applying the algorithm on the available data. For Higher Education Institutions [10–13] and schools [14], this allows finding an approach to each student or pupil and understanding what favors his studying level. It will be able to improve the teaching and learning process by data processing, namely Big Data [15], using machine learning [16, 17] and intelligent data analysis methods [18–21]. If a large amount of data about each student is collected and it is processed in the right way, we can get new knowledge and information about the learning process and forming an informational image of a higher education institution [22–25], the education needs and ways of how to improve it. It helps to increase the success level of certain student and/or entrant groups as well, which will form a conscious young generation that can influence on a variety of social processes. The research was carried out using methods of pupil group achievement statistical and mathematical modeling. The data was collected through an online sociological survey, and an expert evaluation was used as well, which contributed to the increase in the number of informationally important dataset features. For this purpose, the R programming language was used. Some packages (tidyverse, caret) were used to promote the pupils classification, taking into account most of the available characteristics. The algorithm of actions is shown on Fig. 1. The algorithms purpose is to gradually highlight the main steps to achieve the research aim, allow understanding the machine learning algorithms application process, their order to determine the most important features of the dataset.

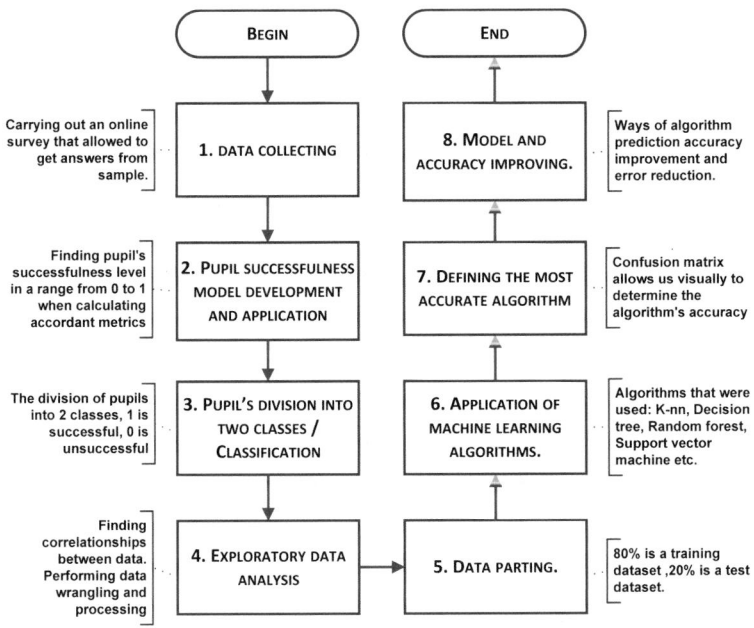

Fig. 1. Block of the algorithm deployment/implementation for the research purpose.

The algorithm of actions beginning is data collection process through an online sociological survey of 10–11 grade students who answered on about 20 questions. Their replies will give the main dataset characteristics.

Data processing was done using R programming language. It allows processing a large amount of information quite quickly, and to carry out qualitative visualization. The equation below is implemented and classification of the sample is carried out according to the results of equation application. In a certain field of the dataset, students with a high success level (P (Successfulness) >= 0.7) were marked as 1 and, accordingly, with a low level (P (Successfulness) < 07) as 0 (Fig. 2). The pupil classification on successful (1) and unsuccessful (0) was done.

EDA (Exploratory data analysis) has been conducted, in order to acquire new knowledge, improve data understanding. This step helps us draw conclusions by identifying correlations in the dataset, defining specific features, finding out the most important characteristics. This, in turn, allows us to create a machine learning model that will provide high prediction accuracy.

Data was randomly divided into training (80%) and test set (20%). The algorithm's accuracy depends on the input data amount. The higher the training set, the more accurate the result may be. Machine learning algorithms have been applied and the most accurate of them is determined. The main purpose is pupils' classification, which is why algorithms are used for this process. The most common are decision tree, logistic regression, k-nn, SVM (Support vector machine), random forest algorithms. This is not a complete list of all existing classification algorithms.

Defining the algorithm results is an important component of the research, which will further improve the entire system. Confusion matrix will detect an error in the classification process, find problems algorithms have encountered. As a result of confusion matrix, the most accurate algorithm will be defined.

Further system improvement is a key aspect of classification process development, its re-application/ reusage on new large amount of data. Model improvement can take place in the following ways: equation and algorithm model improvement for determining the pupil's/ student's success; definition of the most important features and characteristics of the dataset; increasing the amount of information, data.

Based on the research of Chiara, Johnes, Agasisti [26], which has described the process of finding the likely student's score on examinations, using machine learning, an equation for calculating the student's success has been developed. It has been modified and customized for the existent sample. The equation below, which is based on existing metrics, has been used to determine pupils success level at studying.

$$P(Successfulness) = A^{-1}n + \left(\left(\frac{\gamma_i}{r_i}\right) \cdot \gamma_{max} \cdot k_1\right) + k_2 + e_i \cdot k_3 + h_i + \vartheta_i \cdot k_4 \qquad (1)$$

where A is pupil's age; γ_i is average pupil's score [1:12]; r_i is pupil's productivity level (expert evaluation) [1:10]; γ_{max} is maximum average score of the sample [1:12]; e_i is English level [1:9]; h_i is possession of another foreign language (if yes, the $h_i = 0.01$, if no, then $h_i = 0$); ϑ_i is the number of additional lessons (in hours); n is value that corresponds for the age significance coefficient and is determined by the equation below:

$$n = \ln(A_{max}) - A_i \qquad (2)$$

where A_{max} is maximum pupil's age of the sample; k_1 is a coefficient for establishing the correlation between the maximum possible values $A_{max}:\gamma_{max}$ as 1:1, is equal to 0.1; k_2 is a coefficient for calculating the average score and productivity ratio, is calculated by the equation given below:

$$k_2 = \gamma_i \cdot k_5 \qquad (3)$$

where k_5 is equal to 0.05; k_3 is a coefficient for calculating the English level, and is equal to 0.01; k_4 is a coefficient, concerning the number of additional lesson hours, which is equal to 0.015.

4 Results and Discussion

The equation allows us to find values in the range from 0 to 1 inclusive ([1: 1]). If P (Successfulness) > 1, then it is rounded to 1. Before using machine learning algorithms, a pre data analysis has been conducted. According to its results, the most important features have been determined.

The histogram with P (Successfulness) value is shown on Fig. 2. This model is, to some extent, subject to Gauss's law. The red line depicts the arithmetic mean of this sample, it is 0.69. From this we can state the hypothesis that if P (Successfulness) >= 0.7, then the pupil is assigned a feature - successful (1), if not, then - less successful or unsuccessful (0). The histogram shows that the highest percentage of pupils($\sim 10\%$) have a success rate probability within [0.60; 0.69].

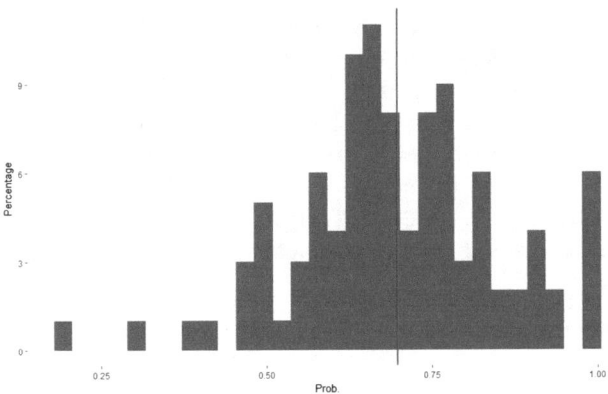

Fig. 2. Histogram P (Successfulness) value distribution among the entrants/pupils sample.

Most of the sample is female gender (65.3%), while the minority is male (34.7%). Figure 3 shows a definite studying direction for entrants by gender.

As we can see, women prefer non-technical (25%), humanitarian (8%) and natural science (12%) areas, whereas male gender prefer technical (16%). Obviously, this is due to the male predisposition to exact sciences such as mathematics, physics, etc.

A correlation between productivity and average score has been found during EDA. Females evaluate their productivity higher than males. Women have higher average score as well. However, men, who have quite high average score, nonetheless, estimate their productivity a bit lower, than women.

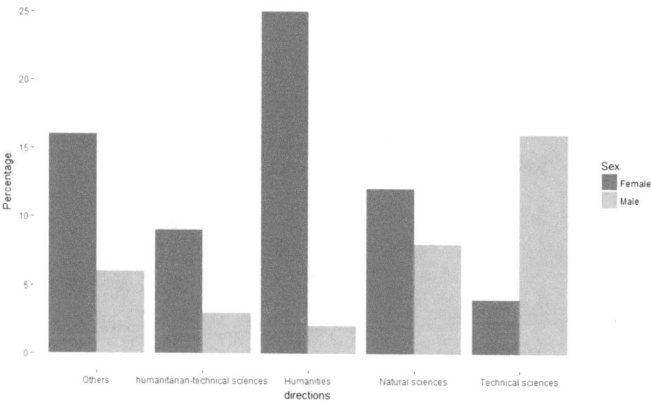

Fig. 3. Distribution diagram of the desired study direction.

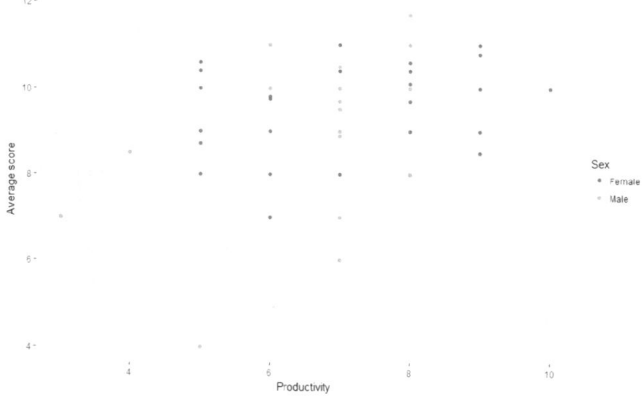

Fig. 4. Productivity and average score correlationship chart divided by gender (male/female).

Correlationship matrix of data features among themselves has been used (Fig. 5) to find out what factors are most conducive to certain indicators of pupil's success. The value of probable successfulness (Prob.) is most influenced by the English level (English_level). The value of this factor is 0.36, which is the highest value across the ma trix. English level and additional lessons (Additional_lessons) factors have influence on productivity (Productivity) as well.

We have filtered out the field "Studying abroad" in order to receive from the sample only those who wish to continue studying abroad, and have found their English level. The next step was to convert the resulting numerical data into a percentage format and change the order of English levels from Beginner to C1 (Advanced). It should be noted that no student from the dataset has had a C2 level. On Fig. 4 it is shown that C1 (Advanced) entrants are the most likely to continue studying abroad, while those with Level A1 (Elementary), A2 (Pre-intermediate), B2 (Upper-intermediate) want to study in Ukraine. The Beginner level answers were rather extreme, because with the

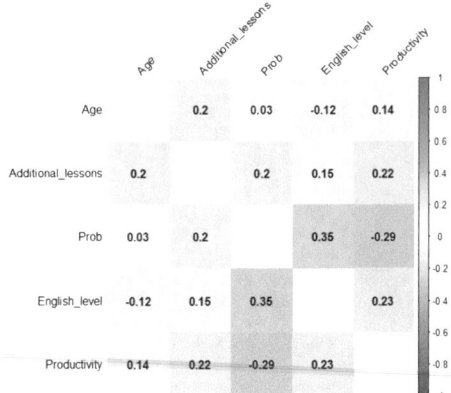

Fig. 5. Data correlationship matrix.

minimum knowledge of foreign languages it is difficult to study outside the country. The average is B1 (Intermediate), because only 18.5% are ready to continue studying abroad (Fig. 6).

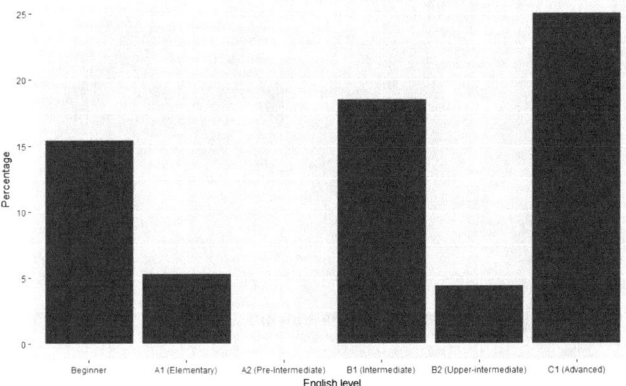

Fig. 6. The diagram of the pupil distribution by English level and desire to study abroad.

It turned out that this year's school and gymnasium students/pupils are more likely to choose history as one of desired subjects, while in lyceums – mathematics (Fig. 7).

This is explained by the fact that Ukrainian lyceums have a focus on mathematics and physics, and most Ukrainian lyceums are included in the list of the best secondary education institutions, while schools and gymnasiums can be profiled (free choice of certain subjects/the school's management determines which subject is profile).

The necessary stage of the research was the application of machine learning algorithms. We used classification algorithms to get the Successfulness column value, which consists of two levels: 1 (successful) and 0 (less successful or unsuccessful).

The code for cross-validation processes, model training based on the k-nn algorithm, and forecast is below. Sentences starting with "#" are comments.

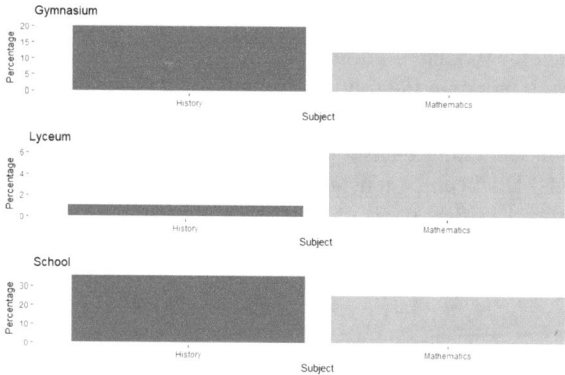

Fig. 7. Percentage advantage diagram of a certain subject.

```
### Cross validation.

trControl <- trainControl(method='repeatedcv',

                          number=15,

                          repeats = 3)
### Model training.

set.seed(222)

fit <- train(formula, data=training, method='knn',
tuneLength=20, trControl = trControl,

             preProc = c("center", "scale"))
### The model's output.

fit
### Making predictions

pred_train <- predict(fit, training)

pred_train
```

Showing the results. The accuracy is 0.9518 (95%) on the training dataset.

```
confusionMatrix(pred_train, training$successfulness)
```

In the "Model training" section, we specified the formula according to the following scheme:

Output ~ Input$_1$ + Input$_2$ + Input$_3$ ... + Input$_n$

Output is a feature of which class we want to define, while Inputs are those parameters on which the model will train and make predictions. The most accurate was the K-nn algorithm. It works in the following way: a certain element acquires the value that is most distributed on the graphic X, Y, among the nearest k-neighboring elements. During the research, a training dataset cross-validation with 15 numbers and 3 repetitions (45 iterations) has been conducted as well, and it is determined that we obtain the best forecast for classification from 21 k numbers of the nearest neighbors. The advantage of the algorithm is that it works well on numerical data, which turned out to be quite a lot to provide high-quality algorithm process. The forecast accuracy for the training sample is 95%, on the test sample −82%, which is a great result.

5   Conclusions

This research can be used to improve the learning and educational processes, to find the individual pupil and student characteristics. The machine learning algorithms will help to determine the successfulness level, interests of students, their foreign language proficiency level, and many others. The input data amount increasing, algorithm model formula and classification process improving are proposed for model and accuracy improving. Prediction accuracy can be improved with accomplishing the P (Successfulness) value equation, adding new features and factors to the dataset. Regression analysis, apart from classification, can be used for predicting discrete values, which will lead to data insights and new knowledge about the sample. Most of the pupil's (more than 60%) have foreign language proficiency (English) less than B2 (Upper-intermediate) but that has not a great influence on the performance metrics. The Hypothesis 1 is rejected; interests have no impact on student's performance. The Hypothesis 2 is confirmed using machine learning models; the described input factors give high accuracy (above 85%). The main successfulness pupil features were defined, the most significant factors, which facilitate to certain processes, were analyzed. The research will help to improve the educational process and pupil's successfulness level.

References

1. Samuel, A.L.: Some studies in machine learning using the game of checkers. IBM J. Res. Dev. **44**(1.2), 206–226 (2000)
2. Zhou, L., Pan, S., Wang, J., Vasilakos, A.: Machine learning on big data: opportunities and challenges. Neurocomputing **237**, 350–361 (2017)
3. Shakhovska, N., Vovk, O., Hasko, R., Kryvenchuk, Y.: The method of big data processing for distance educational system. In: Advances in Intelligent Systems and Computing II. Advances in Intelligent Systems and Computing, vol. 689, pp. 461–473. Springer (2018)
4. Shakhovska, N., Kaminskyy, R., Zasoba, E., Tsiutsiura, M.: Association rules mining in big data. Int. J. Comput. **17**(1), 25–32 (2018). Research Institute of Intelligent Computer Systems Pages
5. Zhuang, Y., Gan, Z.: A machine learning approach to enrollment prediction in Chicago Public School. In: 8th IEEE International Conference on Software Engineering and Service Science (ICSESS), Beijing, pp. 194–198 (2017)

6. du Boulay, B.: Artificial intelligence as an effective classroom assistant. IEEE Intell. Syst. **31** (6), 76–81 (2016)
7. Mukesh, K., Singh, A.J., Handa, D.: Literature survey on student's performance prediction in education using data mining techniques. Int. J. Educ. Manag. Eng. **6**, 40–49 (2017)
8. Kumar, V., Chaturvedi, A., Dave, M.: A solution to secure personal data when Aadhaar is linked with DigiLocker. Int. J. Comput. Netw. Inf. Secur. (IJCNIS) **10**(5), 37–44 (2018). https://doi.org/10.5815/ijcnis.2018.05.05
9. Korzh, R., Fedushko, S., Trach, O., Shved, L., Bandrovskyi, H.: Detection of department with low information activity. In: XIth International Scientific and Technical Conference Computer Sciences and Information Technologies CSIT-2017, Lviv, pp. 224–227 (2017).
10. Korzh, R., Peleshchyshyn, A., Syerov, Yu., Fedushko, S.: University's information image as a result of university web communities'. In: Advances in Intelligent Systems and Computing: Selected Papers from the International Conference on Computer Science and Information Technologies, CSIT 2016, vol. 512, pp. 115–127, Springer, Lviv (2016)
11. Korzh, R., Peleshchyshyn, A., Fedushko, S., Syerov, Yu.: Protection of university information image from focused aggressive actions. In: Advances in Intelligent Systems and Computing: Recent Advances in Systems, Control and Information Technology, SCIT 2016, Warsaw, vol. 543, pp. 104–110. Springer (2017)
12. Korzh, R., Fedushko, S.: Methods for forming an informational image of a higher education institution. Webology **12**(2) (2015). www.webology.org/2015/v12n2/a140.pdf
13. Korzh, R., Peleschyshyn, A., Syerov, Yu., Fedushko, S.: Principles of University's Information Image Protection from Aggression Proceedings of the XIth International Scientific and Technical Conference (CSIT 2016), Lviv, pp. 77–79 (2016)
14. Syerov, Y., Shakhovska, N., Fedushko, S.: Method of the data adequacy determination of personal medical profiles. In: Advances in Artificial Systems for Medicine and Education II. Proceedings of the International Conference of Artificial Intelligence, Medical Engineering, Education (AIMEE 2018) (2018, submitted for publication)
15. Monjurul, B.M.A., Courtney, M.: Educational data mining: a case study perspectives from primary to university education in Australia. Int. J. Inf. Technol. Comput. Sci. (IJITCS) **10** (2), 1–9 (2018)
16. Shakhovska, N., Shakhovska, K., Fedushko, S.: Some aspects of the method for tourist route creation. In: Advances in Artificial Systems for Medicine and Education II. Proceedings of the International Conference of Artificial Intelligence, Medical Engineering, Education (AIMEE 2018) (2018, submitted for publication)
17. Babichev, S., Korobchynskyi, M., Mieshkov, S., Korchomnyi, O.: An effectiveness evaluation of information technology of gene expression profiles processing for gene networks reconstruction. IJISA **10**(7), 1–10 (2018). https://doi.org/10.5815/ijisa.2018.07.01
18. Ali, M.M., Hani, M.I., Mohamed, F.T.: Identity verification mechanism for detecting fake profiles in online social networks. Int. J. Comput. Netw. Inf. Secur. (IJCNIS) **9**(1), 31–39 (2017)
19. El Haji, E., Azmani, A., El Harzli, M.: Using the FAHP Method in the Educational and Vocational Guidance. IJMECS **10**(12), 36–43 (2018)
20. Kaviyarasi, R., Balasubramanian, T.: Exploring the high potential factors that affects students' academic performance. Int. J. Educ. Manag. Eng. (IJEME) **8**(6), 15–23 (2018). https://doi.org/10.5815/ijeme.2018.06.02
21. Ehimwenma, E.K., Crowther, P., Beer, M.: Formalizing logic based rules for skills classification and recommendation of learning materials. Int. J. Inf. Technol. Comput. Sci. (IJITCS) **10**(9), 1–12 (2018)

22. Osaci, M.: Numerical simulation methods of electromagnetic field in higher education: didactic application with graphical interface for FDTD method. Int. J. Modern Educ. Comput. Sci. (IJMECS) **10**(8), 1–10 (2018)
23. Korobiichuk, I., Fedushko, S., Jus, A., Syerov, Y.: Methods of determining information support of web community user personal data verification system. In: Automation 2017. Advances in Intelligent Systems and Computing, vol. 550, pp. 144–150. Springer (2017)
24. Syerov, Y., Fedushko, S., Loboda, Z.: Determination of development scenarios of the educational web forum. In: 2016 XIth International Scientific and Technical Conference Computer Sciences and Information Technologies (CSIT), Lviv, pp. 73–76 (2016). https://doi.org/10.1109/stc-csit.2016.7589872
25. Fedushko, S., Syerov, Y., Korzh, R.: Validation of the user accounts personal data of online academic community. In: 13th International Conference on Modern Problems of Radio Engineering, Telecommunications and Computer Science (TCSET), Lviv, pp. 863–866 (2016). https://doi.org/10.1109/tcset.2016.7452207
26. Chiara, M., Johnes, G., Agasisti, T.: Student and school performance across countries: a machine learning approach. Eur. J. Oper. Res. **269**(3), 1072–1085 (2018)

Modeling of a Cooperative Distance Learning Environment: The Case of Optimal Size of Training Groups

Yurii Koroliuk[(⊠)]

Chernivtsi Institute of Trade and Economics KNUTE, Chernivtsi, Ukraine
yu_kor@ukr.net

Abstract. There has been made an attempt to build a formalized model of learning environment that would take into account endogenous and exogenous parameters of the e-learning process and foresaw the prognostication of its effectiveness. The conceptual model was formalized on the basis of statistic information about the learning procedure of 15 academic groups. As the formalization methods there have been chosen the method of self-organized Kohonen maps, Artificial Neural Network modeling and Group Method of Data Handling.

In the result of clustering with the help of Kohonen maps method there has been found a cluster that encloses academic groups with high performance rate and average number of students – 7. Artificial Neural Network model has proven that performance of students depends greatly on the group size. But, unlike the Kohonen maps, a recommended minimal group size is 10 students. Artificial neural network model and Group Method of Data Handling, though, show a somewhat different result within the scope of the learning course (European Credit Transfer and Accumulation System credits). There has also been discovered that the restriction of the applied methods is the impossibility to estimate the information overload.

Keywords: Cooperative distance learning · Data Mining · Group size · Students' progress · Model of the learning environment

1 Introduction

Learning environment is complex in its content, has active elements, needs organizing and technical supply. Consequences of organizing such environment are long-termed, and cover more than 5 years, and more often even 10 years. That is why formation the educational systems of different levels demands working out adequate models with the possibility to forecast the consequences of the leading parameters influence.

Moreover, modern means of collecting information allow accumulating Big Data, which need to be analyzed and taken into account in the mentioned models of the learning environment. Besides that, implementing network information support of the learning process, not speaking about the system of distance learning, gives a chance to save and monitor a lot of parameters peculiar to different participants of the learning process.

© Springer Nature Switzerland AG 2020
Z. Hu et al. (Eds.): ICCSEEA 2019, AISC 938, pp. 637–647, 2020.
https://doi.org/10.1007/978-3-030-16621-2_59

Learning is a system phenomenon, the structure of which has a number of principal elements. Thus, modern learning technologies foresee the interaction between the participants. It is considered that the group interaction reveals the participants' synergy and improves the performance. Though, the peculiarities of such interaction conditions of appearing and management remain open. The mentioned question was put into the basis of the present research.

The number of students in an academic group is the most attractive parameter when considering the dynamic management of the learning environment effectiveness. Though today, there are few adequate models able to show the dependency of the performance criteria on the number of students in a group. Building the model is complicated due to the big number of statistic research methods of the learning environment Big Data that make the choice of the most adequate method, which can be put into practice, rather difficult. The above mentioned stipulates for the necessity of learning information statistic analysis with the help of different methods in order to find the optimal number of students in a group, for it to have the highest performance rate.

2   Related Work

After the e-learning had appeared in scientific circles, there were started the discussions as to its effectiveness, advantages and disadvantages and peculiarities of its implementation.

Other advantages of distance learning are believed, according to T. Bender, to be the potential of online interaction, the possibility of learning on the Internet, supporting different learning styles and methods [1].

Cooperative learning differs from collaborative one, and as M. Neo states, in the process of such learning students should play and take an active part in the learning process, unlike the methods when students are passive and the teacher is the only authority in class and the only knowledge distributor [2]. More than that, such knowledge distribution leads to that the opportunity to work cooperatively is minimal as students focus on completing ascribed tasks by themselves [3].

In online environment, every student gets broad possibilities to critique, link, reformulate, and combine ideas [4]. Group interaction in e-learning becomes very significant. Many scientists concretize in different ways the advantages of cooperative and collaborative learning [5]. For example, they single out the importance of realizing the common aim when working in a group [6]. Undoubtedly, a common core of the mentioned features of the group interaction efficiency is an open discussion in online environment that should take place in all defined types of inner interaction of the cooperative learning participants [7]. An important fact is admitting the thing that participation measured by interaction has positive influence on learning achievement and gradation [8].

In the theories of cooperative learning the most dynamic and the easiest to control parameter is the number of people in a training group. The optimal number of members in a group may influence the results and the procedure of cooperative e-learning. Though, the size of the training group has been the object of discourse for the last decades.

Speaking about a large-sized group, we consider the problems of information overload of e-learning participants [9]. More specific, but not unanimous are the following suggestions as to the group size:

- 5 members - Abuseileek [10];
- 12 - Tomei [11];
- 13–15 - Qiu [12];
- from 8 to 30 - Rovai [13];
- 15–20 - Khan [14];
- 20 - Roberts and Hopewell [15];
- 25–30 - Arbaugh and Benbunan-Finch [16];
- to 30 - Aragon [17].
etc.

The main point in such suggestions is finding the maximum limit number of participants. A big amount of the participants will lead to the formation of more messages that will overload the discussion with reading materials. So, there is obvious a tendency to support the minimization of the group size. Nye and others substantiated experimentally for the school education that the point estimates of the small-class advantage were larger for lower achieving students in reading, the small-class advantage in mathematics was actually larger for higher achieving students [18]. Finn and Achilles [19] support the effectiveness of small groups only, if the task is to improve reading skills of the pupils. M. Qiu proves experimentally the effectiveness of dividing big online audiences into small discussion groups [12]: "In larger classes, participants were more likely to experience information overload and students were more selective in the notes that they read. A significant positive correlation was found between class size and total notes written. Students' note size and grade-level score were negatively correlated with class size.".

That is why, today organization of effective cooperative learning meets the problem of uncertainty as to the number of students in a training group. There exists the necessity to build a formalized model that would take into account endogenous and exogenous parameters of e-learning process and prognosticate its effectiveness assessment.

3   Methods and the Model

Modeling and learning environment prognostication tasks are solved with the help of application different approaches, in particular, regressive analysis and Data Mining [20], k-means algorithm [21] etc. Undoubtedly, the quality of the model will be defined by the corresponding statistic selection and results validation methods. But it is also believed, that in small-sized groups it is difficult to find the dependency between the number of students and their performance [18]. The main reason of the mentioned may be the lack of the corresponding statistic selection as well as the fact of influencing the progress by other factors. To overcome the stated disadvantage and get the accurate result we need a multiparametric model (1), that would take into account the main factors of the performance:

$$Y(a,b) = F(x_1, x_2, x_3, x_4, x_5) \tag{1}$$

where a – is the index of material mastering (% of the read learning material and colleagues' notes), b – performance index (high quality performance %) of mastering the course by the students, x_1 – number of students in a group, x_2 – number of European Credit Transfer and Accumulation System (ECTS) credits according to the curriculum, x_3 – number of discussions started by the tutor according to the curriculum, x_4 – the size of learning material including the tutor's and students' posts (text is measured in kilobytes), x_5 – averaged indices of students performance for the previous courses (high quality performance %).

As the object of investigation there was chosen the procedure of the learning course «N» in Chernivtsi Institute of Trade and Economics of KNUTE (CHITE KNUTE). The course is being taught for students in Year 3 and upper using full-time and distance forms of learning. When working in class, students listen to the tutor and get the task for independent work. The latter, as well as the additional theoretical material mastering is done in the environment Moodle on CHITE KNUTE (http://www.dist.chtei-knteu.cv.ua:8080/) website, where students form discussion groups and make notes. Moodle is a free and open-source learning management system (LMS) written in PHP and distributed under the GNU General Public License. Developed on pedagogical principles, Moodle is used for blended learning, distance education, flipped classroom and other e-learning projects in schools, universities, workplaces and other sectors [22–24].

On the basis of Moodle CHITE KNUTE system selection on the period 2005–2018 there was made the output table (Table 1). To the selection there were chosen the groups with the number of students 14–18, number of credits for the course from 1,5 to 4,5. The selection was formed on the basis of 15 training groups and total number of students 171. An important parameter was the number of discussions planned by the tutor from 10 to 30 in the case of different groups. The selection was made on the basis of all notes taken by the tutor and students in the system Moodle. These notes were measured in kilobytes with the help of transporting into txt format. As, the final performance is not statistically reliable, there were chosen average indices of students' high performance rate of the group for the whole period of their learning. The percentage selection of the notes read by the students in a group was taken separately. It should be mentioned that the system Moodle is unable to estimate full and thoughtful reading of the posts made by students and the tutor. So, as an estimation point there was taken the fact of opening a post by a student in the group. And, finally, the column "Group number" was informational.

With the aim of interactive distance learning environment modeling there was used the Data Mining technology. The information from Table 1 corresponds to the multidimensional object of investigation – a training group. The above mentioned gave the opportunity to search for groups with similar features.

In particular, there was built the Kohonen map [25] for revealing the group clusters with high and low performance. Besides that, there was also built Artificial neural network (ANN) model of interactive distance learning environment.

Table 1. Data set of interactive distance learning environment

Group number	x_1	x_2	x_3	x_4	x_5	a	b
001	4	1,5	10	982	60	100	65
002	5	1,5	10	1090	62	96	63
003	7	1,5	10	1254	58	98	60
004	7	3	15	1540	52	92	51
005	8	1,5	15	1321	54	90	55
006	9	4,5	30	2010	60	91	60
007	10	3	12	1420	61	89	63
008	13	3	17	1680	55	90	56
009	13	1,5	11	1700	56	88	60
010	14	4,5	25	2450	71	84	68
011	15	3	15	2100	53	81	55
012	15	3	14	1980	55	79	56
013	15	3	16	2050	62	79	61
014	18	1,5	9	2060	64	86	60
015	18	3	13	2350	58	78	59

Where a – is the index of material mastering (% of the read learning material and colleagues' notes), b – performance index (high quality performance %) of mastering the course by the students, x_1 – number of students in a group, x_2 – number of ECTS credits according to the curriculum, x_3 – number of discussions started by the tutor according to the curriculum, x_4 – the size of learning material including the tutor's and students' posts (text is measured in kilobytes), x_5 – averaged indices of students performance for the previous courses (high quality performance %).

The potential of using ANN modeling of complex processes and systems is acknowledged by numerous successful objects under investigation in medicine, biology, sociology, economics, etc. The basis of the success was the connectionist architecture of ANN and its universal approximation ability that emerges from the interactions among its neurons. Taken separately, a single neuron has no real abilities; it is capable only of performing extremely simple operations such as addition and multiplication [26].

As the applied environment for Data Mining, there was chosen the academic version of the program Deductor Studio Lite 5.1 released by Base Group Labs. Taking into account the fact that today there are no generally accepted methods of neural networks architecture formation, the given process is to great extent subjective and dependent on the skills of a researcher as well as on the given task conditions. So, in the process of architecture selection, there were tested more than 15 types with different number of hidden layers and neurons to reach the acceptable exactness of the analysis. The final architecture of the built neural network consists of 5 inputs, 2 hidden layers, that have 5 and 2 neurons accordingly and 2 outputs. The limit number of hidden elements is agreed with the investigation approach [27].

The entries were divided into learning and test sets in the proportion 70 and 30% accordingly. We used sigmoid as an activation function of neurons. As the learning algorithm there was used the method of back propagation [28]. The learning error for the whole data set was 0,5–15%. In the learning process we have identified 100% of test and learning selection that is the evidence of the satisfactory chosen neural network and confirms the adequacy of the model.

To assess the adequacy of the ANN model there was used Group Method of Data Handling (GMDH) when solving the functional (1). The GMDH was first developed by Ivakhnenko [29] as a multivariate analysis method for complex systems modeling and identification. The main idea of the GMDH is to build an analytical function in a feed forward network based on a quadratic node transfer function whose coefficients are obtained using the regression technique [30]. As the applied environment of GMDH realization we have chosen the program GMDH Shell DS.

4   Results Analysis

As it was already stated, clustering of training groups was held with the help of Kohonen maps method. Clustering parameters: number of epochs through which the lines are mixed - 20; method of initial map parameterization – of own vectors; neighboring function is realized step by step; the initial learning radius is 4; the final learning result – 0,1.

The clusterization result is shown in Table 2. With the aim of adequate estimation of performance increase, as its parameter, there was chosen the difference between the current performance of students for the group mastering course "N" and average quality performance of the same group for the previous periods of learning (b-x_5).

So there are got 4 cluster groups: the first one with high performance indices and average number of students about 7 people; the second one with worse performance indices and average number of students – 16; the third one with worse performance and average number of students - 10; the fourth one with the worst performance and 12 students in an averaged group.

The next result was gained in the process of scenario use of the learned ANN model (Fig. 1).

As scenarios, average indices of parameters x_3 (14,8 units), x_4 (1732,5 kilobytes), x_5 (58,7%) were put to dependency (1). According to ANN model the excess of performance indices higher from average ones (b-x_5) depends greatly on the group size. So, in the case of groups with less than 10 members, we see low performance. Except this, when a learning course has 3 credits ECTS, the performance indices are lower for groups with less than 14 members and more than 17.

Table 2. Training groups clusterization

Group number	x_1	a	b	Cluster number	$b-x_5$
001	4	100	65	0	5
002	5	96	63	0	1
003	7	98	60	0	2
005	8	90	55	0	1
009	13	88	60	0	4
011	15	81	55	1	2
012	15	79	56	1	1
013	15	79	61	1	-1
014	18	86	60	1	-4
015	18	78	59	1	1
004	7	92	51	2	-1
007	10	89	63	2	2
008	13	90	56	2	1
006	9	91	60	3	0
010	14	84	68	3	-3

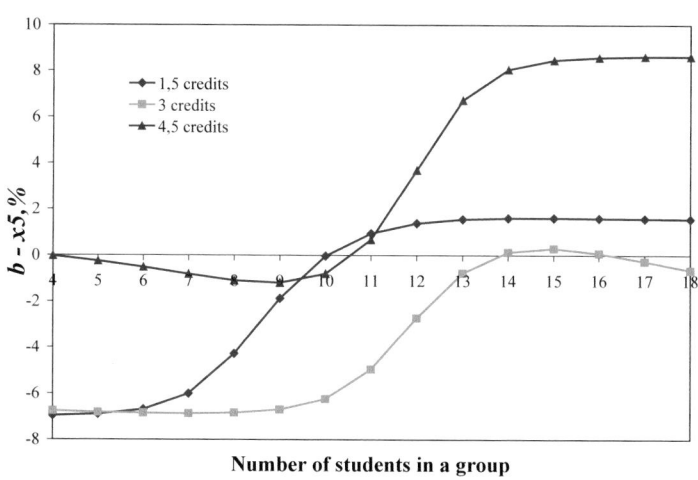

Fig. 1. Performance dependency on the number of students according to ANN model

To substantiate the adequacy of the ANN model, functional (1) was solved with GMDH. In the process of solving, there were input the following parameters: the observations were mixed – odd and even; the relations of learning/check was 60 to 40%; maximum number of neuron layers was 5; the initial width of the layer is 500. The result of using GMDH was dependency (2).

$$b = 0,028 + (1,39 - (2,19485 - x_1^2 \cdot 0,008) \cdot 1,685 +$$
$$(-0,053 - x_2 \cdot (0,201 + (7,043 - x_4 \cdot x_5 \cdot 0,00009 + x_4^2 \cdot 7,039e - 07) \cdot 1,82012 -$$
$$x_1^2 \cdot 0,0079)^2 \cdot 0,6) \cdot 0,65 + (0,201 + (7,043 - x_4 \cdot x_5 \cdot 0,00009 +$$
$$x_4^2 \cdot 7,04e - 07) \cdot 1,82 - (2,19 -$$
$$x_1^2 \cdot 0,008)^2 \cdot 0,6) \cdot 2,518) \cdot 1,606) * 1,065 - (-0,085 + (-0,54 -$$
$$x_3 \cdot (0,52 + x_1 \cdot (7,043 - x_4 \cdot x_5 \cdot 8,49e - 05 + x_4^2 \cdot 7,04e - 07) * 0,088) * 0,26 +$$
$$(0,52 + x_1 \cdot (7,04 - x_4 \cdot x_5 \cdot 8,48e - 05 + x_4^2 \cdot 7,04e - 07) \cdot 0,09) \cdot 4,87) \cdot 04 +$$
$$(0,16 + (0,21 - x_2(7,04 - x_4 \cdot x_5 \cdot 8,48e - 05 + x_4^2 \cdot 7,04e - 07) \cdot 0,565 +$$
$$(7,04 - x_4 \cdot x_5 * 8,49e - 05 + x_4^2 * 7,04e - 07) \cdot 2,08) \cdot 2,2 -$$
$$(-0,03 + x_5 \cdot (7,04 - x_4 \cdot x_5 \cdot 8,49e - 05 +$$
$$x_4^2 * 7,04e - 07) \cdot 0,02)^2 \cdot 0,576) \cdot 0,65)^2 \cdot 0,03$$

$$(2)$$

The coefficient of model (R^2) determination functional (2) is equal to 0,8 that confirms its moderate credibility. Substituting the conditions of the above described scenarios in (2) there will be obvious the performance dependency on the number of students in a group (Fig. 2).

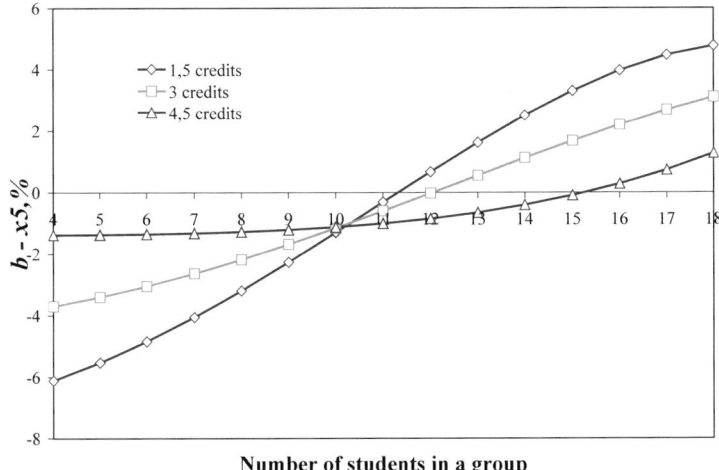

Fig. 2. Performance dependency on the number of students according to GMDH

As it is seen from Fig. 2, GMDH confirms the performance increase in the case of the number of students in a group – more than 10–11. Though, GMDH shows a somewhat different result within the scope of the learning course ECTS credits number.

So, with the increase in the course volume for better performance, bigger groups should be formed: 1,5 credits – 11–12 students, 3 credits – 12 and 4,5 credits – 15 students.

All three methods of statistic investigation confirm the possibility to find the performance rate dependency on the number of students in a group and other accuracy criteria. But from the practical point of view, the obtained results reveal some restrictions as to the used analysis methods, in particular, the impossibility to estimate the information overload that is proven by the results on Figs. 1 and 2.

5   Conclusions, Limitations and Future Work

There was made an attempt to build a learning environment formalized model that would take into account endogenous and exogenous parameters of the e-learning process and foresaw the prognostication of its effectiveness. As the concept of such model there was offered the dependence of mastering the material and students' performance index in the course: number of students in a group; number of ECTS credits according to the curriculum; number of discussions planned by the tutor; volume of learning material with an account of the tutor's and students posts; averaged performance indices for the previous years.

Conceptual model was formalized on the basis of statistic information about the learning procedure of 15 academic groups. As the formalization methods we have chosen the method of self-organized Kohonen maps, ANN modeling and GMDH.

In the result of clustering with the help of Kohonen maps method there was found a cluster that encloses academic groups with high performance rate and average number of students – 7. ANN model has proven that performance of students depends greatly on the group size. But, unlike the Kohonen maps, a recommended minimal group size is 10 students. ANN model and GMDH, though, show a somewhat different result within the scope of the learning course (ECTS credits).

The restriction of the applied methods is the impossibility to estimate the information overload that is proven by the results on Figs. 1 and 2. With the aim to take into account the latter, there is a need to complicate functional (1) by parameters of information notes exchange, messages between the course audience and its formalization in the form of model imitation.

References

1. Bender, T.: Discussion-Based Online Teaching to Enhance Student Learning. Stylus Publishing, Sterling (2003)
2. Neo, M.: Developing a collaborative learning environment using a web-based design. J. Comput. Assist. Learn. **19**, 462–473 (2003)
3. Sharon, S., Shachar, H., Levine, T., Westport, C.T.: The Innovative School: Organization and Instruction. Bergin & Garvey (1999)
4. Smith, M., Winking-Diaz, A.: Increasing students' interactivity in an online course. J. Interact. Online Learn. **2**(3), 1–25 (2004)

5. Njenga, S., Oboko, R., Omwenga, E., Maina, E.: Use of intelligent agents in collaborative M-learning: case of facilitating group learner interactions. Int. J. Mod. Educ. Comput. Sci. **10**, 18–28 (2017)
6. Gillies, R., Ashman, A.: Behaviour and interactions of children in cooperative groups in lower and middle elementary grades. J. Educ. Psychol. **90**, 746–757 (1998)
7. Damon, W., Phelps, E.: Critical distinctions among three approaches to peer education. Int. J. Educ. Res. **13**, 9–19 (1989)
8. Patel, J., Aghayere, A.: Students' perspective on the impact of a web-based discussion forum on student learning. In: Proceedings of the 36th ASEE/IEEE Frontiers in Education Conference, Sandiego, CA (2006)
9. Hewitt, J., Brett, C.: The relationship between class size and online activity patterns in asynchronous computer conferencing environments. Comput. Educ. **49**, 1258–1271 (2007)
10. Abuseileek, A.F.: The effect of computer-assisted cooperative learning methods and group size on the EFL learners' achievement in communication skills. Comput. Educ. **58**, 231–239 (2012)
11. Tomei, L.A.: The impact of online teaching on faculty load: computing the ideal class size for online courses. J. Technol. Teach. Educ. **14**(3), 531–541 (2006)
12. Qiu, M.: A mixed methods study of class size and group configuration in online graduate course discussions. Open library published doctoral dissertation, University of Toronto, Toronto, Ontario, Canada (2009)
13. Rovai, A.P.: Facilitating online discussions effectively. Internet High. Educ. **10**, 77–88 (2007)
14. Khan, B.: Managing e-learning Strategies: Design, Delivery. Implementation and Evaluation. Idea Group Inc. (2005)
15. Roberts, M.R., Hopewell, T.M.: Web-based instruction in technology education. Council on Technology Teacher Education. 52nd Yearbook: Selecting Instructional Strategies for Technology Education. McGraw Hill, Glencoe (2003)
16. Arbaugh, J.B., Benbunan-Finch, R.: Contextual factors that influence ALN effectiveness. In: Hiltz, S.R., Goldman, R. (eds.) Learning Together Online. Research on Asynchronous Learning Networks, pp. 123–144 (2005)
17. Aragon, S.R.: Creating social presence in online environment. New Dir. Adult Contin. Educ. **100**, 57–68 (2003)
18. Nye, B., Hedges, L.V., Konstantopoulos, S.: Do low-achieving students benefit more from small classes? Evidence from the tennessee class size experiment. Educ. Eval. Policy Anal. **24**(3), 201–217 (2002)
19. Finn, J.D., Achilles, C.M.: Answers and questions about class size: a statewide experiment. Am. Educ. Res. J. **27**, 557–577 (1990)
20. Zacharis, N.Z.: Classification and Regression Trees (CART) for predictive modeling in blended learning. Int. J. Intell. Syst. Appl. (IJISA) **10**(3), 1–9 (2018). https://doi.org/10.5815/ijisa.2018.03.01
21. Rawat, B., Dwivedi, S.K.: An architecture for recommendation of courses in e-learning system. Int. J. Inf. Technol. Comput. Sci. (IJITCS) **9**(4), 39–47 (2017). https://doi.org/10.5815/ijitcs.2017.04.06
22. Costello, E.: Opening up to open source: looking at how Moodle was adopted in higher education. J. Open Distance e-Learning **28**(3), 187–200 (2013)
23. Muhsen, Z.F., Maaita, A., Odah, A., Nsour, A.: Moodle and e-learning Tools. Int. J. Mod. Educ. Comput. Sci. (IJMECS) **5**(6), 1–8 (2013). https://doi.org/10.5815/ijmecs.2013.06.01
24. Marikar, F.M.M.T., Jayarathne, N.: Effectiveness of MOODLE in education system in Sri Lankan University. Int. J. Mod. Educ. Comput. Sci. (IJMECS) **8**(2), 54–58 (2016). https://doi.org/10.5815/ijmecs.2016.02.07

25. Kohonen, T.: Exploration of very large databases by self-organizing maps. In: Proceedings of International Conference on Neural Networks (ICNN 1997) (1997)
26. Basse, R.M., Omrani, H., Charif, O., Gerber, P., Bódis, K.: Land use changes modelling using advanced methods: cellular automata and artificial neural networks. The spatial and explicit representation of land cover dynamics at the cross-border region scale. Appl. Geogr. **53**, 160–171 (2014)
27. Nielsen, R.H.: Kolmogorov's mapping neural network existence theorem. In: Proceedings of The IEEE First International Conference on Neural Networks, San Diego, pp. 11–13 (1987)
28. Caruana, R., Lawrence, S., Giles, L.: Overfitting in neural nets: backpropagation, conjugate gradient, and early stopping. In Advances of Neural Information Processing Systems, vol. 13, pp. 402–408 (2001)
29. Ivakhnenko, A.G.: Polynomial theory of complex systems. IEEE Trans. Syst. Man Cybern. **SMC-1**(4), 364–378 (1971)
30. Amanifard, N., Nariman-Zadeh, N., Borji, M., Khalkhali, A., Habibdoust, A.: Modelling and Pareto optimization of heat transfer and flow coefficients in microchannels using GMDH type neural networks and genetic algorithms. Energy Convers. Manag. **49**(2), 311–325 (2008)

Ontology-Based Approach for E-learning Course Creation Using Chunks

Sergiy Syrota[ID], Sergii Kopychko[ID], and Viacheslav Liskin[✉][ID]

Department of Applied Mathematics, National Technical University of Ukraine
"Igor Sikorsky Kyiv Polytechnic Institute", 37 Peremohy Avenue, Kiev, Ukraine
`sergiy.syrot@gmail.com`, `kopuchko.sn@gmail.com`,
`liskinslava@gmail.com`

Abstract. The article is devoted to the chunk-based approach for course ontology creation and didactics of E-learning. Concept of chunk-based approach for course ontology creation will be proposed. Also described what benefits can be obtained using chunk-based approach to accomplish the analysis of student knowledge. A new model of monitoring the interaction between students and chunk-based content is proposed. Development tools of this software engine are explained. Implementation is presented and grounded. As a result, an ontology-based plugin for MOODLE was developed and examples of use are shown.

Keywords: E-learning · Ontology Engineering · Chunk based ·
Ontology-driven E-learning

1 Introduction

More and more people are known to choose online education. The distance courses can be completed at any time and without reference to a specific location. The use of cloud technology has made education more integrated, networked and composite. This makes e-learning as highly effective as the conventional method of learning delivery [1].

Among the e-learning systems most common are LMS (Learning Management Systems) and MOOC (Massive Open Online Courses), Such systems are characterized by high rates of emergence of new educational resources and a large amount of educational data.

At the same time, each new electronic course, as a rule, exists independently, and its authors create all the necessary content from beginning to end. There is no reuse of already existing materials of electronic courses. This leads to significant duplication of information and makes it difficult to analyze both the courses themselves and the results of e-learning. The difficulty of sharing educational materials and facilitate access to content with the help of semantic web to make them accessible and available to different users is one of modern problems [2].

One of the reasons for these problems is the lack of formal links between electronic courses and descriptions of subject areas. Existing systems of e-learning provide data on the revision of certain educational resources and the facts of the implementation of

Z. Hu et al. (Eds.): ICCSEEA 2019, AISC 938, pp. 648–657, 2020.
https://doi.org/10.1007/978-3-030-16621-2_60

independent, control and test work, rather than the complex study of a particular subject area. Also, when monitoring the results of learning, information can be obtained only on the results of passing a certain control measure without a detailed understanding of the learning material at the level of understanding of the subject area.

To date, Internet technologies are actively developing on the basis of ontologies and Web 3.0. Their main purpose is the publication of semantic data for Internet resources, which allows you to create a new, search and logical analysis of Internet content. However, these technologies are not used in systems of computer training and distance education, since there is no ontology of educational disciplines [3].

Recently, the idea of creating Learning Object/Content Repositories (LORs) based on metadata has begun to develop [4]. But there is no common metadata that allows to retrieve similar content from different LORs [4, 5]. The approach based on meta-ontology was proposed in previous works of the authors [6]. In this paper developed the chunk based approach to ontology driven LMS, described e-learning process based on chunks. Also, this paper will demonstrate how an individual learning pathway can be built based on the chunk based approach. All this technologies will give new possibilities in educational field and increase the quality of education. Additionally, the subsystem of monitoring student's knowledge has already been implemented as an example of a plug-in to the LMS MOODLE.

2 The Chunk Based Approach to Ontology Driven LMS

Different kinds of educational content are used in e-learning, such as texts, slides in presentation, video, sound, quizzes, tests. There is a need to link this content among themselves and to allocate a certain "portion" of knowledge, that is, a set of logically related learning material that can be perceived as a unit that is generalized by the topic of learning.

According to the technology offered in [5], the educational material is organized according to meta-ontology approach. The basic essence of such approach is meta-ontological system.

$$Z = <O^{Meta}, P, E>$$ (1)

where O^{Meta} is ontology of the upper level (meta-ontology) which contains general concepts and relations that do not depend on the subject domain and are common for all courses; P is a set of ontologies of subject domain and certain courses; E is rules model of e-learning engine which provide study process.

Meta-ontology O^{Meta} consists of two parts, $O^{Content}$ is a content ontology and $O^{Didactic}$ is a didactic ontology.

The unit of knowledge is facts, information, and skills acquired through experience or education, also the theoretical or practical understanding of a subject. A chunk (a unit of information in one's memory) is an array of letters, words or numbers.

To indicate a certain "portion" of knowledge, that is, a set of logically related learning material, the terms have been used: model, notion, image, quantum and chunk [8–11].

A review [4] of the terms for determining a certain set of logically related material has been allowed to find out that it is more appropriate to use the term "chunk".

The chunking methodology allows us to expand the borders of short-term memory through unification of information into a smaller number of units. Short-term memory can effectively process four blocks, give or take one. Chunking is often used as a general method for simplification of processing [12].

2.1 The Didactic Ontology

The base element (entity) of the didactic ontology is a chunk, while content mapping, and relations conform a didactic model of a discipline.

$$O^{Didactic} = <C, L, R> \tag{2}$$

where $C = \{c_i\}$ is a set of chunks which compose the didactic ontology; $L_i = \{l_{m_i}\}$ is a set of content mappings; $R = \{r_{ij}\}$ is a set of relations, there are considered two types of relations R_1 and R_2. $R_1 \subset C \times C$ is relations among chunks; $R_2 \subset C \times L$ is relations among chunks and content.

Content mappings connect content blocks within the framework of one discipline and varying types of content, whether it is a text file, presentation, video file, or quiz. This allows us to display synchronously different types of content in frames of one browser or even on different devices (Fig. 1).

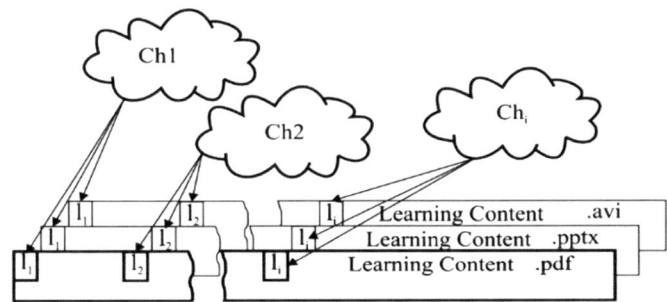

Fig. 1. Chunks to content mapping

The didactic idea is to organize the multi-flow study process. When the different types of content are displayed in few windows simultaneously. For example, one window contains text information, next window shows presentation, another window displays a quiz. This approach gives us the opportunity to organize interactive quizzes. And create a personal learning path depends of student answers (Fig. 2).

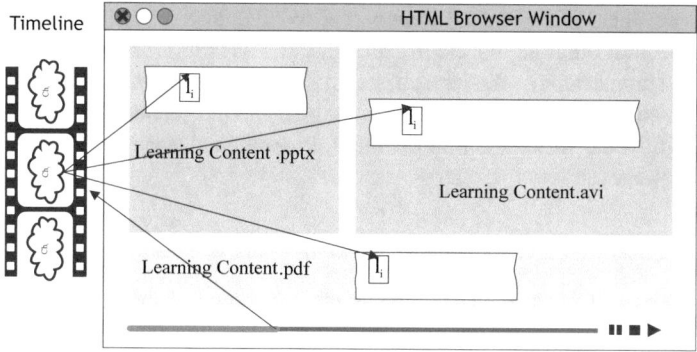

Fig. 2. Multiframe learning.

It can be distinguished that there are relations r_{ij} of different types among the chunks of subject domain. One of that types is "need to understand", for example, to understand what the decision-making process is, student must to understand what criteria, alternatives, scales. and decision-making body are. These relations form a semantic graph of discipline. $<C_i, R>$ whose vertices (C_i) are chunks, and the edges R are the set of relations, "need to understand".

An example of the ontograph fragment of the discipline "Models and Methods of Supporting Decision Making" (Fig. 3).

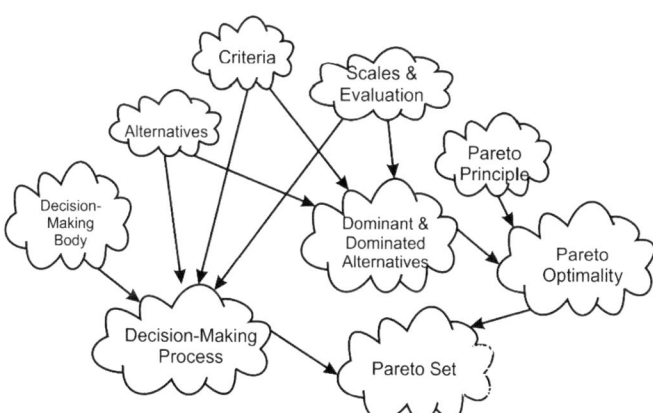

Fig. 3. An example of the ontograph fragment of the discipline "models and methods of decision support"

The vertices of this graph correspond to the chunks, and the edges show the sequence of learning the study material. The topological sorting of the vertices of the graph allows to construct the initial pathway of learning.

In this manner, for example, a chunk called "Pareto Set" follows after all others because it has not outgoing edges. In fact, edges of this kind semantic graph mean the relation "is studied after". As we can see this relation forms directed acyclic graph. Topological sorting of vertexes of this graph gives us the sequence of chunks (topics) to be studied. At the same time impossibility of topological sorting tells us that didactic mistakes present.

2.2   Test Questions and Quizzes Generated Using the Chunks

The content part of meta-ontology reflects the other type of the connections between chunks. In contrast to the traditional approaches to building ontologies of educational disciplines, which in fact consisted in constructing an ontology of the subject domain, it is proposed to construct an ontology by extracting from the text predicates of the form:

notation – essence – link(relation) – description,

For example, the chunk «Pacific States» from USA natural history can relate to next predicate:

"The nick name of State of Oregon is The Beaver State."

In this case *"The nick name of State of"* corresponds to notation, *"Oregon"* – to essence, *"The Beaver State"* – to description, *"is"* – to link(relation). It can be formalized to consider a chunk as a set of chains:

$$C: <N, E, L, D > \qquad (3)$$

where N is notation $E = \{E_i\}$ is a tuple of essences of subject area, L are links between essences and descriptions, $D = \{D_{ij}\}$ is a tuple of descriptions.

Should be noted that every E_i can be substituted for each other (i.e., E_i are in the same class) have paradigmatic relation, this is also true for every D_{ij} with same j. On other hand E with L. However, L are in syntagmatic relation with D. That means L are related semantically with D (can be combined each with other).

Formally the predicates of domain, examples above, can be written like:

"The nick name(s) of the State of E_i is(are) D_{ij}".

Then the set of true predicates must be written:

"The nick name(s) of State of Oregon is(are) {The Beaver State};"
"The nick name(s) of State of Washington is(are) {The Evergreen State};"
"The nick name(s) of State of California is(are) {The Golden State; El Dorado State};"
"The nick name(s) of State of Alaska is(are) {The Last Frontier; Land of the Midnight Sun};"
"The nick name(s) of State of Hawaii is(are) {The Aloha State, Paradise of the Pacific};"

It should be noted that some states may have several nicknames. This fact is reflected by set of D_i in curly brackets. Test questions of closed type one or multiple answer can be generated by combining chains $\langle N, E_i, L, D_{ij} \rangle$ as a correct answers? and chains $\langle N, E_i, L, D_{kj} \rangle$ where $k \neq i$ is destructors.

Using one of supported by MOODLE test question format GIFT [13] the question may look like:

The nick name(s) of State of Alaska is(are) {
~ The Beaver State
~%50% The Last Frontier
~%50% Land of the Midnight Sun
~ The Evergreen State}

In case of wrong or not complete answer student will be directed to WEB page linked with chunk "Alaska nickname" had containing information about origin of nickname.

"Alaska is commonly known as the Last Frontier and sometimes the land of the midnight sun. The frontier reference is due to Alaska being admitted as the 49th state of the union and its location far from the contingent 48 states. Moreover, the rugged landscape and frontier spirit of Alaskans supports this nickname."

For each chunk, a test questions bank Q has been generated for its content ontology.

$$Q = \bigcup_{i=1}^{n} Q_i \qquad (4)$$

where Q_i is the set of generated questions for C_i, $i = 1, \ldots, n$.

In this case, each student has a vector of grades $-\bar{v}$, such that:

$$\bar{v} = (g_1, \ldots, g_n) \qquad (5)$$

where $g_i \in \{-2, -1, \tau, 1, 2\}$

In this case, $g_i \leftrightarrow C_i$, i-th grade corresponds to the i-th chunk.

3   E-Learning Process Based on Chunks

The initial condition of the marks is uncertain, that is, $\bar{v} = (\tau, \ldots, \tau)$.

In the process of learning, the student answers Q_i test questions in accordance with the constructed initial learning pathway.

The grade changes after each student's answer to the question of the corresponding chunk. A graph of possible transitions between grades is shown on Fig. 4.

The arrows marked with the $+$ sign means the correct answer; the rest is not correct. It is assumed that -1 is the lowest grade for a chunk that a student can get, and 2 is the highest.

The ontology-driven system collects information about the student activity during his interaction with LMS using his logs. This information are a history of content attendance and history of grades changes. On the basis of information received system provides recommendations for the possibility of further study or the need for re-studying certain material related with a specific chunk.

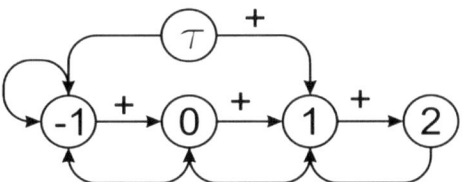

Fig. 4. Graph of conversions between grades

After completing the study of the credit module, each student has his own vector of marks, which system analyzes.

If $\forall g_i = 2$, then the ontology driven e-learning system will move to the next credit module.

If $\exists g_i = \tau$, then the ontology driven e-learning system cannot move to the next credit module, since remaining chunk for which the grade is undefined.

If $\exists g_i = -1$, then the ontology driven e-learning system cannot go to the next credit module, because there are chunk grades for which are unsatisfactory, and re-examination of the material is required.

With the help of the evaluation vector, an ontology driven e-learning system provides recommendations for further education and builds an individual pathway of learning in accordance with the given didactic part of the meta-ontology of the discipline.

Thus, the work with the content and content of the content ontology of the discipline corresponds to the process of knowledge management. Didactic ontology of the discipline implements the sequence of learning, that is, the formalization of the individual learning pathway.

4   Ontology-Based Plugin for MOODLE

The results provide using the e-learning education in MOODLE platform is highly satisfactory [14]. Due to this the ontology-based plugin was developed for MOODLE.

To implement a developed ontology-based plugin for MOODLE, the LAMP web technology stack is used, which represents the main components for building a web service (Linux, Apache, MySQL, PHP).

For a more convenient and quick development of an ontology-based plugin, the special Yii 2 PHP framework is used.

For the separation of users in an ontology-based plugin, the RBAC (Role Based Access Control) module is used to divide users into administrators, teachers, and students.

In order to maintain a permanent HTTP connection between the client and the server in order to implement a synchronized display of educational content, students receive real-time responses using the COMET web application model, namely AJAX long polling technology. As a data format, AJAX uses XML. Unlike regular AJAX queries that, after a certain period of time, send requests to the server and check whether the data has changed, the long pooling browser sends one request to the server and keeps the open connection until the data is changed. When scaling a system, developing new modules that require a permanent connection to the server, such as chat, it is possible to use WebSocket technology. With AJAX, the work on the content of educational disciplines in multi-mode mode is realized.

The screenshot with results of work of ontology-based plugin for MOODLE is presented on Fig. 5.

Fig. 5. The screen form displaying the student activity

Every student can check his results being logged on, using the MOODLE menu commands:

Profile− > Course details− > ⟨Check Course⟩− > Reports− > Status overview.

The first column of a table indicates a list of topics to be studied in appropriate sequence. Second column indicates a grade for topic. The bars indicate part of learning content to have been reviewed by student. Last column keeps the links to log of student's activity on each topic. This screen form contains a list of links to resources that need to be refined in the sequence determined by the individual learning pathway.

The interactive studying module provides the following features: students may review lecture materials in few-framed mode (text and presentation mode, video-lectures, interactive quiz) due to detection of the main objects, terms, chunks, subjects and their structure and referencing, the learning materials are provisioned through few windows simultaneously.

In a dedicated pathway of training, each student is provided with links to the educational content with which he works. After completing the credit module, the material assimilation is checked, that is, the initial control in order to determine whether all the chunks have been learned. And if a student successfully passes test control, he can go to the next credit module, and if not, then the system identifies the chunk on which question he did not respond and rebuilds the pathway of training in such a way as to repeat the necessary material again.

The teacher, using the plugin, receives a rating profile and a status of a pathway for each student. If there are no estimates for a particular chunk or module, the student does not study, and if the points are negative, then you need to change the pathway of training.

Since each discipline has a credit and modular structure, in order to obtain a score for discipline, the student must pass all credit modules and study all chunks from this discipline.

The plugin provide feedback between all participants in the learning process, enable them to control the quality of learning and activities in the E-learning environment, and most importantly adapt the training content by analyzing the student's actions.

The teacher has a possibility to create of a holistic understanding of the subject area understanding among students throughout the educational process.

5 Conclusions

An ontology-based plugin for MOODLE was developed, which allows to expand the main functionality of the LMS. The developed plugin collects information about the student activity and his learning progress. Collected information, gives recommendations to the student and the teacher on further movement in the learning.

The work of the plugin is based on the didactic ontology of the discipline, which forms the basis of the chunk-oriented approach to learning; this provides the opportunity to monitor the student's successes.

Subsequently, the possibilities of this plugin will be expanded and a module developed which, based on available training materials and test results, will generate individual recommendations for student to review appropriate topics before the next lecture, as well as the ability of tracking the presence at the lecture by checking student's activity in the system during the class.

References

1. Adejo, O.W., Ewuzie, I., Usoro, A., Connolly, T.: E-Learning to m-Learning: framework for data protection and security in cloud infrastructure. Int. J. Inf. Technol. Comput. Sci. (IJITCS) **10**(4), 1–9 (2018). https://doi.org/10.5815/ijitcs.2018.04.01
2. Benyahia, K., Lehireche, A., Latreche, A.: Semantic annotation of pedagogic documents. Int. J. Mod. Educ. Comput. Sci. (IJMECS) **8**(6), 13–19 (2016). https://doi.org/10.5815/ijmecs.2016.06.02
3. Dominic, M., Francis, S., Pilomenraj, A.: E-Learning in Web 3.0. Int. J. Mod. Educ. Comput. Sci. (IJMECS) **6**(2), 8–14 (2014). https://doi.org/10.5815/ijmecs.2014.02.02

4. Kurilovas, E.: Learning content repositories and learning management systems based on customization and metadata. In: 2009 Computation World: Future Computing, Service Computation, Cognitive, Adaptive, Content, Patterns. https://doi.org/10.1109/computation-world.2009.23

5. Ferran Ferrer, N., Minguillón Alfonso, J., SpringerLink: Content Management for e-learning. Springer, New York (2011). https://www.springer.com/cda/content/document/cda_download-document/9781441969583-c1.pdf?SGWID=0-0-45-1008047-p174015460. Accessed 21 Nov 2018

6. Syrota, S., Liskin, V., Kopychko, S.: Information technology based on chunk approach for ontology driven E-Learning engine. In: 2018 IEEE First International Conference on System Analysis & Intelligent Computing (SAIC), Kyiv, pp. 1–4 (2018). https://doi.org/10.1109/saic.2018.8516862, https://ieeexplore.ieee.org/document/8516862

7. Liskin, V., Syrota, S.: E-learning information technology based on an ontology driven learning engine. Int. J. Comput. Sci. Inf. Secur. **5**(8), 258–263 (2017)

8. Ganter, B., Wille, R.: Formal Concept Analysis: Mathematical Foundations. Springer (1999)

9. Valkman, Yu.R.: Geshtalty i metafory v kognytivnoj semy`oty`ke. XV mizhnarodna naukova konferenciya IAI-2015 im. T.A. Taran: tezy dopovidej, pp. 31–39. Prosvita, Kiev (2015)

10. Langacker, R.W.: Introduction to Concept, Image, and Symbol. Cognitive linguistics: basic readings, vol. 34, pp. 29–67. Mouton de Gruyter, Berlin, New-York (2006). (Cognitive linguistics research)

11. Miller, G.: The magical number seven, plus or minus two: some limitson our capacity for processing information. Psychol. Rev. **63**, 81–97 (1956)

12. Lidwell, W., Holden, K., Butcher, J.: Universal Design Principles (2010)

13. Syrota, S.V.: V.O. Liskin Rozrobka heneratora testiv dlja "Moodle" na bazi ontolohiji/Sxidno-Jevropejs′kyj zurnal peredovyx texnolohij. T. 5, #2(77), pp. 44–48 (2015). https://doi.org/10.15587/1729-4061.2015.51334

14. Marikar, F.M.M.T., Jayarathne, N.: Effectiveness of MOODLE in education system in Sri Lankan University. Int. J. Mod. Educ. Comput. Sci. (IJMECS) **8**(2), 54–58 (2016). https://doi.org/10.5815/ijmecs.2016.02.07

Batch Size Influence on Performance of Graphic and Tensor Processing Units During Training and Inference Phases

Yuriy Kochura[✉], Yuri Gordienko[✉], Vlad Taran,
Nikita Gordienko, Alexandr Rokovyi, Oleg Alienin,
and Sergii Stirenko

National Technical University of Ukraine "Igor Sikorsky Kyiv
Polytechnic Institute", Kyiv, Ukraine
iuriy.kochura@gmail.com, yuri.gordienko@gmail.com

Abstract. The impact of the maximally possible batch size (for the better runtime) on performance of graphic processing units (GPU) and tensor processing units (TPU) during training and inference phases is investigated. The numerous runs of the selected deep neural network (DNN) were performed on the standard MNIST and Fashion-MNIST datasets. The significant speedup was obtained even for extremely low-scale usage of Google TPUv2 units (8 cores only) in comparison to the quite powerful GPU NVIDIA Tesla K80 card with the speedup up to 10x for training stage (without taking into account the overheads) and speedup up to 2x for prediction stage (with and without taking into account overheads). The precise speedup values depend on the utilization level of TPUv2 units and increase with the increase of the data volume under processing, but for the datasets used in this work (MNIST and Fashion-MNIST with images of sizes 28×28) the speedup was observed for batch sizes >512 images for training phase and >40 000 images for prediction phase. It should be noted that these results were obtained without detriment to the prediction accuracy and loss that were equal for both GPU and TPU runs up to the 3rd significant digit for MNIST dataset, and up to the 2nd significant digit for Fashion-MNIST dataset.

Keywords: Deep learning · Tensor processing unit · GPU · TPUv2 ·
Performance · Accuracy · Loss · Inference time · MNIST · Fashion-MNIST

1   Introduction

Now the current trend of hardware acceleration for deep learning applications is largely related with graphics processing units (GPU) as general purpose processors (GPGPU). But recently the interest to alternative platforms, like variety of alternative hardware including GPUs, GPPs, field programmable gate-arrays (FPGA), and digital signal processors (DSP), and application-specific integrated circuit (ASIC) architectures like NVIDIA TCU (Tensor Core Units) and Google Cloud TPU (tensor processing units), become more and more popular [1]. Despite availability of some performance tests, like Google TPU vs. NVIDIA GPU K80 [2] and Google Cloud TPUv2 vs.

© Springer Nature Switzerland AG 2020
Z. Hu et al. (Eds.): ICCSEEA 2019, AISC 938, pp. 658–668, 2020.
https://doi.org/10.1007/978-3-030-16621-2_61

GPU NVIDIA V100 [3], the systematic studies on the scaling their performance (accuracy, loss, inference time, etc.) in relation to datasets of different sizes and hyper-parameters (for example, different batch sizes) are absent. This especially important in the view of the great interest to the influence of hyper-parameters of deep neural networks (DNN) on their training runtime and performance [4, 5], especially with regard to the batch size, learning rate, activation functions, etc. [6–9].

The main aim of this paper is to investigate scaling of training and inference performance for the available GPUs and TPUs with an increase of batch size and dataset size. The Sect. 2 *Background and Related Work* gives the brief outline of the state of the art in tensor processing and equipment used. The Sect. 3 *Experimental and Computational Details* contains the description of the experimental part related with the selected datasets, networks, and metrics used. The Sect. 4 *Results* reports about the experimental results obtained, the Sect. 5 *Discussion* is dedicated to discussion of these results, and Sect. 6 *Conclusions* summarizes the lessons learned.

2   Background and Related Work

Google's Tensor Processing Unit (TPU) has recently gained attention as a new and novel approach to increasing the efficiency and speed of neural network processing. According to Google, the TPU can compute neural networks up to 30x faster and up to 80x more power efficient than CPU's or GPU's performing similar applications [2]. It is possible, because the TPU is specifically adapted to solve inference problems with the much higher number of instructions per cycle in comparison to CPU, CPIU with advanced vector extensions, and GPU (Table 1).

Table 1. Comparison of the number of instructions per cycle for CPU, GPU and TPU

Type	Instructions per cycle
CPU	$\sim 10^0$
CPU (with vector extensions)	$\sim 10^1$
GPU	$\sim 10^4$
TPU	$\sim 10^5$

The TPU has a systolic array that contains $256 \times 256 =$ total 65,536 arithmetic logic unit (ALUs), and it can process 65,536 multiply-and-add operations for 8-bit integers every cycle. As far as the TPU runs at 700 MHz, it can compute $65,536 \times 7 \times 10^8 = 46 \times 10^{12}$ multiply-and-add operations or 92×10^{12} per second [2].

One of the ways to achieve the highest performance in GPU computing is to hide the long latency and other computational overheads by high data-level parallelism to achieve a high throughput, for example by the high batch size values [10, 11]. In addition to tests on GPU [4–9], recently the thorough performance analysis of the Google TPU was performed with some attempts to estimate influence of hyper-parameters on performance for TPU also [2, 12]. In addition to it this work is aimed to give the answer to some questions, namely, when it could be more efficient to use GPU or TPU during training and inference phases for datasets of various sizes and batch sizes. In the next section the short description of the used datasets, network, equipment, and measurement methods is given.

3   Experimental and Computational Details

Datasets. The MNIST database (Modified National Institute of Standards and Technology database) is a large database of handwritten digits (28 × 28 images) that become a standard benchmark for learning, classification and computer vision systems [13]. It was derived from a larger dataset known as the NIST Special Database 19 which contains digits, uppercase and lowercase handwritten letters.

Fashion-MNIST, a new dataset comprising of 28 × 28 grayscale images of 70,000 fashion products from 10 categories, with 7,000 images per category [14]. The training set has 60,000 images and the test set has 10,000 images. Fashion-MNIST is intended to serve as a direct drop-in replacement for the original MNIST dataset for benchmarking machine learning algorithms, as it shares the same image size, data format and the structure of training and testing splits.

The subsets of these datasets were used with the maximally possible batch size (for the better runtime) starting from 8 images and up to 130 000 images (with duplication of some images to increase the range of the datasets).

Equipment: GPU and TPU. The GPU and TPU computing resources were used to investigate the influence of hardware-supported quantization on performance of the DNNs. NVIDIA Tesla K80 was used as GPU cards during these experiments as Google Collaborative cloud resources (https://colab.research.google.com). Google TPUv2 are arranged into 4-chip modules with a performance of 180 TFLOPS, and 64 of these modules are then assembled into 256 chip pods with 11.5 PFLOPS of overall performance. TPU 2.0 has an instruction set optimized for executing Tensorflow and capable of both training and running DNNs. A cloud TPUv2 version was used as a TPU-hardware during these experiments, where 8 TPU cores were available as Google Collaborative cloud resources also.

Metrics. Accuracy and loss values are calculated for training, validation, and inference phases, then receiver operating characteristic (ROC) curves are constructed and the area under curve (AUC) is calculated per class and as their micro and macro averages (see below in the next section). To emphasize the contribution of initialization phase for GPU and TPU, the following two runtimes (both for GPU and TPU) per image were calculated for each run:

- Time with overheads = the wall time of the 1st epoch/number of images;
- Time without overheads = the wall time of the 2nd epoch/number of images.

The speedup values were calculated as GPU runtimes divided by TPU runtimes.

Deep Neural Network. The following deep convolutional neural network (Fig. 1) was used for this stage of research. The idea behind it was to use simple DNN to get results for reasonable period, but DNN should be complex enough to get the high accuracy and low loss comparable with available results on these standard datasets by other researchers.

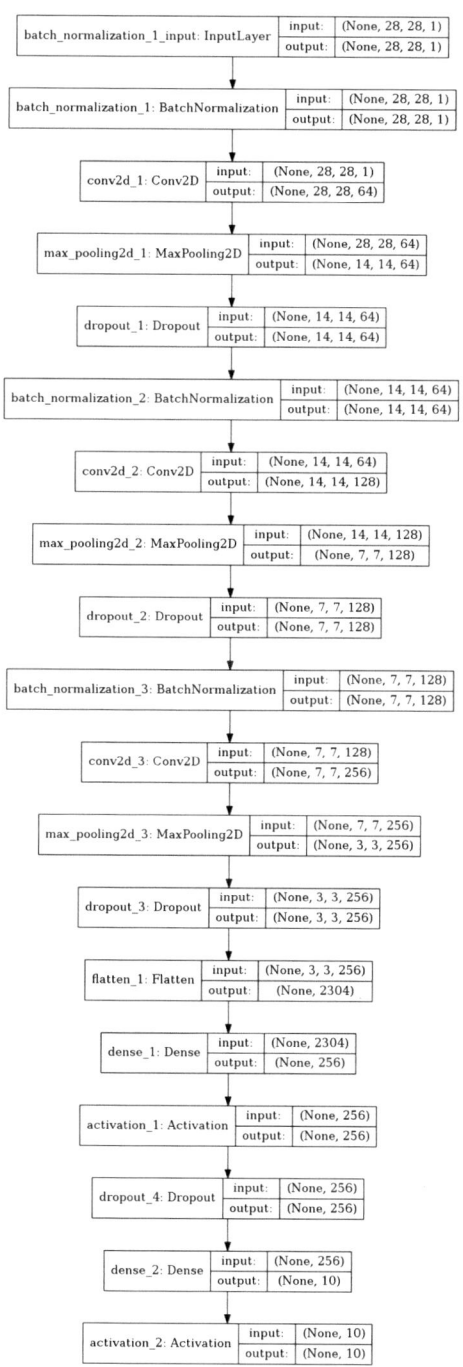

Fig. 1. The structure of the deep neural network used in the work.

4   Results

4.1   GPU

Below the training and validation history is shown for GPU for MNIST dataset (Fig. 2) and the similar plots were obtained for Fashion-MNIST (they are not shown here because of the shortage of space, but will be published elsewhere [12]).

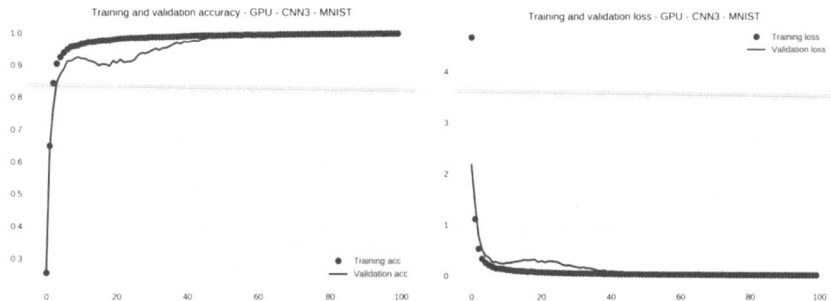

Fig. 2. Accuracy (left) and loss (right) during training and validation on GPU K80 (for the whole training part of MNIST dataset 60000 images)

The ROC-curves and AUC-values (Fig. 3) demonstrate the excellent prediction accuracy, which is used for comparison with the similar experiments on TPUv2.

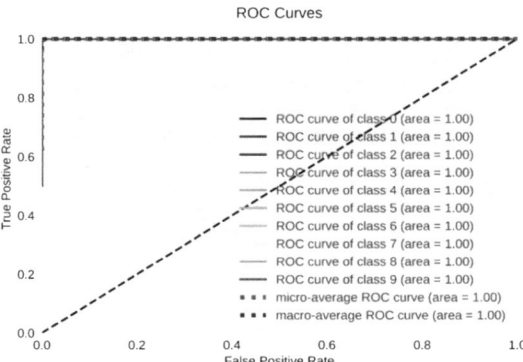

Fig. 3. ROC-curves and AUC-values for 10 classes on GPU K80 (for the testing part of MNIST dataset - 10000 images)

4.2   TPU

Below the similar training and validation history is shown for TPUv2 and MNIST dataset (Fig. 4) and the similar plots were obtained for Fashion-MNIST (again, they are not shown here because of the shortage of space, but will be published elsewhere [12]). These results demonstrate the principally similar evolution to the high values of accuracy and loss as for GPU before.

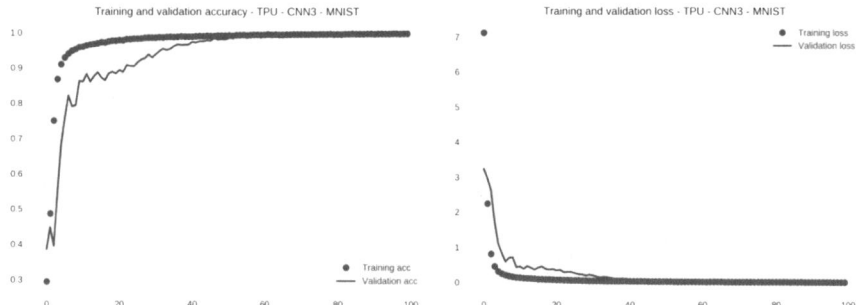

Fig. 4. Accuracy (left) and loss (right) during training and validation on Google TPUv2 (for the whole training part of MNIST dataset 60000 images)

The ROC-curves and AUC-values (Fig. 5) also demonstrate the excellent prediction accuracy, which is similar to the results obtained in experiments on GPU.

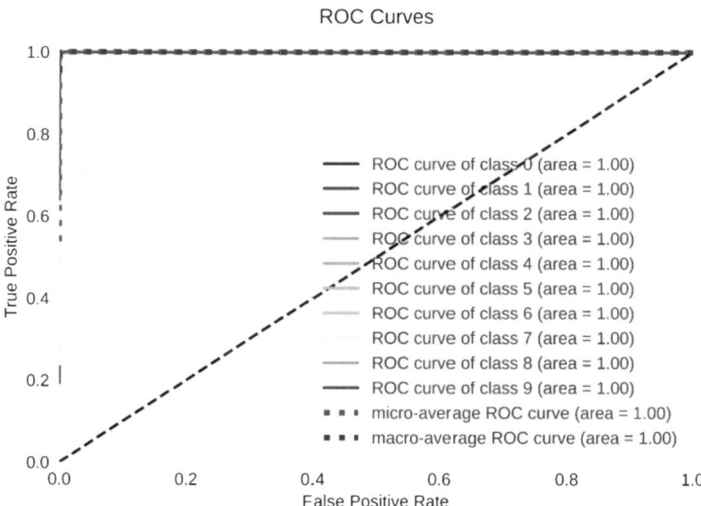

Fig. 5. ROC-curves and AUC-values for 10 classes on Google TPUv2 (for the testing part of MNIST dataset - 10000 images)

The saturation of speedup can be observed for training from the raw timing data (Fig. 6) and speedup plot (Fig. 7) in the point where the number of images equal to ∼ 8000.

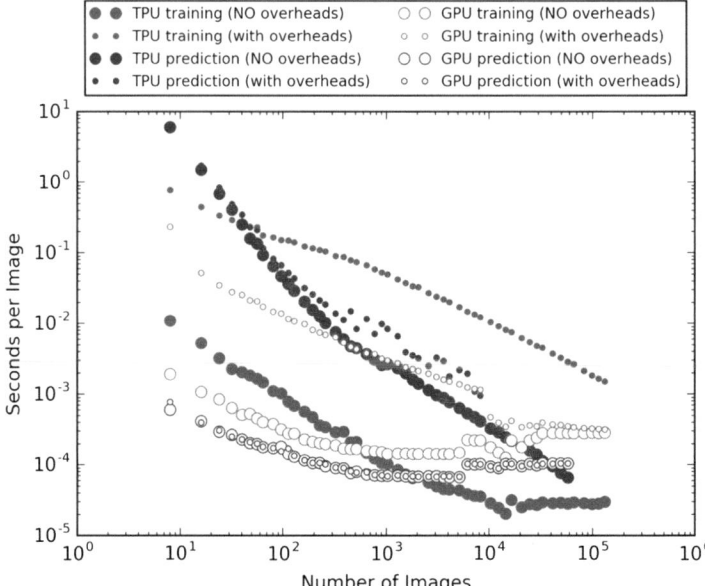

Fig. 6. Comparison of GPU K80 and TPUv2 run times per image.

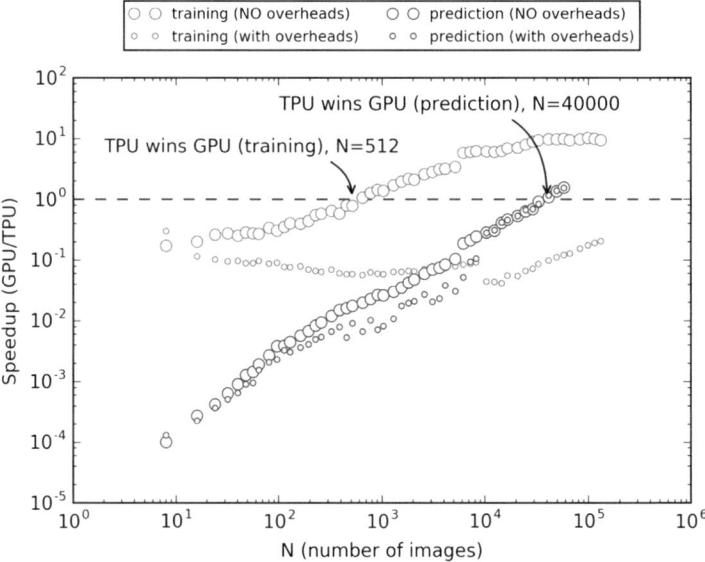

Fig. 7. Speedup of GPU K80 and TPUv2.

It is explained by the inability to use batch size > 8192 for the current datasets with images of sizes 28 × 28. For the bigger images the maximally possible batch size will be lower and the speedup will be saturated earlier, but it will be investigated elsewhere [15].

5 Discussion

These results demonstrate that usage of Google TPUv2 is more effective (faster) than GPU for the large number of computations under conditions of low overhead calculations and high utilization of TPU units. Moreover, these results were obtained for the simple CNN-like deep learning network without detriment to the accuracy and loss that were equal for both GPU and TPU runs up to the 3rd significant digit for MNIST dataset, and up to the 2nd digit for Fashion-MNIST dataset (Table 2).

Table 2. The accuracy and loss or the testing part (10000 images) of the datasets used.

Hardware	Accuracy	Loss
MNIST		
GPU	0.9944	0.02236
TPU	0.9937	0.02214
Fashion-MNIST		
GPU	0.9255	0.2354
TPU	0.9279	0.2434

The prediction accuracy values were equal for both GPU and TPU up to the 3rd significant digit for MNIST, and up to the 2nd significant digit for Fashion-MNIST. The loss values were equal for both GPU and TPU regimes up to the 2nd significant digit for both MNIST and Fashion-MNIST datasets. The significant speedup (Fig. 7) was reached even for extremely low-scale usage of Google TPUv2 units (8 cores only) in comparison to the quite powerful GPU unit (NVIDIA Tesla K80):

- Speedup > 10x for training stage (without taking into account overheads);
- Speedup > up to 2x for prediction stage (with and without overheads).

The speedup values depend on the utilization level of TPUv2 units and increase with the data volume (batch size) under processing, but for the MNIST and Fashion-MNIST datasets with images of sizes 28 × 28 the speedup was started (i.e. speedup becomes > 1) after 512 images (for training) and 40 000 images (for prediction) even.

It should be noted that these results were obtained without detriment to the accuracy and loss for the relatively simple DNN and small images (28 × 28). The current investigations of network size impact and image size impact are under work now and their results will be published elsewhere [15]. In addition to Google TPU architecture, the specific tensor processing hardware tools are available in the other modern GPU-cards like Tesla V100 and Titan V by NVIDIA based on the Volta microarchitecture with specialized Tensor Cores Units (640 TCU) and their influence on training and prediction speedup are under investigation and will be reported elsewhere [15].

As far as the model size limits the available memory space for the batch of images other techniques could be useful for squeezing the model size, like quantization [15, 16] and pruning [17, 18]. These results can be used for selection of the optimal parameters for applications where a large batch of data needs to be processed, for example, for monitoring real-time road condition in advanced driver assistance systems (ADAS), where such specialized architectures like TPU, TCU, and FPGA-based solutions can be deployed [19]. For example, it is especially important for various complex ADAS-related tasks like real-time obstacle and lane detection [20, 21], traffic video analysis under impact of noise [22], and monitoring driver behavior even [23].

6   Conclusions

In this work the influence of dataset size with the maximally possible batch size (for the better runtime) was investigated with regard to the performance of graphic and tensor processing hardware during training and inference phases. During these experiments GPU NVIDIA Tesla K80 was used as a GPU-hardware and Google TPUv2 was used as a TPU-hardware (both were available as Google Collaborative cloud resources). The practical efficiency of the both hardware types was investigated during the numerous runs of the selected DNN on the standard MNIST and Fashion-MNIST datasets. The significant speedup was obtained even for extremely low-scale usage of Google TPUv2 units (8 cores only) in comparison to the quite powerful GPU NVIDIA Tesla K80 card with the speedup up to 10x for training stage (without taking into account overheads) and speedup up to 2x for prediction stage (with and without taking into account overheads). It was shown that the precise speedup values depend on the utilization level of TPUv2 units and increase with the increase of the data volume under processing, but for the datasets used in this work (MNIST and Fashion-MNIST with images of sizes 28×28) the speedup was observed for datasets with >512 images for training phase and >40 000 images for prediction phase. In general, the usage of tensor processing architectures like Google TPUv2 will be very promising way for increasing performance of inference and training even, especially in the view of availability of the similar specific tensor processing hardware architectures like TCU in Tesla V100 and Titan V provided by NVIDIA, and others.

References

1. Lacey, G., Taylor, G.W., Areibi, S.: Deep Learning on FPGAs: Past, Present, and Future. arXiv preprint arXiv:1602.04283 (2016)
2. Jouppi, N.P., et al.: In-datacenter performance analysis of a tensor processing unit. In: International Symposium on Computer Architecture, vol. 45, no. 2, pp. 1–12 (2017)
3. Haußmann, E.: Comparing Google's TPUv2 against Nvidia's V100 on ResNet-50. RiseML Blog, 26 April 2018. Accessed 29 Aug 2018

4. Masters, D., Luschi, C.: Revisiting small batch training for deep neural networks. arXiv preprint arXiv:1804.07612 (2018)
5. Devarakonda, A., Naumov, M., Garland, M.: AdaBatch: adaptive batch sizes for training deep neural networks. arXiv preprint arXiv:1712.02029 (2017)
6. Smith, L.N.: A disciplined approach to neural network hyper-parameters: Part 1 - learning rate, batch size, momentum, and weight decay. arXiv preprint arXiv:1803.09820 (2018)
7. Kochura, Y., Stirenko, S., Alienin, O., Novotarskiy, M., Gordienko, Y.: Performance analysis of open source machine learning frameworks for various parameters in single-threaded and multi-threaded modes. In: Advances in Intelligent Systems and Computing II. CSIT 2017. Advances in Intelligent Systems and Computing, vol. 689, pp. 243–256. Springer, Cham, September 2017
8. Kochura, Y., Stirenko, S., Gordienko, Y.: Comparative performance analysis of neural networks architectures on H2O platform for various activation functions. In: Young Scientists Forum on Applied Physics and Engineering (YSF), 2017 IEEE International, pp. 70–73. IEEE (2017)
9. Kochura, Y., Stirenko, S., Alienin, O., Novotarskiy, M., Gordienko, Y.: Comparative analysis of open source frameworks for machine learning with use case in single-threaded and multi-threaded modes. In: 2017 12th International Scientific and Technical Conference on Computer Sciences and Information Technologies (CSIT), vol. 1, pp. 373–376. IEEE (2017)
10. Zhu, H., Zheng, B., Schroeder, B., Pekhimenko, G., Phanishayee, A.: DNN-Train: benchmarking and analyzing DNN training. Training **8**, 16GBs (2018)
11. Jäger, S., Zorn, H.P., Igel, S., Zirpins, C.: Parallelized training of deep NN: comparison of current concepts and frameworks. In: Proceedings of the Second Workshop on Distributed Infrastructures for Deep Learning, pp. 15–20. ACM (2018)
12. Jouppi, N., Young, C., Patil, N., Patterson, D.: Motivation for and evaluation of the first tensor processing unit. IEEE Micro **38**(3), 10–19 (2018)
13. LeCun, Y., Cortes, C., Burges, C.J.: MNIST handwritten digit database. AT&T Labs (2018). http://yann.lecun.com/exdb/mnist. Accessed 30 Aug 2018
14. Xiao, H., Rasul, K., Vollgraf, R.: Fashion-MNIST: a novel image dataset for benchmarking machine learning algorithms. arXiv preprint arXiv:1708.07747 (2017)
15. Gordienko, Yu., Kochura, Yu., Gordienko, N., Taran, V., Alienin, O., Rokovyi, O., Stirenko, S.: Specialized Tensor Processing Architectures for Deep Learning Models (2018, submitted)
16. Cheng, J., Wang, P.S., Li, G., Hu, Q.H., Lu, H.Q.: Recent advances in efficient computation of deep convolutional neural networks. Front. Inf. Technol. Electron. Eng. **19**(1), 64–77 (2018)
17. Zhu, M., Gupta, S.: To prune, or not to prune: exploring the efficacy of pruning for model compression. arXiv preprint arXiv:1710.01878 (2017)
18. Gordienko, Yu., Kochura, Yu., Taran, V., Gordienko, N.: Adaptive Iterative Channel Pruning for Accelerating Deep Neural Networks (2018, submitted)
19. Li, Y., Liu, Z., Xu, K., Yu, H., Ren, F.: A GPU-outperforming FPGA accelerator architecture for binary convolutional neural networks. ACM J. Emerg. Technol. Comput. Syst. **14**(2), 1–16 (2018)
20. Singh, Y., Kaur, L.: Obstacle detection techniques in outdoor environment: process, study and analysis. Int. J. Image Graph. Signal Process. (IJIGSP) **9**(5), 35–53 (2017). https://doi.org/10.5815/ijigsp.2017.05.05
21. Katru, A., Kumar, A.: Improved parallel lane detection using modified additive hough transform. Int. J. Image Graph. Signal Process. (IJIGSP) **8**(11), 10–17 (2016). https://doi.org/10.5815/ijigsp.2016.11.02

22. Ata, M.M., El-Darieby, M., Elnaby, M.A., Napoleon, S.A.: Traffic video enhancement based vehicle correct tracked methodology. Int. J. Image Graph Signal Process. (IJIGSP) **9**(12), 30–40 (2017). https://doi.org/10.5815/ijigsp.2017.12.04
23. Isong, B., Khutsoane, O., Dladlu, N.: Real-time monitoring and detection of drink-driving and vehicle over-speeding. Int. J. Image Graph. Signal Process. (IJIGSP), **9**(11), 1–9 (2017). https://doi.org/10.5815/ijigsp.2017.11.01

Author Index

© Springer Nature Switzerland AG 2020
Z. Hu et al. (Eds.): ICCSEEA 2019, AISC 938, pp. 669–671, 2020.
https://doi.org/10.1007/978-3-030-16621-2